IPTV Dictionary

**Published By
Althos Inc.**

ALTHOS
SIMPLIFYING KNOWLEDGE

404 Wake Chapel Road
Fuquay-Varina, NC 27526 USA
Telephone: 1-800-227-9681
1-919-557-2260
Fax: 1-919-557-2261
email: Success@Althos.com
web: www.Althos.com

Althos

Copyright © 2006 By Althos Publishing

Printed and Bound by Lightning Source Printing, La Vergne, TN USA

International Standard Book Number: 1-932813-34-9

About the Editor

Mr. Harte is the managing director of Althos, an expert information provider that covers the communication industry. Mr. Harte has over 29 years of technology and business experience an has worked for leading communications technology companies including Ericsson/General Electric, Audiovox/Toshiba and Westinghouse. Lawrence holds degrees of Executive MBA from Wake Forest University (1995) and a BSET from the University of the State of New York, (1990). Mr. Harte has instructed at and received numerous  certificates from many non-university courses including IPTV Systems, VoIP/Internet Telephony, 3G wireless, wireless billing, Bluetooth technology, Internet billing, communications, cryptograph, microwave measurement and calibration, radar, nuclear power, Dale Carnegie, 360 leadership, and public speaking.

Mr. Harte has appeared on television as an industry expert and has been referenced in over 75 telecommunications related articles in industry magazines. He has been a speaker and moderator at numerous industry seminars and trade shows. His magazine publications include Popular Science, Wireless Week, RCR, Cellular Business, Cellular Marketing and others. Between 1993-1995, Mr. Harte wrote a monthly column in Cellular

Marketing called "Techniques." The monthly Techniques column explained the business related issues behind key technology innovations that were developing in the telecommunications industry.

In 2002, Mr. Harte joined Althos as the managing director and in 2003, Mr. Harte became president of Althos. At Althos, Mr. Harte is responsible for the research, training, and publishing divisions. His core focus at Althos is researching, analyzing, testing, and educating companies on alternative communication technologies.

Between 1995 and 2002, Mr. Harte was the president of APDG, a telecommunications research and publishing company. At APDG, he provided telecommunications information to leading companies including Sprint PCS, Delphi (GM), Vodafone, British Telecom, LHS, Radiocommunication Agency, DST, Sony, Ericsson, AT&T, and many others.

Between 1994 and 1995, Lawrence was Vice President of product management for ReadyCom, a cellular voice paging equipment provider. At ReadyComm, Mr. Harte managed product development for cellular voice paging subscriber products. At ReadyComm, he performed market and technical feasibility analysis for new product development and application areas. He designed a product management process for tracking the development of programs and ensured products could met industry and government requirements.

From 1992 to 1994, Lawrence was the New Products and Applications Manager for Ericsson General Electric. As a product manager, he created and managed a core team of line organization managers to develop new product plans. His responsibility included cellular telephone development for PCS, cellular data modems, CDPD, OEM product management, and other new product concepts. Lawrence worked closely with customers such as AT&T, Southwestern Bell, Cantel, and Cellular One during the new product launch of TDMA digital cellular service.

Between 1989 and 1992, Lawrence was a Digital Cellular Development Engineer at Audiovox Corporation. In this position, he was the liaison between digital industry standards committees, engineering at Toshiba Japan, and customers. During this period, He was a voting member of the

Telecommunications Industry Association and an editor of the US Digital Cellular Specification (IS-54, now IS-136). Lawrence supported the sales staff with customers including BMW, BellSouth, Nynex, PacTel (now Airtouch), Bell Atlantic, Southwestern Bell, Bell Mobility (now Bell Canada) and others for technical and marketing issues.

From 1986 to 1989, Mr. Harte was an Automatic Test Equipment Engineer for Test System Associates on site at Westinghouse. At the Westinghouse facility, he converted technical performance requirements into electronic interface devices and created software programs for automatic testing of FB-111 avionics communication equipment.

Between 1977 and 1985, Lawrence was an Electronics Technician in the US Navy which included calibration laboratory supervisor for microwave and RF up to 18 GHz and UHF communication supervisor for the USS Independence.

Lawrence has authored many books and several industry research reports on telecommunications technology including "IP Television Basics" (2006), Introduction to MPEG" (2006), "Introduction to DRM" (2006)," "Introduction to WiMax" (2006), Introduction to IPTV Billing" (2006), "Introduction to Optical Communication" (2005), "Optical Communication Installation Basics" (2005), "Wireless Systems" (2005), "Introduction to WCDMA" (2004), "Introduction to EVDO" (2004), "Introduction to GPRS and EDGE" (2004), "Introduction to GSM" (2004), "Introduction to IS-95 CDMA" (2004), "Introduction to 802.11 Wireless LAN" (2004), "Introduction to Bluetooth(tm)" (2004), "Introduction to SIP IP Telephony" (2004), "Introduction to IP Telephony" (2003), "Voice over Data Networks for Managers" (2003), "Internet Telephone Dictionary" (2003), "Internet Telephone Basics" (2003), "Animated Telecom Dictionary" (2002), "Telecom Basics, 2nd Edition" (2002), "SS7 Basics, 2nd Edition" (2002), "Telecom Made Simple" (2002), "3G Wireless Demystified" (2002), "Delivering xDSL" (2000), "Public and Private Land Mobile Radio Telephone Systems" (2000), "The Comprehensive Guide to Wireless Technologies" (1999), "IS-95 CDMA" (1999), "IS-136 TDMA" (1999), "GSM Superphones" (1998), "Cellular and PCS, The Big Picture" (1997), "Wireless Resale Market Report and Forecast 1997-2002," (1997), "Cellular and PCS/PCN Telephones and Systems", (1996), "Digital Cellular: Economics and Comparative Technologies", (1993), and "Dual Mode Cellular (IS-54/IS-136)", (1991). Lawrence is an inventor of several patents including "Cellular Standby Power Saving Systems" (5,224,152 - issued June 29, 1993), 5,568,513, 5,701,329, 5,794,137.

Lawrence has consulted and been an expert witness for companies including Ameritech, AT&T, Nokia, Hughes Network Systems, Qualcomm, Campbell-Ewald, Casio, Ericsson, Samsung, Sony, AMD, VLSI, Siemens and others.

Contributing Editors and Advisors

Bob Giddy is a professionally qualified electronics engineer with over 30 years experience in manufacturing and design with systems and semiconductor companies. Prior to joining Amino, Giddy worked as a senior manager for National Semiconductor, based in Munich and the northern European headquarters. While at National Semiconductor, he ran the northern Europe sales and marketing operation for the telecommunications market, working with key customers such as Alcatel, Ericsson and Nokia. Giddy later joined NEC Electronics, working for 14 years, first as a director and then as general manager of the northern Europe business. During this time, he established the NEC multimedia design centre in Milton Keynes, UK and oversaw the development of the first single-chip MPEG-2 video decoder. After leaving NEC, Giddy established the European sales and marketing operation for inSilicon, a silicon IP vendor.

Daniel Marcus is the corporate marketing manager for IPTV and broadband solutions at UTStarcom, Inc. based in Alameda, CA. Daniel has worked primarily for startups in the telecommunications industry for eight years in a variety of capacities including marketing, sales and operations, enabling him to understand the unique challenges operators face in the evolving broadband marketplace. He holds a bachelor's degree in political science from the University of California at San Diego.

Les Wyatt is Vice President of Harris Corporation and General Manager of Harris Software Systems, a division of Harris' Broadcast Communications Division based in Denver, Colorado. Mr. Wyatt oversees global operations of the Software Systems business, which includes software product lines in advertising, automation, broadband, digital asset management and media management. Mr. Wyatt has more than 28 years of management experience in leading technology companies. Prior to Harris, he was the Group Vice President and General Manager of EnterpriseOne at PeopleSoft, an enterprise software business focused on the manufacturing industrial sector. Before his position at PeopleSoft, Mr. Wyatt was Chief  Marketing Officer for J.D. Edwards. Prior to that Mr. Wyatt held executive management positions with Harbinger and Texas Instruments. Mr. Wyatt also serves on the Board of directors of the Colorado Software and Internet Association, Colorado's leading technology trade association, where he was Chairman in 2005. Mr. Wyatt holds a bachelor's degree in Mathematics and a master's degree in Computer Science both from Arizona State University.

Danny Wilson is the founder and President of Pixelmetrix Corporation, a manufacturer of preventive monitoring solutions for digital broadcasters. Previously, Mr. Wilson was an executive with Hewlett-Packard's Communication Measurement Division. In this role, he was responsible for numerous telecom and broadcast products including the world's first ATM/B-ISDN Test System. A native of Edmonton, Canada, Mr. Wilson holds a degree in Computer Engineering from the University of Alberta.

Kelly Anderson joined IPDR in 2005 and serves as it's President and COO. Kelly is a widely recognized expert on back office billing and order management processes. Her career spans more than a decade in the communications industry where she promoted industry collaboration for VoIP services, wireline and wireless billing, and provisioning processes. Anderson's was previously at AT&T, Birch Telecom, Daleen Technologies, and Intrado Inc.

J.D. Zeeman is responsible for enabling business to exploit the exciting new technologies of Digital Media (i.e. audio, video, image) at IBM. His industry focus is Telecommunications, along with Media & Entertainment and Energy & Utilities. His global work encompasses strategy, offering development, business development, and leading-edge engagement support. Mr. Zeeman's concentration is on innovative implementations of digital media to address new market opportunities and to fulfill leading-edge client needs. Areas of recent focus include content services (e.g. IPTV, online games), wireless computing, and e-business on demand.

Mr. Zeeman has served as a keynote and panel speaker at industry events including the Broadband and Triple Play Services Conference, Broadband World Forum, Digital Hollywood, Future Image Summit, TelcoTV, and WiFi Planet. Mr. Zeeman holds a Masters of Business Administration from Northwestern University's Kellogg School and a Bachelor of Science in Engineering from Princeton University.

Patrick Christian, founder and managing director of Packet Vision, has more than 25 years' experience of hardware and software design in the data communications and telecoms industries. In 1993 he founded the consultancy company, Blue C Technologies Limited, which worked with many leading technology businesses developing advanced telecommunications systems. In 1998 Patrick co-founded VegaStream Limited to develop high performance voice-over-IP gateways. VegaStream was sold in 2000 to Pace Micro Technology plc for £20m, returning an IRR of 149% to its institutional investors.

In May 2004 Patrick founded Packet Vision which is developing pioneering, network-based addressable IPTV advertising delivery systems. The unique platform combines multiple functions in a slim-line, 'pizza box' style unit that does not require any changes in end user equipment. Patrick has an MA and First Class Honours degree in Electrical Sciences from Cambridge University. He has contributed articles to a number of leading IPTV publications.

Ms. Julia Mooney has extensive experience in product management and R&D within the IP and telecommunications sectors. Prior to joining Irdeto Access in April 2003, she worked as Product Manager for Interactive TV at UPC Media where she particiapted in the launch of UPC Digital TV Interactive Services in December 2001. Prior to this, Julia worked at EUnet International B.V as Product Manager for IP Connectivity Services and handled the Commercial management of EUnets global Internet roaming service as well as managing joint technical implementation projects with a range of industry partners. Julia holds a BSc Hons in Pure and Applied Mathematics from the University of Liverpool and certification in Information Management (Siemens Nixdorf) and Network Engineering (Novell).

William S. Kish is the founder and CTO of Video54. Mr. Kish has extensive experience with C, C++, and Java software development in BSD, VxWorks, Windows, and Linux environments. His expertise is in system and software architecture of 802.11, ATM, Ethernet, IP, SONET, and Optical DWDM networking equipment. He is also knowledgeable in a wide range of networking architectures and protocols. He has proven ability to deliver robust embedded software systems. Mr. Kish works closely with marketing and customers to define and implement innovative product capabilities. He also has experience evaluating the technical and business merits of product offerings including financial ROI analysis.

Mr. Kish has worked for Apple Computers where he defined and developed management software for the Apple Internet Router product. He has also worked at FORE Systems as their lead engineer. Mr. Kish received his Bachelor of Science degree in Electrical and Computer Engineering from Carnegie Mellon University in 1993. Where he graduated with both University and Carnegie institute honors.

Geoff Burke has recently joined Calix as their Director of Video Solutions Marketing. Geoff is a recognized industry expert in all aspects of entertainment services delivery by telecommunications service providers. He has been involved with over 50 North American telco video services deployments and is often asked to share his expertise with the national media, as well as at national and regional industry conventions. His responsibilities include the development and implementation of the company's video solutions strategy, as well as triple play deployment strategy assistance for Calix customers.

Walter Megura is the General Manager for Broadband Networks where he is responsible for the Global business and portfolio strategy for Nortels Broadband Solutions. Previously, Mr. Megura was Regional Vice President Sales for the Americas for Nortel. Prior to that Mr. Megura has held the multiple Vice President leadership positions with responsibilities for Emerging Service Provider (CLEC) segment, High Speed Access Sales for North America, and the Broadband Wireless Access (BWA) business unit with focus on sales strategies for Nortel's broadband technologies in the broadband segments for delivery of data and voice end-to-end networking. Mr. Megura has been an instrumental part of Nortel since March of 1998 when he joined the BWA team.

Mr. Megura's professional background includes 20 years of leadership in both domestic and international business. In the past, Mr. Megura has been a member of the board of directors of the Wireless Communications Association (WCA) as well a member of its Government Regulations Committee. Mr. Megura holds a B.S. degree in Business Administration from Northeastern University, Boston. In 2003 he attended Stanford Executive Management Program.

Philip W. Schuman is the founder of HighView Media. Prior to forming HighView Media, Mr. Schuman was Co-President, Universal Studios Television Distribution where he was responsible for the successful management of the global theatrical and television product distribution organization for Universal Studios, Inc. Mr. Schuman started at Universal as VP Business Affairs in 1995. Previously, Mr. Schuman was Senior Transactions Counsel, at a division of General Electric.

Mr. Schuman began his career obtaining extensive relevant experience specializing in mergers and acquisitions and securities at the law firm of Shearman & Sterling in New York. Mr. Schuman holds a Bachelor of Arts in American History from State University of New York (magna Cum Laude, Phi Beta Kappa) at Buffalo - Buffalo, NY and J.D. from Emory University School of Law (Order of the Coif, Law Review) - Atlanta, GA.

Table of Contents

Numbers

*,G-Any Source and Group Pair

.Com Company-Dot Com Company

.MPG-MPEG

.Net Compact-.NET Compact is an application development platform for the Microsoft .NET program to provide interoperability between programs and devices using XML Web services.

.Sig File-A signature file (.sig) contains text and/or images that are added to the end of electronic mail (email) when they are sent. Signature files usually contain contact information along with company logos and/or tag lines.

.Wav-A program extension code for waveform that is used by Windows for sound files. WAV files are computer digital samples of the analog waveform and they require a relatively large amount of memory compared to compressed sound file types.

.Z-A filename extension that identifies to a Unix system that the file is stored in compressed form by the gzip or the compress utility.

.ZIP-A filename extension that identifies that a file is compressed by the WinZip or PKZip utilities.

μ - Law-Mu

μ, Mu-Greek letter lower case Mu. (1-metric prefix) One millionth or 0.000001. (2-length unit) One micrometer, 0.000001 meter (formerly called one micron). (3-digital code designator) The algorithm for compression or mapping of measured voltage amplitude to binary code value used in conjunction with the DS-1 (T-1) digital PCM encoding of telephone channel waveforms is designated Mu-law or μ-law, in contrast to a similar but distinct algorithm called A-law and used in conjunction with the 2.048 Mbit/s PCM system.

Note that the English letter u is often substituted for μ when the Greek character cannot be produced due to limited typographic capability. Understandable, but often ambiguous!

1 BASE5-An implementation of the 1 Mbps StarLAN IEEE standard network. The maximum segment length for the 1Base5 system is 500m.

1 Pass Encoding-1 pass encoding is a compression or data formatting process that analyzes media or data information a single time to produce a compressed or encoded file or media output.

10 BASE F-A generic designation for a family of 10 Mb/s baseband Ethernet systems operating over optical fiber.

10 Base2-An Ethernet data communication standard that was created by the IEEE for communication over coaxial cable. 10 Base2 has a maximum distance of 185 meters. 10 Base2 is also known as Thin Ethernet.

10 BASE-FB-A baseband Ethernet system operating at 10 Mb/s over two multimode optical fibers using a synchronous active hub.

10 BASE-FP-An Ethernet that transfers data at 10 Mbps over two optical fibers. The 10 Base-FP uses a passive hub to connect data communication devices.

10 BaseT-A data communications system primarily used for computer networks based on the Ethernet IEEE standard 802.3. The 10BaseT system media is twisted pair wire at a data rate of 10 Mbps.

10 BROAD36-A broadband Ethernet system operating at 10 Mb/s over three channels (in each direction) of a private CATV system. Segment length is 3600 meters.

10 GE-10 Gigabit Ethernet

10 Gigabit Ethernet (10 GE)-10 Gigabit Ethernet (10 GE) is a data communication system that combines Ethernet technology with fiberoptic cable transmission to provide data communication transmission at 10 Gbps (10,000 Mbps). The specifications for 10 GE are being developed by the Gigabit Ethernet Alliance. The Gigabit Ethernet Alliance is a group of companies that was formed in January 2000.

10 GE is an extension to the IEEE 802.3 protocol that enables data communication at speeds of approximately 10 billion bits per second (Gbps). The IEEE 802.3ae standard supports two Physical Layer protocols: LAN PHY and WAN PHY. Both of these physical layer protocols use the same Media Access Control (MAC) layer and frame formats.

The LAN PHY supports existing Gigabit Ethernet applications at ten times the bandwidth using a cost-effective solution. For compatibility with existing wide area networks based on SONET/SDH, the WAN PHY has been standardized. The WAN PHY includes a simplified SONET/SDH framer that operates at a line-rate of OC-192/STM-64 resulting in a payload rate of approximately 9.29 Gbps while providing compatibility with the existing Ethernet packet format.

\#

10 Gigabit Ethernet Alliance-An alliance between a group of companies that was formed in January 2000 that is assisting in the creation of a 10 Gigabit Ethernet system.

100 BASE-FX-A baseband Ethernet system operating at 100 Mbps over two multimode optical fibers.

100 BaseT-A data communications system primarily used for computer networks based on the Ethernet IEEE standard 802.3. This system media is twisted pair wire and its data rate is 100 Mbps.

100 BASE-T2-An Ethernet system operating at 100 Mbps over two pairs of Category 3 or higher unshielded twisted pair (UTP) cable.

100 BASE-T4-An Ethernet system operating at 100 Mbps over four pairs of Category 3 or higher using unshielded twisted pair (UTP) cable.

100 BASE-TX-A baseband Ethernet system operating at 100 Mbps over two pairs of STP or Category 5 UTP cable.

100 BASE-X-A generic designation for 100 Base Ethernet systems independent of their underlying physical transmission medium (T- twisted pair or F - fiber).

100 VG-AnyLan-A local area network (LAN) systems that use a demand priority access method. It is standardized by IEEE 802.12.

100/5-A contract stating rights to payment of an additional 100 percent royalties to the instated author at which time the first five episode broadcasts are repeated. The predetermined amount will be paid in installments of 20% according to time of re-run.

100/50/50-Formula used to calculate bonuses to a service provider in correlation to a film subsequent its theatrical debut. (100 percent before first television broadcast, 50 percent after, and an additional 50 percent if the movie is exhibited in theaters overseas before or after first television broadcast.)

1000 BASE-CX-A baseband Ethernet system operating at 1000 Mbps over two pairs of 150 shielded twisted pair (STP) cable.

1000 BASE-LX-A baseband Ethernet system operating at 1000 Mb/s over two multimode or single-mode optical fibers using longwave laser optics.

1000 BASE-SX-An Ethernet system that transmits at 1000 Mbps shortwave laser optics over two multimode optical fibers.

1000 BASE-T-An Ethernet that transmits at 1000 Mbps using four pairs of Category 5 unshielded twisted pair (UTP) cable.

1000 BASE-X-A generic name for an Ethernet system that transmits at 1000 Mbps.

1024 (K)-Widely used but unofficially, the number 1024 (equal to 2 raised to the 10th power) is represented by the capital letter K. This number normally occurs only when describing file size, memory size or other numbers that are integral powers of 2 because of internal use of the binary number system. Take care to distinguish capital K from small k, which represents 1000.

16 Bit Color-Sixteen bit color is the ability to represent up to 65,536 unique colors.

16 Level Quadrature Amplitude Modulation (16-QAM)-16 level QAM (16-QAM) is a combination of amplitude modulation (changing the amplitude or voltage of a sine wave to convey information) together with phase modulation. The use of 16 levels allows each symbol (specific combination of amplitude and phase) to represent 4 bits of information (16 different states).

16mm Film-16mm film is 16 mm wide material that is used to store moving images. 16 mm film was commonly used for educational films.

16-QAM-16 Level Quadrature Amplitude Modulation

1996 Telecommunications Act-Legislation designed to spur competition in the telecommunications industry. Resulting deregulation affects local, long distance, and wireless carriers. Direct competition between these suppliers and other 'non-traditional' competitors (cable, utilities, etc.) should result in better prices and more services for the consumer. Signed into law by President Clinton Feb.8, 1996.

1D Barcode-1 Dimension Barcode

1FB-One Flat Business Line

1G-First Generation

1O BASE5-An implementation of the Ethernet IEEE 802.3 standard on Thicket (RG-6/8) coaxial cable, a baseband medium, at 10 Mbps. The maximum segment length is 500m.

1XEV-1X Evolution

1xEVDO-One Channel Evolution Version Data Only

1xEVDV-One Channel Evolution Version Data and Voice

1xRTT-One Channel Radio Transmission Technology

2 1/2G-Second And A Half Generation

2 Pass Encoding-2 pass encoding is a compression or data formatting process that analyzes media or data information two times to produce a compressed or encoded file or media output. The use of 2 pass encoding allows for the selection of the optical type of compression and potentially offers the ability to compress information more than 1 pass encoding.

2.5G-A term commonly used to describe an interim generation of technology that provides more services and features than second generation (2G) technology but less than the third generation (3G).

2:2 Pulldown-2:2 pulldown is a process of converting film that operates at 24 frames per second to video at 50 fields per second. The 2:2 process operates by repeating a film frame image 2 times to produce 2 fields. This increases the frame rate to 48 fps, which is close enough to 50 fps for most television applications.

20/60/10/10 Formula-A set budget formula based on paying fees at fixed percentages of 20,60,10 and 10 percent of the negotiated price over a period of time. This usually means 20 percent for pre-production, 60 percent during period of production,10 percent is received for the delivery of the directors cut, and the last 10 percent is received for the delivery of the final print.

21CN-21st Century Network

21st Century Network (21CN)-The 21st century network is a communication system proposed by British Telecom that enables advanced multimedia services that standardizes network components through the use of multi-protocol label switching (MPLS).

2-2 Pulldown-2-2 pulldown is a process of converting 24 frames per second film to 50 fields per second video by copying the same image to 2 interlaced fields.

23B+D-Representation of ISDN's primary rate interface (PRI). A PRI is composed of 23 bearer (B) channels at 64Kbps and 1 data (D) channel at 64Kbps.

24/7-Service that is available 24 hours a day, 7 days a week.

2B+D-The combination of two 64 kbps basic rate DS0 (B) and one 16 kbps (D) signaling channel that forms the ISDN basic rate interface (BRI).

2B1Q-Two Binary, One Quarternary

2B4Q Line Coding-Two Binary to Four Quaternary line coding method used for ISDN subscriber lines (the U interface) in North America. In this method of line coding, each two binary bits are mapped into one pulse symbol having one of four

distinct voltage levels on the transmission wires. A sequential method of encoding is used to ensure that the same voltage level will not be transmitted in consecutive symbols, so that the waveform has constant alternation and will be accurately transmitted via coupling transformers and coupling capacitors.

2D Barcode-2 Dimension Barcode

2G-Second Generation

2xEVDO-Two Channel Evolution Version Data Only

3.58 MHz-The approximate frequency that is used as a sub carrier in the NTSC video signal to carry color information.

3:2 Pulldown-3:2 pulldown is a process of converting film that operates at 24 frames per second to video at 60 fields per second. The 3:2 process operates by repeating a film frame image 3 times, repeating the next film frame image 2 times and repeating this process 3 repeats and 2 repeats to create 60 images (fields) per second. 3:2 pulldown is sometimes called Telecine because Telecine was a machine that performed the pulldown conversion.

35mm Film-35mm film is 35 mm wide material that is used to store moving images. 35 mm film was commonly used for cinemas.

3G-Third Generation

3G Wireless-The third generation of technology of the mobile wireless industry. Third generation (3G) systems use wideband digital radio technology as compared to 2nd generation narrowband digital radio. For third generation cordless telephones, 3G wireless describes products that use multiple digital radio channels and new registration processes allowed some 3rd generation cordless phones to roam into other public places.

3G-324M-3G-324M is an application layer protocol that uses text format messages to setup, manage, and terminate multimedia communication sessions on mobile telephone systems.

3GPP-3rd Generation Partnership Project

3GPP File-A 3GP file media format that is defined in the 3GPP standards that are used for transferring multimedia to mobile telephones.

3GPP2-3rd Generation Partnership Project 2

3GSM-3rd Generation GSM

3rd Generation Partnership Project (3GPP)-The 3GPP oversees the creation of industry standards for the 3rd generation of mobile wireless communication systems (W-CDMA). The key members of the 3GPP include standards agencies from Japan, Europe, Korea, China and the United

States. More information about 3GPP can be found at www.3GPP.org.

3rd Generation Partnership Project 2 (3GPP2)-The 3GPP2 is a collaborative group that is working on the creation of 3rd generation industry global standards that provide for high-speed multimedia wireless services. The key members of the 3GPP2 include standards agencies from North America and Asia. The 3GPP2 is similar to the 3GPP project which is developing 3rd generation WCDMA specifications. More information about 3GPP2 and its standards can be found at www.3GPP2.org.

3WC-Three Way Calling

3xEVDO-Three Channel Evolution Version Data Only

3xRTT-Three Channel Radio Transmission Technology

4.43361875 MHz-The frequency of the sub carrier signal that contains the color information that is used in the PAL video standard.

4:2:0 Digital Video-4:2:0 digital video is a CCIR digital video format specification that defines the ratio of luminance sampling frequency as it is related to sampling frequencies for each color channel. For every four luminance samples, there is one sample of each color channel that alternates on every other horizontal scan line.

This figure the format of a 4:2:0 digital video on a display. In this example, a small portion of the video display has been expanded to show horizontal lines and vertical sample points of luminance (intensity) and chrominance (color). This example

shows that the sample frequency for luminance for 4:2:0. is 13.5 MHz and the sample frequency for color is 6.75 MHz. This format has color samples (Cb and Cr) on every other line and that the color samples occur for every other luminance sample.

4:2:2 Digital Video-4:2:2 digital video is a CCIR digital video format specification that defines the ratio of luminance sampling frequency as related to sampling frequencies for each color channel. For every four luminance samples, there are two samples of each color channel.

4B/5B Encoding-A data transmission encoding process that is used to transform each group of four bits to be represented as a five-bit symbol.

4B3T Line Code-A baseband line code (modulation and signaling structure) that transmits four bits for every three levels of symbols.

4G-Fourth Generation

4onIP-MPEG-4 as carried on Internet Protocol (IP). Part 8 of the MPEG-4 standard specifies how MPEG video is to be carried over IP networks.

50 Interlaced (50i)-Fifty interlaced fields (images) per second.

50i-50 Interlaced

56K Line-A telephone circuit that has a data transmission rate of 56 Kbps. It is sometimes called dataphone digital service (DDS).

5B/6B Encoding-A data transmission encoding process that is used to transform each group of five bits to be represented as a six bit symbol. This process is used 100BaseVG networks.

60 Interlaced (60i)-Sixty interlaced fields (images) per second.

60i-60 Interlaced

64 Clear Channel Capability-A channel that allows the end user to transmit 64 kbps without any other constraints such as a maximum ones density or number of-consecutive-zeros restrictions.

64 Level Quadrature Amplitude Modulation (64-QAM)-64 level QAM (64-QAM) is a combination of amplitude modulation (changing the amplitude or voltage of a sine wave to convey information) together with phase modulation. The use of 64 levels allows each symbol (specific combination of amplitude and phase) to represent 6 bits of information (64 different states).

64K Line-A digital communication line that allows the end user to transmit 64 kbps.

64-QAM-64 Level Quadrature Amplitude Modulation

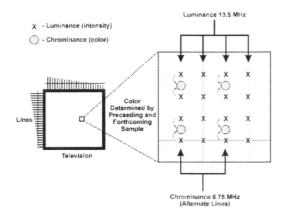

Digital Video 4:2:0 Format

66 Block-A 25 or 50 pair plastic terminal that serves as a RJ21X network interface or telephone system cross connect and mounted at the demarcation point.

8 Bit Color-Eight bit color is the ability to represent up to 256 unique colors.

8 Level PSK (8-PSK)-8 level phase shift keying (8-PSK) is a type of modulation that uses 8 different phase shifts of a radio carrier signal to represent the digital information signal. The use of 8 levels allows a symbol to represent 3 bits of information (8 different states).

8 level Vestigal Sideband (8-VSB)-8 level vestigial sideband modulation is the process of encoding data on a carrier signal using amplitude modulation and removing the redundant sidebands to provide a vestigial (partially developed) sideband signal. Each symbol in an 8-VSB system can have 8 different states (levels) that represent 3 bits of information.

802.11 Access Point (AP)-A radio access point (wireless data base station) that is used to connect wireless data devices (stations) to a wired local area network (WLAN).

802.11 Association-802.11 association is a process of registering a wireless data device (station) with a specific access point (AP) in an 802.11 specified wireless local area network (WLAN) system.

802.11 Basic Service Set (BSS)-The process used by wireless data devices to communicate with other wireless data devices or access points in an 802.11 specified system.

802.11 DeAssociation-802.11 DeAssociation is a process of un-registering (disassociating) a wireless data device (station) with a specific access point (AP) in an 802.11 specified wireless local area network (WLAN) system. This allows the access point to reuse the logical address that was assigned to the wireless data device.

802.11 Direct Sequence Spread Spectrum (DSSS) Mode-A radio access technology used in a 802.11 specified wireless local area network (WLAN) system that uses a code to spread the data signal direct to a much wider bandwidth than the data signal requires. Because each bit of the signal is spread over a wide frequency band, several spread spectrum signals can exist in the same area at the same time with minimal interference.

802.11 Distributed Coordination Function (DCF) Mode-A mode of operation in a 802.11 specified system that allows the independent operation

(distributed access control) of wireless data devices (stations).

802.11 Distribution Service (DS)-The process of distributing packets through access points (APs) and other switching devices in an 802.11 specified wireless local area network system.

802.11 Wireless LAN-The 802.11 wireless local area network (WLAN) is an industry standard developed by the IEEE for wireless network communication. It usually operates in the 2.4 GHz or 5.7 GHz spectrum and permits data transmission speeds from 1 Mbps to 54 Mbps.

802.11A-A version of the 802.11 wireless local area network (WLAN) industry standard that was developed by the IEEE for wireless network communication. It was developed to operate in the 5.7 GHz spectrum and permits data transmission speeds up to 54 Mbps.

802.11B-A wireless local area network (LAN) system that operates in the 2.4 GHz frequency band and has a data transfer rate up to 11 Mbps.

This figure shows the frequency band and radio channel size that is used in the 802.11b system. This example shows that the basic radio channel in the 802.11b system is 25 MHz wide and that the center frequency of the radio channel can be assigned to different points (channels) in the 83 MHz industrial, scientific, and medical (ISM) unlicensed frequency band. This example shows that there can be up to 3 non-interfering (non-overlapping) 802.11b radio channels operating in the same ISM frequency band.

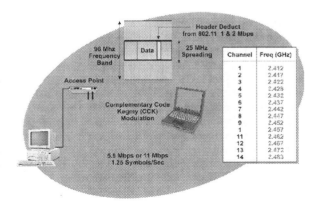

802.11B Radio

802.11d-The 802.11d industry specification defines a wireless LAN (WLAN) system that has the capability to control the channel selection and to restrict the operation of WLAN devices. 802.11d is similar to 802.11b with modifications to define allowed frequencies, power levels, and bandwidth.

802.11e-The 802.11e wireless local area network is an enhancement to the 802.11 series of WLAN specifications that added quality of service (QoS) capabilities to WLAN systems. The 802.11e specification modifies the medium access control (MAC) layer to allow the tracking and assignment of different channel coding methods and flow control capabilities to support different types of applications such as voice, video, and data communication.

802.11g-A wireless local area network (LAN) system that transmits in the 2.4 GHz frequency band and transfers data at up to 54 Mbps.

802.11h-The 802.11h specification is an enhancement to the family of 802.11 WLAN standards that helps to reduce or resolve radio interference issues. The 802.11h specification defines dynamic frequency selection (DFS) and transmit power control (TPC) that can be used to select optimum frequencies and to use only the necessary transmitter power level necessary to provide a WLAN connection reducing the interference from the WLAN system to other wireless devices or systems.

802.11i-An enhanced security protocol that is used in the 802.11 system that uses dynamically changing keys to replace the static security keys used in the original 802.11 system.

802.11k-The 802.11k specification is a proposed wireless LAN (WLAN) industry standard that defines how WLAN systems should support the control of devices within networks including devices that are roaming in other networks.

802.15 Wireless Data Personal Area Network-802.15 is an IEEE working group that specifies wireless personal area communication networks.

802.15.4-802.15.4 is an IEEE wireless technology standard that is used for short range network monitoring and control applications. 802.15.4 is called Zigbee and information on Zigbee can be found at www.Zigbee.org.

802.16 10-66 GHz Wireless MAN-802.16 is an IEEE working group that specifies broadband wireless communication systems. The 802.16 standard defines wireless broadband communication in the 10-66 GHz frequency band.

802.16A 2-11 GHz Wireless MAN-802.16A is an IEEE radio interface specification for fixed broadband wireless access systems that operate in the 2-11 GHz frequency bands.

802.16E Mobile Wireless MAN-802.16E is an IEEE radio interface specification for fixed and mobile broadband wireless access systems that operate in the 2-11 GHz frequency bands.

802.1P-The IEEE 802.1P signaling technique is an OSI Layer 2 standard for prioritizing network traffic at the data link/Mac sublayer. It can also be defined as best-effort QoS at Layer 2. 802.1P traffic is simply classified and sent to the destination without bandwidth reservations being established. 802.1p is a spin-off of the 802.1Q (VLANS) standard. The prioritization field was never defined in the VLAN standard. The 802.1P implementation defines this prioritization field.

802.1Q-The IEEE 802.1Q standard defines the operation of Virtual LAN (VLAN) Bridges that permit the definition, operation and administration of Virtual LAN topologies within a Bridged LAN infrastructure.

802.3-A network standard defined by the IEEE that specifies carrier sense multiple access with collision detection (CSMA/CD) to coordinate access to the network. This is the access method used by Ethernet networks.

802.4-The IEEE standard that defines the MAC layer for Token Bus networks.

802.5-The IEEE standard that defines the MAC layer for the Token Ring networks.

802.6-A network standard adopted by the Institute of Electrical and Electronics Engineers (IEEE) that defines the architecture of metropolitan area networks (MANs). This standard specifies the protocol that is used between high-speed data networks or network nodes.

8-PSK-8 Level PSK

8-VSB-8 level Vestigal Sideband

99.999%-Five Nines

A-Ampere

A Law-A law is the type of non-linear digital voice coding that is commonly used in Europe and other parts of the world, typically with 2.048 Mb/s digital multiplexing. The A Law coding process is used to compress the 13-bit sampling of a digitized audio signal into the equivalent of an 8-bit sample. It does this by assigning a non-uniform voltage difference to each of the consecutive binary code values, except for a small range near zero volts where uniform voltage differences are used between consecutive code values. Another non-linear voice coding system is the Mu Law coding system that is used in North America and Japan with the DS-1 (T-1) digital multiplexing system.

A/B Roll Editing-A/B roll editing is the use of a video switcher or mixer that permits a variety of media segments to be used between the source and output (such as a video recorder).

A/B Switch Box-An A/B switch box is an assembly that allows a user to select from an A or B signal source (e.g. cable television line or antenna line).

A+ Certification-A+ Certification is a certification program that is designed to ensure the competence in basic troubleshooting and repair for computer technicians. The certification is controlled by the Computer Technology Industry Association (CompTIA). A+ certification requires the passing two tests: a core exam and a specialty exam. The core exam tests the general knowledge of PCs technology including installation, configuration, software and hardware upgrading, diagnostics, maintenance, repair, interaction with customers, and safety procedures. The specialty exam tests specific operating system knowledge.

A2DP-Advanced Audio Distribution Profile

AAA-Authentication, Authorization, Accounting

AAC-Advanced Audio Codec

AAL-ATM Adaptation Layer

AAL1-ATM Adaptation Layer 1

AAL2-ATM Adaptation Layer 2

AAL-2-AAL-CU

AAL5-ATM Adaptation Layer 5

AAN-Associated Account Number

AAS-Adaptive Antenna System

AAT-Above Average Terrain

AB-Abbreviated Burst

AB-Access Burst

A-B Box-A-B box is a switching box that allows two or more computers to share a peripheral device such as a printer. It can be switched manually or through software.

Abilene-Abilene is a high-speed communication backbone network used in the Internet 2 system.

ABNF-Augmented Backus-Naur Format

Abort-Abort is a process used in data transmission to discard and ignore all transmitted bits that have been sent by the sender since the preceding flag sequence. The abort process can be invoked by a primary or secondary sending station.

ABR-Available Bit Rate

Absolute Address-Absolute address is a data memory address that is used in a computing device to identify a specific storage location. An absolute address usually requires a larger address word than a relative data address reference.

Absolute Addressing-Absolute addressing is the use of the specific identification codes or address numbers in a LAN or computer system that is permanently assigned to a storage register, location, or device. The antonym for this term is "relative addressing," a method in which the stated relative address must be added to a so-called base number value to obtain the absolute address.

Absolute Measurements-Absolute measurements is the determining of levels or quantities of a signal that are in a standardized measurement unit format.

Absorption-Absorption is the loss of energy (radio or light) that is the result of absorption within the transmission material or transmission medium. This energy is usually converted to heat.

Abstract Service Primitive (ASP)-Abstract service Primitive (ASP) is a description of an interaction between a service-provider and a service-user at a particular service boundary as defined by Open Systems Interconnection (OSI). The ASP operation is independent of the implementation. ASP is defined as part of ATM systems.

Abstract Syntax Notation number 1 (ASN.1)-Abstract syntax notation number 1 is a specification that defines how the format of an object (to the bit level) is described. It is defined in recommendation X.680 of ITU-T, ASN.1. As an example, ASN.1

is the language used in creating and defining SNMP MIBs for network management.

Abstract Test Suite (ATS)-Abstract test suite (ATS) is a set of procedures that are used to evaluate the testing procedures of products supplied by vendors.

AC-Access Carrier

AC-Access Customer

AC-Alternating Current

AC Power-Alternating Current Power

AC Signaling-AC signaling is the insertion of alternating-current signals on a communication circuit to transfer supervisory and address signaling information. Examples of AC signaling include single tone signaling using 2600 Hz or two tone signaling called multifrequency pulsing.

ACADIA Validation-ACADIA validation is a certification process that helps to ensure broadband network equipment and software reliably deliver the services and applications for system operators.

ACAP-Advanced Common Application Platform

ACC-Analog Control Channel

Accelerated Depreciation-Accelerated depreciation is a depreciation method or period of time, including the treatment given cost of removal and gross salvage, that is used when calculating depreciation deductions (asset usage) on income tax returns which is different from the depreciation method or period of time prescribed by the Commission for use in calculating depreciation expense recorded in a company's books of accounts.

Accelerator-An accelerator is a chemical additive which hastens a chemical reaction under specific conditions.

Accelerator Board-Accelerator board is an assembly or printed circuit board that is added to a computer system that enhances the performance of the processor that is installed in the computer.

Acceptable Frame Filter-Acceptable frame filter is a process used in a virtual local area network (VLAN) that qualifies (filters) frames as acceptable. This filter may be set to allow all frames or specific frames that have certain tags (control flags) set.

Acceptance Test-Acceptance test is a test that evaluates the successful operation and/or performance of an electronic assembly or communication system. Acceptance tests usually have specific operation requirements and test measurements.

Acceptance tests are often used as a final product approval and may authorize a product for production or purchase.

Access Channel-Access channel is (1-radio) a radio channel in a wireless system that coordinates the random access of mobile radios to the wireless system. (2-television) A channel on a cable television system that is dedicated for local community use.

Access Charge-Access charge is telecommunications service charges that are approved by the Federal Communications Commission (FCC) that compensate a local exchange carrier (LEC) for connection of local customers to a long distance telephone service company (IXC).

Access Continuum-Access continuum is a flexible network deployment strategy wherein the same services and operations experience can be achieved across all access network elements at the central office, remote terminal, remote node, and the customer premises through common element management, equipment compatibility, and seamless equipment upgrade.

Access Control-Access control are the actions taken to allow or deny use of the services and features of a communication system to individual users.

Access Coupler-Access coupler is a device that allows signals to enter or be extracted (access) from a transmission medium (such as fiber lines).

Access Customer (AC)-Access customer is a user that purchases end user access services (e.g. leased line) from a communication carrier.

Access Discipline-Access discipline is a process that is used by a communication device to gain access to a shared transmission medium. The access discipline can be contention resolution based (all users randomly attempt access) or can be contention free (users must wait for a specific time or token to be given to them.)

Access Domain-Access domain are the groups of communication devices within a network that can directly communicate with each other. These communication devices within a shared system (such as a LAN) operate within the LANs access domain.

Access Failure-Access failure is the termination (failure) of an attempt to access a system in any manner other than through the initiation of the user.

A

Access Fee-Access fee is a fee that is paid for the use of another network to originate, route, or terminate calls. Access fees are commonly paid by a long distance service provider to a local access provider for allowing calls to enter or terminate through the local network.

Access Gateway (AGW)-An access gateway is a device or assembly that transforms data that is received from a device or user into a format that can be used by a network. An access gateway can adjust the modulation, protocols and timing between two dissimilar communication devices or networks.

Access Line-Access line is the physical link (typically a copper wire or fiber) between a customer and a communications system (typically a central office) that allows a customer to access local and toll switched networks. Access lines may include a subscriber loop, a drop line, inside wiring, and a jack.

Access Log File-An access log file is a list of access attempts that have occurred for when potential users request access to software programs or services. The access log file is continually updated (added to) as new access events occur. Access log files can be used to analyze problems or fraudulent activities that have or may occur for applications or services.

Access Method-Access method is a set of rules that when followed allow use of a service. The ability and means necessary to store data, retrieve data, or communicate with a system. FDMA, TDMA and CDMA are examples.

Access Methods-Access methods are the processes used by a communication device to gain access (obtain services) from a system. Some systems allow communication devices to randomly compete for access (contention based) while other systems assign periods of time or setup events (such as token passing). They precisely control (non-contention based) the access times and methods.

Access Minutes-Access minutes is the length of time that telecommunications facilities are used for long distance service (interstate or international service). This is for both originating and terminating calls. Access minute timing stops when one of the parties disconnects.

Access Multiplexing-Access multiplexing is a process used by a communications system to coordinate and allow more than one user to access the communication channels within the system. There

are four basic access multiplexing technologies used in wireless systems: frequency division multiple access (FDMA), time division multiple access (TDMA), code division multiple access (CDMA) and space division multiple access (SDMA). Other forms of access multiplexing (such as voice activity multiplexing) use the fundamentals of these access-multiplexing technologies to operate.

This figure shows the common types of access channel-multiplexing technologies used in wireless systems. This diagram shows that FDMA systems have multiple communication channels and each user on the system occupies an entire channel. TDMA systems dynamically assign users to one or more time slots on each radio channel. CDMA systems assign users a unique spreading code to minimize the interference received and caused with other users. SDMA systems focus radio energy to the geographic area where specific users are operating.

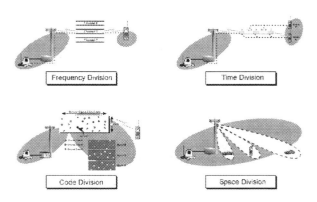

Access Multiplexing Operation

Access Network (AN)-Access network is a portion of a communication network (such as the public switched telephone network) that allows individual subscribers or devices to connect to the core network.

Access Node (AN)-Access nodes are access points or concentration points that allow one or more users to connect to a communication network. Access nodes are usually network devices that allow customers to connect to their communication devices to data networks via wires or wireless connections that use standard transmission protocols such as Ethernet or 802.11 wireless networks.

Access Number-Access number is (1-Wireless) a phone number that can be used to directly dial a cellular or PCS customer in the local wireless system when they are roaming. To contact the roaming customer, the access number is dialed first and when a tone is received, the complete mobile phone number (area code and number) of the roaming customer is entered. The local system will then page the roaming customer. Access numbers were very important in the first few years of cellular service because the systems were not automatically connected. Because most cellular and PCS systems can automatically deliver calls to roaming customers, access numbers are rarely used. (2-calling card) A telephone number of a gateway that allows customer's to enter information that permits the connection of a call or access to information services.

Access Permission-Access permission is the process of assigning permission for a user to gain access to a communication facility.

Access Point (AP)-Access point (AP) is typically, a point that is readily accessible to customers for access to a wireless or wired system. May be called a radio port or access node.

Access Priority-Access priority is a priority level (user priority level) that is assigned to users or devices within a communication network that is used to coordinate access privileges based on network activity or other factors.

Access Probe-An access probe is a process of sending a message or alert signal to an access point or a system to discover if access is possible or to request access to the system.

Access Process Parameter (APP)-Access process parameters (APP) are the parameters used or assigned to a port (logical channel point) in a packet switching, network that coordinate the operation of the port.

Access Protocol-Access protocol is the set of procedures and rules that are used by communication devices to coordinate access to a shared communication media.

Access Provider-Access provider is the company that provides and controls access to communication devices or to other networks.

Access Request Information System (ARIS)-Access request information system (ARIS) is the software system that is used by a customer service center at a carrier that processes access service orders (service activation).

Access Response Channel-Access response channel is a sub channel (logical channel) used for digital cellular systems to coordinate the random access of mobile phones that want to obtain service from the cellular or PCS system.

Access Rights-Access rights are the properties or access right assignment associated with files, directories, or communication services that define the ability of users to change, interact or access services.

Access Router Card (ARC)-An access router card (ARC) is a packet data switching device that allows users to connect different geographically located computer systems and local area data networks. The ARC can adapt between different types of traffic and different protocols integrating a data network into a single enterprise-wide network.

Access Server-An access server is a computer (or computers) that coordinate access for end users who connect to a communication system. The function of access servers can vary from simple access control to advanced call processing services.

Access Service-Access service is a service that connects customers that are located within a local access and transport area (LATA) to interexchange carriers (long distance telephone service providers).

Access Service Area-Access service area is a geographic area established for the provision and administration of communications service. An access service area encompasses one or more exchanges, where an exchange is a unit of the communications network consisting of the distribution facilities within the area served by one or more end offices, together with the associated facilities used in furnishing communications service within the area.

Access Service Group (ASG)-Access service group (ASG) is the group of switching systems associated with the subtending of a tandem switch.

Access Service Request (ASR)-Access service request (ASR) is a process or form by a local carrier to request services from a local network operator. A form for ASR was developed by the alliance for telecommunications industry solutions (ATIS) in 1984 to help define access requests for the long distance service providers.

Access Software-Access software are software programs that are used to transmit and receive information to and from communication systems.

Access Software Provider-Access software provider is a provider of software that provides or enable access control using filtering, modification, or redirecting content.

Access System-An access system is a portion of a communication system that coordinates requests for services and enables the transfer of information when authorized.

Access Tandem (AT)-Access Tandem (AT) is a high level switching system that interconnects low level (local exchange) switching systems. An access tandem can also provide access for nonconforming end offices such as for equal access to other long distance service providers.

Access Tariff-Access tariff is a tariff imposed by federal or state regulators on carriers (typically local exchange carriers) that offer access to telephone exchange services to customers or other companies.

Access Terminal-Access terminals are data input and output devices that are used to communicate with an access point or remotely located computers. Access terminals frequently consist of a keyboard, display monitor, and communication circuitry that can connect the access terminal with the remotely located computer.

Access Time-Access time is the amount of time required to retrieve information from a device or system.

Access Token-Access token is an indicator (flag) that is used in a communication network that identifies a process or user that permits access to a system or service. In Windows 2000, an access token contains the combination of security identifiers along with additional information about the user to ensure reliable and authorized access to systems and services.

Accessibility-Accessibility is the ability of network elements, such as servers, trunks, or ports that can be accessed by users within a group.

ACCH-Associated Control Channel

ACCOLC-Access Overload Class

Account-On LANs or multi-user operating systems, an account is set up for each user. Accounts are usually kept for administrative or security reasons. For communications and online services, accounts identify a subscriber for billing purposes.

Account Administrator-An account administrator is a person that is responsible or authorized to track, change and manage accounts.

Account Average-Account average is the average of assets for all telecommunications company assets associated with a particular account that include maintenance, repair experience, and salvage value expectancies.

Account Inquiry Centers (AICs)-Account inquiry centers (AICs) are call centers that receive and process calls from customers regarding the status of their accounts.

Account Lockout-In network operating systems (Noose's), an account lockout is the result of a count of the number of invalid logon attempts allowed before a user is locked out.

Account Management-Account management is the authorizing, recording, and assignment of costs to users and groups based upon their authorization and use of network resources.

Account Policy-Account policy is the set of rules or processes used by networks and multi-user operating systems that define how users are allowed to access the system based on their predefined account access privilege settings.

Accounting-Accounting is(1-general) the process of recording, assigning, and tracking the usage of resources on a network to specific accounts or users. (2-FCAPS) Accounting is one of the five functions defined in the FCAPS model for network management. The accounting level might also be called the allocation level, and is devoted to distributing resources optimally and fairly among network subscribers. This makes the most effective use of the systems available, minimizing the cost of operation. This level is also responsible for ensuring that users are billed appropriately.

Accounting System-Accounting systems is a set of accounts, rules, processes, software programs, equipment, and other mechanisms that are used to necessary to operate and evaluate a business system for an operations (business decisions), financial (investor reporting), and regulatory (taxes) perspective.

Accreditation-Accreditation is recognition from a perceived or defined authority that a product or service meets specific criteria that is beneficial (e.g. meets UL approval) or required by regulatory laws.

Accredited Standards Development Organization-Accredited standards development organization is an organization composed of members that have been accredited by an institution that is responsible for standards accreditation within an industry.

Accuracy-Accuracy is (1-General) a comparison of an actual signal or measured value as compared to the theoretical or pre-established limits. (2-Network Management) Accuracy is the measurement of interface traffic that does not result in error as used in Network Management. Accuracy can be expressed in terms of a percentage that compares the success rate to the total packet rate over a period of time.

ACD-Automatic Call Distribution

ACE-Advanced Coding Efficiency

ACELP-Algebraic Code Excited Linear Prediction

ACF-Admission Confirm

ACF-Advanced Communications Function

ACI-Access Control Information

ACK-Acknowledgment

Ack-Acknowledgment Message

ACK Channel-Acknowledgement Channel

Ack Message-Ack message is a message that is used to confirm a person or device is willing to participate in a communication session.

Acknowledge Character-Acknowledge character is a control character that is transmitted by the receiving station back to the transmitting station to indicate that the last transmission was successfully received.

Acknowledged Mode-Acknowledged mode is a communication process that requires that a receiver of information to continuously send indications to the sender that indicate it has successfully received information.

Acknowledgement Channel (ACK Channel)-Acknowledgement channel (ACK Channel) is a physical and/or a logical channel that is used to acknowledge the successful reception of data or the completion of a process or step in a communication system.

Acknowledgment Message (Ack)-Acknowledgement message is a message that is responded by a communications device to confirm a message or portion of information has been successfully received. If a communications device is supposed to send ack messages back to the originator and an acknowledgement message is not received, the system will typically resend the message. See also negative acknowledgement message (Nack).

ACL-Access Control List

ACM-Address Complete Message

ACNA-Access Customer Number Abbreviation

ACPI-Advanced Configuration and Power Interface

Acquire-Acquire is the process of obtaining a product or service through competition (e.g. acquire a channel) or assignment (token passing).

Acquisition-(1-General) Acquisition is the process of tuning and decoding a signal from a system. (2-Satellite) The process of a satellite receiver device locking on a satellite's global positioning system (GPS) signal. The acquisition includes tuning to the signal frequency and applying automatic gain control (AGC), synchronization (time alignment), and decoding (processing) of the data signal. (3-customer) Acquisition is the process of acquiring new customers from a population of prospects who were not customers in the past.

Acquisition Cost-Acquisition cost are the combined costs that are associated with marketing and adding new customers to a system or service.

Acquisition Device-(1-digital rights management) An acquisition device is a component, assembly or module that is used to enable the capture and encoding of signals or information. (2-biometrics) A biometric acquisition device is a device or hardware that is used to acquire biometric measurements.

Acquisition Point-(1-general) An acquisition point is a location where information is captured and/or modified. (2-digital rights management) An acquisition point is a location where content is encrypted and usage rights and restrictions are added.

Acquisition Point Certificate-An acquisition point certificate is an authenticated electronic key that is signed to ensure the acquisition point is compliant with required specifications.

Acquisition Time-Acquisition time is the amount of time it takes for an electronic circuit in a communication system to acquire a specific signal. Acquisition time may include radio signal scanning time in addition to the search for a specific pattern of information (synchronization signal).

ACR-Absolute Category Rating

ACR-Anonymous Call Rejection

ACR-Attenuation to Crosstalk Radio

ACS-Advanced Communications Service

Action Safe Area-Action save area is the portion

of display area where action or images will be visible. Approximately 90% of a video frame will be visible on a standard display.

Activate-Activate is the process or action (such as an operator's service screen entry) that starts or reactivates a service.

Activate/Passive Device-In telecommunications systems where a current loop is used to supply power to communications devices an activate device is capable of providing the power for the loop and a passive device is one that can only consume power from the loop for operation.

Activated Channels-Activated channels are those channels engineered at the head end of a cable system for the provision of services generally available to residential subscribers of the cable system, regardless of whether such services actually are provided, including any channel designated for public, educational or governmental use.

Activation-Activation is the process of inputting specific information into a telephone network database to authorize a service account. For a telephone network to provide services, the system must be informed of the services and account codes. Activation of services usually requires that certain customer financial criteria must also be met. After this information is input, the service or telecommunications device will become activated and the customer can request and receive services.

Activation Commission-Activation commission is a commission paid to a retailer or other entity for activating a new customer, or adding additional lines of service to an existing customer. The activation commission is typically paid after a contract with the customer is completed.

Activation Fee-Activation fee is a one-time fee that is charged for the initial setup of communication service. Activation fees are also called "setup fees."

Active-Active is (1-general) a state of a device or circuit where it provides a function. (2-component) A device or circuit that converts a signal through amplification that requires an external source of energy.

Active Hub-Active hub is a device that amplifies transmission signals in a network, allowing signals to be sent over a much greater distance than is possible with a passive hub. An active hub may have ports for coaxial, twisted-pair, or fiber-optic cable connections, as well as LEDs to show that each port is operating correctly.

Active Matrix Display-Active matrix display is a type of image display technology that uses an active matrix to display each picture element where each element has its own control transistor. Because of the active elements of the display, the response time of the display can be faster and sharper than passive types of displays. Active matrix display is also called a thin film transistor (TFT) display.

Active Member Address (AM_ADDR)-An active member address is a unique (usually temporary) address that is used to transfer information to devices which are active in a communication system. For the Bluetooth system, the active member address (AM_ADDR) is a 3 bit code that identifies one of eight devices in a Piconet.

Active Monitor-Active monitor, on a Token Ring network, is the workstation with the highest medium access control (MAC) address, which participates in the monitor contention process. The Active Monitor is responsible for maintaining the token and detecting error conditions, such as a frame that was not removed by its originator. The Active Monitor is the only station on the ring that provides a clock for ring, either 8 MHz for 4 Mbps or 32 MHz for 16 Mbps operation. The Active Monitor's presence is constantly monitored by the Standby Monitor(s) via the Active Monitor Present process, sometimes referred to as Ring Poll.

Active Monitor Present-Active monitor present is a medium access control (MAC) frame that is generated approximately every seven seconds in a token ring network, by a ring's active monitor. The message indicates that an active monitor is operational and alive on the ring. Each standby monitor on the ring monitors for the Active Monitor Present MAC frame. If one is not seen in 14 seconds, the standby monitors will assume the Active Monitor has been removed from the ring and will elect a new active monitor.

Active Server Pages (ASP)-An active server page (ASP) is a script interpreter (real time program operation) that allows the development and use of software program modules to operate on a web server (such as a Microsoft Internet Information Server).

Active Star-An active star is a central node in the centralized network that processes and/or alters information as it passes through the star node.

ActiveX-ActiveX is a development of Microsoft's COM that adds network capabilities by creating a set of component-based Internet- and intranet-oriented applications.

ActiveX Control-ActiveX control is a software control module that requires an ActiveX container (such as a spreadsheet or Web browser) to provide a specific function. ActiveX control allow database access or file accesses that can communicate with another ActiveX containers, ActiveX controls, and to interface with the underlying Windows operating system. ActiveX can directly access files using a security that contain digital certificates that authenticate the source and validity of the control.

Activity Factor-The activity factor is the ratio or percentage of time that a device is performing an operation or service. An example of activity factor is the percentage of time a communication device is actively transmitting data on a communication channel compared to the total time the device spends attached (logically connected) to the communication channel.

Activity Report-Activity report is a report of call records that contain the call identifier, date, usage amount (e.g. transmission time, kbytes of data transferred), destination address (e.g. telephone number or IP address), and other pertinent information.

ACTL-Access Customer Terminal Location

ACTS-Advanced Communications Technologies And Services

Actuator-Actuator is a device or assembly that is used to start the operation, calibration, or switching of a circuit or piece of equipment.

ACU-Alarm Controller Unit

Ad Aggregator-An ad aggregator obtains the rights from multiple content providers to resell and distribute advertising messages through other communication channels. An ad aggregator typically receives and reformats media content, stores or forwards the media content, controls and/or encodes the media for security purposes, accounts for the delivery of media and distributes the media to the systems that provide the media to specific types of customers.

Ad Bidding-Ad bidding is the process of assigning bid amounts that are associated with a specific advertising message along with the criteria that will be used to enable the bidding process. When the criteria is matched (such as matching a search word in an online search), the bid amounts are reviewed, the highest bids are selected and the advertising message(s) are displayed in the order of bid amount.

Ad Compressions-Ad compressions are the number of conversions of advertising messages (selected or automatically compressed) from a larger and/or longer version of an ad to a smaller and/or shorter version of an ad.

Ad Expansions-Ad expansions are the number of conversions of advertising messages (selected or automatically expanded) from a smaller and/or shorter version of an ad to a larger and/or longer version of an ad.

Ad Hoc (Ad-Hoc) Network-Ad hoc (Ad-Hoc) network is a wireless network (typically temporary) comprising only stations without access points that receive and retransmit between stations.

Ad Hoc Query-Ad hoc query is a filtering action or query that is created dynamically or temporarily.

Ad Impression-An ad impression is the presentation of an advertising message or image to a media viewer.

AD Insertion-Ad insertion is the process of inserting an advertising message into a media stream such as a television program. For broadcasting systems, Ad inserts are typically inserted on a national or geographic basis that is determined by the distribution network. For IP television systems, Ad inserts can be directed to specific users based on the viewer's profile.

Ad Insertion Module-Ad insertion modules a process used in cable television or broadcast radio networks that allow the insertion of advertising during pre-determined time segments. This process allows different advertising messages to be inserted in different geographic regions.

Ad Insertion Server-An ad insertion server is an application server that inserts advertising media during pre-determined time segments. This process allows different advertising messages to be sent to specific viewers.

Ad Selection-Ad selection is the process of selecting ("clicking") a link on a link, button or graphic image in an advertising message.

Ad Server-Ad servers are computers that receive requests for advertising media (often from an ad splicer), setup a communication session to the requesting media client and provides the downloading or continuous transmission (streaming) of advertising media.

Ad Space-Ad space is the amount of display space that is dedicated for an ad on a media display (such as an Internet web page).

Ad Splicing-Ad splicing is the process of merging an advertising message into a media stream such as a television program.

Ad Syndication-An ad syndication is the promotion of an advertising message by group of companies in the media industry that agree to work together to promote ads that are accepted by the syndication. Companies in the ad syndication agree to promote ads and share information that is necessary to accomplish advertising objectives.

Ad Window-An ad window is a portion of a screen area that is used to display information associated with a specific advertising message or ad unit.

Adaptation Field-An adaptation field is a portion of a data packet or block of data that is used to adjust (define) the length or format of data that is located in the packet or block of data.

Adaptation Layer (AL)-Adaptation layer (AL) is the processes and protocols that are used to translate information from a layer (such as a transport layer) into a size and format that can be used by another layer (such as an application layer).

Adapter-(1-computer board) An adapter is a printed circuit board that plugs into a computer's expansion bus to provide added capabilities. Common adapters include video adapters, joy-stick controllers, and I/O adapters, as well as other devices, such as internal modems, CD-ROMs, and network interface cards. One adapter can often support several different devices. (2-cable) A cable or connector adapter converts and/or adjusts the physical and/or electrical properties from one connection to another connection.

Adapter Board-Adapter board is an assembly or circuit board that can be inserted into system or equipment to allow added functionality or testing.

Adapter Support Interface (ASI)-Adapter support interface (ASI) is a communication driver specification that was developed by IBM for networking using IEEE 802.5 token-ring systems.

Adaptive Antenna System (AAS)-An adaptive antenna system allows a transmitter to transmit to focused radio beams to increase the transmission range, reduce interference and increase signal quality. When an AAS system is used to allow multiple users to communicate with the same transceiver (multiple beams), it is called spatial division multiple access (SDMA).

Adaptive Channel Allocation-Adaptive channel location is a process of allocating communication channels dynamically based on specific criteria (e.g. system capacity limitations).

Adaptive Differential Pulse Code Modulation (ADPCM)-Adaptive differential pulse code modulation is a process of converting analog voice signals into encoded digital signals through the use of predictive codes that are created by analyzing the previous digital audio signals. ADPCM is derived from the original pulse code modulation (PCM) system that commonly represents an analog signal as 64 kbps (called a DS0). ADPCM systems commonly provide digital signals at 32 kbps and 16 kbps.

Adaptive Equalizer-An adaptive equalizer is a signal processing section in a receiver that can analyze the received signal to adjusts the signal reception to compensate for dynamically changing transmission line conditions (such as the elimination or adding of delayed or multipath signals).

Adaptive Jitter Buffer-An adaptive jitter buffer dynamically adjusts the amount of jitter buffering to take into account the variable delay of a network (such as the Internet) over time. When network conditions allow for low delay transfer of audio signals, the jitter buffer is made very small to reduce any perceived audio delays. When network conditions result in long delays in the transfer of audio signals, the jitter buffer is expanded in size to take this delay into account and provide the best possible audio quality under these conditions. Also known as a Dynamic Jitter Buffer.

Adaptive Modulation-Adaptive modulation is the process of dynamically adjusting the modulation type of a communication channel based on specific criteria (e.g. interference or data transmission rate).

Adaptive Multirate Codec (AMR)-Adaptive multi-rate is a speech compression process that offers multiple speech coding rates used in third generation (3G) wireless systems. The use of AMR allows a lower bit rate (higher compression rates) coding process to be used when system capacity is limited and more users need to be added to the system.

This figure shows the basic adaptive multi-rate speech coding process. This diagram shows that the WCDMA system allows for the use of different speech coding processes and that the system may dynamically control which speech process a used dependent on the needs of the system. This example shows that the speech coding process can change as often as every 20 msec speech frame. The selection of speech coder is primarily determined

Adaptive Predictive Coding (APC)

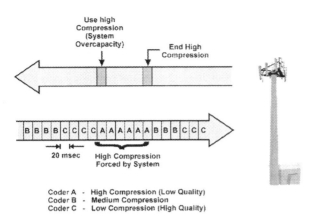

Use high Compression (System Overcapacity)

End High Compression

| B | B | B | B | C | C | C | A | A | A | A | A | B | B | B | C | C | C |

20 msec

High Compression Forced by System

Coder A - High Compression (Low Quality)
Coder B - Medium Compression
Coder C - Low Compression (High Quality)

Adaptive Multi-Rate (AMR) Speech Coder

by the channel quality. However, this example shows that the system can instruct the speech coder to use high compression process (lower audio quality) during periods of system overcapacity.

Adaptive Predictive Coding (APC)-Adaptive predictive coding (APC) is an analog-to-digital conversion process that uses a 1-level or multilevel sampling system for which the system adaptively predicts the future values based on the past values of the quantified signals.

Adaptive Routing-Adaptive routing is a software mechanism that allows a network to reroute messages dynamically, using the best available path, if a portion of the network fails.

Adaptive Spare-Adaptive spare is a spare or standby piece of equipment in a communication network that can replace a failed piece of equipment and adaptively change its configuration or performance to match the characteristics of the failed piece of equipment. Use of adaptive spares reduces the number of backup equipment assemblies that are required in a communications network.

Adaptive Speed Leveling-Adaptive speech leveling is a modem technology that allows a modem to respond to changing line conditions by changing its data rate. As line quality improves, the modem attempts to increase the data rate; as line quality declines, the modem compensates by lowering the data rate. Also known as adaptive equalization.

ADC-Analog to Digital Converter

ADCCP-Advanced Data Communication Control Procedures

Add-Drop Multiplexer (ADM)-Add-drop multiplexers (ADM) are a circuit or assembly that inserts (adds) or extracts (drops) signals from a higher speed communication line such as adding a DS0 channel into a DS1 or E1 communication line. In the Synchronous Optical Network (SONET), it is a network element that provides access to all or some subset Synchronous Transport Signal (STS) line signals contained within an Optical Carrier Level N (OC-N).

Added Channel Framing-Added channel framing is a channel frame process that allows signal elements to occupy consecutive time slots. Added channel framing is also called bunched frame alignment.

Added Digit Framing-Added digit framing is a channel frame process that allows signal elements to occupy non-consecutive digit time slots. Added channel framing is also called distributed frame alignment signal.

Added Main Line Carrier-Added main line carrier is a transmission system that uses two analog carrier signals (frequencies) to provide two telephone channels over one telephone line (2 wires).

Additions (Adds)-Additions (Adds) are the number of customers (subscribers) or devices that are added to a system over a time period.

Additive White Gaussian Noise (AWGN)-Additive white Gaussian noise (AWGN) is a signal that approximates the properties of certain types of "noise" signals normally occurring in nature that has an even distribution of power density across its frequency range. AWGN is random and unpredictable in theory. It is used for mathematical or laboratory analysis of the performance of systems in the presence of undesired signals such as "noise." The term "noise" is used even when the frequency range of the AWGN is not in the audible range to the human ear, or the signal is electrical rather than audible. Audible white noise is used intentionally in certain situations like offices or libraries as light background noise to soothe and placate such as in an office environment, or to prevent people from hearing and being disturbed by other types of noises.

Add-On-Add-on is a telephone call center sales process designed to encourage customers to purchase additional products or services secondary to the primary sale. Occasionally, it is incorrectly referred to as up-selling, which means the quantity or quality of the primary product is offered at a premium and higher cost.

Add-On Conference-Add-on conference is a conference call feature (usually in a PBX system) that allows additional participants to be added to the conference call. To add-on a conference participant, a participant (or moderator) places the existing call or on hold, obtains system dial tone, and connects to the add-on participant. After the new participant has agreed to participate, the originating participant reactivates (re-connects) to the conference call in progress.

Address-Address is a grouping of numbers that uniquely identifies a station in a local area network, a location in computer memory, a house on a street, etc. For a local area network station or a computer memory location, electronic logic can be arranged to ignore messages not bearing the appropriate address and accept messages that do bear the appropriate address.

Address Bus-Address buss is the electronic channel, usually from 20 to 64 lines wide, used to transmit the signals that specify locations in memory.

The number of lines in the address bus determines the number of memory locations that the processor can access, because each line carries one bit of the address. A 20-line address bus (used in early Intel 8086/8088 processors) can access 1MB of memory, a 24-line address bus can access 16MB, and a 32-line address bus can access more than 4GB. A 64-line address bus (used in the DEC Alpha APX) can access 16GB.

Address Classes-Address classes, in a 32-bit IP address, is the number of bits used to identify the network and the host vary according to the network class of the address, as follows:

Class A is used only for very large networks. The high-order bit in a Class A network is always zero, leaving 7 bits available to define 127 networks. The remaining 24 bits of the address allow each Class A network to hold as many as 16,777,216 hosts. Examples of Class A networks include General Electric, IBM, Hewlett-Packard, Apple Computer, Xerox, Digital Equipment Corporation, and MIT. All the Class A networks are in use, and no more are available.

Class B is used for medium-sized networks. The 2 high-order bits are always 10, and the remaining bits are used to define 16,384 networks, each with as many as 65,535 hosts attached. Examples of Class B networks include Microsoft and Exxon. All

Class B networks are in use, and no more are available.

Class C is for smaller networks. The 3 high-order bits are always 110, and the remaining bits are used to define 2,097,152 networks, but each network can have a maximum of only 254 hosts. Class C networks are still available.

Class D is a special multicast address and cannot be used for networks. The 4 high-order bits are always 1110, and the remaining 28 bits allow for more than 268 million possible addresses.

Class E is reserved for experimental purposes. The first four bits in the address are always 1111.

Address Complete Message (ACM)-Address complete message (ACM) is an ISDN User Part trunk signaling message by which a call's destination SSP (switch) acknowledges an initial address message.

Address Field-Address field is a section of a message, generally at the beginning, in which the addresses of the message source and destination are found.

Address Management-Address management is the process of assigning and controlling unique identifying numbers or names for devices in communications system.

Address Management Protocol (AMP)-Address management protocol (AMP) coordinates the assignment, maintenance (changing), and availability of device addresses.

Address Resolution-Address resolution is a process that is used to specifically identify differences between computer addressing schemes. The address resolution process may be implemented by mapping channels on layer 3 (network layer) to specific addresses on layer 2 (data link layer).

Address Resolution Protocol (ARP)-Address resolution protocol is a protocol within TCP/IP (Transmission Control Protocol/Internet Protocol) and AppleTalk networks that allows a host to find the physical address of a node on the same network when it knows only the target's logical or IP address.

Under ARP, a network interface card contains a table (known as the address resolution cache) that maps logical addresses to the hardware addresses of nodes on the network. When a node needs to send a packet, it first checks the address resolution cache to see if the physical address information is already present. If so, that address is used, and network traffic is reduced; otherwise, a normal ARP request is made to determine the address.

Address Signaling-Address signaling is a process of sending the signaling message that includes the dialed telephone number to a exchange carrier. For the public switched telephone network (PSTN), this is performed by dial pulsing (rotary phone) or touch-tone (TM) signals.

Address Signals-Address signals identify call destination information or the digits that are dialed by a caller. The different types of address signaling include Dial Pulse, Multi-Frequency (MF), and Dual-Tone Multi-Frequency (DTMF).

Address Translation Gateway-Address translation gateway is a gateway that can convert the address format (field and/or physical layer structure) from one network to another.

Addressability-Addressability is the ability of communication system to use specific or group addresses to control the distribution of programs from a central location.

Addressable Advertising-Addressable advertising is the communication of a message or media content to a specific device or customer based on their address. The address of the customer may be obtained by searching viewer profiles to determine if the advertising message is appropriate for the recipient. The use of addressable advertising allows for rapid and direct measurement of the effectiveness of advertising campaigns.

This diagram shows how addressable advertising may operate. In this example, a program source is sent with an ad insert period. When the video dis-

tribution point detects the beginning of the ad insert period, it may select and insert ads for specific customers. This allows different ads to be inserted for different viewers, even in the same geographic area.

Addressable IPTV Advertising-Addressable IPTV advertising is the process of enabling separate streams to be sent to different TV sets during the same advertising time slot. Because of this, ads can be tailored according to individual demographics; location; interests; viewing habits; time of day; language and a raft of other factors.

Addressing Authority-Addressing authority is the authority responsible for the unique assignment of Network addresses within a network address domain.

Addressing Domain-Addressing domain is the addresses within each level of network hierarchy where every address is part of a group (domain). Domain addresses are hierarchical. If an addressing domain is part of a hierarchically higher addressing domain (which wholly contains it), the authority for the lower domain is authorized by the authority for the higher domain to assign addresses from the lower domain.

Addressing Space-Addressing space is the amount of memory (range of addresses) available to a computer operating system for software control or processing of data.

Adds-Additions

Adds, Moves and Changes (AMC)-Adds, moves and changes is the process of adding, changing the configuration of equipment and removes (disconnection) of equipment and services for new, existing, and terminated customers.

Adhoc Group-An adhoc group is a temporary group of members or companies that is typically formed to solve a specific problem or function.

Ad-Hoc Network-Ad Hoc

Adjacency-Adjacency is the relationship between neighboring (adjacent) routers that can exchange (swap) routing information.

Adjacent-Adjacent is the term for network devices or communication channels that are electrically or physically located next to each other.

Adjacent Channel-Adjacent channel is a communication channel that has a carrier frequency that is located immediately above or below the frequency of another communication channel.

This diagram shows the relative spacing of adjacent radio communication channels. Additional

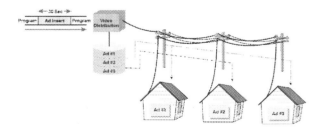

Addressable Advertising

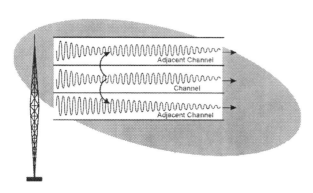

Adjacent Channel System

radio channels are often spaced at bandwidth intervals from the main carrier channel. Adjacent radio channels that are located next to a channel are called adjacent channels.

Adjacent Channel Interference-Adjacent channel interference occurs when the RF power from an adjacent channel overlaps into the frequency band of the channel that is being used by a radio device. This example shows that adjacent channel interference occurs when one radio channel interferes with a channel next to it (e.g. channel 412 interferes with 413). Although each radio channel has a limited amount of bandwidth (10 kHz or 30 kHz wide in this example), some radio energy is transmitted at low levels outside this band. A radio transmitter that is operating at full power can produce enough low-level radio energy outside the channel bandwidth to interfere with radios operating on adjacent channels. Because of alternate channel interference, radio channels that are used

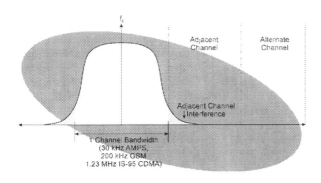

Adjacent Channel Interference Operation

on the same system cannot usually be spaced adjacent to each other in a single radio tower (e.g. channel 115 and 116). A channel separation of 3 channels is typically sufficient to protect most radio channels from adjacent channel interference in mobile communication systems.

Adjudicative Proceeding-Adjudicative proceedings involve future rates or practices, initiated upon the Commission's own motion or upon the filing of an application, a petition for special relief or waiver, or a complaint or similar pleading that involves the determination of rights and responsibilities of specific parties.

ADM-Add-Drop Multiplexer

Administered Roaming-Administered roaming is the process of managing the access rights for customers that move from one service provider's system service area to another service provider's service area and obtain service.

Administrative Distance-Administrative distance is a term used by Cisco Systems, Inc., to express the integrity of a routing-information source. Administrative distance is expressed as a value in the range 0 through 255; the higher the value, the lower the quality of the routing information.

Administrative Unit (AU)-Administrative unit is an information element of a communication system (such as a frame in a SDH system) that contains information that allows the cross-connection and switching of virtual channels (VCs).

Administrator Account-Administrator account is an account on a communications system or software program (such as Microsoft Windows) that is provided with a level of authority and permission that allows the administrator to assign and remove permission to users or groups within the system.

Admission-Admission is the process of a device requesting and receiving authorization to obtain service from a communication system or network.

Admission Confirm (ACF)-Admission confirm (ACF) is the process or message that confirms a device or service is being admitted to a network. An ACF message is defined in H.225.

ADPCM-Adaptive Differential Pulse Code Modulation

ADSI-Analog Display Services Interface

ADSL-Asymmetric Digital Subscriber Line

ADSL Forum-ADSL is a forum that was started in 1994. This forum assists manufacturers and service providers with the marketing and development of ADSL products and services. The ADSL forum has been renamed the DSL forum.

ADSL Modem-ADSL modem is an electronics assembly device that modulates and demodulates (MoDem) asymmetric digital subscriber line (ADSL) signals. ADSL signals are usually transmitted on a twisted pair of copper wires.

An ADSL modem may be in the form of an internal computer card (e.g. PCI card) or an external device (Ethernet adapter). Most ADSL modems have the ability to change their data transfer rates based on the settings that are programmed by the DSL service provider and as a result of the quality of the communication line (e.g. amount of distortion).

ADSL Router-ADSL router is a packet routing device that is commonly used to interfaces a computer or other residential data communication devices with an ADSL telephone line. The use of a router (as opposed to a hub) allows multiple computers within a home to be assigned different Internet protocol (IP) addresses.

ADSL Transmission Unit - Remote (ATU-R)-ADSL transmission unit-remote (ATU-R) is an advanced modem that provides for asynchronous digital subscriber line (ADSL) multi-megabit data rates over unshielded twisted pair (UTP) of copper wires. The ATU-R is usually located at a customer's premises.

The ATU-R can be in various configurations including and internal computer modem (PCI bus), external modem that connects to the Universal Serial Bus (USB), or a bridge device that converts ADSL signals to an 10BastT or 100BaseT Ethernet form.

ADSL2-Asymetric Digital Subscriber Line 2

ADSL-Lite-ADSL-Lite is a limited version of the standard ADSL transmission system. This limited version of ADSL allows for a simpler filter installation that can often be performed by the end user. The limitation of ADSL-Lite is a reduced data transmission rate of 1 Mbps instead of a maximum rate of 8 Mbps.

This figure shows that an ADSL-lite system is similar to the ADSL network with the primary difference in how the end user equipment is connected to the telephone network. The ADSL-lite system does not require a splitter for the home or business. Instead, the end user can install microfilters between the telephone line and standard telephones. These microfilters block the high speed data signal from interfering with standard tele-

phone equipment. The ADSL-Lite end user modem contains a filter to block out the analog signals.

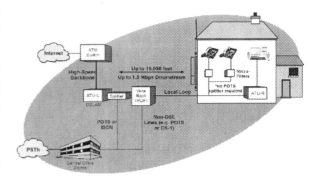

ADSL Lite System

Advanced Audio Codec (AAC)-Advanced audio codec (AAC) is a lossy audio codec standardized by the ISO/IEC Moving Picture Experts Group (MPEG) committee in 1997 as an improved but non-backward-compatible alternative to MP3. Like MP3, AAC is intended for high-quality audio (like music) and expert listeners have found some AAC-encoded audio to be indistinguishable from the original audio at bit rates around 128 kbps, compared with 192 kbps for MP3.

Advanced Coding Efficiency (ACE)-Advanced coding efficiency is a process that provides increased compression specific types of media. ACE is used in the MPEG-4 system to more efficiently characterize and compress arbitrarily shaped objects.

Advanced Common Application Platform (ACAP)-Advanced common application platform is an industry middleware standard used in television systems to allow the user to access additional interactive services such as Internet browsing and electronic programming guides. ACAP development is overseen by ATSC and more information can be found about ACAP at www.ATSC.org.

Advanced Communications Function (ACF)-Advanced communications function is a group of programs developed by IBM that allows the sharing of computer resources by communications links. This allows the interconnection of two or more domains into a single network that can manage multiple domains.

Advanced Core-The Advanced Core visual profile provides scalability for still textures and natural video in the MPEG system.

Advanced Data Communication Control Procedures (ADCCP)-Advanced data communication control procedures (ADCCP) is the American version of HDLC. Bit-oriented, link layer protocol established as a standard by ANSI.

Advanced Encryption Standard (AES)-Advanced encryption standard (AES) is a data encryption standard promoted by the United States government and based on the Rijndael encryption algorithm. The AES standard is supposed to replace the Data Encryption Standard (DES).

Advanced Intelligent Network (AIN)-Advanced intelligent networks (AIN's) are telecommunications networks that are capable of providing advanced services through the use of distributed databases that provide additional information to call processing and routing requests.

In the mid 1980's, Bellcore (now Telcordia) developed a set of software development tools to allow companies to develop advanced services for the telephone network. The advanced intelligent network (AIN) is a combination of the SS7 signaling network, interactive database nodes, and development tools that allow for the processing of signaling messages to provided for advanced telecommunications services.

The AIN system uses a service creation environment (SCE) to created advanced applications. The SCE is a development tool kit that allows the creation of services for an AIN that is used as part of the SS7 network. A service management system (SMS) is the interface between applications and the SS7 telephone network. The SMS is a computer system that administers service between service developers and signal control point databases in the SS7 network. The SMS system supports the development of intelligent database services. The system contains routing instructions and other call processing information.

To enable SCPs to become more interactive, intelligent peripherals (IPs) may be connected to them. IPs are a type of hardware device that can be programmed to perform a intelligent network processing for the SCP database. IPs perform processing services such as interactive voice response (IVR), selected digit capture, feature selection, and account management for prepaid services.

To help reduce the processing requirements of SCP databases in the SS7 network, adjunct processors (APs) may be used. APs provide some of the database processing services to local switching systems (SSPs).

This figure shows the basic structure of the AIN. Companies that want to enable information services use the SMS to interface to SCP databases within the SS7 network. This diagram shows how a prepaid calling card company manages information in a signaling end point (SEP) database. The SEP database communicates to the SS7 network through a SSP. SCE tool kit. The SEP is connected to an interactive voice response (IVR) unit that prompts callers to enter the personal identification number (PIN). The IP then reviews the account and determines available credit remains and informs the SS7 network of the destination number for call routing.

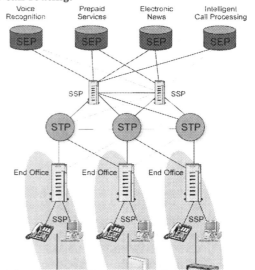

Advanced Intelligent Network (AIN) System

Advanced Peer-To-Peer Networking (APPN)-Advanced peer-to-peer networking (APPN) is IBM's Systems Network Architecture (SNA) protocol, based on APPC (Advanced Program-to-Program Communications). APPN allows nodes on the network to interact without a mainframe host computer and implements dynamic network directories and dynamic routing in a SNA network.

APPN can run over a variety of network media, including Ethernet, token ring, FDDI, ISDN, X.25, SDLC, and higher-speed links such as B-ISDN or ATM.

Advanced Real Time Simple-Advanced real time simple profile is the selection of video compression methods and protocols work well for video conferencing and other real-time multimedia applications.

Advanced Replacement-Advanced Replacement is a process that allows the end user or repair facility to obtain a replacement component (transmitter, receiver, software, etc.) before returning the defective equipment or software. This process usually requires the requesting person or company to obtain a reference number for the advanced replacement component that is requested. The person or company often uses the box from the replacement equipment to return the defective equipment or software.

Advanced Scalable Texture-Advanced scalable texture profile is the selection of video compression and protocols that allows for the sending of image textures in an MPEG system.

Advanced SCSI Programming Interface (ASPI)-Advanced SCSI Programming Interface (ASPI) is the standardized interface for the management and control of SCSI2 devices within a computer system. It is specified by the ANSI standard X3.131-1994.

Advanced Simple-Advanced simple profile is a selection of video compression and protocols in the MPEG-4 system that ad ¼ pel motion compensation, additional quantization tables and global motion compensation to the standard simple profile.

Advanced Streaming Format (ASF)-Advanced streaming format is a Microsoft digital multimedia file format that is used to stream digital audio and digital video. ASF is branded as Windows Media and the file extensions it may use include .WMV or .WMA. ASF files have the ability to synchronize digital audio and digital video along with managing other forms of media.

Advanced Streaming Index (ASX)-Advanced streaming index (ASX) is a listing of the media files, their locations and the necessary parameters to setup multimedia communication sessions.

Advanced Technology Attachment (ATA)-Advanced Technology Attachment (ATA) is an interface system for disk drive interfaces based on ANSI standard x3T10. It is usually called Integrated Drive Electronics (IDE).

Advanced Television (ATV)-Advanced television is television technology that provides audio and video signals that have quality that is better than the quality of existing television broadcast systems (e.g. NTSC or PAL).

Advanced Television Enhancement Forum (ATVEF)-Advanced Television Enhancement Forum (ATVEF) is a group of companies and industry expert that work towards the creation of industry standards for combining Internet content broadcast television.

Advanced Television Services-Advanced Television Services are television services that are provided using alternative technologies (such as digital television). Advanced television services are defined the report: "Advanced Television Systems and Their Impact Upon the Existing Television Broadcast Service", MM Docket 87-268.

Advanced Television Systems Committee (ATSC)-The Advanced Television Systems Committee (ATSC) is an international non-profit organization that assists with the development of advanced television technologies and systems. The membership of the ATSC represents a wide range of broadcast and consumer electronics segments including broadcasting, motion pictures, computers, consumer electronics, and satellite. ATSC was founded in 1982.

ATSC industry standard include digital television (DTV), high definition television (HDTV), standard definition television (SDTV), data broadcasting, surround sound audio, and satellite broadcasting.

Advanced Video Coding (AVC)-Advanced video coding is a video codec that can be used in the MPEG-4 standard. The AVC coder provides standard definition (SD) quality at approximately 2 Mbps.

Advertising-(1-general) The communication of a message or media content to one or more potential customers. (2-Data Networks) The process by which services on a network inform other devices on the network of their availability. Novell NetWare uses the Service Advertising Protocol (SAP) for this purpose. Such a process is used in packet networks to inform switches and router of new routes and route changes.

Advertising Campaign-Advertising campaigns are marketing activities designed to send specific advertising messages to customers about products, services, and options offered by a company.

Advertising Elasticity-Advertising elasticity is the connection between advertising cost or budget to the resulting change of sales or effectiveness of the advertising message.

Advertising on Demand (AOD)-Advertising on

demand is a service that enables service providers and/or end users to interactively request and receive advertising messages. These advertising on demand services are from previously stored media (e.g. advertising video clips) or have a live connection (advertising events or programs that are in real time).

Advertising Plan-An advertising plan is an outline of the objectives, processes and key tasks that a company or person is planning to perform to communicate messages or offers to one or more potential customers.

Advertising Reports-Advertising reports are tables, graphs or images that represent specific aspects of advertising campaigns or the information or data that is created from advertising campaigns.

Advertising Research-Advertising research is the process of gathering and analyzing information that is used to determine the potential or achieved results for advertising campaigns.

Advice Of Charge (AOC)-Advice of charge is the ability of a telecommunications system to advise the user of the actual costs of telephone calls either prior or after calls are made or services are used. For some systems, (such as a mobile phone system) the AOC feature is delivered by short message service.

Advisory Tones-Advisory tones are audio tones that are provided from the telephone system to inform the customer of the call status or change in call status. Advisory tones include busy, dialtone, ringing, fast-busy, call-waiting, and other tones.

Adword Marketing-Adword marketing is the process of selecting key words (adwords) and using them in a marketing program that has advertising messages appear when a user has selected or searched for an adword, portion or an adword or a related word (content match).

AEB-Analog Expansion Bus

Aerial Cable-Aerial cables are cables that can self support itself between attachments on poles, buildings or other fixed objects. Some aerial cable hangs by its own strength while others are attached (lashed) to a supporting guidewire (called a "messenger" wire).

Aerial Lashing-Aerial lashing is the attachment of a cable to a support strand by using helically wrapping materials such as dielectric filament or steel wire to hold the new line to another line.

Aerial Plant-Aerial plant is the physical property and facilities of a telephone, cable or company that uses a transmission system that is mounted above the ground. Aerial plant property includes the telephone poles, transmission cables, amplifiers, and channel multiplexing equipment.

Aerial Telephone Cable-Aerial plant is the physical property and facilities of a telephone company that is mounted above ground. The majority of aerial plant facilities are the 50-600 pair cables that are strung from utility poles.

AERM-Alignment Error Rate Monitor

AES-Advanced Encryption Standard

AES-Aircraft Earth Station

Affiliate-(1-general) A company or person that owns, controls or is owned or controlled by another company or person. Regulations may specify the definition of "affiliate" based on an equity percentage of ownership. Regulations may limit the maximum amount of ownership (e.g. 10%) by an affiliate that is involved in related industries (such as television and radio system ownership) or the amount of ownership by a foreign person or company. (2-marketing) A company or person who markets the products or services of another company or person in return for a fee or commission.

Affiliated Companies-Affiliated companies are the companies that directly or indirectly own or control a company or resource. Regulations may specify the definition of an "affiliated company" based on an equity percentage of ownership. Regulations may limit the types of companies that may participate as an affiliated company based on an amount of ownership (e.g. 10%) and the actions these companies may perform or may be required to perform because they are considered an affiliate company.

Affiliation Through Common Facilities-Affiliation through common facilities is an affiliation that results when a company or person shares resources (such as a communication system or office space) with other companies or people. This affiliation is usually visible when more than one person or company who uses the shared resource can change the shared resources that have potential control of others involved in the use of the resource.

Affiliation Through Common Management-Affiliation through common management is the affiliation between entities (corporation or individuals) that arises where agents of the company serve as a controlling element of the management or board of directors of another entity.

Affiliation Through Contractual Relationships-Affiliation through contractual relationships is the affiliation between entities (corporation or individuals) that arises where the control of agents of an entity is dependent upon contractual terms of another entity.

Affiliation Through Stock Ownership-Affiliation through stock ownership is the affiliation between entities (corporation or individuals) that arises where equity owners (shareholders) own or have the power to control (e.g. vote proxies) of more than 50 percent of the voting stock.

AFH-Adaptive Frequency Hopping

AFP-AppleTalk Filing Protocol

AFS-Andrews File System

Aftermarket-Aftermarket is the market for related hardware, software, and peripheral devices created by the sale of a large number of computers of a specific type.

AGC-Automatic Gain Control

AGCH-Access Grant Channel

Agency-(1-general regulatory) A commission, board of Commissioners, Committee, or other group of commissioners who are authorized to act on behalf of the commission or regulatory department. (2-communications) The Federal Communications Commission (FCC) agency of the U.S. Government as defined by section 105 of title 5 U.S.C., the U.S. Postal Service, the U.S. Postal Rate Commission, a military department as defined by section 102 of title 5 U.S.C., an agency or court of the judicial branch, or and an agency of the legislative branch, including the U.S. Senate and the House of Representatives.

Agency Head-Agency head is the head (chairman) of the Federal Communications Commission agency.

Agent-(1-general) A person or a device that performs tasks for the benefit of someone or some other device. (2-software) A program that performs a task in the background and informs the user when the task reaches a certain milestone or is complete.

A program that searches through archives looking for information specified by the user. A good example is a spider that searches Usenet articles. Sometimes called an intelligent agent.

In SNMP (Simple Network Management Protocol), a program that monitors network traffic.

In client-server applications, an agent is a program that mediates between the client and the server.

Agent Channel-Agent Channel is the distribution channel that makes use of a third party sales force (individual and corporate) to deliver products and services directly to customers.

Aggregate Customer Information-Aggregate customer information is collective data that relates to a group or category of services or customers, from which individual customer identities and characteristics have been removed.

Aggregate Route-Based IP Switching (ARIS)-Aggregate route-based IP switching (ARIS) is a process that is used to establish virtual circuits though switched networks through a network without the need to make switching (routing) decisions at each node through the use of tags (on each packet) that are used to guide the packets through the virtual circuits. Some of the protocols used in the ARIS system include Open Shortest Path First (OSPF) and Border Gateway Protocol (BGP).

Aggregated Link-Aggregated link is a process of combining two or more physical links to provide a single high-speed interface to higher layer protocol layers. This process is also called inverse multiplexing.

Aggregation-Aggregation is the process of combining multiple services or communication circuits into a higher-capacity service or system.

Aggregation Device-An aggregation device is used to combine (concentrate) two or more communication channels onto a higher-speed communication channel.

Aggregator-(1-service provider) A company or service provider that performs the operations required to make multiple physical links function as a combined (aggregated) link. Aggregators typically purchase network services in discounted bulk quantities and pass along the savings or support services to smaller users of the services. (2-billing) A billing aggregator is a company that gathers billing records from one or more companies and posts them to another billing system. An example of a billing aggregator is the gathering of billing of information from many companies that provide information or value added services (e.g. news or messaging services) and transferring these to another basic services carrier for direct billing to a customer.

Aging Margin-Aging margin is the amount of signal loss that can occur over a period of time, usually expressed in decibels, that a signal in a communication path can accept while providing an expected quality level of service.

Aging Process-Aging process, within the Spanning Tree Protocol, is the process that removes dynamic entries from the filtering database when the stations with which those entities are associated have been inactive for a specific time (the aging time).

Aging Time-Aging time, within the Spanning Tree Protocol, is the time after which a dynamic filtering database entry will be removed if its associated station has been continuously inactive.

A-GPS-Assisted Global Positioning System

AGRAS-Air-Ground Radiotelephone Automated Service

AGW-Access Gateway

AI-Artificial Intelligence

AICH-Acquisition Indication Channel

AICs-Account Inquiry Centers

AID-Attention Identification

AIFF-Audio Interchange File Format

AIN-Advanced Intelligent Network

AIOD-Automatic Identification of Outward Dialing

AIP-Abandoned in place

AIP-Application Infrastructure Service Provider

AIR-Allowed Information Rate

Air Space Coaxial Cable-Air spaced coaxial cable is a transmission line that uses air as the dielectric material between conductors. The conductor(s) may be separated from the shield by series of spacers or spiral assembly.

AirTime-(1-radio) Air time is the length a communication service (typically radio transmission time) is used. (2-broadcasting) Time that a person or media source is being transmitted on a broadcast network.

AIX-Advanced Interactive Executive

AK-Authorization Key

A-key-Authentication Key

AL-Adaptation Layer

Alarm-Alarm is an audible or visible indication of a trouble condition. Alarms are classified as minor, major, or critical, depending on the degree of service degradation or disruption.

Alarm Controller Unit (ACU)-Alarm controller unit (ACU) is a control unit that coordinates the reporting of remote alarm information. A common use of an ACU is on a T1-D4 channel bank that used in subscriber loop carrier (SLC) system that interfaces both central office (CO) and remote terminals.

Alarm Indication Signal-Alarm indication signal is an indication signal (such as a code or voltage) that indicate that a failure has been detected on a communication assembly or the failure of a process/program operation.

Alarm Monitoring Service-Alarm monitoring system is a service or company that detects and responds to equipment failures, alarm triggers, or incidents.

Alarm Suppressor Unit (ASU)-Alarm suppressor unit (ASU) is an equipment plug-in module that is used in conjunction with T-carrier systems when fiberoptic lines are used for the transmission line. The use of an ASU eliminates false alarms on a T-carrier span.

Alarmed Equipment-Alarmed equipment is a machine or device that can indicate an alarm condition when some preset level has been exceeded. An example of an alarmed equipment is a cable pulling winch that indicates an alarm when the pulling tension on the cable has exceeded a preset pulling tension limit.

A-law Encoding-A-law encoding is a digital signal companding process that is used for encoding/decoding signals in pulse-code-modulated (PCM) systems. This companding process increases the dynamic range of a binary signal by assigning different weighted values to each bit of information then is defined by the binary system. The A-law encoding system is an international standard. A different companding version is used in the Americas as u-Law.

Alert Tones-Alert tones are the types of alert tones that are available to indicate a particular status of a telecommunications event. An example of an alert tone is a sound that alerts the user that a new short message has been received.

Alert With Info-Alert with info is an alert message that is sent to a communications device (e.g. start ringing) that also contains additional information (e.g. calling name or calling number).

Alerting Sequence-Alerting sequence is a group of symbols or sequences of data that is followed by a control or data message.

Alerting Signals-Alerting signals are signals that are used to alert a user of a service request (e.g. incoming call) or status change (message waiting).

AL-FEC-Application Layer Forward Error Correction

Algorithm-Algorithm is a set of well defined steps or rules that allows for the solution of a problem or processing of information. Commonly the name for a portion of a software program or a function.

Algorithm Specific Signal Processor (ASSP)-Algorithm specific signal processor (ASSP) is an integrated circuit (IC) that is specifically designed for a unique application or series of related applications (e.g. radio signal filtering).

ALI-Automatic Location Identifier

Aliasing-(1-network addressing) Providing temporary or alternative identification codes or names to identify a channel or service. Aliasing allows devices or services to be addressed using a shortened code or allows the user of a name that hides the underlying addressing information. (2-sampled data waveform processing) Signals appearing in the wrong part of the frequency spectrum due to insufficiently frequent sampling of the original waveform.

Aliasing Effects-Aliasing effects are unwanted distortions that result from the conversion of an image where the sampling of the image is at a speed less than half of the most rapid changes in the image. Aliasing effects commonly appear as lines or ripples in the scanned or converted image.

Alignment-Alignment is the process of adjusting a circuit or the status of system components that interact with each other and can be changed so the performance of the system can be enhanced. An example of alignment is the tuning of multiple tuned circuits in a transmitter or receiver assembly so the signal can be amplified in each stage. When all the amplifiers are tuned to the correct frequencies, the system is said to be aligned.

A-Link-Access Link

All-Dielectric Cable-All-dielectric cable is a cable that has the separating materials between the conductors that is made entirely of dielectric (insulating) materials without any metal conductors, armor, or strength members.

Alliance for Telecommunications Industry Solutions (ATIS)-Alliance for Telecommunications Industry Solutions (ATIS) is a North American organization established by Bellcore to develop standards and guidelines for the methods and procedures needed in the telecommunications industry. ATIS has four committees, T1, SONET, Internet Work Interoperability Test Coordination, and Order and Billing Form.

Alligator Clip-Alligator clip is a spring-loaded connector that allows the temporary connection of a test lead to a wire or electronic circuit component. The alligator clip has a long section that contains teeth keep the clip from sliding off the connection (looks like an alligator's mouth).

Allocated Circuit-Allocated circuit is a circuit designed and reserved for the use of a particular customer.

Allocation Group-Allocation group, in automated facility planning, is any message trunk group, specially defined special-service circuit group, or specially defined carrier system group that creates a demand for facilities and equipment.

Allocation Of A Frequency Band-Allocation of a frequency band is the entry in the Table of Frequency Allocations of a given frequency band for the purpose of its use by one or more terrestrial or space radio-communication service or the radio astronomy service under specified conditions. This term shall also be applied to the frequency band concerned.

Allotment of A Radio Frequency Or Radio Frequency Channel-An entry of a designated radio frequency channel in an agreed upon plan, adopted for use by one or more administrations for a terrestrial or space radio-communication service.

Allowed Information Rate (AIR)-Allowed information rate (AIR), in a frame-relay Data Link Connection (DLC), is the maximum data rate is given by the sum of the Committed Information Rate (CIR) and the Excess Information Rate (EIR) of the connection.

All-Routes Explorer Packet-All-routers explorer packet is a packet that is sent in a bridging network to hunt for another device that is attached somewhere in the network.

Alpha Blending-Alpha blending is the combining of a translucent foreground color with a background color to produce a new blended color.

Alpha Channel-An alpha channel is additional information that is associated with each pixel that represents how that pixel is to be blended with background in an MPEG digital video system.

Alpha Paging-Alpha paging are text messages that are sent via operator or computer to an Alpha Pager.

This figure shows how an alpha paging system receives voice, text, or data messages from callers and forwards these messages in text form to an alphanumeric pager. In this example, a sender can access the system by voice or by sending email messages via the Internet. When accessing the system by voice, a caller dials a paging access number and is either connected to an interactive voice response (IVR) unit or to an operator. When connected to an IVR, the user may be given options for specific messages (canned messages) or their voice may be converted to text messages. When connected to an operator, the operator converts (keys in) their messages to text form. When messages are sent via the Internet, their format is changed to a form suitable for the alpha paging system. In any of these cases, the messages are placed in a message queue that holds the message until the system is available (not other messages waiting) before it. When the message reaches the top of the queue (available time to send), it will be encoded (formatted) to a form suitable for transmission on a radio channel. In this example, the message is sent as part of group 4. Sending the messages in groups allows the pager to sleep during transmission of pages from other groups that are not intended to reach the alphanumeric pager. The text message includes the pager address along with the text message in digital form. During the reception of the message, it is stored into the message paging memory area so the pager can display the message after it is received.

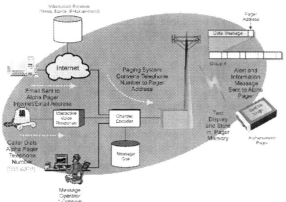

Alpha Paging Operation

Alpha Testing-Alpha testing is the first stage in testing a new hardware or software product, usually performed by the in-house developers or programmers. Alpha testing is the initial internal and possibly limited field testing process used to confirm the operation and performance of new hardware or software products. The key purpose of

Alpha testing is to identify basic problems during typical operating conditions. The typical number of Alpha test participants is 10 to 50.

Alphanumeric-Alphanumeric is a generic term for alphabetic letters, numerical digits, and special characters that can be processed and displayed by a machine. Alphanumeric displays provide a character set that includes letters, numbers, and punctuation marks.

ALS-Alternate Line Service

Alt Attribute-Alternative Attribute

Alt tag-Alternative Text

Alternate Billing Services-Alternate billing services are Calling-cards, collect calls, and third-number-billed calls whose originating party does not pay for the call.

Alternate Buffer-Alternate buffer, in a data communications device, is a section of memory set aside for the transmission or reception of data when the primary buffer is filled. This allows for uninterrupted flow of data in either transmission or reception mode.

Alternate Channel-Alternate channel is a communication channel that has a carrier frequency that is located one channel bandwidth above or below the frequency of another communication channel.

This diagram shows the relative spacing of alternate radio communication channels. Alternate radio channels are spaced at two channel bandwidth intervals from the main carrier channel.

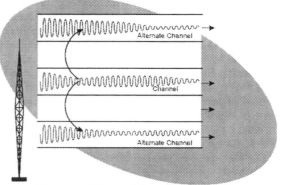

Alternate Channel Frequency Spacing

Alternate Channel Interference-Alternate channel interference occurs when the RF power from an alternate channel (2 channels away from desired channel) overlaps into the frequency band of the channel being used by a radio device.

Alternate DNS Roots-Alternate DNS roots are alternate root domain names where the server is a directory managed by a company rather than that associates top level domain names with their associated IP addresses.

Alternate Gatekeeper-Alternate gatekeeper is a gatekeeper that is assigned to provide backup connections in the event that a primary gatekeeper fails or becomes unavailable.

Alternate Language-An alternate language is an additional track that is a section of a media file (such as a video tape or multimedia file) that can be used to provide audio in different languages.

Alternate Power Supply-Alternate power supply is a backup power supply, which converts commercial ac power to dc to maintain batteries at appropriate charging levels. Alternate power supplies usually include an inverter and regulator for dc/ac conversion to maintain remote site terminals and central office equipment in the event of a commercial power outage.

Alternate Recovery Facility (ARF)-Alternate recovery facility is a backup facility that is equipped and configured to perform the command, control and data communication operations necessary to maintain the orbital parameters of a particular satellite.

Alternate Route-Alternate route is an alternative path or connection of circuits between two points that is used as second or next-choice in the event a primary route is disconnected.

Alternate Routing-Alternate routing is a network switching feature that enables alternate routing of trunk or path assignments. Alternate routing may be enabled if a failure occurs in the primary route (path) or for least-cost routing service.

Alternate Scan-An alternate scan is the process of scanning a matrix of measurements (such as a digital image block) in a predefined scanning pattern to convert the information into a serial data format.

Alternating Current (AC)-Alternating current (AC) refers to the type of electricity that is characterized as a cyclic wave of energy where the voltage varies continuously from positive to negative to positive. When graphed the wave form is sinusoidal. The opposite of AC is DC (direct current).

Alternating Current Power (AC Power)-Alternating current power (AC power) is electrical power that is supplied in a form of alternating current (AC). AC power is usually supplied from an external source such as Utility or generation as the main supply to equipment or assemblies.

Alternative Access Provider-Alternative access provider is a telecommunications service that provides an access connection between the end customer and a telecommunications network. This provider is a different company than the established LEC or PTT company.

Alternative Route-Alternative route is a secondary communications path to a specific destination. An alternative route is used when the primary path is not available.

Alternative Routing Of Signaling-Alternative routing of signaling is the routing of signaling messages (such as in a Signaling System 7 system) through alternative paths as a result in the failure of a primary routing path.

Alternative Text (Alt tag)-Alternative text (alt tag) is text that is displayed in the event an image or other item cannot be displayed. Alternative text is usually displayed when a user positions the mouse cursor over the image or item that can be selected.

ALU-Arithmetic Logic Unit

Always-On-Always-on is a connection to a communications network (such as the Internet) that appears always on to the customer. Although always on connections appear as a dedicated connection to the end user (no need to initiate a dial up sequence), the connection may be temporary and automatically re-established each time the user accesses the network.

AM-Amplitude Modulation

AM Broadcast Band-AM broadcast is the band of frequencies extending from 535 to 1705 kHz.

AM Broadcast Station-AM broadcast is a broadcast station licensed for the dissemination of radio communications intended to be received by the public and operated on a channel in the AM broadcast band.

AM_ADDR-Active Member Address

AMA-Automatic Message Accounting

Ambient Current-Ambient current is a level of electrical current that results from the voltages that are created by random movement of electrons in a circuit without the addition of power.

Ambient Voltage-Ambient voltage are voltages that are created by the random vibration (movement) of electrons in a circuit without the supply of power.

AMC-Adaptive Modulation and Coding

AMC-Adds, Moves and Changes

American National Standards Institute (ANSI)-American National Standards Institute (ANSI) is the US organization that sets the rules and procedures for, and also authorizes specific standards setting organizations. ATIS and EIA/TIA are two ANSI authorized standards setting organizations in the US in the subject area of telecommunications.

American Registry for Internet Numbers (ARIN)-American registry for Internet numbers is a not-for-profit organization that is responsible for the management of Internet protocol (IP) addresses in North America, South America, the Caribbean, and sub-Saharan Africa.

American Society of Composers, Authors and Publishers (ASCAP)-The American society of composers, authors and publishers is an organization that assists and represents people and companies that are involved in the creation and distribution of content.

American Standard Code for Information Interchange (ASCII)-American standard code for information interchange (ASCII) is a widely accepted standard for data communications that uses a 7-bit digital character code to represent text and numeric characters. When companies use ASCII as a standard, they are able to transfer text messages between computers and display devices regardless of the device manufacturer.

This table shows the symbols that can be represented by the 7 bit ASCII code. The columns indicate the upper 3 bits of the ASCII code (b5 - 67) and the rows indicate the lower 4 bits of the ASCII code (b1-b4). For example, to represent the letter A, the ASCII code would be 100 (upper) + 0001 (lower) resulting in ASCII code 1000001. Computers that receive the code 1000001 would display the capital letter A.

	000 (0)	001 (1)	010 (2)	011 (3)	100 (4)	101 (5)	110 (6)	111 (7)
0000 (0)	NUL	DLE	SP	0	@	P	`	p
0001 (1)	SOH	DC1	!	1	A	Q	a	q
0010 (2)	STX	DC2	"	2	B	R	b	r
0011 (3)	ETX	DC3	#	3	C	S	c	s
0100 (4)	EOT	DC4	$	4	D	T	d	t
0101 (5)	ENQ	NAK	%	5	E	U	e	u
0110 (6)	ACK	SYN	&	6	F	V	f	v
0111 (7)	BEL	ETB	'	7	G	W	g	w
1000 (8)	BS	CAN	(	8	H	X	h	x
1001 (9)	HT	EM	)	9	I	Y	i	y
1010 (A)	LF	SUB	*	:	J	Z	j	z
1011 (B)	VT	ESC	+	;	K	[	k	{
1100 (C)	FF	FS	,	<	L	\	l		
1101 (D)	CR	GS	-	=	M	]	m	}
1110 (E)	SO	RS	.	>	N	^	n	~
1111 (F)	SI	US	/	?	O	_	o	DEL

ACK	Acknowledge	FS	Form Separator
BEL	Bell	GS	Group Separator
BS	Backspace	HT	Horizontal Tab
CAN	Cancel	LF	Line Feed
CR	Carriage Return	NAK	Negative Acknowledge
DC	Direct Control	NUL	Null
DEL	Delete Idle	RS	Record Separator
DLE	Data Link Escape	SI	Shift In
EM	End of Medium	SO	Shift Out
ENQ	Enquiry	SOH	Start of Heading
EOT	End of Transmission	STX	Start of Text
ESC	Escape	SUB	Substitute
ETB	End of Transmission Block	SYN	Synchronous Idle
ETX	End of Text	US	Unit Separator
FF	Form Feed	VT	Vertical Tab

American Standard Code for Information Interchange (ASCII)

American Standards Association (ASA)-American Standards Association (ASA) is the predecessor organization to ANSI.

American Wire Gauge (AWG)-American wire gauge (AWG) is a measurement system that provides the diameter of conductors (typically copper wire). The larger the thickness of the wire (higher the gauge), the lower the AWG number and the better the ability of the line to carry electrical signals. Many local telephone access loops use 24 AWG or 26 AWG copper lines.

AMI-Alternate Mark Inversion

AMIS-Audio Messaging Interchange Specification

Amortization-Amortization is the assignment of cost of an asset or acquisition for distribution over the time period that the asset or acquisition will be used.

AMP-Address Management Protocol

Ampere (A)-Ampere (A) is a unit of electric current, equal to a flow rate of approximately 6 250 000 000 000 000 000 electrons per second. The precise definition of an ampere is based on the force between two current-carrying wires. If two straight parallel conductors of infinite length and negligible circular cross-section diameter, are placed 1 meter apart in vacuum, a force equal to 0.000 000 2 newton per meter of length is produced between them when they each carry 1 ampere. Named for André M. Ampère, 19th century French physicist.

Amplified Handset-Amplified handset is a handset (microphone and speaker) that contains an integrated (built-in) amplifier. Amplified handsets may be used by the hearing impaired.

Amplifier-An amplifier is a device or assembly that converts an input signal (usually low level) into another (usually larger) version of itself. Amplifiers increase both the desired signal and unwanted noise signals. Noise signals are any random disturbance or unwanted signal in a communication system that tends to obscure the clarity of a signal in relation to its intended use.

This figure shows the basic process of amplification. An amplifier uses an input signal to control the current or voltage of a device that has an external power supply. The ability of the input signal to control the current or voltage allows the replication of the original signal. The ratio of input signal to the output signal level is called the gain (amount of amplification).

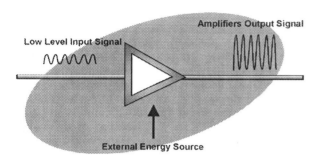

Signal Amplification

Amplifier Nonlinearity-Amplifier nonlinearity is an indicator of distortion that is caused by inconsistent amplification (higher or lower gain to different levels or frequencies of input signals) of a device or system.

Amplitude-Amplitude is a number proportional to the signal voltage or current of a waveform. One measure of amplitude is the peak-to-peak voltage (a measurement of the difference between the largest positive voltage and the most negative voltage of the waveform). Another measure is the maximum positive voltage. Yet another measure is the time-average voltage. The amplitude of a waveform is a signed (positive or negative) quantity.

Amplitude Compandored Single Sideband Modulation (ASSB)-A transmission technique where only one sideband is used for transmission with its amplitude compressed. The amplitude of the sideband signal is expanded at the receiver. No carrier is transmitted and the other sideband is suppressed.

Amplitude Distortion-Amplitude distortion is a variance between the desired or expected amplitude and a signal and the actual received signal. Amplitude distortion can be measured by subtracting the expected test signal from a distorted signal

to allow the measurement of the amount of distortion signal.

Amplitude Equalizer-Amplitude equalizer is a corrective network that is designed to modify the amplitude characteristics of a circuit or system over a desired frequency range. Such devices may be fixed, manually adjustable, or automatic.

Amplitude Modulation (AM)-Amplitude modulation (AM) is the transferring of information onto a carrier wave (such as a radio carrier) by varying the amplitude (intensity) of the carrier signal.

This figure shows that amplitude modulation involves the transferring of information onto a carrier signal by varying the amplitude (intensity) of the carrier signal. This diagram shows an example of an AM modulated radio signal (on bottom) where the high of the radio carrier signal is changed by using the signal amplitude or voltage of the audio signal (on top).

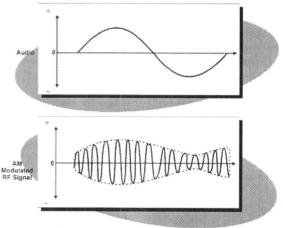

Amplitude Modulation (AM) Operation

Amplitude Modulator Stage-Amplitude modulator stage is the last amplifier stage in a transmitter where the assembly (stage) modulates a radio-frequency signal.

AMPS-Advanced Mobile Phone Service
AMPS-EIA-553
AMR-Adaptive Multirate Codec
AMR-Automatic Meter Reading
AMS-Audience Measurement System
AMSC-American Mobile Satellite Corporation
AMTA-American Mobile Telecommunications Association
AMTS-Automated Maritime Telecommunications

Systems
AMVER-Automated Mutual-Assistance Vessel Rescue System
AMX-Analog Matrix Switch
AN-Access Network
AN-Access Node
Analog-Analog is an information form that is represented by a continuous and smoothly varying amplitude or frequency changes over a certain range such as voice or music. Analog lines allow the representation of information to closely resemble the original information signal.

This figure shows a sample analog signal created by sound pressure waves. In this example, as the sound pressure from a person's voice is detected by a microphone, it is converted to its equivalent electrical signal. Also, the analog audio signal continuously varies in amplitude (height, loudness, or energy) as time progresses.

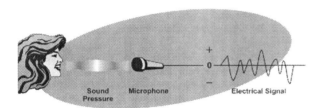

Analog Audio Signal

Analog Access Channel-Analog access channel is a control channel on an analog cellular system that uses frequency shift keying (FSK) modulation to pass digital messaging signals between the mobile phone and cell site. See: access channel.
Analog Bridge-Analog bridge is a circuit or system that allows an analog communication session to be connected (to include) another users.
Analog Cable System-An analog cable system distributes television (and other information services) via a cable television distribution system in analog modulated form.
Analog Capture-Analog capture is the process of receiving and storing analog video images. Analog video capture typically refers to capture of analog video images (e.g. NTSC, PAL or SECAM) into digital form.

Analog Carrier System-Analog carrier system is a transmission system that uses the modulation (amplitude, frequency, or phase) of carrier signal to transport an analog information signal.

Analog Cellular-Analog cellular is an industry term given to first generation (1G) cellular systems that transmit voice information using a form of analog modulation (e.g. FM). Analog cellular systems may have digital control channels. Analog cellular systems primarily provide voice and low-speed data communication services over a wide geographic area.

This figure shows a basic analog cellular system. This diagram shows that there are two types of radio channels; control channels and voice channels. Control channels typically use frequency shift keying (FSK) to send control messages (data) between the mobile phone and the base station. Voice channels typically use FM modulation with brief bursts of digital information to allow control messages (such as handoff) during conversation. Base stations typically have two antennas for receiving and one for transmitting. Dual receiver antennas increase the ability to receive the radio signal from mobile telephones, which typically have a much lower transmitter power level than the transmitters in the base station. Base stations

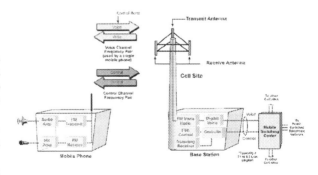

Analog Cellular System (1st Generation)

are connected to a mobile switching center (MSC) typically by a high speed telephone line or microwave radio system. This interconnection must allow both voice and control information to be exchanged between the switching system and the base station. The MSC is connected to the telephone network to allow mobile telephones to be connected to standard landline telephones.

Analog Compression-Analog compression is a process that uses a signal processing components or assemblies (such as a compandor and expandor) to convert the format of analog signals into a form that occupies or requires fewer resources to transfer or store analog signals. Analog compression allows more information to be transmitted on a communication channel.

Analog Facsimile-Analog facsimile is a facsimile that transmits images on an analog communication line through the conversion of images or shades of images to analog signals (tones).

Analog Fiber-Analog fiber is a fiber strand that is used to transmit optical signals in analog (unstructured) form.

Analog Matrix Switch (AMX)-Analog matrix switch (AMX) is a switching system that can interconnect analog telephone lines or analog telephone devices.

Analog Media-Analog media is the format of information that is used to express information (media) that is represented in a form that can have levels or signal composition of any level (analog).

Analog Microwave-Analog microwave is a microwave transmission system that transfers information through the modulation of a microwave carrier signal. The type of modulation used may be amplitude, frequency or phase shift, but the analog input signal is used as the source of modulation information.

Analog Mobile-Analog mobile is a cellular telephone that is limited to utilizing only analog FM radio transmission for voice communications.

Analog Multiline-Analog multiline is a device that uses frequency division to multiplex two telephone lines on a single 2-wire facility (copper pair). Analog multiline is used by telephone companies (telcos) in service areas where access to wire facilities are limited.

Analog Scrambling-Analog scrambling is a process of altering or changing an analog electrical signal (continuously changing signals) to prevent the listening or interpretation of signals by unauthorized users. Scrambling involves the changing of a signal according to a known process (such as inverting parts of the signal) so that the received signal can reverse the process to decode the signal back into its original (or close to original) form.

Analog Set Top Box (Analog STB)-An analog set top box is an electronic device that adapts an analog television broadcast signal into a format that is accessible by the end user. Analog set top boxes are commonly located in a customer's home to allow the reception of video signals on a television or computer.

Analog Signal-An analog signal is a direct representation of a physical process. For instance, an analog electromagnetic signal representing your voice on a telephone line is represented by continuous variations in voltage. Loud sounds are represented by large voltages, soft ones by small. High voices are represented by high frequency variations in the voltage, low ones by low frequencies. Analog signals provide the most nuanced and precise record of a physical process because of this exact representation. However, they are more easily distorted by noise and other factors than are digital signals.

This figure shows how an analog signal exactly matches each portion of the information source. This diagram shows a person who is creating sound pressure waves that are converted into an electrical signal via a microphone. This example shows that each portion of the sound pressure wave is represented by its own instantaneous electrical signal level.

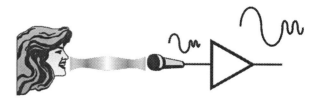

Analog Signal

Analog Signal Processing-Analog signal processing involves the conversion of analog signals into another form using analog (continuous) circuits or systems. Analog signal processing includes filters, shaping circuits, combiners, and amplifiers to change their shape and modify the content of analog signals.

Analog STB-Analog Set Top Box

Analog Subscriber Loop Carrier (ASLC)-Analog subscriber loop carrier (ASLC) is a highly efficient analog transmission system that uses existing distribution cabling systems to transfer analog information between the telephone system (central office) and a users telephone. An analog double sideband carrier system is sometimes used in facilities-starved areas as a permanent or temporary engineering solution. It employs bi-directional transmission over a single exchange grade cable pair used for control. One control pair derives 8 channels using FDM over a 8 to 144 kHz spectrum. These systems are designed for 140 dB with repeater spacing at 35 dB intervals and are successful in low-density suburban applications (e.g. summer home areas).

Analog Switching System-Analog switching system is a switching system in which the connection route uses analog transmission.

Analog Tape-Analog tape is a magnetic tape storage format that changes magnetic information on the tape to represent analog signals.

Analog Telephone Adapter (ATA)-Analog Telephone Adapter (ATA) is a device that converts analog telephone signals into another format (such as digital Internet protocol). These adapter boxes may provide a single function such as providing Internet telephone service or they may convert digital signals into several different forms such as audio, data, and video. When adapter boxes convert into multiple information forms, they may be called multimedia terminal adapters (MTAs) or integrated access devices (IADs).

Analog telephone adapters (ATA) must convert both the audio signals (voice) and control signals (such as touch tone or hold requests) into forms that can be sent and received via the Internet.

Analog Television Adapter (ATVA)-Analog Television Adapter (ATVA) is a device that converts digital multimedia signals (such as MPEG) into analog television signals (such as NTSC or PAL). These adapter boxes may provide a single function such as providing Internet television ser-

vice or they may convert digital signals into several different forms such as audio, data, and video. When adapter boxes convert into multiple information forms, they may be called multimedia terminal adapters (MTAs) or integrated access devices (IADs).

Analog television adapters (ATA) must convert video, audio, and control signals (such as requests for changing channels) into forms that can be sent and received via data networks such as the Internet.

Analog Terminal Adapter (ATA)-Analog terminal adapter (ATA) is a communications adapter that allows analog telephone devices (e.g. a computer modem) to interconnect to digital telephone systems.

Analog to Digital Converter (ADC)-Analog to digital conversion is a process that changes a continuously varying signal (analog) into a digital values. A typical conversion process includes an initial filtering process to remove extremely high and low frequencies that could confuse the digital converter. A periodic sampling section that at fixed intervals locks in the instantaneous analog signal voltage, and a converter that changes the sampled voltage into its equivalent digital number or pulses.

This diagram shows how an analog signal is converted to a digital signal. This diagram shows that an acoustic (sound) signal is converted to an audio electrical signal (continuously varying signal) by a microphone. This signal is sent through an audio band-pass filter that only allows frequency ranges within the desired audio band (removes unwanted noise and other non-audio frequency components).

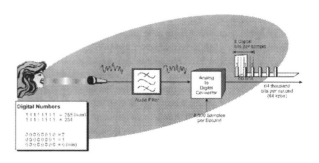

Signal Digitization Operation

The audio signal is then sampled every 125 microseconds (8,000 times per second) and converted into 8 digital bits. The digital bits represent the amplitude of the input analog signal.

Analog Transmission-Analog transmission is a system that is capable of transferring continuously varying signals (analog signals) between points. The system may directly transfer the analog signal or the analog signal may modify another carrier (such as a radio carrier).

Analog Video-Analog video is the representation of a series of multiple images (video) through the use of a rapidly changing signal (analog). This analog signal indicates the luminance and color information within the video signal.

Sending a video picture involves the creation and transfer of a sequence of individual still pictures called frames. Each frame is divided into horizontal and vertical lines. To create a single frame picture on a television set, the frame is drawn line by line. The process of drawing these lines on the screen is called scanning. The frames are drawn to the screen in two separate scans. The first scan draws half of the picture and the second scan draws between the lines of the first scan. This scanning method is called interlacing. Each line is divided into pixels that are the smallest possible parts of the picture. The number of pixels that can be displayed determines the resolution (quality) of the video signal. The video signal television picture into three parts: the picture brightness (luminance), the color (chrominance), and the audio.

Analog Voice Channel (AVC)-Analog voice channel (AVC) is a radio channel on a cellular system that typically uses FM modulation to transfer voice (audio) signals. The analog voice channel also sends digital messages during brief periods by muting the voice signal.

This figure shows that an AMPS voice channel pairs are primarily used to carry FM modulated voice signals. This diagram also shows that a supervisory audio tone (SAT) out-of-band audio control signals are added by the base station and re-transmitted back to the base station by the mobile device. This example shows that control messages can be sent by briefly replacing audio signals with FSK messages.

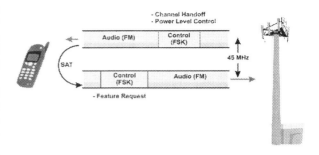

AMPS Voice Channel

Anamorphic-Anamorphic is a process used to reduce the size of a widescreen film or video onto a narrower medium such as screen or DVD. Anamorphic cinema squeezes wide (2.35:1) frames onto 35mm film. Anamorphic widescreen movies stored on DVD at 720x480 are resampled and stretched (by creating new pixels) to fill 1280 pixels horizontally.

Ancillary Services-Ancillary services are additional features or services provided, including call features, detail billing, voice mail, etc.

Andrews File System (AFS)-Andrews file system (AFS) is a protocol that allows communication devices to access file servers in remote locations using TCP/IP protocol. AFS protocol was developed at Carnegie Mellon University. AFS includes a memory cache that helps to the administration of the file transfer process.

Angle Modulation-Modulation of the phase angle of a carrier, as in some forms of phase and frequency modulation.

Angstrom-Angstrom is a measure of the length of light wavelengths. A single angstrom is 0.1nm or $10^\wedge$-10 m,

ANI-Automatic Number Identification

ANI Identification-ANI identification is a software processing function at an end office (local) switch that forwards the billing number to the termination point. ANI is needed for the caller identification service offered by many service providers.

Animated GIF-Animated Graphic Interchange Format

Animated Graphic Interchange Format (Animated GIF)-An animated graphic interchange format (GIF) tool is a sequence of GIF images that are combined in time sequence to simulate movie sequence animations.

Animated Icons-Animated Icons are small and relatively short graphic animations that are linked to a software program or function. The purpose of the animated icon is to allow the user to quickly identify (find) their programs. Animated Icons are commonly used to gain the attention of the user from other static (non-moving) icons.

Animation-Animation is a process that changes parameters or features of an image or object over time. Animation can be a change in position of an image within a video frame to synthetically created images that change as a result of programming commands.

ANM-Answer Message

Announcement Server-An announcement server provides network announcements to callers (e.g. "please wait while your call is being connected").

Announcement Trunk Group-Announcement trunk group is a trunk group used to inform customers or operators about call status or to access announcement services.

Anonymizer-An anonymizer is a company or service that is located between a user or consumer and a supplier of a product or service who isolates the identity of the user (such as an IP address) from the vendor.

Anonymous Call Rejection (ACR)-Anonymous call rejection is a service that allows a telephone customer (wireless or wired) to reject calls from callers who have selected a privacy feature that disables the display of their calling telephone number.

Anonymous FTP-Anonymous FTP is a security process that allows general users to have limited access to file servers without the need for registering for account identification and passwords. The user enters anonymous as an unregistered user and the password is usually your e-mail address.

ANSI-American National Standards Institute

Answer Message (ANM)-Answer message (ANM) is an ISDN User Part trunk signaling message that indicates the called party has answered the call.

Answer Mode-Answer mode is a call processing function that determines how a telephone device (e.g. a fax or a modem) will respond to incoming calls. The answer mode may be set to no answer, number of rings, or immediate answer.

Answer Signal-Answer signal is a control message (signal) that indicates in incoming call (or service request) has been answered or accepted. The answer signal can be a current flow (loop start) or a digital message that is sent on a separate signaling channel.

Answer Supervision-Answer supervision is the sending of an off-hook supervisory signal back to an exchange carrier's point of termination to show that the called party has answered.

Answering Machine-An answering machine is a device that can automatically answer telephone calls, play a prerecorded greeting message, store audio information, and allow retrieval and deletion of messages.

Antenna-An antenna is a device used to convert signals between electrical and electromagnetic form. Antennas are usually designed to operate over a specific frequency range. Directional antennas are designed to focus (concentrate) the transmitted energy in a particular direction to achieve antenna gain.

This diagram shows how an antenna system is used to connect a transmitter and receiver to each other along with the key characteristics of the transmission. The transmitter provides an electrical signal in the RF signal range. This signal is transmitted through a cable or transmission line to the antenna. As the signal travels through the transmission line, some of the energy leaks from the cable and some of the energy is converted to

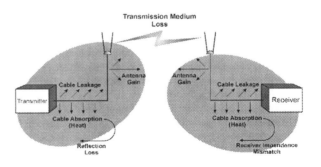

Antenna System

heat in the cable. The antenna converts the electrical energy to electromagnetic energy that transfers through air and other mediums (e.g. buildings and trees). The amount of energy that is converted depends on the impedance match between the transmission line and the antenna. When there is an impedance mismatch, some of the energy is reflected back through the transmission line and is not converted. Some antennas focus energy in a particular direction. Depending on the amount of directional focus, the antenna can appear to have gain in a particular direction (at the expense of reduced gain in other directions). The electromagnetic signal is then transferred between antennas at a loss of approximately 20 dB per decade in free space transmission (through air). This means that for every 10 times the distance (e.g. 10 meters to 100 meters), the signal energy will drop by 99% (decrease to 1%). The receiving antenna converts electromagnetic energy into electrical energy. Receiving antennas can sense energy in a particular direction. Depending on the amount of directional focus, the receiving antenna can appear to have gain in a particular direction (at the expense of reduced gain from signals that are received from other directions). This signal is then transferred through a cable or transmission line to the receiver. As the signal travels through the transmission line, some of the energy leaks from the cable and some of the energy is converted to heat in the cable. The amount of energy that is received from the cable depends on the impedance match between the transmission line and the receiver. When there is an impedance mismatch, some of the energy is reflected back through the transmission line and is not received.

Antenna Array-An antenna array is a group of several antennas linked together to facilitate a specified degree of directivity.

Antenna Bandwidth-Antenna bandwidth is the frequency range (bandwidth) that an antenna can effectively convert electrical signals into electromagnetic waves.

Antenna Combiner-Antenna combiner is a group filters (usually tuned cavity filters) that allow the outputs of multiple transmitters to be connected to the same antenna assembly. Because each tuned cavity in the antenna combiner is tuned to the specific frequency of its transmitter, high power radio energy from other transmitters at a different frequency is blocked (highly attenuated) from entering into the transmitter.

This diagram shows how tuned cavity filters are used to connect the output of four high power transmitters to the same antenna assembly. In this example, each transmitter is tuned to a different frequency along with their associated tuned cavity. Each tuned cavity allows a specific narrow band of frequencies to pass through it. This diagram shows that when other high power signals are applied to the tuned cavity with a different frequency, the signal is highly attenuated (blocked) from entering back into other transmitters.

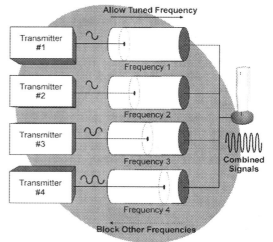

Antenna Combiner Operation

Antenna Electrical Beam Tilt-An antenna electrical beam tilt is the vertical directing of an antenna radiation pattern through the use of electrical means (phase shifting) so that the radiation pattern can be adjusted along the horizontal plane.

Antenna Farm-An antenna farm is a geographic location (area) that is used to group antennas that may have a common impact on aviation as designated by the Federal Communications Commission.

Antenna Frequency-Antenna frequency is the center of a frequency band that an antenna is capable of transmitting. The band of frequencies (antenna bandwidth) the antenna may be capable of transmitting ranges from less than one percent (e.g. waveguide antenna) to over 10%, dependant on the design and type of the antenna.

A

Antenna Gain-Antenna gain is the ratio of energy that is supplied to an antenna to the amount of energy transmitted from the antenna in a specific direction. The antenna gain is usually referenced to the direction of maximum radiation.

This diagram shows antenna systems that have different amounts of gain (directivity). In this example, a handheld telephone has a small amount of gain so they can transmit equally in most directions. The antenna that is mounted on a car can have more gain (directivity) as it will not be tilted (changed angle) as much as the portable telephone.

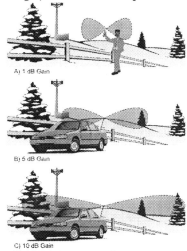

Antenna Gain

Antenna Resistance-Antenna resistance is the total resistance of the transmitting antenna system at the operating frequency and at the point at which the antenna current is measured.

Anti Reflection Coating-Anti-reflecting coating is a coating used to reduce the amount of light reflected from a surface and increase the amount transmitted through it. The coating may be of one or several thin films, typically one quarter wavelength thick each. The material for the coating is selected primarily based on its index of refraction. The combination of film thickness and appropriate index supports constructive interference of the light to be reflected and destructive interference of the light to be transmitted. This technology has application in many optical components where reflected light represents lost signal intensity.

Anti-Aliasing-Anti-aliasing is a process of filtering or processing to avoid the creation of unwanted distortions that result from the conversion of an image where the sampling of the image is at a speed less than half of the most rapid changes in the image. Aliasing effects commonly appear as lines or ripples in the scanned or converted image.

Antitrust Laws-Antitrust law has the meaning given it in subsection (a) of the first section of the Clayton Act, except that such term includes section 5 of the Federal Trade Commission Act to the extent that such section 5 applies to unfair methods of competition.

Antivirus Program-Antivirus program is a program that can detect the presence of and possibly eliminate a computer virus. Anti-virus program usually locate and identify viruses by looking for previously identified patterns or suspicious activity in the system.

Any Source and Group Pair (*,G)-A source and group pair (S,G) is the identification information used for multicast source trees. For multicast shared trees that can have sources at any location in the tree, the notation is changed to (*,G) which indicates the source address (any address) and the group address.

Anycast-Anycasting is the use of a common IP address (a unicast address) to allow the sharing of media transmission services on the Internet. Routers that receive the anycast IP address to the nearest node that offers the anycast service.

AOC-Advice Of Charge

AOD-Advertising on Demand

AOS-Alternate Operator Services

AOSP-Alternate Operator Services Provider

AP-802.11 Access Point

AP-Access Point

AP-Adjunct Processor

AP Address-Access Point Address

APC-Adaptive Predictive Coding

APC Connector-Angled Physical Contact Connector

APCO-Associated Public Safety Communication Officials

Apco 16-Associated Public Safety Communication Officials 16 Standard

Apco 25-Associated Public Safety Communication Officials Project 25 Standard

APD-Avalanche Photodiode

APDU-Application Protocol Data Units

API-Application Program Interface

APM-Advanced Power Management

APM Factor-Attempts Per Message

APN-Access Point Name

APOC-Advanced Paging Operators Code

APON-ATM Passive Optical Network

APP-Access Process Parameter

APPC-Advanced Program-To-Program Communications

Applet-An applet is a small software program that uses the Java programming language to request, transfer and process information in a computer.

AppleTalk-Apple talk is a protocol suite developed by Apple Computer, used in Macintosh computers and other compatible devices.

Applicant-An applicant is the entity that submits a form or application to participate in an event (such as a communication license auction) that may include all holders of partnership and other ownership interests and a percent of stock (equity) interest or outstanding stock in the entity (e.g. more than 5%) and officers and directors of that entity.

Application-An application is a software program that is designed to perform operations using commands or information from other sources (such as a user at a keyboard). Popular applications that involve human interface include electronic mail programs, word processing programs, and spreadsheets. Some applications (such as embedded program applications) do not involve regular human interaction such as automotive ignition control systems.

Application Based Call Routing-Application based call routing is a process that controls call routing of an incoming call based on the type of selected application such as sales, customer service, or order tracking.

Application Class-Application class is a group of client applications that perform similar services, such as voice messaging or fax-back services.

Application Decomposition-Application decomposition is the ability to divide an application into functional parts. This allows for applications to more easily evolve into more advanced programs by the upgrading of the individual functional parts.

Application Environment-Application environment is the processes and configurations of other programs, systems and services that may effect and be affected by running an application.

Application Firewall-An application firewall is a data filtering device that uses and analyzes application protocols to filter (block) unwanted data that is sent between a computer server or data communication device and a public network (e.g. the Internet). Application firewalls continuously look for data packet transmission patterns associated with a specific application (such as IP Telephony) that indicates authorized or unauthorized use to the server.

This figure shows how voice over data network (VoIP) telephone service can work through a firewall. This example describes how IP telephone registration sets up the firewall to allow call control messages (such as an incoming call) to reach the IP telephone device. When an IP telephone first senses it has been connected to a data network, it attempts to register with it's call server. In this example, the call server is at a distant location outside the firewall. Because the IP telephone is part of the local area network (LAN), it is a trusted device and the firewall allows it to request a communication session with the data network (probably the Internet). This creates a communication session and the firewall remembers the details of the communication session. These communication session details include the Internet address of the call server and the IP telephone device address in the LAN. When packets are received in the future from the call server with the Internet address assigned to the IP telephone, these packets will be forwarded to the Internet telephone device.

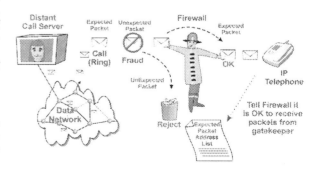

VoIP Firewall Operation

Application Flow-Application flow is a stream of frames or packets among communicating processes within a set of end stations.

Application Infrastructure Service Provider (AIP)-Application infrastructure service provider (AIP) are companies that provide communication application services such as email, web hosting, and voice communication.

Application Interface-An application interface is the messages, processes and/or hardware equipment that allow applications (such as software programs) to communicate and interact with each other.

Application Layer-The application layer coordinates the information interface between the communication device and the end user. The application layer receives data from the underlying protocols and processes this information into a form required or requested by the user or endpoint device. The application layer usually requests or responds to requests for a communication session. The location of the application layer is at the top of protocol stacks. The application layer is layer 7 in the open system interconnection (OSI) protocol layer model.

Application Layer Forward Error Correction (AL-FEC)-Application layer forward error correction is a process that added, changes or provides forward error correction to transferred data for specific types of applications.

Application Management-Application management is the management of the installation, operation, and resource allocation for software and communication applications.

Application Note-Application notes are descriptions or instructions that are provide to assist in the application or design of a device, product, or service into a system. Manufacturers commonly provide application notes to help their customers to use their products or services in their systems.

Application Policy Server (APS)-Application policy server (APS) is a communications server (computer with a software application) that coordinates the allocation of network resources and the priority for service for clients who use application services (such as email and voice communication).

Application Profile-Application profiles are particular implementation of protocols, feature operations, and/or processes that ensure applications operate in a specific manor.

Application Program Interface (API)-Application program interface (API) are defined and documented entry points into a software application where other programs may interact with the application in order to provide customized extensions or perform special processing functions. Typically an API is a public function call that then itself calls on the services of the application. In this way the API hides the underlying details and complexities of the application software making it easier for programmers to add custom functionality.

Application Protocol Data Units (APDU)-Application protocol data is a packet structure that is used in the H.450 (packet voice services) specification series.

Application Protocols-Application protocols are commands and procedures used by software programs to perform operations using information or messages that are received from or sent to other sources (such as a user at a keyboard). Application protocols are independent of the underlying technologies and communication protocols. The use of well-defined application protocols (agreed commands and processes) allows the software applications to interoperate with other programs that use the application protocol independent from the underlying technologies that link them together (such as wires or wireless connections.)

This figure shows that wireless personal area networks (WPANs) can use standard industry protocols to connect to standard communication applications. In this example, a laptop computer is communicating with a wireless mouse, a personal digi-

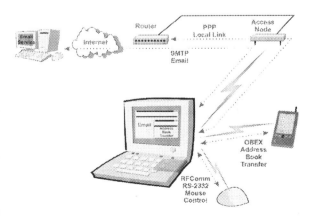

Application Protocols

tal assistant (PDA), and an access node using a single WPAN PCMCIA card. When communicating with the mouse, the laptop uses the standard RS-232 protocol. When transferring (exchanging) items between an address book stored in the laptop and an address book stored in the PDA, it uses the standard Object Exchange (OBEX) protocol. To connect to the Internet, the laptop connects through an access node to a router using standard point-to-point protocol (PPP). This diagram shows that the PPP connection is only part of the communication link that reaches an email server that is connected to the Internet. The computer uses standard simple mail transfer protocol (SMTP) to send and retrieve email messages.

Application Server (AS)-An application server (AS) is a computer and associated software that is connected to a communication network and provides information services (applications) for clients (users). Application servers are usually optimized to provide specific applications such as database information access or sales contact management.

Application servers are a key component of nextgen networks, and an enabler of IP-based enhanced services. These enhanced services will generate much-needed new revenue streams for service providers. Examples include all forms of conferencing, voice mail and unified messaging. Being softswitch-based, application servers have the flexibility to easily offer services that go beyond the feature set of legacy switched telephony. In terms of network configuration, the application server works in tandem with the media server, providing it with business logic and instructions for delivering enhanced services.

Application Service Element (ASE)-Application service element (ASE) is a software program or portion of a communication protocol that is part of an application layer of a protocol stack. Several ASEs may be combined to form a complete application protocol.

Application Service Provider (ASP)-Application service provider (ASP) is a company that provides an end user with an information service. An ASP owns or leases computer hardware and software system that allows one or more users to access information services on or through that computer systems.

Application Specific Integrated Circuit (ASIC)-An integrated circuit (IC) that is designed to provide specific signal processing functions. This is in contrast to general purpose IC's that perform more general signal processing functions. ASICs are often created from a gate array (batches of logical gates) through the use of a custom mask that interconnection the gates so they can perform specific functional operations.

Applications-Applications are the software programs that require voice over data communication technology. Many of the communications applications and services that were available in the year 2000 were designed for the narrow-band applications (below 56 kbps). These applications included limited graphic web browsing, text based on-line shopping, email and word processor file transfer. Low cost broadband services such as xDSL systems provide a tremendous opportunity for the development of richer, more enhanced applications. These applications require services such as streaming video, rapid image file transfer or high speed data file transfer services.

Applications Generator (Apps Gen)-Applications generator (Apps Gen) is a software capability that enables users to develop programmatic coding using high-level input to save time and reduce the programming tasks.

Applications Processor-Applications processor is a computer system or information service provider that is dedicated to processing applications such as voice mail or billing systems.

Applied Metadata-Applied metadata is information that describes how the media may be used. Examples of applied metadata include duration, format and display restrictions.

APPN-Advanced Peer-To-Peer Networking

Apps Gen-Applications Generator

APS-ACCUNET Packet Service

APS-Application Policy Server

Arbitration-(1-general) A process or set of rules that is used to manage conflicts. (2-computers) The process of competing for computer resources such as memory or peripheral devices, made by multiple processes or users.

ARC-Access Router Card

ARCH-Access Response Channel

Architecture-Architecture is the functional design of a network, computer or telecommunications system elements and the relationships between them. The architecture usually includes hardware and software components.

Architecture for Voice, Video and Integrated Data (AVVID)-AVVID is a network structure standard that defines the types of devices used in a voice over data (multimedia) network and how they are interconnected and used within the network. The AVVID structure allows for system expansion, efficient feature deployment, security, and increased reliability. AVVID was developed by Cisco.

Archival-Archival is a process or media storage device that has a long-term minimum life-spans over which the information will not become corrupted.

Archiving Files-Archiving files is a process where the information contained in an active computer file is made ready for storing in a non-active file, perhaps in off-line or near-line storage. Typically when files are archived, they are compressed to reduce their size. To restore the file to its original size requires a process know as unarchiving.

ARDIS-Advanced Radio Data Information Service

Area Code-An area code is a 3-digit number that generally identifies a geographic area of a switch that provides service to a telephone device. In North America, the Numbering Plan Area (NPA) is the area code. In countries other than North America, the area code may have any number of digits depending on the regulation of telecommunication in that country.

Area Exchange-Area exchange are geopolitical areas that are defined as region authorized to provide local telephone services. Small metropolitan areas or a collection of towns often share a single area exchange.

ARF-Alternate Recovery Facility

Argument Separator-Argument separator is a code or structure that separates information elements for records or data within programs such as databases, program languages, and spreadsheets. These separators delimit the information elements to allow the software programs to identify and process each element according to their program's requirements. Examples of an argument separator include a comma, tab, or semicolon.

ARIB-Association for Radio Industries and Business

ARIN-American Registry for Internet Numbers

ARIS-Access Request Information System

ARIS-Aggregate Route-Based IP Switching

Arithmetic Coding-Arithmetic coding is a data compression technique for a sequence of data. Arithmetic coding can compress with more efficiency than Huffman coding. This is because arithmetic coding allows compressed bits to be shared between two (or more) encoded symbols, unlike Huffman coding in which each compressed bit belongs exclusively to a single encoded symbol. However arithmetic coding has requirements on arithmetic accuracy, buffer memory, and computation which Huffman coding can avoid.

Arithmetic Logic Unit (ALU)-Arithmetic logic unit (ALU) is a part of a central processing unit (CPU) that performs arithmetic and logical operations on data.

Arithmetic Overflow-Arithmetic overflow is a condition that occurs in a digital computing device or system when the result of a calculation that is greater than the computing device can store or display.

Arithmetic Register-Arithmetic register is a short-term memory storage location (register) that holds the operands or the results of an arithmetic operation.

Arithmetic Unit-Arithmetic unit is the processing portion of a computing system that performs a majority of the arithmetic operations processed by the computer.

ARJ-Admission Reject

Armor-Armor is a protective layer or mechanical shield that is located around a cable or assembly. When metal armor is used for fiber optic cables, it may require electrical grounding for safety.

Armored Cable-Armored cable is one or more transmission lines (such as copper wire, coax, or optical fibers) that are covered in a sheathed or coated protective outer jacket. The armor protects the cable from accidental or intentional bending or cutting of the inner cable. Armored cable may be used in buried cable or where cables are exposed to stressful environmental conditions. An example of an armored cable is a stainless steel handset cord that is used on public telephones. The steel armor protects the cable from excessive wear from usage and vandalism.

ARP-Address Resolution Protocol

ARP Cache-ARP code is a data structure that provides the current mapping of 32-bit IP addresses to 48-bit MAC addresses.

ARPAnet-Advanced Radio Research Projects Agency Network

ARPC-Average Revenue per Customer

ARPU-Average Revenue Per User

ARQ-Automatic Retransmission Request

Arrears-Arrears is the assessment of charges for services that have been used (e.g. usage).

Arrival Rate-Arrival rate, as related to call centers, is the pattern at which incoming calls are received at the call center. Arrival rates are classified as steady, random or peaked. This term is also used in queuing theory to describe the rate at which entities enter into an ordered list to be processed.

Artifacting-Artifacting is the process of producing unintended, unwanted visual aberrations in a video image.

Artifacts-Artifacts, in general, are results, effects or modifications of the natural environment produced by people. In the processing or transmission of audio or video signals, a distortion or modification produced due to the actions of people or due to a process designed by people. Unintended, unwanted visual aberrations in a video image. In all kinds of computer graphics, including any display on a monitor, artifacts are things you don't want to see. They fall into many categories (such as speckles, in scanned pictures) but they all have one thing in common: they are chunks of stray pixels that don't belong in the image.

Artificial Intelligence (AI)-Artificial intelligence (AI) is a deductive reasoning process that can be applied on an automated basis by computer processing. Artificial intelligence simulates the reasoning capabilities much like the human mind through user input. British mathematician Alan Turning introduced the Artificial intelligence term in the 1950's.

Artificial Line Interface-Artificial line interface is the ability of a piece of transmission equipment to attenuate its output level to meet the required transmission level.

ARTS-Advanced Radio Technology Subcommittee

ARU-Audio Response Unit

AS-Application Server

AS-Autonomous System

As Built Data Log-As built data logs are records of installation routes and measurements for cables and assemblies. As built data logs help technicians to troubleshoot communication lines and systems as they provide locations and expected performance results (such as optical communication line losses) at the time the systems were installed and setup.

ASA-American Standards Association

ASAI-Adjunct Switch Application Interface

ASCAP-American Society of Composers, Authors and Publishers

ASCII-American Standard Code for Information Interchange

ASCII Extended Character Set-ASCII extended character set are the character codes with values from 128 to 255 in the ASCII Character Set. These codes are used to represent special characters and non-printing control commands to computer hardware. The codes in this range are not standardized and may be used by computer hardware manufactures and software developers for special purposes and therefore may not be compatible across different systems. Examples of use are for alternate language symbols, mathematical symbols and line drawing symbols.

ASCII Standard Character Set-ASCII standard character set is the American Standard Code for Information Interchange (ASCII) that uses an 8 bit binary code to represent the letters of the alphabet. The numerical values from 0 to 127 are known as the Standard ASCII set and contain the most commonly used punctuation symbols and non-printing control codes. The values from 128 to 255 are known as the Extended ASCII character set.

ASE-Accredited Systems Engineer

ASE-Application Service Element

ASF-Advanced Streaming Format

ASG-Access Service Group

ASI-Adapter Support Interface

ASI-Asynchronous Serial Interface

ASIC-Application Specific Integrated Circuit

ASLC-Analog Subscriber Loop Carrier

ASN-Autonomous System Number

ASN.1-Abstract Syntax Notation number 1

ASP-Abstract Service Primitive

ASP-Active Server Pages

ASP-Application Service Provider

Aspect Ratio-Aspect ratio is the ratio of the number of items (such as pixels on a screen) as compared to their width and height. The aspect ratio determines the frame shape of an image. The aspect ratio of the NTSC (analog television) standard is 4:3 for conventional monitors such as home television sets, and 16:9 for HDTV.

This figure shows how aspect ratio is the relationship between width and height expressed as width:height. This diagram shows that wide screen television has an aspect ratio of 16:9 and that standard television and computer monitors have an aspect ratio of 4:3.

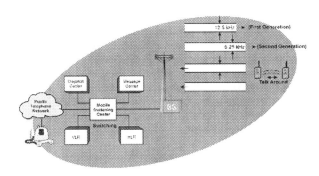

Aspect Ratio

ASPI-Advanced SCSI Programming Interface
ASR-Access Service Request
ASSB-Amplitude Compandored Single Sideband Modulation
Assemble Edit-Assemble edit is a process of combining (assembling) a series of media segments (clips) to produce a program.
Assembler-Assembler is an executable program that translates a set of statements written in assembly level programming language into the machine specific code that executes the instructions specified in the statements. Assembly language is very closely related to the actual instructions and architecture of the microprocessor being targeted and therefore assembly language instructions typically map one-for-one to machine level instructions. In a high-level language such as C, Fortran or Pascal; one statement may create dozens of machine level instructions.
Assembly Language-Assembly language is a low-level programming language that each statement corresponds to a single machine language instruction that a processor can execute. Assembly language is very closely related to the actual instructions and architecture of the microprocessor being targeted and therefore assembly language instructions typically map one-for-one to machine level instructions. Assembly languages are specific to a given microprocessor.
Assembly Line-Assembly line is a group of machines or assembly areas that form a sequence of operations that are used to assemble circuit boards or products.

Asset Management-Asset management is the process of acquiring, maintaining, distributing and the elimination of assets. Assets may be in the form of hardware (e.g. equipment), software (e.g. applications) or information content (e.g. media programs).
Asset Tracking-Asset tracking is the process of determining the location and or status of an item or asset.
Assigned Frequency Band-Assigned frequency band is the frequency band where emission of a mobile station (transmitter/receiver) is authorized at specific levels.
Assigned Pairs-Assigned pairs are communication lines (wire pairs) that are assigned for customer service. These lines may be working or idle.
Assignment Of A Radio Frequency Or Radio Frequency Channel-The sending of a channel assignment or authorization message that instructs a mobile station (mobile radio) to use a specific radio frequency, time slot, and/or channel code for a communication session (voice or data).
Assignment Of Authorization-Assignment of authorization is a transfer of authorization to provide communication services from one party to another.
Associate Session-Associate session is a communication session that is associated with another session. These sessions may be synchronized (such as video and audio).
Associated Broadcasting Station-Associated broadcasting station is a broadcasting station or stations that are part of a system or network that is licensed as an auxiliary broadcast location for which it is principally used.
Association-Association is the recognition or registration of a device (such as a wireless data terminal in an 802.11 WLAN system) by a node or access point that is part of the network. Association allows communication links to be established with a device that may be temporarily connected to specific points within a network.
Association for Radio Industries and Business (ARIB)-Association for Radio Industries and Business (ARIB) is an association in Japan that oversees the creation of telecommunications standards.
ASSP-Algorithm Specific Signal Processor
Assurance-Assurance is the process of making sure that customers receive the levels of service that they have purchased or agreed to purchase.

Assurance may include commitments to a high-level of overall customer satisfaction of quality of service.

Assurance Level-Assurance level is the level of probability that a service or product will meet a specific criteria or range of limits. Assurance level is often expressed as a percent. For example, there is 99% assurance (probability) that a dialtone will be available in a subscriber loop.

Assured Delivery-Assured delivery is a protocol said to provide assured delivery if each packet is guaranteed to be delivered. The sender accomplishes this via receiver acknowledgement and retransmission when packets are not acknowledged. Examples of protocols that provide assured delivery are TCP/IP and IEEE 802.2 LLC connection oriented services.

Asterisk-Asterisk is the symbol denoted with *. This symbol is often used as a wildcard within a computer operating and file system. It is typically used to match one or more characters within a specific filename. For example a file system with the files Dog.txt, Hog.txt and Fog.txt would return all three files when asked for *og.txt.

ASU-Alarm Suppressor Unit

ASX-Advanced Streaming Index

Asymmetric Digital Subscriber Line 2 (ADSL2)-Asymmetric digital subscriber line 2 (ADSLs) is an evolved version of ADSL that uses a more advanced modulation technology over the 1.1 MHz of frequency bandwidth to increase the data transmission rate up to 12 Mbps (downstream).

Asymmetric Protocol-An asymmetric protocol is a set of commands and processes that have differences between reception, initiation and functions performed by the protocol commands as a result of which direction the command travels or based on the type of device that sends or receives the protocol.

Asymmetric-Asymmetric channels are two-way communication channels that allow for transmission rates that can vary by direction. For example, the downlink broadcast channel may be a high-speed channel (e.g. 1.9 Mbps) and an uplink (reverse direction) channel may only be 15 kbps.

Asymmetric Codec-An asymmetric codec is a coder/decoder compression device or process (software program) that uses different compression and decompression processes. Asymmetric codecs often have more complex compression processes and less complex (more efficient) decoding/playback processes.

Asymmetric Coding-Asymmetric coding is the process of changing information into a different form or set of codes (coding) prior to transferring or transmitting the information to a receiver that decodes the information and the coding process on the transmitter side differs from the decoding process on the receiver side. An example of asymmetric coding is the compression of video information where the coding process on the transmitter side may be much more complex than the decoding process.

Asymmetric Compression-Asymmetric compression are techniques where the computational complexity of the decoding process is not the same as that of the compression process. Many compression techniques are asymmetric because only the encoding process requires a search over possible choices of encodings or requires analysis of the input. There are many compression scenarios where asymmetric compression is tolerable, for example: decoding will be done many times for each encoding; decoding needs to be done on a device with much less computational power than the encoding device; encoding and decoding will be done by each device so that the total computing power required on each device is equal.

Asymmetric Cryptography-Asymmetric cryptography is the use of separate mathematical functions for the encoding and decoding of information where one key is used for the encoding of information and a separate key is used for decoding.

Asymmetric Digital Subscriber Line (ADSL)-Asymmetric digital subscriber line (ADSL) is a communication system that transfers both analog and digital information on a copper wire pair. The analog information can be a standard POTS or ISDN signal. The maximum downstream digital transmission rate (data rate to the end user) can vary from 1.5 Mbps to 9 Mbps downstream and the maximum upstream digital transmission rate (from the customer to the network) varies from 16 kbps to approximately 800 kbps. The data transmission rate varies depending on distance, line distortion and settings from the ADSL service provider.

This figure shows that a typical ADSL system can allow a single copper access line (twisted pair) to be connected to different networks. These include the public switched telephone network (PSTN) and the

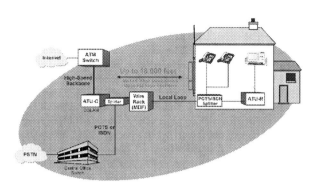

Asymetric Digital Subscriber Line (ADSL) System

data communications network (usually the Internet or media server). The ability of ADSL systems to combine and separate low frequency signal (POTS or IDSN) is made possible through the use of a splitter. The splitter is composed of two frequency filters; one for low pass and one for high pass. The DSL modems are ADSL transceiver units at the central office (ATU-C) and the ADSL transceiver unit at the remote home or business (ATU-R). The digital subscriber line access module (DSLAM) is connected to the access line via the main distribution frame (MDF). The MDF is the termination point of copper access lines that connect end users to the central office.

Asymmetric Encryption-Asymmetric encryption is the process of encoding and decoding of voice or data information so that it cannot be used by unauthorized users through the use of different keys for the encryption and decryption process. Asymmetric encryption involves a pair of matched keys: one key to lock and another key to unlock the data. Usually both keys can perform locking and unlocking of data: Key A can only unlock things locked with key B, and key B can only unlock things locked with key A.

Asymmetric Key Encryption-Asymmetric key encryption is the process of using the different keys for encryption and decryption.

Asymmetrical-(1-General) Asymmetrical transmission is two-way communication that has different data transmission rates in send (forward) and receive (reverse) directions. (2-Bluetooth) Asymmetrical is a type of Asynchronous Connectionless (ACL) link that operates at two different speeds in the upstream and downstream directions. An example of an asymmetrical connection is the Bluetooth ACL link. The Bluetooth specification specifies a maximum data rate of up to 723.2 Kbps in the downstream direction, while permitting 57.6 Kbps in the upstream direction. See also symmetrical.

Asymmetrical Private Virtual Circuit (Asymmetrical PVC)-Asymmetrical private virtual circuit is a virtual circuit that permits uneven (asymmetrical) data transmission rates for each direction of transmission.

Asymmetrical PVC-Asymmetrical Private Virtual Circuit

Asymptotic Coding Gain-Asymptotic coding gain is a processing gain of a coding system that can be obtained when the signal to noise ratio (SNR) approaches infinity.

Asynchronous-Asynchronous channels are dynamically adjusted channels that do not have a fixed synchronization with some other reference signal. The communications on an asynchronous channel is not sequential and may appear random or unbalanced in nature.

This diagram shows the process of data transmission using asynchronous (unscheduled) transmission. In this example, each message or block of data that is transmitted in an asynchronous data communication system must include indicators (delimiters) that identify the start and stop of a block of data. The blocks of data usually include some bits of information that are dedicated for flow control (e.g. routing and/or error protection).

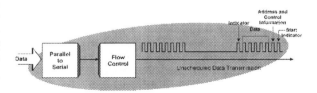

Asynchronous Data Transmission

Asynchronous Communication System- Asynchronous communication system is a data communication system that dynamically can send and receive data blocks where data blocks are individually synchronized.

Asynchronous Devices- Asynchronous devices are devices that transmit communication signals at irregular time intervals. Examples of asynchronous devices include mobile telephones and local area network data communication devices.

Asynchronous Interface Protocol- Asynchronous interface protocol is a protocol that enables network signaling messages to be sent to a data terminal as an ASCII message. This protocol was developed by the Consultative Committee for International Telephony And Telegraphy (CCITT) for access to public packet-switched service.

Asynchronous Serial Interface (ASI)- Asynchronous serial interface is a serial transmission format standard that is used by digital video broadcasting (DVB) to transport MPEG-2 digital video signals.

Asynchronous Time-Division Multiplexing- Asynchronous time-division multiplexing is an asynchronous transmission system that uses time slots to transfer uncoordinated data transfer (non-synchronized).

Asynchronous Transfer Mode (ATM)- Asynchronous transfer mode (ATM) is a packet data and switching technique that transfers information by using fixed length 53 byte cells. The ATM system uses high-speed transmission (155 Mbps) and is a connection-based system. When an ATM circuit is established, a patch through multiple switches is setup and remains in place until the connection is completed. ATM service was developed to allow one communication medium (high speed packet data) to provide for voice, data and video service.

As of the 1990's, ATM has become a standard for high-speed digital backbone networks. ATM networks are widely used by large telecommunications service providers to interconnect their network parts (e.g. DSLAMs and Routers). ATM aggregators operate networks that consolidate data traffic from multiple feeders (such as DSL lines and ISP links) to transport different types of media (voice, data and video).

This figure shows a functional diagram of an ATM packet switching system. This diagram shows that there are three signal sources going through an ATM network to different destinations. The audio signal source (signal 1) is a 64 kbps voice circuit. The data from the voice circuit is divided into short packets and sent to ATM switch 1. ATM switch 1 looks in its routing table and determines the packet is destined for ATM switch 4 and ATM switch 4 adapts (slows down the transmission speed) and routes it to it destination voice circuit. The routing from ATM switch 1 to ATM switch 4 is accomplished by assigning the ATM packet a virtual circuit identifier (VCI) that ATM switch can understand (the packet routing address). This VCI code remains for the duration of the communication. The second signal source is a 384 kbps Internet session. ATM switch 1 determines the destination of these packets is ATM switch 4 through ATM switch 3. The third signal source is a 1 Mbps digital video signal from a digital video camera. ATM switch 1 determines this signal is destined for ATM switch 4 for a digital television. In this case, the communication path is through ATM switches 1, 2, and 4.

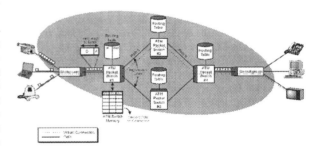

Asynchronous Transfer Mode (ATM) System

Asynchronous Transfer Mode 25 Mbps (ATM25)- A 25 Mbps version of ATM. The ATM25 standard was developed primarily for corporate networks. However, the QoS advantages of ATM and customer needs for switched services for digital video and Internet access has stimulated interest in ATM for the DSL industry.

AT- Access Tandem

AT Command Set- AT command set is a simple communication language (list of commands) that is used to setup and control modulator/demodulator (modem) devices. The AT command set was developed by Hayes Microcomputer Products to control their modems. Some of the commands include telephone dialing (pulse or tone), adjustment the audio volumes, data transmission rates, and programming modem on how to answer incoming calls.

AT&T Consent Decree-AT&T consent degree is the order for AT&T to divest parts (the break-up) of its business and the restriction not to allow AT&T to provide local telephone service.

ATA-Advanced Technology Attachment

ATA-Analog Telephone Adapter

ATA-Analog Terminal Adapter

ATB-All Trunks Busy

ATC-Autotune Combiner

ATG-Air-To-Ground

ATIS-Alliance for Telecommunications Industry Solutions

ATM-Asynchronous Transfer Mode

ATM Adaptation Layer (AAL)-ATM adaptation layer (AAL) is a set of standard protocols that allow for the conversion (translation) of user traffic (such as IP or Ethernet data packets) into a size and format that can be contained in the small payload of an ATM cell. User traffic is converted (segmented) from its original form into packets sizes that can be sent inside small ATM packets. After they have been transferred through the ATM network, they are assembled back into their original form at the destination.

All AAL functions occur at the ATM end-station rather than at the switch. These protocol layers are designed to allow for constant bit rate (CBR), variable bit rate (VBR), unspecified bit rate (UBR), and other types of services.

ATM Adaptation Layer 1 (AAL1)-ATM adaptation layer 1 (AAL1) is the layer within the ATM protocol that converts the 53 byte packets from the network into the form used by the customer for constant bit rate (CBR) services.

ATM Adaptation Layer 2 (AAL2)-ATM adaptation layer 2 (AAL2) is the layer within the ATM protocol that converts the 53 byte packets from the network into the form used by the customer for variable bit rate (VBR) services. AAL2 Supports connection oriented traffic such as compressed voice and data.

ATM Adaptation Layer 5 (AAL5)-ATM adaptation layer 5 (AAL5) is a more efficient class of service for ATM than AAL1. AAL5 was previously called Simple and Efficient AAL (SEAL).

ATM Address-ATM address is a 20 byte address used in asynchronous transfer mode (ATM) systems that identify the country, area, and end-system identifiers. ATM address formats are defined in the user network interface (UNI) specification.

ATM Backbone Switch-ATM backbone switch is a switch that receives and forward standard 53 byte asynchronous transfer mode (ATM) packets and is located in the interconnection backbone networks of carriers to interconnect slower switches or edge (network interface) switches. Because the ATM backbone switch can prioritize packets for different types of services based with varying bandwidth and quality of service (QoS) requirements, ATM multi-service switches are well suited as backbone switches. The processing of standard size packets allow ATM backbone switches to provide bandwidth over 200 Gbps.

ATM Cell-ATM cell is a 53 byte packet of data (called a "cell") that is used in an ATM network. An ATM cell is usually divided into a 5 byte header and 48 byte payload. The ATM header is primarily used for local connection routing information to the next switching point.

ATM Forum-ATM forum is a forum that was started to assist manufacturers and service providers with marketing and development of ATM products and services. The ATM forum was started in 1991.

ATM header-ATM header is the 5 byte portion of the 53 byte ATM cell that contains the addressing information for the ATM cell. There are different ATM cell header structures; user network interface (UNI) and network to network interface (NNI). The header structure for UNI provides for end-to-end addressing and the NNI structure only provides for inter-network addressing.

ATM Packet-ATM packet is a 53 byte packet that contains an 5 byte header and 48 byte payload.

This diagram shows that ATM packets have a fixed length of 53 bytes. Of these, 5 bytes are used for address and control information and 48 bytes are dedicated for data. This diagram also shows that the ATM packet structure varies dependent on its use. When it is used for user to network interface (UNI), it contains additional generic flow control (GFC) bits. When it is used for network to network interface (NNI) communication, it uses the additional bits for additional virtual path indicator (VPI) channels.

ATM Passive Optical Network (APON)-An ATM passive optical network (APON) combines, routes, and separates optical signals through the use of passive optical filters and ATM protocol. The APON distributes and routes signals without the need to convert them to electrical signals for routing through switches.

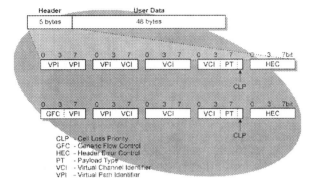

ATM Packet Structure

ATM Switch-ATM switch is a packet switch in an ATM network that receives and forwards fixed length 53 byte cells. The ATM switch is connection oriented so the paths for packets are pre-established at the beginning of a communication session. The ATM switch can prioritize the routing of packets based on the routing address or content of the packet.

ATM25-Asynchronous Transfer Mode 25 Mbps

Atom-An atom is a segment of content within a media file. Atoms may have characteristics such as time duration and index pointers and they can be structured in a hierarchy to allow for rapid development and coordination of complex movies.

Atomizing-Atomizing is the process of breaking down the average size of a media product or program into its smallest usable segments (smallest chunks).

ATP-AppleTalk Transaction Protocol

ATS-Abstract Test Suite

ATSC-Advanced Television Systems Committee

Attach-Attach is the process of establishing a connection (a physical and/or logical connection) between a client (e.g. workstation) and a network server (e.g. a file server).

Attach Terminal-Attach terminal is the process that is used by a software application program that assigns a computer terminal for exclusive use by the application.

Attachment-An attachment to an electronic mail is non-text data included in an e-mail using Multipurpose Internet Mail Extensions (MIME). An e-mail message may contain any number of attachments. Each attachment has a "MIME type" property that suggests to the user's e-mail application the data type of the attachment, for example HTML text or a JPEG image. Depending on the software, in some cases the user's e-mail application will display attachments within the e-mail application (like JPEG images) and in other cases an attempt will be made to open an external application, sometimes requiring assistance or permission from the user, or the user may save the attachment as a file.

Attack-Attack, in network and computer security parlance, is an attempt to disable or gain unauthorized access to a computer system or network by exploiting weaknesses within the operating system or implemented security measures.

Attempt-Attempt is any process of requesting or demanding service from a communications system.

Attendant-An attendant is a person who answers, screens, or directs calls in a communication system.

Attendant Call Waiting Indication-Attendant call waiting indication is the indication light or message on the attendant console that indicates that one or several calls are in queue to be answered. The indication may change (e.g. flash or ring) when additional thresholds (e.g. maximum number of waiting calls) are reached.

Attendant Conference-Attendant conference is a PBX system feature that permits an attendant to establish a conference call between the public telephone lines and PBX extensions.

Attendant Console-An attendant console is a specialized telephone or computer device that is used by the operator or attendant to answer incoming calls and allow the call attendant to reroute the calls to other extensions.

Attendant Forced Release-Attendant forced release is an attendant or operator activated release of a call that will "disconnect" all parties on that circuit entered by the attendant.

Attendant Switchboard-An attendant switchboard is a communication device (phone and display) that provides the ability for a receptionist (attendant) to identify and answer incoming calls, interact with callers, and redirect (transfer) calls to the proper extension. Attendant switchboard consoles in an IPBX system are software programs that typically operate on a standard multimedia computer. Attendant switchboards display incoming Caller ID information, have graphical call status indications (hold, in-use), allow quick access to company directories, and permit the simple transfer of calls through the use double-clicks.

Attention Identification (AID)-Attention identification (AID) is an access control attention identifier that is used for initiating communication between a data terminal and a host system.

Attenuation-Attenuation is the amount of signal energy (power or level) caused by a transmission channel. Attenuation can be caused by absorption (conversion of energy to heat), mismatching of transmission lines (causing signal energy to be reflected), or other forms of distortion that reduces the received signal level energy.

Attenuation to Crosstalk Radio (ACR)-Attenuation to crosstalk radio (ACR) is a comparison between signal attenuation and the amount of crosstalk. ACR is usually measured in decibels.

Attenuator-An attenuator is a device that reduces the signal energy (e.g. power) that is transferred through it. Attenuators are normally rated as a ratio in decibels (dB) as compared between the input and output signal levels.

Attributes-Attributes are the characteristics of an object or information element (e.g. file) that can be used to identify, qualify, or assist in the control of that object or element.

ATU-C-ADSL Transmission Unit - Central Office

ATU-R-ADSL Transmission Unit - Remote

ATV-Advanced Television

ATVA-Analog Television Adapter

ATVEF-Advanced Television Enhancement Forum

AU-Administrative Unit

AuC-Authentication Center

Audibility-Audibility is the quality of being able to be heard. For most humans, the frequency range of audibility extends roughly from 20 Hz to 20 kHz.

Audible-Audible is the range of sound that can be heard by a normal human.

Audible Sound-Audible sound spans a range of frequencies from around 20 hertz (cycles per second) to 20 kilohertz. To be audible, the power intensity of the sound must be at least 1.0E-12 watt/square meter. This acoustic power intensity level is also denoted as 0 acoustic decibels. (Note that 0 dB in electric power measurement is 1 milliwatt of power, a different base unit that does not involve an area but only a power measurement.)

Audience Measurement System (AMS)-An audience measurement system is a combination of equipment, protocols and transmission lines that are used to monitor, record and potentially analyze the selections and habits of television viewers.

Audio-Audio is a signal that is composed of frequencies that can be created and heard by humans. The frequency range for an audio signal typically ranges from 15 Hz to 20,000 Hz.

Audio Bridge-Audio bridge, in telecommunications, is a device that mixes multiple audio inputs then provides the composite audio back to each communication device, less that devices audio input. An example of an audio bridge is a conference call.

Audio Broadcast Services-Audio broadcast services are the transmission of program material (typically audio) that is typically paid for by advertising. Most commercial stations receive the bulk of their ad revenues from local advertising, as opposed to television, which gets most of its revenue from network advertising.

Audio Broadcasting-Audio broadcasting is the transmission by program material (typically audio) that can be simultaneously received by receivers that are capable of receiving and decoding the radio signal to recover the original audio or information signal.

Audio Chat Room-Audio chat rooms are real-time communication services that allow several participants (typically 10 to 20) to interact act much like an audio conference session. Audio conference chat rooms may be public (allow anyone to participate) or private (restricted to those with invitations or access codes.)

Audio Clipping-(1-signal) Signal distortion that results from the non-linear (non-equal) processing of a signal through a device or circuit (such as an amplifier) where some of the signal is lost due to the maximum or minimum processing ability of the device or circuit. (2-voice activity) A short time period of audio distortion (primarily muting) that occurs at the beginning of a speech or audio signal in a system that uses voice activity detection. This time period results from the lag of the voice activity detection (VAD) circuit to allow audio to pass through when audio signals have been detected.

Audio Conference-Audio conferencing (also called teleconferencing) is a process of conducting a meeting between two or more people through the use of telecommunications circuits and equipment.

Audio Digitization-Audio digitization is the conversion of analog audio signal into digital form. To convert analog audio signals to digital form, the analog signal is digitized by using an analog-to-digital (pronounced A to D) converter. The A/D converter periodically senses (samples) the level of the analog signal and creates a binary number or series of digital pulses that represent the level of the signal. The typical sampling rate for conversion of analog audio ranges from 8,000 samples per second (for telephone quality) to 44,000 samples per second (for music quality).

Audio Dubbing-Audio dubbing is the copying of audio information from one source to another without affecting the other forms of media (such as a video tape). An example of audio dubbing is the insertion of narration audio onto a video program.

Audio Frequency-Audio frequency is the frequency in Hertz (Hz) of an audio signal. The typical adult can hear audio frequencies between 20 Hertz and 20000 Hertz. Audio compact discs can play frequencies up to 22050 Hertz.

Audio Messaging Interchange Specification (AMIS)-An industry specification that defines how messages are exchanged on a network between voice mail systems.

Audio Programming Services-Audio programming services are the providing of information content (programs) by a communication system operator (such as a radio broadcast station).

Audio Quality-Audio quality is the ability of an audio device or transfer system to recreate the key characteristics of an original audio signal or sound.

Audio Response Unit (ARU)-Audio response unit is a device or system that can translate data files (usually stored on a computer) into audio voice messages.

Audio Segment-An audio segment is a portion of a media file (such as a video or multimedia file) that contains the audio media.

Audio Streaming-Audio streaming is a real-time system for delivering audio, typically over the Internet. Upon request, a server system will deliver a stream of audio (usually compressed) to a client. The client will receive the data stream and (after a short buffering delay) decode the audio and play it to a user. Internet audio streaming systems are used for delivering audio from 2 kbps (for telephone-quality speech) up to hundreds of kbps (for audiophile-quality music).

Audio Synchronization-Audio synchronization is a process that adjusts the timing of the audio signals to match the presentation of other media (such as video or a slide presentation).

Audio Track-An audio track is a section of a media file (such as a video tape or multimedia file) that is associated with a video signal. Multiple audio tracks may be used to provide stereo or additional independent soundtracks (such as for other languages).

Audio Video Distribution Transport Protocol (AVDTP)-Audio video distribution transport protocol is the communication messages and processes that are used for audio and video distribution.

Audio Video Interleaved (AVI)-Audio video interleaved (AVI) is a Microsoft multimedia digital video format that interleaves digital audio and digital video frames into a common file. AVI files contain an index file of the media components.

Audio Video Transport (AVT)-Audio video transport (AVT) is an IETF working group that is responsible for audio and video transport protocols.

Audio Watermarking-Audio watermarking is a process of adding or changing information in an analog or digital audio media tape, streaming media or other form of audio media to uniquely identify the media and/or its authorized uses. Audio watermarking may be performed by adding audio tones above the normal frequency or by modifying the frequencies and volume level of the audio in such a way that the listener does not notice the watermarking information.

This figure shows how digital watermarks can be added to digital audio to provide identification information. The digital watermark is added as a

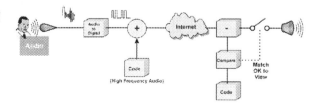

Audio Watermarking

A

high frequency audio component that is typically not perceivable to the listener of view of the media.

Audiovisual Objects-Audiovisual objects are parts of media images (media elements). Media images or moving pictures may be analyzed and divided into audiovisual objects to allow for improved media compression or audiovisual objects may be combined to form new images or media programs (synthetic video).

Audit Trail-An audit trail is the availability of information elements (such as financial transactions and usage events) that can be linked to determine the origin and history of the usage of a product, service or information (content).

Augmented Backus-Naur Format (ABNF)-Augmented backus-naur format (ABNF) is a metalanguage syntax used in defining and utilizing the session initiation protocol (SIP) as defined in RFC2543. An example ABNF construct used to describe a SIP message is as follows: SIP-message = Request | Response.

Authentication-Authentication is a process of exchanging information between a communications device (typically a user device such as a mobile phone or computing device) and a communications network that allows the carrier or network operator to confirm the true identity of the user (or device). This validation of the authenticity of the user or device allows a service provider to deny service to users that cannot be identified. Thus, authentication inhibits fraudulent use of a communication device that does not contain the proper identification information.

This figure shows the operation of a basic authentication process used in a radio communication system. As part of a typical authentication process, a random number that changes periodically (RAND) is sent from the base station. This number is regularly received and temporarily stored by the mobile radio. The random number is then processed with the shared secret data that has been previously stored in the mobile radio along with other information in the subscriber to create an authentication response (AUTHR). The authentication response is sent back to the system to validate the mobile radio. The system processes the same information to create its own authentication response. If both the authentication responses match, service may be provided. This process avoids sending any secret information over the radio communication channel.

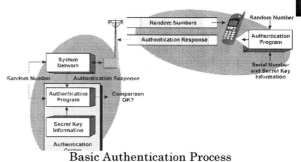

Basic Authentication Process

Authentication Center (AuC)-Authentication center (AuC) is a part of a network that manages the encryption keys that validate the identity of customers and enable voice privacy services. A single authentication center may process validation requests using different keys, random numbers and encryption algorithms.

Authentication Credentials-Authentication credentials are the information elements that are used to identify and validate the identity of a person, company or device. Authentication credentials may include identification codes, service access codes and secret keys.

Authentication Key (A-key)-Authentication key (A-key) is a secret key that is stored in the cellular or PCS system and entered into a mobile phone to create the shared secret data (SSD) that is used to validate the identity of the subscriber. See also shared secret data (SSD).

Authentication Procedure-Authentication procedure is the sequence of steps carried out by two end points of a communication system to exchange the information necessary to insure that some aspect of the communication session are valid. This may include user validation, data validation or service validation.

Authentication URL-Authentication URL is a URL or address for a device (such as an IP telephone) where the authentication information is maintained to allow the proper security for incoming information.

Authentication via Challenge-Response -A method of authentication used by digital cellular telephones, military communication systems, Microsoft Windows 2000 and other systems. This authentication process responds with a "challenge" number value (different for each instance of authentication) when a user contacts a server, upon which the user then performs a cryptographic operation, and then returns the result to the server. The user equipment and the server both contain a copy of an internal secret number, used in the cryptographic operation. The server also performs the same cryptographic operation, and if the two results are the same, the user is considered authentic. Although the transactions can be intercepted and observed by an eavesdropper, they cannot be imitated or used to produce a false authentication, because different numbers are exchanged in each instance of the authentication process.

Authentication, Authorization, Accounting (AAA)-Authentication, Authorization, and Accounting (AAA) are the processes used of validating the claimed identity of an end user or a device, such as a host, server, switch, or router in a communication network. Authorization is the act of granting access rights to a user, groups of users, system, or a process. Accounting is the method to establish who, or what, performed a certain action, such as tracking user connection and logging system users.

Authoring Tools-Authoring tools are software and/or hardware tools to allow creation of a digital media presentation from multiple source media. For example, a video authoring tool may allow editing of video clips into a single video presentation, while a web authoring tool may allow creation of graphical and text content for pages on a web site. Authoring tools may also function as encoding tools.

Authorization-Authorization is the enabling of services to a device or customer that requests services. Authorization is often part of the billing and customer care (BCC) system and is maintained in a customer database service profile.

Services are initially enabled in a network as a result of provisioning. Provisioning is a process within a company that allows for establishment of new accounts, activation, termination of features, and coordinating and dispatching the resources necessary to fill those service orders. Provisioning is usually part of customer care systems.

Authorized Billing Agent-An authorized billing agent is a person or company that is used by a telecommunications service provider to perform billing and collection services.

Authorized Domain-An authorized domain is a group of users who are provided access to and possibly allowed to distribute media or information to other users within their authorized domain. Examples of an authorized domain are users or devices within a household.

Auto Attendant-The auto attendant feature is used to route incoming telephone calls based on selections or information provided by the incoming caller. The auto attendant feature may use interactive voice response (IVR) to prompt the caller to select the call routing based on category choices or it may use the calling number identification to determine the destination (e.g. a telephone number for a specific sales group).

Auto Configuration-Auto configuration is the process of allowing a device or system to detect, initialize, update or program features and parameters without the direct assistance of a user or technician.

Auto Dial Number in Short Message-Auto dial in short message is a feature that allows a telephone (typically a mobile telephone) to automatically call a telephone number that is part of a short message.

Auto Discovery-Automatic discovery is a process where a network manager automatically searches through a range of network addresses and discovers specific types or all types of devices present in that range of addresses. The auto discovery process may be manually initiated by the network administrator or it may be initiated after a new device automatically registers after it is connected to the network.

Auto Redial-Audio redial is the ability for a telephone to repeatedly dial a telephone number in the event the first dialing attempt is unsuccessful. In a wireless system, the maximum number of attempts for redialing may be limited to ensure the system does not become overloaded with requests for service.

Auto Responder (Autoresponder)-Autoresponders are software programs that automatically create response messages when incoming messages are received. Autoresponder capabilities vary from simple responses to incoming email messages such as "I'm currently out of the office, will be back," to personalized letters such as "Dear Tom, Thanks for requesting more information on our ABC product, from the information you gave use, we...".

Auto Sensing-Auto sensing is a process during

which a network device automatically senses the speed of another device. For example, Ethernet hardware with auto-sensing 10 base-T / 100 base-T ports will automatically determine which of those two network speeds is supported by a device connected to the port, then use the faster choice if possible.

Autodialer-An autodialer is a machine or assembly that is programmed to dial a list of telephone numbers that dialed automatically. In some cases, the autodialer may be used to dial numbers and automatically detect voice activity (e.g. someone answers the call) and then connect the call to an available customer service representative.

AUTODIN-Automatic Digital Network

Auto-Discovery-Auto-discovery is a method of discovering devices or services that may be attached or disconnected from a system or network. Auto-discovery is typically performed via layer 3 network connectivity and there are many vendors who have written network auto-discovery tools. Cisco's Cisco discovery protocol (CDP) is an example of a discovery tool Cisco uses to identify it's equipment connections and parameters.

Automated Attendant System-Automated attendant system is a processor control system that performs telephone console attendant functions such as answering a call, transferring callers to specific user stations, directing callers to voice mail, or performing other related call-routing functions without the assistance of a live attendant. The caller's activation's of these features occurs through pressing keys that activate DTMF signaling.

This diagram shows how computer telephony systems can be used to create virtual (simulated) call attendants. In this diagram, a call is received to the main telephone number of the company to the computer telephony board. The automated telephony call processing software detects a ring signal, answers the phone (creates an off-hook signal) and plays a pre-recorded message informing the caller of options they may choose to direct the call to a specific extension. In this example, the automated call attendant software decodes DTMF tones or limited list of voice commands to determine the routing of the call. The automated call attendant software then determines if the destination choice is within the option list and if the extension is available. If the extension is available, the automated attendant will send a command to the com-

puter telephony board switching the call to the selected extension. If the extension is not valid or not available, the automated attendant will provide a new voice prompt with updated information and additional options.

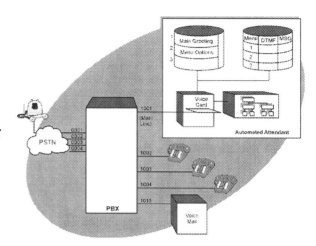

Automated Attendant Operation

Automated Playout-Automated playout is the process of streaming or transferring media to a user or distributor of the media at a predetermined time, schedule or when specific criteria have been met (such as when user registration and payment).

Automated Voice Response Systems (AVRS)-Automated voice response systems is a system that will automatically answer an incoming telephone call and provide voice instructions or information to the caller. The caller's response to these instructions may be keypad tones or even spoken words and will be used by the system to route the call to the appropriate extension or to other sources of additional information.

Automatic Busy Redial-Automatic bust redial is a feature for automatically redialing a busy number at certain intervals.

Automatic Call Distribution (ACD)-ACD is a system that automatically distributes incoming telephone to specific telephone sets or stations calls based on the characteristics of the call. These characteristics can include an incoming phone number or options selected by a caller using an interactive voice response (IVR) system. ACD is the process of management and control of incoming calls so that the calls are distributed evenly to attendant positions. Calls are served in the approximate order of their arrival and are routed to service positions as positions become available for handling calls.

This figure shows a sample automatic call distribution (ACD) system that uses an interactive voice response (IVR) system to determine call routing. When an incoming call is initially received, the ACD system coordinates with the IVR system to determine the customer's selection. The ACD system then looks into the databases to retrieve the customers' account or other relevant information and transfer the call through the PBX to a qualified customer service representative (CSR). This diagram also shows that the ACD system may also transfer customer or related product information to the CSR.

(caller's phone carrier) informs the remote (distant) switch of automatic callback request. This reserves (blocks) the called number from receiving additional calls until the automatic callback service is completed. When the called number becomes available, the remote switch sends a message to the local switch and this rings the original caller's number (possibly with distinctive ring feature.)

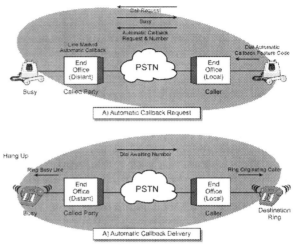

Automatic Callback

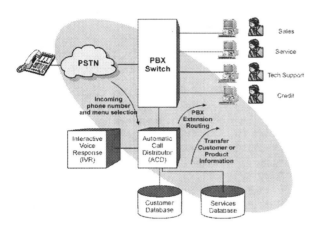

Automatic Call Distribution (ACD) Operation

Automatic Callback-Automatic callback is a CLASS service feature that allows a caller to complete a call to a busy station by dialing an activation code (usually a single digit) and hanging up. The system automatically rings both parties when the lines are available.

This figure shows the basic operation of automatic callback. To activate automatic callback service, after a call has dialed a number that is busy, the customer dials an automatic callback feature code and hangs up the telephone. The local switch

Automatic Gain Control (AGC)-Automatic gain control (AGC) is an assembly or circuit that is part of a communications receiver that automatically adjusts the received signal level so that its level is approximately the same regardless of the received radio signal level. AGC is often used to supply a constant level signal to a demodulator assembly.

This diagram shows how a varying level signal that is supplied to a communication receiver can be adjusted to a fairly constant signal by an automatic gain control (AGC) system. This diagram shows that a varying radio signal is supplied to a signal level detector (diode) and a variable gain amplifier. The output of the detector is used to inversely vary the gain of the amplifier. As a result, the amplifier produces a near constant signal level that can be provided to the demodulator assembly.Control (AGC) Operation

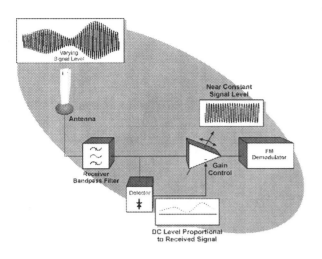

Automatic Gain Control (AGC) Operation

Automatic Ingestion-Automatic ingestion is the process of automatically transferring media into a storage or content management system through a system that has established rules, schedules or procedures.

Automatic Location Identifier (ALI)- Automatic location identifier (ALI) is a number (such as a telephone number or MAC address) that identifies a location of a device or assembly.

Automatic Message Accounting (AMA)- Automatic message accounting (AMA) is an automatic system for recording data describing the origination time of day, dialed number and time duration of a call for purposes of billing. The earliest systems used punched paper tape, later replaced by magnetic computer tape and then later magnetic computer disk. AMA is a term mostly used in the public network, and similar terms, some used in private, PBX, or inter-carrier systems are Call Detail Recording (CDR), Station Detail Message Recording (SMDR), and Automatic (calling) Number Identification (ACNI or ANI).

Automatic Number Identification (ANI)- Automatic number identification (ANI) is the providing of the originating telephone number, including an extension number in a Centrex system or PBX system. The ANI is an administrative number provided by the telephone system and may not be the actual originating number.

Automatic Retransmission Request (ARQ)- Automatic retransmission request is an acknowledgment process whereby the sending device can retransmit blocks of data that were received incorrectly at the receiving device.

Automatic Roaming-Automatic roaming is the ability of a customer (such as a mobile telephone user) to make and receive calls automatically outside of the customer's home area.

Autonomous Cells (system)-Autonomous cells sit in a wireless network that are not directly controlled by the mobile switching center for the normal assignment of calls. These autonomous cells may be used as private systems and are not listed as a potential candidate for call transfer for neighboring cells that are part of the public system.

Autonomous Registration-Autonomous registration is a process where a mobile radio independently transmits information to a wireless system that informs it that it is available and operating in the system. This allows the system to send paging alerts and command messages to the mobile radio. The mobile radio may be stimulated to register with the system when it detections it has entered into a new radio coverage area or it detects a registration request message.

Autonomous System Number (ASN)- Autonomous system number (ASN) is a unique number that identifies autonomous systems connected to the Internet. ASNs are assigned by the InterNIC and are used by routing protocols such as Border Gateway Protocol (BGP) to identify the autonomous system.

Autoresponder-Auto Responder

AUTOSEVOCOM-Automatic Secure Voice Communications Network

AUTOVON-Automatic Voice Network

Auxiliary Joint-Auxiliary joint is a lead sleeve installed over polysheath telco cable that will provide a soldering surface to facilitate the subsequent soldering of a main sleeve as a splice closure.

Avail-An 'avail' is the time slot within which an advertisement is placed. Avail time periods usually are available in standard lengths of 10, 20, 30 or 40 seconds each. Through the use of addressable advertising, which may provide access to hundreds of thousands of ads with different time lengths, it is possible for many different advertisements, going to different audiences to share a single avail.

Availability-Availability is a measurement that indicates the connection status or a commitment to provide a minimum amount of connection status of a network during a period of time. Availability may be measured by a connection time or by a minimum performance measurement (e.g. at a minimum data transfer rate). Availability is often tied to reliability.

Available Bit Rate (ABR)-Available bit rate (ABR) is a communications service category that provides the user with a data transmission rate that varies dependent on the availability of the network resources. ABR service may provide the user with feedback as to the changed data transfer rate and may have established minimum and maximum levels of data transmission rates.

Avalanche Photodiode (APD)-Avalanche photodiode is a semiconductor diode device that has a barrier region that is very sensitive to photons and the reception of photons causes an avalanche of electrons across p/n junction. This causes a rapid increase in current with the detection of light (photon) energy. APDs are used in optical networks to detect light, typically at the receiver. It is a semiconductor pn junction diode that is heavily biased so that each time a photon is absorbed by the device, the resulting conduction electron has so much energy that it can knock loose many other electrons and each of those electrons can generate others, resulting in an avalanche of electrons and therefore a large current. This causes a rapid increase in current with the detection of light (photon) energy. APDs can detect faint optical signals. However, APDs require higher voltages than other types of semiconductor devices.

This figure shows the basic operation of an optical APD photodetector. This example shows that an avalanche photodetector is composed of a p-type and n-type semiconductor material that is separated by an intrinsic material. A reverse bias is applied to the PIN causing a small amount of leakage current flows between the anode and cathode of the PIN detector. This example shows an APD is designed to allow the reception of a photon to cause an avalanche of many electrons making an APD many times more sensitive than a PIN detector.

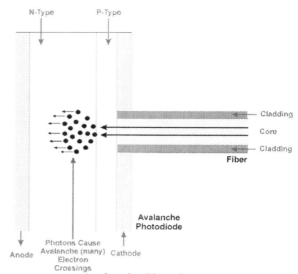

Avalanche Photodetector

AVC-Advanced Video Coding
AVC-Analog Voice Channel
AVDTP-Audio Video Distribution Transport Protocol
Average Power-Average power is the measure of power or energy that occurs over a period of time.
Average Revenue Per User (ARPU)-Average revenue per user (ARPU) is an indicator of a service operators business's operating performance. ARPU measures the average monthly revenue generated for each customer unit, such as a cellular phone or cable television customer that a carrier has in operation. Severely declining ARPU typically is a negative sign that may indicate a carrier is adding too many low-revenue generating customers to its subscriber base.
Average Transfer Delay-Average transfer delay is the average time that occurs between the sending of packets from their originating point to complete delivery at the destination station.
AVI-Audio Video Interleaved
AVI 2.0-Audio Video Interleaved 2
AVL-Automatic Vehicle Location
AVRCP-Audio Video Remote Control Profile
AVRS-Automated Voice Response Systems
AVSD-Analog Simultaneous Voice and Data
AVT-Audio Video Transport
AVVID-Architecture for Voice, Video and Integrated Data
AWG-American Wire Gauge

AWGN-Additive White Gaussian Noise

AWOS-Automatic Weather Observation Station

Axial Splice Enclosure-An axial splice enclosure is a container that holds splices and the cables enter in one end and leave from same end.

Axis-Axis is the straight line (center line) through the center of an object or objects (such as an optical fiber).

Azimuth-Azimuth is an angle measured (usually in degrees) in the horizontal plane between a reference direction (such as geographic north, or the strongest beam direction of an antenna) and a specific direction.

B

b-Bit

B-Byte

B 911-Basic 911

B Battery-Part of a telephone system power supply that supplies direct current for operating relays and other components. The B battery is typically 20 volts.

B Channel-A logical channel in an Integrated Services Digital Network (ISDN) that provides a 64 kbps connection to a switching system or to the non-switched portion of a network. This channel is used for transferring voice or data information between customer premises equipment and an end-office switching system.

B Connector-A connector that is used for splicing twisted wire pairs. B connectors are sometimes called beans. B connectors are shaped like a plastic tube that is approximately 2.5 cm (1 inch) long. They include metal teeth inside to help ensure a good connection is made when the B connector is crimped (compressed) and to ensure that the connector will not slide off the splice. Some B connectors contain a water-retardant jelly inside to reduce the effects of corrosion on the wire splice.

B Frame-Bidirectional Frame

B Wire-European term that corresponds to North American Ring wire.

B2BUA-Back to Back User Agent

B8ZS-Bipolar with 8 Zero Substitution

B8ZS Line Code-Bipower With 8-Zero Substitution

Babble-The aggregate crosstalk from many interfering channels.

Babbling Tributary-A communication device (station) that continuously transmits unnecessary messages.

Back Channel-A back channel is a communication channel that can transfer information (such as the channel quality information of the received channel) back from a receiver to the sender of the information.

Back Door-In an otherwise secure system, an intentional way for a trusted party to circumvent the security with secret knowledge such as the nature of a designed-in security flaw. Since back doors might be found and exploited by malicious parties, they weaken the security of what may appear to be a well-protected system.

Back End-A system or database server function that processes data via a network connection.

Back Haul-Back haul is the process of extending the use of communication facilities or more efficient circuits by using communicating routing lines that are longer than would be typical for a specific type of service. Back haul allows for cost effective sharing of facilities by either sharing network facilities (such as a switching system) or communication circuits (sharing long haul lines with many more users). An example of using back haul to reduce cost is the use of long distance lines to connect cellular radio towers to distant mobile cellular switching facilities. The use of back haul lines reduces or eliminates the need to install mobile switching equipment in a local system.

This diagram shows how long distance communication lines can be used to back haul communication circuits to a distant switching system to eliminate the need to install a local switching system. In this example, the distant mobile radio communication towers (cell sites) operate as they would if they were connected to a local switching system. The long distance back haul lines carry the control information and the voice communication channels.

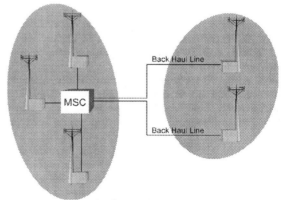

Back Haul System

Back Hoe Fade-A reduction in the ability of a communication system to route calls due to the cutting of a buried communication cable (e.g. fiber optic cable). The reduction in capacity comes from the automatic re-routing of communication circuits through other systems that have a lower capacity than the original communication circuit.

Back Nine-A sum of series episodes ordered by networks used to complete a full season of the required number of episodes.

Back Office Operations-Back office operations are processes and systems that are used to assist with the operation and management of communication systems. Back office operations may include billing and customer care systems, accounting, maintenance services and asset management.

Back Porch-The short time interval in an analog television video wave form immediately following the horizontal synchronizing pulse. The color synchronizing burst is transmitted during the back porch time interval. See the figure appearing with the term Horizontal Scan.

Back Pulling-Back pulling is the process of pulling a cable through a duct or pipe from a point that is between the end points. Back pulling may be performed to minimize the stress (pulling tension) on the cable or because it is not possible to feed the cable at one or both of the endpoints.

Back to Back User Agent (B2BUA)-A user agent (UA) that is part of a SIP system that provides some advanced call processing features.

Back Wrapping-Back wrapping is the unwinding of cable from a cable spool (cable reel) due to the continued rotation of the cable reel after the cable pulling has stopped.

Backbone-A network backbone is the core infrastructure of a network that connects several major network components together. A backbone system is usually a high-speed communications network such as ATM or FDDI.

Backbone Network-A communications network that connects the primary switches or nodes within the network. The backbone network is usually composed of high-speed switches and communication lines.

Backbone Route-A high-capacity transmission line or combination of circuits that interconnects various multiplexing or distribution points. Backbone routes are efficient transmission facilities to allow communication traffic aggregation at selected points to maintain the circuit loading (percentage of user of the backbone transmission line).

Backcharging-A process of charging for usage of a service when the service request is initiated rather than when the service is connected. An example of backcharging is the process of starting the billing

time on a mobile telephone when the user initiates the call, not when the call is actually connected.

Backdoor Attack-A backdoor attack is an attempt to disable or gain unauthorized access to a computer system or network by exploiting a designed-in (backdoor) security flaw. Since back doors might be found and exploited by malicious parties, they may be difficult to detect.

Back-End Processor-This is another computer system or dedicated microprocessor that is optimized to perform a specialized task in order to offload work from the main processing resources.

Backend System-Backend systems perform supporting functions for business operations. Examples of backend systems include billing, customer service and inventory management.

Background-Background class is the delivering media or data that operates concurrently with other media transfers without interfering with the normal operation of the other media transfers (transferring in the background). Background class may be used for applications such as email or web browsing that do not require real time interaction.

Background Communication-Communication that is simultaneously performed on another channel than the communication channel that is in use. An example of background communication is the operation of a white board during a video conference call.

Background Noise-Electrical signal energy in a channel or frequency bandwidth that is not part of the desired received signals.

Background Noise Regeneration-Speech codecs like G.729 may use a silence suppression technique to avoid sending unnecessary bits during periods of no speech, however the lack of any sound at the decoder during these times sounds odd to a listener. For a more natural sound during these moments, a background noise regeneration technique may be used to play audio that sounds like the normal background noise.

Background Processing-The process by which lower priority programs are scheduled to use computer system resources only when no other higher priority tasks are available for execution. This allows the computing resources to be made available to higher priority programs that must perform an immediate action.

Background Program-A separate program or processing thread that runs at a lower priority and therefore only uses the resources of the computer system when no other higher priority tasks (foreground tasks) are available for processing.

Background Sprite-A background sprite is a graphic object that is located behind foreground objects. Background sprites usually don't change or change relatively slowly.

Background Task-A processing job (task) that is performed concurrently with a primary job where the background processing steps are usually performed during periods of low activity or idle periods during the processing of the primary task.

Backhaul-Backhaul is the process of transferring packets or communication signals over relatively long distances to a separate location for processing.

Backhauling-The linking of cell sites to a mobile switching center (MSC) through telephone lines. The term backhauling is often used when referring to cellular or PCS systems that provide service by backhauling their cell sites to a MSC that is located in a different geographic region. In some rural areas, it is not economically viable to install and connect a switching center to PSTN to service the cell sites. Cellular or PCS carriers in adjacent markets sometimes enter into relationships to provide the switching services via backhauling.

Backing Out-The process of returning the state of a system to its prior known state after a transaction fails to complete. This process guarantees the integrity of the information contained within the system. Used primarily for database and distributed systems.

Backlight-A backlight illuminates an object end from behind the object or from behind the viewing point.

Backoff-A process that is used in a data communications network to prevent repeated collisions during data access attempts. Using the backoff procedure, when a failed access attempt occurs, the communication device will delay its next access transmission request (backoff) before a second attempt is made to access the network.

Backoff Algorithm-A backoff algorithm is a program or process that increases the wait time before a device reattempts to access a system or resource. Backoff algorithms are used to help ensure systems are not overwhelmed with repeated attempts for system access requests from devices that are not provided system access on their first request.

Backplane Bus-A collection of electrical wiring that interconnects the slots where modules in a computer system are inserted. The backplane bus wiring connects the same signal to identically numbered pins of each module in the computer system.

For example, the "power on" signal might be connected to pin 3 of each module via the backplane bus wiring.

Backshell-A backshell is a portion of a connector that guides and attaches to the cable.

Back-To-Back Connection-A connection between the output of a receiver and the input of a transmitter. Also may refer to private line circuits that are connected via modems or CSU connected back to back using the appropriate roll-over cable.

Backup-(1-data) A copy of data information, usually on a storage device that is stable over relatively long periods of time. (2-process) The process of transferring files or information to a storage device or media.

Backup Domain Controller-A backup domain controller is a server (computer) that maintains a copy of company's or person's domain information. This includes lists of authorized users, account information, and lists of other primary and backup servers it may communicate with.

Backup Link-A communication link that is used for communication in the event a primary or specified link becomes inactive or disabled.

Backup Power Supply-A redundant power supply that takes over if the primary power supply fails. This may or may not be automatic although when supplied as part of the overall system it usually is.

Backup Program-This is an application program that is used to make archival copies of the contents of a computer file system. Most backup programs provide methods for both full and incremental backups. Incremental backups only archive files that have been modified since the last full backup. Most commercial backup programs provide a means of also restoring the archived information if the need arises.

Backup Server-A program that administers the copying of users' files so that at least two up-to-date copies exist.

Backward and Forward Compatibility-Backward and forward compatibility is the ability to use existing and future modes of operation of a system or service.

Backward Channel-In channel in a data transmission system that is opposite that of the primary channel. The backward channel may be a low capacity channel that is used for transmission control or supervisory signals.

Backward Compatibility-Ability of new hardware or software to operate effectively with older versions of the same equipment or programs.

Backward Congestion Notification-Backward congestion notification indicates to sending (downstream) switching devices in a data communication network that congestion is occurring and packets that are received may be discarded. The sending switch can then change the priority of packet discarding and send and indication to other switches indicating network congestion. This should eventually reduce the amount of data end users are sending into the network.

Backward Error Correction (BEC)-Error correction techniques in which the receiver detects any errors and requests retransmission of the erroneous data. When error rates are low or zero, BEC can be very efficient. However if the same error occurs repeatedly BEC techniques can never transmit the data properly. Compare to forward error correction (FEC).

Backward Explicit Congestion Notification (BECN)-A control bit carried within the overhead of a data packet in a frame-relay network that indicates network congestion exists in the backward (opposite) direction of its flow. This control bit allows higher-level protocols within the data communications equipment (DCE) and data terminal equipment (DTE) to take appropriate bandwidth allocation action if necessary.

Backward Indicator Bit (BIB)-A bit in an SS7 signaling message, by its status change at the remote end, requests retransmission because of messages received out of sequence.

Backward Prediction-Backward prediction is the process of estimating the likely changes or occurrences that may have occurred within previous media, images or media components in a sequence of media (such as audio packets or video frames).

Backward Sequence Number (BSN)-A field in an SS7 signal unit sent that contains the forward sequence number of a correctly received signal unit being acknowledged.

Backwards Compatible-Backward compatibility is the ability to use future modes of operation of a system or service on existing (legacy) systems.

BACP-Bandwidth Allocation Control Protocol

Bad Block-Within a mass storage device the storage medium is divided up into blocks that each contain a fixed amount of data, typically 512 or 1024 bytes. If a particular block on the medium is defective it is marked as a bad block so that the operating system will not use it to store information.

Bad Track Table-A list of the unusable areas on a storage disk, usually defective tracks.

Badged-An English term that represents that a product manufactured by one firm that is sold by another. The Original Equipment Manufacturer (OEM) produces the product with identification and/or brand of the selling firm.

BAIC-Barring of All Incoming Calls

Balance condition-The condition in which two signals are effectively separated or equal levels.

Balance Test Line-Equipment in a central office that provides a proper point to terminate a line for echo-balance and noise testing. Access to such equipment through a dial-up telephone line facilitates testing of trunks by measurement systems and by field service personnel.

Balanced Line-A transmission line that uses two conductors where each of the conductors have equal voltage but opposite polarity levels to form a balanced line.

Balanced link access procedure (LAPB)-The data link layer protocol that connects terminals and computers to a packet switched network. It can be viewed as the same as the asynchronous balanced mode of HDLC.

Balanced Modulator-A modulator that combines the information signal and the carrier so that the output contains the two sidebands without the carrier.

Ballistics Test-An outside plant cable test that measures the charging and discharging capacity in a pair of conductors. This meter deflection allows a technician to approximate the length of the conductors.

BAN-Billing Account Number

Band-A range of frequencies defined specified upper and lower frequency limits.

Band Elimination Filter (BEF)-A filter that has an attenuation band for a specific range of frequencies. See band reject filter.

Band Reject Filter-A band reject filter rejects a specific band of frequencies. A band reject filter is sometimes called a notch filter.

Band Splitter-A band multiplexer or filter assembly that is used to split a frequency band into several smaller frequency bands.

Band Stop Filter (BSF)- A device that blocks (stops) a specific range of frequencies and allows all other frequencies to pass through with minimum attenuation. See band reject filter.

Band Wipe-A band wipe is a video transition that replaces an outgoing clip with expanding bands of incoming clip until the incoming clip completely covers the outgoing clip.

Bandpass-A range of frequencies (between upper and lower limit frequencies) that allow signals to pass through without low signal loss (low attenuation).

Bandpass Filter-A filter designed to reject or block all frequencies not within a given bandwidth. Such filters may be used to reject noise or other signal bands close in frequency to that of the desired signal. Filters typically require a trade-off among how much signal they pass (the amount of loss), how strongly they reject the undesirable signal, and how sharp the dividing line between passed and unpassed signals is.

This diagram shows a banpass filter that is used to block high and low frequency component parts of an input signal. In this example, both the high frequency noise and low frequency noise are attenuated by the bandpass filter. This example shows the desired frequency is allowed to pass through the filter with minimal attenuation.

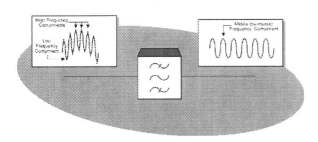

Bandpass Filter Operation

Bandwidth-A term that defines the signals that occupy a portion of a frequency spectrum, particularly a radio system or their data transmission rate. Analysis or measurement of the signals or signal waveforms of such a system will show that most (or substantially all) of the power contained in that signal can be found in a designated portion of the frequency spectrum. The difference between the highest and lowest frequency describing that portion of the spectrum is the bandwidth of the signal. Frequency (radio spectrum) bandwidth is measured in units of hertz or cycles per second and data transmission bandwidth is measured in bits per second.

Bandwidth Allocation Control Protocol (BACP)-Bandwidth allocation control protocol (BACP) is a multi-channel bonded channel management protocol that can be used to negotiate and manage the bandwidth of IP communication channels. BACP is used during an Internet engineering task force (IETF) standard that can be used to combine multiple 64 kbps sessions on an ISDN connection.

Bandwidth Awareness-Bandwidth awareness is the process and/or protocols that are used to allow for the recognition of the bandwidth that is available to devices that are connected to a network. For example, a media server needs to discover the connection speed of a multimedia computer that has requested to view a digital video stream. If the server has bandwidth awareness capability, it can determine an optimum compression and data transmission rate for the requested digital video stream.

Bandwidth Confirm (BCF)-The process or message that confirms the bandwidth assigned to a communication connection. A BCF message is defined in H.225.

Bandwidth Control-Bandwidth control is the process of detecting and adjusting the bandwidth that is available or assigned to a particular device or service.

Bandwidth Cost-Bandwidth cost is the combination of transmission or proportioned system equipment usage cost(s) between entry and exit points of a path on a communications network along with associated equipment usage and data processing costs divided by the amount of bandwidth used or reserved. For example, a leased line of 45 Mbps that costs $5,000 per month, this equates to a bandwidth cost of $111 per Mbps per month ($5,000/45).

Bandwidth On Demand (BoD)-A system that allows different data transmission rates based on requests from the customer, their application (e.g. voice or video), and the data transmission capability of the system.

Bandwidth Reservation-Bandwidth reservation is a process that is used to reserve bandwidth capacity through devices or communication line for specific communication sessions or services.

Bandwidth Scalability-Bandwidth scalability is the ability of a system or program to provide a service to a number of users at different bandwidth rates.

Bandwidth Stealing-Bandwidth stealing is the use of bandwidth that has been allocated to another device, system or service.

Bandwidth Target-Bandwidth target is the desired or assigned data transmission rate for use with a particular service or application.

Bandwidth Variation-Bandwidth variation is the changes in the available data transmission rate for a communication service. Bandwidth variation may occur as the data transmission activity for a number of users who share a communication network changes resulting in the available data transmission rates for other users who are connected and sharing the network resources.

BAOC-Barring of All Outgoing Calls

Bar Code-A standardized sequence of typically black vertical bars separated by white spaces that may be read with an optical decoder. The decoder reads the sequence of bars and interprets them as alphanumeric characters. The pattern of adjacent thick and thin bars is located on a contrasting background. Each decimal digit of an identification number is represented by a binary bit group in a bar code, and parity check digits are typically appended. A bar code can be scanned by an optical scanner device for input to a computer system for purposes of inventory control or the like.

Bare Board-A particular form factor of a circuit board with the necessary system connectors, power and ground conductors and standardized parts layout but without components. These boards are typically used for building prototypes and specialized functions that are not commercially available.

Bare Wire-An electrical conductor that does not contain any type of shielding over its surface area.

BARG-Billing and Accounting for Roaming Group

Barn Door Wipe-A barn door wipe is a video transition that slides the view of a new media segment across the screen during the transition (like sliding a door).

Barrel Connector-A small cylindrical connector made with two female ends so that it may be used to connect between two cables both terminated with male connectors.

Base Address-A numeric value used as a reference in the calculation of addresses during the execution of a computer program. Usually defined by the beginning real address of a partition or other segment of computer memory.

Base Layer-A base layer is one of multiple layers that can be combined to produce higher resolution images or video. A base layer may be decoded separately to provide a low resolution preview of the image or video and to reduce the decoding processing requirements (reduced complexity).

Base Rate-A fixed amount charged for a service per a specific time period. Normally this term is associated with a service that has a fixed charge (for a basic amount or type of service) and variable charge (applied only when more than the basic services are used).

Base Station (BS)-The radio part of a mobile radio transmission site (cell site). A single base station usually contains several radio transmitters, receivers, control sections and power supplies. Base stations are sometimes called a land station or cell site.

A base station contains amplifiers, radio transceivers, RF combiners, control sections, communications links, a scanning receiver, backup power supplies, and an antenna assembly. The transceiver sections are similar to the mobile telephone transceiver as they convert audio to RF signals and RF to audio signals. The transmitter output side of these radio transceivers is supplied to a high power RF amplifier (typically 10 to 50 Watts). The RF combiner allows separate radio channels to be combined onto one or several antenna assemblies without interfering with each other. This combined RF signal is routed to the transmitter antenna on top of the radio tower via low energy-loss coaxial cable.

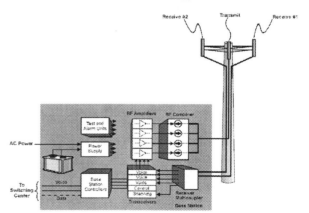

Base Station (BS) Functional

Base Station Controller (BSC)-A base station controller is an automatic coordinator (controller) that permits one or more base transceiver stations (BTS) in a wireless network to communicate with a mobile switching center.

Base Station Transceiver-A combined transmit-

B

ter and receiver assembly (transceiver) that is used as the radio processing part of a cellular system radio transmission site (cell site). The base station transceiver consists of all radio transmission and reception equipment, it provides coverage to a geographic area, and is controlled by a base station controller.

Base Traffic Load-In trunk forecasting, the average load offered on the first route available between two identified areas. Base load is found by averaging the traffic measured during the same 1-hour period each day over a period of several days. Base loads often are used to forecast future loads.

Base Transceiver Station (BTS)-The radio part of a wireless network (typically cellular or PCS) that includes the transmitters and receivers, antennas and tower that is used to communicate with mobile radios. A BTS is connected to a base station controller (BSC).

Base Year-Any 12 consecutive months for which data is collected for determining base loads.

BASE64-A binary-to-text encoding mechanism specified in RFC 1521, used to convert arbitrary sequences of binary data into a strictly limited subset of printable ASCII characters. Each set of 3 bytes (that is, 24 bits) is converted to a set of 4 characters from the ASCII subset. One application is for converting binary attachments into printable text for MIME-formatted e-mail attachments.

Baseband-A digital signal that occupies the entire bandwidth of a channel. Used in short range networks such a LAN's.

Baseband Channel-An information content (channel) that is used to modulate or encode a transmission medium. When used with radio signals, the high frequency component is called the broadband channel.

Baseband LAN-A local area network (LAN) that uses a single carrier frequency on the communication channel that represents the actual information that is transferred (no conversion required). Ethernet, Token Ring and ArcNet LAN's are examples of baseband transmission systems.

Baseband Signal-The original form of an information signal. Baseband signals can be applied to a modulator to impose its information on a carrier signal (radio wave or optical) to produce a broadband signal.

Baseband Signaling-The direct transmission of signal information without any conversion to another transmission medium.

Baseline-The process of determining and docu-menting network throughput and other performance information when the network is operating under what is considered a normal load. Measured performance characteristics might include error-rate and data-transfer information, along with information about the most active users and their applications.

Basic Animated Texture-The basic animated texture profile is the combination of processes and protocols used in the MPEG-4 system that uses wavelet-compressed shapes to create texture images.

Basic Input/Output System (BIOS)-The software or processes that controls the input and output of information to devices connected to a computer. The BIOS contains the set of instructions that sets up and tests the hardware when the computer is first turned on (booted). It starts the loading of the operating system and coordinates the operation of computing devices, such as floppy drives, hard disks, CD ROMs, video cards, mouse, and keyboards. The BIOS program is stored in non-volatile memory and is rarely changed (if ever). It is pronounced "bye + Ose."

Basic Rate Interface (BRI)-Basic rate interface is a data interface that is used in the integrated services digital network (ISDN) system for end user devices or low speed connections. The BRI interface provides up to 144 kbps of information that is divided into two 64 kbps channels (voice or data) and one 16 kbps control channel (data). The 64 kbps channels are referred to as the B channels and the 16 kbps channel is called the D channel.

Basic Trading Area (BTA)-A geographic region in the United States where area residents do most of their commerce activities. The United States has been divided into 493 BTAs. Some PCS licenses issued in the United States in the mid 1990's (bands C through F) were awarded based on BTA values.

Batch Processing-A process that does not occur in a real time mode. In a batch process, events are collected and then forwarded in batches to the processor that is "idle" until required to begin.

Battery-A battery is a device that stores electrical energy for use at a later time. Some batteries are designed and used for a single use (primary cell) and other batteries are designed for repetitive use (rechargeable).

Battery, Overvoltage, Ringing, Supervision, Coding, Hybrid, Test (BORSCHT)-A group of functions provided in Subscriber Line Circuits (LCs). It stands for:

B: Battery supply to subscriber line.

O: Over voltage protection (prevents damage to equipment or hazard to subscriber in case of lightning pulses or a "cross" connection with a high voltage power line).

R: Ringing current supply.

S: Supervision of subscriber set (determining busy/idle or off hook vs. on hook status).

C: Coder and decoder for analog-digital conversion when an analog telephone is used with a digital switch or transmission system.

H: Hybrid coils or "induction" coil (2 wire to 4 wire conversion) when audio frequency voltage waveforms flow in both directions on the subscriber lines but are separated in the telephone set for connections to the microphone and earphone. Also, a similar separation/combination of two opposite unidirectional signal flows in the switch or trunk transmission system vis-à-vis the two-wire analog subscriber lines.

T: Test.

BORSCHT is a group of functions provided to an analog line from a line circuit of a digital central office switch. An analog electronic switch can omit C and possibly H. A line circuit on a switch with metallic matrix (SXS, Xbar, 1,2,3ESS) only detects call origination and disconnects itself. The term BORSCHT is attributed to John W. Iwerson of AT&T Bell Laboratories in the 1960s.

Baud (Bd)-Name for the unit of data symbols per second. This name is taken from the name of the 19th century French teletypewriter machine innovator Emiel Baudot. For a method of modulation or encoding in which there is a choice of only two symbol values per symbol interval, or one bit per symbol (such as two-level pulse voltages) the baud rate is equal to the bit rate (bits per second). For a method of modulation or encoding in which there are more than two symbol values per symbol interval (and thus 2 or more bits per symbol) the bit rate is higher than the baud rate. For example, QPSK phase modulation and 2B4Q pulse coding both have 4 symbol values per symbol interval and thus the bit rate (bits per second) is twice the symbol (baud) rate. (Please do not make the error of writing "baud per second.")

Baud Rate-The number of signaling elements (symbols) per second on a transmission medium. For some line codes, such as bipolar, baud rate is the same as bit rate. However, in many applications, the baud rate is below the bit rate. For exam-ple, in 2B1Q coding, each quaternary signaling element conveys 2 bits of information, so the baud rate is one-half the bit rate. The spectral characteristics of a line signal depend on the baud rate, not the bit rate. For high-speed digital communications systems, one state change can be made to represent more than one data bit.

This diagram shows that the baud rate is not always the same as the bit rate as each baud (symbol) can have several states that represent multiple binary bits.

Baudot-Murray Code-Baudot-Murray (ITU

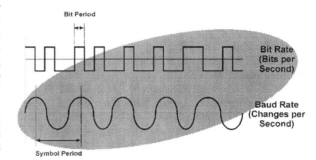

Baud Rate

alphabet No. 2) is a 5-bit character code used by older teletypewriters and telex machines, and also by most teletypewriters for the deaf. Frequently called Baudot code. It is typically transmitted with an added start bit and a lengthened stop bit (7 bits total) in an asynchronous sequence.

Bay-An equipment casing that can hold electronic assemblies such as transmission or call processing assemblies. Commonly called an "Equipment Bay."

Bayonet Neill Concelman (BNC)-Bayonet Neill Concelman is a set of BNC connectors that are used to transfer component video signals.

Bayonet-Neill Concelman Connector (BNC Connector)-A Bayonet-Neill Concelman connector is a type of connector that is used for coaxial and RF systems and cables.

BBS-Bulletin Board System

BC-Billing Center

BCC-Billing and Customer Care

BCC-Blind Carbon Copy

BCCH-Broadcast Control Channel

BCF-Bandwidth Confirm

BCH-Broadcast Channel

BCH Code-Bose, Chaudhuri, And Hocquenhem

B

BCI-Billing Correlation Identifier
BCS-Block Check Sequence
Bd-Baud
BDA-Bi-Directional Amplifier
Beacon Channel-A channel used in the Bluetooth system to support parked slaves where the master establishes a beacon channel when one or more slaves are parked. The beacon channel consists of one beacon slot or a train of equidistant beacon slots, and is transmitted periodically with a constant time interval. When parked, the slave will receive the beacon parameters through a Link Management Protocol (LMP) command. See also Park Mode.
Beaconing-In a token-ring network, a process used by all nodes on a faulting ring to isolate the failure. When a node determines that it cannot transmit and receive reliably, the device will "beacon". This involves transmitting a special Beacon MAC frame every few milliseconds. Any station receiving a Beacon MAC frame must repeat the frame. Within a few seconds, the exact location of a cable fault can be isolated and the MAC addresses of the network interface cards between the fault discovered.
When a ring is beaconing, user communication cannot take place.
Beam Forming-Beam forming is a process of directing the energy of a transmitted signal to specific areas. Some beam forming systems allow for the dynamic creation of beam patterns to maximize the performance to specific end points (receivers) and to minimize the effects of interference to and from other devices.
Beamsplitter-An assembly or device that receives an input light wave and divides it into two or more separate beams.
Bearer Services (BS)-Bearer services are telecommunication services that are used to transfer user data and control signals between two pieces of equipment. Bearer services can range from the transfer of low speed messages (300 bps) to very high-speed data signals (10+ Gbps).
Bearer services are typically categorized by their information transfer characteristics, methods of accessing the service, inter-working requirements (to other networks) and other general attributes. Information characteristics include data transfer rate, direction(s) of data flow, type of data transfer (circuit or packet) and other physical characteristics. The access methods determine what parts of

the system control could be affected by the bearer service. Some bearer services must cross different types of networks (e.g. wireless and wired) and the data and control information may need to be adjusted depending on the type of network. Other general attributes might specify a minimum quality level for the service or special conditional procedures such as automatic re-establishment of a bearer service after the service has been disconnected due to interference. Some categories of bearer services available via the telephone system include synchronous and asynchronous data, packet data and alternate speech and data.
BEC-Backward Error Correction
BECN-Backward Explicit Congestion Notification
BEF-Band Elimination Filter
Behavioral Score-A measure of a customer's credit worthiness that is based on past payment history. Sometimes termed "Internal Score", or "Payment Behavior Score".
Bel-A unit of power measurement where each positive unit represents a tenfold (10x) increase in power and each negative unit represents a tenfold decrease in power ratio (1/10th). This unit of measurement is named in honor of Alexander Graham Bell.
The commonly used version of the Bel unit is one tenth of a bell, or a decibel (dB). One bell equals 10 db. The Bel unit is a logarithmic value. If an amplifier increases the power of a signal by 100 times, the power gain of the amplifier is equal to 2 Bel, or 20 dB.
Bellcore-A telecommunications research and development consortium that was developed by the seven regional companies that resulted from AT&T's divestiture of the Bell Operating Companies. These were Ameritech, Bell Atlantic Corporation, BellSouth, NYNEX, Pacific Telesis, Southwestern Bell Corporation, and US West. Bellcore was sold in the late 1990s to Science Associates, and is now called Telcordia.
Benchmark-Benchmarking is a process of establishing targets or objectives as compared to (benchmark to) existing systems or performance standards.
Bend Loss-Bend loss is the attenuation of an optical signal that results from the bending of an optical fiber. The bending causes the traversing pattern (transmission path) to change so that some of the light energy is reflected outward from the cable instead of being reflected and channeled through the center core of the fiber.

Bend Radius-Bend radius is the maximum degree an optical fiber cable can be bent before significant or specific signal attenuation or cable breakage occurs. The typical bend radius is approximately 5 to 10 times the outside diameter of the cable.

Bent Pipe-Bent pipe satellite transmission is a process of transferring radio signals through a satellite by redirecting the signal (bending) to a new direction rather than receiving, processing, and re-transmitting the signal. The bent pipe method reduces the complexity of the transferring device as it is only retransmits the original signal in a new direction without analyzing or altering its information content.

BER-Bit Error Rate

BER Threshold-Bit Error Rate

Berkeley Internet Name Domain (BIND)-A Domain Name Server (DNS) implementation available for server operating systems. BIND is a widely deployed DNS implementation on the Internet.

BERT-Bit Error Rate Test

BES-Best Effort Service

Best Effort Service (BES)-Best effort is a level of service in a communications system that doesn't have a guaranteed level of quality of service (QoS).

Best Mode-In accordance with United States law, an application for patent must include a description of the best mode contemplated by the inventor, as of the time of filing the application for patent, of practicing or implementing the invention which is claimed.

The purpose of the best mode requirement is to force the inventor to not only disclose his or her invention, but also the best way in which the invention can be used. The best mode requirement is intended to prevent an inventor from obtaining a patent on an invention while keeping secret the best way to exploit the invention.

Best Path-The selection of an optimal route between source and destination and stations through a wide area network. The best path can be determined through routing protocols such as RIP and OSPF, best path can be based on lowest delay, cost or other criteria. As a result, the path chosen in one direction is not necessarily the same as the opposite direction.

Beta-(1-general) Second letter of the Greek alphabet. Used as a mathematical symbol for the current amplification factor of a junction transistor. (2- testing) A second stage of product testing (following alpha testing), typically allowing selected potential consumers use the product in the field. (3- video) A now-obsolescent videotape format for consumer magnetic tape video recording, and a shortened form of a broadcast quality video recording system, named Betacam, that uses metallic particle video recording tape in a cassette having the same dimensions as the consumer Beta product. Betacam is used by several television network news broadcasters.

Beta-Betacam

Beta Testing-Beta testing is the field testing process used to confirm the operation and performance of new hardware or software products before a product is officially released. Beta testing is the second stage for testing a new hardware or software product, usually performed by friendly customers or affiliates of the manufacturer or developer. The key purpose of Beta testing is to identify problems and the reliability of operation during normal field operating conditions. The typical number of Beta test participants is 50 to several hundred.

Beta Videotape-Beta videotape is a video tape storage format that is used to store video recording images on magnetic tape. Betacam format stores images in color difference components (Y, R-Y, and B-Y). The Beta consumer videotape recording format is completely different from the professional Betacam format.

Betacam (Beta)-Betacam™ is a video tape storage format that is used to store video recording images on 1/2 inch magnetic tape. Betacam uses the set of color difference signals Y (luminance), R-Y, B-Y.

Betacam SP-Betacam superior performance (Betacam SP) is an improved version of Betacam. Betacam SP uses a recording system that has a wider bandwidth.

BETRS-Basic Exchange Telecommunications Radio Service

B-Frame-Bidirectional Frame

BFSK-Binary Frequency Shift Keying

BG-Border Gateway

BGMP-Border Gateway Multicast Protocol

BGP-Border Gateway Protocol

BGP4-Border Gateway Protocol Version 4

BHM-Busy Hour Minutes

BHMC-Busy Hour Minutes of Capacity

BHT-Busy Hour Traffic

Bias-(1-mathematic) A deviation from a reference value. (2-electrical) A force (electrical, mechanical, or magnetic) that is used to establish a level that is necessary to operate a device or assembly.

BIB-Backward Indicator Bit

BICSI-Building Industry Consulting Service International

Bid Gap-Bid gap is the difference in bid amounts between two advertisers who are competing for a position (typically the top spot) on a pay-for-placement search engine.

Bid Management-Bid management is the process of monitoring and adjusting the bid amounts for pay for placement sponsored ad listings.

Bidirectional-Bidirectional transmission is the transfer of information in two directions.

Bi-Directional-Bi-directional communication is the ability to send and receive information on a communication channel. While bi-directional connections may allow two-way connectivity, the connections may not transmit and receive at the same time.

This diagram shows how a single fiber can be used to allow optical signals to flow in two different directions. This example shows that a light source on one end of the optical system uses a different wavelength than the light source on the opposite end. The signals are combined and split by an optical splitter.

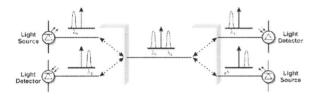

Optical Bidirectional Transmission on a Single Fiber

Bidirectional Carrier-A telephone company (telco) carrier system that utilizes 1 cable runs. This cable run uses 100 pair separation between transmit and receive pairs.

Bidirectional Frame (B Frame)-A bidirectional frame is an image in a motion video sequence that is coded using difference and motion compensation from previous and future images.

Bidirectional Frame (B-Frame)-Bidirectional frames (B-Frames) are images (pictures) within a sequence of images (such as in a video sequence) that are created using information from preceding images and images that follow (such as from intra frames (I-Frames) and from predicted frames (P-

Frames). Because B-Frames are created using both preceding and images that follow, B-Frames offer more data compression capability than P-Frames

Bi-Directional Prediction-Bi-direction prediction is a process that uses past reference and future reference frames to build frames in-between other frames.

Bidirectional Repeater-A repeater that regenerates a signal in both directions of transmission. It can be used in single pair cable operation. Bi-directional repeaters are placed within T1 spans at 3,000-6,000 feet intervals.

Bi-directional Services-Bi-directional services are communication connections that allow the operation of applications require two-way communication. An example of bi-directional services include video conferencing, chat rooms, and video mail.

Bi-directonial Frame-Bi-directional frames are images whose data contents are created from both previous and future frames in a sequence of images.

BIFS-Binary Format for Scenes

Big Firewire Jack-Big firewire jack is the large connector version for the Firewire personal area network system.

Bilateral Peering-Bilateral peering is the process of inquiring or exchanging information among network elements (such as routers) between two companies.

Bill Cycle Code-An identifier assigned to all customers who are to be billed on the same run of the billing process.

Bill Of Materials (BOM)-A list of parts (components and other assemblies) that are the materials that are used to produce a product, quantity of product, or an assembly. The listing of parts in a BOM is often assigned a cost to estimate the construction cost of the product or assembly.

Bill Period-The period for which the recurring charges apply. In monthly billing for instance, the Bill Period typically corresponds to the period between the current bill and the next bill for charges billed in advance, and between the previous bill and the current bill for charges billed in arrears (e.g. April 5th to May 4th).

Bill Pool-A bill pool is a group of billing records that have been processed by billing system to include the necessary charging rate information. The bill pool usually contains records that are ready for the final stage of bill processing.

Bill Posting-The process of recording an invoice payment into a billing system.

Bill Rendering Fees-Bill rendering fees are charges for the production of billing records in specific formats such as paper or other forms of stored media (e.g. CD ROM). The fees for bill production (bill rendering) may vary based on the detail level that is included in the bill format.

Bill Run-The process of triggering the start of the billing process that gathers and processes billing charges associated with specific customers or accounts. Also sometimes referred to as "Bill Round", or "Billing", or "Month End", etc. Typically, one or more cycles are assigned to a Bill Run; i.e. all customers who have been assigned those cycles will be processed during this Bill Run.

Billed Minutes-The actual time, in reported minutes, for messages of a duration equal to or greater than an initial period. When the reported minutes are less than the initial period, the initial period minutes are shown as billed minutes. For example, a one-minute conversation on a connection having an initial period of three minutes would be billed as three minutes.

Billing-The process of grouping service or product usage information for specific accounts or customers, producing and sending invoices, and recording (posting) payments made to customer accounts.

Billing Aggregator-A billing aggregator is a company that gathers billing records from one or more companies and posts them to another billing system. An example of a billing aggregator is the gathering of billing information from many companies that provide information or value added services (e.g. news or messaging services) and transferring these to another basic services carrier for direct billing to a customer. The customer pays the carrier, the carrier pays the aggregator, and the aggregator pays the value added service provider company. The billing aggregator performs services similar to a clearinghouse.

Billing and Customer Care (BCC)-A set of functions and processes related to generating customer invoices. These generally include Event (network usage) Rating, Invoicing, and tools to establish and maintain the customer profile.

Billing and custom care systems convert the transfer of bits and bytes of digital information within the network into the money that will be received by the service provider. To accomplish this, billing and customer care systems provide account activation and tracking, service feature selection, selection of billing rates for specific calls, invoice creation, payment entry and management of communication

with the customer.

Billing and customer care systems are the link between end users and the telecommunications network equipment. Telecommunications service providers manage networks, setup the networks to allow customers to transfer information (provisioning), and bill end users for their use of the system. Customers who need telecommunication services select carriers by evaluating service and equipment costs, reviewing the reliability of the network, and comparing how specific services (features) match their communication needs. Because most network operations have access to systems with the same technology, the billing and customer care system is one of the key methods used to differentiate one service provider from another.

This figure shows an overview of a billing and customer care system that is used for communication services. This diagram shows the key steps for billing systems. First, the network records events that contain usage information (for example, connection time) that is related to a specific call. Next, these events are combined and reformatted into a single call detail record (CDR). Because these events only contain network usage information, the identity of the user must be matched (guided) to the call detail record and the charging rate for the call must be determined. After the total charge for the call is calculated using the charging rate, the billing record is updated and is sent to a bill pool (list of ready-to-bill call records). Periodically, a bill is produced for the customer and as payments are

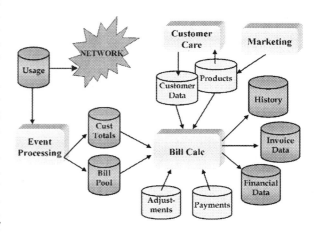

Billing and Customer Care (BCC) System

received, they are recorded (posted) to the customer's account.

Billing Block-A billing block is the listing of credits (in a block format) for a program in public media such as newspaper ads, billboards, etc.

Billing Center (BC)-A billing center is a facility and is the associated equipment that is responsible for the billing of communication and/or information services.

Billing Company-A company that bills customers for services such as collect calls or long distance changes. The billing company may or may not be the same as the company that provides the service.

Billing Correlation Identifier (BCI)-A code (16-octets) that is used to identify a particular call that is made within the PacketCable system.

Billing Cycle-To distribute the bill processing requirements, billing systems divide customers into groups that allows their bills to be processed in specific cycles (or "billing cycles.") The billing cycles are different for groups of customers. This allows the billing system to only batch a portion of the billing records each time. These billing records must be forwarded for delivery (to a bill printer or for electronic distribution).

This diagram shows how a billing system can divide the invoicing process into billing cycles. In this example, each month is divided into 4 groups of billing cycles. At the end of each billing cycle, the billing records for each customer in a particular billing group are gathered, processed, and converted into invoices.

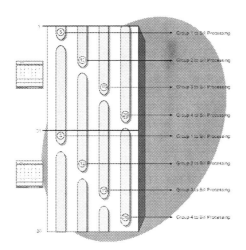

Billing Cycle Operation

Billing Data-All billing data collected during a telephone call. Billing data usually includes call time, originating port number, calling telephone number, call class, dialed number, connection indicator, home/roam indicator (wireless), answer time, disconnect time, timeout, and other billing related information. Billing data is usually transferred on an AMA tape or transferred electronically.

Billing Event-A measurable condition in a network that represents the usage of a network resource by a customer. Several billing events usually occur for each communications session (e.g. telephone call). Billing events are often supplied to a mediation device that combines the billing events into a single call detail record (CDR).

Billing Increment-The smallest amount of time or resource that can be billed or charged. Billing increments include units of time (e.g. 1 minute or 6 second increments), number of messages, thousand bytes of data (kBytes), or other amount that can be calculated into a charge or criteria for a billing system.

Billing Interconnection Percentage (BIP)-The percentage of interconnection charge that is calculated based on the percentage of a communication route or amount of resources used when there are multiple carriers that are providing a service.

Billing Media Converter (BMC)-A device that transfers automatic message accounting (AMA) data from end offices to regional accounting offices using magnetic tape.

Billing Name and Address-The name and address provided to carrier (such as a local exchange company) by each of its customers to which the company directs bills for its services.

Billing Service Bureau-A billing service bureau is a company that provides billing services to other companies. The services provided can range from billing consolidation to complete billing operations that include gathering billing records, processing invoices, mailing or issuing the invoices, and posting payments.

Billing System-A system (usually a combination of software and hardware) that receives call detail and service usage information, grouping this information for specific accounts or customers, produces invoices, creating reports for management, and recording (posting) payments made to customer accounts.

This figure shows a standard billing process. In this diagram, the customer calls customer care or works with an activation agent to establish a new wireless account. The agent (customer care) enters the customer's service preferences into the system, checks for credit worthiness, and provides the customer with a phone number so that the customer may make and receive calls through the telephone network. As the customer makes calls, the connections made by the network (such as switches) create records of their activities. These records include the identification of the customer and other relevant information that are passed onto the billing system. The billing system also receives records from other carriers (such as a long distance service provider, or a roaming partner). The billing system now guides and updates these call detail records (CDRs) to their correct customer and rating information. As information about the customer is discovered (e.g. rate plan), the updated billing records are placed in a billing pool so that they may be combined into a single invoice that is sent to the customer. The customer then sends his payment to the telecom service provider. Payments are recorded in the billing system. History files are then updated for the use of customer service representatives (CSRs) and auditing managers.

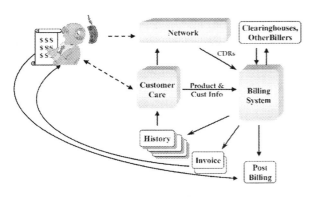

Standard Billing Process Operation

Billing Tape-Monthly call detail records that are generated by the carrier. Resellers often use the billing tape to generate their own bills. (see Call Report).

Billing Telephone Number (BTN)-A number recorded by the switch on a Call Detail Record (CDR) identifying the party to be billed for the call.

This number is not always necessarily a telephone number, it could be a Calling Card Number or any other number used to identify the party responsible for payment.

Billing Unit-A billing unit is a value or characteristic that is used in the measurement of service or product usage. Examples of a billing unit include minutes, megabytes transferred or program viewings.

Bin-A bin is a location for storing and organizing video or audio clips in a media editor.

Binary-A numbering system based upon the powers of 2 used to represent data in digital computer systems. Binary data consists of a positional sequence of 1s and 0s. Each position is used to represent a specific power of 2. For example, the decimal number 19 is written in binary as 10011. This corresponds to (2 to the 4th power) + (2 to the 1st) + 2 to the 0th) which is 16 + 2 + 1 = 19.

Binary Digit-A number in the binary system. Also called a bit of information.

Binary Format for Scenes (BIFS)-Binary format for scenes is part of the MPEG-4 standard that deals with synchronizing video and audio.

Binary Frequency Shift Keying (BFSK)-A digital modulation process where each binary digit is represented as a different (unique) frequency.

Binary Message-A string of binary digits that are defined using a message code table to determine the meaning of the message. The use of binary messages results in many different messages being defined in a limited number of data bits.

Binary Phase-Shift Keying (BPSK)-Binary phase shift keying (BPSK) is a modulation process that converts binary bits into phase shifts of the radio carrier without substantially changing the frequency of the carrier waveform. The phase of a carrier is the relative time of the peaks and valleys of the sine wave relative to the time of an unmodulated "clock" sine wave of the same frequency. BPSK uses only two phase angles, corresponding to a phase shift of zero or a half cycle (that is, zero or 180 degrees of angle).

Binary Protocol-Binary protocols are the languages, processes, and procedures that perform functions used to send control messages and coordinate the transfer of data using binary (non-text) based messages.

BIND-Berkeley Internet Name Domain

Binding-The relating of an identifier (such as a telephone number) to another identifier or resource (such as an IP address or service).

Binding Post-A binding post is an attachment point for an electrical or mechanical connection within a device or assembly.

Biometric Access Control-Biometric access control is the process of authenticating access through the use of human measurements. Examples of biometric access control include voiceprinting, fingerprinting, handwriting analysis, and eye retinal scanning.

Biometrics-Biometrics is the measurement of identifying life characteristics or behavioral traits such as fingerprints, retina patterns or voice prints

BIOS-Basic Input/Output System

BIOS Enumerator-Basic Input/Output System Enumerator

BIP-Billing Interconnection Percentage

BIP-Bit Interleaved Parity

Bi-Polar Violation (BPV)-In digital telephone multiplexing systems, two level logic digital signal pulses in the electronic multiplexer are converted into three level pulses for purposes of transmitting the pulses via transmission paths that incorporate transformers or coupling capacitors. These devices cause waveform distortions if the transmitted waveform does not have a zero average voltage over the time interval corresponding to several pulses. Before transmitting, a binary 1 signal is converted into a zero volt pulse (an interval with no pulse voltage). A binary zero signal is converted into either a positive pulse (of about 3 volts amplitude) or a negative pulse (with negative 3 volts amplitude). A simple memory device in the coder causes each zero value pulse output to be set to the opposite polarity from the previous zero value pulse output. This line coding method is called Alternate Mark Inversion (AMI), and is sometimes called "pseudo-ternary" since it has three voltage pulse levels (+3, 0, -3 volts) but the three levels do not represent three independent code values but are instead dependent on two underlying code values. At the receiving location, the three different pulse levels are converted back into just two levels. Because of the design, no two distinct consecutive 3 volt pulses should ever have the same polarity. If they appear to do so, we call this a bi-polar violation. It indicates that either a pulse has been "lost" in the transmission, or that an extra pulse has been produced (perhaps due to lightning or some other source of electrical interference). Certain other types of line coding such as HDB3 or B8ZS intentionally produce BPVs for certain all-zero binary data strings.

Bipolar with 8 Zero Substitution (B8ZS)-A technique specified in the ITU-T G.703 recommendation, B8ZS is used in some North American DS-1 (T-1) links to permit transmission of unrestricted binary data strings, including consecutive zero binary values, and thus provide clear channel digital transmission. B8ZS substitutes one of two predefined special bi-polar violation (BPV) pulse sequences for strings of eight consecutive zero voltage values, with the intentional bipolar violation codes being inserted in bit positions 4 and 7 of the eight bit group. The bipolar violation codes are of opposite polarity, thus producing a necessary average voltage of zero over the entire eight bit group. The specific bipolar violation code of the two is chosen depending on the polarity of the last preceding pulse. B8ZS, which generally is used on newer DS-1 (T-1) and ISDN PRI circuits, offers clear channel communications of 64 kb/s on each channel. AMI (Alternate Mark Inversion) the older technique, suffers from a loss of timing recovery when 8 consecutive zeros are transmitted. The 2.048 Mb/s E-1 primary rate multiplexing standards utilize a similar (but different in detail) pulse substitution method, HDB3 (High Density Bipolar 3) line coding, which also is specified in G.703.

Birds of Feathers (BOF)-A group of interested people and companies who form a temporary BOF group to assess the interest and need for a new Internet protocol specification.

Biscuit-A term sometimes used for external wall mounted RJ-45/RJ-48 jack.

B-ISDN-Broadband Integrated Services Digital Network

B-ISUP-Broadband Integrated Services Digital Network User Part

Bit (b)-In digital signals, a bit is a single unit of data. Generally, a bit has a value of either one or zero corresponding to on or off for an optical or electrical signal. Bits are typically clustered into groups of eight, called bytes or octads. Different coding schemes, such as ASCII, have been developed to assign meaning to the bytes.

Bit Duration-The time it takes to transmit a single encoded bit on the physical transmission medium.

Bit Error-A bit error occurs when one or more encoded bits received by the receiver were corrupted at some point during transmission.

Bit Error Rate (BER)-BER is calculated by dividing the number of bits received in error by the total number of bits transmitted. It is generally used to denote the quality of a digital transmission channel.

Bit Error Rate (BER) Threshold-The maximum rate of bit errors (usually in percentage form) for which the number of bits received in error exceeds the error correction capability of the digital communication system.

Bit Error Rate Test (BERT)-Bit error rate testing is the measurement of how many bits are received in error for a transmission under specific transmission conditions. BERT is performed by setting up a communication channel with the appropriate conditions (e.g. signal levels), sending (transmitting) a known pattern of bits and comparing the received bits to the original bits. The number of bits received in error compared to the total number of bits sent is the bit error rate (BER).

Bit Interleaved Parity (BIP)-A simple parity check mechanism. SDH implements two BIP parity checks, BIP-8 and BIP-24. BIP-8 is used in the RSOH, for regenerator section error control, and in the VC overheads, of any level. BIP-24 is used in the MSOH, for regenerator section error control.

Bit Oriented-Term used to denote that fields consisting of one or more bits are used to convey control information within the communications protocol.

BIT Rate-A measurement of the transfer rate of digital signals through a channel. The bit rate is the number of bits transmitted in a specified amount of time. Bit rate is usually expressed as bits per second (bps).

Bit Slice Processor-A microprocessor that has its control circuits and arithmetic and logic units on separate integrated circuits. For example, a 16-bit microprocessor function might be sliced into four 4-bit functions.

Bit Slippage-A phenomenon that occurs in a parallel digital data bus when one or more bits of the bus become mistimed in relation to the rest. The result is that when the data on the parallel bus is next latched, erroneous data is generated. The most common cause for bit slippage is differential cable length for one or more bits of the parallel bus.

Bit Stream-(1-data flow) The sequence of encoded bits that are flowing over a digital communications channel. (2-humor) A place where engineers go to find bitty fish that byte in channels. The engineers often throw back the bitty ones.

Bit Stream Generator-A bit stream generator is an electronics assembly device that creates and receives digital signals between communication devices. A bit stream generator may be in the form of an internal computer card (e.g. PCI card) or an external device (Ethernet adapter).

Bit Stream Transmission-The transmission of characters at fixed time intervals without stop and start elements. The bits that make up the characters follow each other in sequence without interruption.

Bit String-A linear sequence of bits.

BIT Stuffing-Bit stuffing is the insertion of additional bits in a data stream maintain a constant data transmission rate. Bit stuffing may be used to correct for frequency differences between the clock of unsynchronized inputs. It is a method for changing the rate of a digital signal in a controlled manner so that it can match a rate different from its own inherent rate, usually without loss of information. Bit stuffing may also be called pulse stuffing.

Bitrate-Bit rate is the number of bits that pass (transfer) between corresponding equipment in a data transmission network.

Bits per Hertz (BPH)-Bits per Hertz is a measure of the number of bits that are transferred for each Hertz of bandwidth. BPH is commonly used to help describe the spectral efficiency of a communication channel or system.

Bits Per Inch (BPI)-The number of bits that a magnetic tape or computer tape cartridge can store per linear inch of length.

Bits Per Second (BPS)-The common measurement for data transmission that indicates the number of bits that can be transferred to or from a communications device in one second.

Bit-Sliced Arithmetic Coding (BSAC)-Bit sliced arithmetic coding is the process of offering differing levels of media quality through increasing the data rates (in increments).

Bitstream-The stream of compressed data that is the input to a decoder, especially for compressed audio or video.

BitTorrent-A bit torrent is a rapid file transfer that occurs when multiple providers of information can combine their digital data (bits) transfer into a single stream (a torrent) of file information to the receiving computer.

BLA-Bridged Line Appearance

Blade-A blade is a card or module that can be placed into a backplane of a communication system.

Blank And Burst Signaling-Blank and burst signaling is a process of sending control messages between telecommunications devices where control data temporarily mutes voice or user data and replaces it with a control message. This is also known as in-band signaling.

This figure shows the basic process of blank and burst signaling used in the AMPS system. This diagram shows that the audio is muted when a control message is sent. This diagram shows that the control message contains a preamble and a sync field to allow the mobile device to determine that it is receiving a control message and that it should mute the audio.

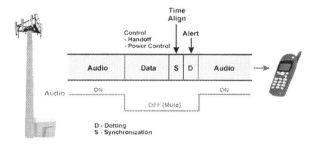

AMPS Blank and Burst Signaling

BLEC-Broadband Local Exchange Carrier
BLEC-Building Local Exchange Carrier
BLES-Broadband Loop Emulation Service
Blind Transfer (BT)-The process of transferring a call to another extension or phone number without telling the person who's calling that they are being transferred. Blind transfers are sometimes called cold transfer or unsupervised transfer.
B-Link-Bridge Link
Block-(1-data) An information unit or group of data bits that usually include a header, information element, and an ending error-checking code. (2-image) In encoding of an image or one frame of video, the image is divided into blocks of pixels like a checkerboard, for example each block is a set of 8 by 8 pixels. Each block is then processed separately.

Block Call-Block call is a call processing feature that allows a telephone user to block incoming calls from specific telephone numbers.
Block Check Character-A character added at the end of a message or transmission block to facilitate error detection. Generally block check characters are derived via an algorithm that is applied to the block of information being transmitted.
Block Check Sequence (BCS)-A block check sequence is a calculated code that is used to determine (check) if the bits within a block of data have been received correctly during transmission.
Block Code-A series of bits or a number that is appended to a group of bits or batch of information that allows for the detecting and/or correcting of information that has been transmitted. Block codes use mathematical formulas that perform an operation on the data that will be transmitted. This produces a resulting number that is related to the transmitted data. Depending on how complex the mathematical formula is and how many bits the result may be, the bock code can be used to detect and correct one or more bits of information.
Block Convertor-A device or assembly that shifts a band of frequencies to another band of frequencies.

This diagram shows that a frequency block converter shifts an entire band or group of frequency channels to a new frequency band. This diagram shows that the incoming frequency band A is shifted up to a new frequency band B. All the channels that are received from frequency band A are shifted up by the same amount.Block Convertor Operation

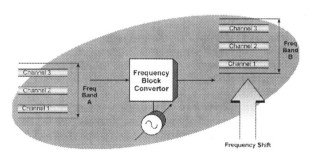

Block Convertor Operation

Block Diagram-A high level abstraction of the various functions or components of a network, system or circuit depicted using simple geometric figures. Lines and arrows between the geometric figures denote an interaction of some type between the connected blocks. Block diagrams are typically used to simplify the understanding of a complex system.

Block Error Probability-The ratio of incorrect blocks to the total number of successful block deliveries during a measurement period.

Block Matching-Block matching is the process of matching the images in a block (a portion of an image) to locations in other frames of a digital picture sequence (e.g. digital video).

Blockage-The temporary lack of access because of high traffic in a switching system or in a subscriber line concentrator.

Blocked Attempt-An attempted call that cannot be further advanced toward its destination because of an equipment shortage or failure in a network.

Blockiness-In video and image compression, a common artifact of low-quality compression when too few bits are used. Blockiness is an obviously perceptible contrast of color at the boundaries of the encoding blocks with a codec like JPEG or MPEG video. Blockiness can be partially concealed by post processing the image to blur the block boundaries.

Blocking-Blocking is a condition that occurs in a communication network when all permitted trunk paths or circuits are in use or busy that prevents a user from accessing services. In the case of telephone call blocking, a message may be returned to the customer that the system is unavailable.

Blocking Capacitor-A capacitor included in a circuit to stop the passage of direct current.

Blocking Factor-The likelihood of no free channels in a cellular telephone system being available when dialed.

Blocking IPR-An intellectual property right (IPR), patent, copyright or other right proprietary to an individual, group or company, which precludes someone else from making, using or selling that invention.

Blocking Objective-The design objective for the maximum ratio of allowed unsuccessful access compared to total access attempts (average blocking ratio) for which a group of switches, servers or networks is engineered or administered.

Blocking Probability-The percentage of calls that cannot be completed within a one hour period due to capacity limitations. For example, if within one hour 100 users attempt accessing the system, and 10 attempts fail, the blocking probability is ten percent.

Blocking Ratio-For a group of servers, the ratio of unsuccessful access attempts compared to total access attempts within a specified time interval.

Blowing Cable-Blowing cable is the use of compressed air or gas (such as Nitrogen) to force a cable through a tube or conduit. A float is attached to the front of the fiber cable so compressed air can push it through the tube or conduit.

Blown Fiber-Blown in Fiber

BLSR-Bidirectional Line Switched Ring

Blue Signal-In telecommunications applications, an alarm signal composed of ones and zeros (101010, etc.) substituted for loss of a valid input signal to indicate loss of the signal to downstream equipment.

Bluetooth-Bluetooth is a wireless personal area network (WPAN) communication system standard that allows for wireless data connections to be dynamically added and removed between nearby devices. Each Bluetooth wireless network can contain up to 8 active devices and is called a Piconet. Piconets can be linked to form Scatternets. Information about Bluetooth technology and wireless data can be found at www.Bluetooth.com.

Bluetooth was named after Harald Blatand, King of Denmark. King Blatand was head of Denmark from 940 to 985 A.D and he is known for uniting the Danes and Norwegians. It seems appropriate to name the wireless technology that unifies communication between diverse sets of devices after King Blatand.

This diagram shows the basic radio transmission process used in the Bluetooth system. This diagram shows that the frequency range of the Bluetooth system ranges from 2.4 GHz to 2.483 GHz and that the basic radio transmission packet time slot is 625 usec. It also shows that one device in a Bluetooth piconet is the master (controller) and other devices are slaves to the master. Each radio packet contains a local area piconet ID, device ID, and logical channel identifier. This diagram also shows that the hopping sequence is normally determined by the master's Bluetooth device address. However, when a device is not under control of the master, it does not know what hopping sequence to use so it listens for inquiries on a standard hopping sequence and then listens for pages using its own Bluetooth device address.

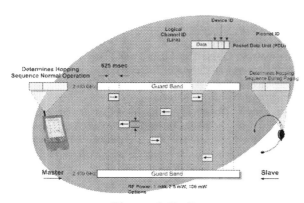

Bluetooth Radio

This diagram shows how a Bluetooth system can connect multiple devices on a single radio link. This diagram shows that a laptop computer has requested a data file from a desktop computer. When this laptop computer first requests the data file, it accessed the Bluetooth radio through a serial data communication port. The serial data port was adapted to Bluetooth protocol (RFComm) and a physical radio channel was requested from the local device (master) to the remote computing device (slave). The link manager of the master Bluetooth device requests a physical link to the remote Bluetooth radio. After the physical link is created,

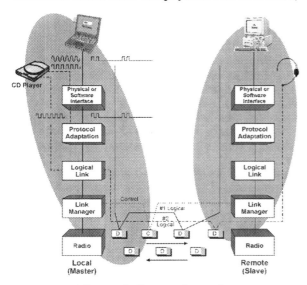

Bluetooth System Operation

the logical link controller sends a message to the remote device requesting a logical channel be connected between the laptop computer and the remote computer. The logical link continually transmits data between the devices. In this diagram, the user then requests that a CD Player send digital audio to a headset at the remote computer. Because a physical channel is already established, the logical link controller only needs to setup a 2nd logical link between the master Bluetooth device and the remote Bluetooth device. Now data from the CD ROM will be routed over the same physical link between the two Bluetooth devices.

Bluetooth Address-A Bluetooth address is a unique non-changeable 48 bit address that identifies a specific Bluetooth device. A portion of the Bluetooth address identifies the equipment manufacturer and a portion is a sequence number assigned by a manufacturer. The Bluetooth address is also known as the "BD_ADDR".

Bluetooth Host-A computing device such as a mobile telephone or data access point that communicates and provides services with other Bluetooth communication devices.

Bluetooth Inquiry-Bluetooth inquiry is a process used in a Bluetooth system to determine the address (and possibly the name) of other Bluetooth devices that are operating in the same area. Bluetooth inquiry is the process that requests specific information from a computer or communication device to determine its access code or availability for a communication session.

This diagram shows a simplified diagram of how the Bluetooth system inquiry process discovers nearby Bluetooth communication devices. This diagram shows each Bluetooth transmitter and receiver has a standard preprogrammed frequency tuning sequence called the general inquiry access code (GIAC). The GIAC access code causes the transmitter and receiver frequency sequence to change from one channel frequency to channel frequency. Because the transmitter and receiver do not know exactly when each will transmit, to initially capture information from a transmitter to a receiver, the transmitter changes frequency often (3200 steps per second). The receiver changes frequency using the same hopping sequence, but at a slower hop rate (1 step every 1.28 seconds). Each time the transmitter stops on a specific frequency, it transmits two short inquiry identification packets (they contain the GIAC access code). Eventually, the transmitter and receiver wind up on the same channel and the receiver will capture at least one of these inquiry packets. After waiting a random amount of time to avoid collisions with other Bluetooth devices that may have also received the inquiry packet, the receiver will respond to the transmitter informing it of its address. The transmitter has now discovered that another device is operating near it and it can use that address to contact (page) the other Bluetooth device and setup a communication session.

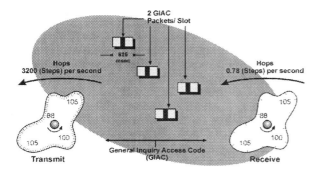

Bluetooth Inquiry Process

Bluetooth Paging-Bluetooth paging is the process used in a Bluetooth system to create logical connections between devices. Paging involves getting the attention of a device by using a hopping sequence of the device that it wants to connect to. Once the device discovers it has been paged and if the device desires to allow a connection to be made ("is connectable"), the paging sequence is completed when the hopping sequence is changed to the paging device's hopping sequence.

Bluetooth Qualification Administrator (BQA)-The Bluetooth qualification administrator (BQA) is responsible for overseeing the administration of the qualification program. The BQA ensures that qualified products can be listed on the Bluetooth website.

Bluetooth Qualification Body (BQB)-The BQB is authorized by the Bluetooth qualification review board (BQRB) to be responsible for the checking of declarations and documents against requirements,

reviewing product test reports, and listing conforming products in the official database of Bluetooth qualified products.

Bluetooth Qualification Test Facility (BQTF)-BQTFs are test labs that are recognized and certified by the BQRB as being capable of testing and qualifying Bluetooth devices.

Bluetooth Radio-Bluetooth radio systems use radio 1 MHz wide radio channels that hop over 79 channels. A Bluetooth radio transceiver transmits and receives modulated electrical signals wirelessly between peer Bluetooth devices.

BMC-Billing Media Converter

BMC-Broadcast/Multicast Control Protocol

BMDP-Broadcast Media Distribution Protocol

BMI-Broadcast Music International

BNC-Bayonet Neill Concelman

BNC-British Naval Connector

BNC Connector-Bayonet-Neill Concelman Connector

Board-To-Board Test-A test performed before a cutover from an existing switching system to a replacement switching system. The test ensures the proper connection and assignment of lines on the new system by physical comparison to the lines on the existing switching system.

BoD-Bandwidth On Demand

BOF-Birds of Feathers

Boilerplate-Boilerplate is a sample structure or document that is used to assist in the creation and formatting of other documents.

BOM-Bill Of Materials

Bonded ADSL-Bonded DSL

Bonding-Bonding is a procedure that creates a relationship between two devices based on a common link key. The link key is created and exchanged during the bonding procedure and is stored by both devices for use in future authentication. Bonding is used in the Bluetooth system to create a trusted association between Bluetooth devices.

Bong-A unique tone that telephone carriers add to an audio signal while a call is in progress to alert the user that another action on the users part is required (such as to dial an access code number for a calling card).

BooP-Bootstrap Protocol

Boot-(1-computer) Booting is the process of starting a computer and/or loading an operating system into memory. (2-cable) A boot is a protective case that covers the physical attachment of a connector to a cable. A boot may only serve a cosmetic (appearance) or it may provide some mechanical isolation of cable to the connector minimizing the effects of pulling or wiggling of the cable.

Boot Loader-A boot loader is a small program, typically in ROM, that executes after the central processing unit is reset or powered on that takes care of loading the operating system and necessary system drivers from disk.

Bootstrap Protocol (BooP)-BooP is a protocol that is used to acquire initializing information from a data communications network that a device is connected to. The use of BooP allows a communication device to obtain a network address and to initially determine where to initially start transmitting packets of data.

Border-A perimeter of a system or network. Borders can be physical or logical boundaries that can be crossed at locations known as border gateways.

Border Gateway (BG)-A router or data processing device that connects one network (such as the Internet) to another network (such as a private LAN). Because the border gateway is a single point of entry, it filters all data going between the networks.

Border Gateway Multicast Protocol (BGMP)-Border gateway multicast protocol is a protocol that is used to communicate between routers to help determine the selection of routes for multicast packets over multiple network domains.

Border Gateway Protocol Version 4 (BGP4)-An inter-domain routing protocol that is used to span (link) autonomous systems on the Internet. BGP4 is capable of aggregating routes listed within a router's memory and the use of BGP4 results in a reduction of the size of routing tables.

BORSCHT-Battery, Overvoltage, Ringing, Supervision, Coding, Hybrid, Test

Bose, Chaudhuri, And Hocquenhem (BCH) Code-A class of cyclic redundancy codes used to provide error detection and correction capabilities to a communication link.

Bot-Robot

Bottleneck-A point in a data communications path or computer processing flow that limits overall throughput or performance.

BPDU-Bridge Protocol Data Unit

BPH-Bits per Hertz

BPI-Bits Per Inch

BPL-Broadband Over Powerline

BPON-Broadband Passive Optical Network

BPS-Bits Per Second

BPSK-Binary Phase-Shift Keying

BPV-Bi-Polar Violation

BQA-Bluetooth Qualification Administrator

BQB-Bluetooth Qualification Body

BQRB-Bluetooth Qualification Review Board

BQTF-Bluetooth Qualification Test Facility

BR-Base Radio

Branch Splice-A cable splicing method whereby cable feeder or distribution counts are separated within a splice closure. The total incoming cable count splits into 2 or more separate paths for final distribution via drop-off at various fixed or ready access terminals along the way. For example cable 10 pairs 1-600 separates into cable 101 pairs 1-300 and cable 101 pairs 301-600 in a Y type of arrangement.

Brand Equity -Brand equity is the term used to describe the extra value that a company or product can command in the marketplace because of the branding activity associated with it.

Branding-Branding is the process of creating an awareness of a company (corporate brand), product (product brand), or service (service brand) to the purchasers, users, or influencers of products or services.

Branding System-A branding system is the equipment and hardware that can be used to add branding logos and identification into or along with the delivery and display of content.

Breadboard-A board used for prototype assembly. The term came from the use of low cost wooden breadboards that allowed inventors to easily attach sockets and other electrical or electronic components so they could be interconnected by temporary wires. A breadboard is often used to describe circuits or systems that are assembled to test if a design or technology will work.

Breakage-A final balance on a prepaid service that is never used.

Break-In-Busy Override

Break-Out-Break Out Cable

Breakout Cable-A breakout cable is a wire assembly that allows the communication lines to be easily separated (breakout) and connected. Breakout cables usually contain tight buffer tubes and have separate strength members for each inner jacket assembly.

Breakpoint-A point, measured as a distance from a central office, beyond which service is not provided without a mileage charge. A technology breakpoint is one beyond which service cannot be providing by an existing transmission system. An economic breakpoint is one at which it is more costly to provide service on one technology than on another.

BREW-Binary Runtime Environment for Wireless

BRI-Basic Rate Interface

Brickwall Filter-An ideal type of security filter that is used in a data communications network. A brick wall filter has rapid response with no loss to passband data and infinite loss stopband.

Bridge-A bridge is a data communication device that connects two or more segments of data communication networks by forwarding packets between them. Bridges extend the reach of the LAN from one segment to another.

Bridge devices operate at layer 2 of the OSI reference model, used to connect two or more LAN segments. Bridges provide more intelligence and fault isolation than repeater, which operate at Layer 1. However, unlike routers, which operate at Layer 3, bridges do not provide broadcast domains.

Bridges operate by inspecting each packets MAC header and forwarding the packet based only on this information. Each bridge keeps a table which enables it to determine the egress port to which each packet should be forwarded.

In general, there are two classes of bridges: Transparent bridges are commonly found in Ethernet-like networks and make use of only MAC addresses. Source Route bridges are commonly found in Token-Ring networks, and utilize a Routing Information Field in the MAC header to forward packets.

The terms Routing Information Field and Source Routing are slightly misleading. While these terms have the word "route" in them, these are purely Layer 2 protocols and should not be confused with Layer 3 routing protocols such as RIP or OSPF.

This figure shows the basic operation of a bridge that is connecting 3 segments of a LAN network. Segment 1 of the LAN has addresses 101 through 103, segment 2 of the LAN has addresses 201 through 203, and segment 3 of the LAN has addresses 301 through 303. The table contained in the bridge indicates the address ranges that should be forwarded to specific ports. This diagram shows a packet that is received from LAN segment 3 that contains the address 102 will be forwarded to LAN segment 1. When a data packet from computer 303 contains the address 301, the bridge will receive the packet but the bridge will ignore (not forward) the packet.

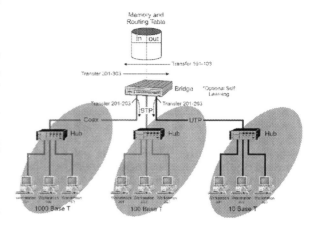

Data Bridge Operation

Bridge Identifier-In the Spanning Tree Protocol, a catenation of a Bridge Priority value and the bridge's MAC address. The bridge with the numerically lowest bridge identifier value will become the Root Bridge in the spanning tree.

Bridge Number-A locally-assigned, ring unique number identifying each bridge in a source routed catenet.

Bridge Port-A network interface on a bridge.

Bridge Protocol Data Unit (BPDU)-A type of packet data that is used in an ATM system to exchange management and control information between other systems (bridge).

Bridge Tap-A bridge tap is an extension to a communication line that is used to attach two (or more) end points (user access lines) to a central office. Bridge taps provide connection options to the telephone company on connecting different communication lines to a central office without having to install new pairs of wires each time a customer requests a new telephone line.

This figure shows how reflections from bridge line tap can cause distortion. This signal shows that some of the energy from the bridge tap is reflected back to the communications line. This reflected signal is a delayed representation of the original signal. Typically, bridge taps must be removed from communications lines that use DSL technology.

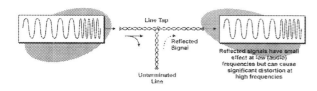

Bridge Tap Reflections

Bridge Tap Reflections-Bridge tap reflections are signals that are redirected from a line or assembly that is connected (tapped) into a transmission line. Bridge tap reflections may occur from the impedance mismatch of the tapped line or from the impedance mismatch of the device that the tap is connected to.

Bridge Transit Delay-The delay between the receipt of a frame on one port and the forwarding of that frame onto another port of a bridge.

Bridged LAN-A network that is formed my connecting two or more LAN segments with a bridge. See bridge.

Bridged Line Appearance (BLA)-A communication line that is designed to appear as a bridged communication line.

Bridged Tap-A communication line (tap) that is connected to another communication line between a receiver and a transmitter. The bridged tap appears as a stub or side branch on the main line. Bridged taps allow communication lines to have other possible termination points (possibly to allow connection to different customers in the future). At normal audio signal frequencies (300Hz to 3300Hz), a bridged tap does not significantly affect the electrical transmission characteristics. However, at high frequencies (such as those used in DSL technologies), bridged taps can distort the transmission of electrical signals (commonly called a shunting affect).

Bridge-Router (Brouter)-When the functionality of a router is combined with a bridge function, it is called a BRouter. The BRouter can route one or more specific protocols, such as TCP/IP, and bridge other protocols. This allows a BRouter to operate at either the data-link layer or the network layer of the OSI Reference Model.

Bridging-The shunting or paralleling of one circuit with another.

Bridging Connection-A parallel connection that draws some of the signal energy from a circuit, often with imperceptible effects on the normal operation of the circuit.

Bridging Loss-The loss at a given frequency resulting from connecting an impedance across a transmission line.

Bridging Repeater-A specially designed central-office repeater used for patching pulse code modulation (PCM) lines so that service will not be interrupted when a patch is removed.

British Naval Connector (BNC)-A twist-locking bayonet connector that is used to connect coaxial cable. This connector was originally developed for the British Navy.

This diagram shows several types of British Naval Connectors (BNC). A BNC connector is a twist-locking bayonet connector that is used to connect coaxial cable. This connector was originally developed by the British Navy. BNC connectors are common connectors in a network environment because of their versatility and universal acceptance. Most network peripherals will have a port that allows a cable fitted with a BNC connector to be attached.

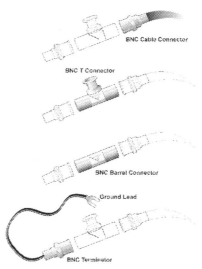

British Naval Connector (BNC)

BRJ-Bandwidth Reject

Broadband-(1-data transfer) A term that is commonly associated with high-speed data transfer connections. When applied to consumer access networks, broadband often refers to data transmission rates of 1 Mbps or higher. When referred to LANs, MANs, or WANs, broadband data transmission rates are 45 Mbps or higher. (2-radio bandwidth) A frequency bandwidth that is much larger than the required bandwidth to transfer the information signal. For example, using a 1 MHz wide radio channel to transmit a 4 kHz limited audio signal.

Broadband Access System-A broadband access system is a portion of a high-speed communication system (over 1 Mbps) that coordinates requests for services and enables the transfer of information when authorized.

Broadband Communications-Broadband communication service is the transfer of digital audio (voice), data, and/or video communications at rates greater than wideband communications rates (above 1 Mbps).

Broadband Integrated Services Digital Network (B-ISDN)-Usually refers to the portions of a digital network operating at data transfer rates in excess of 1.544 or 2.048 Mbps. The B-ISDN network often uses ATM to enables transport and switching of voice, data, image, and video over the same network equipment.

Broadband Integrated Services Digital Network User Part (B-ISUP)-An SS7 protocol that defines the signaling messages that are used to control ATM broadband connections and services.

Broadband Local Exchange Carrier (BLEC)-A service carrier (provider) that offers broadband services locally. Broadband services are usually defined as the ability to transmit a large amount of information, including voice, data, and video, that has a combined data transmission rate that exceeds 1 Mbps.

Broadband Network-A network that is capable of transmitting a large amount of information, including voice, data, and video that have a combined data transmission rate that exceeds 1 Mbps. Sometimes called wideband transmission, it is based on the same technology used by cable television.

Broadband Over Powerline (BPL)-Broadband over powerline is the transmission of broadband data signals (over 1 Mbps) over power lines. BPL can refer to sending broadband over high voltage or low voltage power lines.

Broadband Router-A router that is able to receive and transfer packets in a broadband network (1 Mbps+).

Broadband Signal-A term that is used to describe high-speed data communication signals. For consumer data communication, broadband signals have a data transmission rate above 1 Mbps are usually considered Broadband signals. For telecommunication or data networks, data transmission rates above 45 Mbps (OC1) are considered broadband.

Broadband Telephony-The use of a broadband data connection (such as high-speed DSL or cable modem Internet connection) to make telephone calls.

Broadband Television (Broadband TV)-Broadband television is the delivery of digital television services over broadband data connections. Broadband television systems may be able to control and guarantee the quality of television services if the underlying broadband connections have enough bandwidth.

Broadband TV-Broadband Television

Broadband Video-Broadband video is the transfer of digital video information through a broadband communication system. Broadband video is typically delivered at a data rate above 1 Mbps.

Broadband War-The broadband war is the conflict that is occurring between multiple types of communication service provider companies who can each provide broadband (1 Mbps+) communication services. Key tactics in the broadband war include increasing the performance of broadband connections (higher data transmission rates) and declining costs of broadband service. Broadband competitors include telephone companies (DSL), cable television (Cable Modems), mobile telephone (EVDO, HSPDA), fiber systems (FTTH or FTTC) and fixed wireless (WiMax).

Broadband Wireless-Broadband wireless is the transfer of high-speed data communications via a wireless connection. Broadband wireless often refers to data transmission rates of 1 Mbps or higher.

Broadband Wireless Access (BWA)-Broadband wireless access is a term that is commonly associated with wireless high-speed data transfer connections. When applied to consumer access networks, broadband often refers to data transmission rates of 1 Mbps or higher.

Broadcast-Transmission of an information signal to a specified geographic area or network. This allows the same information to be received by all customers in that geographic area that can successfully receive (demodulate) and decode the information.

This diagram shows two broadcast examples: radio broadcast and network broadcast. Part (a) shows a radio broadcast tower that is sending an audio broadcast to all radios that are within its radio signal coverage area. Part (b) shows a network broadcast system that sends a data message that is coded to indicate the message is a broadcast message. This message contains an address that indicates it is a broadcast message. When routers or other data distribution devices receive this message, each distribution device forwards the data broadcast message to the other network parts for which it is connected to. All communication devices that are connected to the network can receive the broadcast message.

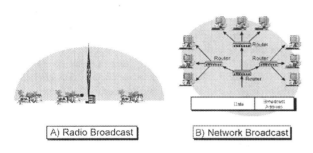

Broadcast Communication Operation

Broadcast Address-A well-known multicast address signifying the set of all stations.

Broadcast Advertising-Broadcast advertising is the sending of the same advertising message to all the recipients (such as television viewers) who can receive the broadcast signal.

Broadcast and Unknown Server (BUS)-A process used for LAN Emulation (LANE) in an asynchronous transfer mode (ATM) system to handle broadcast and multicast data. The BUS communicates with the LAN emulation server (LES) to register and resolve addresses between a 48-bit Ethernet address and ATM addresses. The BUS

system labels each device transmission with both addresses.

Broadcast Encryption-Broadcast encryption is a process of protecting voice or data information from being obtained by unauthorized users in a broadcast system. Broadcast encryption involves the use of a data processing algorithm (formula program) that uses one or more secret keys that both the sender and receiver of the information use to encrypt and decrypt the information. Without the encryption algorithm and key(s), unauthorized listeners cannot decode the message. When the encryption and decryption keys are the same, the encryption process is known as symmetrical encryption. When different encryption and decryption keys are used (such as in a public encryption system), the process is known as asymmetrical encryption.

Broadcast Fax-A process or service that broadcasts a fax message to a list of pre-defined recipients.

Broadcast Media Distribution Protocol (BMDP)-Broadband media distribution protocols are the commands and processes that control and enable broadcast distribution systems (such as IPTV) to adjust storage and distribution heuristically according to trends in subscriber behavior. BMDP also provides a buffer against network jitter and greater tolerance for peak bursts in traffic. And example of how BDMP operates is the setup of temporary storage of a popular digital television program as it is streamed through a multicast distribution system so that viewers can watch the program from a local media server rather than having to reconnect to a distant media server.

Broadcast Messaging-A service that broadcasts the same message to a group of recipients that are connected to a network or who can receive the broadcast message via a communications medium. This diagram illustrates how broadcast SMS messages are delivered to users in a cellular (wireless) system. Like the point-to-point messaging service, the message first goes to a message center (step 1) where the message center determines that this message is designated for all communication devices. The message is stored with a broadcast message code identifier (step 2). The cellular system then transmits the broadcast message on designated message channels (step 3). These broadcast channels may be part of a control channel in every cell site where the broadcast message is designated to be received or may be on other types of channels (e.g. voice/traffic channels). Unlike point-to-point and point-to-multipoint messages, communication devices do not acknowledge receipt of broadcast messages. If a communication device is off or is not tuned to the message channel, it misses the broadcast message. To address this limitation, messages may be broadcast several times, and mobile telephones that have already received message may ignore repeats (steps 4-5).

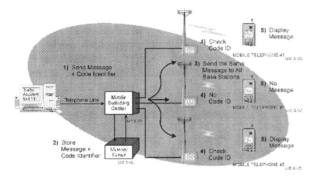

Broadcast Messaging Operation

Broadcast Monitoring-Broadcast monitoring is the process of receiving and reviewing media that is transmitted on a broadcast channel to determine if a particular media item has or has not been broadcasted. Broadcast monitoring may be performed to ensure an advertisement has been inserted on a broadcast television system as defined in an advertising agreement or broadcast monitoring may be used to ensure some media is not broadcast (e.g. licensed content).

Broadcast Music International (BMI)-Broadcast music international is an organization that assists and represents people and companies that are involved in the creation and distribution of film, theater and television content.

Broadcast Radio-Broadcast radio technology allows a single transmitted signal to be heard by many radio receivers. The first radio broadcast transmission systems used amplitude modulated (AM) radio transmission to provide audio service. Radio broadcast technologies have evolved to allow for enhanced audio (stereo) and data signals (such as radio broadcast data system).

Broadcast Storm-A broadcast storm is the congestion that is created on a communication network

when a large quantity of packets is broadcasted.

Broadcast Television (Broadcast TV)-Broadcast television is the sending of video and audio signals to devices (such as television sets) that can receive broadcast signals. Television broadcast technology was developed in the 1940s. The success of the television marketplace is due to standardized, reliable, and relatively inexpensive television receivers and a large selection of media sources. The first television transmission standards used analog radio transmission to provide black and white video service. These initial television technologies have evolved to allow for both black and white and color television signals, along with advanced services such as stereo audio and closed caption text. This was a very important evolution as new television services (such as color television) can be on the same radio channel as black and white television services.

Broadcast Television Systems Committee (BTSC)-The broadcast television system committee stereo system is a recommended process that is used to provide multiple channels (stereo), television sound transmission and audio processing. BTSC stereo is defined in FCC Bulletin OET 60.

Broadcast Transmissions-Sending the same signal to many different places, like television broadcasting station. Broadcast transmission can be over optical fibers if the same signal is sent to many subscribers. For LAN's broadcast messages are usually associated with Ethernet that uses CSMA/CD at its MAC layer.

Broadcast TV-Broadcast Television

Broadcast/Multicast Control Protocol (BMC)-Broadcast and multicast control (BMC) protocol coordinates the transmission and reception of broadcast and multicast services for the WCDMA system. The main functions of BMC include the reception, storage, and distribution of broadcast and multicast messages such as broadcast short messages to the appropriate applications.

Broadcaster-A broadcaster is a company that transmits or provides information to users that are connected or able to access signals on the broadcast network.

Broadcasting-Broadcasting is a process that sends voice, data, or video signals simultaneously to group of people or companies in a specific geographic area or who are connected to the broadcast network system (e.g. satellite or cable television system). Broadcasting is typically associated with radio or television radio transmission systems that send the same radio signal to many receivers in a geographic area. Broadcasting can also be applied to distribution or point-to-point networks where all users that are connected to the network can receive the same information signal.

Broadcasting Satellite Service-Broadcasting satellite service is the transmission of audio and video signals to devices or individuals by satellite broadcasting.

Broadcasting Service-Broadcasting service is the transmission of information signals (audio, video and/or data) that is intended for direct reception by multiple receivers (e.g. consumers).

Broadcatching-Broadcatching is the process of consumers selecting, capturing and consuming the media they want from an endless array (broad selection) of choices.

Brouter-Bridge-Router

Brownout-A reduction in the servicing of customers as a result of the demand for service exceeding the service processing capability of the service provider's equipment or staff. Brownout usually occurs during a peak period. Because service access attempts by customers increase during a brownout period (customers repeatedly attempt to get service), service providers may discontinue services to groups of customers during brownout.

Browser-A software program or module (called a client) that is used to convert information that is available on the Web portion of the Internet into forms usable by a person (text, graphics and sound). Also called a web browser.

Browser Plug In-A browser plug-in is a software program that works with a web browser application to enhance its capabilities. An example of a browser plug-in is a media player plug-in. The media player decodes and reformats the incoming media so it can be displayed on the web browser.

Browsing-Browsing is the process of searching through an information storage system to identify and/or acquire information without the need to know the specific name or format of information that is being searched.

BRQ-Bandwidth Request
BS-Base Station
BS-Bearer Services
BSA-Basic Service Area
BSAC-Bit-Sliced Arithmetic Coding
BSC-Base Station Controller
BSCS-Business Support And Control System

BSE-Basic Service Element
BSF-Band Stop Filter
BSN-Backward Sequence Number
BSP-Backbone Service Provider
BSS-802.11 Basic Service Set
BSS-Base Station Subsystem
BSS-Basic Services Set
BSSID-Basic Service Set Identifier
BT-Blind Transfer
BTA-Basic Trading Area
BTN-Billing Telephone Number
BTS-Base Transceiver Station
BTSC-Broadcast Television Systems Committee

Buffer-(1-memory) An allocation of memory storage that set aside for temporary storage of data. The use of memory buffers allow data transfer rates to vary so that differences in communication speeds between devices do not interrupt normal processing operations. (2-optical) A layer placed between a fiber and its jacket to provide additional protection to the fiber. The layer is often made of a thermoplastic material.

Bug-(1-problem) A problem that occurs in software or hardware that can be fixed by changes to the software or design changes to the hardware. The term seems to have started from a problem that occurred with an early model computer system when a moth got caught inside the machinery resulting in a bug problem. (2-audio) A microphone or listening device that is concealed for audio surveillance.

Building Local Exchange Carrier (BLEC)-A building owner, real estate owner, or corporation that provides local exchange carrier services to the tenants or customer's of their building(s).

Bulk Basis-Billing system that is based on the billing name and address information for all the local exchange service subscribers of local exchange carrier.

Bulk Billing-A method of billing telephone customers in which the charges for all messages of a given type, chargeable to a particular account for a billing period, are combined and billed as a single amount. Message-unit billing is the most frequent application of the bulk-billing procedure.

Bulk Encryption-Bulk encryption is the combined encryption of some or all channels that are transferred in a communication system to many receivers.

Bulkhead Receptacles-Bulkhead receptacles are the connectors on a patch panel that are used to connect to lines and equipment. The network configuration can be easily changed by moving cables on the bulkhead connectors instead of cutting and splicing lines.

Bullet Proof (Bulletproof)-Bullet proof is the ability of a system or process to deter or remain unchanged to the effects of unauthorized users or processes.

Bulletin Board System (BBS)-A bulletin board system is a data communication service that allows users to connect to it (usually via a dial-up modem connection) and share and/or exchange messages. Many users of bulletin board systems have changed to using Internet Newsgroups.

Bulletproof-Bullet Proof

Bumper Ad-A bumper ad is an advertising clip that is played before or after a media or video segment or clip.

Bundling-The process of combining different products and services into a "package" offer, and then offering it to a customer at a separate, combined price.

Bundling Services-The combining of different services into one service offering so the customer can communicate with one company for several different services. An example of bundling is the combination of cellular, PCS, local and long distance services as one service package.

Buried Cable-Buried cable is the location of a cable under the surface of the ground in such a manner that it cannot be removed without disturbing the soil.

Burst-(1- general) A short transmission of information (data). (2- GSM) A time slot of information that lasts 577 usec. There are several burst types including normal, synchronization, frequency correction, access and dummy. (3-CDMA) A time slot of information that lasts 1.25 msec).

Burst Collisions-The overlapping of transmission bursts between time slots at a receiver. This may occur from varying amounts of propagation time in a mobile communication time where some mobile radios are close and others are far away. Dynamic time alignment (variable transmission time periods) was created to solve this challenge. Dynamic time alignment allows mobile radios at a distance to start the transmission early so their transmission bursts will be received in the correct time slot period.

Burst Errors-Burst errors are the distortion or failure of a digital receiver to correctly decode groups of digital bits. Burst errors typically have a high bit error ratio (BER) compared to the overall BER of a communication link or channel.

Burst Switching-The techniques of switching packets in bursts through a network.

Bursty Data-An attribute of a communications channel where the data comes in unpredictable bursts instead of a continuous stream. Typical voice communications is an example of bursty data because there are long periods of silence followed by periods of speech.

BUS-Broadcast and Unknown Server

Bus (Signal Path)-A central conductor for the primary signal path. The term bus also may refer to a signal path to which a number of inputs may be connected for feed to one or more outputs.

Bus Network-A network where each communication device (typically computers) are connected to a common bus. Each computer that is connected to the bus network uses a transmission control process to sense availability and contention (simultaneous access) on the network.

Bus Segment-A section or piece of bus that is electrically continuous, with no intervening components such as repeaters. Electrical continuity may be maintained using fixtures such as coaxial cable connectors, but no signal boosting equipment (repeaters) are used, the piece of cable on the other side of such equipment is defined as another bus segment.

Bus Topology-In a local area network (LAN), a topology consisting of a single shared medium, typically coaxial cable, to which multiple devices are connected. These devices monitor signals on the medium and selectively copy the data addressed to them. Modern day Ethernet hubs implement a logical bus topology while providing a star-wired network structure. (See also: bus topology, ring topology, tree topology.)

This figure shows a bus network. This diagram shows that data communication devices can hear each other on the common (shared) bus.

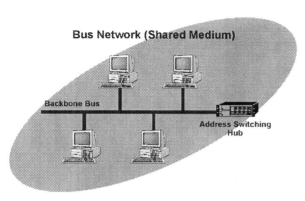

Physical Bus Topology

Business Group-A collection of lines that serve a single business location, share a common number dialing plan, and are assigned other business features. It is commonly called a Centrex group.

Business Services Data Base-An application residing in a service control point (SCP) that provides call-handling instructions to a service switching point (SSP).

Busy Hour-The hour in a day when the total usage of the network, trunk connection, or the switching system is greater than at any other hour during the day. Telephone systems and networks are typically designed to meet a specific quality level that can be provided during the busy hour (e.g. maximum number of blocked calls).

Busy Hour Minutes (BHM)-The number of minutes during "busy hours" ordered by a customer. Other terms: BHMOT represents originating usage, BHMTT represents terminating usage.

Busy Hour Minutes of Capacity (BHMC)-The maximum amount of access minutes that an interconnecting carrier (such as an IXC) expects to route through end office switch at peak activity.

Busy Hour Traffic (BHT)-Busy hour traffic is the amount of call traffic in hours that occur during the busiest time of the day.

Busy Override (Break-In)-A feature that allows access to a busy number for an emergency.

Busy Tone-An audible tone or signal that indicates to a caller that their call cannot be received because a called line is unavailable (in use).

Busy/Idle Flag-A busy/idle flag used in cellular radio transmission is an indicator that is transmitted by the Mobile Data Base Station (MDBS) periodically to indicate whether the reverse channel is currently in the busy state or the idle state.

BWA-Broadband Wireless Access

BWDP-Bandwidth Distance Product

Bye Message-A message that is used to inform a person or device that a communication session is ending. The Bye message is defined in session initiation protocol (SIP).

Bypass-Bypass service is the routing of calls or communication sessions around any other networks facilities to avoid toll charges. Bypass is commonly discussed with the ability to bypass local exchange carriers (LEC) to save on interconnection charges.

Bypass LAN-The capability of a station to be electronically or optically isolated from a network while maintaining the integrity of the network ring. This is found in the newer hub-based token ring networks.

Bypass Relays-Relays in a ring network that permits message traffic to travel between two nodes that are not normally adjacent. Usually, such relays are arranged so that any node can be removed from the ring for servicing and the two nodes on either side of the removed node are now connected via the bypass relay.

Byte (B)-A group of bits, typically eight, used to represent data values from 0 to 255. These values may represent alphabetic or control characters or numbers less than 255. A single byte is typically the smallest data value manipulated by a computer. In order to represent larger values, multiple bytes are grouped together into words that are 16, 32, 64 or 128 bits in length.

Byte Multiplexing-A form of time-division multiplexing (TDM) in which the whole of a byte from one subchannel is sent as a unit, interlacing in successive time slots with complete bytes from other subchannels.

C

C-coulomb

C Wire-European term corresponds to North American Sleeve wire.

C/A Interference Ratio-Carrier to Adjacent Channel

C/I-Carrier To Interference Signal Ratio

C/I-Carrier-To-Interference Ratio

C/I-Co-Channel Interference

C/N-Carrier to Noise Ratio

CA-Call Agent

CA-Certificate Authority

CA-Conditional Access

Cabinet-In communication systems, an enclosure that is used to hold equipment or electronic assemblies.

Cable Assembly-A cable assembly is composed of one or more cables (multiple transmission lines) with connection points (usually connectors or terminations).

Cable Binder- A device that holds multiple cable pairs.

Cable Box-(1-optical or electrical) A cable box is a container assembly that holds wires, cables and devices. (2-television) An electronic device that adapts a communications medium to a format that is accessible by the end user. Set top boxes are commonly located in a customer's home to allow the reception of video signals on a television or computer.

Cable Cap-A cable cap or water seal is a covering that is applied to the end of a cable to seal the end of the cable from the entry of water and other unwanted substances.

Cable Card-A cable card is a portable credit card size device that can store and process information that is unique to the owner or manager of the cable card. When the card is inserted into a cable card socket (such as into a set top box), electrical pads on the card connect it to transfer information between the electronic device and the card. Cable cards are used to identify and validate the user or a service and may contain decryption keys that allow for the descrambling of cable transport channels.

Cable Clamp-A cable clamp is a device that is used to hold a cable in a particular physical position.

Cable Color Code-Cable color codes are unique identifying colors on or inside cables that are used to uniquely identify wireless and/or fibers within a cable or group of cables.

Cable Converter-Cable converters, commonly called a "set top" box are electronic devices that convert an incoming cable television signal into a form that can be displayed on a video device typically a television or computer. The set-top box is typically located in a customer's home to enable the reception and/or interaction with services on the customer's television or computer.

Cable Crimper-A cable crimper is a tool that can compress a connector or crimp sleeve that it typically used to hold a connector to a cable end. A crimp sleeve is a tube, usually constructed of metal, that can be compressed (crimped) to surround and hold a cable or wire to a connecting device or assembly.

Cable Duct-Cable ducts are pipes or conduits that are installed underground or in a building, whose purpose is to protect the cables installed within them.

This figure shows cable ducts that are inside a duct. These cable ducts are hollow and the allow cables to be pulled through the cable after the duct has been installed.

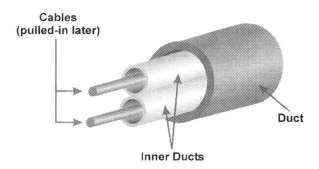

Cables (pulled-in later)

Duct

Inner Ducts

Cable Ducts

Cable Enclosure-A cable enclosure is a plastic or metal container that is used to cover and protect wires or cables. Cable enclosures may be used to protect the ends of cable or to protect splices. Cable enclosures may contain multiple container shells. An outer shell may be used to provide mechanical and environmental protection and the inner shell may be used to hold the cables or a splice tray.

Cable End Box-A cable end box is a container assembly that is used to provide access for pulling cables and wires through conduits that are connected to the cable end box.

Cable Entrance Facility (CEF)-The steel, wooden, tile, or plastic ductwork or pathway that allows for cabling to enter a building.

Cable Fire Rating-Cable fire rating is the classification of the temperature ranges and exposure times that cause emissions from a substance or cable. Cable fire ratings are specified in the national electrical code (NEC).

Cable Flange-A cable flange is a guiding mechanism that assists in the spooling (pulling) of cable on a cable spool or reel.

Cable Grounding-Cable grounding is the connecting of a device or circuit to an electrical ground point or to a conductor that is grounded.

Cable Hanger-A cable hanger is a device that is used to hold a cable and to attach a cable to a structure by hanging. Cable hangers may be designed for specific types of cables.

Cable Head End-The portion of a cable television communication system that receives, formats, and transmits the carrier signals to the distribution system. Because cable television companies have started to offer two-way telecommunication services such as Internet and telephone service, the head end equipment has been expanded to include voice and data gateways.

This figure shows a diagram of a simple head-end system. This diagram shows that the head-end allows the selection of multiple video sources. Some of these video sources are scrambled to prevent unauthorized viewing before being sent to the cable distribution system. The video signals are supplied to video modulators that convert the low frequency video signals into their radio frequency television channel. The output of each modulator is combined and connected to the distribution trunk.

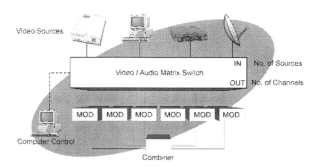

Head End System

Cable Input Selector Switch-A cable input selector is a switch that allows a user to select between different cable lines such as a cable television system or antenna.

Cable Installation-Cable installation is the process of identifying and/or creating routing paths for cable lines, handling cable rolls and pulling through conduits or attaching cable to supporting structures.

Cable Jacket-A cable jacket is a layer of material surrounding wires or fibers inside a cable. A jacket may be used to insulate the conductors and/or to hold conductors and materials together inside a cable assembly.

Cable Jacket Slitter-A cable jacket slitter is a tool that is used to create a cut along the side of cable jackets without penetrating the cut into the inner conductors. To operate the cable jacket slitter, the tool is located where the slit will begin, the handle is compressed so the cutting edge enters the side of the jacket and the cable jacket slitter is pulled toward the end of the cable. The cutting blade of a cable jacket slitter may be adjustable to avoid cutting into the inner conductors.

Cable Label-Cable labels are unique identifying information (typically numbers and letters) that are used to identify specific cables. Cable identifiers usually a unique number for the specific cable along with other information that relates to the type of cable and information about its routing and connection points within a communication system. Cable labels are usually attached within inches of each end of the cable. Cable identifiers are used by designers and technicians during design, installation and troubleshooting of the system.

Cable Laboratories, Inc. (CableLabs)- Cablelabs® is a non profit consortium of members from the cable television industry founded in 1988 that oversees and assists in the development of technologies used in cable television systems.

Cable Locating Equipment-Cable locating equipment is a test instrument that can be used to locate buried or covered cables, lines, or pipes. Cable locating equipment typically operates by injecting a radio frequency signal onto the cable, lines, or pipes to allow a receiver to detect the RF signal located within or near the cable, lines, or pipes.

Cable Loop (Cableloop)-A cable loop is a bundle (loop) of cable that is part of a communication line (such as a pole mounted cable TV or telephone lines) that provides additional cable length that may be necessary to perform a splice or cable path reconfiguration at a later time. Cable loops are sometimes called service loops.

Cable Loss-Cable loss is the signal loss through the cable (insertion loss), often expressed in decibels (dB) per unit length.

Cable Markings-Cable markings are identifying marks that are located on or in a cable. Cable markings may identify the type of cable along with its manufacturing information.

Cable Modem (CM)-A communication device that Modulates and demodulates (MoDem) data signals to and from a cable television system. Cable modems select and decode high data-rate signals on the cable television system (CATV) into digital signals that are designated for a specific user.

There are two generations of cable modems; First Generation one-way cable modems transmit high speed data to all the users into a portion of a cable network and return low speed data through telephone lines or via a shared channel on the CATV system. First generation cable modems used asymmetrical data transmission where the data transfer rate in the downstream direction was typically much higher than the data transfer in the upstream direction. The typical gross (system) downstream data rates ranged up to 30 Mbps and gross upstream data rates typically range up to 2 Mbps. Because 500 to 2000 users typically share the gross data transfer rate on a cable system, cable modems also have the requirement to divide the high-speed digital signals into low speed connections for each user. The average data rates for a first generation cable modem user rage up to 720 kbps. Until the late 1990's, most cable modems used first generation technology.

Second generation cable modems offered much data transmission rates in both downstream and upstream directions. Second generation cable television systems use two-way fiber optic cable for the head end and feeder distribution systems. This allows a much higher data transmission rate and many more channels available for each cable modem. As of the year 2000, approximately 35% of the total cable lines in the United States had already been converted to HFC technology.

This figure shows a basic cable modem system that consists of a head end (television receivers and cable modem system), distribution lines with amplifiers, and cable modems that connect to customers' computers. This diagram shows that the cable television operator's head end system contains both analog and digital television channel transmitters that are connected to customers through the distribution lines. The distribution lines (fiber and/or coaxial cable) carry over 100 television RF channels. Some of the upper television RF channels are used for digital broadcast channels that transmit data to customers and the lower frequency channels are used to transmit digital information from the customer to the cable operator. Each of the upper digital channels can transfer 30 to 40 Mbps and each of the lower digital channels can transfer data at approximately 2 Mbps. The cable operator has replaced its one-way distribution amplifiers with precision (linear) high frequency bi-directional (two-way) amplifiers. Each high-speed Internet customer has a cable modem that can communicate with the cable modem ter-

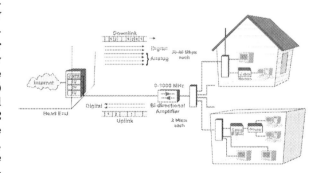

Cable Modem Overview

mination system (CMTS) modem at the head end of the system where the CMTS system is connected to the Internet.

Cable Modem Termination System (CMTS)-A cable modem termination system is a system located in the headend of the cable television system that coordinates the overall operation of the cable modem system. The CMTS controls the gateways (Internet to data) and end user cable modems. The CMTS not only manages the data paths to allow end users to connect to the Internet, but also provides cable modem authentication, IP address assignment, billing functions and is responsible for the majority of Media Access Control (MAC) functionality in a cable modem network. A single CMTS typically controls hundreds or even thousands of end-user cable modems.

Cable Pay-Off Stand-A cable pay-off stand is a fixture that is used to hold a cable reel and allow it to spin and release the cable during cable pulling.

Cable Plant-Cable plant is the physical property and facilities of a telephone, cable or company that uses a cable based transmission system. Cable plant property includes the transmission cables, amplifiers, and channel multiplexing equipment.

Cable Plowing-Cable plowing is the pushing of a blade or other earth splitting apparatus to create a narrow trench that is followed by the insertion of cable into the plowed hole. After inserting the cable, the cable plow may fill in or re-compact the earth burying the cable.

Cable Programming Service-Cable programming service is the providing of information content (programs) by a cable television system operator.

Cable Pullback-Cable pullback is the movement of a cable back into a conduit, cable tray or source of the cable (such as a cable reel). Cable pullback can occur due to expansion or stretching of the cable during installation or to thermal contraction.

Cable Puller-A cable puller is a device a pulling device that is used to pull wire or cables through conduits or other cable channel guides.

Cable Pulling-Cable pulling is the process of pulling a cable through a duct or pipe.

Cable Ready Consumer Electronics Equipment-Cable ready consumer electronics equipment is television, video recorders, or other communication devices that are designed to be capable of receiving the channels and media that are commonly sent on cable television systems.

Cable Reel Trailer-A cable reel trailer is a rolling assembly that holds and can dispense (uncoil) cable from a role as it is installed.

Cable Retention-Cable retention is the securing of a wire or cable within through the use of pressure connectors, straps, or adhesive.

Cable Sag-Cable sag is the distance a cable moves from its straight line axis between its end connection points.

Cable Seal-A cable seal is a device or material that is used to close or isolate a cable end from the environment.

Cable Service-Cable service is the transmission of video programming and other media services to customers (subscribers) via a video and high-speed media distribution system (typically over fiber and/or cables).

Cable Spacer-Cable spacer is a device that holds one or more cables at a distance from each other.

Cable Strap-A cable strap is a device that is used to hold a cable and to attach a cable to another structure by applying pressure to the jacket of the cable. Cable straps come in different sizes and it is important to use a cable strap that is the correct size for the cable and its applications so that the cable does not become bent, crushed or stressed.

Cable Stripper-A cable stripper is a tool that is used to remove the jackets of cable and their inner conductors. The types of cable strippers include a cable knife, snips and a cable sheath cutter.

Cable Stripping-Cable stripping is the process of removing the outer and inner jackets of a cable. Cable stripping is performed to allow for cable splicing or the attachment of connectors to inner conductors.

Cable Stub-A short length of cable, usually less than 25 feet long, that is terminated and shipped with telephone company (telco) terminal hardware, left slack and readied for splicing at the time of activating the terminal with "live" working facilities.

Cable System Operator-A cable system operator is a company or organization who provides cable service (video and other media distribution) over a high-speed communication system who performs the operation the distribution system and controls access to the media that is distributed on the system.

Cable Tag-A cable tag is an identifying plate, sticker or label that is attached to a cable or located somewhere near the cable (such as on a telephone pole).

Cable Telephony-A process of providing telecommunications services through the use of community access television (CATV) systems. Cable telephony services usually combine voice telephone, Internet access, digital cable television (TV), and analog cable TV.

This diagram shows a CATV system that offers cable telephony services. This diagram shows that a two-way digital CATV system can be enhanced to offer cable telephony services by adding voice gateways to the cable network's head-end CMTS system and media terminal adapters (MTAs) at the residence or business. The voice gateway connects and converts signals from the public telephone network into data signals that can be transported on the cable modem system. The CMTS system uses a portion of the cable modem signal (data channel) to communicate with the MTA. The MTA converts the telephony data signal to its analog audio component for connection to standard telephones. MTAs are sometimes called integrated access devices (IADs).

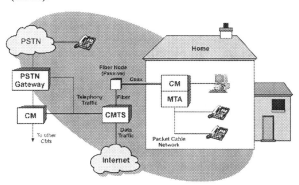

Cable Telephony System

Cable Television (CATV)-Cable television is a distribution system that uses a network of cables to deliver multiple video and audio channels. CATV systems typically have 50 or more video channels. In the late 1990's, many cable systems started converting to digital transmission using fiber optic cable and digital signal compression.

This figure shows a cable television network that has both television distribution and cable modem transmission capability. This diagram shows that the head-end of a cable television (CATV) system is the initial distribution center for the CATV system. The head end is where incoming video and television signal sources (e.g., video tape, satellites, and local studios) are received, amplified, and modulated onto TV carrier channels for transmission on the CATV cabling system. The cable distribution system is a cable (fiber or coax) that is used to transfer signals from the head end to the end-users. The cable is attached to the television through a set-top box. The set-top box is an electronic device that adapts a communications medium to a format that is accessible by the end-user.

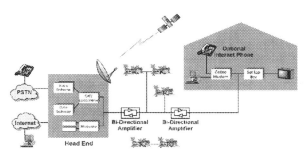

Cable Television Network

Cable Tension-Cable tension is the tensile (pulling) load that exists on a cable.

Cable Termination-Cable termination is the process of connecting or the assembly that is used to connect the end of a cable assembly (wires or fibers) to devices or connectors.

Cable Tie-A cable tie (also called a tie wrap) is a flexible strap that contains a self locking eyelet at one end that allows the other end of the strap to enter and latch so it cannot be pulled back out. The cable tie is looped around the cable and another object (such as a hanger or conduit pipe) and the end is pulled through the cable tie eyelet so a snug wrap is formed.

Cable Transfer-Cable transfer occurs when working cables are spliced and the cable count is changed per engineering documents.

Cable Transmission Loss-The amount of signal or radio frequency (RF) energy that is lost while it travels on or through a cable. Signal loss during cable transmission may result from the cable's electrical characteristics (e.g. shape), the type of conductors, and materials. High frequency signal energy transmitted through twisted pair cable usually radiates (leaks) energy from the conductors. For coaxial cable, much of the high frequency energy is kept within the coaxial cable. Coaxial cables generally have greater signal loss at higher frequencies than lower frequencies due to signal absorption (conversion to heat). As a result, cable losses are usually calculated for the highest frequency that the cable is rated for.

Cable Vault-A concrete enclosure that is designed for routing, splicing and connecting communication cable lines. Cable vaults are also known as manholes.

This diagram shows cable vault duct numbers. This diagram shows that cable vault walls that face to the north and east directions are numbered from left to right and from top to bottom. Cable vault walls that face to the south and west directions are numbered from right to left and from top to bottom. This example shows that this numbering scheme allows cables to easily bend and travel straight through to ducts with the same number.

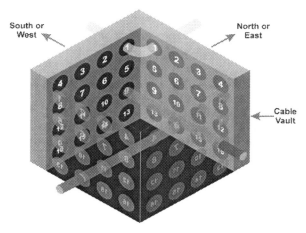

Cable Vault Duct Numbers

Cable Weight-Cable weight is the amount of weight a cable has per unit of measure (usually in feet or meters). Cable weight requirements may be specified to determine how strong a messenger wire must be to support a cable span between poles or mounting points.

Cablecasting-Cablecasting is the providing or retransmission of television programs on a cable broadcast system.

CableLabs-Cable Laboratories, Inc.

Cableloop-Cable Loop

Cabling Safety-Cabling safety are the procedures used to ensure the safety of installers and users of systems that use cables.

CABS-Carrier Access Billing System

CAC-Call Admission Control

CAC-Circuit Access Code

Cache-A memory storage area that temporarily stores information that is repeatedly accessed several times. The cache memory can be accessed more quickly than other memory storage areas such as a CD ROM or hard disk. This allows computers to process information faster. A cache is also used to temporarily store files from other locations (such as Internet Web pages) so it is possible to return to this information without having to transfer the information again.

Caching-Caching is a process by which information is moved to a temporary storage area to assist in the processing or future transfer of information to other parts of a processor or system.

Caching Server-A caching server is a computer that can receive, process, and respond to an end user's (client's) request information or information processing and temporarily store (cache) information that may be used again during a server's communication session.

CAD-Computer Aided Dispatch

CAFC-Court of Appeals of the Federal Circuit

CAI-Common Air Interface

CALEA-Communications Assistance For Law Enforcement

CALEA Server-Communications Assistance for Law Enforcement Act Server

Call Accounting-The processing call routing information that tracks the services used and their amount of use associated with the call.

Call Admission Control (CAC)-(1-ATM) A traffic management feature of ATM networks that ensures that virtual channel connections are not offered unless enough bandwidth is available on its network. (2-voice) Call Admission Control (CAC) is a concept that applies to voice traffic. CAC is a deterministic and informed decision that is made before a voice call is established and is based on whether the required network resources are available to provide suitable QoS for the new call.

Call Agent (CA)-In a communication network, the call agent is responsible for call processing functions related to setup, maintenance, and teardown communication sessions (e.g. voice calls).

Call Center (Call Centre)-A call center is a place where calls are answered and originated, typically between a company and a customer. Call centers assist customers with requests for new service activation and help with product features and services. A call center usually has many stations for call center agents that communicate with customers. When call agents assist customers, they are typically called customer service representatives (CSRs).

Call centers use telephone systems that usually include sophisticated automatic call distribution (ACD) systems and computer telephone integration (CTI) systems. ACD systems route the incoming calls to the correct (qualified) customer service representative (CSR). CTI systems link the telephone calls to the accounting databases to allow the CSR to see the account history (usually producing a "screen-pop" of information).

Call Centre-Call Center

Call Clearing-The process used (e.g. signaling messages) to clear a call connection. Call clearing involves informing the network elements of the release of connection ports and call processing (e.g. echo canceling) activities.

Call Congestion-The ratio of calls lost due to a lack of system resources to the total number of calls over a long interval of time.

Call Control (CC)-The processes associated with establishing, maintaining, and releasing a voice or data call.

Call Delay-The time delay that is experienced when a call arrives at a switching device or system and resources are not immediately available.

Call Detail Record (CDR)-A call detail record that holds information related to a telephone call or communication session. This information usually contains the origination and destination address of the call, time of day the call was connected, added toll charges through other networks, and duration of the call.

This figure shows the basic structure of a call detail record (CDR). This diagram shows that a usage data report (UDR) contains a unique identification number, the originator of the call, the called number, the start and end time of the call. This diagram also shows an additional charge for operator assis-

tance and that a UDR dynamically grows as more relevant information becomes available.

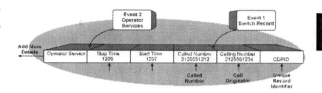

Call Detail Record (CDR) Structure

Call Detail Record (CDR) Processing-The process of creating a call billing record through the gathering and manipulating of billing event and service record information related to the call. Call record processing involves receiving the call record, guiding the record to the specific billing account, rating the record, and routing the record to the appropriate database for collections or account settlement.

This figure shows the general process that is used to identify and rate (bill) a call. This diagram shows that a call detail record evolves as it passes through the rating process. In the first step, a rate band is determined. Then, the identification information on the CDR is used to identify a specific

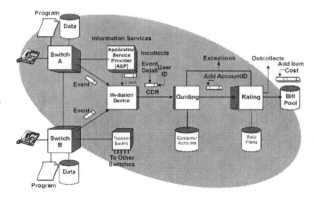

Call Detail Record (CDR) Processing Operation

customer's account (guide the record). The customer's rate plan is discovered and the unit (usage) and fixed (per event) charging rates are gathered and calculated. The new information (rate band, call charge amount) is added to the call detail record and it is moved to the bill pool, as it is ready to be billed.

Call Detail Record Server (CDR Server)-A server that collects and processes call usage information for the creation of call detail record (CDR) accounting and billing information.

Call Diversion-The automatic forwarding of telephone calls to a programmable number if the called telephone number is non-operational. This number is chosen and changed by the user.

Call Forwarding-A call processing feature allows a user to have telephones calls automatically redirected to another telephone number or device (such as a voice mail system). There can be conditional or unconditional reasons for call forwarding. If the user selects that all calls are forwarded to another telephone device (such as a telephone number or voice mailbox), this is unconditional. Conditional reasons for call forwarding include if the user is busy, does not answer or is not reachable (such as when a mobile phone is out of service area).

This diagram shows how call forwarding can be used to automatically redirect telephone calls based on specific conditions. This example shows that a call may be redirected by the switching system to one extension if the user is busy and to a different extension if the user has programmed the extension as "Do Not Disturb" or if it is busy. When the call is received by the system destined for extension 1001, the call processing system uses the

indication of busy along with a redirection table to determine the call must be automatically transferred to extension 1003.

Call Hold-A feature that allows a user to temporarily hold and incoming call, typically to use other features such as transfer or to originate a 3rd party call. During the call hold period, the caller may hear silence or music depending on the network or telephone feature.

This diagram shows how a call can be temporarily placed on hold so the call can stay connected without the user having to continue conversation with the caller. During hold, the audio from the user is muted. For an analog line, the call hold feature involves placing a load (connection) across the line so that current may continue to flow through the circuit. For digital systems, the call hold feature may send a call hold message back to the system (such as a signaling message on the signaling channel) so the system can know that the status of the telephone station has changed to "hold."

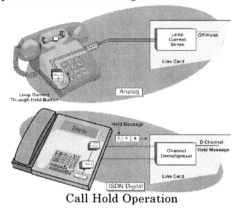

Call Hold Operation

Call Forwarding Operation

Call Intrusion-The unannounced or un-requested connection of another caller to a call that is in progress.

Call Log-The recording of information that is associated with a communication device or service.

Call Management Record (CMR)-Records relating to a call that was processed by a call manager in an Enterprise communication system (such as an iPBX system that uses AVVID protocols).

Call Me Later-Call me later is a service that allows an Internet user or television viewer to select an option that requests that someone from a company or offered service call them later.

Call Me Now-Call me now is a service that allows an Internet user or television viewer to select an option that requests that someone from a company or offered service call them immediately.

Call Origination-Call origination on a wireless system is the initiation of a call or communication session by a mobile or fixed radio. The call origination process begins with a communication session request (such as the mobile radio user dialing a number and pressing send). This is followed by an attempt to access the communication system on an access channel. After attempting to access the system, the mobile radio listens to see if a response to its' communication request has been processed (called "contention resolution). If it does not hear a response in a pre-determined maximum time, it will wait a random amount of time before attempting access again to prevent repeated collisions with other users. If the system is not busy, it will send out an acknowledgment message and assign the mobile radio to a communication channel.

Call Pickup-A telephone call processing (switch control) feature that enables a telephone user to answer a telephone from another telephone station. When a call is received, a key sequence is entered from specific groups or any telephone (dependent on how the system is setup) and the call is redirected to the extension or line that has picked up (entered the code) the line.

This diagram shows the operation of a telephone system that has a call pickup group feature. In this example, an incoming call is received and is directed toward the main extension 1001. When the call is received for extension 1001, the system uses the call pickup group list to determine that extensions 1001, 1002, 1003, and 1004 are programmed for call pickup. This allows any of these extensions to answer the call by pressing a key sequence (key "3" in this example).

Call Processing Language (CPL)-A language that is based on extended markup language (XML) that is used to describe and control Internet (VoIP) telephony services. CPL is used by end users to create telephone services that integrate VoIP with existing Email, WEB, and other applications. It can be used with other VoIP protocols such as session initiation protocol (SIP).

Call Progress Tones-Signaling control tones that are sent to inform the device or system of the progress of a call. Examples of call progress tones include dial tone, ringback, and busy tone.

Call Rate (CR)-The number of calls that are received over a defined period of time (such as during an hour or day.) The call rate may be characterized by a specific period of time such as the busy hour (BH) so its usage often include enough qualifying information to assure that it will be properly understood.

Call Rating-A process of assigning a value or cost to a telephone call or communication session.

This figure shows the basic call rating process used in a billing system. This diagram shows that a call detail record evolves as it passes through the rating process. In the first step, a rate band is determined. Then, the identification information on the CDR is used to identify a specific customer's account (guide the record). The customer's rate plan is discovered and the unit (usage) and fixed (per event) charging rates is gathered and calculated. The new information (rate band, call charge amount) is added to the call detail record and it is moved to the bill pool as it is ready to be billed.

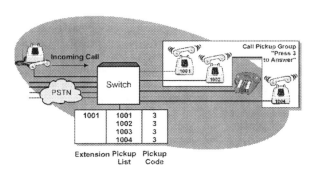

Call Pickup Operation

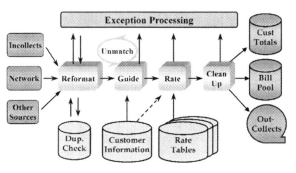

Billing System Call Rating

Call Reference-In the Integrated Services Digital Network (ISDN) Q.931 protocol, an information element that identifies the call that corresponds to a Layer 3 message.

Call Reference Flag-In the Integrated Services Digital Network (ISDN) Q.931 protocol, a flag that indicates which side of an interface allocates the call reference value.

Call Report-A cellular telephone record stored on a data acquisition system (DAS) tape containing the overall timing information, mobile number, dialed digits, and appropriate indicators to ticket a call for every call completed or attempted through the system. (See billing Tape).

Call Restrictions-In call-processing feature that restricts telephone call origination to specific groups of authorized numbers (typically local phone numbers or no-international calls).

Call Routing-(1-circuit switching) The process of determining the path of a call from point of origination to point of destination. (2-packet switching) The steps taken to ensure a connection can be made from an origination point to its destination point. The packets that transfer through this connection may actually take different paths.

Call Routing Tree-A graphic display (looks like a tree) of the different call routing paths that can be taken by call routing decision logic.

Call Screening-A process that allows a call recipient to review the incoming telephone number (and name if available) to determine if they desire to answer the incoming call or to transfer the call to another person.

Call Server-A call server is a particular form of application server that manages the setup or connection of telephone calls. The call server will receive call setup request messages, determine the status of destination devices, check the authorization of users to originate and/or receive calls, and create and send the necessary messages to process the call requests.

Call Setup-Call setup are the call processing steps (events) that occur during the time a call is being established, but not yet connected.

Call Signaling-The process of sending control information during a call. Call signaling may be in band (muting the audio while sending control information) or out of band (on a separate signaling channel (such as SS7) during the call.

Call Supervision-The process of monitoring a communication line or trunk for changes in call status. These changes can include start-dial (off-hook), feature change requests, or call termination (on-hook).

Call Termination-The process of terminating a call or communication session. Call termination can be initiated by any of the users of the call or the system that is connecting the call.

Call Timers-Clock timers that are used to count time or events that occur with a telecommunications device. This may be the airtime or usage of a telecommunications service or how many calls were made during a particular time period.

Call Topology-The interconnection relationships between switches or routers in a communication network. Interconnection types can vary from circuit switched TDM to packet switched VoIP.

Call Trace-Call trace allows a subscriber to initiate a call trace request message that allows the dialed digits of a caller to be stored for investigation. The activation of the call trace service alerts the telephone service operator to "tag" the originator's number to allow authorities to investigate the originator of the unwanted or unauthorized call. Some of the call trace activation codes include *57 on a touchtone phone or the dialing of 1157 on a rotary (pulse) phone. If the call trace of the last call was completed successfully, an announcement should be heard. The service operator will usually release the call trace information to law enforcement agencies and a signed authorization from the subscriber may be required.

This diagram shows the call trace operation. The call trace operation starts with the reception of an unwanted call. When the call is received, an event record is created in the switch that records the connection through the switch (resources used) and this record usually contains the calling party number associated with the incoming call. This example shows that the customer dials a call trace feature code to inform the carrier to trace the last call. The local switch (recipients phone carrier) reviews the call detail information stored within the switch (or billing system) to determine if a calling number was provided. If the number was provided and is available, the information will be stored in the carriers system. The customer then calls the carrier and requests that the number be sent to local authorities for further action. This example shows that the customer completes a call trace release form that authorizes the transfer of the number to authorities.

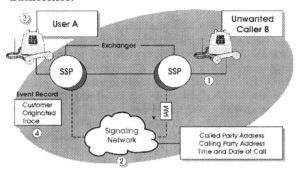

Call Trace Operation

Call Transfer-A call connection routing feature that transfers a call from one telephone or extension to another. Call can occur at various stages of the call conversation through the use of a system call request feature. The system special service request feature is often called a "Flash" feature. The flash feature is created to indicate a desire to recall a service function or to activate a custom calling feature (such as a call transfer request).

A flash feature service request can be created when the user initiates a short on-hook interval or through the sending of a special service request message. The short on-hook interval is created by a momentary operation of the telephone switch hook, during a prolonged off-hook period. The special service request message can be sent by a button on a telephone (such as a PBX telephone) or by pressing the SEND key on a mobile telephone.

This diagram shows a typical call transfer process used in a private telephone system. In this example, a caller is connected from the public telephone network to extension 1001 via incoming line 0001 on a telephone system (PBX). The user on extension 1001 desires to transfer the call to extension 1011. The first step involves extension 1001 sending a hold signal to the telephone system that allows it to place incoming line 0001 on hold. This is followed by the user at extension 1001 sending the digits indicating the destination of the call transfer (1011). The system then connects extension 1001 with 1011 and the user at 1001 may inform the user at 1011 that a call is being trans-

ferred to them. When the user at 1001 hangs up, the connection between 1001 and 1011 is disconnected and a connection is made between incoming line 1001 and 1011.

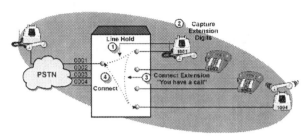

Call Transfer Operation

Call Waiting (CW)-A telephone call processing feature that notifies a telephone user that another incoming call is waiting to be answered. This is typically provided by a brief tone that is not heard by the other callers. Some advanced telephones (such as digital mobile telephones) are capable of displaying the incoming phone number of the waiting call.

After the service provides the subscriber with the notification of an incoming call while the subscriber's call, controlling subscriber can either answer or ignore the incoming call. If the controlling subscriber answers the second call, it may alternate between the two calls.

This diagram shows how call-waiting service may be provided on an analog telephone line. In this example, a call is in process with caller 1. During the call, a second caller (caller 2) dials the telephone number of the user that has call waiting. The system discovers that the line or extension is busy on another call. The system also determines that this user has the call waiting service processing feature available so it sends a call waiting message tone to the user (only heard by the user). If the user desires to answer the call, the user sends a flash message (a momentary open on the line) that indicates to the telephone system to place the current call in progress on hold and switch to the other incoming call (caller 2). Each time a flash message is sent, the line alternates between each incoming caller.

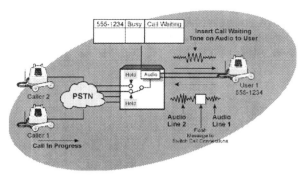

Call Waiting Operation

Callback-A call processing service that reverses the connection of calls. This process is divided into the call setup (dial-in) and callback stages. The caller dials a number that provides access to the callback service. The callback gateway receives the call and prompts the caller to say or enter (e.g. by touch tone) the number they desire to be connected to and the number they want the callback service to connect to. The callback center then originates calls to both numbers and connects the two individuals to each other.

Called Line Identification-The identification information carried by an SS7 packet that provides the destination receiver to identify the source of the call.

Called Party-A called party is the person who receives a telephone call.

Caller ID-An optional telephone service that provides a receiving telephone device with the phone number of the originating caller, which can be displayed to the destination person prior to receiving the call. The caller ID is transmitted as a data parameter in the SS7 Initial Address Message from the originating end switch to the destination end switch in the process of setting up the call. Some caller ID services can also provide directory name listing information, derived separately from the LIDB database. Caller ID information is typically transferred as a type-202-modem-compatible data signal between the first two ringing cadence cycles of the alerting tone.

Calling Card-An identifying number or code unique to the individual, that is issued to the individual by a common carrier and enables the individual to be charged by means of a phone bill for charges incurred independent of where the call originates.

This figure shows how a calling card can be used to initiate calls through a telephone network. In this example, the customer uses the telephone number on the pre-paid calling card to initiate a call to a switching gateway. The gateway gathers the calling card account information by either prompting the user to enter information or by gathering information from the incoming call (e.g. prepaid wireless telephone number). The gateway sends the account information (dialed digits and account number) to the real time rating system. The real-time rating system identifies the correct rate table (e.g. peak time or off peak time) and inquires the account determine the balance of the account associated with the calling card. Using the rate information and balance available, the real time rating system determines the maximum available time for the call duration. This information is sent back to the gateway and the gateway completes (connects) the call. During the call progress, the gateway maintains a timer so the caller cannot exceed the maximum amount of time. After the call is complete (either caller hangs up), the gateway sends a message to the real time rating system that contains the actual amount of time that is used. The real time rating system uses the time and rate information to calculate the actual charge for the call. The system then updates the account balance (decreases by the charge for the call).

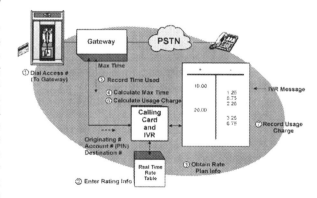

Calling Card Operation

Calling Line Identification (CLI)-A service, which displays the calling number prior to answering the call that allows telephone customers to determine if they want to answer the call. The calling number may be used by the telephone device to look-up a name in memory (e.g. mom) and display the name along with the phone number.

This figure shows the calling number identification operation. Calling number identification operation starts with the reception of a call. When the call is received, the initial address message (IAM) contains the calling party number of the incoming call. The IAM may contain additional information such as the text name of the calling party. This example shows that the local switching system extracts this information and combines this information with the ring signal (using different frequencies and amplitudes) and sends it to the customer during the alerting (ringing) process. If the customer has the appropriate display equipment, the calling number information is display as the telephone rings.

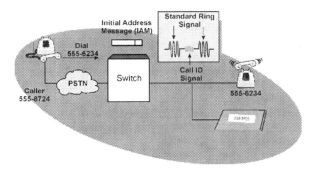

Calling Line Identification (CLID) Operation

Calling Line Identification Display (CLID)- Calling line identification display (CLID) identifies for display the originating or calling phone number.

Calling Line Identification Presentation (CLIP)- The transmission of a calling number to the receiver of a call. The calling number may originate from the caller's equipment or it may be created by the network.

Calling Name Delivery- A feature that enables a customer to view the name of the calling party as well as the date and time of the call on a display device.

Calling Party- A calling party is the person who originates a telephone call.

Calling Party Pays (CPP)- A communication service that bills the calling party for the delivery of their call through a network (such as a mobile communication network or freephone service) or for the providing of information (such as a news service.)

CAM- Conditional Access Message

CAMA- Centralized Automatic Message Accounting

Camcorder- A camera recorder (camcorder) is a portable video and audio recording device. The storage formats used for camcorders include VHS, Betamax, cassettes, and miniature digital video disks (DVDs).

CAMEL- Customized Applications For Mobile Enhanced Logic

Camera Phone- A camera phone is a telephone (usually a mobile telephone) that is combined with a digital camera.

Campaign Management- The process usually utilized for direct marketing (i.e. direct mail or phone sales), to manage the conceptualization, prioritization, planning, execution, and post campaign measurement of activities.

Campus Switch- A switch used within a campus backbone. Campus switches are generally high-performance devices that aggregate traffic streams from multiple buildings and departments within a site.

Cancel Message- A message that is used to inform a person or device that a previous message or service that has been requested has been cancelled. The cancel message is defined in session initiation protocol (SIP).

Candelas (CD)- A unit of measure for luminous intensity (photons per second).

CAP- CAMEL Application Part

CAP- Carrierless Amplitude And Phase

Capacitance- Capacitance is the measure of the capability of a device or assembly to store electrical energy.

Capacity- (1-communication) The maximum information carrying ability of a communications facility or system. The unit of capacity measurement for the facility or system depends on the type of services or information content that are provided by the facility or system. (2-energy) The amount of electrical energy that a device such as a capacitor or battery can store. The unit of measure for capacity of a capacitor is the Farad.

Capacity Limit-A capacity limit is the maximum amount of service (such as data transmission rate) or number of customers that a system can provide services to at a defined level of service.

Capacity Management-Capacity management is the use of processes and system for the assignment and optimization of network bandwidth through the monitoring of performance metrics, workload analysis, application resources, and overall network demand.

Capacity Planning-Capacity planning applies to proactively assessing systems, whether network or computer based systems, and forecasting the needed future growth of the system. Capacity planning in network management can be a function of the Configuration or Performance Management level of the FCAPS model.

CAPCS-Cellular Auxiliary Personal Communication Services

CapEx-Capital Expenses

Capital Cost Model-A financial cost model that recognizes the differences in regulatory accounting versus tax accounting for telecommunications networks and other systems.

Capital Cost per Subscriber-Capital cost per subscriber is the total cost of network investment required to provide services to subscribers (customers). Capital costs include the purchase of network equipment, installation costs, initial license fees, development costs, and other non-operational costs.

Capital Expenses (CapEx)-CapEx is the capital expenses that include the purchase of land, buildings and, most importantly, the build out of network capacity in a telecommunications system.

CAPs-Competitive Access Providers

Capstan-(1-recording system) A motor-driven shaft or spindle used to control (and establish) a constant speed of the magnetic media as it moves through the recording path. A servo circuit controls the speed of the capstan relative to a reference signal. Any error between the capstan velocity (as determined by sensors) and the reference produces a correction signal that is fed back by the servo to increase or decrease capstan rotation. (2-cable reel) A rotating drum or cylinder used for pulling cables by exerting traction upon a rope or pull line passing around the drum.

Caption-Text or titles that are inserted in a video program.

Caption Windows-Caption windows are the invisible area of a screen display that defines the top and bottom limits of a roll-up caption. The lowest row of the caption window is called the base row.

Capture Effect-An effect associated with the reception of frequency-modulated signals in which, if two signals are received on the same frequency, only the stronger of the two will appear in the output. The complete suppression of the weaker carrier occurs at the receiver limiter, where it is treated as noise and rejected.

Card-(1-electronics) The term card is commonly used to represent a printed circuit board or electronics assembly that is designed can to be inserted (plugged-in) to a computer or electronic device. (2-software) A card is a block of data that holds information for the screen of a portable communication device. A deck of cards may be maintained to allow the rapid display and management of information used in portable devices.

CARE-Customer Account Record Exchange

Carosel Content-Carosel content is the moving of media or video to a receiving device through the sequential and repeating delivery of segments of media. If a receiving device cannot successfully obtain a portion of the media content, it simply waits for the media crosel to circle around again.

CAROT-Centralized Automatic Reporting On Trunks

Carriage-The established procedures of a cable TV system regarding the carrying of signals of local TV stations on its channels.

Carried Load-The average volume in a traffic system. Carried load equals offered load when all calls are served. Carried load is less than offered load when some calls are denied service.

Carried Traffic-That part of the traffic offered to a group of servers that successfully seizes a server. Carried traffic equals offered traffic less the overflow traffic.

Carrier-(1-signal) A sine wave that can be modulated in amplitude, frequency or phase for the purpose of carrying information. (2-company) A business organization providing telecommunications service, such as a radio common carrier or a telephone service provider. (3-frequency) The frequency of the radio carrier signal. (4-level) The radio energy (power) of a carrier signal, typically expressed in decibels in relation to some nominal (reference) level.

Carrier Access Billing System (CABS)-A system that is used by network access providers to bill carriers for their customer's access to the network facilities.

Carrier Detect (CD)-Carrier detect is a control signal that indicates the device has detected a carrier (modem) signal.

Carrier Identification Code (CIC)-A 3-digit code that uniquely identifies a telecommunications carrier within the North American Numbering Plan (NANP). The CIC is indicated by an XXX in a Carrier Access Code where X can be any digit, 0 through 9. After an XXX code has been assigned to a carrier, the code is retained for use with either Feature Groups B (95OA-0XXX, 950-1XXX) or D (10XXX) throughout the area served by the NANP. (See also: Carrier Access Code, pre-subscription, primary interexchange carrier.

Carrier Interconnection Plan-A long-term plan for inter-connecting interexchange and international carriers to exchange carrier intraLATA networks. This plan also can be called the Exchange Access Plan. The term equal access commonly is used to refer to the features provided by this plan.

Carrier Noise Level-The noise level resulting from undesired variations of a carrier in the absence of any intended modulation.

Carrier Sense-In Ethernet, the act of determining whether the shared communications channel is currently in use by another station.

Carrier Sense Multiple Access (CSMA)-A system access process that allows multiple radios or stations to gain the attention (service) from a system. During the CSMA process, each station "listens" to a channel to sense whether it is busy prior to attempting access to the system. If the channel is not busy, the station will randomly attempt access. An access request collision detection process is used to allow the requesting station to recognize if the system response is from its access request as a result of a request from another station. If the station determines that the response is not resulting from its service request, that station will wait for a random period before it attempts to transmit a service request again.

Carrier Sense Multiple Access/Collision Avoidance (CSMA/CA)-A network access method in which each device signals its intent to transmit data before it actually does so. This keeps other devices from signaling information, thus preventing collisions between the signals of two or more devices. CSMA/CA is used at Home Radio Frequency (HomeRF) networks, which operates in the same unlicensed 2.4 GHz ISM band as the Bluetooth specification.

Carrier Sense Multiple Access/Collision Detection (CSMA/CD)-A carrier sense multiple access transmission scheme in which transmission resulting in collisions are followed by the transmitting stations backing off the network a random amount of time before attempting to retransmit. CSMA/CD is used as the basis of Ethernet networks. When multiple collisions occur for the same packet, Ethernet stations typically back off in exponentially increasingly large random time amounts to further reduce the probability of collision. Collision avoidance algorithms are also common in Ethernet, whereby a station will listen for silence on the media before transmitting.

Carrier Sensitive Routing (CSR)-Selecting the call routing of a communication session based on carrier preferences. The preferences used may include cost, quality, and congestion.

Carrier Shift-(1-modulation) A method of keying a radio carrier for transmitting binary data that consists of shifting the carrier frequency in one direction for a mark signal and in the opposite direction for a space signal. (2 - envelope) A condition resulting from imperfect amplitude modulation whereby the positive and negative excursions of the envelope pattern are unequal, thus effecting a change in the power associated with the carrier. Carrier shift may be positive or negative.

Carrier System-A system that carries many individual signals over a shared facility by multiplexing, that is, combining those signals for transmission. The most common transmission technique is time-division multiplexing, in which each information channel uses the transmission medium for assigned time intervals. Frequency-division multiplexing, in which each information channel occupies an assigned portion of the frequency spectrum, also has been used. Carrier systems range in size from two to many thousands of individual channels.

Carrier to Adjacent Channel (C/A) Interference Ratio-The amount of interference level from an adjacent signal in comparison to the desired carrier signal. The C/A ratio is commonly expressed in dB. Different types of systems can tolerate different levels of adjacent channel interference dependent on the modulation type and error protection systems.

Carrier To Interference Signal Ratio (C/I)-The amount of interference level from all unwanted interfering signals in comparison to the desired

Carrier to Noise Ratio (C/N)

carrier signal. The C/I ratio is commonly expressed in dB. Different types of systems can tolerate different levels of interference dependent on the modulation type and error protection systems.

Carrier to Noise Ratio (C/N)-The amount of interference level from noise signal energy in comparison to the desired carrier signal. The C/N ratio is commonly expressed in dB. The C/N measurement is commonly used at the edges of mobile communication systems as noise is the only interference with the reception of a radio signal. This is called a "noise limited system."

Carrierless Amplitude And Phase (CAP)-Carrierless amplitude and phase (CAP) modulation is very similar to QAM modulation. The difference is the continuous shifting of phase (or signal mix) of the carrier signal level. CAP modulation was designed to help reduce the effects of crosstalk and to simplify the signal processing of modulated signal. CAP transmits data signals on a single high bandwidth modulated carrier.

Carrier-To-Interference Ratio (C/I)-The ratio of the desired radio carrier signal with respect to the combined interference due to other radio channel interfering signals.

CARS-Community Antenna Relay Service

Carterfone Decision-A Federal Communications Commission (FCC) decision in 1968 that allowed customers to directly connect their own customer-provided equipment (CPE) to telephone company networks. Prior to this decision, some telephone companies prohibited the attachment of such equipment to their networks. Equipment that is to be connected to telephone networks must conform to industry standards and FCC regulations to ensure the protection of the telephone network from potential damage that may result from defective CPE.

CAS-Centralized Attendant Service
CAS-Channel Associated Signaling
CAS-Communicating Application Specification
CAS-Conditional Access System
CAS-CPE Alerting Signal

Cascade-(1-circuit) To connect the output of a device or system to the input of another device. (2-Windows) The arranging of program windows so they overlap with each other.

Cascaded-An arrangement of two or more circuits in which the output of one circuit is connected to the input of the next circuit.

Cascaded Codecs-The connection of one coder/decoder through another coder/decoder. The cascading of coders (such as speech coders used in mobile communication systems) usually results in significant degradation of voice quality.

CASE-Computer Aided Software Engineering

Case Sensitive-Case sensitive is the process of differentiating between characters that are upper case and lower case. A computer system that is case sensitive would identify Case and case as two different words or text blocks.

Cash Cost Per User (CCPU)-Cash cost per user is the amount of cash that is required to provide services and/or equipment to a user.

Cassette-A self-contained package of reel-to-reel blank or recorded magnetic tape that is continuous and may be rewound on demand.

CAT-Conditional Access Table
CAT5-Category 5

Cataloging-Cataloging is the process of identifying media and selecting groups of items to form a catalog.

Category 5 (CAT5)-Category 5 unshielded twisted-pair wiring commonly used for 10baseT and 100BaseT Ethernet networks and rated by the EIA/TIA

Catenet-A collection of networks (typically LANs) interconnected at the Data Link layer using bridges. Also known as a bridged LAN.

Cathode Ray Tube (CRT)-An electronic vacuum tube in which a beam of electrons is directed to a phosphor-coated screen that glows where struck by an electron. CRTs are used to display information, as in computers, radar systems, television receivers, monitors and test equipment.

CATV-Cable Television
CATV-Community Access Television
CAV-Component Analog Video

Cavity-A resonant device, usually drum-shaped or cylindrical, that acts as a filter in a radio frequency system.

CB-Citizens Band
CB Radio-Citizens Band Radio
CBC-Cipher Block Chaining
CBCH-Cell Broadcast Channel
CBD-Central Billing District
CBQ-Class Based Queuing
CBR-Constant Bit Rate
CBS-Common Base Station
CBT-Central Buffer Tube
CBT-Computer Based Training

CBT-Core Based Trees
CBU-Caribbean Broadcasting Union
CC-Call Control
CC-Calling Channel
CC-Carbon Copy
CC-Conference Calling
CCBS-Customer Care And Billing System
CCC-Clear Channel Capability
CCC-Copyright Clearance Center
CCCH-Common Control Channel
CCD-Charge Coupled Device
CCH-Common Channel
CCIR-Comite' Consultatif International de Radiocommunications
CCIR-International Radio Consultative Committee
CCIrl-Comite' Consultatif International Telegraphique et Telephonique
CCITT-Consultative Committee for International Telephony And Telegraphy
CCK-Complementary Code Keying
CCPU-Cash Cost Per User
CCS-Centum Call Seconds
CCS-Common Channel Signaling
CCSRL-Control Channel Segmentation and Reassembly Layer
CCTV-Closed-Circuit Television
CD-Candelas
CD-Carrier Detect
CD-Commitment Date
CD-Compact Disc
CD-Compact Disk
CDDD-Customer Desired Due Date
CDDI-Copper Distribution Data Interconnection
CDG-CDMA Development Group
CD-I-Compact Disk Interactive
CDL-Coded Digital Locator
CDM-Code Division Multiplexing
CDMA-Code Division Multiple Access
CDMA-IS-95
CDMA2000-Code Division Multiple Access 2000
CDN-Content Delivery Network
CDP-Cisco Discovery Protocol
CDP-Customer Demarcation Point
CDPD-Cellular Digital Packet Data
CDR-Call Detail Record
CD-R-Compact Disk Read Only
CDR Processing-Call Detail Record
CDR Server-Call Detail Record Server
CD-RW-Compact Disk Read/Write
CDVCC-Coded Digital Verification Color Code

CE-Consumer Electronics
CEBus-Consumer Electronics Bus
CEEFAX-A service developed in England by the BBC engineers launched a teletext service in 1974 called CEEFAX. These text broadcasts appeared and continue today (though there are many services available from other broadcasters) as pages of information on the TV one can access by punching in a three-number code on one's remote control and use of the Fast Keys (as we mentioned earlier). The "home page" (for want of a better term) of CEEFAX is 100.
CEF-Cable Entrance Facility
Ceiling-The highest available price, maximum.
Cell-(1-cellular system) A radio coverage area associated with a fixed-location cellular or PCS radio tower that is interconnected with other cells to provide radio coverage to a larger geographic area. The term cell is often visualized as a hexagon as a relative building block depicting the ability of a cellular system to continually split so the system capacity can continually increase as new customers are added to the system. (2 - battery) A primary (disposable) or secondary (rechargeable) unit that stores energy for the supply to electrical or electronic equipment. (3- fuel) An electrochemical cell that produces electricity from the chemical energy of a fuel and an oxidant.
Cell Broadcast-The broadcasting of messages to one or a group of cell sites so that mobile radio receivers that are located anywhere within the coverage area of those cell sites can receive the broadcast messages.
Cell Loss Priority (CLP)-A field contained in the header of an ATM packet (cell) that identifies the cell's discard priority in the event that network congestion occurs. A value of 1 indicates the cell can be discarded and a value of 1 indicates the cell has the highest priority.
Cell Rate-The cell rate is the number of cells per second (53 byte ATM packets) that are required to be passed through a switch for a particular call or communication session.
Cell Rate Margin (CRM)-The difference between an assigned cell rate (bandwidth allocation) and the sustained cell rate in an ATM system.
Cell Site-A transmitter-receiver tower, operated by a wireless carrier (typically cellular or PCS), through which radio links are established between a wireless system and mobile and portable units.

Cellular-Cellular radio is wireless telephone system that divides geographic areas into small radio areas (cells) that are interconnected with each other. Each cell coverage area has one or several transmitters and receivers that communicate with mobile telephones within its area.

Cellular Carrier-A radio common carrier that provide cellular telephone service.

Cellular Intercarrier Billing Exchange Roamer (CIBER)-A billing standard designed to promote inter-carrier roaming between cellular telephone systems. The CIBER format is developed and maintained by CiberNet. Cibernet is owned by the Cellular Telecommunications Industry Association (CTIA).

This figure shows some of the information (fields) contained in a type 22 CIBER record. This example shows that the type 22 Ciber record field structure has been updated from the previous type 20 record structure to include additional fields that allow for telephone number portability (enabling telephone number transfer between carriers). This list shows that fields in the Ciber record primarily include identification of airtime charges, taxes, and interconnection (toll) charges.re

Type 22 Record - Sample of Fields

- Home Carrier SID/BID
- MIN/IMSI
- MSISDN/MDN
- ESN/IMEI
- Serving Carrier SID/BID
- Total Charges and Taxes
- Total State/Province Tax
- Total Local Tax
- Call Date
- Call Direction
- Call Completion Indicator
- Call Termination Indicator
- Caller ID
- Called Number
- LRN
- TLDN
- Time Zone Indicator
- Air Connect Time
- Air Chargeable Time
- Air Rate Period
- Toll Connect Time
- Toll Chargeable Time
- Toll Carrier ID
- Toll Rate Class

Cellular Intercarrier Billing Exchange Roamer (CIBER) Structu

Cellular Network-Cellular systems are comprised of a set of radio towers that are strategically distributed over a geographical area in order to provide a continuous service coverage area. A mobile switching center (MSC) provides the switching and control functions necessary to connect calls from the public switched telephone network (PSTN) to the individual mobile telephones. The MSC also manages the radio resources within the entire network.

Cellular Operator-The owner and/or operator of a cellular network

Cellular Radio-A mobile radio system having two distinguishing properties compared to other mobile radio systems: 1) Re-use of the same radio frequency/frequencies in different cells of the system. 2) Handoff (or handover) of a call in progress, typically between two adjacent cells. Each cell has a distinct radio coverage area (aside from some intentional overlap of radio coverage at the boundaries between adjacent areas) and base antennas. Property 1 distinguishes cellular radio from earlier systems like Improved Mobile Telephone Service (IMTS), that use the same frequency throughout an entire city or large service area. Property 2 distinguishes cellular radio from trunked radio, which has reuse (property 1) but not handover. Some people use the term Personal Communication System (PCS) for certain types of cellular radio systems because they operate on different radio frequency bands (1.9 GHz vs. 800 MHz, for example), or because they are digital vs. analog, or because of the particular service provider who owns and operates the base infrastructure equipment.

Cellular System- A fully automatic, wide-area, high-capacity RF network made up of a group of coverage sites called cells. As a subscriber passes from cell to cell, a series of handoffs maintains smooth call continuity.

This figure shows a basic cellular system. The cellular system connects mobile radios (called mobile stations) via radio channels to base stations. Some of the radio channels (or portions of a digital radio channel) are used for control purposes (setup and disconnection of calls) and some are used to transfer voice or customer data signals. Each base station contains transmitters and receivers that convert the radio signals to electrical signals that can be sent to and from the mobile switching center (MSC). The MSC contains communication controllers that adapt signals from base stations into a form that can be connected (switched) between other base stations or to lines that connect to the public telephone network. The switching system is connected to databases that contain active customers (customers active in its system). The switching system in the MSC is coordinated by call processing software that receives requests for service and processes the steps to setup and maintain connections through the MSC to destination communication devices such as to other mobile telephones or to telephones that are connected to the public telephone network.

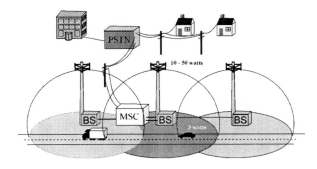

Basic Cellular System

CELP-Code Excited Linear Prediction

CELP-Code Excited Linear Predictive

Center Frequency-The center frequency is the central point that a frequency band/channel or radio carrier occupies. For a modulated signal, it is the initial frequency (or resting frequency) of the carrier before it is modulated.

Center Pulling-Center pulling is the process of pulling a cable through a duct or pipe from a point that is between the end points. Center pulling may be performed to minimize the stress (pulling tension) on the cable or because it is not possible to feed the cable at one or both of the endpoints.

Center Tap-A connection made at the electrical center of a coil.

Center Wavelength-The center wavelength is the central point that an optical spectrum band (bandwidth) that an optical channel or optical carrier occupies. For a modulated signal, it is the initial frequency (or resting frequency) of the carrier before it is modulated.

Central Exchange (Centrex)-Centrex is a service offered by a local telephone service provider that allows the customer to have features that are typically associated with a private branch

exchange (PBX). These features include 3 or 4 digit dialing, intercom features, distinctive line ringing for inside and outside lines, voice mail waiting indication and others. Centrex services are provided by the central office switching facilities in the telephone network.

Central Office (CO)-The name commonly used in North America to identify the switch in a telephone network that connects local customers. The international name is "public exchange" or "telephone exchange." CO actual refers to the building that contains the switching and interconnection equipment.

Central Processing Unit (CPU)-The electronic components of a computer that interpret instructions, performs calculations, moves data in main computer storage, and controls input/output operations. Often this expression refers to a mainframe or midrange machine.

Centralized Architecture-Centralized architecture is a system that is designed to use a central intelligence to coordinate the overall operation of the system. The use of a centralized communication system allows one point to coordinate the operation of all other points and devices within the network.

Centralized Attendant Service (CAS)-A centralized group of customer service operators (attendants) who answer incoming calls for multiple telephone systems.

Centralized Automatic Message Accounting (CAMA)-A system for the recording of detailed billing information at a central location other than an end office, usually at a tandem.

Centralized Automatic Reporting On Trunks (CAROT)-A system that automatically and routinely reports on the performance of trunk lines to ensure high availability and to avoid potential failures. The reporting system may check for return loss, line noise, timing precision, transmission and call processing performance criteria.

Centralized Control-A system that relies on a central intelligence to coordinate the overall operation of the system.

Centralized Message Data System (CMDS)-A system that allows interexchange of billing and usage data between telephone exchange networks. CMDS has four parts: (1) Centralized Message Data System (CMDSI) which provides for collect services (collect, third party number, and calling card); (2) Carrier Access Billing System (CABS) that bills interexchange carriers (IXCs) for local exchange access services; (3) 800 Service Usage; and (4) Meet Point Billing which involves the billing of access services (via CABS) provided to two or more interexchange carriers.

Centralized Routing-Centralized routing uses a coordinated network control to determine the routing of packets of data through a communication network. Centralized routing switches are not very intelligent (e.g. "dumb") and the system switching can be more vulnerable to central system failures.

Centralized Security-Centralized security is the coordination of access control, authorization and encryption of data through a communication network from a centralized authority.

Centralized Trunk Test Unit (CTTU)-A test system that test trunks through a data link on a switch from a centralized location.

Centrex-Central Exchange

Centric-Centric is the central part or core focus of a process or system.

Centronics Printer Connector-A parallel wire connector traditionally used for computer printers, named for the first manufacturer to use this particular connector for a printer.

Centum Call Seconds (CCS)-A measurement of communication trunk usage that equals 100 seconds of continuous usage. Because the standard time interval for communication network engineering is based on activity over an hour, the system load is expressed in hundred call seconds (CCS) for one hour. A single hour has thirty-six hundred call seconds (equal to one erlang).

CEO-CABS End Office

CEPT-Conference Of European Postal And Telecommunications Administrations

Certificate-A certificate is an authenticated document or electronic key that is signed, typically by some higher authority, and that usually contains useful information in an encapsulated unit that provides a higher level of assurance of a request or service.

Certificate Authority (CA)-A certificate authority is an authorized company that validates and provides secure socket layer (SSL) certificates that enable SSL authentication and encryption between communication devices. There are several certificate authority companies and are sometimes referred to as a "trusted authority."

Certificate Revocation-Certificate revocation is the process of sending a command to a user or device that updates, disables or removes or changes the usage access rights for its associated media or data.

Certificate Revocation List-A certificate revocation list is a group of users or devices that have been identified as having expired or invalid certificates.

Certified Output Protection Protocol (COPP)-Certified output protection protocol is a set of commands and processes (protocol) that is used to assist in the protection of video content.

Certifying-Certifying is a process of reviewing, testing or monitoring a product, system or service by a qualified person or measuring device using calibrated test equipment to validate that the performance is within defined characteristics (such as a test specification).

CEV-Controlled Environmental Vault

CFB-Call Forwarding-Busy

CF-End-Contention Free End

CFNR-Call Forwarding No Response

CFP-Contention Free Period

CFRP-Carbon Fiber Reinforced Plastic

CFV-Call For Votes

CG-Character Generator

CG-Charging Gateway

CGF-Charging Gateway Function

CGI-Common Gateway Interface

CGSA-Cellular Geographic Service Area

Chain Broadcasting-Chain broadcasting is the simultaneous transmission of a program by two or more connected transmission devices.

Chain of Title-A chain of ownership based on history of a literary property, which can be treaced back to the time of creation.

Challenge Handshake Authentication Protocol (CHAP)-Challenge handshake authentication protocol is a security protocol that is used to authenticate (validate the identity) a user or device that is requesting a service before they are given access to service. CHAP protocol uses previously known information to calculate an authenticated response and only the calculated response is transferred during the authentication process rather than the key or other secret data.

Change Point-The point at which a cable run changes from one type of cable to another.

Channel Acquisition-Channel acquisition is the process of finding and acquiring access to a communication channel.

Channel Associated Signaling-A signaling method in which the signaling information necessary for the traffic carried by a single channel is transmitted in the channel itself or in a signaling channel permanently associated with it.

Channel Associated Signaling (CAS)-Channel associated signaling (CAS) is the transmission of signaling information within the voice channel. With CAS, circuit state is indicated by one or more bits of signaling status sent repetitively and associated with that specific circuit. E1 CAS uses the ABCD signaling bits of channel (timeslot) 16. T1 CAS robs a bit from every 6th frame of a multiframe. In both cases, these signaling bits set conditions such as On Hook, Off Hook, Coin Control, and Dial Pulsing.

Channel Bandwidth-(1-data) Channel bandwidth the maximum or fixed data transmission rate of a communication channel. (2-radio) The difference between the upper frequency limit and lower frequency limit of allowable radio transmission energy for a channel.

Channel Banks-Channel banks are groups of communication channels that allow a single higher speed communication channel to be divided into multiple communication channels.

Channel Block-Channel blocks are groups of channels that are assigned or used together.

Channel Bonding-Channel bonding is the process of combining two or more communication channels to form a new communication channel that can use and manage the combined capacity of the bonded transmission channels.

This figure shows how multiple transmission channels can be bonded together to produce a single channel with higher data transmission rates. This diagram shows how two transmission channels can be combined using a bonding protocol. This diagram shows that a bonded session is requested and negotiated on a single communication channel. Once the bonded session has been setup, the bonding protocol is used to monitor and manage the bonded connection.

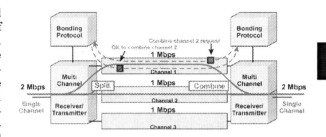

Channel Bonding

Channel Branding-Channel branding is the process of creating an awareness of a company (corporate brand), product (product brand), or service (service brand) to the users or viewers of specific channels (such as a television channel).

Channel Busy Tone-An audible signal indicating that a call cannot be completed because all switching paths or toll trunks are busy, or that equipment is blocked. The tone is applied 120 times per minute. The channel busy tone also is called fast busy, all trunks busy, or reorder tone.

Channel Capacity-The amount of data or channel transmission capability of a communication channel.

Channel Change Delay-Channel change delay is the period of time that occurs from when a user presses the channel change button on the remote until a stable picture appears for the requested channel. Channel changing time is the sum of the individual delays associated with processing in the set top box, communication with the network, IGMP join & leave, channel rights validation, and finally, acquisition and synchronization with the video stream itself. Channel change delay can be a significant amount of time associated with changes in media stream signals (such as IPTV).

Channel Codes-Channel codes are unique patterns or codes that are combined with, mixed into, and used to modify information that is sent on a communicate channel to identify each channel that are sent on a common transmission channel.

Channel Coding-Channel coding is a process where one or more control and user data signals are combined with error protected or error correction information. After a sequence of digital data bits has been produced by a digital speech code or by other digital signal sources, these digital bits are processed to create a sequence of new bit patterns that are ready for transmission. This processing typically includes the addition of error detection and error protection bits along with rearranging of bit order for transmission.

Channel Collision-The result of two cellular phones trying to use the same channel at the same time. Cellular networks provide safeguards to inhibit this.

Channel Combiner-A device or filter assembly that allows several modulated carrier signals (physical channels) to be grouped on to the same transmission channel or antenna system.

Channel Descriptor-Channel descriptor is a message or information parameters that describe the characteristics or parameters associated with a communication channel. The use of a channel descriptor can permit more accurate and successful reception and decoding of information that is sent on a communication channel.

Channel Encoder-A device that converts a signal into a form suitable for transmission over the communications channel.

Channel Equalizer-A channel equalizer is a device, system or process that adapts a receiver to counteract unwanted signal changing effects of a transmission channel to improve the quality of the received signal.

Channel Equipment-Channel equipment is an interface assembly that allows end user devices to send and receive control signals. For telephone line channel equipment, these control signals typically include tip and ring along with supervisory signaling leads.

Channel Establishment-(1-general) The process of connecting a communication channel (physical and/or logical) (2-Bluetooth) Procedure for establishing a Bluetooth channel (a logical link) between two Bluetooth devices using the Bluetooth File Transfer Profile Specification. Channel establishment starts after link establishment is completed and the initiator sends a channel establishment request.

Channel Expansion-Channel expansion is the process of adding channels to a system or network.

Channel expansion may be performed by adding new communication channel frequencies (if they are available), adding lines or sub-dividing or enhancing the capacity of existing channels so they can transfer or share additional channels on the same resources or lines.

Channel Extension-A system that enables peripheral equipment, such as high-speed tape spoolers, printers, and terminals, to be connected to mainframe computers many miles away.

Channel Frequencies-Channel frequencies are the frequencies assigned for specific channels in a communication system.

Channel Interleaving-Channel interleaving is the process of offsetting radio channel frequency spacing at nearby radio towers in a wireless communication system. This allows radio service provides to place more radio channels in each tower, reducing the total number of towers needed to provide a number of communication channels in a geographic area. Channel interleaving reduces the interference between nearby channels because the radio energy for both the nearby channel and interleaved channel is primarily concentrated at the center of the each radio channel frequency band. There is a reduced occurrence (minimized interference) of both modulated carriers fully occupying their frequency bands at the same time. In essence, it is acceptable to overlap in to each other's channel frequency band, just not at the same time.

This diagram shows how radio channels can be interleaved (offset in frequency) at nearby radio towers to allow mobile radio service providers to place more radio channels in each tower (reducing the total number of towers needed). This example shows that the radio signal energy is primarily concentrated at the center of the radio channel and

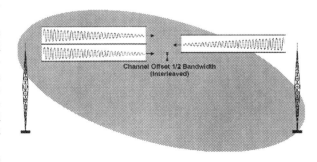

Channel Interleaving System

that the signal energy decreases at the signal is propagated from the tower. The use of frequency offsetting (interleaving) radio channels at nearby radio towers minimizes the amount of interference between channels compared to channels located at the same frequency.

Channel Listing-A channel listing is a directory of the available channels on a communication system.

Channel Loading-Channel loading is a ratio of the number of users authorized to operate on a particular channel or system compared to the number of users that actively transmit on a system. An example of channel loading is a private telephone system that may have 5 extensions for every telephone line (loading of 5:1) because the average business telephone user only talks for 1-2 hours per day.

Channel Management-Formal process utilized to manage the creation, staffing, tasking, and measurement of sales and customer support channels.

Channel Mapping-A relationship that allows multiple communication channels to be assigned to other transmission channel formats. An example is mapping multiple DS3 channels to the frames within a synchronous optical network (SONET) transmission channel.

This diagram shows how multiple communication channels can be mapped (related to) specific portions of another communication channel (payload). Mapping allows several communication channels to share the same physical or logical data transmission medium.

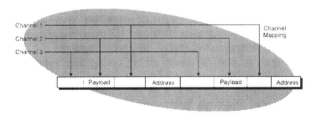

Channel Mapping Operation

Channel Modulator-A channel modulator is used to convert video signals into television broadcast channels. Channel modulators are used in cable-TV networks to convert a video program signal (such as CNN or MTV) and converts it with an RF carrier frequency for a television channel that is distributed through the CATV network. The modulator converts both video and audio signals. The frequency of this channel modulator carrier determines the television channel number (I.e., 2 to 99) that the program will be received on by subscribers.

Channel Multiplexing-Channel multiplexing is a process that divides a single transmission path into several parts that can transfer multiple communication (voice and/or data) channels. Multiplexing may be frequency division (dividing into frequency bands), time division (dividing into time slots), code division (dividing into coded data that randomly overlap), or statistical multiplexing (dynamically assigning portions of channels when activity exists).

Channel Normalization-Channel normalization is the adjustment of frequency, time, and/or power characteristics of a communication signal to compensate for changes or distortions that have occurred during the transmission of the signal through a communication channel.

Channel Occupancy-The time that a radio channel is used to carry traffic.

Channel Pair-Channel pairs is the grouping of channels together to form a bi-directional (two-way) communication circuit.

Channel Quality Message (CQM)-Messages sent on a communication system (such as a mobile communication system) that provide the remote connection (e.g. base station) with channel quality information. Channel quality information may include carrier level received signal strength indication (RSSI) and bit error rate (BER).

Channel Rate-The data rate at which information is transmitted through the channel or communications media, typically stated in bits per second (BPS).

Channel Reliability-The percentage of time a channel is available for use in a specific direction during a specified period.

Channel Re-use-The re-use of the same radio channel frequency at a distant radio tower in a wireless network. The minimum separation distance for the towers is determined by several factors including the power level and ability of the modulation type to reject interference. See also: Co-channel Interference.

Channel Service Unit (CSU)-A channel service unit is the hardware that is used to assign communication channels from one or more data terminal equipment (DTE) devices to logical channels on a multi-channel communication circuit. The CSU can be customer premises equipment (CPE) or provided by the telecommunications service provider. The CSU is often combined with a data service unit (DSU).

Channel Spacing-The spacing in a radio frequency band (in Hertz) between adjacent radio carrier signals. It is measured from the center of one channel to the center of the next adjacent channel.

Channel Splitter-A device or assembly that can split the signal energy into multiple outputs (taps) or separate communication channels (de-multiplex).

Channel Stream-A channel stream is a data communications channel (stream of data) between a transmitter (such as a radio base station) and a receiving device (such as a mobile telephone).

Channel Structure-Channel structure is the division and coordination of a communication channel (information transfer) into logical channels, frames (groups) of data, and fields within the frames that hold specific types of information.

Channel Surfing-Channel surfing is the rapid changing of selected viewing channels (channel flipping).

Channel Switching-The switching or rearrangement of dedicated channels that may use electronic equipment, such as digital cross-connect systems or circuit switches.

Channel Time Slot-A repetitive period of time or number of bits related to a frame or group of information on a physical channel that is used for a specific channel.

Channel Translator-A device or system that is used to change the channel frequency of one channel to another channel. Channel translators are commonly used to convert cable system upstream channels (from the customer to the system) into downstream channels (from the system to the customer).

Channel Unit-Equipment that provides transmission and signaling functions for one voice or data channel out of the channels that are available. A channel unit plugs into a carrier channel bank. (See also: CSU/DSU.)

Channelization-Channelization is the allocation of communication circuits to channels and the forming of these channels. Channelization divides the bandwidth of a communications link or communication circuit into smaller increments that can be assigned to multiple communication channels.

Channelized Carrier-A communication line (carrier) that is divided into multiple logical communication channels. Users may be allowed to access some or all of the logical channels.

CHAP-Challenge Handshake Authentication Protocol

Character Generator (CG)-A character generator is a device or program that used to generate graphics, letters, or numbers in a video format. The characters are subsequently keyed over program video or background video.

Character Interval-The total number of unit intervals, including synchronizing, information, error checking, and control bits, required to transmit any given character in a communications system.

Character Reader-A device capable of reading typescript and producing a machine-readable output without operator assistance.

Character Set-A set of different characters such as letters, numbers, and symbols that are agreed upon to represent characters in a computing system.

Charge-(1-electric) The product of electric current and time. The SI unit of electric charge is the coulomb, named for a 19th century French physicist. One coulomb is the product of one ampere of current and one second of time. There are approximately 6,250,000,000,000,000,000 electrons in one coulomb of electric charge. To state that in another way, the electric charge of one electron is negative and its magnitude is 1.6 E-19 coulomb.(2-verb) The process of replenishing or replacing the electric charge in a capacitor, secondary cell or storage battery. (3-cost) Charge is the name for a price or fee or cost of some goods or services. The verb form of this word is the action of computing or applying a cost.

Charge Coupled Device (CCD)-A charged coupled device is an electronic device that contains light-sensitive element arrays. When light is sensed by the CCD, it translates the intensity of the light into digital data. CCD devices can have several thousand horizontal and vertical elements which can create several million pixels of information.

Charge Number-A charge number is an identification code that is used to identify specific accounts for billing purposes.

Charges-Charges are the rates or costs associated with a usage of a service or product.

Charging Attributes-Charging attributes are the characteristics of a service object (e.g. data transfer rate) that can be used to identify, qualify, or assist in the billing of that service.

Charging Gateway (CG)-A charging gateway is a device or processing system that combines, processes, and reformats billing detail records (CDRs, IPDRs, etc) into a format that can be used by a billing system.

Charging Gateway Function (CGF)-A charging gateway function is a device or processing system within a GSM, GPRS, or WCDMA system that combines, processes, and call detail records (CDRs) into a format that can be processed by a billing centre.

Chassis Ground-A connection to the metal frame of an electronic system that holds the components in place. The chassis ground connection serves as the ground return or electrical common for the system.

Chat (Chatting)-Chat is a name for "instant messaging" between users that is typically performed over the Internet.

Chat Room-Chat rooms are real-time communication services that allow several participants (typically 10 to 20) to interact act through the use of text messaging. Chat rooms may be public (allow anyone to participate) or private (restricted to those with invitations or access codes.)

Chatting-Chat

Cheapernet-A slang term for thin-wire coaxial Ethernet (10BASE2).

Checksum-A number that is calculated from a block of data or sequence of data bits that is sent along with the data and used by the receiver to detect if transmission errors occurred. The receiver will use the same calculation process on the received data to calculate an compare the appended check sum with the value it has calculated. In some cases, the checksum can be used to correct some of the bit errors.

Cherry Picking-Cherry picking is the process of selecting media or calls based on preferred criteria. Examples of cherry picking include the selection of network television programs that are distributed to a cable television system or the selection of incoming sales calls from a specific region in a call center.

Chip Manufacturer-A chip manufacturer is a company that manufactures the semiconductor wafers that are used inside integrated circuits (chip manufacturer).

Chip On Board (COB)-A group of electronics components in very small plastic packages that are directly mounted on a printed circuit (PC) board. Unlike integrated circuits (ICs) that have a sturdy protective shell, COB circuits use the structure of the PC board for durable mounting and connection. Because COB circuits eliminate a mounting case and COB assemblies can be placed on both sides of a PC board, the density of components can high and the thickness of the PC board assembly can remain small.

Chipset-Chip Set

Choke-An inductor that passes direct current but attenuates or blocks alternating current. A choke often is installed on a direct-current power lead to prevent undesirable alternating currents from reaching and interfering with the powered equipment.

Chroma Key Filter-A chroma key filer is a device or software program that can replace a color with another color or image.

Chroma Keying-Chroma keying is the use of a single color (such as a green or blue screen behind a weather person) so that the chroma can be replaced by another image (weather reports).

Chrominance-Chrominance is the portion of the video signal that contains the color information (hue and saturation). Video picture information contains two components: luminance (brightness and contrast) and chrominance (hue and saturation).

Chrominance Subcarrier-Chrominance sub carrier is a carrier within a television signal that contains (is modulated with) the chrominance information.

cHTML-Compact Hypertext Markup Language

Churn-The process of customers disconnecting from one telecommunications service provider. Churn can be a natural process of customer geographic relocation or may be the result of customers selecting a new service provider in their local area.

CIBER-Cellular Intercarrier Billing Exchange Roamer

CIC-Carrier Identification Code

CIC-Circuit Identification Code

CID-Conference Identifier

CID-Connection Identifier

CIDR-Classless Inter-Domain Routing

CIDR-Classless Internet Domain Routing

CIF-Common Interchange Format

CIF-Common Intermediate Format

Cinematography-Cinematography is the production of moving pictures and it translates to writing in light. Cinematography usually refers to all aspects of the capturing and production of moving pictures.

Cinepak-Cinepak is a cross-platform computer digital video compressor (codec) that was developed in the early 1990's.

CIOA-Classical IP over ATM

Cipher-An algorithmic transformation of data typically used to prevent unauthorized viewing or alteration of the original data. There are many different types of ciphers. One of the most common is based upon the Digital Encryption Standard, DES, algorithm.

Cipher Key (Kc)-(1-general) A secret key that is used in the authentication process. (2-GSM) A code key created by the GSM encryption algorithm that results from the key code Ki and a random number that is sent by the system. Kc is used by the GSM network as part of the authentication process.

CIR-Committed Information Rate

Circuit-(1 - communication) Any communication path through which any information can be transferred. (2 - electronics) A combination of electrical processing components that perform a process (such as signal amplification) or function (clock display processor).

Circuit Access Code (CAC)-A computer-generated code identifying a trunk group.

Circuit Assignment-A process that identifies, reserves, or designates a partially or wholly inventoried equipment item to a circuit.

Circuit Bonding-The combining of multiple circuits to one or more communication channels to increase reliability through circuit redundancy or to share data services.

Circuit Data-The continuous flow (or continuous connection) between two data devices.

Circuit Identification Code (CIC)-An information code that identifies a circuit between a pair of SS7 exchanges, for which signaling is being performed (14 bits in the ANSI version and 12 bits in the ITU version ISDN User Part).

Circuit Miles-The route miles of circuits in service, determined by measuring the length in miles of an actual path followed by a transmission medium.

Circuit Noise-The noise that is measured or heard across an audio pair between the tip and ring conductors in a telephone system.

Circuit Provision-The process used by carriers to determine the need for trunks and special-service circuits.

Circuit Provision Center-A center that assigns facilities and equipment for message trunk circuits and designs special-service circuits and carrier systems. A circuit provision center also generates and maintains circuit records and inventory and assignment records for all interoffice facilities and equipment.

Circuit Provisioning-The process used by a service provider (network operator) that provides a circuit connection between two (or more) points.

Circuit Reliability-The percentage of time a circuit is available to the user during a specified period of scheduled availability.

Circuit Switch-A circuit switch is a device or assembly that receives signals on input ports and provides a continuous connection (may be a physical or logical connection) to output ports.

Circuit Switched Data-Circuit switched data is a data communication method that maintains a dedicated communications path between two communication devices regardless of the amount of data that is sent between the devices. This gives to communications equipment the exclusive use of the circuit that connects them, even when the circuit is momentarily idle.

To establish a circuit-switched data connection, the address is sent first and a connection (possibly a virtual non-physical connection) path is established. After this path is setup, data is continually transferred using this path until the path is disconnected by request from the sender or receiver of data.

This figure shows the basic operation that uses circuit-switched data. In this example, a laptop computer is sending a file to a company's computer that is connected to the public switched telephone network (PSTN). The laptop computer data communication software requests the destination phone number from the user to connect to the remote computer. This telephone number (the address) is used to connect a path through the PSTN switches until the call reaches the destination computer. The dialed number is first connected through local switch #1, port number 4236. This port number is assigned to a memory location in the switch that routes the data connection through a high-speed line, time slot 6 to an Inter-eXchange Carrier's (IXC) switch. The IXC switch then assigns a memory location in its switch to a high-speed line, time slot 3 that connects to local switch #2. Local switch #2 assigns a memory location in its switch to port number 1249. This port connects to the remote computer. Once this path through the network is setup, it remains constant throughout the data communications session regardless of how much data is transferred between the laptop computer and the company's computer.

ports and each switch only adds a small amount of transfer time between ports. After all the switching connections are made, an audio path can be connected between. Throughout the connection, this path will be maintained through the initial path (the same switch ports) without any changes.

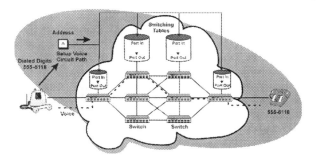

Circuit Switching Operation

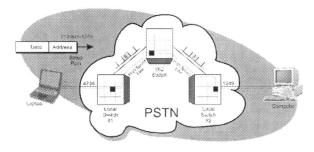

PSTN Circuit Switched Data

Circuit Switching-A process of connecting two points in a communications network where the path (switching points) through the network remains fixed during the operation of a communications circuit. While a circuit switched connection is in operation, the capacity of the circuit remains constant regardless of the amount of content (e.g. voice or data signal) that is transferred during the circuit connection.

This figure shows how circuit switching is used for voice communication. In this example, a telephone is dialing a telephone that is connected to a distant switch. When the user dials the telephone, the dialed digits are captured and used to program the circuit switches between the two telephones. Each switch then has assigned input ports and output

Circular MIL-The measurement unit of the cross-sectional area of a circular conductor. A circular mil is the area of a circle whose diameter is one mil, or 0.001-inch.

Circular Polarization-Circular polarization is the process of rotating the orientation of electromagnetic wave propagation along the direction of propagation.

Cisco Discovery Protocol (CDP)-The Cisco Discovery Protocol or CDP is a proprietary discovery protocol developed by Cisco Systems, Inc. CDP is implemented on most of Cisco's network devices by default. CDP is a layer 2 discovery protocol. Cisco devices send out CDP packets every 30 seconds and are Layer 2 multicast packets (MAC address: 01-00-0c-cc-cc-cc).

Cladding-Cladding is the material that surrounds the core of an optical fiber. The cladding redirects (reflects or refracts) light so that it can travel through the fiber instead of escape from the fiber core. Cladding also protects the fiber core against scattering due to surface-contaminant impurities. In plastic-clad silica fibers, the plastic cladding also can serve as the coating.

Claims-Claims are the specially written sentences at the end of patent documents which set forth the metes and bounds of the patent monopoly. The claims describe the scope of protection of the patent. Claim drafting is subject to a myriad of specific requirements for form and style which vary among countries and regions. Claims of identical language may mean different things in different countries.

Clamp (Clamping)-(1-circuit) The device or process that restores the dc component of a signal. A video clamp circuit, usually triggered by horizontal synchronizing pulses, reestablishes a fixed dc reference level for the video signal. Some damp circuits clamp sync tip to a fixed level, and others clamp back porch (blanking) to a fixed level A major benefit of a clamp is the removal of low-frequency interference, especially power line hum. (2-mechanical) A device used to couple parts together mechanically.

Clamping-Clamp

CLASS-Custom Local Access Signaling Services

CLASS (TM) Services-A group of telecommunication services that provide selective-call screening, alerting, and calling-identification delivery functions. Bellcore owns the trademark for CLASS (TM) services.

Class 4-A voice communications switching system used to interconnect local telephone switching systems. The class 4 switching system was one level above the class 5 end office switching system. Class 4 switches are also known as "Tandem Switches."

Class 5 Electronic Switching System-A classification of a switching system that is used by local telephone service providers. A class 5 switch is the last point in the network prior to the customer. Class 5 switches usually can handle anywhere from 10,000 to 100,000 customers.

Class A Amplifier-An amplifier that is operating in class A (fully linear).

Class AB Amplifier-An amplifier that is operating in class AB. Class AB amplifiers can amplify between 180 degrees to 360 degrees of the input signal (somewhat non-linear).

Class B Amplifier-An amplifier that is operating in class B. Class B amplifiers can amplify a signal for 180 degrees of its input cycle (non-linear amplification for particular portions of the input signal).

Class Based Queuing (CBQ)-A priority scheduling method that gives preferential treatment traffic communication sessions that have different priority levels for each class of communication session (such as real-time audio traffic and near-real time

video). CBQ allocates variable bandwidth for each class of traffic so all applications continue to operate under heavy traffic conditions.

Class C Amplifier-An amplifier that is operating in class C. Class C amplifiers can amplify a signal for less than 180 degrees of its input cycle (non-linear amplification for a majority of portions of the input signal).

Class D IP Address-Class D IP addresses are IP addresses that range from 224.0.0.0 to 239.255.255.255 which are used to identify multicast groups.

Class Of Service-(1-multimedia) The type of service associated with a particular application or communication session. The class of service usually requires a specific quality of service (QoS) level. (2-telecommunications) Categories of services that are provided by tariff for charging customers for the particular service they select. Examples of class services include flat rate, coin, toll free (800) service, and PBX. (3-SS7) Services provided by the Signaling Connection Control Pant (SCCP) to its users.

Class of Service (CoS)-Class of Service is a way of managing traffic in a network by grouping similar types of traffic. 802.1p, Type of Service (ToS), and DiffServ are three main CoS technologies.

Classical IP-Classical IP is a set of industry specifications that were developed by the Internet Engineering Task Force (IETF) that allow ATM networks to connect LAN systems.

Classical IP over ATM (CIOA)-Classical IP over ATM is the process of sending IP packets (connectionless packets) over ATM (connection-oriented) networks.

Classless Addressing-An addressing system used in an IP communication system that uses a network suffix to depict the address range.

Classless Inter-Domain Routing (CIDR)-A routing solution that does not require the use of the upper parts of a routing address to be used to determine the specific ("class") of network. The CIDR address includes the number of bits used in the address mask. CIDR is pronounced "cider."

Classless Internet Domain Routing (CIDR)-In response to the limitations of class A, B, and C IP addresses, the Internet community implemented CIDR (pronounced "cider"). CIDR allows IP addresses to be broken down into subnets on arbitrary bit boundaries. CIDR networks are described with a "slash". For example, 192.168.128.0/17 represents a subnet specified by the high-order 17 bits

C

of the given IP address. When a router adds this route to its forwarding table, any packet with a destination IP address that matches the high-order 17 bits of that address will be treated as members of the same subnet and forwarded accordingly.

Clear Channel-(1-radio, especially radio broadcasting) A carrier frequency that is licensed to only one transmitter in the entire nation, thus ensuring no radio interference. (2-digital telephone systems) A digital channel, typically 64 kb/s, for the subscriber that can carry any binary bit stream without restriction, including a string of binary zeros.

Clear Channel Capability (CCC)-A characteristic of DS1 transmission in which the 192 information bits in a frame can carry any combination of zeros and ones.

Clear Defective Pair-A facility modification that requires the repair of a defective wire pair for a specific service order or related line and station transfer.

Clear Field-In reference to blank spaces found on screen before and after the scroll of credits found at the end-title of a film.

Clear Text Authentication-An authentication process that relies on a user identification name and password information without encrypting the information (relatively low security).

Clear To Send (CTS)-Clear to send is a control signal that indicates that the transmission of communication information can proceed. CTS is a control signal used on an RS-232 serial bus communication line.

Clearing House (Clearinghouse)-A clearinghouse is a company or association that transfers billing records and/or performs financial clearing functions between carriers that allow their customers to use each other's networks. The clearinghouse receives, validates and accounts for telephone bills for several telephone service providers. Clearinghouses are particularly important for international billing because they convert different data record formats that may be used by some service providers and convert for the currency exchange rate.

Clearinghouses provide a variety of services including processing proprietary records (e.g. switch records) into formats understandable by the member carriers' billing systems, validate charges from carriers with intersystem agreements, and extract unauthorized or un-billable billing records. Clearinghouses transfer messages in a standard format such as exchange message record (EMR),

cellular inter-carrier billing exchange roamer (CIBER), or transferred account process (TAP) format. The EMR format is often used for billing records in traditional wired telecom networks and the CIBER and TAP formats are used for wireless networks. The records may be exchanged by magnetic tape or by other medium such as electronic transfer or CD ROM.

Clearinghouse-Clearing House

CLEC-Competitive Local Exchange Carrier

CLI-Calling Line Identification

CLI-Cellular Line Interface

CLI-Command Line Interface

Click Through Rate (CTR)-Click through rate (usually in percentage form) is a ratio of how many clicks a link or advertising message receives from visitors compared to the number of times the link or advertising message is displayed. An example of click through rate is a link that is clicked 5 times out of 100 displays to visitors is 5%.

Click to Call (CTC)-Click to call is a combined voice and data service that allows a user who is viewing a web page or IP television program to click a link or image on the display to initiate a voice call. The link contains an embedded address (URL or IP address) that connects to a call server along with the necessary software (such as SIP) that allows for the setup and connection of the call. Click to dial service is similar in concept to the 'mailto:' link that can launch a user's email software when selected.

Click to Dial or Click to Call-Click to Dial is a combined voice and data service that allows a user who is viewing a web page to click a link on that web page to initiate a voice over Internet call. The link contains an embedded address (URL or IP address) that connects to a call server along with the necessary software (such as SIP) that allows for the setup and connection of the call. Click to dial service is similar in concept to the 'mailto:' link that can launch a user's email software when selected.

Click-Through-The process of selecting ("clicking") a link on an Internet web page.

CLID-Calling Line Identification Display

Client-A computer, hardware device or software program that is configured to request services from a network. Client also may refer to the codec or terminating device located at one end of a network node.

Client Software-Software that is controlled by a user that requests processing services from another computer or system. An example of client software is an Internet browser that requests information from a web site.

Client/Server Architecture-A form of distributed computing in which each application is viewed as a series of inter-dependent tasks accomplished by several different computers linked on a network, for example, a personal computer can be a client while it is linked to a remote processor acting as a server. The various tasks performed by the client is considered to be front end operations and those performed by the server are backend operations

This diagram shows the configuration of a client/server network. A client/server network is a form of distributed computing in which each application is viewed as a series of inter-dependent tasks being accomplished by several different computers linked on a network. For example; a personal computer can be a client while it is linked to a remote processor while it is acting as a server for another computer. The various tasks performed by the client are considered to be front end operations and those performed by the server are backend operations.

may be isolated video segments (such as the launch of a space shuttle) or they may be a short portion of a larger movie or video (a video clip of a famous movie scene).

CLIP-Calling Line Identification Presentation

Clipper-A limiting circuit which ensures that a specified output level is not exceeded. It does this by restricting the output waveform to a maximum peak amplitude.

Clipping-Signal distortion that results from the non-linear (non-equal) processing of a signal through a device or circuit (such as an amplifier) where some of the signal is lost due to the maximum or minimum processing ability of the device or circuit.

This diagram shows how a signal may experience clipping if its level exceeds the maximum capabilities of an amplifier. In this example, an input signal varying from 0 to 1 Volt is applied to an amplifier that has a gain of 10. Because this amplifier has a maximum output voltage of 5 volts, when the input signal exceeds 0.5 Volts, the output signal reaches a maximum (clipped) to 5 Volts.

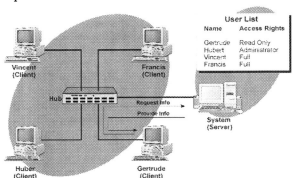

Client/Server Network

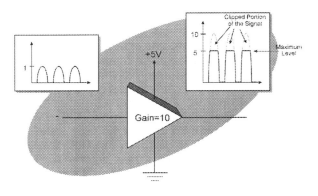

Clipping Operation

C-Link-Cross Link

Clip-(1-insert) The trigger point or range of a key source signal at which the key or insert takes place. (2-control) Clip also may refer to the control that sets this action. To produce a key signal from a video signal, a clip control on the keyer control panel is used to set the clip code. (3-test) A flexible connector that can self-attach to an electronic component or assembly test point. (4-media) Clips are short movie, video or audio segments. Video clips

Clipping Level-An electronic limit beyond which unwanted distortion will occur in the audio or video signal

CLIR-Calling Line Identification Restriction

CLLI-Common Language Location Identification

CLLI Code-Code Common Language Location Identifier Code

Clock-A clock is a reference source of timing information that is used to coordinate the operation of equipment, machines or systems. A clock can be an electronic circuit that produces timing pulses that is used or shared by several components, circuits or assemblies.

Clock Frequency-The master frequency of periodic pulses that are used to synchronize the operation of equipment.

Clock Jitter-Undesirable random changes in clock timing signal (phase deviation.)

Clock Reference-A clock reference is a repetitive accurate signal that is used to synchronize or coordinate the operations of a system or communication line.

Cloning-Duplicating certain information necessary for the connection of communication (such as wireless telephones) devices in a fraudulent manner. Cloning the ESN's of existing customers is an example of cloning.

Closed Captioning-Closed captioning is a service that adds additional information (e.g. text captions) to television programs to enabled people with disabilities (such as deaf people). Closed captioning information is typically provided by an additional signal that requires a closed captioning decoder to receive and display the captions on the television screen.

Closed Numbering Plan-A numbering plan in which all local numbers comprise the same number of digits, all area or zone codes comprise a fixed number of digits, etc. The North American Numbering Plan (NANP) is nominally a closed numbering plan. All local numbers are 7 digits in length. All area codes are 3 digits in length. There are some few exceptions to this rule, since certain short codes are used, such as 0 for the operator/attendant, 911 for emergency service, etc. See also the antonym Open Numbering Plan.

Closed Source Software-Closed source software is files, programs, and applications that do not include original source codes from which the programs were created. Closed source programs do not allow (or easily allow) changes to be made to the application software.

Closed Standard-Closed standards are operational descriptions, procedures or tests that are part of an industry standard document or series of documents that are available to a limited number of companies. Closed standards are commonly created by one company or through the participation of a limited number of companies.

Closed User Group (CUG)-(1-access restriction) A group of directory numbers sharing an access restriction such that any directory number can reach others in the group but cannot access outside numbers. (2-cellular system) Advanced features such as 4-digit dialing authorized for a closed group of users of the service. (3-X25 protocol) In the X.25 packet-switching protocol, a facility indicating a virtual grouping of terminals that can communicate only with other members of that group. The feature can be extended to a closed user group with outgoing access, or a closed user group with incoming access.

Closed-Circuit Television (CCTV)-A private (closed circuit) network of security that display images on one or more television (video) monitors.

CLP-Cell Loss Priority

CLTP-Connectionless Transport Protocol

Cluster-(1-data communications) A group of storage devices, processors, or computers that share a common bus or network and function as a single system or sub-system within a larger network. These devices are often coordinated by a cluster controller. (2-memory device) The amount of memory storage capability within a sector on a diskette or hard disk. (3-SS7) In the Signaling System 7 protocol, a set of signaling points that are identifiable as a group within the signaling-point code-address spare.

Clustering-Clustering is the grouping of resources or servers within a network to increase system reliability and/or to distribute the processing requirements of the system.

CM-Cable Modem

CM-Circuit Merit

CM-Code Multiplexing

CM-Configuration Management

CMAC-Control Mobile Attenuation Code

CMC-Cellular Mobile Carriers

CMDS-Centralized Message Data System

CML-Chemical Markup Language

CMOS-Complementary Metal Oxide Semiconductor

CMR-Call Management Record

CMR-Common Mode Rejection

CMRS-Cellular Mobile Radiotelephone Service

CMRS-Commercial Mobile Radio Service

CMS-Content Management System

CMTS-Cable Modem Termination System

CMYK-Cyan, Magenta, Yellow and BlacK

CN-Core Network

CND-Called Number Delivery

CNET-The first generation analog cellular radio system used in Germany and other parts of the world.

CNG-Comfort Noise Generator

CNI-Calling Number Identification

CO-Central Office

Coating-A protective material (usually plastic) applied to an optical fiber immediately after drawing to preserve its mechanical strength and to cushion it from external forces that can induce microbending losses.

Coax-Coaxial cable

Coaxial cable (Coax)-A coaxial cable (coax) is a multi conductor cable comprising a central wire conductor surrounded by a hollow cylindrical insulating space of air, or solid insulation, or mostly air with spaced insulating disks, finally surrounded by a hollow cylindrical outer conductor. Invented by Lloyd Espenschied in the 1930s for transmission of wideband television and radar signals, it is also used for other purposes. The hyphenated form of the name is preferable to avoid confusion with the English word "coax." Versions using more than one internal conductor are called "tri-ax," "quad-ax" for three total conductors or four total conductors respectively. Co-ax exhibits lower radiated electromagnetic power losses because the electromagnetic field is confined inside the outer conductor. The form with the hyphen is preferable to avoid confusion with the English word coax.

This figure shows a cross sectional view of a coaxial cable. This diagram shows a center conductor that is surrounded by an insulator (dielectric). The insulator is surrounded by the shield. This diagram shows that during transmission, electric fields extend perpendicular from the center conductor to

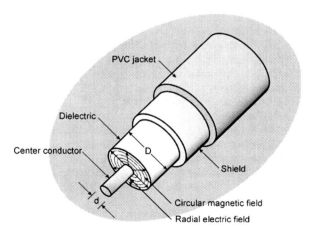

Coaxial Cable Diagram

the shield and magnetic fields form a circular pattern around the center conductor.

Co-axial Lightning Suppressor-A device or assembly that protects a coaxial cable by shorting surge voltages on the inner conductor to ground (shield).

Coaxial Switch-A coaxial switch is a device or an assembly that can connect and disconnect signals for coaxial cables.

COB-Chip On Board

Co-Channel-A communication channel that is operating on the same carrier frequency at a nearby transmission site. The ability to use channels on the same frequency at other transmission sites is made possible due to the attenuation of the signal because of distance. This is referred to as frequency reuse.

Co-Channel Interference (C/I)-Radio signal interference that results when two radio channels, that are operating in the same geographic area on the same frequency, are not separated by enough distance.

COD-Connection Oriented Data

Code Character-The representation of a discrete value or symbol in accordance with a specified code.

Code Common Language Location Identifier Code (CLLI Code)-CLLI codes are used to provide unique identification of facilities (equipment and cables) between any two interconnected CLLI coded locations. The CLLI code is mnemonic code that can be a maximum of 38 characters. An example of a CLLI code is 115T3NYCNY20DALTX. This example shows a T-3 carrier is connected between New York City, New York and Dallas, Texas.

Code Conversion-In address signaling, the alteration of leading received digits in order to direct a call through subsequent switching offices.

Code Division Multiple Access (CDMA)-A system that allows multiple users to share one or more radio channels for service by adding a unique code to each data signal that is being sent to and from each of the radio transceivers. These codes are used to spread the data signal to a bandwidth much wider than is necessary to transmit the data signal without the code.

Code Division Multiplexing (CDM)-The process of multiplexing several signals by allowing separate signals to be applied to a common transmission physical channel or line by using specific code sequences to uniquely identify each signal.

This diagram shows how IS-95 CDMA radio channels can provide multiple communication channels through the use of multiple coded channels. This diagram shows that a code pattern mask is used to decode each communication channel. The channel mask is shifted along the radio channel until the code chips (or a majority of the code chips) match the expected code pattern. When a match occurs, this produces a single bit of information (a logical 1 or 0). This example shows that the use of multiple code patterns (multiple masks in this example) allow multiple users to share the same radio channel.

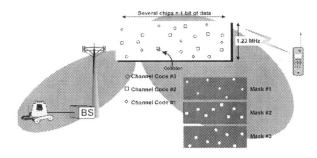

Code Division Multiplexing (CDM) Operation

Code Domain-The process of examining the power level, phase etc of each of the Walsh Codes in a CDMA system.

Code Excited Linear Prediction (CELP)-A lossy compression technique for speech, sometimes pronounced like "kelp" or "selp". In general, a CELP decoder produces a segment of audio samples by filtering an excitation signal through a "linear predictor" filter, also sometimes called the synthesis filter. The excitation is one of several possibilities as indicated by a code in the bitstream, hence the term "code excited". The linear predictor filter typically has an order of about 10 coefficients, the values of which are sent in the bitstream in quantized form. At the CELP encoder, the linear predictor coefficients may be derived using the Levinson-Durbin algorithm, after which the encoder does a search over some possible excitation codes for a best match. CELP codecs are a type of vocoder. CELP-based standard speech codecs

include G.728 (16 kbps), the GSM CELP codec (13 kbps), and the Federal Standard 1016 codec (4.8 kbps).

Code Excited Linear Predictive (CELP)-Code excited linear predictive coding is a rapid analog-to-digital conversion technique that employs a level or multilevel sampling system in which the value of the signal at each sample time is predicted to be a particular linear function of the past values of the quantized signal. The function is identified through the use of codebooks.

Code Multiplexing (CM)-Code multiplexing is the process that allows multiple users to share a communication channel for service by adding a unique code to each data signal that is being sent to and from each of the communication transceivers. These codes are used to spread the data signal to a bandwidth much wider than is necessary to transmit the data signal without the code.

Code of Practice #3 (COP3)-Code of practice number 3 specifies working guidelines for the deployment of television services over telecom lines, notably over digital subscriber line (DSL) and fiber transmission systems. The most significant attribute of COP3 is the use of forward error correction (FEC) on the packet stream to compensate for the high transmission error rate found on DSL circuits.

Code Rates-Code rates are the ratio of information bits to a coding process to the total number of bits created by the coding process. A coding rate of ¼ indicates for each information bit into the coding process (such as 8 kbps) there will be 4 bits created for transmission (output of 32 kbps). The higher the code rate, the higher percentage of error detection/correction overhead.

Code Schemes (CS)-Code schemes are a mixture of coding processes used in communication systems. Code schemes may include different modulation and channel coding types. The selection of code scheme that is used usually depends on the type of communication medium. Communication channels that have a high percentage of errors may use strong error protection coding and robust modulation types. For communication channels that have a high quality link, more efficient modulation technology and minimal error protection may be used to increase the data transmission rate.

Code Set-The complete set of representations defined by a code or by a coded character set.
Codec-Coder/Decoder
CODEC-H.261 Video Coding and Decoding

Codec Negotiation-Codec negotiation is the process where two (or more) communication devices negotiate for which coder/decoder (CoDec) devices they will use during a communication session. Communication devices use codec negotiation when they have access to multiple audio, data, and video codecs (data compression devices). The codec negotiation process may include the use of customer preferences (e.g. use of low-bit rate codecs due to bandwidth costs), preferred codecs due to quality of service (QoS) requirements (e.g. good audio or video quality), or availability of compatible codecs.

This diagram shows how two voice gateways negotiate for codec selection in a voice over data network using user assigned preferences and options determined by equipment availability. In this example, the calling gateway sends a connection request message to the remote gateway. This connection request indicates that the calling gateway prefers to use a G.711 64 kbps speech codec because it has enough bandwidth and the caller prefers a high quality audio signal. Unfortunately, the receiving gateway cannot accept the request for the G.711 codec because its access bandwidth is low speed (28 kbps). The receiving gateway sends back a request to use the G.722 codec. When the calling gateway receives this request, it declines the request because it does not have a G.722 codec available. It then requests the G.729 codec (industry standard minimum) and the receiving gateway confirms the request and both devices use the G.729 codec.

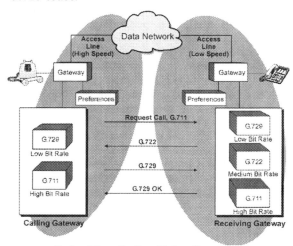

Codec Negotiation Voice Operation

Coded Orthogonal Frequency Division Modulation (COFDM)-Coded orthogonal frequency division multiplexing (COFDM) is a process of modulating a digital communication channel into multiple independent frequency carrier channels that do not interfere with each other.

This figure shows how COFDM divides a single radio channel into multiple coded sub-channels. This example shows that a high-speed digital signal is divided into multiple lower-speed sub channels. Part of the COFDM process is to code each channel of bits in such as way that bits from other channels have a relationship to them (error protection). Each COFDM sub channel transmission is independent from each other. In this example, a portion of a sub channel is lost due to a frequency fade. Due to the COFDM encoding process, the missing bits from one channel can be re-created from the bits received from the other channels.

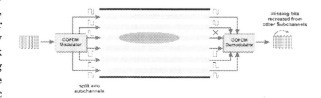

Code Orthogonal Frequency Division Modulation (COFDM) Transmission

Coded Speech-A speech signal that has been changed into a standard digital code form.

Coded Trunks-Trunks that are coded for identification so they can be identified as they hunt (search) for connections.

Codepoint-In the Signaling System 7 protocol, the single coding of a value within an information element.

Coder/Decoder (Codec)-Codecs are devices or software that are used to compress (code) or expand (decode) information to a fewer number of bits for more efficient transmission and storage. Normally the term codec applies only to compression of human-perceived signals such as speech, audio, images, or video.

Coding-(1-digital) A process of changing digital bits to include error protection bits and/or signaling bits prior to the sending or storing of the information. (2-software) The process of writing instructions or commands for software programs.

Coding Scheme-A coding scheme is a combination of modulation, error protection, and signaling control in a communication system.

Coefficient of Friction-Coefficient of friction is the ratio of force that results from sliding an object across a surface. The coefficient of friction can be used to determine how much tensile load (pulling load) a cable will experience as it is pulled through a duct or conduit. The coefficient of friction is different for dry conditions and lubricated conditions.

COFDM-Coded Orthogonal Frequency Division Modulation

Coherence-The correlation between the phases of two or more waves.

Coherent-Coherent is a condition that is characterized by a fixed phase relationship among points on an electromagnetic wave. With regard to light, it is a condition where all of the rays of a transmitted signal are aligned on the same axis of transmission. Lasers produce coherent light.

Coherent Network-A network in which inputs, outputs, signal levels, bit rates, digital bit stream structures, and signaling information all are inherently compatible throughout the network.

Coherent Pulse-The condition in which a fixed phase relationship is maintained between consecutive pulses during pulse transmission.

Coin Telephone-A telephone that allows the collection of coins. William Gray invented the coin telephone.

Coinless Public Telephone Service-A service for use of a coinless public telephone for originating calls only. The phone is served by a single-party, loop-start line with no extensions and no custom calling features.

Cold Boot-The startup process that begins when equipment (such as a computer) or device is first powered on. A cold boot is used to start an operating system from a known good point (before other programs are activated).

Cold Joint-An inadequately heated soldered connection, leaving the wires held in place by rosin flux, not solder. A cold joint sometimes is referred to as a dry joint.

Cold Shrink-Cold shrink is a material that has its size contract without the application of heat. Cold shrink tubing contains a removable inner core that expands the cold shrink until it is installed. When the inner core is removed (pulled out), the cold shrink contracts surrounding the inner object.

Collaboration-The process of two or more people working together on a shared medium (such as a word processor file or sketch whiteboard). Collaboration usually enables a group of people to work together in real-time to share screen displays, documents, and video images.

Collaborative Filtering-Collaborative filtering is the process of reviewing, analyzing and selecting information (data) through the combined actions of individuals or companies.

Collapsable Ads-Collapsible ads are images or video clips that can be reduced or shortened in time and/or content. Collapsible ads may have been initially expanded or they may be interstitial (inserted) ads between other media content.

Collapsed Backbone-A collapsed backbone method of interconnecting networks by using a switch or router as a central relay device rather than using multiple switching or concentration points.

Collection-The methods and procedures used by the service provider to receive payments due from a customer (also known in Europe as "chasing up".) Sometimes also refers to the methods used to minimize the risk that a customer will end up in arrears.

Collections-Collections are activities that a service provider performs to receive money from their customers. Ideally, all customers will receive their bills and pay promptly. Unfortunately, not all customers pay their bills and service providers must have a progressive collection process in the event a customer does not pay their bill.

When customers are first added to a system, they are rated on the probability that they will pay their bills. This is accomplished by using information on their application and reviewing the credit history as provided by an independent credit reporting agency.

The collection process for delinquent customers usually starts by sending a reminder messages to the customer be mail or recorded audio message. If initial attempts to collect are unsuccessful, more aggressive collection activities will progress that include restricted calling, service disconnection and sending or selling the uncollected invoice to a collection service.

Collective Action-Collective action is a process or activity that is performed by or for the benefit of a group of companies and/or people who share a common interest.

Collective Licensing System-A collective licensing system is a process that allows a collective group of technologies or intellectual property to be licensed as a complete group instead of identifying and negotiating licenses for each part separately.

Collimation-Collimation is the adjustment of the direction of light beams so they become more parallel with each other.

Collision-Data collisions occur when two or more data stream attempt to occupy a telecommunications media at the same time causing each stream to be corrupted. Typically, each transmitting host that detects the collision will wait for some period of time and try again. This situation is most often associated with Ethernet and its access method carrier sense multiple access with collision detection (CSMA/CD).

Collision Detection-Collision detection is a process that detects simultaneous interfering transmission has taken place on a transmission channel. Typically, each transmitting station that detects the collision will wait some period of time before attempting to transmit again.

Collision Domain-The set of stations among which a collision can occur. Stations on the same shared LAN are in the same collision domain; stations in separate collision domains do not contend for use of a common communications channel.

Collision Fragment-The portion of an Ethernet frame that results from a collision. On a properly configured and operating LAN, collision fragments are always shorter than the minimum length of a valid frame.

Co-location-The location of equipment from systems or multiple carriers at the same facility. Colocation commonly refers to the location of a competing telephone service provider's equipment at a local telephone company's switching facilities. This enables providers of interstate or competing telecommunications service providers to connect their facilities directly to those of a local exchange carrier (LEC).

Co-location-Colocation

Color Adjust Filter-A color adjust filter is image editing software that allows for the selection and changing of color aspects of images, objects or individual channels.

Color Code-(1-general) The word "color" is used in both the literal and figurative sense in telecommunications. The literal meaning occurs in component identification methods using visible colors. The figurative meaning is used for situations in which different component frequencies are not descriptive of actual visible light. One figurative meaning is used for different wavelengths of infrared light (different "colors") used in WDM and DWDM fiber optics. Electromagnetic radiation in the infrared frequency range is not actually visible to the human eye. A noise or an artificial random signal waveform that has uniform power density across the audio frequency or radio frequency spectrum or bandwidth of interest is called "white" noise . A similar signal that has higher power density at the low frequency (longer wavelength) part of its spectrum is called "pink noise" by analogy to pink visible light. (2-cable colors) the twisted-pair cable color code (PIC) using various body and stripe colors on wire insulation to distinguish various different wire pairs in a cable. (3-fiber optic) Fiber-optic color code using distinct jacket colors to distinguish different optical fibers in a multi-fiber cable. (4-Resistors and other component values) The resistor/capacitor color code using a distinct color stripe to represent each decimal digit of a component's resistance or capacitance value. (5-channel code) A special case is the entirely figurative meaning in terms such as Digital Verification Color Code, used in IS-136 cellular systems to distinguish different radio signals that use the same carrier frequency but are intended for use in different cells. The term "color" is used there in an entirely figurative sense meaning "identification."

Color Depth-In uncompressed digital images, the number of bits per pixel. For example, a color image that uses 8 bits for each color uses 24 bits per pixel. For a low color depth like 8 bits per pixel, a color palette may be used to map each value to one of a small number of more precisely represented colors, for example to one of 256 24-bit colors.

Color Field-In the NTSC system, the color subcarrier is phase-locked to the line sync so that on each consecutive line, subcarrier phase is changed 180 degrees with respect to the sync pulses. In the PAL system, color subcarrier phase moves 90 degrees every frame. In NJSC, this creates four different field types; in PAL there are eight To make clean edits, alignment of color field sequences from different sources is crucial.

Color Frame-In NTSC, a 4-field sequence that exists between the occurrence of exact coincidences of the color subcarrier signal. This span results from the uneven number of cycles of subcarrier per line (227.5) and per 525-line frame (119437.5). In PAL video, a similar sequence of eight fields exists because of the more complex phase structure of the PAL signal.

Color Gamut-The entire range of values a component signal or a combination of component signals may take on that are reproducible at the display device. Some component formats are interdependent (Y, R-Y, and B-Y), and the valid color gamut cannot be evaluated by looking at a single component alone.

Color Palette-The hard-coded colors that a video card can display on a computer.

Color Resolution-Color resolution is the ability to select and display the correct chrominance (color) expressed on a display.

Color Space-A set of components of a color, such as RGB or YUV. Any visible color can be represented as a set of values corresponding to the components of a color space.

Color Space Conversion-The process of converting a color from the representation in one color space to a representation in a different color space. A color conversion can be computed as a constant linear relationship (a matrix multiplication) followed by clipping to ensure the resulting color values lie within the valid ranges. Color space conversion is required for encoding or decoding images or video in many cases. For example, a JPEG image or a frame of MPEG video is decoded into a pixel-map of YUV color values, which need to be color-converted to RGB color values if the image is to be displayed on an RGB display device. In some operating systems, a graphics display API allows specifying display colors using an application-specified color space so that the color space conversion can be done by the video display hardware when supported.

Color Space Transform-Color space transform is the conversion (transformation) of a color image into different forms that represent the original color image. Examples of color transforms include red, green and blue (RGB) or cyan, magenta, yellow and black (CMYK) formats.

Color Subcarrier-The frequency that carries the color information in the baseband composite video signal. in NTSC, the color sub carrier is 3,579,545 Hz, usually rounded off in text to 3.58 MHz.

Color Timing-The synchronization of the burst phase of two or more video signals. Proper color timing ensures that no color shifts occur in the picture when the signals are mixed in a switcher or other video device.

Color Video Baseband Signal (CVBS)-A color video baseband signal is a modified black and white analog video signal that includes chrominance signals.

Colorplus-Colorplus is the process of inverting and adding interlaced fields in a video signal to cancel the effects of cross luminance (intensity signals causing changes in color levels) and cross color (color signals causing increased intensity levels).

Colour-British English spelling of color.

Co-marketing-The joint marketing of products or services. Co-marketing of products or services commonly involves the manufacturer or service provider giving marketing allowances or incentives to retailers or other sales focused companies for specific marketing promotions in their market areas.

Combination Trunk-A trunk line that can operate as a direct inward dial (DID) or direct outward dial (DOD) trunk. This allows them to receive (ring-in) and send calls.

Combiner-(1-radio frequency) A device that is used to combine several channels on to a common transmission line or antenna system. (2-video manipulator) In a video or image processing system, a device or assembly that controls the interaction between video channels or images. The combiner prioritizes the display of video sources or images and controls the type of graphic transitions that can occur between them.

COMET-PreConditions Met

Commerce Revenue-Commerce revenue is a source of money that is created through the produce sales activity of customers. Commerce revenue is different than service revenue. An example of commerce revenue is the purchasing of a pizza through an interactive television advertisement message. The commerce revenue is the money generated from the pizza sale and not the money that is generated from the television service.

Commercial Radio Operator-A commercial radio operator is a person or company who holds a license to provide commercial radio services.

Commercial Use-The providing of communication and/or media transmission on a fee basis.

Committed Information Rate (CIR)-Committed information rate is a guaranteed minimum data transmission rate of service that will be available to the user through a network. Applications that use CIR services include voice and real time data applications. CIR can be measured in bits per second, burst size, and burst interval.

Some service providers allow users to transmit data above the CIR level. However, when data is transmitted above the CIR level, some of the data may be selectively discarded if the network becomes congested.

Committee For European Electrotechnical Standardization-Committee For European Standardization

Committee For European Standardization (Committee For European Electrotechnical Standardization)-A European standards organization also known collectively as the Joint European Standards Institute.

Common-A point that acts as a reference for circuits, often equal in potential to the local ground.

Common Air Interface (CAI)-The specifications for a radio communication system that standardize the communication between radio transmitter(s) and radio receiver(s) within that system. If the radio equipment is designed to meet the common air interface (CAI) specifications, theoretically equipment or software developed by one company should be capable of operating with other equipment or software designed by a different company. It's defined in terms of Access Method, Modulation Scheme, Vocoding Method, Channel Data Rate and Channel Data Format.

Common Carrier-(1-general) A company that carries goods, services, or people from one point to another for the public. (2-telecommunications) A company that provides communications services and typically is subject to a regulatory agency.

Common Channel (CCH)-Common channels are accessible and shareable by a variety of communication devices. Common channels are commonly used to send commands or instructions to a group of communication devices (such as system or device identification information).

Common Channel Signaling (CCS)-A signaling technique in which signaling information relating to a multiplicity of circuits (trunks), is conveyed over a separate single channel by addressed messages. Common channel signaling system #7 ("SS7") is the primary system used for interconnec-

tion of telephone systems. SS7 sends packets of control information between switching systems.

The SS7 network is composed of its own data packet switches, and these switching facilities are called signal transfer points (STPs). In some cases, when advanced intelligent network services are provided, STPs may communicate with signal control points (SCPs) to process advanced telephone services. STPs are the telephone network switching point that route control messages to other switching points. SCPs are databases that allow messages to be processed as they pass through the network (such as calling card information or call forwarding information).

This diagram shows the basic structure of the SS7 control signaling system. The SS7 network is composed of its own data packet switches, and these switching facilities are called signal transfer points (STPs). In some cases, when advanced intelligent network services are provided, STPs may communicate with signal control points (SCPs) to process advanced telephone services. STPs are the telephone network switching point that route control messages to other switching points. SCPs are databases that allow messages to be processed as they pass through the network (such as calling card information or call forwarding information). Messages originate and terminate at a service switching point (SSP.) A SSP is a part of the end office (EO) switching system. End offices are sometimes called central offices (COs).

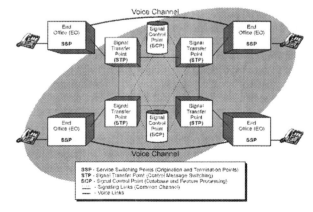

Common Channel Signaling System

Common Control-A system in which items of control equipment are commonly shared (such as network control signaling).

Common Equipment-Any apparatus used by more than one unit of equipment, or any equipment used by more than one channel; that is, equipment common to two or more channels.

Common Gateway Interface (CGI)-A standard software program interface that allows Web servers to interact with user specified applications such e-commerce programs and databases. CGI describes and uses variable control tags that are inside an HTML file to allow information to transfer from user applications (programs and web pages) to the Web server host program. Because the CGI defines specific information variables and how these variables (information elements) are processed, CGI scripts can be created and used on any web based operating system.

Common Interchange Format (CIF)-Common interchange format (CIF) is an image resolution format that is 360 pixels across by 248 pixels high (360x248). The CIF standard is defined in the ITU H.261 and H.264 compression standards. CIF coding includes interframe prediction (using key frames and difference frames), mathematical transform coding and motion compensation.

Common Intermediate Format (CIF)-A video format, adopted by the Specialists Group of the International Telegraph and Telephone Consultative Committee (CCFIT), for transmission of video signals on Integrated Services Digital Networks (ISDNs). The standard CIF has 288 lines of 360 pixels each for luminance and half as many lines and pixels for each of two chrominance, or color, components. A second version, 1/4 CIF, has one-fourth that resolution.

Common Language Location Identification (CLLI)-A standard code used by telecommunications systems to identify specific locations of a switching office or network element. A CLLI code is composed of 11 alphanumeric characters. The first four characters of the CLLI code are an abbreviated place name. Characters 5 and 6 are state abbreviations. Positions 7 and 8 identify a specific building, and 9,10, and 11 represent a particular piece of equipment.

Common Mode-Signals identical with respect to amplitude, frequency, and phase that are applied to both terminals of a cable and/or both the input and reference of an amplifier.

Common Mode Hum-Power line interference (usually 60 Hz) that appears on both terminals of a cable with the same phase, amplitude, and frequency.

Common Mode Rejection (CMR)-A measure of how well a differential amplifier rejects a signal that appears simultaneously and in phase at both input terminals. As a specification, CMR usually is stated as a decibel ratio at a given frequency.

Common Open Policy Service (COPS)-The COPS protocol allows a system to implement policy decisions by allowing a client to obtain system configuration and parameter information from a policy server. COPS is defined in RFC 2748.

Common Return-A return path that is common to two or more circuits, and that returns currents to their source or to ground.

Common Scrambling Algorithm (CSA)-A common scrambling algorithm is an encryption process that can be used by multiple companies and systems. For digital transmission systems, the digital video broadcasting (DVB) standards organization has defined a CSA for encrypting digital video streams.

Common System-A system that shares power, interconnections, or environmental support for network elements associated with transmission and switching.

Common Trunk-Trunks within a telephone systems that are accessible to all groups of service (trunk) grades.

Commonwealth Broadcasting Association-An association of public broadcasting authorities of Commonwealth countries, numbering about 50.

Communication Center-A facility responsible for the reception, transmission, and delivery of information..

Communication Channel-A medium that transfers information from a source to a destination. A communication channel may transport one or many communication circuits.

Communication Link-A communication link is a transmission system associated switch that allow information to pass between two points.

Communication Server-A communication server is a computing system that can receive, process, and respond to an end user's (client's) request for communication services.

This figure shows the different types of servers used in some SIP based communication systems. This example shows that a call manager (proxy server) receives and processes call requests from communication units (IP telephones). The administrator server coordinates accounts to the system. A unit manager (location server) functions as a location server by tracking the IP address assigned to the communication units. The gateway manager identifies and coordinates communication through the available gateways. The system manager coordinates the communication between the different servers and programs available on the system.

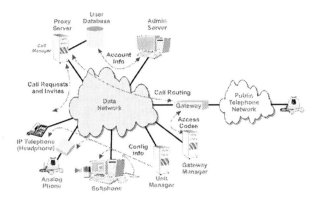

SIP Communication Servers

Communication Services-Communication services are the processes that transfer information between two or more points. Communication services may involve the transfer of one type of signal or a mix of voice, data, or video signals. When communication services only involve the transport of information, they are called bearer services. When communication services involve additional processing of information during transfer (such as store and forward), they are known as teleservices.

Communications-The conveyance of information, including voice, images, and/or data, through a transmission channel without alteration of the original message.

Communications Assistance For Law Enforcement (CALEA)-A statute that was enacted by the U.S. congress in 1994 to define requirements of telephone service providers to provide

wiretap capabilities to law enforcement agencies. To attach a listening device to a communication line, the law enforcement agency must have a surveillance order from a court of competent jurisdiction.

Communications Channel-The medium and Physical layer devices that convey signals among communicating stations.

Communications Medium-The physical medium used to propagate signals across a communications channel (e.g., optical fiber, coaxial cable, twisted pair cable).

Communications Processor-A processor that manages and monitors the flow of voice calls and coverts (interfaces) to different communications protocols. A call processor may be part of an advanced intelligent network for call delivery.

Communications Satellite-An orbiting object that relays signals between ground-based communications stations.

Communications Satellite Corporation (COMSAT)-A corporation that provides the US portion of international satellite services between the United States and other countries. COMSAT is a part owner of the International Telecommunications Satellite Organization (INTELSAT) and the International Maritime Satellite Organization (INMARSAT) satellite systems. COMSAT was established by Congress in 1962.

Communications System-A collection of individual communications networks, transmission systems, relay stations, tributary stations, and terminal equipment capable of interconnection and interoperation to form an integral whole. The individual components must serve a common purpose, be technically compatible, employ common procedures, respond to some form of control, and, in general, operate in unison.

Communications Workers Of America (CWA)-A labor union established for the telecommunications and printing fields.

Community Access Television (CATV)-A system or process of delivering quality television reception by taking signals from a well-situated central antenna and delivering them to people's homes by means of a coaxial cable network.

Community Antenna Relay Service (CARS)-The 12.75 to 12.95 GHz microwave frequency band that the FCC has assigned to the CATV industry for use in transporting TV signals.

Community String-A password used with the SNMP protocol, SNMP community strings are used for both read only and read/write privileges. A community string is case sensitive, and may include some punctuation characters. The community string is sent along with SNMP operations like "get" and "set-request" packets. Read/write community strings are needed when issuing "set-requests".

Commutator-A circular assembly of contacts insulated one from another, each leading to a different portion of the circuit or machine.

Compact Disc (CD)-A compact disc is an optical storage system that has a total storage capacity of over 600 Mbytes on a 5-inch diameter disc. For digital audio applications, it can store 74 minutes of audio using a sampling rate of 44.1 kHz.

Compact Disk (CD)-A compact disk is a data storage medium that represents information in the form of changes in optical properties of a light signal as it is reflected from the surface of the disk. The disk is modified (burned) using a strong optical source (a laser light) to change its optical properties. Some compact disks can only be recorded once (CD-R) and others can be re-written (CD-RW).

Compact Disk Interactive (CD-I)-Compact disk interactive is a data format that is used to store, manage and allow interaction with digital video and software applications on a standard compact disk (650 MB or 700 MB).

Compact VHS (C-VHS)-Compact video home system (C-VHS) is a smaller version of the standard VHS video tape storage format that is used to store video recording images on 1/2 inch magnetic tape. VHS offers 240 lines of resolution. The C-VHS cassette typically holds 30-40 minutes of analog video.

Companding-Companding is a system that reduces the amount of amplification (gain) of an audio signal for larger input signals (e.g., louder talker). The use of companding allows the level of audio signal that enters the modulator to have a smaller overall range (higher minimum and lower maximum) regardless if some people talk softly or boldly. As a result of companding, high-level signals and low-level signals input to a modulator may have a different conversion level (ratio of modulation compared to input signal level). This can create distortion so companding allows the modulator to convert the information signal (audio signal) with less distortion. Of course, the process of companding must be reversed at the receiving end, called expanding, to recreate the original audio signal.

This figure shows the basic signal companding and expanding process. This diagram shows that the amount of amplifier gain is reduced as the level of input signal is increased. This keeps the input level to the modulator to a relatively small dynamic range. At the receiving end of the system, an expanding system is used to provide additional amplification to the upper end of the output signal. This recreates the shape of the original input audio signal.

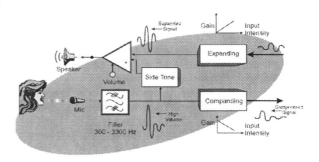

Analog Signal Companding and Expanding Operation

Compatibility-The ability of different systems to exchange information in usable form.

Compensation-(1-signal) The process of passing a signal through an element or circuit with characteristics the reverse of those in the transmission line, so that the net effect is a received signal with an acceptable level/frequency characteristic. (2-financial) The providing of financial or other rewards for performance of services or products.

Competitive Access Providers (CAPs)-A telecommunications service provider that offers competing services to an established (incumbent) telephone service provider. CAPs typically compete with a local exchange carrier (LEC). CAPs can provide service by reselling local service from the LEC.

Competitive Local Exchange Carrier (CLEC)-A telephone service company that provides local telephone service that competes with the incumbent local exchange carrier (ILEC).

Compile-To process a program from one language (usually a high-level language) into a language (machine language) that can be used by a computer.

Compiler-An executable program that converts a specific grammar of a computer programming language into the necessary machine level instructions needed to carry out the operations specified in the statements on a specific type of microprocessor. These grammars are typically called high-level programming languages and the compiler will insure that the language constructs obey the rules specified by the particular grammar. Compilers are also capable of linking together multiple files containing machine level instructions into a single executable program.

Most compilers do much more than this, however; they translate the entire program into machine language, while at the same time, they check your source code syntax for errors and then post error messages or warnings as appropriate.

Completing Field-A term designating those end offices at which a tandem can terminate, that is, complete calls.

Complex Tone-An aural stimulus whose free-air pressure is not sinusoidal over time. The waveform may or may not repeat periodically.

Complex Wave-A waveform consisting of two or more sine wave components. At any instant of time, a complex wave is the algebraic sum of all its sine wave components. Examples include voice and music signals.

Compliant Chip-A compliant chip is an integrated circuit that is validated to conform to a specification.

Compliant Device-A compliant device is a component or assembly that is validated to conform to a specification.

Component-(1-general) An assembly, or part thereof, that is essential to the operation of some larger circuit or system. A component is an immediate subdivision of the assembly to which it belongs. (2-SS7) in the Signaling System 7 protocol, the portion of the Transaction Capabilities Application Part that identifies a component type, provides correlation between components, specifies operations to be performed, and contains the parameters relevant to that operation.

Component Analog Video (CAV)-Signals, which represent the luminance and color information of a video picture. Each signal contains an analog voltage that varies with picture content. CAV also is referred to as analog component.

Component Connections-Component connections are the separate electrical signal connection points on a video capture card that carry the intensity and color components of video signals.

Component Video-Component video consists of three separate primary color signals: red, green, and blue (RGB). The combination of component video can produce any color and intensity of picture information. Some component video systems are converted into a luminance (brightness) signal and two color difference signals.

Component Video Signals-A set of signals, each representing a portion of the information needed to generate a full color image, for example RGB; Y, I, 0: or Y, R-Y, B-Y

Composite-(1-signal) A composite signal is the combination of multiple signals. Composite signals may be created by using a signal multiplexer that adds or multiplies the information from the multiple input signals. (2-video) Refers to a type of video signal or color monitor where color signals are carried by a single input signal and electrically separated (processed) inside the monitor.

Composite Analog-A complex signal (such as video) that is a composite of several analog signals.

Composite Blanking-The complete TV blanking signal, composed of both line rate and field rate blanking signals. (See also: line blanking, field blanking.)

Composite Color Signal-A signal consisting of combined luminance and chrominance information formed through frequency domain multiplexing. NTSC and PAL video signals are of this type.

Composite Signal-Composite signals use multiple information signals to produce a new combined (composite) signal. An example of a composite signal is a composite audio and video television signal.

Composite Sync (CS)-A video synchronizing signal that contains horizontal and vertical synchronizing information. Composite sync often is referred to simply as sync.

Composite Triple Beat-An intermodulation distortion often present in multichannel AM systems such as CATV and MMDS; characterized by horizontal streaking in a television picture.

Composite Video-Composite video is a single electrical signal that contains luminance, color, and synchronization information. NTSC, PAL, and SECAM all are examples of composite video formats.

Compositor-A video compositor is a device or system that can take two or more video inputs or graphic images and combine them into one composite video signal.

Compound Error-Two or more errors on a data element or record.

Compound Mailbox-A compound mailbox is a media storage system or service that holds multiple types of media and messages such as fax, voice mail, e-mail, and video mail.

Compress-(1-general) A process of converting information into a smaller form through the elimination of redundancy or the use of mathematical modeling of the information. (2-digital image) digital picture manipulator effect by which the picture is squeezed or made proportionally smaller.

Compressed Air Canister-A compressed air canister (can) is a container that holds compressed air that can be directed to a particular area through a nozzle or tube. Compressed air canisters are commonly used by technicians to remove unwanted dirt and particles from cables, connectors and electronic assemblies.

Compressed Real-time Transport Protocol (CRTP)-Defined in RFC2508: Compressing IP/UDP/RTP Headers for Low-Speed Serial Links. Compressed RTP indicates that the RTP header is compressed along with the IP and UDP headers. The size of this header may still be just two bytes, or more if differences must be communicated. This packet type is used when the second-order difference (at least in the usually constant fields) is zero. It includes delta encoding for those fields that have changed by other than the expected amount to establish the first-order differences after an uncompressed RTP header is sent and whenever they change.

Compressed Video-Compressed video is a sequence of images or moving picture segment that has been processed using a variety of computer algorithms and other techniques to reduce the amount of data required to accurately represent the content.

Compression-(1-digital) The processing of digital information to a form that reduces the space required for storage. (2-distortion) The altering of signal quality caused by the non-linearly conversion (compression) process in audio and video compression systems.

Compression Algorithm-The program or process that converts information into a smaller form (such as a smaller data file size).

Compression Layer-A compression layer is the portion of a system (such as an MPEG system) that is related to the reduction of data storage or transmission rate of information or data signals.

Compromise Network-A network consisting of resistance and capacitance that is adjusted for an average impedance value for all circuits terminated on a switching system. It is used as a balancing network in conjunction with a hybrid, a test termination, or an idle circuit termination.

Compulsory Licensing-Compulsory licensing is the requirement imposed by a governing body that forces a holder of intellectual property (e.g. a patent) to allow others to use, make or sell a product, service or content. Compulsory licensing usually requires the user of the intellectual property (licensee) to pay the owner (licensor) a reasonable license fee along with non-discriminatory terms.

Computer-A device, circuit, or system that processes information according to a set of stored instructions. Typically, an electronic computer consists of data input and output circuits, a central processing unit, memory storage, and facilities for interaction with a user, such as a keyboard and a visual monitor or readout.

Computer Aided Dispatch (CAD)-A computerized communication system that can coordinate and/or track mobile vehicles. CAD systems can be automated messaging devices to complex computer systems that display maps and vehicle positions on a computer monitor.

Computer Aided Software Engineering (CASE)-A process that uses software tools to assist the development of software programs. The use of CASE tools allows for the efficient use of functional requirements to develop software programs. CASE tools may simply manage the development process, validate software operation, or even automatically produce the code from functional specifications.

Computer Based Training (CBT)-The use of computers to assist the providing of training. Early CBT systems used text based screens and CD-ROM technology to provide for self paced learning experiences. CBT systems commonly provided training in small sections followed by a test section to ensure student learning prior to allowing additional sections to be presented. CBT is sometimes called computer-assisted instruction (CAI).

Computer Graphics-Computer graphics is the creation of images on a display screen or image formats in a file through the processing of information by a computer.

Computer Language-The words and rules of construction for phrases and sentences that direct the operation of a computer.

Computer Peripheral-An auxiliary input, output, or storage device under the control of a computer.

Computer Telephony (CT)-Computer telephony (CT) systems are communication networks that merge computer intelligence with telecommunications devices and technologies.

This figure shows a sample CTI system computer that contains a voice card. This voice card is connected to a multiple channel T1 line. The voice card connects digital PBX stations through the voice card to individual DS0 channels on the T1 line when calls are in progress. Several software programs are installed on this system that provide for call processing, IVR, ACD, voice mail, fax, and email broadcasting. The monitor shows a directory of extensions. The advanced call processing feature shows text names along with the individual extensions to allow callers to automatically search through a company's directory without the need to use an operator.

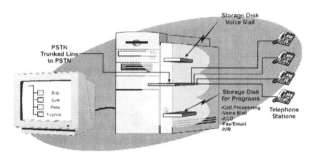

Computer Telephony (CT) System

Computer Telephony Integration (CTI)-CTI is the integration of computer processing systems with telephone technology. Computer telephony provides PBX functions along with advanced call processing and information access services. These services include, pre-paid telephony access control, interactive voice response (IVR), call center management, and private branch exchanges (PBX).

This diagram shows how a telephone system can be integrated with a computer system to provide for advanced call processing and information services. This diagram shows the core of the system is a voice card or PBX switch that is controlled by call processing software. The call processing software is customized with information about the system equipment it is operating with. This diagram shows that the computer telephone integration (CTI) system also interfaces to a company database to allow the call center to receive and update information based on both telephony commands (automatic number identification and user selected options) and customer service representative (CSR) screen commands.

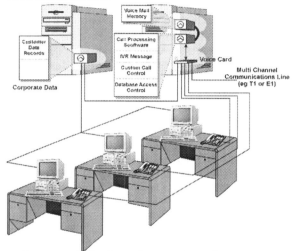

Computer Telephony Integration (CTI) System

Computer Terminal-Computer terminals are input, output, and processing devices that interface human operators to data communication systems. Computer terminals frequently consist of a keyboard, video display monitor, central processor unit (CPU), and communication circuitry that can connect the computer terminal with a data communication network.

The term "computer terminal" is often used to describe multiple types of devices including dedicated data "dumb" terminals, scientific workstations, and other types of computers that can communicate with other computers or a host computer.

Computer-Aided Design/Computer-Aided Manufacture-A design system that uses computers to assist in the creation of diagrams and images.

Computer-Aided Dialing-The use of a computer system to assist in dialing telephone numbers. Computer aided dialing systems are often used for telemarketing services.

COMSAT-Communications Satellite Corporation

Concatenation-Concatenation is the process of joining two or more items together. Concatenation is used to combine multiple communication channels together to form data transmission systems that have higher capacity than the available individual communication channels.

Concentration-The process of combining traffic from many low-usage communication lines to a lesser number of high-usage communication lines (trunks).

Concentrator-In a communication system, a concentrator combines multiple communication sources on to a higher capacity communication path. A concentrator usually provides more efficient communication capability between multiple low-speed channels to one or more high-speed channels.

Concurrent Streams-Concurrent streams are the simultaneous sending or receiving of multiple streams. Concurrent streams may be used to increase the overall data transmission rate where the sources or reception points for each stream may be more limited at one end compared to the other end.

Conditional Access (CA)-Conditional access is a system or service access control process that is used in a communication system (such as a broadcast television system) to limit the ability of users to obtain or use media or services. Conditional access systems can use uniquely identifiable devices (sealed with serial numbers) and may use smart cards to store and access secret codes.

Conditional Access Message (CAM)-Conditional access messages are commands and data that enable the descrambling and decoding of media for one or more users of media. CAMs are composed of control messages, service keys and user keys. For the MPEG systems, there are two types of CAMs; entitlement control messages (ECM) and entitlement management messages (EMM).

Conditional Access Provider-A conditional access provider is a company that provides software, security process and/or services that can be used in a communication system (e.g. broadcast television system to limit the access of media to authorized users.

Conditional Access System (CAS)-A conditional access is a security process that is used in a communication system (such as a broadcast television system) to limit the access of media to authorized users. Conditional access systems can use uniquely identifiable devices (sealed with serial numbers) and may use smart cards to store and access secret codes.

Conditional Access Table (CAT)-A conditional access table holds information that is used by an access device (such as a set top box with a smart card) to decode programs that are part of a conditional access system (e.g. on-demand programs). The CAT table provides the packet identifier (PID) channel code that provides the entitlement management messages (EMM) to the descrambler assembly.

Cone Of Protection-In reference to lightning, the space enclosed by a cone formed with its apex at the highest point of a lightning rod or protecting tower, the diameter of the base of the cone having a definite relationship to the height of the rod or tower. When overhead ground wires are used, the space protected is referred to as a protected zone.

Conference Bridge-(1-telephone) A telecommunications facility or service which permits callers from several diverse locations to be connected together for a conference call. The conference bridge contains electronics for amplifying and balancing the loudness of each speaker in a conference call so everyone can hear each other and speak to each other. Background noises are suppressed and typically only the current two or three loudest speakers' voices are retransmitted to other participants by the bridge, while a speaker's own voice audio is not sent back to that speaker to avoid audio feedback, echo or "squealing" self-oscillation. (2- text) A facility to receive the character codes from the keyboards of multiple participants and retransmit this text to the display or printer of all participants in the conference. First used with electromechanical teletypewriters for private networks of automobile parts suppliers. Modern implementations on the Internet are typically called a "chat room."

Conference Call-The connection of three or more telephones to a telephone conversation.

This diagram shows how a conference call can use a conference bridge to allow several users to effectively communicate in a conference call (3 or more users). This example shows that this conference bridge uses audio level detectors to determine the level of the microphone audio level for each conference call participant that is talking. As a person begins to talk, the conference bridge increases the gain on the microphone and decreases the gain on the speaker line. This process effectively dynamically reduces the background noise from non-participating members while providing good sound quality to participants that are talking.

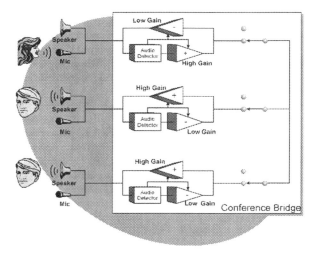

Conference Call Operation

Conference Server-A conference server manages the connection and removal of users for multi-party conference calling (e.g. 3-way calling).

Conferencing-The process of adding additional callers to join into a phone conversation or data conference.

Configuration-(1-system) A relative arrangement of interconnected equipment or software in a system. (2-file) The information used to adapt an equipment or software program to its environment (configuration).

Configuration Management (CM)-Configuration management is one of the five functions defined in the FCAPS model for network management. With configuration management network operation is monitored and controlled. Hardware and programming changes, including the addition of new equipment and programs, modification of existing systems, and removal of obso-

lete systems and programs, are coordinated. An inventory of equipment and programs is kept and updated regularly.

Configuration Message-In the Spanning Tree Protocol, a BPDU that carries the information needed to compute and maintain the spanning tree.

Conformance-(1-general) The ability or certification of a device or system to perform or act as defined to an agreed specification or process. (2-Bluetooth) When conformance to a Bluetooth profile is claimed by a vendor all mandatory capabilities for that profile must be supported in the specified manner (process-mandatory). This also applies for all optional and conditional capabilities for which support is indicated. All mandatory, optional, and conditional capabilities for which support is indicated are subject to verification as part of the Bluetooth certification program. See also Bluetooth Qualification Review Board.

Conformance Points-Conformance points are a combination of profiles and levels in a system (such as an MPEG system) where different products can interoperate (by conforming to that level and profile). An example of conformance points is if ability of a mobile video server to support creation and playback of a simple visual profile at level 0, any mobile phone that has these conformance points should be able to play a video with these profiles and levels.

Conformance Test-Conformance testing is a group of operational and performance tests that help to determine if a product or system satisfies the requirements of an industry standard or product and/or service specification.

Conformance Testing-Conformance testing is the process of testing a particular product or system to determine if it meets specific requirements or if it will be interoperable with other products or systems of the same type. Conformance testing is often performed by an independent company or laboratory to ensure accurate and unbiased results.

Congestion-A condition that exists when the demands for service on a communications network exceed its capacity to deliver that service.

Congestion Notification-Congestion notification is a control flag signaling system that is used in a data transmission system to indicate the status of network congestion. Congestion notification allows data communication devices that are connected to the data network to send or delay packet transmission.

Forward congestion notification indicates to upstream data switching devices in a communication network that data that is being transmitted through congested switches and it is likely that some of the remaining data or packets may be discarded. The upstream switch can then change the discard priority accordingly.

Congestion Time-The time or probability that a system is congested over any time period.

CONN-Connect

CONN ACK-Connect Acknowledgement

Connect (CONN)-In the Integrated Services Digital Network (ISDN) Q.931 protocol, the layer 3 CONNect message that indicates call acceptance by a called party.

Connect Day-The day of the week on which a connection is made.

Connect Time- (1-cellular) That period of time a cellular phone is in radio contact with a cell site. The connect time is not the same as the duration of the conversation. (2 - telco) The local time at a calling party's location when a connection was made.(3 - general) The amount of time that a given circuit is in use.

Connected State-A connected state is the logical or physical connection of a device to a communication circuit. While devices in a connected state are capable of transferring voice or data information, they may not be actually transferring data at all times.

Connecting-A phase in the communication between devices when a connection between them is being establishment phase is completed.

Connecting Link-A channel and/or facility used to connect a communication device (terminal) to a transceiver (transmitter/receiver) or communication device.

Connection Establishment-(1-general) The procedures used to establish a connection between communication devices. This may involve the establishment of physical connections, assignment of logical channels, and the selection and allocation of communication protocols. (2-Bluetooth) A procedure for establishing a connection between the applications on two Bluetooth devices. Connection establishment starts after channel establishment is completed. At that point, the initiating device sends a connection establishment request. The specific request used depends upon the application.

Connection Handle-A connection handle is a temporary name or number that assigned to a communication channel within a system to allow information to transfer through the system using the connection handle name.

Connection Identifier (CID)-A CID is a unique name or number that is used to identify a specific logical connection path in a communication system. This diagram shows how a connection identifier (CID) is used in a data communications network to subdivide frames that are transmitted through a communication network so the portions of data can reach their specific destination channels. This example shows that the CID is used to de-multiplex the frames (divide channels) that pass through an access device. In this example, 4 telephone channels are multiplexed onto frames that are sent through access device 1. These frames are sent on a permanent virtual circuit (PVC) to access device 2. Access device 2 uses the data link connection identifier 7 to route the frames to channel demultiplexer. The channel demultiplexer uses the channel identifier (CID) code to route the data to each specific digital telephone device.

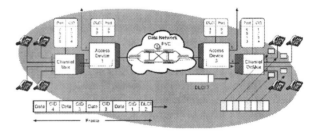

Connection Identifier (CID) Operation

Connection Layer-A connection layer is a processing function that is responsible for sending, processing, and receiving commands that setup, manage, and terminate (end) communication connections.

Connection Oriented Data (COD)-A communication connection that allows the sequential delivery of its component parts (packets or frames). COD assures that a supported application such as voice or video will receive data with a minimum amount of delay.

Connectioniess Network Service-A network service that transfers information between end users without establishing a logical connection or virtual circuit.

Connectionless-A communication connection that allows the delivery of its component parts (packets or frames) by a variety of network connection paths. Connectionless service helps assure that packets of data will arrive even if a portion of the network is disabled or disconnected. This interconnection allows data information to start transferring without first establishing a connection and without immediate acknowledgment of receipt. Sometimes it is (imprecisely) called datagram. Internet protocol is an example of connectionless service.

Connectionless Data Service-Service at a given layer of the OSI Reference Model in which there is no connection setup phase.

Connectionless Protocol-A packet-switching format that requires that each packet in a series carry complete addressing information so that it can find its own way through a network without using logical channels.

Connectionless Switching-Connectionless switching is a process of allowing portions of information (packets) to independently find its way to its destination without the setup of a predetermined route.

Connectionless Transport Protocol (CLTP)-A protocol that manages end-to-end transport data addressing and error correction to allow routing of packets through a network. CLTP is only a routing protocol and does not guarantee the delivery of data. It is the open systems interconnect (OSI) protocol equivalent to user datagram protocol (UDP).

Connection-Oriented-A communications model in which stations establish a connection before proceeding with data exchange and in which the data constitutes a flow that persists over time.

Connector-A connector is a device or assembly that mounted on a cable or device that permits the acoustic, electrical, or optical connection and disconnection to other cables or assemblies.

Connector Block-A block of insulating material supporting terminals that can be used to connect electric conductors.

Consideration-Consideration is the item or value (such as money or services) that is provided for the exchange of assets or rights to the use assets (such as the right to view a movie).

Console-The computer access device that allows an operator to communicate with or control a network. A console commonly consists of a display monitor and keyboard.

Consolidated Billing-Consolidated billing is the process of combining all of the customer's communication charges and credits for multiple types of services (e.g. wireless and data lines) on one bill or invoice.

Consolidated Carrier-A communications service provider that integrates several communication services. Also known as an integrated carrier.

Constant Bit Rate (CBR)-Constant bit rate is a class of telecommunications service that provides an end user with constant bit data transfer rate. CBR service is often used when real time data transfer rate is required such as for voice service.

Constant Voltage Source-A source with low, ideally zero, internal impedance, so that voltage will remain constant, independent of current supplied.

Constant-Current Source-A source with infinitely high output impedance so that output current is independent of voltage, for a specified range of output voltages.

Constant-Voltage Charge-A method of charging a secondary cell or storage battery during which the terminal voltage is kept at a constant value.

Constellation-A method of displaying multiple information parts of an signal. A constellation may display phase and amplitude on an x-y axis. Using the final parameter of time where the signal has a cycle of one rotation around the x-y axis, points on the constellation may display information elements (logic 1's and 0's).

Consultant-A person with special skills. Also a person who borrows your watch, tells you what time it is, then charges you for the time of day. See also Guru.

Consultative Committee for International Telephony And Telegraphy (CCITT)-Original French language name "Comité Consultatif International Télégraphique et Téléphonique" abbreviated CCITT. A part of the United Nations Economic Scientific and Cultural Organization (UNESCO), based in Geneva, that develops worldwide telecommunications standards. Standards (called recommendations) were published every four years in complete sets. Each section of the standards appeared in a book having a letter code such as G for coding/decoding, Q for signaling, I for ISDN, V for modems and multiplexers, etc. Individual standards within each book are designated by a number following the letter, as in Q.931 or I.451 for ISDN call processing signals (appears in two different books for historical reasons), V.32 for certain modems and error correcting codes, X.25 packet networks, T.30 for facsimile, etc. In 1993, after a reorganization, the organization's name was changed to International Telecommunication Union-Telecommunications Sector (ITU-T or just ITU), and even though ITU now creates recommendations and standards, you will still hear the CCITT standards mentioned. ITU standards are still identified by letter-number codes of the same form, but they are no longer published in a full book, nor are they issued on a rigid four year schedule. Instead, each standard appears when a revision is deemed necessary. ITU standards can be purchased as a downloaded document via the ITU Internet wet site www.itu.int

Consumer-A consumer is a person or company who uses a product or service.

Consumer Demand Curve -Economic model that defines the different level of service that an individual customer will demand at a different price/service level.

Consumer Electronics (CE)-Consumer electronics is the devices or products that are purchased or used by the general public (consumers).

Consumer Electronics Bus (CEBus)-A communications transmission system that transfers data on local (typically residential) power lines. The standard for CEBus is EIA IS-60. CEBus works similar to an Ethernet data network system.

Contact Connector-A physical contact (PC) connector is a fiber optic connector that provides physical contact between the fibers at the connection. Contact connectors can have very low loss because of the minimal spacing between fibers results in low loss. Contact connectors can have difficulty in applications that require multiple connections (disconnecting and reconnecting) or in environments with high vibration or mechanical stresses due to the potential of scratching or deforming the fiber ends within the connector.

Contact Noise-A noise resulting from current flow through an electrical contact that has a rapidly varying resistance, as when the contacts are corroded or dirty.

Contact Resistance-The resistance at the surface when two conductors make contact.

Container File-A container file is a digital media file format that holds multiple types of media types. A container file may contain a header (begin-

ning portion) that describes the types of media, their location and their characteristics that are contained within the container file.

Container Format-A container format is the organization of objects (such as digital audio and digital video) that are located within a file or streaming media source.

Contaminated Trunk Group-A trunk group that carries both local exchange and interexchange carrier traffic.

Content-Content is the information contained within a message, call, media program or web site display.

Content Aggregation-Content aggregation is the process of combining multiple content sources for distribution through other communication channels.

Content Aggregator-A content aggregator obtains the rights from multiple content providers to resell and distribute through other communication channels. A content aggregator typically receives and reformats media content, stores or forwards the media content, controls and/or encodes the media for security purposes, accounts for the delivery of media and distributes the media to the systems that sell and provide the media to customers.

Content Control-(1-service) Content control is the process of authorizing and provisioning of access of content to users. (2-interaction) Content control is the ability of a user to review, select and interact with content sources.

Content Conversion-Content conversion is the process of modifying or changing data or media (content) into a different form. An example of content conversion is the changing of an analog videotape format into a digital video disk (DVD) format.

Content Creation-Content creation is the production or origination of information that is contained within a message, call, media program or web site display.

Content Creator-A content creator is a person or company that has created an original form of intellectual property.

Content Delivery-Content delivery is the transfer of content from one location to another location.

Content Delivery Network (CDN)-A content delivery network is a network of service that is designed or used for the distribution of content. CDNs may have redundant links and multiple streaming servers to ensure file downloads occur efficiently and reliably.

C

Content Delivery Platform-A content delivery platform is the combination of system hardware and software that is used to process and distribute content.

Content Delivery System-A content delivery system is the combination of equipment, protocols and transmission lines that are used to transfer content between addressable locations.

Content Developer-A content developer is a person or company who creates images or video.

Content Distribution-Content distribution is the process of transferring content to one or more persons, companies or points.

Content Distributor-A content distributor is a person or company that receives content from one or more companies, stores and manages the content and transfers content to one or more companies.

Content Ingestion-Content ingestion is the process of transferring media into a storage or content management system.

Content License-A content license is a contract that grants specific rights to use of content. An example of a content license is permission to distribute a television program to users in a cable television system.

Content Licensing-Content licensing is the defining, authorizing and compensating for the rights to develop, use or sell content (media or information).

Content Management System (CMS)-A content management system identifies, categorizes, and manages the storage and distribution of content.

Content Owner-A content owner is a person or company that owns the rights to intellectual property (content).

Content Producer-Content producers are companies or developers of media content. Content producers may directly provide distributors or network providers with access to content.

Content Proliferation-Content proliferation is the process of copying and redistribution of content.

Content Protection (CP)-Content Protection is the end-to-end system preventing content from being pirated or tampered with in a communication network (such as in a television system). Content protection involves uniquely identifying content, assigning the usage rights, scrambling and encrypting the digital assets prior to play-out or storage (both in the network or end user devices) as well as the delivery of the accompanying rights to allow legal users to access the content.

Content Protection For Recordable Media (CPRM)-Content protection for recordable media is a specification that defines the hardware and software that can be used to identify and protect the usage rights for media that can be transferred and stored. The CPRM specification includes a cryptographic process (media encryption) that is renewable.

Content Provider-A content provider is a person or company that provides content or intellectual property to distributors or users of content.

Content Repository-A content repository is a storage system that holds content. Content repositories are commonly located inside a network that is protected from unauthorized users.

Content Rights-Content rights are the authorized uses and the allowable distribution methods that can be used for content (typically data or media).

Content Rules-Content rules are the properties or access right assignments associated with files, directories, or communication services that define the ability of users to change, interact or access content.

Content Scamble System (CSS)-A content scrambling system is a method of encryption that is used for stored media (e.g. DVDs) to prevent the use or copying of the media information. To decode and play the information, a player with a decoding algorithm must be used. The CSS process is designed to prevent access to the media from devices that do not have the appropriate decoding algorithm so the media cannot be copied or manipulated by unauthorized users and devices (e.g. via DVD burners).

Content Scheduling-Content scheduling is the process of selecting content and establishing a time that the content will be played, transferred, made available (accessible) or streamed.

Content Scrambling System (CSS)-Content scrambling system is a content protection system that combines hardware and software programs to prevent unauthorized use of digital versatile discs (DVDs).

Content Segment-A content segment is a portion of content or media such as a television program, show or event.

Content Segment License (CSL)-A content segment license is a command or message that authorizes the decoding and usage rights (such as viewing) of a content segment (such as a television program).

Content Server-A content server is a computer system that provides content or media to devices that are connected to a communication system (such as through the Internet). The content servers' many function is to process requests for media content, setup a connection to the requesting device and to manage media transfer during the communication session.

Content Source-A content source is the system or provider of content. Examples of content sources include satellite systems, video servers and live network feeds.

Content Tracking-Content tracking is the monitoring or following of the flow or usage of information content.

Content User-A content user is a person, company, or group that receives, processes or takes some form of action on information services or intellectual property (content).

Content Workflow-Content workflow is the process of assigning, creating and managing the creation or distribution of content.

Content Wrapper-A content wrapper is data or information that is added to media (such as a video program). Content wrappers may provide descriptive and content protection information.

Contention-A condition that exists when two or more devices attempt to transmit at the same time using a shared channel.

Contention Based Access Control-Contention based access control is the independent operation (distributed access control) of communication devices (stations). In contention-based system, communication devices randomly request service from channels within a communication system. Because communication requests occur randomly, two or more communication devices may request service simultaneously. The access control portion of a contention-based session usually involves requiring the communication device to sense for activity before transmitting and listen for message collisions after its service request. If the requesting device does not hear a response to its request, it will wait a random amount of time before repeating the access attempt. The amount of random time waited between retransmission requests increases each time a collision occurs.

Contention Control-A system for allocating channels within a communication system (such as a Ethernet data network or cellular telephone system) where requests for service from two or more communication devices may occur simultaneously. The control portion usually involves the sensing of collisions and assignment of random delays for repeated access attempts.

Contention Free Access Control-Contention free access control is the coordinated operation (assigned access control - Infrastructure Mode) of communication devices (stations). In contention free systems, communication devices wait until they receive a polling message or for their assigned transmission period before they transmit any information. Because a master host coordinates the transmission of all the devices within its networks, no device will transmit at the same time (contention free). The access control portion of a contention free session usually involves requiring the communication agreeing to listen to a single host before transmitting any data. To confirm transmitted data has been successfully received; the polling message will usually include information about the status of packets that have been received. If the sending device does not receive a confirmation of transmission in the polling message, it will retransmit the data again after it receives another polling message.

Contention Interval-Contention interval is the amount of time that is assigned before a communication device is allowed to attempt access again in a random access contention based communication system (such as a mobile telephone or Ethernet system) after an access transmission attempt has failed.

Contingency Planning-Planning of a course of action such that the plan will be invoked only if the contingency materializes.

Contingent Compensation-Dependent on event(s) where as goals of a film are reached, the sum that is due to be paid, or that has been paid.

Continuation Application-An application for patent claiming priority to a pending application for patent which is identical to the co-pending application, but claiming a different invention.

Continuation-in-Part Application-An application for patent claiming priority to a pending application for patent to which new matter (such as an improvement to the invention) has been included.

Continuity-(1-circuit) A continuous path for the flow of current in an electric circuit.

Continuity Check-A physical electrical check made to a circuit or circuits requested by a network (such as the SS7 network) to verify that an acceptable path (for data, speech, etc.) exists for transmission.

Continuity Test-A continuity test is a quick, general test that is used to measure the continuous path of a cable by using a ohm meter, circuit tester or tone and probe kit. For copper cables, the leads of an ohm meter are connected both conductors to confirm a low or near zero ohms reading. In addition both individual conductors should be tested in relation to a ground source. Simply attach one meter lead to electrical ground and the other to either the tip or ring side. A high or infinite ohms (resistance) reading should be achieved.

Continuous DTMF-Continuous Dual Tone Multi-Frequency

Continuous Dual Tone Multi-Frequency (Continuous DTMF)-A process for some telephones (especially mobile telephones and PBX telephones) that allows a dual tone multifrequency (DTMF) signal to be sent as long as the key is depressed. Some telephones send DTMF for a pre-defined time limit (such as 100 msec). Short DTMF tones may not allow access to services such as voice mail and answering machines. However, long DTMF tones that are sent over a poor communication line (such as an analog radio channel) may be recognized as multiple digits if temporary lapses or distortion occurs during DTMF transmission.

Continuous Transmission-A mode of operation where the mobile does not cycle its power level down when the modulating signal amplitude is low or off.

Continuous Wave (CW)-An electromagnetic signal in which successive oscillations of the waves are identical and have a constant or unvarying frequency and amplitude. Continuous wave usually refers to the output of a device (e.g., an optical fiber laser) which is turned on, but which is not modulated with a signal. Historically, in the first half of the 20th century, this term was used in radio to describe a radiotelegraph signal typically produced by an electronic (vacuum tube) oscillator, in distinction to the irregular, non-constant amplitude and non-constant frequency signal produced by an electric spark or arc oscillator.

Continuous Wave Laser-A laser in which the coherent light beam is generated continuously, as is normally needed for optical fiber communications systems.

Continuously Variable Slope Delta (CVSD)-Continuously variable slope delta is a form of analog to digital conversion coding which converts analog signals into digital signal that represent the change (slope) of the analog signal. The conversion process involves sampling the analog signal and creating digital pulses that indicate which way the analog signal is changing. The same consecutive digital signals (digital levels) progressively increase or decrease the approximated signal.

Contrast-The range of light-to-dark values of an image that are proportional to the voltage difference between black and white voltage levels of the video signal. The contrast control adjusts video gain (white bar, white reference).

Contrast Ratio-In video applications, the ratio of the highlight output level divided by the low light output level. In theory the contrast ratio of a TV system should measure as much as 300:1. In reality, there are several limitations, including the CRT itself. Light from adjacent elements contaminate the area of each element, and room ambient light further contaminates the image emitted from the CRT. Well-controlled viewing conditions should yield a practical contrast ratio of 30:1 to 50:1.

Control-(1-general) The supervision that an operator or device exercises over a circuit or system. (2-comparison) A comparison of record counts or attribute totals against invoice-type data.

Control Bits-Bits contained within a sequence of data bits ("bit stream") that are used to control the transmission of information. Control bits are non-data transmission overhead bits.

Control Bus-In routing switchers, the interconnecting communications path between control panels or devices and the routing matrices.

Control Busy Hour-In a telecommunications network, the hour with the highest usage. (See also: busy hour.)

Control Center-A centralized work location from which managers and support staff administer the bulk of central office daily work requests. Examples include network terminal equipment centers, switching control centers, and frame control centers.

Control Channel-A communication channel in system (such as a radio control channel) that is dedicated to the sending and/or receiving of controlling messages between devices (such as a base station and a mobile radio). On a mobile radio system, the control channel sends messages that include paging (alerting), access control (channel assignment) and system broadcast information (access parameters and system identification).

Control Channel Segmentation and Reassembly Layer (CCSRL)-Control channel segmentation and reassembly layer (CCSRL) is a process of dividing blocks of control information into smaller size blocks (segmentation) and reassembling these smaller blocks into their original large block form when they are received.

Control Channels-An allocation of frequencies in a communication system (such as a cellular telephone network) that are used to convey control information rather than voice traffic.

Control Character-A character in a computer program whose occurrence in a particular context initiates, modifies, or stops an action that affects the recording, processing, or interpretation of data.

Control Keys-Keys on a computer keyboard that, when pressed simultaneously, invoke control functions, for example, the use of "Control-D" to terminate a computer work session. One of the keys generally is marked Control or CTRL, and the other is an alphabetic character.

Control Mobile Attenuation Code (CMAC)-Used in a message sent to the mobile from the base which assigns the mobile to an absolute (specified) power level. This is important in small diameter cells where the mobile must access the system at a low power level to prevent co-channel interfering with other control channels.

Control Office-An exchange carrier center or office responsible for the installation and maintenance of a given access service furnished to an access customer.

Control Packet-A packet of data that is sent in a communication system that is used to control the transmission (link layer) or higher- level communication layer.

Control Plane-The portion of a network system that is involved in the setup, management, and termination of communication sessions and services.

Control Unit-A part of the microprocessor whose function is to read and decode instructions from the memory.

Control Word-A control word is a message or command that initializes, changes or terminates (ends) a process.

Controlled Environmental Vault (CEV)-A controlled environmental vault (CEV) is an underground room that is used to contain electronic and/or optical equipment under controlled thermal and humidity conditions.

Controlled Load Service-Controlled-load service provides a variable bandwidth for each communication session that varies based on factors including the amount of network activity (e.g. heavy traffic) and quality of service requirements (e.g. real-time compared to non-real time communication application).

Controlled Rerouting-In the Signaling System 7 protocol, the controlled transfer of signaling traffic from an alternative signaling route to the normal signaling route, when it becomes available.

Controller/Sequencer-An electronic assembly that receives and executes instructions from memory to allow the coordination of other electronic assemblies or processing communications data flow.

Controlling Rights-Controlling rights are authorized terms that allow a person or company to control the delivery or use of services, products or intellectual property (content).

Convention-A generally acceptable symbol, sign, or practice in a given industry.

Convergence Billing-An all-inclusive bill that combines charges for: local, long distance, data, Internet, and possibly utilities (water, gas, electric) and cable TV. Convergent billing generally implies one all-encompassing view of the Customer for all subscribed services, and a unified view of the product portfolio to enable cross-product packages.

Convergence Time-The amount of time it takes an echo canceller to train (learn) how to cancel echoes from the incoming audio signal.

Convergent-A process of integrating services such as local telephone service, mobile communications, cable television, and long distance together onto a single system. These services and/or systems may be truly integrated or loosely tied together (sometimes called "stapled") so allow a customer to have a single billing and customer care access point.

Convergent Billing-Convergent billing is the combining of billing information for multiple types of services such as television, telephone service and data communication services. Convergent billing or systems may be tightly integrated or loosely tied together (sometimes called "stapled") so allow a customer to have a single billing and customer care access point.

Convergent Market Cross Mapping -Technique used to compare the populations of customers "owned" by different product line managers, in order to determine the potential value that sharing of those customers can have for each product manager.

Convergent Network-A network incorporating wireless, optical and copper transmission media and multiple protocols. In the past, networks have often been built for a specific purpose and/or based on a specific technology. In the future, more and more interconnections among networks are anticipated to maximize communications options.

Conversation-As used in Link Aggregation, a set of traffic among which ordering must be maintained.

Conversation Class-Conversation class is the providing of communication service (typically voice) through a network with minimal delay in two directions. While conversation has stringent maximum time delay limits (typically tens of milliseconds), it is typically acceptable to loose some data during transmission due to errors or discarding packets during system overcapacity.

Conversation Minute Miles-The product of the total number of message minutes carried on a trunk or circuit group and the average route miles of the trunk or circuit group.

Conversation Minutes-The actual time during which a customer uses a connection for voice transmission.

Conversation State-A state (mode) of a mobile telephone phone when it tuned to a voice or traffic channel and is providing a voice path for a call. The mobile telephone may provide additional information (such as MAHO0 during the conversation state. When the conversation ends, the phone will enter the control channel scanning and locking state.

Conversion Factor-A factor applied to known data to adjust it and make it appear as other data. For example, reported minutes can be converted to conversation minutes through the use of a conversion factor.

Conversion Rate (CR)-(1-Internet Marketing) Conversion rate is a measure of the people who log on to a web page and select a process or purchase via that web site. A high conversion rate percentage usually indicates how much more valuable a web site is to its visitors.

Conversion Timing-An agreed-upon scheduled date and time to begin a cutover in a coordinated conversion.

Convolutional Coding-Convolutional coding is an error correction process that uses the input data to create a continuous flow of error-protected bits. As these bits are input to the convolutional coder,

an increased number of bits are produced. Convolutional coding is often used in transmission systems that often experience burst errors such as wireless systems. Convolutional coding systems are represented by the ratio (rate) of input bits to output bits. A ½ rate convolutional coder creates (outputs) 2 bits for each 1 bit (input) it receives.

This figure shows a convolutional coder that continuously receives the data bits in sequence to create a new digital signal that combines both the original information and new error protection bits. This example shows a 1/2 rate convolutional coder that generates two bits for every one that enters.

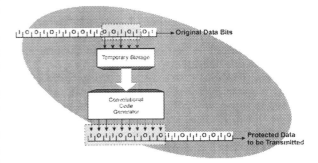

Convolutional Coding

Cookie-A cookie is a small amount of information that is stored on a web user's computer (a client) that is used by a web site (web server) to help control the content and format of information to the user during future visits to the web site.

Cooling Stand-A cooling stand is a device or fixture that is used to hold hot devices, connectors or parts to allow them to cool before handling them.

Co-Op-Co-Operative Advertising

Co-Operative Advertising (Co-Op)-An amount of funds, percentage of sales, or marketing allowance that is provided to a distributor or retailer for their advertising of specific products or services. To receive the co-operative advertising funds, the distributor or retailer may be required to provide proof that the advertising was performed and paid for.

Coordinated Restoral-An equipment or service restoral process that is coordinated. Coordinated restoral systems can sequentially or simultaneously restore equipment and or services.

Coordinated Universal Time (UTC)-Coordinated universal time is a reference time scale that is maintained by the Bureau International de l'Heure (BIR). UTC is commonly used as a basis of a coordinated dissemination of standard frequencies and time signals used in communication systems.

COP3-Code of Practice #3

COPP-Certified Output Protection Protocol

Copper Cross Connect-A copper cross connect system allows access lines (copper lines) to be connected to several different DSL modems. There are two key reasons to use a copper cross connect system. The first reason is to allow a copper wire access line to be connected to different digital subscriber line modems. This could be because the customer may upgrade to a new type of modem (e.g. ADSL to VDSL) or if a DSL modem fails, a spare DSL modem could be connected to the customer's line. The second reason is to allow the access line to be connected to a DSL modem only when a connection is required. This would allow a DSL service provider to install lesser number of modems in a system than they have customers for.

Copper Distribution Data Interconnection (CDDI)-CDDI is an adaptation of the FDDI protocol for use on copper cable.

COPS-Common Open Policy Service

Copy Port-Synonymous with Mirror Port.

Copy Protection-Copy protection is the use of technologies and/or processes that are developed to prevent or reduce the abiltiy to copy data or media.

Copyright-A copyright is a monopoly which may be claimed for a limited period of time by the author of an original work of literature, art, music, drama, or any other form of expression - published or unpublished. Copyrights are Intellectual Property Rights which give the owner, or assignee, the right to prevent others from reproducing the work or derivates, including reproducing, copying, performing, or otherwise distributing the work. Copyrights can, in many countries be claimed without registration. Most countries, or regions, however, have a copyright office where copyrights can be officially registered.

Copyright Clearance Center (CCC)-The copyright clearance center (www.Copyright.com) is a licensing system that is used to obtain permissions for the use of copyrighted materials and to collect and distribute royalties to the owners of the copyrighted materials.

Cordless-Cordless systems are short range wireless telephone systems that are primarily used in residential applications. Cordless telephones typically use radio transmitters that have a maximum power level below 10 milliwatts (0.01 Watts). This limits their usable range of a 100 meters or less.

Cordless Telephone System-A system consisting of two transceivers, one a base station that connects to the public switched telephone network and the other a mobile handset unit that communicates directly with the base station. Transmissions from the mobile unit are received by the base station and then placed on the public switched telephone network. Information received from the switched telephone network is transmitted by the base station to the mobile unit.

This figure shows the basic operation of cordless telephones. This diagram shows that initiating (dialing) a call on a cordless telephone typically starts with the cordless telephone scanning the available channels to determine if they are busy before initiating a call. After the cordless phone has found an unused radio channel, it will send an access request message to the cordless base station. The cordless base station continuously scans all of the available channels to determine if it is receiving a request from its cordless handset(s). After the cordless base station has received the request message (or an indication of a call request), it will

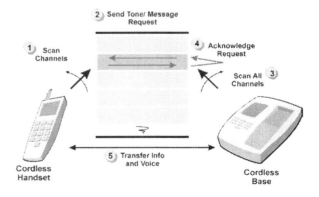

Cordless Telephone Operation

respond to the cordless handset with a message or tone of its own. This allows the cordless handset and base station to determine which radio channel to use during the call. This example shows that additional information may be exchanged on this radio channel to continue the setup (such as sending the dialed digits) and management of the call (such as sending special features such as call hold).

Cordless Telephony, Second Generation (CT2)-Cordless telephony generation 2 (CT2) is an industry standard that allows cordless telephone technology to access small coverage area public radio systems in places such as airports and railway stations. The CT2 system does not normally allow the receiving of calls in a public area.

This diagram shows that a CT-2 system includes radio devices (cordless portable part - CPP) and radio base stations (cordless fixed part - CFP). The CT-2 system radio channel has a 100 kHz bandwidth with a gross data transfer rate of 72 kbps. The radio channel is divided into 2 msec frames and each frame is divided into 1 msec transmit and a 1 msec time slots. The CT-2 system uses time division duplex (TDD) multiplexing so that one slot is used in the forward direction and one slot is used in the reverse direction to provide full duplex (simultaneous) voice communication.

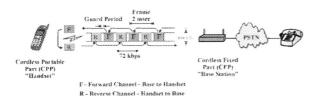

Cordless Telephony, Second Generation (CT2) Operation

Cordless Telephony, Third Generation (CT3)-Cordless telephony generation 3 (CT3) generally refers to an early version of digital European cordless telephony (DCT) that was developed to operate in the 900 MHz frequency band. CT3 allows for both initiating and receiving of calls in public areas.

Core-(1-general) The central part of a device or system. (2-magnetic) A magnetic core is a memory system that uses magnetic field levels to store information. (3-optical) Fiber optic cores are the portion of an optical cable that conducts (transfers) light from the entry to exit points.

Core Area-The urban center of a municipality such as the downtown region. Core area also can refer to the portion of an urban wire-center area characterized by its large buildings.

Core Based Trees (CBT)-Core based trees are a distribution structure for sparse mode multicasting in a data network. For CBT systems, new group members send join messages to a designated core router. Any routers that the join message passes along the way also begin to relay the multicast. This builds a tree up as the packets move to the core.

Core Metadata-Core metadata is information that describes the core (identifying and physical) information about media or data. Examples of core metadata include the title, author and color.

Core Network (CN)-The core network is the central network portion of a communication system. The core network primarily provides interconnection and transfer between edge networks.

Core Size-A primary specification for an optical fiber, stated in microns. The core size does not include cladding. The core size determines the end surface area that accepts and transmits light.

Core Specification-A core specification is a document or set of documents that are intended primarily to define the operation and essential technical requirements for items, materials, or services.

Core Switch-A backbone switch that interconnects to other core switches and edge switches. Core switches maintain information about virtual paths that are connected through the network.

Corner Pullbox-A corner pullbox is a rectangular or square steel box installed at a location where two conduits or ducts meet at a 90 degree angle. The use of a corner pullbox removes a 90 degree bend in a cable or duct run and providers for an interim pulling location to minimize the length of cable that is pulled at one time. This reduces the tensile load stress on the cable.

Corporate LAN-A corporate LAN is a private data communication network that uses high-speed digital communications channels for the interconnection of computers and related equipment in geographic areas that are managed by a corporation.

Corporation For Open Systems International-A user and vendor forum of computer and communications companies whose goal is to ensure a selection of interoperable, multi-vendor products and services operating under Open Systems Interconnection (OSI), Integrated Services Digital Network (ISDN), and related international standards.

CoS-Class of Service

COSMOS-Computer System For Mainframe Operations

Cost Of Clearing Blockage-In multiple outside plant, the average cost of rearranging a network to fulfill an inward service order each time a blockage occurs.

Cost of Ownership-Cost of ownership is the acquisition cost (purchase price) of hardware and software along with the costs for operation and maintenance.

Cost Of Service-The total cost of providing utility service and includes operating expenses, depreciation, taxes, and a rate of return adequate to service investment capital.

Cost Per Plant Unit-An average cost for performing work on a plant unit within a plant classification, such as placing a telephone pole.

Cost Per Thousand (CPM)-The cost for each thousand units associated with media such as TV time, radio spots, Internet ad impressions, print ads, etc. For example, a television ad might sell for $200 CPM. An ad for television show with 2,000,000 viewers will therefore cost $200 x 2,000 or $400,000 dollars.

Cost Per Work Unit-An average cost for performing a work unit within a plant classification, such as placing a telephone pole.

Costa Model -The Costa Model is a representation used to help define the way that telephone companies (telcos) use different departments and functions to optimally attract and keep customers. Key layers of the Costa Model include awareness, familiarity, attraction, preference and selection (sales).

Cost-Based Pricing-A method of setting rates for telephone services based on the costs for providing those services to the exchange carrier.

COT-Customer Originated Trace

Couch Potato-A couch potato is a person who spends an above average amount of time on a couch watching television programming.

Country Code-A 1-, 2-, or 3-digit number that identifies a country or numbering plan to which international calls are routed. The first digit is always a world zone number. Additional digits define a specific geographic area, usually a specific country.

Couple-The process of linking two circuits by inductance, so that energy is transferred from one circuit to another.

Coupler-A device that is used to transfer energy from one transmission medium to another transmission medium. A coupler is typically used to provide a sample of a transmission signal to another device. A coupler may also be used to insert a signal from one transmission line to another transmission line.

Coupling-Coupling is the transferring of a signal or energy between components or systems. Coupling may be performed by direct connection or through other forms of transference such as through the capacitance between adjacent assemblies.

Coupling Factor-A figure representing the combined effect of the various couplings between a power circuit and a communications circuit in producing noise in the communications circuit.

Coupling Loss-Coupling loss is the power loss that results when a signal from a source (such as an input line) is coupled (transferred) to another point (such as a coupled or adjacent output line). An example of coupling loss is the 10% reduction in power (from input to output port) that results from a 10 dB coupling device where the coupled output port is 10dB (90%) below the input line.

Cover Shots-Shots used to replace unfit footage in a film, used to meet necessary censorship requirements.

Coverage-(1-general) The geographical area over which the signal strength of a given radio frequency is available for service. (2-broadcasting) The "footprint" of a broadcast station's signal, usually measured in terms of percentage of homes covered in a designated geographic area or in terms of the signal strength. (See also: penetration.)

Coverage Area-A geographical area that has a sufficient level of radio signal strength from a transmitting tower to provide an acceptable level of signal reception. An acceptable level of signal reception may be determined by signal to noise ratio (for analog systems) or bit error rate (for digital systems).

Coverage Holes-Coverage holes are portions of radio coverage in a wireless system where the customer cannot receive radio signals due to low radio signal levels or distorted radio signals. Coverage holes are commonly caused by multipath fading where multiple signals are combined in such a way that the signal levels cancel causing reduced signal levels.

Covert-Adjective used to describe undercover operations by government agents. "Covert" communications are generally encrypted.

COW-Cell Site On Wheels

CP-Content Protection

CPA-Combined Paging And Access

CPA-Cost Per Acuisition

CPAGCH-Compact Packet Access Grant Channel

CPC-Cost Per Click

CPCCCH-Compact Packet Common Control Channel

CPCCH-Compact Packet Paging Channel

CPCH-Common Packet Channel

CPE-Customer Premises Equipment

CPI-Computer to PBX Interface

CPICH-Common Pilot Channel

CPL-Call Processing Language

CPM-Call Progress Message

CPM-Cost Per Thousand

CPN-Calling Party Number

CPNI-Customer Proprietary Network Information

CPP-Calling Party Pays

CPRM-Content Protection For Recordable Media

CPS-Calls Per Second

CPU-Central Processing Unit

CPU Space-A protected memory space addressed only by the CPU itself. The CPU space is used for processor internal functions or vectored exception processing.

CQM-Channel Quality Message

CR-Call Rate

CR-Conversion Rate

Crack Program-A crack program is a software application that can be used to defeat or change the security settings of another program or service.

Cracking-Cracking is the process of attempting to obtain access to a computer or information system without authorization of the owner or operator of the system.

Cramming-Cramming is the fraudulent addition of charges for services that were not agreed to by the end customer.

Crapplet-A Java applet that does not operate as desired.

Crash-A complete or partial failure of a hardware device or a software program or operation.

Crawl-Crawling is the movement of text across a video display.

CRC-Cyclic Redundancy Check

CRC-Cyclical Redundancy Check

Credit Score-A measure of a customer's credit worthiness that is based on information obtained from an independent credit bureau. Sometimes also termed "External Score".

Crimp Ring-A crimp ring is a circular band, usually constructed of metal, that can be compressed (crimped) to hold a pipe or assembly to a fitting or mounting device.

Crimp Sleeve-A crimp sleeve or crimp ring is a circular band, usually constructed of metal that can be compressed (crimped) to hold a connector assembly to a cable jacket.

Crimp Termination-A crimp termination is a connector or terminal device that is attached to a conductor or fiber by applying pressure (crimping) of a lug or connector onto a wire or fiber cable.

Crimp Tool (Crimper)-A crimp tool (crimper) is a tool that can compress a connector or crimp tube that it typically used to hold a connector to a cable end.

Crimper-Crimp Tool

Critical Report Date-(1-testing) The date on which point-to-point testing is completed. (2-service order) The date on which implementation groups must report that all documents and material have been received to carry out a service order.

Critical Section-The feeder section with the shortest lifetime in a route.

CRM-Cell Rate Margin

CRM-Customer Relationship Management

CRNC-Controlling RNC

Cross Bar Switch-A cross bar switch is a switching system that uses mechanical connections (switching relays) to connect input and output ports (lines) to connected telephone calls.

Cross Border Compatilbity-Cross border compatibility is the ability of communication systems and protocols of one system to be able to communicate and control communications and protocols of another system.

Cross Color-Cross color is the unwanted process of the luminance (intensity) portion of a signal influencing or contributing to the color signals.

This can often be seen around high-intensity white images that display color artifacts (e.g. fine color lines).

Cross Compiler-A compiler that runs on a host computer and outputs machine level instructions and generates executable programs for a target microprocessor / computer that is different than that running on the host. This is typically done to support new architectures or smaller micro-controller based systems that have insufficient resources to host their own compilation software.

Cross Connect Switch-A switch (typically telephone channel) that connects pre-designated channels in different links on a semi-permanent basis. Cross connect switches do NOT set up a different switched channel path for each individual call dialed by the originating subscriber.

Cross Luminance-Cross luminance is the unwanted process of the color portion of a signal contributing to the luminance (intensity) signal. This can often be seen around color saturated images vertical edges.

Cross Modulation-The interference experienced when a carrier signal becomes modulated by an unwanted signal, as well as being modulated by its desired signal.

Cross Polarization-Cross polarization is the electromagnetic relationship between two radio waves where one is polarized horizontally and the other is polarized vertically.

Cross Sell-Marketing activity that is designed to encourage customers to buy different products from the same company. In telecommunications, cross sell could include selling wireless or ISP service to wireline customers.

Cross Talk-A problem where the audio from one communications channel is imposed on another channel.

Crossover Cable-Another name for a null modem cable. A cross over cable reverses the transmit and receive communication lines to allow the direct connection of computers without the need for hubs, switches, or routers.

Crosstalk-Crosstalk is the undesired leakage of a signal from one communications channel to another.

This diagram shows that crosstalk occurs when signal energy from one transmission line is transferred (coupled) to the other. This diagram shows that crosstalk can be divided into two categories: near end crosstalk (NEXT) and far end crosstalk

(FEXT). NEXT results when some of the energy that is transmitted in the desired direction seeps into one (or more) adjacent communication lines from the originating source. FEXT occurs when some of the digital signal energy leaks from one twisted pair and is coupled back to a communications line that is transferring a signal in the opposite direction. Generally, NEXT is more serious than FEXT as the signal interference levels from NEXT are higher.

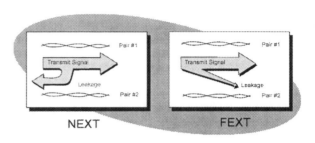

Operation Crosstalk Operation

Crowbar-A short-circuit or low-resistance path placed across the input to a circuit.

CRT-Cathode Ray Tube

CRTP-Compressed Real-time Transport Protocol

Crush Rating-Crush rating is the specification of the maximum amount of force or pressure that a cable or assembly can experience before it may begin to experience a reduction in physical size or distort in shape potentially causing damage to its inner components.

Crush Resistance-Crush resistance is the maximum amount of force or pressure that a cable or assembly can experience before it may begin to experience a reduction in physical size or distort in shape potentially causing damage to its inner components.

Cryptanalysis-Cryptanalysis is a process of decoding an encoded message without knowing the value of the key(s) used in the encryption process.

Crypto-Crypto is a term that is used to describe the encryption or scrambling of data or information.

Crypto Period-A crypto period is a period of time that content has a particular encryption code. Crypto periods can be relatively short (in seconds) and a content segment (such as a movie) may be divided into many crypto periods.

Cryptography-Cryptography is a process that data manipulation that protects the identity, integrity, and privacy of information that is transferred through a communication system. Cryptography uses keys and mathematical formulas to convert information into a form that requires the recipient of the information to use a key to decode the information back to its original form. Encrypted information cannot be decoded (or not easily decoded) by a recipient of the information without the key.

Cryptology-The science of hidden, disguised, or encrypted communications. Cryptology embraces communications security and communications intelligence.

Crystal-(1-general) A solidified form of a substance that has atoms and molecules arranged in a symmetrical pattern. (2-piezoelectric) A crystal, such as quartz, that will generate a voltage when excited. (3-quartz) A piezoelectric crystal cut from natural quartz. (4 - X-cut) A crystal with its major flat surfaces cut so that they are perpendicular to the electrical (X) axis of the original quartz crystal.(5-XY-cut) A cut crystal that has characteristics similar to those of the X-cut and the Y-cut crystals. (6-Y-cut) A crystal with its major flat surfaces cut so that they are perpendicular to the mechanical (y) axis of the original quartz crystal.

Crystal Controlled Oscillator-An oscillator that couples a crystal's piezoelectric-effect to a tuned oscillator circuit so the crystal frequency adjusts (pulls) the oscillator frequency to its the crystals natural resonant frequency.

Crystal Filter-A filter that uses piezoelectric crystals to create resonant or anti-resonant circuits that allow desired frequencies (signals) to pass through.

CS-Code Schemes

CS-Composite Sync

CSA-Common Scrambling Algorithm

CSC-Customer Service Center

CSE-Cellular Signal Enhancer

Cseq-Command Sequence

CSL-Content Segment License

CSMA-Carrier Sense Multiple Access

CSMA/CA-Carrier Sense Multiple Access/Collision Avoidance

CSMA/CD-Carrier Sense Multiple Access/Collision Detection

CSO-Chemical Safety Officer

CSR-Carrier Sensitive Routing

CSR-Customer Service Record

CSR-Customer Service Representative

CSS-Cascading Style Sheets

CSS-Cellular Subscriber Station

CSS-Content Scramble System

CSS-Content Scrambling System

CSTD-Cable System Terminal Device

CSU-Channel Service Unit

CT-Computer Telephony

CT0-Cordless Telephony Generation 0

CT1-Cordless Telephony Generation 1

CT2-Cordless Telephony, Second Generation

CT2+-Cordless Telephony, Second Generation, Enhanced

CT3-Cordless Telephony, Third Generation

CTC-Click to Call

CTCH-Common Traffic Channel

CTCSS-Continuous Tone-Controlled Squelch System

CTI-Computer Telephony Integration

CTIA-Cellular Telecommunications Industry Association

CTR-Click Through Rate

CTS-Clear To Send

CTTU-Centralized Trunk Test Unit

Cue Tones-Cue tones are signals that are embedded within media or sent along with the media that indicate an action or event is about to happen. Cue tones can be a simple event signal or they can contain additional information about the event that is about to occur. An example of a cue tone is a signal in a television program that indicates that a time period for a commercial will occur and how long the time period will last.

CUG-Closed User Group

Curie Point-The temperature at which piezoelectric properties cease.

Curing Threshold-The amount that a customer is required to pay in order for the collections process to be suspended. This amount is not necessarily the full amount due to-date and is dependent on: the type of customer, collectible amount, and behavior score.

Current Amplifier-A low output impedance amplifier capable of providing high current output.

Current Carrying Capacity-A measure of the maximum current that can be carried continuously without damage to components or devices in a circuit.

Cursor-A cursor is a small icon or spot that appears on a display. The cursor has a particular image display location associated with it (horizontal and vertical position). This location may be used to select items on a screen or to be used as a reference point (mark a spot) to be used for future reference (such as length calculations between cursor marks).

Custom Call Routing-Custom call routing is the processing and redirecting of calls based on the specific needs or desires of a user or application (such as a call center).

Custom Calling Services-Custom local area signaling services (CLASS) are telephone service features available in a local access and transport area (LATA) that are primarily based on information that can be processed inside the telephone network. CLASS features include call forwarding, caller identification, and three-way calling.

Custom Local Access Signaling Services (CLASS)-A set of telephone services and enhanced features available in a local access customers that may include calling number delivery or calling name delivery (CND), message waiting, and other features.

Customer-An entity that has the ultimate responsibility for: signing contracts, making payments, allowing use of the services. Sometimes, this term also refers to prospects, or potential prospects.

Customer Account Record Exchange (CARE)-Procedure used for the exchange of customer records between the local exchange carrier (LEC) and the long distance (LD) carrier primary interexchange carrier selected (PIC'ed) by the customer.

Customer Acquisition Cost-Customer acquisition cost is the average cost to a carrier of signing up an individual subscriber. Some of the factors included in the cost of acquiring customers include handset subsidies, activation commissions, sales, marketing, advertising, and promotional campaigns.

Customer Care And Billing System (CCBS)-A system that provides customer account tracking, service feature selection, billing rates, invoicing and details.

Customer Control And Management-The capability of a customer to remotely reconfigure special-service circuits using control circuits, software, and remotely configurable network elements.

Customer Data Rate-The data transmission rate that the customer receives.

Customer Desired Due Date (CDDD)-The date the customer desires delivery or operation of a product or service.

Customer Group-A common term for a group of customers with similar characteristics (e.g. business customers.)

Customer Interconnection Record-A record that is created at a control point for inter-connection.

Customer Loop-A dedicated communications channel, usually a pair of wires and/or a digital loop carrier channel, between a customer's telephone and a serving central office.

Customer Network Management-All activities associated with the planning, operation, administration, and maintenance of the communications network of a corporate customer. This term also refers to an arrangement that enables customers to manage their own networks, for example, to move Integrated Services Digital Network (ISDN) access channels from circuit-switched to packet-switched service.

Customer Premises Equipment (CPE)-All telecommunications terminal equipment located on the customer's premises, including telephone sets, private branch exchanges (PBXs), data terminals, and customer-owned coin-operated telephones.

Customer Problem Handling -Customer problem handling is the organizational unit responsible for taking trouble reports from customers and making sure that the problems are resolved.

Customer Proprietary Network Information (CPNI)-Customer proprietary network information is business and/or technical information that is unique or only made available to a customer that relates to the quantity, configuration, routing, or usage of a telecommunications service subscribed to by a customer.

Customer Provisioning-Customer provisioning is the process of delivering products and/or services to the customer. It is the operational aspects of delivering the service, from setting up accounts, user authentication, billing/return policies, channel deployment, packaging and customer support.

Customer Rating-The process of rating the credit worthiness of a customer to determine what services they are authorized to use and the amount of deposit the customer may be required to pay.

Customer Record Information System-An electronic data processing system for keeping customer records and billing information, generally using magnetic tape.

Customer Relationship Management (CRM)- Customer relationship management is the process or system that coordinates information that is sent and received between companies and customers. CRM systems are used to schedule activities, allocate resources, and help control the sales activities within a company.

Customer Retrial-A customer's subsequent attempt, within a measurement period, to complete a call or a request for a service.

Customer Self Care-Customer self care is the process of a allowing the customer to review and/or activate and disable services without the direct assistance of a customer service representative (CSR). Customer self care can be as simple as providing account billing information to the customer by telephone through the use of an interactive voice response (IVR) system to providing interactive service activation menus on an Internet web site.

This figure shows that a customer self-care system can be used to review product or service options, check billing records, and change customer feature options (called "provisioning"). This diagram shows that the customer can contact the billing and customer care system via a gateway. The gateway may contain Internet web access (for graphic displays) or interactive voice response (IVR) systems to allow the customer to select their account, receive billing and customer care information, and possibly change feature options.

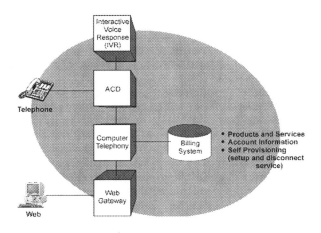

Customer Self Care Operation

Customer Service Class-The customer category of telecommunications services, including business, residence and public.

Customer Service Representative (CSR)-A company representative who manages customer communication for account inquiries, complaints, follow-up support, or other service related issues.

Customer Trouble Report Analysis Plan-An administrative procedure for recording trouble reports, furnishing a database from which analyses can be performed, and summarizing closed trouble data for exchange carrier administrative reports.

Customer Value Assessment -The analysis of how much value a customer provides to a company.

Customized Applications For Mobile Enhanced Logic (CAMEL)-Customized applications for mobile network enhanced logic is an intelligent network service specification that allows service providers to create custom service applications for mobile telephone systems. CAMEL operates on a "services creation node" in a GSM or WCDMA network. Examples of CAMEL applications include time of day call forwarding, multiple telephone extension service, and automatic call initiation on special conditions (trigger).

Cut-(1-video) A transition between two pictures that is instantaneous, without any gradual change. (2-crystal) The orientation of a crystal with respect to its electrical and mechanical axes.

Cut Back-Cut back is the process of transferring a video or multimedia source to the original video source or feed.

Cut Off Date-A date after which major redesign or reprogramming of a project cannot be achieved without serious effects on costs or completion date.

Cut Off Frequency-(1-general) The frequency above or below which the output current in a circuit is reduced to a specified level. (2-radio frequency) The frequency below which a radio wave fails to penetrate a layer of the ionosphere at the angle of incidence required for transmission between two specified points by reflection from the layer.

Cut Off Wavelength-(1-general) The shortest wavelength at which only the fundamental mode of an optical signal is capable of propagation. (2-theoretical) The shortest wavelength at which a single mode can propagate in a single-mode fiber. At wavelengths below cutoff, several modes propagate, and the fiber is no longer single-mode but multimode.

Cut Point-A cut point is a position on film or a

time marker on a video segment where a clip or segment will be connected or spliced to an overlying image or video.

Cut-Off Call-A call that has been disconnected by any means other than user intended.

Cutoff Date-The last date that transactions will be included in the current billing cycle.

Cuts-Cuts are the final versions of a film that have completed editing processing.

CVBS-Color Video Baseband Signal

C-VHS-Compact VHS

CVSD-Continuously Variable Slope Delta

CVSD-Continuously Variable Slope Delta Modulation

CVT-Circuit Validation Test

CW-Call Waiting

CW-Continuous Wave

CWA-Communications Workers Of America

CWDM-Coarse Wave Division Multiplexing

CWID-Call Waiting Identification

CWTS-China Wireless Telecommunication Standard Group

Cyan-Also called "process blue," one of the three subtractive primary colors.

Cyan, Magenta, Yellow and BlacK (CMYK)-The three primary subtractive colors used for printing ink or photographic color transparencies and prints. Black is used with these three subtractive primary colors because available inks or pigments do not inherently produce a solid black when all three are combined. Instead they produce a muddy brown, and the black ink is used to supplement the desired black areas. For the additive primary colors, see RGB.

Cyberspace-A term that is commonly used to describe an interconnected public data network (such as the Web) that has tools available to allow users to find and retrieve data from the network.

Cyclic Code-Cyclic code is an error correction code that uses a codeword from a particular cyclic code.

Cyclic Redundancy Check (CRC)-An error-checking technique in which bytes at the end of a packet are used by the receiving node to detect transmission problems. The bytes represent the result of a calculation performed on the data portion of the packet before transmission. If the results for the same calculation on the received packet are not equal to the transmitted results, the receiving node can request that the packet be resent. (See also: cyclic redundancy check code.)

An error detection and/or correction method that is used to determine if a series of data bits were received correctly during transmission. To setup a CRC error checking process, the original bits of data are supplied to a CRC generator. The CRC generator uses a specific mathematical formula to create a new group of data bits called the CRC check sum (bits). The CRC check bits are typically appended to the data bits that are being sent. The receiver of the message compares the result of the CRC generator on the receiver end to determine if the bits were received without error. In some cases, CRC check bits can be used to help correct some bits that were received in error during transmission, this is called Forward Error Correction (FEC) or Error Correction Code (ECC).

Cypher Key-A key used for the encryption process of data information.

D

D Channel-This is the out-of-band control channel associated with ISDN bearer channels, (see B Channel), and is used to establish, monitor, terminate and provide enhanced telephone services. The D channel provides a 16kbps or 64kbps packet-mode connection between a serving switch and a customer's premises. The channel carries signaling and control information for B channel activity and also can carry user data in the form of packets.

D/R-Distance to Reuse Ratio

D-1-The generic name used to describe a digital component video recording system. The D-1 system records digital component video to the CCIR-601 standard on 19 mm magnetic tape.

D1 Video Format-D1 is an uncompressed digital video format standard that is overseen by the Society of Motion Picture and Television Engineers (SMPTE). The D1 format is a component signal format, which has a relatively high bandwidth requirement as compared to combined (composite) color formats.

D2 Video Format-D2 is an uncompressed digital video format standard that is overseen by the Society of Motion Picture and Television Engineers (SMPTE). The D2 format is a composite signal format, which has a reduced bandwidth requirement as opposed to having data for each video component. The D2 format includes four audio channels and an analog cue channel.

D2-MAC-Duobinary Data MAC

D-3-A digital composite video recording format for NTSC or PAL. A 1/2-inch medium is used. The concept is similar to the D-2 format.

D4 Connector-Deutche Institut Normung 4 Connector

DA-Destination Address

DA-Distribution Amplifier

DAB-Digital Audio Broadcast

DAC-Digital To Analog Converter

DAC-Dual Attached Concentrators

DACS-Digital Access Cross-Connect System

DAI-Delivery Application Interface

Dailup-Dialup is a data connection that requires the user to initiate (dial up) a connection.

DAK-Deny Any Knowledge

DAL-Dedicated Access Line

DAM-Digital Answering Machine

DAM-Digital Asset Management

DAMA-Demand Assigned Multiple Access

DAML-Digital Added Main Line

D-AMPS-Digital Advance Mobile Phone Service

Dark Fiber-Dark fiber is a fiber strand in a cable that is unused by other light sources. Dark fiber may be used by customers to send any form or optical transmission (analog or digital.).

DARPA-Defense Advanced Research Projects Agency

DARS-Digital Audio Radio System

Dart Leader Stroke-The initial discharge that largely determines the path taken by a lightning flash. A dart leader develops continuously, and a stepped leader develops in short steps. Both are followed by a high-current discharge flash, usually in the reverse direction to the leader.

DAS-Dual Attached Stations

DAS Tape-Data Acquisition System

Dashboard-A dashboard is a screen or an online location where you can get all the information you need for an account or service. A dashboard usually offers the user the ability to manage account information (such as changing features or preferences for an account).

DAT-Digital Audio Tape

Data-Data is any representation of information that has an assigned meaning and is suitable for processing, transmission, communication, or interpretation by humans, systems, or computers.

Data Base Administration System-An operations support system that maintains a database of billed-number screening data. The system also handles auditing and fraud reporting as well as other duties, including calling card, collect, and bill-to-third number information.

Data Bus-An electrical or optical path through which data is exchanged between different systems or parts of systems. In fiber optics, an optical waveguide is used as a common trunk to which several terminals can be interconnected through optical couplers.

Data Channel-A data channel is a transmission path for data from a transmitter to a receiver. A data channel may be composed of one or more logical channels.

Data Circuit Terminating Equipment (DCE)- Data circuit terminating equipment is the communication channel terminating equipment that allows data terminal equipment (DTE) to be connected to a network.

Data Communications-Data communication is the transmission and reception of binary data and other discrete level signals that can be represented by a carrier signal that can represent the discrete (usually on-off levels) for signal transmission. There are two basic types of data communications: circuit-switched data and packet-switched data. Circuit-switched data provides for continuous data signals while packet-switched data allows for rapid delivery of very short data messages.

Data Communications Equipment (DCE)-The equipment that establishes, maintains, and terminates a data connection, as well as performs the signal conversion and coding required for communication between data terminal equipment (DTE) and a data circuit.

Data Communications Exchange-The hardware that enables data to be transmitted, received, switched, in real time to several different destinations over several different routes and channels, simultaneously.

Data Compression-Data compression is a technique for encoding information so that fewer data bits of information are required to represent a given amount of data. Compression allows the transmission of more data over a given amount of time and circuit capacity. It also reduces the amount of memory required for data storage.

Data Coupler-An interface (interconnection) device that connects a data device (such as a computer) via a telephone device (such as a telephone or mobile radio) with a telecommunications network. A coupler typically includes circuitry that protects the telecommunications network from damage that may result from failure of the coupler or data device that is connected to the data coupler.

Data Element-A data element is the lowest or smallest common denominator in a database management system or media file. A data element is considered indivisible, representing a specific element such as customer names, phone numbers, and addresses.

Data Encapsulation-Data encapsulation is a process of inserting the entire contents of a data packet into the payload (data portion) of another packet.

Data Encoding Scheme-The data encoding scheme is the process used to format information on a communication channel or how information is located onto a storage device such as a hard disk or CD ROM.

Data Encryption Standard (DES)-The data encryption standard is an encryption algorithm that is available in the public domain and was accepted as a federal standard in 1976. It encrypts information in 16 stages of substitutions, transpositions and nonlinear mathematical operations.

Data File-A data file is a group of information elements (such as digital bits) that are formatted (structured) in such a way to represent information. While the term data file is commonly associated with databases (tables of data), a data file can contain any type of information such as text files or electronic books.

Data Integrity-Data integrity is the accuracy of data information as compared to its original information source. Data integrity can be verified through the use of error detection codes that are sent along with the original data information. Error detection codes relate to the original data information through the use of mathematical processes. To verify data integrity, the received data information is processed using the same error detection coding mathematical process and comparing the output to the received error detection bits. If they match, the data integrity has been verified.

Data Line Interface-An interface, assembly, or connection point where a data line is connected to a telephone network.

Data Line Monitor-A device that connects to a data line and measures the performance of the data. This includes the data signal levels, the addressing accuracy, and protocol performance. The data line monitor only measures the data, it does not modify the information that is being transmitted.

Data Link (DL)-A transmission path between communications devices. The data link includes all equipment and signals in the connection.

Data Link Connection Identifier (DLCI)-A temporary channel identifier used in a communication system to identify a specific circuit along with its required communication parameters (such as peak data rates). The DLCI in a frame relay system is 10 bits. It is pronounced ("dill-see").

This diagram shows how a data link connection identifier (DLCI) is used in a data communications network to route information packets through a network and local data system to reach their destination. This example shows that the DLCI uses access devices to determine which virtual circuit is being used. This diagram shows how a computer that is connected to access device 3 is sending a print file to the printer connected to access device 1 using permanent virtual circuit (PVC) 1. Access device 2 is coordinating a digital voice connection to access device 1 using switched virtual circuit (SVC) 2. This example shows that the DLCI is only used by the access device. The data network uses its own routing tables to provide a virtual path connection through the network.

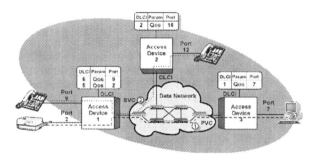

Data Link Connection Identifier (DLCI) Operation

Data Link Layer-The data link layer is the second layer (layer 2) of the seven-layer Open Systems Interconnection (OSI) protocol model that facilitates the detection of and recovery from transmission errors and a single data link. A data link connection is built on one or more physical connections.

Data Local Exchange Carrier (DLEC)-A service provider that specializes in transferring data. A DLEC is a competitive local exchange carrier (CLEC) that competes with other carriers in the local area such as the local exchange carrier (LEC). DLECs commonly use DSL for data transmission.

Data Mart-A specialized form of data warehouse. Data marts are typically smaller and more focused in purpose than the larger, more generalized data warehouses.

Data Mining-Data mining is the process of reviewing and analyzing information (data) for the purpose of identifying common characteristics that may be useful for other purposes. Data mining is commonly used in marketing programs to identify people or customers that have specific types of needs or buying patterns.

Data Mining Model-A particular type of statistical model constructed with the use of a statistical analysis product that uses data mining technology.

Data Model-An abstract construct defining the classes, types, and relationships between the data to be stored in a data base. Data Models are typically used to define how a database will be constructed in order to guarantee builders the best possible organization of data and indexes.

Data Multicasting-Data multicasting is the process of transmitting media channels to a number of users through the use of distributed channels (copying media channels) as they progress through a network.

Data Network-Data networks is a system that transfers data between network access points (nodes) through data switching, system control and interconnection transmission lines. Data networks are primarily designed to transfer data from one point to one or more points (multipoint). Data networks may be composed of a variety of communication systems including: circuit switches, leased lines and packet switching networks. There are predominately two types of data networks, broadcast and point-to-point.

This figure shows the basic types of data networks. This diagram shows several types of local area networks (LANs) including Ethernet, Token Ring and FDDI. It also shows that small networks can be interconnected to form wide area networks. Data networks can be private networks or public networks. It is also possible to encrypt (protect) data information that is transmitted on public networks to form virtual networks.

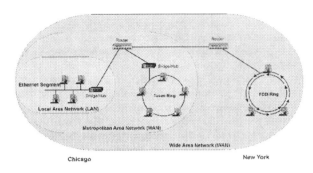

Data Networks

Data Over Cable Service Interface Specification + (DOCSIS+)-A version of Data Over Cable Service Interface Specification (DOCSIS) that is used for wireless broadband.

This figure shows an example of how the DOCSIS+ system can be used to provide wireless broadband service on MMDS frequency band. This diagram shows that standard DOCSIS CMTS equipment has been installed at the head-end of a cable system and standard DOCSIS cable modems are installed at the end user location. To allow the DOCSIS system to transmit on frequency bands, a transverter is located at both the head end and at the end user's location to convert the DOCSIS television data channels to and from MMDS radio channels.

A standard used by cable television systems for providing Internet data services to users. The DOCSIS standard was developed primarily by equipment manufacturers and CATV operators. It

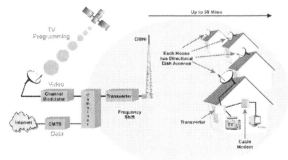

Data Over Cable Interface Specification for Wireless (DOCSIS+)

details most aspects of data over cable networks including physical layer (modulation types and data rates), medium access control (MAC), services, and security.

Data Packet-In Internet Protocol (IP) networks, the smallest amount of data that can packaged and transmitted from source to destination. In more general communications terms, any group of data which has been organized and labeled for transmission on a network or serial line. Each networking technology has minimum and maximum packet sizes which it supports.

Data Pipe-A data pipe is a single communication channel that is used to transfer data and this data may be used for one or many services.

Data Port-A connection point (usually a logical channel) that allows a computer or device to connect to other devices.

Data Processing Equipment-Data processing equipment is the equipment, assemblies, or peripherals (e.g. magnetic or other data storage media) that is used to receive, process, and output data information.

Data Rate-Data Transmission Rate

Data Service Unit (DSU)-A data service unit is a device that interfaces between data terminal equipment (DTE) and a data communication network. A DSU is the digital equivalent of the analog modem. DSUs are commonly used or combined with channel service units (CSUs) to allow specific channels of a communication line to communicate with specific DTE devices.

Data Service Unit/Channel Service Unit (DSU/CSU)-Devices that combine the functionality of data service units (DSU) and channel service units (CSU) to adapt data from user communication systems to communication lines with multiple channels. The DSU portion as an interface between a customer's data terminal equipment and a data communication network. DSU are the digital equivalent of the analog modem and are translation codecs (COde and DECode) coupled with a network termination interface (NTI). The CSU portion is used to coordinate communication from one or more data terminal equipment (DTE) devices to logical channels on a multi-channel communication circuit.

Data Services-Data Services are communication services that transfer information between two or more devices. Data services may be provided in or outside the audio frequency band through a communication network. Data service involves the establishment of physical and logical communication sessions between two (or more) users that allows for the non-real time transfer of data (binary) type signals between users.

Data Set Ready (DSR)-Data set ready is a control signal that indicates a device is ready to communicate.

Data Sink-The equipment that stores data after broadcast.

Data Speed-The data transfer rate of a signal. Data speed is also called baud rate or bit rate.

Data Storage-Data storage is the retaining of information over a period of time (typically in digital form).

Data Storage Fees-Data storage fees are charges for the allocation and/or usage of storage media. Storage fees may be charged for raw data storage or for the storage of particular types of media (such as movies, pictures and voice mailboxes). Data storage fee rates may vary based on the reliability (such as backup) and data transfer performance available from the data storage device.

Data Stream-A continuous flow of digital information (data).

Data Telephone-A data telephones is a telecommunication device that integrates analog telephone functions with a data communication interface. Because many data telephones use voice over Internet (VoIP) protocols, they are often referred to as an Internet telephones.

Data Terminal-Data terminals are data input and output devices that are used to communicate with a remotely located computer or other data communication device. Data terminals frequently consist of a keyboard, video display monitor, and communication circuitry that can connect the data terminal with the remotely located computer.

Data Terminal Equipment (DTE)-In a data communications network, the data source, such as a computer, and the data sink, such as an optical storage device. (See also: data sink, data source, network channel terminating equipment, channel service unit, data service unit.)

This diagram shows data terminals that are connected through data communication equipment (DCE) to allow a user to receive and send communication through a network to a remote computer. In this diagram, the data terminal allows the user to view information on a monitor and enter infor-

mation through the keyboard. This data terminal is connected through data communication equipment (a modem) that converts the data terminal's digital signals into an analog form (audio signal) that can be sent through the telephone line to a remote computer.

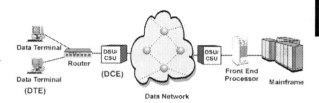

Data Terminal Equipment (DTE)

Data Terminal Ready (DTR)-Data terminal ready is a control signal on a communication line that indicates a device is ready to receive communication information.

Data Throughput-Data throughput is the amount of data information that can be transferred through a communication channel or transfer through a point on a communication system.

Data Transfer Adapter (DTA)-A data transfer adapter (DTA) converts the data bits from a computing device into a format that is suitable for transmission on a communication channel that has a different data transmission format.

Data Transfer Cost-Data transfer cost is the fee paid for the transferring of data into or out of a data network.

Data Transfer Rate-Data transfer rate is the number of bits, characters, or blocks of time passing between corresponding equipment in a data transmission network.

Data Transmission-Data transmission is the transfer of data (information) from one point to another point. The data may be transferred using one or more types of communication channels. The physical transfer of the information can be sent in analog or digital form.

Data Transmission Rate (Data Rate)-Data transmission rate is the amount of digital information that is transferred over a transmission medium over a specific period of time. Data transmission rate is commonly measured in the amount of bits that are transferred per second (e.g. bps, Mbps).

Data Warehouse-(1-information management) An information management service that stores, analyzes, and processes information that is derived from transaction systems. (2-system) A specialized database system, dedicated to the storage and retrieval of information for purposes of analysis and business intelligence investigation.

Data Word-In data communications, a character string, binary element string, or bit string that is considered as an entity.

Database-Databases are collections of data that is interrelated and stored in memory (disk, computer, or other data storage medium). Database systems are typically accessed and controlled by computer terminals that are connected to the same data network as the database system.

Database Administrator (DBA)-A person who is responsible for the organization, design, and implementation of a company's databases.

Database Dip-A process of an information search within a database. An example of a database dip used in a telecommunication system is the searching in a signaling control point (SCP) database to find the actual telephone associated with a toll free (800) or freephone (0800) telephone number.

Database Grooming-The modification of an existing database to sort information and remove redundant or unwanted information.

Database Management System (DBMS)-A system that is controls access to, organization of, security, and application interfaces to information data.

Datacasting-Datacasting is the transmission of ancillary or related material in combination with program content that enhances or supplements the original program. Examples of datacast services include program information, web pages, or news services.

Datagram-In packet switching, a self-contained data packet representing a portion of a message. A datagram is sent independently through a network to its destination. It is reassembled with other packets into the original message.

Datagram Service-Service at the network layer in which successive packets may be routed independently from end to end. There is no call setup phase. Datagrams may arrive out of order. In Internet Protocol parlance, datagram service generally implies the use of UDP or non-assured delivery.

Dataport-(1 - mobile radio) A connection port on a mobile radio that allows data devices such as fax machines to be connected to the mobile radio. (2 - channel unit) One channel unit from a digital channel bank that connects data signals from customers to a digital data network without the use of a digital multiplexer.

Date and Time Stamps-Date and time stamps are time references that indicate when an event occurred. Date and time stamps may be stored with, appended to or entered into a separate data file. An example of data and time stamps is storing the date and time information when voice mail systems have received a voice message.

Date of Conception (Invention)-Date of Conception is relevant in the United States where patents are awarded on a first-to-invent basis. Date of Conception has no meaning where patents are awarded on a first to file basis.

Contrary to common belief, the date of conception is not the moment when an invention was first thought of, rather the date of conception occurs when a complete, working concept of the invention is completed. The date of conception is that date when a person skilled in the art would be able to reduce the invention to practice without undue experimentation.

Date of Invention-Date of Invention is only relevant where patents are awarded on a first-to-invent basis. Date of Invention has no meaning where patents are awarded on a first to file basis. The date of invention is the date when the invention is actually, or constructively, reduced to practice. The filing of an application for patent is considered a "constructive" reduction to practice.In certain circumstances, the date of invention can be the same as the date of conception, if it can be shown that the inventors worked diligently and without undue delay to reduce the invention to practice.

DAVIC-Digital Audio Video Council

Day Rate Period-The period of time for which day rates apply. It may vary for different services, including interstate, intrastate, and overseas services. (See also peak time).

dB-Decibel (See also: Bel, decibel)

dB-decibel

DB-Dummy Burst

DBA-Database Administrator

dBd-Decibels dipole
dBi-Decibels isotropic
dBm-decibel milliwatt
dBm-Decibels Relative to One Milliwatt
DBMS-Database Management System
dBmW-Decibel Milliwatt
DBS-Direct Broadcast Satellite
DC Power-Direct Current Power
DC Subcarrier-A DC subcarrier is a separate modulation portion of a carrier signal (sub-carrier) that is located in the center of the carrier frequency bandwidth.
DC/MA-Dynamic Channel Multiple Access
DCA-Dynamic Channel Allocation
DCCH-Dedicated Control Channel
DCCH-Digital Control Channel
DCD-Downlink Channel Descriptor
DCE-Data Circuit Terminating Equipment
DCE-Data Communications Equipment
DCF-Distributed Coordinated Function
DCF Mode-802.11 Distributed Coordination Function
DCH-Dedicated Channel
DCS-Digital Cross Connect System
DCS-1800-Digital Cellular System 1800 MHz
DCS-1900-Digital Cellular System 1900 MHz
DCT-Digital Cordless Telephone
DCT-Discrete Cosine Transform
DCTI-Desktop Computer Telephone Integration
DDD-Direct Distance Dialing
DDR-Digital Disk Recorder
DDS-Digital Data Storage
DE-Discard Eligibility
Dead Spot-An area within a wireless system (typically cellular or PCS) where, for one reason or another, signals do not have a sufficient level to an acceptable level of communications. See also: Signal Fading.
Dead Spots-Dead spots are portions of radio coverage in a wireless system where the customer cannot initiate or receive calls or radio signals due to low radio signal level.
Dead Zone-(1-potentiometer) The dead zone is the portion of a potentiometer through which a change in wiper position causes no change in output. (2-measurement) A range of values outside the range of a measuring instrument. (3-TDR) A dead zone is an area on a TDR or OTDR trace that cannot be seen or analyzed due to the launching of the initial pulse. A dead zone is caused when the receiver of the TDR is overwhelmed (blinded) by the initial pulse level.

DeAuthentication-DeAuthentication is a process that is used to remove the association between authenticated devices.
Debugging-Debugging is the process of finding, analyzing, and fixing software errors (bugs) in a software program.
Decapsulation-The process of removing protocol headers and trailers to extract higher-layer protocol information carried in the data payload. See also Encapsulation.
Decentralized Architecture-Decentralized architecture is a system that is designed to use a distributed intelligence to coordinate the operation of the system.

decibel (dB)-A measurement that expresses the ratio of two amounts of power by use of the logarithm. The decibel is 1/10th the amount of a Bel, which was a measurement named after Alexander Graham Bell. The formula for expressing a number related to the power ratio, in decibels, is "10 log (P2/P1)" where "log" represents the common or base 10 logarithm. P2 is a second power level, and P1 is a first power level. P1 and P2 are both expressed in the same power unit, typically in watts.
The decibel is a non-linear measurement. Examples using power: 3 decibels (.3 Bels) indicates that P2 is approximately 2 times P1; 10 decibels (1 Bel) indicates that P2 is exactly 10 times as large as P1; and 20 decibels (2 Bels) indicates that P2 is exactly 100 times as large as P1.
Since power can also be expressed as voltage squared divided by resistance, the logarithmic decibel measure of the ratio of two voltages V2 and V1, measured at two points in a system where the local ratio of voltage to current are R2 and R1 respectively, then the formula for expressing a power ratio in decibels is "20 log (v2/v1) - 10 log (R2/R1)". An equivalent formula is "20 log (V2/V1) + 10 log (R1/R2)". For situations in which R2 and R1 are equal, these last two formulas reduce to "20 log (V2/V1)".
The decibel unit was introduced into the telephone industry about the year 1910. The power ratio of the input power to the output power for a 1 kHz test tone, for 1 mile (1.6 km) of number 19 (American wire gauge) copper wire, is almost exactly 1 dB. At that time, 19 gauge wire was widely used in the telephone industry (but was later almost completely replaced with 22, 24 or 26 gauge wire), and it was customary for many engineers and technicians to describe the power ratio of one

mile of 19 gauge wire by the inappropriate name "1 mile of signal loss". Use of the decibel gave a formal basis and a mathematically consistent formula for this practice. Use of a logarithmic method of expressing the power ratio allows the engineer or technician to add the logarithmic dB units for multiple "miles" of wire, rather than multiplying the actual numeric power ratios. The dB unit has also been applied extensively to audio power ratios and light power ratios. Remember that a dB is an expression of a ratio, and not an absolute power unit. When the denominator power term is taken to be a pre-agreed unit of power, then dB can be used to express the logarithm of an absolute power level. Examples of such absolute expressions are dBm, dBW, and acoustic decibels.

decibel milliwatt (dBm)-The power in dB as referenced to one milliwatt. For example, 0 dBm is 1 milliwatt, 20 dBm is 100 milliwatts, and -10 dBm is 0.1 milliwatt.

Decibels Relative to One Milliwatt (dBm)-The power of a signal referenced to 1 mW (milliwatt).

Decipherment-The decoding of data that is encrypted with a cipher key.

Decoding-Decoding is the process of converting encoded words into the original signal. For example, in pulse code modulation (PCM), decoding is the conversion of 7-bit (D1 type) or 8-bit (including D2, D3, D4, and D5 type) pulse code modulation (PCM) words to analog signals. Decoding is the inverse of encoding.

Decompression-Decompression is the processing of compressed digital information to convert it to its original uncompressed format.

Decompressor-A decompressor is a device, software or assembly that decodes and expands compressed information (such as a compressed digital media file).

Decryption Content Scramble System (DeCSS)-DeCSS is a software program that was created to remove the CSS protection on DVDs so the media could be manipulated and copied.

DeCSS-Decryption Content Scramble System

DECT-Digital Enhanced Cordless Telephone

Dedicated Access-Dedicated access is the permanently assigned (non-switched) channel or circuit for the exclusive use of a particular subscriber.

Dedicated Access Line (DAL)-A communication line that has its data transmission capability reserved (dedicated) for a specific user.

Dedicated Bandwidth-A configuration in which the communications channel attached to a network interface is dedicated for use by a single transmitter or receiver and does not have to be shared.

Dedicated Channel (DCH)-A dedicated channel is an uplink or downlink communication channel that is only accessible by one device.

Dedicated Media-A configuration in which the physical communications medium used to connect a station to either a hub or another station constitutes a point-to-point link between the devices.

Dedicated Plant Assignment Card-An outside plant record that gives the particulars of a distribution plant served by a dedicated outside plant control point, or serving area interface. The card lists the address of every housing unit served from the control point or interface, as well as the telephone number, central office equipment, type of service, feeder and distribution pair numbers, and the address of the distribution terminal. The dedicated plant assignment card generally has been replaced by mechanical loop assignment systems, such as the Facilities Assignment and Control System (FACS).

Dedicated Trunk-A communication trunk that is connected directly through to a particular phone or hunt group. Dedicated trunks bypass attendant consoles.

De-Emphasis-De-emphasis is the reduction of the high-frequency components of a received signal to reverse the pre-emphasis that was placed on them to overcome attenuation and noise in the transmission process.

Defacto Standards-Defacto standards are specifications that are widely used or accepted which are not officially recognized or controlled by standards organizations. An example of a defacto standard is Microsoft Windows operating system.

Default-Default is the initial value(s) or process(es) for a particular service or application.

Default Parameters-Default parameters are the specific values of the variables that are used for a particular service or application. An example of a default parameter is an "Include on Mailing List" option on a web page that is default set to "include" on the mailing list.

Default Port-A physical or logical port that is preconfigured or defined to be the communication port for traffic when a port assignment is not provided or when a process fails. Applications and protocols in communication systems such as the Internet use default ports so they can start specific applications or protocol communication when a packet is first received. For example, when an IP packet is received with the port number 21, it specifies file transfer protocol (FTP) should be used.

Deferment-Deferment is the delaying or postponing of something, usually referring to payment.

Deferred compensation-A sum of money to be paid that is delayed in transaction due to non-occurrence of events required to make payment possible.

Deflection-The control placed on electron direction and motion in CRTs and camera tubes by varying the strengths of electrostatic (electric) or electromagnetic fields.

Defocus Effect-A digital picture manipulation term meaning a controlled blurring of the image.

Degeneration-(1-recording) The loss of quality on a videotape, typically resulting from multiple generations of copying the material. (2-amplifier circuit) The process of reducing the gain of an amplifier stage by applying negative feedback (feedback that is 90 degrees out of phase) to the input.

Deglitcher-A circuit used to limit the duration of switching transients in a digital system.

Delay-The amount of time it takes for a signal to transfer or for the time that is required to establish a communication path or circuit.

Delay Circuit-A circuit designed to add time delay to a signal passing through it by a specified amount.

Delay Equalizer-A network that adjusts the velocity of propagation of the frequency components of a complex signal to counteract the delay distortion characteristics of a transmission channel.

Delay Line-A device that delays a signal by a very accurate amount of time. They are used in synchronizing electronic signals for computing, telecommunications and radar applications. SAW devices are often used for delay purposes.

Delay Spread-A product of multipath propagation where symbols become distorted and eventually will overlap due to the same signal being received at a different time. It becomes a significant problem in mountainous areas where signals are reflected at great distances.

Delay Time-The sum of waiting time and service time in a queue.

Delay Variation-The difference, in microseconds, between the maximum and minimum possible delay that a packet will experience as it goes out over a channel. This value is used by applications to determine the amount of buffer space needed at the receiving side in order to restore the original data transmission pattern.

Deliberate Churn -Deliberate churn is a type of voluntary churn. Deliberate churn occurs when the customer decides to terminate service because they have found a more attractive service elsewhere.

Delinquency Threshold-The amount owed by the customer that will trigger the collection process to be initiated.

Delinquent-Delinquent is status of a claim or debt which has not been paid by the date specified in the agency's written notification or applicable contractual agreement, unless other satisfactory payment arrangements have been made by that date, or, at any time thereafter, the debtor has failed to satisfy an obligation under a payment agreement with the agency.

Delivery Multimedia Integration Framework (DMIF)-Delivery multimedia integration framework is a structure that allows a multimedia system (such as MPEG) to identify the sources of media and the transmission characters for that media source (such as from a high bandwidth low error rate DVD or through a limited bandwidth mobile telephone system). The use of DMIF allows the playback system to become independent from the sources and their transmission limitations.

Delivery Multimedia Integration Framework (DMIF) Application Interace (DAI)-DMIF application interface is the messages and processes that allow applications (such as media player programs) to receive, decode and time sequence media streams. The DAI allows a media player to obtain and display multimedia independent of the underlying sources of media.

Delta-In a fiber communication system, delta is the normalized refractive index difference of a fiber. Delta is approximately equal to the difference in the indices of refraction of the core and the cladding divided by the index of the core.

Delta Frame-A type of video image frames used in digital video applications. The delta frame contains only differences between the current frame it represents and the frame that precedes it. A delta frame tends to be considerably smaller than an intra-frame. A stream of temporarily compressed data must be interspersed with key frames (also called intra-frames, or stand-alone frames) in addition to delta frames, so that audio/video synchronization can occur.

Delta Modulation (DM)-A form of digital modulation where voltage changes of an analog signal are changed into a fixed difference of value (or delta), and a +/- sign. For adaptive delta modulation, differences in the voltages are not fixed, and vary depending on past history.

Demand Analytics-Collection of data mining models that can be utilized to help define how changes in operational expense (OPEX) spending (advertising, marketing, customer service etc.) will impact the customers demand for telecommunications services.

Demand Assigned Multiple Access (DAMA)-Demand assigned multiple access, is a process that allows the sharing of the capacity of a communication channel or groups of communication channels through the assignment of unused channels as demand for capacity increases.

Demand Based Build Out-Demand based build out is the process of installing and adding equipment in areas as customers (subscribers) are added to the system. This allows service providers to invest in their systems as their revenue grows rather than installing large and complex networks with the anticipation that customers will eventually subscribe to their services.

Demarc-Demarcation Point

Demarcation Point (Demarc)-The physical and electrical boundary between an end user's telecommunication equipment and the telecommunications network. The demarcation point establishes point of ownership and accountability.

DeMilitarized Zone (DMZ)-A demilitarized zone (DMZ) is trusted part of a communications network, typically behind a firewall, that allows unrestricted access and transfer of information between devices. Information that passes in or out of a DMZ may be delayed and filtered. This may cause challenges with real-time communication applications such as IP telephony and media streaming.

Demodulation-Demodulation is the process of recovering an information signal that has been previously modulated on a radio, optical or other electromagnetic carrier signal.

Demodulator-A demodulator is a circuit or a device that recovers (extracts) an information signal from a carrier (transport) signal. The output from a demodulator is usually the original baseband information signal format.

This diagram shows how a demodulator converts a modulated carrier signal into an information signal. This diagram shows that the demodulator compares the modulated carrier (carrier with the changes) to an unmodulated carrier (pure carrier signal) to produce the information signal (representing only the changes of the carrier signal).

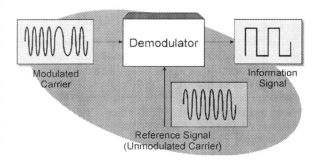

Demodulator Operation

Demultiplexer (DeMux)-(1-general) A device used to separate two or more signals that were previously combined by a compatible multiplexer and are transmitted over a single channel. (2-digital) A circuit or device for separating two or more signals received sequentially from a single communications channel into individual bit streams, and sending the bit streams to their respective destinations. In analog systems, a demultiplexer separates two or more signals received in a specified frequency spectrum into individual channels.

This diagram shows how demultiplexing can extract two or more low speed channels from a higher speed communication channel. In this dia-

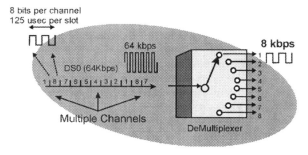

Demultiplexing Operation

gram, there is a single 64 kbps communication signal that is supplied to a demultiplexer circuit. The demultiplexer gathers and routes 8 bits data to a specific port during each 125 usec time slot. This example shows that the data on each port is sent out (clocked out) at 8 kbps.

Demultiplexing-Demultiplexing is a process that separates individual channels from a common transport or transmission channel.

DeMux-Demultiplexer

DEN-Directory Enabled Network

Denial Of Service (DoS)-A denial of service attack is a process that inhibits or reduces the ability of authorized users from gaining access to communications systems through the continual transmission of service requests or messages that disable communication sessions.

Dense Mode Multicast-Dense mode multicasting is the distribution of media to multiple users within a data network where many or most of the users that are connected to the network are part of the multicast group.

Dense Wave Division Multiplexing (DWDM)-Dense wave division multiplexing is a version of fiber optic communication that combines many optical channels on a single fiber, typically used to increase the data transmission capacity of previously installed fiber. Dense wave division multiplexing provides a significant increase in capacity compared to wavelength division multiplexing (WDM) that combined up to four different optical wavelengths on a single fiber. As of year 2005, DWDM systems up to 160 different wavelengths on one fiber, providing the capability of transferring over 1 trillion bits of data per second. Each optical channel (wavelength) is capable of providing approximately 10 Gb/s (OC-192 9.3 Gb/s data rate). The wavelengths on a DWDM system can be separated by as little as 0.4 nm.

Deny Any Knowledge (DAK)-A claim by a customer that a call was not made (as in "I deny any knowledge of this call").

Department Of Communications (DOC)-A government agency of countries throughout the world that sets policies and rules regarding telecommunications within their country. In the United States, the FCC is the equivalent of the DOC in some other countries.

Depiction Release-A Contract releasing rights to portray and often fictionalize and dramatize elements of ones life story in television or film form.

Depolarization-Depolarization is the changing or randomizing of a desired electromagnetic wave orientation (horizontal or vertical) that occurs when the electromagnetic wave passes through a medium.

Depth of Field-Depth of field of the distance between the closest object that is in focus to the object furthest away that is in focus.

Derating Factor-An operating safety margin provided for a component or system to ensure reliable performance. Typically, a derating allowance also is provided for operation under extreme environmental conditions, or under stringent reliability requirements.

Derivative Work Right-Derivative work right is the authorization to modify, extract or embed materials to create a new work.

DES-Data Encryption Standard

DES Encryption-The data encryption standard is an encryption algorithm that is available in the public domain and was accepted as a federal standard in 1976. It encrypts information in 16 stages of substitutions, transpositions and nonlinear mathematical operations.

Descrambler-A descrambler is a device or circuit in a receiver that rearranges a received pattern of digital signals, or frequency bands in an analog system, to compensate for scrambling that took place in the transmitter.

Descrambling-Descrambling is a process of converting an encoded or encrypted signal back into its original form through the use of keys and/or descrambling algorithms.

Design-Design is the configuration of components systems or programs so they perform specific functions or services.

Design Error-Design error is the configuration of components and/or systems so that they do not or cannot perform specific functions and/or services.

Desktop-The screen layout on a computer display that typically includes program icons and toolbars.

Desktop Collaboration-Desktop collaboration is the process of interconnecting desktop computers and their programs to each other so the desktop users can share and edit the same documents or use the same programs.

Desktop Computer Telephone Integration (DCTI)-Desktop computer telephony integration is the merging of telephony services with computer technologies.

Desktop Management Interface (DMI)-The process of managing desktop workstation hardware and software components automatically, often from a central location. The DMI is an application programming interface (API) that enables a software program to collect information about a computer environment.

Destination Address (DA)-The address for whom a message is intended.

Detaching-Detaching is the process of removing a connection (a physical and/or logical connection) between a client (e.g. workstation) and a network server (e.g. a file server).

Detailed Billing-A type of billing in which the details of each message, such as the date of call, number called, and charge, are listed as separate items on a toll statement that is included with a customer's bill.

Detariffed-A billing and regulatory term designated by the FCC and state public utility commissions that often refers to deregulated rate structures on subscriber-owned inside premise wiring and related CPE.

Detection-(1-General) Detection is a process of sensing that a signal is present or an event has occurred. (2-Demodulation) The process of separating a modulated wave from a carrier signal.

Detector-A device that senses the presence or level of a signal. A detector senses the presence of electromagnetic or optical energy and creates an electrical signal that represents the level of the sampled signal.

Developed Portion Of A Wire Center-In long-range outside plant planning, land that is either completely built up, built up with scattered vacant lots, vacant but surrounded by developed land, or a cluster of more than 200 living units in an otherwise undeveloped part of a wire center. Such areas are also called developed areas.

Development Activities-During the production of a film, all actions taken in interest of that film.

Development Fee-A development fee is an amount of money paid to one in order to compensate them for talent or services rendered in connection with the development stage of a project.

Development Kit-A development kit is the combination of software development tools and possibly hardware that allow a company or person to develop products, communication applications and/or services. A development kit usually includes code editors, compilers, debuggers, utility libraries, along with technical information that instructs the developer on how to use the kit.

Development Plan-A development plan is a document that defines the key steps and processes that are likely to be used during the development of a product or service.

Development Procedures-Development procedures or documents and supporting materials that define the actions and procedures that are used for the development of products or services.

Development Stage-Development stage is the period of time it takes to complete a project from beginning to end.

Deviation-(1-general) A departure from a standard or specified value. (2-modulation) The peak difference between the instantaneous frequency of an AM signal and the carrier frequency. (3-phase) The peak difference between the instantaneous angle of a phase-modulated wave and the angle of the carrier.

Device Certificate-A device certificate is information contained within a device (usually in digital form) that can uniquely identify its identity along with other characteristics of the device. Device certificates are usually issued by an authority (a trusted party) that guarantees the information in the certificate is correct.

Device Class-Device class is a parameter that indicates the capabilities of a device. Device class capabilities may include maximum and minimum transmitter power levels, available modulation and coding types, and which services the device supports.

Device Configuration-Device configuration is the assigning of the parameters or changing of adjustable elements of a device to enable the device work within a system and/or to perform specific features.

Device Descriptor-A device descriptor contains general information about a data communication device. It usually includes information that applies to the device configuration and communication settings. Each device in a USB system has only one device descriptor.

Device Discovery-Device discovery is the processes used to request and receive the identification address, name, and services of other devices. For the Bluetooth system, device discovery information includes address, clock, class of device, used page scan mode, and names of devices.

Device Management-Device management is the process of identifying, adding, and configuration devices that are part of a system. Device management may be a manual process or it may be automatically performed through the use of protocols that can discover and configure equipment that is added to a network and remove registrations from devices that are removed from a network.

Device Pairing-Device pairing is the process of associating two devices with each other. During the pairing process, identifying information that is unique to each device is stored in the paired device. After devices have been paired, they can automatically identify each other during future communication sessions.

Device pairing may be pre-established at the time of manufacture, it may occur during the first attempted use of the device with another product, or it may be manually performed through the selection of codes or switches located on or in the device. To initiate the pairing process, one or both devices must enter a pairing mode of operation. This is usually performed locally, and not performed automatically to avoid the potential of other unauthorized devices from pairing. Pairing mode may be entered into by a manual key sequence or by holding a hidden button for a few seconds.

Device Under Test (DUT)-A device under test is a device that is attached or communicating with a test system for the purposes of testing, performance measurements and/or diagnostic purposes. A DUT may be placed in a special test mode to permit test commands to be received and processed.

DFA-Doped Fiber Amplifier

DFB-Distributed Feedback Laser

DFE-Discrete Feedback Equalizer

DFS-Dynamic Frequency Selection

DFT-Discrete Fourier Transform

DGPS-Differential Global Positioning Service

DHCP-Dynamic Host Configuration Protocol

DHTML-Dynamic hypertext markup language

Diagnostic Records-Records relating to a call or communication session that was processed by a call manager or a call server.

Diagnostics-A program, often built into a system, that tests the functionality of the system and reports the results. Diagnostic systems that are separate and simply monitor the operation of the subject system are considered to be non-intrusive.

Dial By Name-The ability to dial a person by spelling their name out on the telephone or communication device keypad.

Dial Map-A dial map is the systematic use of certain prefix digits to dial a destination via user selected routing. An example is the use of the dialed prefix "9" from within a PBX to first select an outside local telephone line so that the originator can then dial a (typically 7 digit) local city telephone number. Similarly, a PBX may use the dialed prefix "8" to select a tie line to another PBX. This figure shows how a basic dial map operates. This diagram shows that there are several dial plan rules that are used each time a number is dialed. The first step in the dial map is to determine if the first digit is a 0, 3, 8, or 9 are dialed. This first rule allows the system to determine if the caller desires to reach the attendant (0), is calling an internal number (3+), long distance (8), or outside line or emergency services (9). This rule changes how the next digit is processed. If the first digit is a 3, it is an internal call (4 digits for this system) and the system will wait for 3 more digits before attempting to connect the call to another unit in the system. If the first digit is 8, the system will capture multiple digits and analyze the call as a long distance public telephone number (country code, city code, exchange code, and extension). If the first digit is a 9, the system will analyze the following digits to determine if it is an emergency call (for example 911) or a local telephone call. If it is a local telephone call, the system will wait until it has sufficient digits and connect the call to a local gateway. If the next 3 digits were 911, it would connect the call to a local gateway and route to the emergency services number.

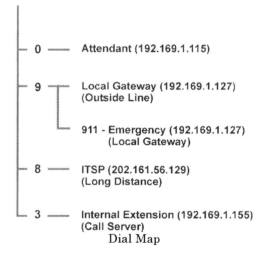

Dial Map

Dial Plan-A dial plan (also called a dialing scheme) is the numbering system that is used by a company to identify devices within their network by unique numbers. After a system has been setup, a dialing plan is developed for each communication unit (or groups of communication units).

Dial Plan-Dialing Plan

Dial Pulse (DP)-Dial pulse (DP) signaling senses and counts the changes in current flow, such as from a rotary dial telephone, to allow the user to send address information (dialed digits) to the telephone system.

Dial Tone-A signal tone provided from a local telephone service provider that indicates that the telephone network is ready to send a call (to receive dialed digits). The dial tone signal is usually a combination of 350 Hz and 440 Hz signals.

Dial Up Line-(1-general) A communication line that is established by sending the dialed digits to the network that allows the connection to be made. The term dial up is commonly used with the process of manually connecting a computer to the Internet through a standard telephone line. (2-humor) An ancient method of connecting to the Internet using MODEMS that signal over a telecom system known as the plain old telephone service or POTS. This method of communication was outlawed in the mid 2000s as part of the Computers With Disabilities Act.

Dialed Number Identification Service (DNIS)-A call identification service typically provided by a toll free (800 number) network. The DNIS information can be used by the PBX or automatic call delivery (ACD) system to select the menu choices, call routing, and customer service representative information display based on the incoming telephone number.

Dialing Plan (Dial Plan)-A dial plan (also called a dialing scheme) is a systematic use of certain prefix digits to dial a destination via user selected routing. An example is the use of the dialed prefix "9" from within a PBX to first select an outside local telephone line so that the originator can then dial a (typically 7 digit) local city telephone number. Similarly, a PBX may use the dialed prefix "8" to select a tie line to another PBX. A dialing plan differs from a numbering plan by being used inside a particular private telephone system, and also the specifics of different dialing plans are different in different PBXs or different private networks in the same country, while a numbering plan is uniform throughout an entire country.

Dictionary Attack-A process of attempting to gain access to a communications resource by sequentially cycling through a list of words (such as a dictionary) that are likely to be used for passwords or account information.

DID-Direct Inward Dialing

DID Assignments-Direct inward dialing (DID) assignments are the mapping (connecting) of extensions in a private telephone system to the incoming calls on a common telephone line that identifies the incoming direct dial telephone numbers.

Dielectric-A dielectric is any material that resists the flow of electricity. A dielectric is an insulator.

Dielectric Nature-A dielectric nature is the characteristic of a material or environment to act as an insulator against electrical currents.

Differential Encoding-Differential encoding is a process of transmitting information that is produced by a logical combination of the current data bit and the previous data bit.

Differential Motion Vector (DMV)-A differential motion vector is a line that describes the direction (angle) and how much (length) that an object has moved as compared to another video frame or sequence.

Differential Phase Shift Keying (DPSK)-Differential phase shift keying is a modulation technique for transmitting digital information by representing the information as discrete phase changes of a carrier signal. The phase is changed if the current data bit is different from its predecessor. At a receiving end, phase changes are detected by comparing the phase of each signal element with the phase of a preceding signal element.

Differential Pulse Code Modulation (DPCM)-(1-general) A modulation method that uses pulse coding which represents the difference in successive voltages samples of an analog signal. Each individual sample is approximated by a discrete set of values. (2-video) A modulation technology used for compressing video signals so they can be transmitted in the available capacity of an Integrated Services Digital Network (ISDN) channel.

Differential Quadrature Phase Shift Keying (DQPSK)-A type of modulation that alternates between 4 different phase shifts to represent the digital information symbol signal. During one symbol, these shifts are +/- 45 and +/- 135 degrees and during the alternate symbol, these shifts are +/- 90 and +/- 180 degrees.

Differentiated Services (DiffServ)-A differentiated service is a protocol that identifies (tags) different types of data with data transmission requirement flags (e.g. priority) so that the routing (switching) network has the capability to treat the transmission of different types of data (such as real-time voice data) differently.

DiffServ-Differentiated Services

DIFS-Distributed Interframe Space

Digest Authentication-An authentication process that processes a user identification code and password identification without sending the raw data through the communication network to validate the true identity of the user. The digest authentication process begins by the challenger (usually the service provider) sending a random number (nonce value) that is used by the user (usually the client) along with other previously known information (secret key) to calculate a response using an encryption algorithm. The end result is sent back to the challenger who also has access to the previously known information and random number that is used to calculate the same result. If the result matches, the authentication passes.

Digital Video 25 (DV25)-Digital video 25 is a standard digital video format that stores digital video signals at 25 Mbps (4 Mbytes per second).

DigiBeta-Digital Beta

Digital-Digital electronic devices use electric currents or voltages that are intentionally restricted to take on a limited set of values for their intended use, rather than allowing continuous variation of the current or voltage. Typically, only two voltage values are used, for example, having values of approximately zero or five volts. Because undesired signals (noise, interference) are much smaller (typically less than 0.1 volt, for example) the digital signals often can be transmitted or recorded without errors because the presence of small deviations of the signal do not confuse the device from correctly interpreting the voltage as definitely representing one of the two intentional voltage levels. When a digital coded representation of an analog signal is used, and the digital part of the system does not introduce any errors, the only degradation of the signal is due to the inherent inaccuracy of the initial encoding device (codec) that converts the signal from analog to digital representation. This inaccuracy can be controlled by the design of the codec.

This figure shows a digital signal that is in the form of a series of bits and these bits are combined into groups of 8 bits to form Bytes (B). In this example, the bits 01011010 are transferred in 1 second. This results in a bit (transmission) rate of 8 bps.

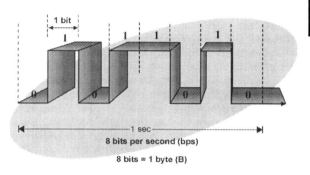

Digital Signal

Digital 8mm-Digital 8mm is a small self-contained video cassette package of 8mm reel-to-reel magnetic tape that is used for video signals and may be played or rewound on demand.

Digital Access Cross-Connect System (DACS)-A digital switching system that interconnects specific communication channels (time slots) between digital multiplexed lines (usually t-carrier lines).

Digital Answering Machine (DAM)-A digital answering machine is a device that contains digital memory which can automatically answer telephone calls, play a prerecorded greeting message, store audio information, and allow retrieval and deletion of messages.

Digital Asset-A digital asset is a digital file or data that represents a valuable form of media. Digital assets may be in the form of media files, software (e.g. applications) or information content (e.g. media programs).

Digital Asset Management (DAM)-Digital asset management is the process of acquiring, maintaining, distributing and the deletion of information (electronic) assets.

Digital Audio-Digital audio is the representation of audio information in digital (discrete level) formats. The use of digital audio allows for more simple storage, processing, and transmission of audio signals.

Digital Audio Broadcast (DAB)-Digital audio broadcasting is a communication system that transmits voice, music and other types of information using digital transmission. The DAB signal is typically shared with additional digital information on a single digital radio channel.

Digital Audio Radio System (DARS)-Digital audio radio services (DARS) is a radio system that provides audio programming via satellite transmission. It is similar to Direct Broadcast System (DBS) TV systems. The initial application of DARS was by CD Radio in 1990. It is also called Digital Audio Broadcasting (DAB) outside the U.S.

Digital Audio Tape (DAT)-A digital audio tape (DAT) is a magnetic tape storage format that uses a cartridge that has a 4 mm wide metal coated tape that stores audio information in digital form. The standard sample rate for DAT is 44.1 kHz and a single DAT can provide storage over 3 GB of storage capacity.

Digital Audio Video Council (DAVIC)-The digital audio video council was established in 1994 to develop audio-visual industry specifications. The applications for these specifications ranged from broadcasting to video on demand (VOD). In 1999, DAVIC determined that it had completed its required tasks and the council was disbanded and new work is now followed by TV Anytime Forum (www.TV-anytime.org).

Digital Beta (DigiBeta)-DigiBeta is a 1/2" video tape storage format that is used to store video recording images on magnetic tape.

Digital Broadcasting-The process of transmitting the same digital data signal to all users that are connected to the digital broadcast network. Digital broadcast signals may be encoded in a way that only some of the users may be capable of decoding digitally broadcast messages (e.g. a specific pay-per-view movie channel).

Digital Cable Television System-A digital cable system distributes television (and other information services) via a cable television distribution system in digital modulated form.

Digital Cellular-An industry term given to the new cellular technology that transmits voice information in digital form. This differs from Analog cellular in that the method of transmission for voice/data information is by means of digital signals.

This figure shows a basic digital cellular system. This diagram shows that there typically is only one type of digital radio channel called a digital traffic channel (DTC). The digital radio channel is typically sub-divided into control channels and digital voice channels. Both the control channels and voice channels use the same type of digital modulation to send control and data between the mobile phone and the base station. When used for voice, the digital signal is usually a compressed digital signal that is from a speech coder. When conversation is in progress, some of the digital bits are usually dedicated for control information (such as handoff). Similar to analog systems, digital base stations have two antennas to increases the ability to receive weak radio signals from mobile telephones. Base stations are connected to a mobile switching center (MSC) typically by a high-speed telephone line or microwave radio system. This interconnection may allow compressed digital information (directly from the speech coder) to increase the number of voice channels that can be shared on a single connection line. The MSC is connected to the telephone network to allow mobile telephones to be connected to standard landline telephones.

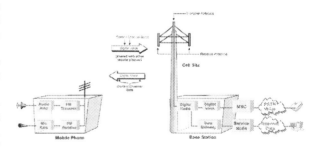

Digital Cellular System (2nd Generation)

Digital Certificate-A digital certificate is information in binary (digital) form that is used to identify the identity and possibly the capabilities of a device, system or software application. Digital certificates are usually issued by an authority (a trusted party), which ensures that the information in the certificate is correct.

Digital Compression-Digital compression is a process that uses a computing device (such as a digital signal processor) to analyze a digital signal and create a new data signal that represents of the original signal using a lesser number of digital bits. Digital compression allows more information to be transmitted on a communication channel.

Digital compression devices use mathematical formulas and codebook tables to compress the data. Mathematical formulae transform the original signal into it characteristic parts such as frequency and amplitude. Codebook tables contain blocks of high occurrence information (such as particular tones used in fax machines). When transmitting digital information that has been compressed, only the parameters (such as the frequency, amplitude and code book word) are sent on the transmission channel. When the digital information is received, the compression process is reversed by a decoder to produce the same (or similar) initial signal.

Digital Console-A digital console is a communication system access device that allows an operator to communicate with or control a network and or other devices. A digital console commonly consists of a display monitor and keyboard whose display and functions can be dynamically changed through the use of software.

Digital Content-Digital content is information or media that is stored in digital form.

Digital Copy Protection-Digital Copy protection is the mechanism and/or information that is sent or embedded within content to help ensure digital content is used in conformance with its usage rules and it can provide an auditable trail of how an asset is used within a consumer device. Copy protection defines the following rules for use of the content.

1.Copy Freely. User is free to make as many copies as they want.

2. Copy Never. The user is not able to make any copies.

3. Copy Once. Any device that makes a copy of a piece of content, will set the copy bits on the content to copy never on the copy and where possible on the original as well.

For video signals copy protection bits may be embedded in content in a number of ways including: 1. Within the video blanking interval (VBI) lines of the analogue output (CGMSA) 2. Within the media stream format (such as MPEG)

Digital Cordless Telephone (DCT)-Digital cordless telephony (DCT) typically refers to the use of digital transmission for cordless telephony.

Digital Data Storage (DDS)-Digital data storage is a digital information storage format that is used to store information on magnetic tape. The DDS format for data storage using 4 mm DAT tape cartridges were created in 1989 by Hewlett Packard and Sony corporations. There are several DDS storage formats. DDS-1 can store up to 2 GB of data, DDS-2 can store up to 8 GB of data, DDS-3 can store up to 24 GB of data and DDS-4 can store up to 40 GB of data. Over time, the information stored on DDS tapes begins to degrade and the tapes will need to be recreated. The number of uses of DDS tapes is also limited to approximately 2000 short recording sessions or 100 full recording sessions.

Digital Desktop-A desktop workplace for the employee that consists primarily of digital devices. The digital desktop digital devices are usually a computer, printer, and digital telephone.

Digital Device-A digital device is an electronic device that uses digital logic to perform its operational functions.

Digital Disk Recorder (DDR)-A digital disk recorder is a memory device that can store (record) digital video directly into memory storage (typically a hard disk).

Digital Dolby 5.1®-Digital Dolby (also known as Dolby 5.1) is an audio compression and channel multiplexing format that allows for the transmission of up to 6 channels (5 audio and 1 sub-audio).

Digital Enhanced Cordless Telephone (DECT)-The DECT system is a digital cordless and WPBX system. DECT was originally developed by the European Telecommunications Standards Institute (ETSI) technical standards committee in the late 1980s and the specification was released in 1992 and commercial equipment was available by 1993. The number of DECT handsets in use by 2004 is in excess of 50 million. It was first intended that the use of the DECT system be for wireless office. After its release, it has been adapted to allow home cordless, public cordless, and radio local loop (RLL).

The DECT system includes three key parts; the mobile radio portable part (PP), the radio base station fixed part (RFP), and the interconnecting system fixed part (FP). There are two version of DECT; the European version and the American version. The European version uses a very wide radio channel to allow up to 12 simultaneous wireless telephones to share each channel. The American version uses a slightly more narrow radio channel and allows up to 8 users to share a single radio channel. Personal Wireless Telecommunication (PWT) is an adaptation of DECT for the North American market.

DECT technology is managed and promoted by the DECT forum. The DECT forum helps to promote DECT technology worldwide, assists in the allocation of radio frequencies for DECT systems, provides forums that allows developers and providers to share information, and to manage the evolution of DECT technology to ensure reasonable migration from older legacy equipment to improved versions of DECT. More information about DECT forum and technology can be found at www.DECT.org.

This figure shows a DECT radio system. It shows that a DECT system includes radio devices (portable part - PP), radio base stations (radio fixed part - RFP), and interconnection equipment (fixed part - FP). The DECT system radio channel has a 1.728 MHz bandwidth with a gross data transfer rate of 1.152 Mbps. The radio channel is divided into 10 msec frames and each 10 msec frame is divided into 5 msec transmit and a 5 msec receive frames that contain 12 time slots each. The DECT system uses time division duplex (TDD) multiplexing so that one slot in the 5 msec transmit group is used in the forward direction and one slot in the 5 msec receive group is used in the reverse direction to provide full duplex (simultaneous) voice communication. This example shows that the DECT system can be used in a home environment or in an office system.

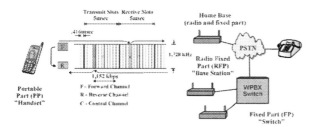

Digital Enhanced Cordless Telephone (DECT) Operation

Digital Headend-Digital headend are the network components that are used to mange and distribute digital media content in a cable television network. Digital headends can range from the simple conversion of analog video to digital form for transmission to the interactive control, delivery, and management of digital content.

Digital Ingest-Digital ingest is the process of transferring digital media into a computer storage system or network.

Digital Loop Carrier (DLC)-A highly efficient digital transmission system that uses existing distribution cabling systems to transfer digital information between the telephone system (central office) and a users telephone and/or computer equipment. A DLC system usually includes a high-speed digital line (e.g. T1) from a central office and a remote digital terminal (RDT). The RDT converts the high-speed digital line to low speed lines (analog or digital) for routing to the end customers.

Digital Media-Digital media is the format of information that is used to express information (media) which is represented in a form that can have levels or signal composition of specific discrete levels (digital).

Digital Microwave-A microwave transmission system that transfers digital information through the modulation of a microwave carrier signal. The type of modulation used may be amplitude, frequency or phase shift, but the digital signal is used as the source of modulation information.

Digital Millennium Copyright Act (DMCA)-The digital millennium copyright act is a regulation (statute) that covers the technological methods to protect copyrights. A key aspect of the DMCA is restrictions on the creation or selling of products or systems that are designed to allow users to get around copy protection. The DMCA became law in October 1998.

Digital Modulation-A modulation process where the amplitude, frequency or phase of a carrier signal is varied by the discrete states (On and Off) of a digital signal.

This figure shows different forms of digital modulation. This diagram shows ASK modulation that turns the carrier signal on and off with the digital signal. FSK modulation shifts the frequency of the carrier signal according to the on and off levels of the digital information signal. The phase shift modulator changes the phase of the carrier signal in accordance with the digital information signal. This diagram also shows that advanced forms of modulation such as QAM can combine amplitude and phase of digital signals.

D

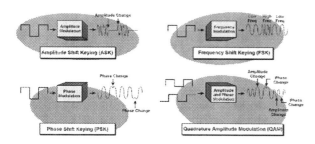

the same while the publisher information can change.

This figure shows digital object identifier (DOI) structure. This example shows that a DOI number is composed of a prefix that is assigned by the registration agency (RA) of the international DOI foundation and suffix that is assigned by the publisher of the content. This example shows that the first part of the prefix identifies the DOI directory that will be used and the second part identifies the publisher of the content.

Digital Modulation Operation

Digital Multiplex System (DMS)-The trade name of a line of digital telephone office switches from Nortel Networks (formerly Northern Electric, then Northern Telecom). There are DMS-10 and DMS-100 end subscriber switches, DMS-200 and DMS-250 transit/tandem switches, DMS-MTX cellular/wireless switches and DSM-300 international gateway switches (used to connected between North America and other countries).

Digital Object-A digital object is a group of data (digital) bits that represent data, information or images.

Digital Object Identifier (DOI)-A digital object identifier is a unique number that can be used to identify any type or portion of content. DOI numbers perform for long term (persistent) and locatable (actionable) identification information for specific content or elements of content. This content can be in the form of bar codes (price codes), book or magazine identification numbers or software programs. The DOI system is managed by the International DOI foundation (IDF) that was established in 1998. More information about DOI numbering can be found at www.DOI.org.

DOI numbers point to a DOI directory which is linked to specific information about a particular object or information element. The use of a DOI directory as a locating mechanism allows for the redirecting of information about identification information as changes occur in its identifying characteristics. For example, a book identification number may belong to the original publisher until the copyright of the work is sold to another publisher. At this time, the owner of the item content changes. The item number on the book can remain

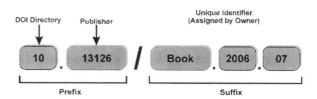

Digital Object Identifier Structure

Digital Power Line-Digital power line is a term that refers to the sending of digital information through electric power lines.

Digital Pricing Extortion-Digital pricing extortion is a practice that is used by television service providers to artificially inflate the price of analog programming tiers to create small incremental pricing differences between analog and digital tiers, thereby encouraging digital upgrades by subscribers

Digital Program Insertion (DPI)-Digital program insertion is the process of splicing media segments or programs together. Because digital media is typically composed of key frames and difference pictures that compose a group of pictures (GOP), the splicing of digital media is more complex than the splicing of analog media that has picture information in each frame which allows direct frame to frame splicing.

Digital Programming Extortion-Digital programming extortion is a practice of lining up or transferring highly valued program content that is delivered on analog channels so they are only available on digital channels to help motivate viewers to upgrade their service to include digital programming.

Digital Property Rights Language (DPRL)- Digital property rights language is a set of instructions and procedures that are used to define rights of digital media. It was invented by Mark Stefik of Xerox's Palo Alto research center in the mid 1990s and has transformed into extensible rights markup language (XrML). The initial version of DPRL was LISP based and 2nd version (2.0) was based on XML to allow the flexibility of describing new forms of data and processes.

Digital Rights Management (DRM)- Digital rights management is a system of access control and copy protection used to control the distribution of digital media. DRM involves the control of physical access to information, identity validation (authentication), service authorization, and media protection (encryption). DRM systems are typically incorporated or integrated with other systems such as content management system, billing systems, and royalty management. Some of the key parts of DRM systems include key management, product packaging, user rights management (URM), data encryption, product fulfillment and product monitoring.

Digital Service Unit (DSU)- A device that interconnects the customer's digital telephone equipment to a telephone network.

Digital Service, Level 0 (DS-0)- A 64,000 b/s channel, the worldwide standard public telephone industry bit rate for digitizing one voice conversation. There are 24 DS-0 channels in a 1.544 Mb/s DS-1 digital multiplex bit stream, and 30 DS-0 traffic channels plus two additional DS-0 channels used for synchronization and signaling (a total of 32 DS-0 channels) in a 2.048 Mb/s E-1 digital multiplex bit stream.

Digital Set Top Box (Digital STB)- A digital set top box is an electronic device that adapts a communications medium to a format that is accessible by the end user. Digital set top boxes are commonly located in a customer's home to allow the reception of digital video signals on a television or computer.

Digital Signal- Digital signals consist of a series of ones and zeros, most often represented in telecommunications signals by two different voltages. For example a +5 Volt level could represent a logical 1 (one) and 0 Volt level could represent a logical 0 (zero). The ones and zeros are called bits. Several bits (usually eight) are grouped into a byte and each byte is defined to have a specific meaning, such as a specific letter on a keyboard. Digital signals are used to represent specific levels on an analog signal. While a digital signal cannot represent every point on an analog wave, they can come close enough to be almost indistinguishable. Digital signals are much easier to process by computer systems and they are able to resist the effects of noise better than analog signals.

Digital Signal 1 (DS-1)- The primary rate telephone industry digital multiplexing system used in North America and Japan. It combines 24 DS-0 (64 kb/s) channels and a single 8 kbit/s synchronizing bit stream for a total of 1.544 Mb/s. A different primary rate multiplexing system that combines 30 DS-0 channels with a 64 kb/s channel for signaling and another 64 kb/s channel for synchronization and control, for a total of 2.048 Mb/s, is used elsewhere. The 2.048 Mb/s system is sometimes named E-1, or MIC, or CEPT Primary Rate Multiplexing. Both the 1.544 Mb/s DS-1 system and the 2.048 Mb/s system are recognized ITU standards. The 2.048 Mb/s system was designed to have some improvements and a slightly larger channel capacity than the 1.544 Mb/s DS-1 system, and was intentionally incompatible, a result attributed by some industry observers to a motive of protecting the European market from imported product competition. (Later, manufacturers in different countries changed their objectives from the former strategy of setting intentionally incompatible standards in different countries to the present strategy of setting internationally compatible standards in all countries.) DS-1 is not a trade name of any manufacturer. T-1 is effectively a synonym of DS-1 in North America, and was originally a trade name of just one manufacturer, but today it is widely used for all compatible products regardless of manufacturer.

Digital Signal 3 (DS-3)- A standard digital transmission line that is divided into twenty eight DS1 (T1) channels. The gross transmission rate for a DS3 channel is 44.736 Mbps. A single DS3 provides for 672 standard (64 kbps) voice channels.

Digital Signal Level (DSx)- Digital signal (DS) transmission is a hierarchy of digital communication channels and lines that range from 64 kbps to 565 Mbps. Lower level DS structures are combined to produce higher-speed communication lines. There are different structures of DS levels used throughout the world with significant variations between North American and European systems. DSx has been used to represent the digital transmission standards where the "x" denotes which service is under discussion.

Digital Signal Processing-Digital signal processing is the manipulation of digital signals into other forms using computing circuits or systems. Digital signal processors use software programs to allow them to perform complex signal processing operations such as filtering, modulation, data compression, and shaping of the information (such as digital audio signal) that are represented by digital signals.

Digital Signal Processor (DSP)-An integrated circuit (chip) that is designed specifically for high-speed manipulation of digital information. DSP chips operate using software programs to allow them to perform complex signal processing operations such as filtering, modulation, data compression, and information processing. The use of DSPs in communication circuits allows manufacturers to quickly and reliably develop advanced communications systems through the use of software programs. The software programs (often called modules) perform advanced signal processing functions that previously complex dedicated electronics circuits. Although manufacturers may develop their own software modules, DSP software modules are often developed by other companies that specialize in specific types of communication technologies. For example, a manufacturer may purchase a software module for echo canceling from one DSP software module developer and a modulator software module from a different DSP software module developer.

This figure shows typical digital signal processor that is used in a digital communication system. This diagram shows that a DSP contains a signal input and output lines, a microprocessor assembly, interrupt lines from assemblies that may require processing, and software program instructions. This diagram shows that this DSP has 3 software programs, digital signal compression, channel coding, and modulation coding. The digital signal compression software analyzes the digital audio signal and compresses the information to a lower data transmission rate. The channel coding adds control signals and error protection bits. The modulation coding formats (shapes) the output signal so it can be directly applied to an RF modulator assembly. This diagram also shows that an optional interface is included to allow updating of the software programs that are stored in the DSP.

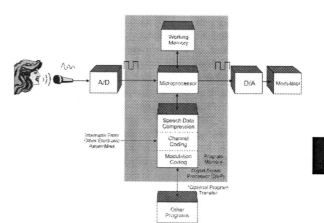

Digital Signal Processor (DSP) Operation

Digital Signal Regeneration-Digital signal regeneration is the process of reception and restoration of a digital pulse or lightwave signal to its original form after its amplitude, waveform, or timing have been degraded during transmission. The resultant signal is virtually free of noise or distortion.

Digital Signaling Tone (DST)-A tone that is sent on the analog radio channel to indicate a change in status (e.g. end call).

Digital Signature-A number calculated from the contents of a file or message using a private key and appended or embedded within the file or message. The inclusion of a digital signature allows a recipient to check the validity of file or data by decoding the signature to verify the identity of the sender.

Digital Sound Broadcasting (DSB)-Digital sound broadcasting is the process of transmitting audio broadcast information using digital transmission. The digital sound broadcast signal is typically shared with additional digital information that describes or supplements the audio program content.

Digital Speech Interpolation (DSI)-In addition to multiplexing through channel division, statistical multiplexing can also be used by distributing transmission of a communications channel over idle portions of multiplexed channels. An example of statistical multiplexing is digital speech interpolation (DSI). DSI is a technique that dynamically allocates time slots for voice or data transmission to a user only when the have voice or data activity. This increases the system capacity as transmission for other users can occur when others are silent. This figure shows the process of multiplexing using DSI. This diagram shows a communication circuit that has 96 independent communication channels (one communication link that has 96 time slots). The DSI system monitors the activity of each voice conversation (a voice channel) using a voice activity detector (VAD). The VAD is an electronic circuit that senses the activity (or absence) of voice signals. This is used to inhibit a transmission signal during periods of voice inactivity.

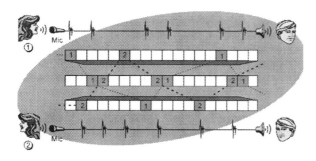

Digital Speech Interpolation (DSI) Operation

Digital STB-Digital Set Top Box

Digital Storage Medium (DSM)-A digital storage medium is a form of material that can be used to store information in digital levels (1s and 0s).

Digital Subscriber Line (DSL)-Digital subscriber line is the transmission of digital information, usually on a copper wire pair. Although the transmitted information is in digital form, the transmission medium is usually an analog carrier signal (or the combination of many analog carrier signals) that is modulated by the digital information signal.

This figure shows a simplified ADSL communication system that consists of a digital subscriber line access multiplexer (DSLAM), local distribution lines that start from a main distribution frame (MDF) wire cabinet that brings the connection to

the digital subscriber line (DSL) modem at the customer's location. Modems in the DSLAM convert the digital signals from the Internet to high frequency signals that travel down the telephone line to the DSL modem. The DSL modem converts the RF signals back to its original digital form so it can be provided to the customer's computer. Most DSL technologies (such as ADSL shown in this example) transmit the data information on frequencies above the audio channel. This allows for the simultaneous transmission of analog and data signals on the same telephone line. The highest frequencies are used transmission from the DSLAM to the DSL modem and frequencies just above the audio band are used to transmit from the data from the customer to the DSLAM. Typical DSL technology allows up to 6 Mbps to be transmitted to the customer and up to 640 kbps can be received from the customer.

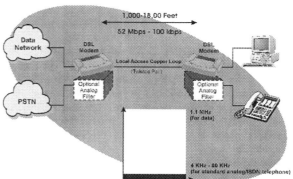

Digital Subscriber Line (DSL) System

Digital Subscriber Line Access Multiplexer (DSLAM)-A digital subscriber line access multiplexer is an electronic device that usually holds several digital subscriber line (DSL) modems that communicate between a telephone network and an end customer's DSL modem via a copper wire access line. The DSLAM concentrates multiple digital access lines onto a backbone network for distribution to other data networks (e.g. Internet).

This figure shows that a digital subscriber line access multiplexer (DSLAM) concentrates multiple digital subscriber lines onto a high-speed backbone network (e.g. ATM or Ethernet) for distribution to other data networks (e.g. Internet). In this diagram, the DSLAM contains a backbone assembly (multiple sockets) that allow for the insertion of DSL modem line cards and that each DSL modem line card can provide service to more than one customer. This allows the service provider to add DSL modem line cards as the number of DSL lines increase. This DSLAM also contains the simple network management protocol (SNMP) communication capability that can be used to control the DSLAM and the DSLAM configuration information is stored in a management information base (MIB).

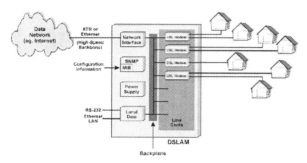

Digital Subscriber Line Access Multiplexer (DSLAM)

Digital Subscriber Line Modem (DSL Modem)-A DSL modem is an electronics assembly device that modulates and demodulates (MoDem) digital subscriber line (DSL) signals. DSL signals are usually transmitted on a twisted pair of copper wires. A DSL modem may be in the form of an internal computer card (e.g. PCI card) or an external device (Ethernet adapter). Most DSL modems have the ability to change their data transfer rates based on the settings that are programmed by the DSL service provider and as a result of the quality of the communication line (e.g. amount of distortion).

Digital Tape-Digital tape is a magnetic tape storage format that changes magnetic information on the tape to represent digital (discrete level) signals.

Digital Telephony-Digital telephony is a communication system that uses digital data to represent and transfer analog signals. These analog signals can be audio signals (acoustic sounds) or complex modem signals that represent other forms of information.

Digital Television (DTV)-Digital television is a process or system that transmits video images through the use of digital transmission. The digital transmission is divided into channels for digital

video and audio. These digital channels are usually compressed. Digital television systems commonly use one of the motion picture experts group (MPEG) standards to reduce the data transmission rate by a factor of 200:1.

Digital Terrestrial Television (DTT)-Digital terrestrial television is the broadcasting of digital television signals using surface based (terrestrial) antennas. DTT is also called digital video broadcasting terrestrial (DVB-T).

Digital Transmission-Digital transmission is the process of sending information in digital (discrete level) form.

Digital Transrating-Digital transrating is the process of converting digital information from one transmission rate to another transmission rate. An example of digital transrating is the conversion of a high-speed digital video signal that is received from a satellite into a medium-speed digital video signal that is transferred through the Internet.

Digital Tuner-A digital tuner is a processing device that can receive and decode digital media channels.

Digital Turnaround Devices-Digital turnaround devices convert or distribute digital signals or media that are in one format into digital signals or media in another format.

Digital Video-Digital video is a sequence of picture signals (frames) that are represented by binary data (bits) that describe a finite set of color and luminance levels. Sending a digital video picture involves the conversion of a scanned image to digital information that is transferred to a digital video receiver. The digital information contains characteristics of the video signal and the position of the image (bit location) that will be displayed.

This figure shows the basic process used by digital video to compress the video signal. This example shows that the first frame in a video sequence is a key frame. The next sequence of image data sent is the changes from the key frame. This diagram shows a person who is waving. Because they are sitting still, only the hand changes are sent in frames after the key frame. The information (data) sent for the changed images are much smaller than the full image information. This allows digital video to be compressed by substantial amount for video that does not have rapid changes.

Digital Video Broadcast (DVB)

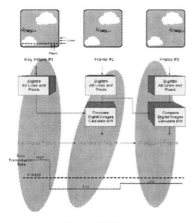

Digital Video

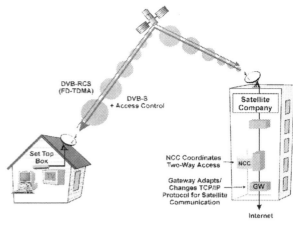

Satellite DVB-RCS System

Digital Video Broadcast (DVB)-Digital video broadcasting is the sending of television signals over digital transmission channels. DVB transmission can be over different types of systems including broadcast radio, satellite systems, cable television systems and mobile communications. DVB industry standards that are published by the joint technical committee (JTC) of the European Telecommunications Standards Institute (ETSI). These standards can be obtained at www.ETSI.org.

Digital Video Broadcast Return Channel via Satellite (DVB-RCS)-Digital video broadcast with return channel via satellite is an industry standard for high-speed two-way Internet communications via GEO satellite communication.

This figure shows how a satellite system can provide two-way communication capabilities. This diagram shows that the forward direction (satellite to end user) has higher data transmission rates than is available from the user to the satellite. This example also shows that the broadband satellite communication system includes an Internet gateway that can optimize the IP communication session to compensate for challenges created by the added transmission delay time.

Digital Video Broadcasting Cable (DVB-C)-Digital video broadcasting cable is the broadcasting of digital television systems using cable television based systems. DVB-C is also called digital cable television.

Digital Video Broadcasting Common Interface (DVB-CI)-A digital video broadcasting common interface is a defined connection (physical and electrical) that allows for detachable modules to be connected to television equipment (such as a set top box). These detachable modules may perform different types of functions including channel decoding, media reformatting or storage of user features and preferences.

Digital Video Broadcasting Common Scrambling Algorithm (DVB-CSA)-Digital video broadcasting common scrambling algorithm is a standard DVB encryption process that can be used by multiple companies and systems.

Digital Video Broadcasting Data (DVB-Data)-Digital video broadcasting data is a standard developed by the DVB to allow data broadcasting on DVB transmission channels.

Digital Video Broadcasting Handheld (DVB-H)-Digital video broadcasting handheld is the broadcasting of digital television systems using surface based (terrestrial) antennas to portable handheld devices. DVB-H uses a low bandwidth portion of the DVB-T system to transfer the video signal.

Digital Video Broadcasting Internet Protocol (DVB-IPI)-Digital video broadcasting Internet protocol is a standard developed by the DVB to allow IP data transmission on DVB transmission channels.

Digital Video Broadcasting Multimedia Home Platform (DVB-MHP)-Digital video broadcasting multimedia home platform is an industry standard for a digital television system that allows interactive applications.

Digital Video Broadcasting Satellite (DVB-S)-Digital video broadcasting satellite is the broadcasting of digital television systems using satellite transmission systems. DVB-S is also called digital satellite television.

Digital Video Broadcasting Teletext (DVB-Txt)-Digital video broadcasting teletext (DVB-Txt) is a service that transfers data information along with television signals to allow the simultaneous display of text and video on the television.

Digital Video Broadcasting Terrestrial (DVB-T)-Digital video broadcasting terrestrial is the broadcasting of digital television systems using surface based (terrestrial) antennas. DVB-T is also called digital terrestrial television.

Digital Video Cassette (DVC)-A digital video cassette is a self-contained package of reel-to-reel magnetic tape that is used for digital video signals and may be played or rewound on demand.

Digital Video Compression (DVC)-Digital video compression is the reduction of the number of digital bits required to represent a video signal by digital coding techniques. When compressed, a digital video signal can be transmitted on circuits with data rates relatively 50 to 200 times lower than their original uncompressed form.

Digital Video Home System (D-VHS)-Digital video home system (D-VHS) is a video tape storage format that is used to store digital video recording images 1/2 inch magnetic tape. Digital VHS stores the actual digital media allows for much higher resolutions than the 240 lines of resolution offered by standard VHS tape or 400 lines of resolution offered by S-VHS tapes. D-VHS is backward compatible with VHS and S-VHS so analog VHS tapes can typically be played in a D-VHS tape player..

Digital Video Interactive (DVI)-A digital video transmission system that allows real time (or near real time) interaction to allow changing of the information content. DVI combines digital video and audio and allows the computer user to control the operation of the media display. DVI is a registered trademark of Intel.

Digital Video Processing-Digital video processing is the process of converting and/or modifying digital video signals from one format (such as an encrypted form) into another form using digital signal processing.

Digital Video Quality (DVQ)-Digital video quality is the ability of a display or video transfer system to recreate the key characteristics of an original digital video signal. Digital video and transmission system impairments include tiling, error blocks, smearing, jerkiness, edge busyness and object retention.

Digital Video Recorder (DVR)-A digital video recorder is a device that stores video images in digital format.

Digital Video Tape (DV)-A digital video tape is a small self-contained package of reel-to-reel magnetic tape that is used for digital video signals and may be played or rewound on demand.

Digital Videotape (DV)-Digital videotape is a magnetic tape storage format that changes magnetic information on the tape to represent digital video signals. There are several industry formats for digital video.

Digital Visual Interface (DVI)-Digital visual interface is a standard protocol that was created by the digital display working group (DDWG) is used to transfer high-speed digital video information to monitors. The DVI system uses a protocol called transition minimized differential signaling between the digital video source and display device. The standard includes a connector that is capable of providing the new high-speed digital display signal along with legacy VGA interfaces.

Digital Voice Coding-Digital voice coding is the processing of digital audio information into a specific format. Digital voice coding is typically used to assist in the transmission and/or compression of the signal in a communication system.

This diagram shows the basic process that is used for digital voice compression process. In this diagram, a digital audio signal (64 kbps PCM signal) is continuously applied to a digital signal analysis device. The analysis portion of the speech coder extracts the amplitude, pitch, and other key parameters of the signal and then looks up related values in the code book for the portion of sound it has analyzed. Only key parameters and code book values are transmitted. This results in data compression ratios of 4:1 to over 16:1.

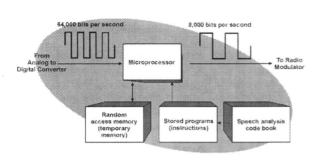

Digital Voice Compression Operation

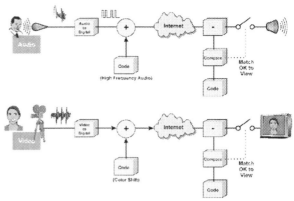

Digital Watermarking

Digital Voltmeter-A voltmeter that displays its readings in a digital format, either by LCDs or by a digital output signal supplied to another instrument.

Digital Watermark-A digital watermark is a signal that is hidden (typically is imperceptible to the user) in a digital signal (such as in the digital audio or a digital image portion) that contains identifying information. Ideally a digital watermark would not be destroyed (that is, the signal altered so that the hidden information could no longer be determined) by any imperceptible processing of the overall signal, for example high-quality lossy compression, slight equalization, or digital-to-analog-to-digital conversion.

This figure shows how watermarks can be added to a variety of media types to provide identification information. This example shows that digital watermarks can be added to audio or video media. The digital watermark is added as a code that is typically not perceivable to the listener of view of the media.

Digital Wrapper-A digital wrapper is data or information that is added to media (such as a video program). Digital wrappers may provide descriptive and content protection information.

Digital8-Digital8 is a video camcorder format that can record digital video onto standard 8mm and Hi8 tapes.

Digital-To-Digital Transfer (D-t-D)-The process of transferring digital information (audio, video, or data) from one machine to another in the digital domain. The advantage in D-to-D transfers is that no signal degradation occurs because this process bypasses all of the analog circuitry which can degrade overall performance.

Digitization-Digitization is the conversion of analog into digital form. To convert analog signals to digital form, the analog signal is digitized by using an analog-to-digital (pronounced A to D) converter. The A/D converter periodically senses (samples) the level of the analog signal and creates a binary number or series of digital pulses that represent the level of the signal.

The common conversion process is Pulse Code Modulation (PCM). For most PCM systems, the typical analog sampling rate occurs at 8000 times a second. Each sample produces 8 bits digital that results in a digital data rate (bit stream) of 64 thousand bits per second (kbps).

Digital bytes of information are converted to specific voltage levels based on the value (weighting) of the binary bit position. In the binary system, the value of the next sequential bit is 2 times larger. For PCM systems that are used for telephone audio signals, the weighting of bits within a byte of information (8 bits) is different than the binary system. The companding process increases the dynamic range of a digital signal that represents an analog signal; smaller bits are given larger values that than their binary equivalent. This skewing of weighing value give better dynamic range. This companding process increases the dynamic range of a binary signal by assigning different weighted values to each bit of information than is defined by the binary system.

Two common encoding laws are Mu-Law and A-Law encoding. Mu-Law encoding is primarily used in the Americas and A-Law encoding is used in the rest of the world. When different types of encoding systems are used, a converter is used to translate the different coding levels.

This figure shows the basic audio digitization process. This diagram shows that a person creates sound pressure waves when they talk. These sound pressure waves are converted to electrical signals by a microphone. The bigger the sound pressure wave (louder the audio), the larger the analog signal. To convert the analog signal to digital form, the analog signal is periodically sampled and converted to a number of pulses. The higher the analog signal, the larger the number of pulses. The number of pulses can be counted and sent as digital numbers. This example also shows that when the digital information is sent, the effects of distortion can be eliminated by only looking for high or low levels. This conversion process is called regeneration or repeating. This regeneration progress allows digital signals to be sent at great distances without losing the quality of the audio sound.

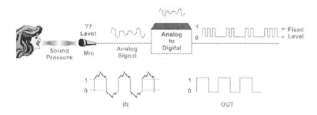

Digitization Process

Diplex Filter-A diplex filter is a device or assembly that combines two bandpass filters with different frequency bands to allow a transmitter to pass through the filter in one frequency range (transmit frequency band) and allow receiver signals to pass through on another frequency range (receive frequency band). The bandpass filters block signals from the transmitter from entering into the receiver.

Diplexer-(1-mobile communication) A device that enables the outputs of two wireless base stations or antenna to be combined onto a single feeder cable, which allows co-sitting of mobile communication systems. (2-television) A device that combines the transmission of audio and video signals over a common channel.

Direct Broadcast Satellite (DBS)-Direct broadcast satellite is a satellite with enough range and power to be received by small dish antennas suitable for consumer home use. DBS can be sent to both direct individual homes as well as received by communities by means of retransmission over a small TV station or cable TV system. In the late 1990's, the DBS marketplace became a formidable competitor to the traditional cable industry. DBS systems provide digital-quality pictures and have the potential to offer high speed interactive services. By using digital compression technology, DBS systems can offer a greater number of channels than analog cable systems to both PCs and TVs. DBS systems can also be customized to provide unique services for limited video on demand (VOD), near video-on-demand (NVOD) and interactive pay-per-view channels.

Direct Connect-(1-mobile radio) Direct connection is the communication between two or many mobile radios or telephones. While the users may be directly communicating with each other, the radios may be actually connected through a relay site or cell site. (2-local telephone bypass) The direct connection of a customer to an Inter-exchange carrier (IXC) via copper, wireless, or optical connections. This connection bypasses the local telephone company's switching and transmission facilities.

Direct Distance Dialing (DDD)-The automatic completion of customer dialed long distance or toll calls in response to signals from a customer's telephone where no operator assistance is required.

Direct Inward Dialing (DID)-Direct Inward Dialing (DID) connections are trunk-side (network side) end office connections. The network signaling on these 2-wire circuits is primarily limited to 1-way, incoming service. DID connections employ different supervision and address pulsing signals than dial lines. Typically, DID connections use a form of loop supervision called reverse battery, which is common for 1-way, trunk-side connections. Until recently, most DID trunks were equipped with either Dial Pulse (DP) or Dual Tone Multifrequency (DTMF) address pulsing. While many wireless carriers would have preferred to use Multifrequency (MF) address pulsing, a number of LEC's prohibited the use of MF on DID trunks.

This figure shows the basic operation of direct inward dialing in a PBX system. This diagram shows how a caller has dialed a person in a company through the public telephone number. When this call is received by the end office (EO) switch in the public telephone company, the public telephone operator connects the call to one of the available incoming trunk lines between the telephone company switch and the PBX switch. This example shows that in addition to connecting the call and sending an alert (ringing) signal, the called number is also sent to the PBX. This allows the PBX system to lookup the called number to determine which extension the call should be connected to. The PBX system then connects the incoming trunk line to the correct telephone extension.

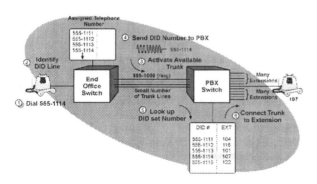

Direct Inward Dialing (DID) Operation

Direct Modulation-Direct modulation is the changing of a carrier signal (such as a laser or LED light source) by the controlling (modulating) its power source and therefore, controlling the amount of light it emits.

Direct Outward Dialing (DOD)-A feature that allows private telephone systems users (PBX or Centrex) to directly call public telephone numbers without the need to use an attendant or operator. The use of "dial 9" to get an outside line is a direct outward dialing feature.

Direct Sequence Spread Spectrum (DSSS)-A transmission technique in which multiplies an information signal with a sequence code (or information signal) to allow the modulation signal to occupy a frequency bandwidth that is much wider than is necessary to represent the information signal alone. The DSSS system then uses the multiplying code to decode the received signal.

Direct To Home (DTH)-Direct to home is satellite service that provides broadcast signals direct to end users.

Directional Antenna-A directional antenna focuses the transmitted or received signal in a specific direction. Directional antennas are used to provide signal gain in a specific direction while reducing the signal levels (that may cause interference) in other directions.

This diagram shows how the energy of an antenna can be focused (directed) to a particular area. This diagram shows the focusing of the beam into a main lobe and that the transmission patterns of directional antennas usually result in the creation of unwanted side lobes.

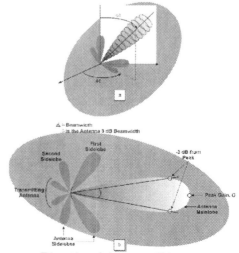

Directional Antenna Diagram

Directional Coupler-A device that provides a sample of transmitted energy as the signal passes through in a particular (forward direction). Signals that pass through the directional coupler in the opposite direction (e.g. reflected signals) are isolated from the sample port.

This diagram shows how a directional coupler allows a signal to pass through while it directs (samples) a portion of an input signal to a coupled output port. This example shows how a 20 dB directional computer (20 dB = factor of 100 times) allows 99% of the RF signal to pass through the main output while directing 1% (10 mWatts) of the signal to the coupled output.

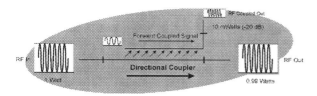

Directional Coupler

Directional Differences-Directional differences are variances in performance measurements that result with a transmission line or system is tested in the opposite direction. Directional differences can result in optical systems due to the use of direction sensitive components such as angled connectors.

Directional Filter-A device or assembly that combines high-pass and/or low-pass filters that are used to separate a frequency range in a portion of transmission bands in a specific direction of transmission. For example, a directional filter can be used in a bi-directional system to extract a particular signal in a specific bandpass frequency range in a single direction.

Directories-(1-computer storage) File directories are portions of a computer storage area that are dedicated to the listing of file names and location pointers (starting addresses and file size) for the files contained on the computer storage system. (2-Internet) A web site directory is a listing of web pages that have specific characteristics (e.g. subjects). Web site search visitors enter key words (category search words) to find web pages that contain the characteristics associated with the search

words. Directories are typically different than search engines because URL submission and categorization process involves a human review process.

Directories URL-An address that is preprogrammed into or used by devices (such as an IP Telephone) where information is kept regarding direct listings.

Directory-(1-computer) A list of all the files on a floppy diskette, hard disk, or web site. A directory may also contain other information such as the size of the files and the amount of free space remaining. (2-telephone listing) A telephone directory. (3-Internet listings) A web site portal (search engine or listing) that identifies companies, products, or listings of specific category types.

Directory Enabled Network (DEN)-A directory enabled network is the coordination of communications within a network through the use of a central information database that contains information about users, applications and network resources.

Directory Gatekeeper-A gatekeeper that is used in a large voice over data communications network as a central information point for other gatekeepers. A directory gatekeeper allows an administrator to manage the configuration of the network without having to configure many local area systems.

DISA-Direct Inward Service Access

Disassociation-A process of de-registering a wireless data device (station) with a specific access point (AP) in an 802.11 specified wireless local area network (WLAN) system.

Disaster Recovery-The processes that are used to restore services after a significant interruption (disaster) in communications systems. Disaster recovery processes usually occur after events such as fires, floods, or earthquakes. However, disaster recover may also occur after critical equipment failures or information corruption that occurs from software viruses.

Discard Eligibility (DE)-A control flag system to indicate the essential nature of the packet's data that is transmitted through a packet data network. The DE flag(s) allow systems to selectively discard data packets or frames that are non-essential. This process allows some data transmission systems to send more data than is agreed to (dynamic bandwidth). If the network is not congested, it may allow the extra packets of data to reach their destination.

Disconnect for Non-Payment (DNP)-A transaction transmitted to the Network Operations system requesting that a customer's service be disconnected due to non-payment.

Discontinuous Reception (DRx)-Discontinuous reception (DRx) is a process of turning off a radio receiver when it does not expect to receive incoming messages. For DRx to operate, the system must coordinate with the mobile radio for the grouping of messages. The mobile radio (or pager) will wake up during scheduled periods to look for its messages. This reduces the power consumption which extends battery life. This is sometimes called: sleep mode.

This shows how paging groups can be used to provide for discontinuous reception capability. This diagram shows that the paging channels can be divided into 200 msec groups and that paging groups are typically associated with the last digit of the mobile devices telephone number. This provides for 10 groups with a typical maximum delay of 2 seconds.

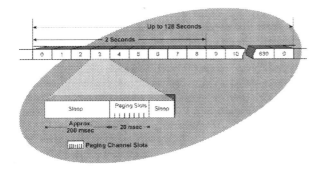

Discontinuous Reception (DRx) Sleep Mode
Operation

Discontinuous Transmission (DTx)-The ability of a communications system to inhibit transmission when no or reduced activity is present on a communications channel. DTx is often used in mobile telephone systems to conserve battery life.

Discount-Discounts are reductions of pre-established fees or tariffs that are given for specific reasons. Discounts may be in the form of a specific amount or they may be based on a percentage of an item price or invoice amount. Discount types may be coded using specific identification codes and dis-

count rates may be applied based on the specific type of sale or customer category using a discount schedule.

Discoverable Device-A discoverable device is a communication device that is within range of another communication device that will respond to an inquiry message. For the Bluetooth system, there are two types of discoverable modes: limited and general. In the first case, a device may be available for discovery for a limited period of time, during temporary conditions, or for a specific event. In the second case, a device may be available for discovery on a continuous basis.

Discovery-Discovery is the process and protocols that are used to allow for the recognition of devices that are operating within a network or radio coverage area and what services they are capable of using. For example, a palmtop needs to discover the home network and find a service that will provide palmtop-to-PC synchronization capabilities.

Discovery Metadata-Discovery metadata is information (data) that describes the content and attributes of the content that is contained in a collection of media (such as a media program).

Discrete Cosine Transform (DCT)-Discrete cosine transform, is a form of frequency analysis that is applied to discrete signals (e.g. binary data) to produce an output that is composed of the frequency components and the levels (coefficients) that represent the original digital signal. A DCT output is composed of a DC component (basic intensity) and a series of increasing frequency components that reflect the complexity of the underlying data.

Discrete Fourier Transform (DFT)-A Fourier transform applied to a periodic sequence of complex values (known as samples) at discrete times. The result of this transform is a periodic sequence of complex values at discrete frequencies. Because the input and output of the DFT are periodic sequences, each can be represented by only one period of samples. Since the DFT is a mapping of one finite sequence to another finite sequence, it can be implemented by a computer in a straightforward way, unlike a general Fourier transform of an arbitrary continuous function.

Discrete Multi-Tone (DMT)-A data communications process that transfers a high speed data communication channel by dividing it into several narrow sub-channels and sending them independently through frequency divided channels. When the sub-channels are received, the low speed parts are recombined to create the original high-speed data transmission signal.

The advantage of sending several sub-channels is the ability to independently adjust the transmission levels of each sub-channel signal. Because the frequency response of the line can vary and distortions can occur on specific frequencies (where only a few sub-channels may be affected). DMT is used in DSL systems as it adapts well to the hostile environment of copper wire transmission.

This figure shows a discrete multitone transmission system (DMT) system. In this diagram, a high-speed data signal is divided into several low speed data signals. Each low speed data signal modulates a sub-channel. The sub-channels are combined and supplied to the copper wire. At the receiving end, each sub channel is received and decoded. The sub-channel data signals are re-combined to recreate the original high-speed data signal.

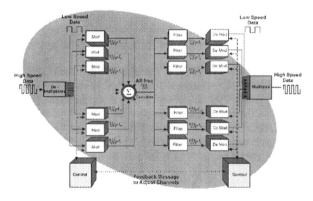

DMT Transmission System

Disk Mirroring-A data protection strategy that uses redundancy of information on two (or more) storage devices (disks) to allow for real time backup and recovery of information. The process of disk mirroring is the storage of the same information on two disks. One disk is used as the primary source of information and the other disk is used when a failure is detected on the primary disk.

This figure shows the process of mirroring information on two disk storage devices to increase the reliability of information storage. Disk mirroring is also known as redundant array of inexpensive disks (RAID). This diagram shows that disk Mirroring is performed by a RAID controller card that controls both primary and secondary disk stor-

age devices. Information is stored to both hard disks simultaneously. If the primary hard drive fails, the controller will automatically begin to use the secondary hard disk and alert the user that one of the hard disks has failed.

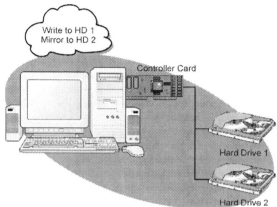

Disk Mirroring Operation

Dispersed Control-A system that coordinates the distribution of the call processing or switching intelligence to multiple control parts of a system or network.

Dispersion Rate-Dispersion rate is a reference value that is used to calculate the bandwidth capacity of a single mode fiber over a specific distance. The dispersion is determined by the physical construction of the fiber or cable and is expressed in terms of picoseconds per nanometer of wavelength per kilometer of distance (ps/nm/km). Using the dispersion rate as opposed to BWDP, this allows the maximum distance to be calculated to vary based on the width of the optical source. The more narrow the optical source (e.g. different laser types), the longer the maximum distance.

Dispersion Shifted Fiber (DSF)-Dispersion shifted fiber is a fiber that is constructed to shift the typical zero dispersion wavelength (lowest attenuation loss) of 1310 nm to another wavelength (usually near 1550 nm). The materials in a dispersion shifted fiber are selected so that the chromatic dispersion cancels out waveguide dispersion.

Display Formatting-Display formatting is the positioning and timing of graphic elements on a display area (such as on a television or computer display).

Displayable Character-Any letter, number or symbol which can be displayed on-screen display.

Disruptive Technology-A new technology that significantly reduces or eliminates the value of existing technology implementations by performing or providing the service faster and at lower cost.

Distance Learning-Distance learning is the process of providing educational training to students at locations other than official learning centers (schools). Distance learning has been available for many years and is now used in elementary education (grades K-12), higher education (college), professional (industry), government training and military training. In the early years, distance learning was provided through the use of books and other printed materials and was commonly referred to as "correspondence courses".

Distance learning has evolved through the use of broadcast media (e.g. televisions) and moved onto individual or small group training through the availability of video based training (VBT) or computer based training (CBT). These systems have developed to interactive distance learning (IDL) as the computer allowed changes in the training.

Distance Vector-A distance vector is a measurement of the length of a path between points and the direction the path has (from its origin to its end). When used in networks, a distance vector may represent the length of a connection or the number of routers between connection points.

Distance Vector Multicast Routing Protocol (DVMRP)-Distance vector multicast routing protocol are the commands and processes that assist a router in the selection of routes based on the distance between connections. DVMRP keeps track of the distance of other connections between routers in a network. DVMRP uses Internet Group-Management Protocol (IGMP) to transfer routing information with neighboring routers.

Distance-Vector Routing Protocol-A routing protocol which mathematically computes routes using a measurement of distance. This measurement is known as the distance vector and can be based on a link's speed or other characteristic of the link. Each router utilizing a D-V protocol periodically transmits all or a subset of its routing table, in a routing-update message, at a regular interval to each of its adjacent routers. As this routing information proliferates through the network each router recomputes the distance to other routers by adding each link's D-V weight. The routing update messages identify new destinations as they are added to the network, convey link failures information, and calculate distances to all known destinations.

Distance vector routing protocols are often contrasted with Link-State routing protocols, such as Open Shortest Path First (OSPF) which requires each router to send only its local connection information, not too its neighboring routers, but instead to all routers in the internetwork.

In short, link-state algorithms require much more intelligence at each router but only require small update messages to be sent to each router. Distance Vector algorithms send large updates, but only to adjacent routers.

D-V protocols are easier to implement than Link-State protocols, however, D-V protocols are less resilient and take longer to converge. Routing Information Protocol (RIP) is the most common D-V routing protocol in use. More information about D-V, sometimes called Bellman-Ford algorithms can be found in RFC 1058, the RIP protocol specification.

Distinctive Ringing-A service feature that alerts a customer via a special ring (usually short, long or rapid ring) that an incoming call is received that has a different purpose or priority from others that are received on that same telephone line. Distinctive ringing is used for sharing multiple phone numbers on a single line or for priority ringing.

This diagram shows the operation of a telephone system that has distinctive ringing feature. This diagram shows a single telephone that is assigned two different telephone numbers even though the telephone operates on one telephone number (one switch port). In this example, when an incoming call is received for the registered number 555-6234, it is re-directed (forwarded) towards the actual destination number 555-1234 along with information that allows the system to uniquely identify the call with a dual ring (2 rings in the 2 second ring period). When calls are received to 555-1234, the ring is a single 2 second/4 second cadence. This allows the receiver of the call to determine which telephone number was dialed by the distinctive ring sound.

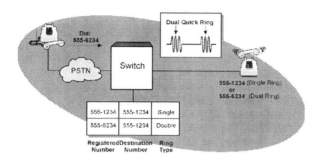

Distinctive Ringing Operation

Distortion-(1-general) The inaccurate reproduction of a signal caused by changes in the signal waveform. The difference between the wave shape of an original signal and the signal after it has traversed the transmission circuit. (2-delay) The distortion caused by the later arrival of higher-frequency components of a complex waveform as a result of the slower travel speed of higher-frequency components. (3-envelope delay) The distortion caused by a delay of the envelope or group of signals passing through a network. (4 - frequency) The changes in the relative amplitudes of different frequency components of a complex wave form. (5 - harmonic) The distortion caused by the creation of harmonics of a fundamental frequency. (6 - intermodulation) The distortion produced when two or more waves (or a complex waveform involving two or more frequencies) pass through a nonlinear device that produces sum-and-difference modulation product frequencies. (7 - linear) A distortion that is independent of the signal amplitude. (8 - nonlinear) A distortion that is dependent on signal amplitude. (9 - optical waveguide) The signal distortion caused by three primary mechanisms: waveguide dispersion, material dispersion, and profile dispersion. In addition, the signal may suffer degradation from intramodal distortion and multimode distortion. (10 - phase-frequency) The distortion resulting from the difference between phase delay at a given frequency and at a reference frequency.

Distress Stream-A distress stream is a low bit rate (low resolution) media stream that is used to display media to the viewer when the error rate or availability of the standard or high-resolution media stream becomes unusable.

Distributed Billing-Distributed billing is a network that is designed to receive and process call detail and service usage information where the reporting and reconciling of billing records may be performed at various billing concentration (distribution) points within the billing system. Distributed billing transfers some of the intelligence (bill processing functions) to various points (distributed intelligence) to reduce the number of billing records that reach the central billing system.

Distributed Control-A system that distributes the call processing or switching intelligence to multiple parts of the system or network.

Distributed Routing-Distributed routing uses intelligent routers or switches that forward packets toward their destination rather than using centrally coordinated routing paths. This can provider for a more robust network as each switch or router makes its own routing decisions without the need for a centralized control center.

Distribution-(1-switching network) The capability in a switching system of connecting an input to any of several outputs. (2-traffic network) The separation of calls on incoming trunk groups at a toll or tandem office and their recombination on outgoing trunk groups. (3-data network) The process of distributing information from one station to one or more stations.

Distribution Amplifier (DA)-A distribution amplifier is an active device used to replicate an input signal, typically providing several outputs, each of which is identical to the input. A DA also may include delay and/or cable equalization capabilities.

Distribution Cable-A cable or cabling system that is used to transfer signals from a central location (e.g. a central office or the head end of a CATV system). to end customers.

Distribution Channel-A distribution channel is the route that a product service uses to get from the original manufacturer or supplier to the customer or end user.

Distribution Channels-Distribution channels are the organizational units (internal and external) that are dedicated to the process of distributing products and services directly to customers through retail or interpersonal facilities.

Distribution Fee-Distribution fee is an amount of money charged by most film and television distributors in order to be compensated for selling or licensing of programming on behalf of producers.

Distribution Network-(1-general) A system of cables and terminals that interconnect communication devices. (2-cable television) The portion of a cable television system that links the head end to the end customers televisions. (3-telephone) Cables from a main telephone switching or distribution junction that usually contain from 25 to 200 pairs of wires.

Distribution Service (DS)-A service of distributing the same information to multiple receivers that are connected to a network.

Distribution System (DS)-(1-communication) A system or network that is composed of transmission lines and switching or hub equipment that allows signals are messages to be transferred from one point to one or many other points in the network. (2-802.11 WLAN) The process or system that is used by access points (APs) to communicate with each other.

Distribution Tree-A distribution tree is the data packet transmission routes that are taken by multicast packets in a data communication network. A distribution tree has an origination point (a route) that connects to branches in the distribution tree.

Distributor-(1-signal) The module within a link aggregator responsible for assigning frames submitted by higher-layer clients to the individual underlying physical links. (2-Product) A company or individual that sells products they receive from other companies to retailers or customers.

Diversity-Diversity is the process of combating the effects of path fading in a radio communications system by combining two or more received signals.

DivX-DivX is a digital video format that evolved as an alternative to the standard MPEG digital media format to allow the use of AVI file containers. DivX uses MPEG-4 digital video compression technology and MP3 digital audio compression technology.

DL-Data Link

DLC-Digital Loop Carrier

DLCI-Data Link Connection Identifier

DLEC-Data Local Exchange Carrier

D-Link-Diagonal Link

DLL-Dynamic Link Library

DLMR-Digital Land Mobile Radio

DM-Delta Modulation

DMA-Differential Mode Attenuation

DMAT-Digital Music Access Technology

DMCA-Digital Millennium Copyright Act

DMI-Desktop Management Interface

DMIF-Delivery Multimedia Integration Framework

DMS-Digital Multiplex System

DMT-Discrete Multi-Tone

DMV-Differential Motion Vector

DMZ-DeMilitarized Zone

DND-Do Not Disturb

DNIS-Dialed Number Identification Service

DNP-Disconnect for Non-Payment

DNS-Domain Name Server

DOA-Dead On Arrival

DOC-Department Of Communications

DOCSIS-Data Over Cable Service Interface Specifications

DOCSIS+-Data Over Cable Service Interface Specification +

Documentation Server-A documentation server is a computer that stores reference materials, brochures or the locations of documents. A documentation server is typically attached to a network so users, managers or order processing systems can select, access and distribute documentation.

DOD-Direct Outward Dialing

DOI-Digital Object Identifier

Dolby AC-3®-Dolby AC-3 is a digital compression process that was developed by Dolby® laboratories that is commonly used in movie theaters and on DVDs.

Dolby Digital®-Dolby digital is the process that is used to code, compress and structure digital audio.

Dolby E®-Dolby E is an audio compression and transmission format that allows the transmission of up to 8 channels of audio and its descriptive metadata to be transmitted on a single high-speed data connection (such as a single AES-3 connection).

Dolby(r) Noise Reduction (Dolby NR)-Dolby is an audio signal processing system that is used to reduce the noise or hiss that was invented by Ray Dolby. The original Dolby noise reduction process that was developed in 1960s used companding and expanding to adjust the dynamic range of the audio into a range that was more suitable for stored or transmitted medium. Since its original development, various enhancements to the Dolby system have been developed including Dolby A, Dolby B, Dolby C, Dolby S, Dolby SR and Digital Dolby.

Domain Name-A domain name is the unique text or sequence of characters that is used to identify an address where information can be accessed on a data communication network. A domain name is associated with one or more IP addresses through the use of Domain Name Service (DNS).

Domain Name Server (DNS)-A domain name server is a data processing device (e.g. a computer) that translates text and numeric names for an Internet addresses. A DNS uses a distributed database containing addresses of other DNS servers that may contain the Internet address.

Domain Name Service-The process of converting a domain name into its corresponding IP address.

Domestic Satellite (DOMSAT)-A satellite system used for domestic communications generally within the continental United States, Alaska, Hawaii, Puerto Rico, and the Virgin Islands.

Domestic Satellite Carrier-A common carrier that provides communications services within the United States via that owned or leased satellite facilities.

Dominant Carrier-A dominant carrier is a service provider that has relatively strong market power (e.g. has the ability to control or influence industry prices).

DOMSAT-Domestic Satellite

Dongle-A dongle is a small assembly or device that attaches to another object or system. Dongles can contain active electronic circuits that allow them to be used as part of an encryption or digital rights management (DRM) system to enable access or decoding to information.

Doorway Page-An Internet web site that is developed to attract users or visitors. Doorway pages are typically optimized to have high ranking in search engines and they act as a portal for visitors to be transferred to other web sites that sell related products or services.

Doppler-An offset frequency that is the result of a moving antenna relative to a transmitted signal.

DoS-Denial Of Service

Dots Per Inch (dpi)-Dots per inch is the number of dots that appear per inch on the horizontal axis of a display or printing device.

Double Play-Double play refers to providing of two main services such as voice and data, data and video, or video and data on one network. For cable MSOs, this usually means building out the next generation network to DOCSIS 2.0 specifications, for Carriers this often means building out fibre or VDSL (very fast DSL) networks. Usually it is the larger MSOs and Telecom Carriers that roll out triple play services, and the advantage is that they can sign customers to a bundle of two services,

thereby increasing revenue and customer loyalty.

Double-Banger-Dressing rooms for actors or members of cast, usually in the form of a double wide trailer.

Downlink-(1-Satellite) The portion of a communication link used for transmission of signals from a satellite to a mobile or fixed receiver. (2-cellular system) The radio link from the base station to the mobile station.

Download and Play-Download and play is a process of downloading a media program (an audio or video file) and then playing it after the file has completely downloaded.

Downloading-Downloading is the transferring of a program or of data from computer server to another computer. Download commonly refers to retrieving files from a web site server to another computer.

Downsampling-Downsampling is the process of sampling one or more channels (segments) of information at a rate or resolution that is lower than the primary sample rate. A common example of downsampling is 4:2:2 where for each 4 bits that represent the intensity information in a video signal, 2 bits are used to represent the color (chrominance) components.

Downstream-The direction of transmission, usually from a network to an end customer.

Downtime-An amount of time that a communication network or computer system is not available to users. Downtime usually occurs from hardware failure, software crashes, or operator errors.

DP-Dial Pulse

DPBX-Digital PBX

DPC-Destination Point Code

DPCCH-Dedicated Physical Control Channel

DPCM-Differential Pulse Code Modulation

DPDCH-Dedicated Physical Data Channel

DPI-Digital Program Insertion

dpi-Dots Per Inch

DPM-Defects Per Million

DPRL-Digital Property Rights Language

DPSK-Differential Phase Shift Keying

DQDB-Distributed Queue Dual Bus

DQPSK-Differential Quadrature Phase Shift Keying

DRAM-Dynamic Random Access Memory

Drift-Drift is a change in a desired fixed characteristic such as wavelength or frequency that occurs over a period of time.

Driver-(1-general) An electronic circuit that supplies an isolated output to drive the input of another circuit. (2-fiber optic) The electric circuit that drives the light-emitting source, modulating it in accordance with an intelligence bearing signal. (3-software) A software module that controls an input/output port or external device, such as a keyboard or a monitor.

DRM-Digital Rights Management

DRM Controller-A DRM controller is the software and/or hardware that allows users to access content through a digital rights management system. DRM controllers receive requests to access digital content, obtain the necessary information elements (e.g. user ID and key codes), performs authentication (if requested) and retrieves the necessary encryption keys that allows for the decoding of digital media (if the media is encoded).

DRM Packager-A DRM packager is a program or system that is used to combine content (digital audio and/or video), product information (e.g. Metadata) and security codes to a media format or file that is sent from a content provider to a user or viewer of the content.

DRM Wrapper-A digital rights management (DRM) wrapper is a block of information contained in a media file that provides access to the content inside the media file.

Drop Reel-A drop reel is a spool that is used to hold and feed drop wire during cable installation.

Drop Shadow-A drop shadow is an image that represents a shadow and is offset from an object or text in an image or video. Drop shadows add dimension to images or video.

Drop Wire-A drop wire is the wire or pairs of wires that are connected between a customer's premises and a nearby network line. Although the first drop wires were connected from a telephone pole to a building, drop wires can be buried or aerial.

Drop/Add-A microwave link or other form of communication system whereby signals are dropped to customers along the transmission path. Conversely, a channel can accept and add new signals at intermediate points.

Dropped Calls-Cellular telephone calls that are inadvertently disconnected from the system because of interference, inadequate coverage or lack of capacity.

Dropped Frame-Dropped frames are images in a moving image sequence of images that are lost or intentionally dropped (removed) from the sequence.

DRx-Discontinuous Reception

Dry Circuit-A communication line that does not provide electrical power with the information signal. The use of a dry line requires the terminating equipment to supply its own power.

DS-802.11 Distribution Service

DS-Distribution Service

DS-Distribution System

DS-0-Digital Service, Level 0

DS-1-Digital Signal 1

DS-1C-Digital Signal, level 1 Combined

DS-3-Digital Signal 3

DS-4-Digital Signal 4

DSAP-Destination Service Access Point

DSB-Digital Sound Broadcasting

DSC-Digital Selective Calling

DSCH-Downlink Shared Channel

DSCP-DiffServ Code Point

DSE-Data Switching Exchange

DSF-Dispersion Shifted Fiber

DSI-Digital Speech Interpolation

DSL-Digital Subscriber Line

DSL bonding-The combining of multiple DSL communication lines to provide for higher data rate. For example, if eight 1.5 Mbps DSL lines are combined, the data transmission rate is 12 Mbps. This is also called inverse multiplexing.

DSL Bridge-A device that translates the protocol between a DSL modem and a DSL network. A DSL only translates the protocol and does not assign a separate address to the end user.

DSL Concentrator-An interface that allows more local loop telephone lines to share a digital subscriber line access multiplexer (DSLAM) that is allowed by the number of DSL modems that are installed in the DSLAM. The DSL concentrator acts as a mini-switch connecting the local loop to the DSL modem when data service is requested.

DSL Forum-A forum that was started in 1994 to assist manufacturers and service providers with the marketing and development of DSL products and services. The DSL forum was previously called the ADSL forum.

DSL Microfilter-A DSL microfilter is a blocking filter device that attaches to a telephone jack that blocks unwanted high-speed data signals from entering into the telephone.

DSL Modem-Digital Subscriber Line Modem

DSL Splitter-A circuit, device or component that divides a DSL signal into separate voice and data outputs. A DSL splitter is typically used for ADSL and VDSL systems.

The DSL splitter separates the existing telephone signal from the high speed data signal. In the United States, the standard telephone signal (POTS) frequency band extends up to 8 kHz. In Europe, standard telephone signals include additional high frequency components such as 12 kHz billing increment impulses that extend up to 12 kHz. When the DSL splitter is used to allow ISDN signals, the frequency band for the ISDN signal extends up to 80 kHz (120 kHz for ISDN in Germany).

DSLAM-Digital Subscriber Line Access Multiplexer

DSM-Digital Storage Medium

DSn-Digital Service Hierarchy

DSP-Digital Signal Processor

DSR-Data Set Ready

DSS-Direct Station Select

DSSS-Direct Sequence Spread Spectrum

DSSS Mode-802.11 Direct Sequence Spread Spectrum

DST-Digital Signaling Tone

DSU-Data Service Unit

DSU-Digital Service Unit

DSU/CSU-Data Service Unit/Channel Service Unit

DSx-Digital Signal Level

DTA-Data Transfer Adapter

DTC-Digital Traffic Channel

DTCH-Dedicated Traffic Channel

DTD-Document Type Definition

D-t-D-Digital-To-Digital Transfer

DTE-Data Terminal Equipment

DTH-Direct To Home

DTM-Dual Transfer Mode

DTMF-Dual Tone Multi-Frequency

DTMF Decoder-A device or process that converts dual tone multi-frequency (DTMF) signals into another form (such as data digits).

DTMF Receiver-A device or process of receiving dual tone multi-frequency (DTMF) signals and converting them into another form (such as data digits).

DTR-Data Terminal Ready

DTS-Digital Termination Service

DTT-Digital Terrestrial Television

DTV-Digital Television

DTx-Discontinuous Transmission

Dual Band-A wireless device that is capable of accessing radio channels on two bands of frequencies (such as cellular and PCS).

Dual Fiber Cable-A fiber cable that contains two separate single-fiber cables.

Dual Mode-Dual mode is the ability of a device or system to operate in two different modes (not necessarily at the same time). For wireless systems, it refers to mobile devices that can operate on two different system types.

This figure shows how the IS-95 CDMA system can provide both digital and analog "dual mode" operation. This diagram shows that a dual mode CDMA mobile radio can originate and receive calls in two different system types; CDMA and AMPS analog. The mobile device typically searches for CDMA channels first. If it cannot find the CDMA channels, it will then scan for analog channels.

Dual Seizure-The condition which occurs when

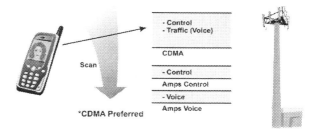

IS-95 CDMA Dual Mode Operation

two exchanges (switches) attempt to seize the same circuit at approximately the same time. (See also: glare.)

Dual Tone Multi-Frequency (DTMF)-DTMF signaling is a means of transferring information from a user to the telephone network through the use of in-band audio tones. Each digit of information is assigned a simultaneous combination of one of a lower group of frequencies and one of a higher group of frequencies to represent each digit or character. There are 8 tones that are capable of producing 16 combinations; 0-9, *, #, A-D. The letters A-D are normally used for non-traditional systems (such as the military telephone systems).

This diagram shows how dual tone multi-frequency (DTMF) tones can be used to send dialing information from a telephone to a telephone system. There are 8 different frequencies that can be combined to represent 16 keys. The keys A-D are not usually included on standard telephone sets. To represent each button, two tones are combined. In this example, the button 3 is pressed, followed by a pause, then button 2 is depressed. Button 3 is represented by the combined tones 1477 Hz and 697 Hz. Button 2 is represented by the combined tones 1336 Hz and 697 Hz. To determine if the user is finished dialing, a timer is used. When the user has stopped dialing, the digits can be sent to the call processing section of the telephone system to initiate the call.

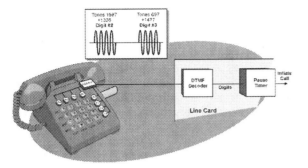

DTMF Dial Operation

DualCam-A digital camera that can take pictures and be used as a WebCam to provide still or continuous images to the Internet.

Dub-(1-copy) To copy information on one videotape or audiotape to another. (2-a copy) A copy of a videotape or audiotape.

Duct-(1-general) A pipe or conduit, installed underground or in a building, whose purpose is to protect the cables installed therein. (2-height) The height above the Earth of the lower boundary of an elevated propagation duct. (3-metal floor wiring) One of various proprietary schemes of cable ducting to provide flexibility in equipment installation in office areas. (4-nest) A number of cable ducts provided for and laid in one trench. (5-propagation) A layer of cold air under warm air, experienced in some areas, that causes microwave signals to propagate further than normally possible. (6-surface) A radio duct whose lower boundary is the surface of the earth. (7-thickness) The difference in height between the upper and lower boundaries of a tropospheric radio duct.

Due Date - Billing-The date by which payment due must be received before the collections process is triggered. This date is typically dependent upon the invoice date, and is typically different for each cycle.

Dumb Switch-A term commonly applied to a time division multiplexed (TDM) telecommunications switch that only can connect lines to each other based on instructions a controller or other computer.

Dumb Terminal-A computer display terminal that serves as a slave to a host computer. A dumb terminal has a keyboard for data entry and a video display, but no computing power of its own.

Dummy Burst (DB)-A burst of information that contains no user data information. Dummy bursts are used to fill a time slot or frame with information to ensure a continuous flow of data is being sent to a channel or time slot.

DUN-Dial-Up Network

Dunning-Unique treatment of customers for the purpose of collection of service charges or account balances.

Duobinary Data MAC (D2-MAC)-Duobinary data MAC is a European hybrid video format that combines digital audio with analog video to produce high definition television (HDTV).

DUP-Data User Part

Duplex Channels-Duplex channels are the combining of 2 one-way communication channels (one forward and one reverse) to allow for simultaneous communication in two directions. To create simultaneous two-way communication, channel duplex can be in the form of time division duplex (TDD) or frequency division duplex (FDD).

Duplex Transmission-Duplex transmission is the simultaneous transmission of two information signals that allows simultaneous 2-way communication.

This figure shows the basic operation of FDD and TDD systems. In example A (FDD), device 1 (on the left side) is transmitting audio on frequency 1 that is received on device 2 (on the right side). Device 2 is transmitting audio on frequency 2 that is received by device 1. Because this communication operates on 2 different frequencies, audio can simultaneously sent in both directions (full duplex). Example B (TDD) shows a time division duplex (TDD) system. In a TDD system, device 1 transmits information for brief period while device 2 listens on frequency 1. Device 2 then transmits information while device 1 listens on the same frequency 1. This process repeats so a continual transfer of information occurs in both directions. If the audio of each device is compressed before sending and the audio is expanded after receiving, the audio appears to be simultaneously sent in both directions. In example C (combined FDD & TDD), device 1 transmits information for brief period while device 2 listens on frequency 1. Device 2 then transmits information while device 1 listens on frequency 2. This process repeats so a continual transfer of information occurs in both directions.

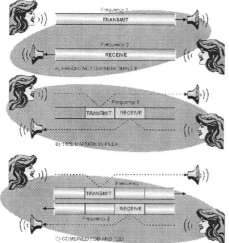

FDD and TDD Duplex Systems

Duplexer-A combined filter device that permits a transmitter and receiver to share the same antenna assembly by using filters with different frequency bands. The use of a duplexer prevents transmitter power output to the antenna from transferring to sensitive receiver assembly.

This diagram shows how a duplexer allows one antenna to be connected to a transmitter and receiver at the same time. This example shows that the transmitter and receiver use different frequencies and the duplex filter (duplexer) contain two bandpass filter assemblies. This diagram shows that the bandpass filter allows the transmitter frequency to reach the antenna and that the receiver filter blocks the high-energy transmitter frequency from entering into the receiver. However, the duplexer's receiver bandpass filter does allow for receiver band frequencies (receiver channels) to enter into the receiver.

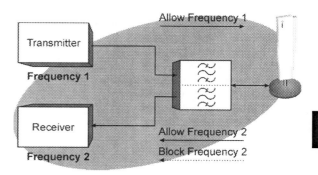

Duplexer Operation

Duplication Rights-Duplication rights is the permission given to duplicate content or portions of content in specific formats (such as duplicating an article for the employees of a company).

Duration-Duration is the time that elapses between the start of an event (such as answering a call) and the termination of the event (such as the ending of a call).

Duration Based Charging-Duration based charging is the rating of billing cost that is determined by the duration (start to end) time of the service regardless of the amount of data or service used.

Dust Cap-A dust cap is a cover or an assembly that covers a cable end or opening in an assembly to prevent contamination (such as dust) from entering into the cable or assembly.

DUT-Device Under Test

DV-Digital Video Tape

DV-Digital Videotape

DV25-Digial Video 25

DVB-Digital Video Broadcast

DVB-C-Digital Video Broadcasting Cable

DVB-CI-Digital Video Broadcasting Common Interface

DVB-CSA-Digital Video Broadcasting Common Scrambling Algorithm

DVB-Data-Digital Video Broadcasting Data

DVB-H-Digital Video Broadcasting Handheld

DVB-IPI-Digital Video Broadcasting Internet Protocol

DVB-MHP-Digital Video Broadcasting Multimedia Home Platform

DVB-RCS-Digital Video Broadcast Return Channel via Satellite

DVB-S-Digital Video Broadcasting Satellite
DVB-T-Digital Video Broadcating Terrestrial
DVB-Txt-Digital Video Broadcasting Teletext
DVC-Digital Video Cassette
DVC-Digital Video Compression
DVCC-Digital Verification Color Code
D-VHS-Digital Video Home System
DVI-Digital Video Interactive
DVI-Digital Visual Interface
DVMRP-Distance Vector Multicast Routing Protocol
DVQ-Digital Video Quality
DVR-Digital Video Recorder
DWDM-Dense Wave Division Multiplexing

Dynamic-A process, item, or information element that has parameters that can change at unplanned times.

Dynamic Address-An address that is assigned to a device or service, usually at the beginning of a communication session.

Dynamic Allocation-Dynamic allocation is the assignment of a resource on an unscheduled basis. Dynamic allocation is a process of sensing the need for a communication resource (such as a channel) and the assignment (allocation) of the resources that are required by that communication need.

Dynamic Frequency Selection (DFS)-Dynamic frequency selection is a process that allows devices or users to request, select or change an operating frequency at various times.

Dynamic Host Configuration Protocol (DHCP)-Dynamic host configuration protocol is a process that dynamically assigns an Internet Protocol (IP) address from a server to clients on an as needed basis. The IP addresses are owned or controlled by the server and are stored in a pool of available addresses. When the DHCP server senses a client needs an IP address (e.g. when a computer boots up in a network), it assigned one of the IP addresses available in the pool.

This figure shows how a computer uses DHCP to obtain a temporary IP address when it requires an Internet communication session. In this example, the computer requests a connection with an Internet service provider (ISP) via a modem that is connected to a universal serial bus (USB) line. When the Internet service provider receives the request for connection, it assigns an IP address from its list of available IP addresses. The computer will then use this IP address for all of its communications with the Internet until it disconnects the connection to the ISP.

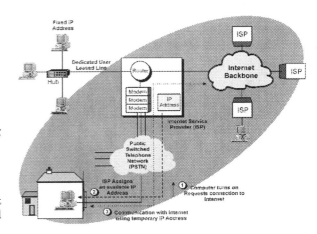

Dynamic Host Configuration Protocol (DHCP) Operation

Dynamic hypertext markup language (DHTML)-Dynamic hypertext markup language is an evolved version of HTML that ads scripts and features that allows for the creation of web pages that have content that is processed after the user has selected some options.

Dynamic IP Addressing-Dynamic IP addressing is a process of assigning an Internet protocol address to a client (usually and end user's computer) on an as needed basis. Dynamic addressing is used to conserve on the number of IP addresses required by a server and to provide an enhanced level of security (no predefined address to use for hackers).

Dynamic Link Library (DLL)-A feature of an operating system (e.g. Windows) that allow executable software code modules to be loaded on demand and linked when the applications begin to operate (run time). This enables the software code to access the latest parameters in its related software applications. DLL files are unloaded when they are no longer needed (when the application is closed).

Each DLL applications that is initiated is copied into the working memory of the computer. A key benefit of using dynamic linked libraries is that the executable program files are not as large because the frequently used routines can be put into DLL files.

Dynamic Load Balancing-A process that is used to evenly distribute incoming calls to customer service agents. Dynamic load balancing can be implemented in Automatic Call Distributor (ACD) systems.

Dynamic Loading-Dynamic loading is the process of obtaining files or media as the need for the data is determined. An example of dynamic loading is the obtaining of a digital video clip file and loading it into a media player after a person has selected the link to that file.

Dynamic Page-Dynamic Web Page

Dynamic Power Control-The combination of self regulated power control and system power control. Self controlled power level is performed by sensing the received power level and increasing the transmitter power level as the received level decreases due to increased distance from the base station transmitter. System controlled power control is a process of controlling the power level in a cellular system where the base station receiver monitors the received signal strength of a mobile telephone and control messages are transmitted from the base station to the mobile telephone commanding it to raise and lower its transmitter power level as necessary to maintain a good radio communications link.

Dynamic Random Access Memory (DRAM)-A type of memory that temporarily stores information and requires continual refreshing of the information. DRAM information is completely lost when the power and refreshing is removed.

Dynamic Range-(1-analog system) The range or extremes in amplitude, from the lowest to the highest points that a system or radio is expected to operate. The dynamic range is typically expressed in decibels against a reference level. (2-digital) The number of bits used to define the range of a given signal.

Dynamic Routing-Dynamic routing is the process of automatically re-routing communication paths or circuits as the network traffic levels (e.g. levels of congestion) change. Dynamic routing is sometimes called adaptive routing.

Dynamic Time Alignment-Dynamic time alignment is a technique that allows a radio system base station to receive transmitted signals from mobile radios in an exact time slot, even though not all mobile telephones are the same distance from the base station. Time alignment keeps different mobile radio's transmit bursts from colliding or overlapping. Dynamic time alignment is necessary because subscribers are moving, and their radio waves' arrival time at the base station depends on

their changing distance from the base station. The greater the distance, the more delay in the signal's arrival time. Transmission delay is approximately 3 microseconds per km (or 5 microseconds per mile).

This diagram shows how the relative transmitter timing in a mobile radio (relative to the received signal) is dynamically adjusted to account for the combined receive and transmit delays as the mobile radio is located at different distances from the base station antenna. In this example, the mobile telephone uses a received burst to determine when its burst transmission should start. As the mobile radio moves away from the tower, the transmission time increases and this causes the transmitted bursts to slip outside its time slot when it is received at the base station (possibly causing overlap to transmissions from other radios.) When the base station receiver detects the change in slot period reception, it sends commands to the mobile telephone to advance its relative transmission time as it moves away from the base station and to be retarded as it moves closer.

Dynamic Traffic Routing-Dynamic routing is the automatic selection of alternative communication routes or systems. The choice of alternative routes may be dependent on cost of service, traffic congestion, line failure, or other criteria that may change the choice of routing path.

E

E&M-Ear and Mouth Signaling

E&O-Errors And Omissions

E.164 International Public Telecommunications Numbering Plan-The International Telecommunications Union (ITU), a division of the United Nations, has defined a world numbering plan recommendation, "E.164." The E.164 numbering plan defines the use of a country code (CC), national destination code (NDC), and subscriber number (SN) for telephone numbering. The CC consists of one, two or three digits. The first digit identifies the world zone. The number of digits used for telephone numbers throughout the world varies. However, no portion of a telephone number can exceed 15 digits. There are several "E" series of ITU numbering recommendations that assist in providing unique identifying numbers for telephone devices around the world.

This diagram shows the world (telephone) numbering plan recommendation, "E.164" developed by the International Telecommunications Union (ITU). This diagram shows the numbering plan divides a telephone number into a country code (CC), national destination code (NDC), and subscriber number (SN) for telephone numbering. The CC consists of one, two or three digits and the first digit identifies the world zone. This diagram shows that the local number can be divided into an exchange code (end office switch identifier) and a port (or extension) code.

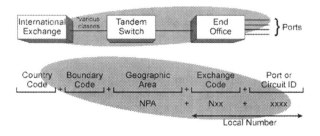

E.164 Telephone Numbering System

E/O-Electronic to Optical Conversion

E1-A communication line that was developed by European standards that multiplexes thirty voice channels and two control channels onto a single communication line. The E1 line uses 256 bit frames and transmitted at 2.048 Mbps.

E411-Enhanced 411

E911-Enhanced 911

EAP-Extensible Authentication Protocol

Earnings Before Interest, Taxes, Depreciation, and Amortization (EBITDA)-EBITDA is used by finance and investment analysts to gain a more accurate appraisal of the true net worth and/or cash flow of a company. EBITDA reflects the ability of a business to achieve profitability without the influence of financial positions that may vary between different companies.

EAS-Emergency Alert System

Easement-Easement is the right of use of another person's land that is granted by the owner of the land or by an authorized government agency.

Eb/No-The ratio of bit energy to a noise signal.

E-BCCH-Extended Broadcast Channel

EBCDIC-Extended Binary Coded Decimal Interexchange Code

EBITDA-Earnings Before Interest, Taxes, Depreciation, and Amortization

eBook-Electronic Book

EBPP-Electronic Bill Presentation and Payment

EBTS-Enhanced Base Transceiver System

e-Business-Electronic Business

EC-Exchange Carrier

E-Carrier-A digital carrier system adopted by the international telecommunications union (ITU). Each digital signaling level supports several 64 kbps (DS0) channels. E-carrier was initially used in Europe and now is used throughout the rest of the world. The different digital signaling levels include;

- DS-1, 2.048 Mbps with 30 channels + 2 control channels
- DS-2, 8.448 Mbps with 120 channels
- DS-3, 34.368 Mbps with 480 channels
- DS-4, 139.264 Mbps with 1920 channels
- DS-5, 565.148 Mbps with 7680 channels

ECC-Error Correcting Code

Echo-A type of transmission impairment in which a signal is reflected back to the originating source.

In the transmission, the reflected signal often is attenuated and delayed, resulting in an echo.

Echo Cancelling-Echo cancellation is a process of extracting an original transmitted signal from the received signal that contains one or more delayed signals (copies of the original signal). Echoes may be created in a baseband or broadband signal. When echoes occur on an audio baseband signal, it is usually through acoustic feedback where some of the audio signal transferring from a speaker into a microphone. When echoes occur on a broadband signal, it is usually the result of the same signal (such as a radio signal) that travels on different paths to reach its destination. In either case, echoed signals cause distortion and may be removed by performing via advanced signal analysis and filtering.

This figure shows how echoes can be removed by an echo canceling system. In this example, the transmission of the words: "Hello, is Susan there" experience the effects of echo. When the signal is supplied to an echo canceller (a sophisticated estimating and subtraction machine), the echo canceling device takes a sample of the initial audio and tries to find echo matches of the input audio at delayed periods (the amount of echo time). In this example, it does this by creating various delayed versions the audio signal and different (reduced) amplitude (echo volume usually decreases as time increases), and comparing the estimate the audio that contains the echo. When it finds an exact match at a specific audio level, the echo canceller can subtract the echo signal. This produces audio without the echo.

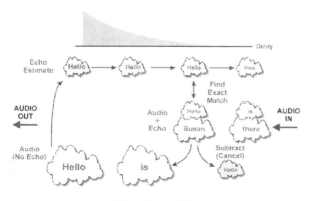

Echo Cancelling

Echo Canceller-A signal processing device or circuit that reduces the effects of echo signals. This is performed by calculating an estimate of the expected echo signal and subtracting this estimate from the signal in which the echo appears. Echo cancellers are essential for communication systems that have long signal processing delays such as long distance voice, satellite and digital mobile telephony lines.

ECL-Emitter Coupled Logic

ECM-Entitlement Control Messages

E-commerce or ECommerce-Electronic Commerce

Economic Bypass-A form of bypassing communication services where the cost of the bypass service is lower than that of an equivalent telephone company service.

Economic Conditions Churn -Economic conditions churn is a type of voluntary, incidental churn that occurs when customers loose their jobs or face other economic crisis's, and must terminate service.

Economy of Scale-Economy of scale is the process of decreasing item or service cost as the volume of items produced or services provided increases.

ECPM-Effective Cost Per Thousand

ECSA-Exchange Carriers Standards Association

ECSD-Enhanced Circuit Switched Data

EDGE-Enhanced Data Rates For Global Evolution

EDGE-Enhanced Data Rates For GSM Evolution

EDGE Compact-Enhanced Data Rates For Global Evolution Compact

Edge Network-An edge network is a system that is used to connect a private or enterprise network to a backbone or other interconnecting network. These networks are usually capable of providing low to medium speed connections.

Edge Router-A device used to connect a private or enterprise network to a Service Provider's network. These devices are usually capable of more advanced routing protocols, such as BGP-4 and OSPF, and usually provide services such as tunneling, authentication, filtering, billing, traffic shaping and rate policing and network address translation. Depending on the service provider, the device may be owned and managed by the service provider or by the customer.

Edge Switch-A switch located at the boundary between two networks. The edge switch usually provides access to an end users system to a carrier's core or backbone network. Edge switches are sometimes called access nodes and service nodes.

EDI-Electronic Data Interchange

EDSS1-European Digital Subscriber Signaling no. 1

EE-End To End

EFCI-Explicit Forward Congestion Indication

Effective Resistance-The increased resistance of a conductor to an alternating current resulting from skin effect, relative to the direct-current resistance of the conductor. Higher frequencies tend to travel only on the outer skin of the conductor, whereas DC flows uniformly through the entire area.

Effective Transmission-Effective transmission is the actual performance of transmission. It can be measured based upon precise or subjective tests of transmission repetition rates.

Effects-(1-video) The process of combining two or more video images to create a new composite image. (2-audio) The process of adding special audio elements, such as echo, to a program to enhance the overall aural impact.

Effects Memory-The capability of a video production switcher to store and recall effects created on the system through the use of computer control techniques.

Effects System-The portion of a video production switcher that performs mixes, wipes, and cuts between background and/or special effects key video signals.

EFM-Eight To Fourteen Modulation

EFM-Ethernet in the First Mile

EFR-Enhanced Full Rate

EGP-External Gateway Protocol

EGPRS-Enhanced General Packet Radio Service

Egress-(1-signal) A process where a strong signal inside a communication system leaves the transmission line or system. (2-network) The process of data or signals exiting from a communication network.

Egress Filtering-Egress filtering is the process of restricting the types of connections or packets that are transferred out from a network.

Egress Firewall-An egress firewall is a device or software that is installed between a computer server or data communication device and a public network (e.g. the Internet) to restrict the flow of packets out of the network.

Egress Packet Filtering-Egress packet filtering is the process of decoding and searching packets that are exiting a network to determine and alter its contents or change it's routing or contents based on filtering information (e.g. type of service).

EHR-Enhanced Half Rate

EIA-Electronic Industries Association

EIFS-Extended Interframe Space

EIGRP-Interior Gateway Routing Protocol

EIR-Equipment Identity Register

EIRP-Effective Isotropic Radiated Power

EIT-Event Information Table

EKTS-Electronic Key Telephone System

Elastic Buffer-A storage device that has capacity to store data and add variable amounts of delay before retransmitting the data. An elastic buffer smoothes out the variable transmission delay that digital audio signals may experience when transmitted through a packet switching network (such as the Internet).

Electromagnetic Coupling-The process of coupling two signals using electromagnetic energy.

Electromagnetic Energy-Energy with both electronic and magnetic components. In telecommunications, this energy is typically in the form of a wave of varying electronic and magnetic fields. The electronic component is usually established by applying a varying voltage to a conductor or semiconductor. The resulting motion of electrons generates a magnetic field.

Electromagnetic Interference (EMI)-Electromagnetic interference (EMI) is the radiation (transmission) of undesirable electro- magnetic waves from an electronic circuit or device. These EMI waves may interfere with the normal operation of other circuits or devices.

Electromagnetic Spectrum (EMS)-Electromagnetic spectrum is the range of frequencies of electromagnetic waves. Radio and microwave are at the low frequency, long wavelength end of the spectrum, followed by the infrared light used in optical networks. Visible light represents a small portion of the spectrum at a higher frequency than infrared. Ultraviolet light is higher in frequency and shorter in wavelength than visible light. X-rays and gamma rays are still higher frequency portions of the electromagnetic spectrum.

Electromagnetic Wave-A periodic variation in electric and magnetic fields that travels through a medium, forming a wave. For instance, when a voltage is applied across a conductor, its electrons move. This electron acceleration results in an electric field, which induces a magnetic field. If the voltage is varied systematically, the direction and

E

magnitude of both electric and magnetic fields will also vary systematically. The result is the apparent propagation of a wave through the conductor. An electromagnetic wave travels at the speed of light in a vacuum, or 186,300 miles per second (300,000,000 meters per second) and more slowly through other media.

Electronic Bill Presentation and Payment (EBPP)-Electronic bill presentation and payment is the systems, applications and processes that are used to group, process and display billing information along with the ability for the user to interact and pay invoices and balances that are due on accounts.

Electronic Commerce (E-commerce or ECommerce)-A shopping medium that uses electronic networks (such as the Internet or telecommunications) to present products and process orders.

Electronic Industries Association (EIA)-A trade association that develops standards for electronic components and systems and represents manufacturers of electronic systems and parts.

Electronic Mail (Email or e-Mail)-A process of sending messages in electronic form. These messages are usually in text form. However, they can also include images and video clips.

Electronic Programming Guide (EPG)-Electronic programming guides are an interface (portal) that allows a customer to preview and select from possible lists of available content media. EPGs can vary from simple program selection to interactive filters that dynamically allow the user to filter through program guides by theme, time period, or other criteria.

Electronic Publishing (ePublishing)-The process of developing, converting, publishing and/or distributing books and/or documents through electronic networks such as the Internet.

Electronic Switching System (ESS)-A system that can connect incoming and outgoing digital lines together through the use of temporary memory locations. For an ESS system, a computer controls the assignment, storage, and retrieval of memory locations so that a portion of an incoming line (time slot) can be stored in temporary memory and retrieved for insertion to an outgoing line.

Electronic to Optical Conversion (E/O)-The process of changing an electronic signal into an optical one. Optical signals, or lightwaves, are used to transport information rapidly. However, many critical network processing steps can only be per-

formed on electronic signals. After the electronic processing has been completed, the signal is changed back into an optical signal in order to be transported in the optical network. See also Optical to Electronic Conversion (O/E).

Electronic Voice Mail (EVM)-A system that stores messages in electronic form (usually digital audio) that can be saved, moved, and retrieved by the mailbox owner.

Elektrisches Telescop-An Elektrisches Telescop is a mechanical electric telescope that was invented by Paul Nipkow in 1884. His invention scanned, transmitted and received images through the use of a scanning disk and a sensor composed of selenium. The selenium cell converted the light image into electrical current as the disk rotated. This electrical signal was sent to a receiving device that had a disk spinning at the same rate and synchronized with the sending disk. The receiving disk had holes, which allowed the creation of small images.

Element Management System (EMS)-EMS is becoming more synonymous with Network Management Station (NMS) as telephony and data service providers come together to manage heterogeneous networks. EMS is used mainly in the telephony networks for provisioning versus the data networks.

Elementary Blocks-Elementary blocks are groups of pixels 8 x 8. Display images that can be divided into elementary blocks as a source for an image compression process (such as JPEG compression).

Elementary Stream (ES)-Elementary streams are the raw information component streams (such as audio and video) that are part of a program stream.

Eliminate-Eliminate is a process used to remove media components or segments from an image or a video program.

E-Link-Extended Link

ELT-Emergency Locator Transmitter

ELV-Expendable Launch Vehicle

E-Mail Notification-E-mail notification is the process of sending a subscriber an email message that notifies them that information has changed (such as a new voicemail).

Email or e-Mail-Electronic Mail

EMBARC-Electronic Mail Broadcast to A Roaming Computer

Embed-To insert all of an information element or object directly into another information element or object.

Embedded Device-An embedded device is a component or assembly that is integrated into a larger device, assembly, or system.

Embedded Object-An embedded object is block of information located in a file that is a complete copy of information from another file or application. By embedding an object in a file or document rather than linking it, the object can more rapidly load and it can be manipulated and/or modified directly in the application.

Embedded Operating System-An embedded operating system is a group of software programs and routines that directs the operation of a device in its tasks and assists programs in performing their functions that is not generally accessible by the user. The embedded operating system software is responsible for coordinating and allocating system resources. This includes transferring data to and from memory, processor, and peripheral devices. Software applications use the embedded operating system to gain access to these resources as required.

Embedded Operations Channel (EOC)-A communications channel that is designed to be part of a communications circuit. The EOC allows for commands (usually system step, change and test commands) to be transferred without the need to interfere with an established communications link it is paired with.

Embedded System-An embedded system is a special-purpose computer system built into a larger device. An embedded system is required to meet very different requirements than a general-purpose personal computer.

EMD-Equilibrium Mode Distribution

Emergency Alert System (EAS)-Emergency alert system is a system that coordinates the sending of messages to broadcast networks of cable networks, AM, FM, and TV broadcast stations; Low Power TV (LPTV) stations and other communications providers during public emergencies.

Emergency Service-Emergency service is the providing of communication services for a need that is unforeseen by the service provider. Emergency service users may be given higher access and call control priority than other users on the communication systems.

EMI-Electromagnetic Interference

EMLPP-Enhanced Multi-Level Priority And Pre-Emption

EMM-Entitlement Management Messages

Empirical-A conclusion that is based not on pure theory, but on practical and experimental work.

EMR-Exchange Message Record

EMRP-Equivalent Monopole Radiated Power

EMS-Electromagnetic Spectrum

EMS-Element Management System

EMS-Enhanced Messaging Service

Emulator-A program or device that simulates the operation of another software program or hardware device. The use of an emulator allows developers to simulate the operation of programs or devices quickly without the need to build expensive prototypes or to spend significant resources on the debugging of software.

Encapsulating Bridge-A bridge that encapsulates LAN frames for transmission across a backbone where the bridges networks are dissimilar.

Encapsulation-(1-connection) Encasing an electrical connection in a container with protective material to seal it to the environment (e.g. keep out water and air). (2-software program) The grouping of software code into a single entity, object or program. The encapsulation of a software program isolates the inner workings of the program from the programmers and other programs that use it. (3-layered protocols) The adding of additional protocol layers to a protocol data unit (PDU) to add new control or routing capabilities to the PDU. (4-data packet) A process of inserting the entire contents of a data packet into the payload (data portion) of another packet.

Encipherment-The process of encrypting data with a cipher key.

Encoded Chroma Key-A chroma key that uses an encoded video signal instead of separate RGB or Y, Cr, Cb signals for deriving the key.

Encoded Data-Encoded data is the information that is coded and formatted. The coding and formatting may range simple reorganization of bits to adding control and error protection bits.

Encoder-(1-general) An encoder is a device that processes one or more input signals into a specified form for transmission and/or storage. (2-video) A device used to form a single (composite) color signal from a set of component signals. An encoder is used whenever a composite output is required from a source (or recording) that is in a component format. (3- digital) A device that adds codes and error-correction information to digital data prior to recording or transmission. (4-mechanical) Any device that can be attached to various mechanical systems to sense (measure) and transmit numerical (encoded) information about the system. General motion, rotation and vibration are some attributes which can be so measured.

Encoding-(1-modulation) For digital modulation systems such as pulse code modulation (PCM), encoding is the process of converting the magnitude of a sample to a 7-bit code (DI-type), or an 8-bit code (D2- or later). Encoding is a transmit function. (See also: decoding, quantizing, sampling.) (2-data) In the context of data compression, as a gerund, the process of compressing the source data into compressed form. As a noun, the compressed data which is the result of the encoding process.

Encoding Standard-The encoding standard is the specification of how information is processed and formatted on a communication channel or how information is located onto a storage device such as a hard disk or CD ROM.

Encoding Tools-Software and/or hardware tools to create compressed digital media from source media. Encoding tools may also function as authoring tools.

Encrypted File-An encrypted file is a collection of information or data (such as digital bits) that is coded (encrypted) in a way that does not allow unauthorized users to view or use the file.

Encrypted File System-An encrypted file system is the coding (encryption) of the file directory in a way to prevent unauthorized viewing or access to files stored in computers or on network storage systems.

Encryption-Encryption is a process of a protecting voice or data information from being obtained by unauthorized users. Encryption involves the use of a data processing algorithm (formula program) that uses one or more secret keys that both the sender and receiver of the information use to encrypt and decrypt the information. Without the encryption algorithm and key(s), unauthorized listeners cannot decode the message. When the encryption and decryption keys are the same, the encryption process is known as symmetrical encryption. When different encryption and decryption keys are used (such as in a public encryption system), the process is known as asymmetrical encryption.

This diagram shows how encryption can convert non-secure information (clear text) into a format (cyphertext) that is difficult or impossible for a recipient to understand without the proper decoding keys. In this example, data is provided to an encryption processing assembly that modifies the data signal using an encryption key. This diagram also shows that additional (optional) information such as a frame count or random number may be used along with the encryption key to provide better information encryption protection.

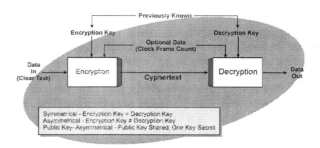

Encryption Operation

Encryption Algorithm-An encryption algorithm is a mathematical process that modifies data or information using keys or private information to prevent unauthorized companies or people from being able to make use of the information, media or data.

Encryption System-An encryption system is the combination of equipment, protocols, coding algorithms and key management that is used secure information or data transfer services.

End Cap (Endcap)-An end cap is a fitting or assembly that is installed on the end of a cable to seal the cable from entry or exit of water, gasses, dust and other materials.

End Of Message (EOM)-A frame that indicates the end of a message that has been transmitting.

End of Procedure frame (EOP)-A frame that indicates the end of a procedure that has been running.

End Office (EO)-An end office is a switching system that interconnects calls between local customers and the telephone network. Each end office switch can usually supply service up to 10,000 customers. In larger areas (such as a city), established LECs may have several EO switches. The EO switches are interconnected using a higher level tandem switch. If is a significant amount of calls regularly processed between end offices, they may be directly connected via high-speed communication lines (trunks).

End Pulling-End pulling is the process of pulling a cable through a duct or pipe from the one end (the cable feed) to the other end of the cable run.

End Station-A PC or host that allows communications between ATM end stations and LAN end stations.

End To End (EE)-End to end refers the connection of a communication from one end user to another end user.

End to End Encryption-End to end encryption is a process that uses encryption keys at each end of a communication session to encrypt the data that passes between the two end points.

End to End Solution-An end to end solution is a product or service that is offered from a company or vendor that includes all the functions or services necessary to provide the service or the solution they offer.

End User-A customer or device that uses communications services.

End User Device-End user devices are communication adapters such as telephones, fax machines, or private telephone systems that adapt signals from a telecommunication system to a format (such as audio or visual) to a form that is suitable for an end user.

End User Licensing Agreement (EULA)-An agreement between an end-user of a product (usually a software product) and the owner of the product (e.g. software developer) that defines the terms of how the end user must abide by when they use a product. The end user is often prompted to enter into an EULA prior to installing and using a software application.

Endcap-End Cap

End-To-End-(1-general) End-to-end is the complete process or responsibility for a system or service. (2-communication) The overall process from the origination of the request or service to the connection or completion of the service or command. End-to-end commonly is used to refer to commands or communication over multiple communication systems and/or processes.

End-to-End Solution-An end-to-end solution is a system or service that can manage and control the necessary equipment and resources to operate the system and/or provide the service. This may include communicating with systems and devices outside the direct control of the company providing the system.

End-User Point Of Termination-The network interface that connects the end-user to the network.

Energized-The condition when a circuit is switched on or powered up.

Energy-Energy is the product of mass and half the square of velocity (for velocity values that are small relative to the speed of light). The SI unit of energy, consistent with electrical units, is the joule (also called the watt second), product of one kilogram of mass, the constant 1/2, and the square of a 1 meter per second velocity. Energy can also be described equivalently as the product of power and time, and the joule (or watt second) is the product of one watt of power with one second of time. The joule is named for James P. Joule, a 19th century Irish physicist (his name rhymes with foul).

In non-technical discourse, the four words force, energy, power and momentum are often unfortunately used as synonyms for each other. In science and technology, each of these words describes a distinct and separate physical quantity.

Energy Dispersal-Energy dispersal is a process that distributes data or information over multiple bits or over a wider bandwidth with the intent of distributing the concentration of information more evenly within a communication channel. Energy dispersal may process (XOR) a data signal using a pseudo-random binary sequence (PRBS) to more evenly distribute bits that are sent on a communication channel to avoid the potential of sending (concentrating) multiple 1s or 0s in sequence. Some applications of energy dispersal is called randomizing.

Enhanced 911 (E911)-E911 is an emergency telephone calling system that provides an emergency dispatcher with the address and number of the telephone when a user initiates a call for help. The E911 system has the capability of indicating the contact information for the local police, fire, and ambulance agencies that are within a customers calling area.

Enhanced AMPS-IS-94

E

Enhanced Service Provider (ESP)-A provider of communication services that enhance the services provided by existing carriers. ESPs often provide information processing services such reformatting data or information management services. Because ESPs use existing communication services, ESPs do not typically file tariffs.

Enhanced Telecom Operations Map (eTOM)-Enhanced telecom operations map (eTOM) is a business system structure that allows communication service providers to define, develop, and procure business and operational support systems. The eTOM framework was created by the telemanagement forum.

Enhancement Layer-An enhancement layer is a stream or source of media information that is used to improve (enhance) the resolution or appearance of underlying (e.g. base layers)..

Enhancing-The process of electronically adjusting the quality and sharpness of a signal or image. Enhancing also may refer to sweetening audio, for example, by adding laugh tracks and sound effects.

ENIAC-Electronic Numerical Integrator and Computer

Enterprise Network-The set of Local, Metropolitan, and/or Wide Area Networks and internetworking devices comprising the communications infrastructure for a geographically-distributed organization.

Enterprise Switch-A switch used within an enterprise backbone. Enterprise switches are generally high-performance devices operating at the Network layer that aggregate traffic streams from sites within an enterprise.

Entertainment Application-Entertainment applications are software programs and/or services that provide a diversion or an amusement experience.

Entitlement Control Messages (ECM)-Entitlement control messages are conditional access system commands commonly used in broadcast systems that contain access parameters and procedures. ECMs typically provide access and decoding information for short time periods (such as every 10 seconds). ECMs are sent continuously to allow the media stream to be decoded as encryption codes change.

Entitlement Management Messages (EMM)-Entitlement Management Messages are information elements that allow the encryption and/or decryption of information (such as a descrambling code for a digital television channel).

Entitlements-Entitlements are rights to use a product, service or media.

Entrance Facilities-The structure or conduit assembly that permits communication cabling to enter a building or facility.

Entropy-A measure of the amount of information in a message, which can be expressed in units of bits. If the probability of each possible variation of the message is known, the entropy can be calculated with a simple formula. The entropy of the message is always no greater than the number of bits used to represent the original message, so entropy coding techniques can generally be used to compress the information in the message into fewer bits.

Entropy Coding-Entropy coding is a general term for lossless data compression techniques that take advantage of non-uniformity in the probability distribution of the data. The theoretically optimal compression performance of an entropy coding method would be, on average, to compress each symbol to a number of bits equal to the entropy of that symbol. Practical entropy coding techniques include Huffman coding and arithmetic coding, which can both achieve performance close to that theoretical optimum.

ENUM-tElephony NUmber Mapping

Enumeration-Enumeration is the process of discovering the communication parameter and configuration state of a data device that has been connected to a computer. When the computer (the host) discovers a new device, it determines what type of data transfers it anticipates transferring. Enumeration is a process used in the many computing systems including the Universal Serial Bus (USB) system.

Environmental Communications-
Environmental communication is the broadcasting of information about the environmental conditions that may cause hazardous conditions (e.g. sea conditions and weather status) in which vessels operate.

Environmental Hazards-Environmental hazards are the surrounding conditions or situations in an area that can be a source of danger resulting in the loss of health, life or property.

EO-End Office
EOC-Embedded Operations Channel
EOM-End Of Message
EOM-End Of Message Frame
EOP-End of Procedure frame

EPG-Electronic Programming Guide

EPIRB-Emergency Position Indicating Radio Beacon Station

EPON-Ethernet Passive Optical Network

Epoxy-Epoxy is a liquid material that solidifies upon heat curing, ultraviolet light curing' or mixing with another material. Epoxy is sometimes used for fastening fibers together or for fastening fibers to joining hardware.

ePublishing-Electronic Publishing

EQ-Equalization

Equal Access-A telephone service that provides a communication user with an equal choice of long distance carriers.

Equalization (EQ)-(1-general) Equalization is a process of adjusting received radio signal to counter the effects of distortion caused by multiple receptions of the same radio signal that have been delayed in time. (2-audio) The process of improving the sound quality of an audio signal by increasing or decreasing the gain of the signal at various frequencies. (3-video) The process of altering the frequency response of a video amplifier to compensate for high-frequency losses in coaxial cable.

Equalize-The process of inserting in a line a network with transmission characteristics complementary to those of the line, so that when the loss or delay in the line and that in the equalizer are combined, the overall loss or delay is approximately equal at all frequencies.

Equalizer-(1-general) A network that adjusts the frequency characteristics of a transmission circuit to allow it to transmit selected frequencies in a uniform manner. (2-device) A device installed in a signal transmission path to compensate for differences in time delay or in amplitude occurring in the transmission path for different frequency components of the signal. The purpose of the equalizer is to make the amplitude of all frequency components equal after the equalizer process (see: amplitude equalizer, graphic equalizer), or to make the time delay equal for all frequency components (see: Dispersion), or both. Some equalizers are self-adjusting adaptive equalizers. (3-absolute delay) A circuit or device within a network that is adds time delay to frequency components or circuits so the signals can be received from multiple sources at approximately the same time. (4-adaptive) An equalizer that constantly readjusts the amount of equalization (frequency or time delay) to adjust a distorted input signal to produce an equalized output signal. (5-delay) An equalizer that can delay the transmission of some frequencies so that the delay of all frequency components is approximately the same.

Equipment Cabinet-An enclosure assembly that allows the installation of electronic assemblies or components. Common equipment racks are 7 feet tall by 24 to 26 inches wide. The inside mounting area is often 19 or 22 inches wide (the rack). Some equipment cabinets come with power supplies and cooling fans.

Equipment Configuration-Equipment configuration is the process of sending information to a device that is used to adapt the equipment or software program to its environment (configuration).

Equipment Identity Register (EIR)-An equipment identity register is a database in a mobile telecommunications network (e.g. GSM) that contains the identity of telecommunications devices (such as mobile phones) and the status of these devices in the network (such as authorized or not-authorized). The EIR is primarily used to identify mobile phones that may have been stolen or have questionable usage patterns that may indicate fraudulent use. The EIR has three types of lists; white, black and gray. The white list holds known good IMEIs. The black lists holds invalid (barred) IMEIs. The gray list holds IMEIs that may be suspect for fraud or are being tested for validation.

Equipment Rack-A container that holds and interconnects electronic or electrical assemblies. A common size for an equipment rack is 19 inches (48.26 cm) wide. Electronic assemblies (called "modules") typically slide or are guided into the equipment rack where connectors located in the back of the equipment rack make contact with the modules.

This diagram shows an equipment rack that is used for mounting and interconnecting network communication equipment. This diagram shows that several modules (line cards) can be mounted in each equipment rack. This diagram shows that the back (call the "backplane") of the equipment rack contains connectors that can interconnect the modules to each other and to cables from other equipment racks or devices.

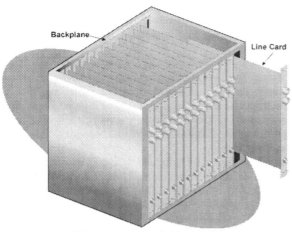

Equipment Rack Diagram

Equipment Room (ER)-A room or space that is dedicated for communication equipment and cable connection points. Equipment rooms may be centralized points that allow one or many companies to access telephone equipment, data communication equipment, and communication line connection points.

ER-Equipment Room

Erase-The process of discarding, obliterating, or marking information as deleted from a storage medium.

ERL-Echo Return Loss

Erlang-A measurement unit of the average traffic usage of a telecommunications facility during a period of time (normally a busy hour) with reference to one hour of continuous use. The capacity of Erlangs is the ratio of time during which a facility is occupied to the time the facility is available for occupancy with reference to one hour. For example, a 12 minute call is 0.2 Erlangs. 1 Erlang equals 36 centum call seconds (CCS).

ERLE-Echo Return Loss Enhancement

ERMES-European Radio Messaging System

ERP-Effective Radiated Power

Error Blocks-Error blocks are groups of image bits (a block of pixels) in a digital video signal that do not represent error signals rather than the original image bits that were supposed to be in that image block.

Error Concealment-Error concealment is a process that is used by a coding device (such as a speech coder) to create information that replaces data that has been received in error. Error concealment is possible when portions of the signal output of the coder has some relationship to other portions of the signal output and that the relationship can be used to produce an approximated signal that replaces the lost information period (lost bits).

Error Correcting Code (ECC)-Error correcting codes are additional information elements (codes) that are sent along with an information (data) signal that can be used to detect and possibly correct errors that occur during transmission and storage of the media. Error correction codes conform to specific rules or formulas to create the code from the data information that is being sent. Error correction codes require an increase in the number of signal elements than are transmitted increasing the required data transmission rate.

This figure shows how block coding error detection and correction bits are added to the data to be transmitted (e.g. digital speech) by supplying blocks of data to a block code generator. To create an error protection code, the block cyclic redundancy check (CRC) parity generation divides a given block of data by a defined polynomial formula. The quotient and remainder are appended to the data stimulus to allow comparison when received. Using the same polynomial formula, the received data can be compared to the received error protection bits. In some cases, the formula can be used to fix some of the bits that were received in error.

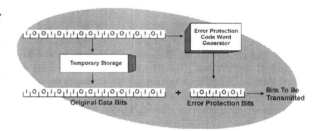

Block Error Coding

Error Correction-Error correction is the process of adding data bits to a transmitted signal that can be used to help identify errors and possibly correct bits that were received in error due to distorted radio transmission.

Error Detection-The process of detecting for bits that are received in error during data transmission. Error detection is made possible by sending additional bits that have a relationship to the original data that can be verified. See also: error correction.

This figure shows the basic error detection and correction process. This diagram shows that a sequence of digital bits is supplied to a computing device that produces a check bit sequence. The check bit sequence is sent in addition to the original digital bits. When the check bits are received, the same formula is used to check to see if any of the bits received were in error.

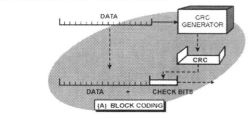

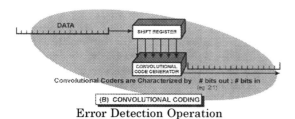

Error Detection Operation

Error Listing-A data set or printout of errors and adjustments. The listing can include totals, comparisons with other data, and a coded description of an error condition. An error listing is sometimes called an exception report.

Error Prone-Error prone is a condition or probability that information will contain inaccurate data or will loose information during transfer.

Error Protection-The process of adding information to a data signal (typically by sending additional data bits) that permits a receiver of information to detect and/or correct for errors that may have occurred during data transmission.

Error Recovery-In the context of decoding compressed bitstreams, the ability of the decoder to recover from catastrophic bit errors or lost bits within the bitstream. For example, a decoder may need to terminate decoding the previous bitstream segment and completely resynchronize with a new synchronization point with the bitstream. The speed of error recovery may depend on both bitstream characteristics and decoder implementation.

Error Resilience-Error resilience is the ability of a processing system to detect, correct or adjust for errors.

Error Vector Magnitude (EVM)-Is a measurement of phase modulation accuracy of the transmitter as compared to predicted vectors.

Errors-(1-general) A collective term that includes all types of edit rejects, inconsistencies, transmission deviations, and control failures. (2-difference) The difference between the measurement of a particular quantity and the true value of the quantity. (3-digital system) Data that has been lost or damaged through either system problems or media defects. The most common general types of errors are hard errors where the errors occur over a long time period and are repeatable and soft errors that are non-reoccurring and may be hard to identify and duplicate for verification.

Errors And Omissions (E&O)-A type of insurance that is provided to production companies and studios to offer security in case the copyright of a film might be jeopardized.

ES-Elementary Stream

ESA-Extended Service Area

Escalation-The process of taking a trouble call or ticket up through increasing levels of management until the problem gets resolved.

ESCON Connector-Enterprise Systems Connections

ESMR-Enhanced Specialized Mobile Radio

ESMTP-Extended Simple Mail Transport Protocol

ESN-Electronic Serial Number

ESP-Enhanced Service Provider

ESS-802.11 Extended Service Set

ESS-Electronic Switching System

ESS-Extended Services Set

Estimated Load-The usage activity of a system during a measurement period, as determined by available data and traffic theory.

ETC-Enhanced Throughput Cellular

ETDMA-Extended Time Division Multiple Access

Ethernet-Ethernet is a packet based transmission protocol that is primarily used in LANs. Ethernet is the common name for the IEEE 802.3 industry specification and it is often characterized by its data transmission rate and type of transmission medium (e.g., twisted pair is T and fiber is F). Ethernet systems in 1972 operated at 1 Mbps. In 1992, Ethernet progressed to 10 Mbps data transfer speed (called 10 Base T). In 2001, Ethernet data transfer rates included 100 Mbps (100 BaseT) and 1 Gbps (1000 Base T). In the year 2000, 10 Gigabit fiber Ethernet prototypes had been demonstrated. Ethernet can be provided on twisted pair, coaxial cable, wireless, or fiber cable. In 2001, the common wired connections for Ethernet was 10 Mbps or 100 Mbps. 100 Mbps Ethernet (100 BaseT) systems are also called "Fast Ethernet." Ethernet systems that can transmit at 1 Gbps (1 Gbps = 1 thousand Mbps) or more, are called "Gigabit Ethernet (GE)." Wireless Ethernet have data transmission rates that are usually limited from 2 Mbps to 11 Mbps.

Originally created by an alliance between Digital Equipment Corporation, Intel and Xerox, Ethernet DIX, is slightly different than IEEE 802.3. In Ethernet the packet header includes a type field and the length of the packet is determined by detection. In IEEE 802.3, the packet header includes a length field and the packet type is encapsulated in an IEEE 802.2 header. Most modern day "Ethernet" devices are capable of using both protocol variation, however, older equipment was not able to do this.

This figure shows several types of Ethernet LAN systems and the approximate distance devices can be connected together in these networks. Thicknet Ethernet uses a low loss coaxial cable to provide up to 500 meters of interconnection without the need for repeaters. Thinnet systems use a relatively thin coaxial cable system and the typical signal loss in this cable restricts the maximum distance to approximately 185 meters. 100 BaseT systems use category 5 UTP cable and the maximum distance is approximately 100 meters.

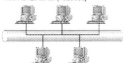

Thicknet Ethernet (10Base5)

Thicknet (10Base5)
Segment up to 500 meters w/o repeaters
Uses the 5-4-3 Rule
Cable access: Transceiver w/AUI interface
Cable spec: RG 6 to 8
Speed: 10Mbps

General:
Specified by IEEE 802.3
Protocol: Carrier Sense Multiple Access w/ Collision Detection
Standards Speeds: 10Mbps & 100Mbps
Experimental Speeds: Above 1Gbps

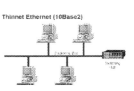

Thinnet Ethernet (10Base2)

Thinnnet (10Base2)
Segment up to 185 meters w/o repeaters
Uses the 5-4-3 Rule
Cable access: BNC T-connector
Cable spec: RG 58
Speed: 10Mbps

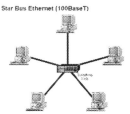

Star Bus Ethernet (100BaseT)

(100BaseT)
Category 5 UTP/STP up to 100 meters
Cable access: RJ45
Cable spec: UTP/STP Category 5
Speed: 10/100Mbps

Ethernet System

Ethernet Address-The unique non-changeable 48 bit address that identifies a specific hardware device in an Ethernet network. This includes routers, network interface cards (NIC's), printers, and bridges. A portion of the Ethernet address identifies the equipment manufacturer and a portion is a sequence number assigned by a manufacturer. The Ethernet address is also call the medium access control (MAC) address or hardware address.

Ethernet Hub-In 10-Base-T and 100-Base-T Ethernet, a specific type of repeater which usually has between four and forty-eight ports. The hub allows Ethernet to be star-wired while still being a logical bus. This enables the IEEE 802.3 MAC protocol, based on collision domain (CSMA/CD), to be utilized while having the added advantage of centralized wiring.

This figure shows an Ethernet hub. This diagram shows that one of the computers has sent a data message to the hub on its transmit lines. The hub receives the data from the device and rebroadcasts the information on all of its transmit lines, including the line that the data was received on. The hub's receiver and transmit lines are reversed from the computers. This allows the computers that are connected to the hub to hear the information on their receive lines. The sending computer uses the echo of its own information as confirmation the hub has successfully received and retransmitted its information. This indicates that no collision has occurred with other computers that may have transmitted information at the same time.

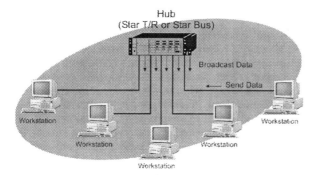

Ethernet Hub Operation

Ethernet in the First Mile (EFM)-Developed as standard IEEE 802.3ah, commonly known as Ethernet First Mile (EFM). This technology seeks to provide high-speed Ethernet-like communication over voice grade copper wire and fiber optic cables. "First mile" refers to the cables running from homes and business to the first device operated by a carrier or service provider outside of the customer premises. The technology leverages the ubiquity of Ethernet in the LAN with existing wire infrastructure already in place.

Ethernet Passive Optical Network (EPON)-A Ethernet Passive Optical Network (EPON) utilizes Ethernet framing and protocols on a passive optical communication system. EPONs allow service providers and customers to be required to own and understand a single networking technology: Ethernet.

EPONs are passive in that all devices in the physical cable plant from the service provider to the end-customer are passive. Passive splitters are used to direct light in fibers to multiple locations, much like coax splitters are used in cable TV networks. Likewise, precise timing and/or frequency control is utilized to allow multiple end-users to access the same upstream fiber. The key advantage of EPONs are their ability to provide high-speed access to many end-users with a single fiber, while

not requiring any advanced electronics devices to be located in environmentally harsh locations such as atop telephone poles. EPONs are very similar in architecture, applicability and deployment challenges as cable modem networks based on DOCSIS.

Ethernet Switch-A device that relies on enhancements to the original Ethernet specification that transfer data directly between ports without the need to re-broadcast the information to all ports of the device. Switched Ethernet incorporates modified layer 2 (Data Link Layer) electronics in order to allow an individual 10 Mbps (for Ethernet) or 100 Mbps (for fast Ethernet) user to transfer data directly to an end segment.

ETL-Extract, Transform, and Load

eTOM-Enhanced Telecom Operations Map

ETS-ETSI Technical Specification

ETSI-European Telecommunications Standards Institute

EULA-End User Licensing Agreement

Eureka 147-Eureka 147 is a digital audio broadcasting (DAB) system that transmits voice and other information using digital transmission. The DAB signal is typically shared with additional digital information on a single digital radio channel.

European Telecommunications Standards Institute (ETSI)-An organization that assists with the standards-making process in Europe. They work with other international standards bodies, including the International Standards Organization (ISO), in coordinating like activities.

EVDO-Evolution Version Data Only

Event-A process or data transfer that has initiated, changed, or completed in a communication system. Event records identify these changes and are commonly used for billing systems. Events that cause other processes to start are called trigger events.

Event - Billing-A change of equipment configuration or the use of a network resource that is recorded, usually for billing purposes. Event information is used by the billing system to determine the rate and amount of usage that will be billed to a customer. Events are sometimes referred to as Usage, CDR, xDR, Ticket, Message.

Event Driven-Event driven is an action or software program operation that starts when a specific event or trigger has occurred.

Event History-Event history is the records and related information of the activities that have been performed or related to a customer or prospect (a customer record). Example of items in an event history include phone calls to customers, items mailed to customers and responses received from questionnaires sent to customers.

Event Information Table (EIT)-An event information table is a group of data elements that is transmitted on an MPEG stream that contains the details of specific TV programs, including the program name, start time, duration, genre and possibly age rating. The EITs are used by the middleware of the set top box to create an on-screen Electronic Program Guide.

Event Management-The recording, organizing, and displaying events that occur in communication systems. These events may include the type of event such as a program activated by a software application or information that is generated by event triggers (some preset level). Event management is used to determine the status and changes (such as increased bandwidth or CPU usage) that are occurring in communication systems to help find potential problem areas before they cause system failures.

Event Record-Event record is a data record that holds information related to a specific event. Event records for communication systems contain information about the specific usage of a network element or service. Examples of event records include the path and connection time used by a specific switch in a communication network. A communication session (such as a voice call) typically involves multiple events (such as multiple switch connections). Call detail records (CDRs) are typically created by combining multiple event records.

Events-(1-general) External information that is received that indicates a change in status within a communication system. (2-Bluetooth) Incoming messages to the L2CA layer along with any timeouts. Events are categorized as Indications and Confirms from lower layers, Request and Responses from higher layers, data from peers, signal Requests and Responses from peers, and events caused by timer expirations.

EventsML-Events Markup Language

EVM-Electronic Voice Mail

EVM-Error Vector Magnitude

Examination-The process of qualifying an application for patent for grant. Also known as "substantive examination," this process includes an analysis of the claims with regard to Prior Art to determine whether or not the claims meet all statu-

tory and regulatory requirements. Examination is performed by qualified patent examiners who have expertise in the technical areas where they work.

Exception Reports-Exception reports are tables, graphs or images that represent data or information that was not able to be processed using existing programs (e.g. could not find or match data records) or had results that were unexpected or which fall outside allowable ranges.

Exchange-(1-switch) The term used in some countries to refer to a network switch (See "Switch"). (2-company) A communication company or companies for the administration of communication service in a specified area, which usually embraces a city, town, or village and its environs, and consisting of one or more central offices, together with the associated plant, used in furnishing communication service in that area. The first exchange was installed in 1878.

Exchange Carrier (EC)-A company that provides local telecommunications services. Exchange carriers are generally regulated by a national regulatory agency for interstate or inter-regional services and are regulated by state or local agencies for internal local access and transport area (InterLATA) services.

Exchange Message Record (EMR)-An exchange message record is a standard data format for the exchange of messages between telecommunications systems that is often used for billing records. The records may be exchanged by magnetic tape or by other medium such as electronic transfer or CD ROM.

Exchange Point-An exchange point is a location, device or facility where several networks may interconnect with each other (a common switch or node).

Executable Short Message-A message that is received by a Subscriber Identity Module card in a wireless system (such as a mobile phone system) that contains a program that instructs the SIM card to perform processing instructions.

This diagram shows how an executable short message can be sent by a system operator to add a new feature into a mobile phone. The executable short message is a program that is stored in the SIM card and interacts with the operation of the mobile phone to allow the new feature to operate. The system simply sends the file as an executable message directly to the mobile phone identification. When the complete executable short message has been received in the SIM card, it is stored in memory and this program can complete (run) instructs that allows the new feature to operate.

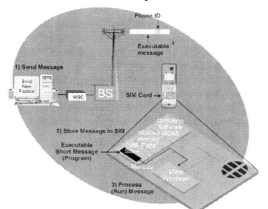

Executable Short Message Operation

Expandability-The ability of a system to supply processing or services without significant changes to its fundamental assemblies. Measures of expandability in communication include the maximum number of customers that can receive service, number of radio transmitter towers that can be connected to a switching system, and maximum data transfer rates on a communication line.

Expandable Ads-Expandable ads are images or video clips that can be enlarged or extended in time and/or content. Expandable ads may first appear as images, links or banner ads.

Expanding-Expanding is a process that increases the amount of amplification (gain) of an audio signal for smaller input signals (e.g., softer talker). The use of expanding allows the level of audio signal that leaves the modulator to have a larger overall range (lower minimum and higher maximum) regardless if some people talk softly or boldly. As a result of expanding, high-level signals and low-level signals output from a modulator may have a different conversion level (ratio of modulation compared to input signal level). This could create distortion so expanding allows the modulator to convert the information signal (audio signal) with less distortion. Of course, the process of expanding must be initiated at the transmitting end, called companding, to recreate the original audio signal.

Expansion Loop-An expansion loop is a short length of cable that is part of a communication line (such as a pole mounted cable TV or telephone

lines) that provides additional cable length that may be necessary to compensate for changes in the length of the cable due to temperature changes or cable loading (wind or ice).

This figure shows how an aerial expansion loop provides protection to a cable for the expansion and contraction of the messenger wire. This diagram shows that the expansion loop is placed at the telephone phone. As the messenger cable moves, the cable will flex taking from or returning cable to the expansion loop.

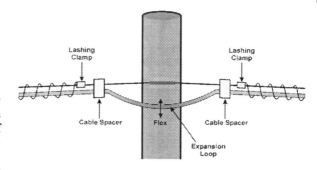

Aerial Cable Expansion Loop

Expansion Slots-A connection point and associated assembly mounting hardware that allows circuit cards or assemblies to be added to a computer or communication equipment. Examples of expansion slots include PCI connection in a desktop computer or a PCMCIA slot in a laptop computer.

Experimental Broadcast Station-A experimental broadcast station is a radio transmitter that is licensed for experimental or developmental transmission of radio telephony, television, facsimile, or other types of telecommunication services intended for reception and use by the general public.

Experimental Period-Experimental period is a time period that radio transmission is used for test, maintenance and experimentation. For radio broadcast radio stations, this is usually the time between 12 midnight local time and local sunrise.

Experimental Radio Service-Experimental radio service is a service that uses radio waves for the purpose of experimentation and research projects.

Explorer Frame-In source routing, a frame used either to perform route discovery or to propagate multicast traffic. There are two types of explorer frames: Spanning Tree Explorers and All Routes Explorers.

Extended Allocation-Extended allocation is the assignment of a resource that exceeds the basic assignment allocation. Extended allocation is a process of determining an increased need or the availability of a resource and the assignment (allocation) of the resources that is in excess of a typical resource assignment.

Extended LAN-An extended local area network is a group of LAN systems that are interconnected by bridges (packet transfer points).

Extended Simple Mail Transport Protocol (ESMTP)-An extended (enhanced) version of the original SMTP protocol to better support text languages, graphics, audio and video files. It is described in RFC 1651.

Extender Board-An adapter board that extends a module outside of its frame to allow easier access to the module's components for troubleshooting and alignment.

Extensibility-The ability to upgrade existing systems or services without significant changes to the existing systems.

Extensible Hypertext Markup Language (xHTML)-Extensible Hypertext Markup Language (XHTML) Basic is a text based software communication standard that is used to allow the web software developer to define new (extensible) elements of a Internet web pages. xHTML was created by the World Wide Web Consortium (WC3) in 1996 to provide a common markup language for wireless devices and other small devices with limited memory. It is a widely supported open technology (i.e. non-proprietary technology) that is used for data exchange between any type of application that can understand XML. The combination of XML with Hypertext Markup Language (HTML) produces a web presentation language that is flexible (extensible) based on the needs of the application it is being used for.

This figure shows an example of how XHTML Basic operation can be used to process information requests from mobile devices. This example shows that a user has sent movie time request from a Wireless device to their preferred entertainment information provider. This request is routed through the cellular tower and to the cellular system that forwards the request to the Internet using the selected Internet address of the information provider. The Internet routes the request to a WAP server. The WAP server determines that this request is for a document or information that is written in XHTML Basic and stored on the WAP server. The XHTML Basic program is accessed, and the requested data is sent via the Internet through the cellular tower to the wireless device for display. The requested data could be any web site that, for example, has been converted to XHTML Basic from HTML so that it can be displayed correctly on any wireless device.

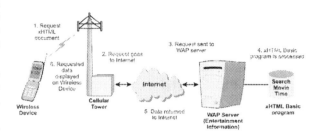

xHTML Basic Operation

Extensible Markup Language (XML)-A software standard that is used to define exchangeable elements of a web (HTML) page. Extensible Markup Language was developed in 1996 by the World Wide Web Consortium (W3C). It is a widely supported open technology (i.e. non-proprietary technology) for data exchange. XML documents contain only data, and applications display that data is various ways. XML permits document authors to create their own markup for virtually any type of information. Therefore, authors can use XML to create entirely new markup languages to describe specific types of data, including mathematical formulas, chemical formulas, music and recipes.

eXtensible Media Commerce Language (XMCL)-Extensible media commerce language is a software standard that is used to define exchangeable elements multimedia content and digital rights associated with the media so the media can be exchanged (interchanged) with networks or business systems.

Extensible Rights Markup Language (XrML)-
Extensible rights management language is a XML
that is used to define rights elements of digital
media and services. Extensible Rights Markup
Language was initially developed by Content
Guard and its use has been endorsed by several
companies including Microsoft. The XrML language provides a universal language and process
for defining and controlling the rights associated
with many types of content and services.

Because XrML is based on extensible markup language (XML), XrML files can be customized for specific applications such as to describe books (ONIX)
or web based media (RDF). For more information
on XrML see www.XrML.org.

Extensible Solution-An extensible solution a system, product or service that is capable of being
changed or updated to allow for added services and
benefits.

Extension-(1-software) A set of commands or protocols that are used to extend the capabilities of
another application or protocol. Software extensions are commonly used to rapidly extend the
capabilities of an existing application or protocol
without changing the underlying application or
protocol. (2-telephone) An additional telephone
connected to a line. Allows two or more locations to
be served by the same telephone line or line group.
May also refer to an intercom phone number in an
office. (3-file name) The optional second part of a
PC computer filename. Extensions begin with a
period and contain from one to three characters.
Most application programs supply extensions for
files they create.

Extension Negotiation-The process of negotiating which software extensions will be used between
devices or systems. Extension negotiation requires
the communication devices to identify which software extensions they can use (and possibly their
preferences for using such extensions) to define a
common set of extensions that can be used between
communication devices.

External Gateway Protocol (EGP)-EGP is an
Internet protocol that is used to exchange routing
information between switches and routers outside
the network domain where it is located. An EGP
can indicate if a given network is reachable.
However does not make routing or priority decisions.

External Interface-A device that adapts internal
networks to external networks. The use of an exter-

nal interface allows devices and applications that
are not directly connected to a network to request
and receive some or all services from the network.

External Modem-An external is a modem is a self
contained (stand-alone) modem that is typically
connected and connected between a transmission
line (such as a phone line) and a communication
device (such as a computer) by a cable. External
modems may have display indicators (LEDs) on the
front of the chassis indicate the current status or
activity of the modem. The connection to the computer is typically a standard interface such as an
Ethernet or USB connector. This allows an external modem can be used with different computers at
different times and also with different types of computers.

External Modulation-Modulation of a carrier signal source, such as a laser or LED, by varying its
emitted signal. In direct modulation, the amount of
light that is emitted is varied. In external modulation, the emission (such as light) is constant and
some mechanism is used to alternately block and
pass it. See also Electro-Optic Modulator and
Electro-Absorption Semiconductor Modulator.

External Network-A network that is not part of a
specific network domain (such as the Internet
when compared to a company data network).
External networks may allow direct or limited
access to network information (routing and
addressing) and control of network resources
(bandwidth and QoS) depending on the interconnection type and network access (security) provisioning.

External Radio Modem-External radio modems
are self contained radios with data modems that
allow the customer to simply plug the radio device
to their USB or Ethernet data port on their desktop
or laptop computer. External modems are commonly connected to computers via standard connections
such as universal serial bus (USB) or RJ-45
Ethernet connections.

External RF Power Amplifier-A device that is
capable of increasing power output when used in
conjunction with, but not an integral part of, a
transmitter.

Extract-A process of removing information or data
from another (usually larger) data source.
Extracting often involves data pointers that indicate the start of the data block with the information data and processing steps for extracting and
confirming the data integrity.

E

Extranet-An extranet is a network (typically an Internet network) that is only accessible to customers, affiliates, or members rather than the general public. Extranets may use the public Internet as its transmission system. However, extranet data connections through the Internet are usually encrypted and accounts are password protected to prevent access by unauthorized users.

Eyedropper-An eyedropper is a cursor icon that appears when a color tool is selected. The eyedropper typically allows the user to select a color from anywhere on the display.

E-Zine or EZine-Electronic Magazine

F

F-farad

F/R-CPHCH-Forward/Reverse Common Physical Channels

FAB-Fulfillment, Assurance, and Billing

FAC-Feature Access Code

FACCH-Fast Associated Control Channel

FACCH/F-Full Rate Fast Associated Control Channel

FACCH/H-Half Rate Fast Associated Control Channel

FACH-Forward Access Channel

Facilities-Facilities are the parts (elements) of a network or system that are owned or leased by a company or person. The term "facilities" is sometimes generalized to describe buildings, utilities, or communication equipment.

Facility-(1-telephone) Any one of the elements of the physical telephone plant needed to provide service, such as switching systems, cables, and microwave radio transmission systems. (2-line) Transmission plant between offices. (3-outside plant) The outside plant from a central office mainframe to a point of termination on a customer's premises, but not including customer equipment. (4-packet switching) A specific X.25 service, such as Closed User Group and/or a protocol element that conveys information about that service.

FACN-Foreign Agent Control Node

Facsimile (Fax)-Facsimile is the representation of optical images by electrical signals for transmission over communication systems. When the electrical signal is received, it is converted back to an optical format for display or printing of the original image. This process is commonly called FAXing.

Fax usually involves the transmission of still images (typically black and white with no intermediate shades of gray). The most widely used type of FAX service today is described by ITU standard T.30, Group 3. Group 3 facsimile uses a digital modem over a voice grade public telephone channel to transmit the FAX signal. Group 4 FAX also permits user-selectable pixel size and typically operates via a 64 kb/s ISDN channel. Group 1 was an early analog fax system, typified by the Xerox Telecopier. Group 2 was an early digital FAX system. Groups 1 and 2 are obsolescent, although some Group 3 FAX machines are backward compatible with Group 2 machines.

Fade Filter-A fade filter decreases the image intensity of the outgoing media segment and increases the intensity of the incoming media segment in a video.

Fade Margin-Fade margin is the amount of signal loss, usually expressed in decibels, that a radio signal in a communication path is anticipated to change (or budgeted to change) due to transmission impairments. By budgeting a fade margin in a communication link, this helps to ensure that typical signal fading periods do not result in a lower than expected quality of service.

Fading-(1-radio signal) The variation in radio signal strength that results from changes in the characteristics of the radio transmission path. Fading can be caused by several different factors including signal reflection, ionosphere, and interference from other radio channels. (2-video) A progressive level of translucence (see-through) from one image or video screen to another.

Fail Safe Operation-A type of system control that allows for the automatic operation of additional equipment or reconfiguration of existing equipment to prevent the improper functioning of communications or systems in the event of circuit loss or impairment.

Failure-A detected cessation of capability to perform a specified function or functions within previously established limits. A failure is beyond adjustment by the operator through means of controls normally accessible during routine operation of the system. (This requires that measurable limits be established to define "satisfactory performance.")

Failure Mode And Effects Analysis (FMEA)-An iterative documented process performed to identify basic faults at the component level and determine their effects at higher levels of assembly.

Fair Reasonable and Non-Discriminatory (FRAND)-Fair, reasonable and non-discriminatory are licensing terms (fees and restrictions) that are offered by owners of intellectual property that allows other companies to build products that incorporate their technology while providing the potential for sustainable profits.

Fairness Scheduling-Fairness scheduling algorithm that coordinates the sequences of processes or information so that the data transmission rate or application processing time is fairly distributed to users of the system or services. Fairness scheduling can be used to overcome the differences in capabilities of the access or systems for each user.

False Colors Effect-A digital picture manipulator effect that permits user adjustment of colors in the picture.

Family of Technologies-A family of technologies is a collection of technologies that share a common technology or industry category.

Fan In-The greatest number of separate inputs acceptable to a single specified logic circuit without adversely affecting performance.

Fan-Out-Fan Out Cable

FAQs-Frequently Asked Questions

Far End-Far end is a reference point that is at the end of a transmission line or circuit (near the exit or final termination point).

Far End Crosstalk (FEXT)-Far end crosstalk (FEXT) is the leakage of signal that is coupled to a nearby cable or electronics circuit (called crosstalk) where the unwanted signal is received on the far end (remote end) of the cable.

This diagram shows how far end crosstalk (FEXT) can cause interference at the distant end of a transmission line. FEXT occurs when some of the transmitted signal energy leaks from one twisted pair and is coupled back to a communications line that is transferring a signal in the opposite direction. Generally, FEXT has a lower level of signal interference as the crosstalk levels at the receiver end are lower then crosstalk at the sending (near) end.

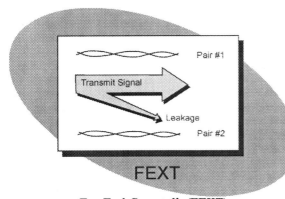

Far End Crosstalk (FEXT)

Fast Busy-An alert tone that indicates communication resources (such as switch trunks) are not available. A fast busy signal operates at twice the normal busy signal rate (120 tones/minute).

Fast Cache-Fast cache is a process that temporarily increases the data transmission bandwidth to allow the rapid transfer of buffering (cache) information which can reduce the initial buffering time for a media streaming session.

Fast Ethernet-The standard Ethernet protocol that transfers data at 100 Mbps capacity. Fast Ethernet is commonly called 100BaseT.

Fast Fourier Transform (FFT)-An alternative implementation of the Discrete Fourier Transform that eliminates significant computational redundancy.

Fast Frequency Shift Keying (FFSK)-A digital modulation process where each digit is represented a different (unique) frequency.

Fast IR-Fast Infrared

Fast Reconnect-Fast reconnect is a process of re-establishing a communication session within a relatively short amount of time. Fast reconnection typically involves the ability to skip over some of the initial steps that occur in the setup of a communication connection because some of the parameters of the session are already known (such as an IP address) and can be used in the re-establishment of the connection.

Fast Recovery-Fast recovery is a process of restarting or continuing a service within a relatively short amount of time. Fast recovery typically involves the ability to skip over some of the initial steps in the setup of a service or software application as some of the parameters of the service are already known (such as the type of media and file pointers) and can be used in to restart the service or application.

Fast Start-Fast start is the process of having a media file to quickly start playing (or perceived as quickly) after the media file is requested. Examples of fast start is the ability of a digital television channel to begin playing a movie as soon as or shortly after a user has selected the television channel.

Fast Stream Start-Fast stream start is the process of having a streaming media source to quickly start playing (or perceived as quickly) after the media file is requested.

Fast Streaming-Fast streaming is the process of streaming media with minimal delays. To overcome the challenges of initialization and buffering, fast streaming systems may use a variety of techniques including temporarily increasing the data transmission bandwidth, stream thinning and variable compression rates to minimize the buffering time.

Fault-A condition that causes a device, a component, or an element to fall to perform in a required manner. Examples include a short circuit, a broken wine, or an intermittent connection.

Fault Finder-A test set or other device that enables faults to be identified and localized.

Fault Location-Fault location is the processes that are used to determine the location (and sometimes the cause) of transmission line or system malfunction.

Fault Management-Fault management is one of the five functions defined in the FCAPS model for network management. Fault management identifies the network problems, failures, events and corrects them. Fault management is the reactive form of network management. SNMP traps, syslog, and RMON typically are used in fault management.

Fault Tolerance-Fault tolerance is the ability of a network or sub-system to continue to operate in the event of a hardware or software failure. Fault tolerant systems are typically able to identify the fault and replace the failed component or sub-system with another equipment.

Fault Tolerant Network-A network designed or engineered to remain in operation in the event of system or component failures. For example, a fault-tolerant network might use alternate routing.

Fault Tree Analysis (FTA)-An iterative documented process of a systematic nature performed to identify basic faults, determine their causes and effects, and establish their probabilities of occurrence.

Fault-management, Configuration, Accounting, Performance, and Security (FCAPS)-Fault-management, Configuration, Accounting, Performance, and Security is a categorical model of the working objectives of network management and is a standard adopted by the International Telecommunications Union (ITU). There are five levels, called the fault-management level (F), the configuration level (C), the accounting level (A), the performance level (P), and the security level (S).

Favored Nations-A verbal or written contract made insuring all parties equal treatment concerning specific issues.

Fax-Facsimile

Fax Back-A service that allows callers to request information that they will be delivered to them by fax. Fax back service usually involves the use of an interactive voice response (IVR) system to provide information to the caller of the fax back service. The user selects appropriate documents they desire to have sent, usually making selections with the keypad. The system then requests the caller to enter the destination fax number so the documents can be sent to the caller.

Fax Mail (FxM)-A communications service that delivers documents in fax format by converting a document (such as a word processor document) to a fax image format delivering it to a fax machine.

Fax Mailbox-A portion of memory, usually located on a computer hard disk that receives and sends fax images. The fax images are often in compressed digital image format.

Fax Modem-An adapter that is capable of both modem and facsimile transmission. A fax modem may be an internal card (such as a PCI card) or an eternal assembly. The fax modem can receive and convert digital fax images from their analog transmission form. The fax modem is capable communicating with both modem and fax protocols. There are several versions of modem and fax protocols and a fax modem may be capable of communicating with some (e.g. slow speed) or all of them.

Fax On Demand (FOD)-A telecommunications transmission service that sends previously stored faxes to a user that has requested a specific fax message. For FOD service, the caller listens to the options available and requests that additional information is delivered by fax. FOD service is often used to deliver previously stored instructions or marketing materials.

This diagram shows how a fax on demand (FOD) system can store and automatically deliver faxes. The manager of this system has previously stored operating manuals in fax mailboxes at the company. Each fax document that is stored is assigned to a fax mailbox. The verbal (audio) name that is associated with this fax mailbox is programmed into the interactive voice response (IVR) system. This allows the user to hear their options and to make a fax mailbox selection. This example shows how a caller dials a telephone number that this company uses for automatic fax back service (555-2345 in this example). When the phone system receives a

F

call on this line, it automatically routes the call to the IVR system at extension 1001. The IVR system prompts the caller to select the fax mailbox and to enter the destination fax number (555-8111). The fax mailbox (usually a storage area on the computer hard disk) retrieves the data and sends it to a computer fax generator. The destination digits (destination fax number) are also sent to the fax generator. This allows the computer to create the fax from the fax mailbox data and send the fax to the destination fax machine.

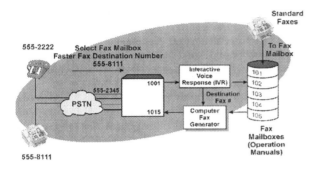

Fax on Demand (FOD) Operation

Fax over Internet Protocol (FoIP)-A process of sending fax communication signals over the Internet. If the fax signal is in analog form (voice or fax, the signal is first converted to a digital form. Packet routing information is then added to the digital fax signal so it can be routed through the Internet.

Fax Profile-(1-general) Protocols or procedures that adapt an information format into a format suitable for facsimile communication. (2-Bluetooth) Defines the protocols and procedures used by Bluetooth devices implementing the fax part of the usage model called "Data Access Points, Wide Area Networks." A Bluetooth cellular phone or modem may be used by a computer as a wireless fax-modem to send or receive fax messages.

F-BCCH-Fast Broadcast Control Channel

FBG-Fiber Bragg Grating

FC Connector-Face Contact

FCAPS-Fault-management, Configuration, Accounting, Performance, and Security

FCB-Frequency Correction Burst

FCC-Federal Communications Commission

FCC Type Approval-A approval from the FCC that identifies the radio equipment manufactured has passed tests certifying it meets the minimum FCC requirements for that type of radio equipment. Most radio devices must meet several FCC specification requirements to receive FCC type approval. Companies typically use an independent testing lab to certify that equipment meets FCC requirements.

FCCH-Frequency Correction Channel

FCS-Frame Check Sequence

FDD-Frequency Division Duplex

FDDI-Fiber Distributed Data Interface

FDDI Connector-A FDDI connector is a two fiber strand keyed contact fiber optic connector. The FDDI connector is pull proof and wiggle proof. The official FDDI connector name is media interface connector (MIC).

FDIS-Final Draft International Standard

FDM-Frequency Division Multiplexing

FDMA-Frequency Division Multiple Access

FDTC-Forward Digital Traffic Channel

Fear Factor-One or more reasons why potential customers avoid deciding to purchase a produce or service that may satisfy their business or personal needs.

Fear, Uncertainty, Doubt (FUD)-A marketing tactic that is used to discourage customers from buying their competitors.

Feasibility Study-A feasibility study is a document that defines the requirements for the development of a product or service, the potential revenue or benefits of developing the product or service, and if the resources required to develop the product match the company capabilities and objectives.

Feathering-Feathering is an image processing effect that gradually (progressively) changes the edges of an image or object so that it appears to have a seamless blend with a background and/or other images.

Feature-(1-general) Features are a specific operations or characteristics of a piece of equipment or a service provided to a user. (2-end office) Features are specific call processing features that are provided by the switching system that is located at the end office (also called the end office).

Feature Access Code (FAC)-The codes assigned (usually numeric codes) that users may use to access the features of communication system.

Feature Group-A switched access service offered by a local exchange carrier to an inter-exchange carrier, in North America.

Feature Race-A feature race is a competitive environment where manufacturers or vendors continually add new features to gain perceived advantages.

FEC-Forward Error Correction

FECN-Forward Explicit Congestion Notification

Federal Communications Commission (FCC)-The federal communications commission (FCC) is a government agency of the United States that establishes and enforces laws and regulations regarding interstate radio and wired communications services. The agency was established by the Communications Act of 1934. The FCC must certify (FCC type approval) radio and computer equipment before it can be sold in the United States.

Feed-(1-radio frequency) The wires, cable, or waveguide used to connect an antenna with its radio transmitter or receiver. (2-broadcast) A video (television) or audio signal source.

Feedback Loop-A feedback loop is a path that connects a signal or information back to a controlling component, device or system.

Feeder-A feeder is a transmission line linking a radio communication transmitter or receiver with its antenna.

Feeder Cable-Feeder cables are one of several large cables that provide a physical connection between a central office and distribution cables that connect to end customers. Feeder cable usually is placed in underground conduit. A main feeder originates at a central office and usually follows a single route through a large, defined area. Branch feeders, or subfeeders, extend from a main feeder to provide facilities in a well defined segment of a main feeder area. Feeder cable often is referred to as feeder plant. When combined with distribution cables, it constitutes a customer loop.

Feeder Relief-The process of making more wire pairs available at a given demand point. This can be done either by adding new cables or by rearranging a network to make unused pairs available where needed.

Feeder Route Schematic-An schematic drawing of the feeder lines in a communication network.

Feeder Sections-Linear segments of feeder routes defined so that the number of pairs in a route can be effectively and economically matched to present and future demand. Ideally, the feeder section should be associated uniquely with each allocation area.

FER-Frame Error Rate

FET-Field Effect Transistor

FEXT-Far End Crosstalk

FFSK-Fast Frequency Shift Keying

FFT-Fast Fourier Transform

FGA-Feature Group A

FGB-Feature Group B

FGD-Feature Group D

FGS-Fine Granularity Scalability

FHMA-Frequency Hopping Multiple Access

FHSS-Frequency Hopping Spread Spectrum

FHSS Mode-802.11 Frequency Hopping Spread Spectrum

FIB-Forward Indicator Bit

Fiber Amplifier-A device made from optical fiber and a pump laser that is designed to produce stimulated emission of photons at the signal wavelength, leading to an amplified signal. There are two major types of fiber amplifiers: Raman amplifiers and doped fiber amplifiers like EDFAs.

Fiber Cable-A cable that is constructed of an inner core of optical fibers (glass or plastic) that are covered by high-density polyethylene. Fiber cable may be wrapped with various types of strengthening materials including steel wire and high-density polyethylene.

This figure shows single mode and multimode fiber lines. This diagram shows that multimode fibers have a relatively wide transmission channel that allows signals with different wavelengths to bend back into the center of the fiber strand as they propagate down the fiber. The diagram also shows that single mode fiber has a much smaller transmission channel that only allows a specific wavelength to transfer down the fiber strand.

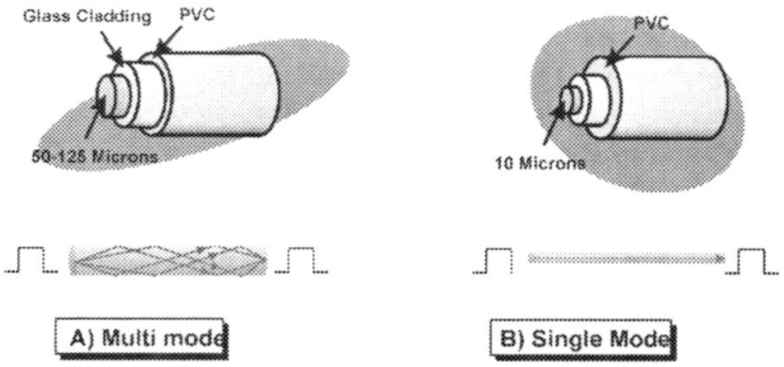

Fiber Optic Cable Diagram

Fiber Distributed Data Interface (FDDI)-Fiber distributed data interface (FDDI) is a computer network protocol that utilizes fiber optic or copper cable as the transmission medium to provide a token-passing, logical ring topology network operating at 100 Mbps. FDDI also provides for a mode of operation whereby two counter rotating rings are used to provide immediate fail-over and ring recover should a fiber cut occur. FDDI was commonly used as a backbone network to interconnect lower speed Ethernet and Token Ring networks within an enterprise. The American National Standards Institute standard X3T12 defines the protocol.

This figure shows FDDI system that uses dual rings that transmit data in opposite directions. This diagram shows one dual attached station (DAS) and a dual attached concentrator (DAC). The DAS receives and forwards the token to the mainframe computer. The DAC receives and token and coordinates its distribution to multiple data devices that are connected to it.

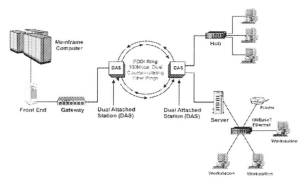

Fiber Distributed Data Interface (FDDI) System

Fiber Hub-A termination point for optical fiber cables. A hub contains multiplexers and equipment that convert optical signals to electrical signals that are then carried to customer locations over physical pairs.

Fiber Optic Receiver-A fiber optic receiver detects a signal carried by an optical fiber and converts it to an electronic signal for processing. A receiver includes at least a photodetector, signal processing circuitry and a decision circuit (for digital signals) or a demodulator (for analog signals). The photodetector is usually a p-i-n diode or an avalanche photodiode. The processing circuitry typ-ically includes amplification and filtering functions. The decision circuit or demodulator is used to recover the original, electronic signal from the transmitted signal.

Fiber Optic Transmission System (FOTS)-A term sometimes used by carriers to descript the generic application of SONET and SDH optical networks.

Fiber Optic Transmitter-A device that takes in an information signal (usually electronic) and converts it into an optical signal that it then launches into a network, typically via an optical fiber. A transmitter consists at least of a light source and a modulator (unless modulation is done directly).

Fiber Strand-Fiber optic cable is a strand of glass or plastic that is used to transfer optical energy between points. The size of most fibers is from 10 to 200 microns (1/100th to 1/5th of a mm). Optical fibers are typically used in a unidirectional mode (e.g., data moves in only one direction). Because of this, every transmission system requires at least two fibers (one for transmission and one for reception).

Fiber To The Curb (FTTC)-A distribution system that uses fiber optic cable to connect telephone networks to nodes that are located near homes or any business environment (near the curb). The fiber optic transmission is used to provide broadband services beyond the central office, all the way to the last 50-100 feet from the subscriber. The service pedestal is said to be "at the subscriber's curb."

Fiber To The Home (FTTH)-A distribution system that uses fiber optic cable to connect telephone networks to nodes that are located in the homes of customers. The fiber optic transmission is used to provide broadband services beyond the central office, all the way through the drop wire to the optical node that is located in the customers home.

Fiber To The Premise (FTTP)-Fiber to the premise is a distribution system that uses fiber optic cable to connect telephone networks to nodes that are located within businesses and homes. FTTP is also known as fiber to the home (FTTH) and fiber to the building (FTTB).

Fiber Transmitter-A device or assembly that converts electrical signals to optical signals for transmission on an optical fiber line. Fiber transmitters can use carrier modulation (on-off) or sometimes use analog modulation (amplitude, frequency, or phase) to transfer the information signal (such as an RF signal) to a fiber communication line.

Fiberoptic Cable-See Fiber Optic Cable

Fidelity-Audio fidelity is the degree to which a system, or a portion of a system, accurately reproduces at its output the essential characteristics of the signal impressed upon its input.

Field-(1-data) A specified number of bits in a data record that is designated for a particular type information. (2-energy) Electric and/or magnetic lines of force in a specific area or region. (3-video) Half of an NTSC picture frame, which has half of the lines, required to produce a picture image. This is to interlace video signals where adjacent lines are contained in alternate fields for each picture.

Field Blanking-The blanking signals occurring at the end of each video field, used to make the vertical retrace invisible. Field blanking also is referred to as vertical blanking.

Field Emission-The emission of electrons from a surface, caused by a large external field rather than a heated filament effect.

Field Frequency-Field frequency is the number of video fields that are transmitted per second.

Field Interlacing-Field interlacing is the process used to create a single video frame by overlapping two pictures where one picture provides the odd lines and the other picture provides the lines. Field interlacing is performed to reduce flicker.

Field Label-The label (name) of a field within a database record.

Field Repeater-The preset pattern of each successive frame of composite 525 and 625 television. Composite video (PAL, NISC, and SECAM) carries the color information on a subcarrier signal whose cyclic pattern repeats.

Field Testing-Field testing is a process of testing a device, assembly, or system at a location that typically involves its normal operation.

Field Upgradeable-Field upgradeable is a product or system that can be updated without the need to return the product to the manufacturer or to a repair facility.

Field Upgrade-A field upgrade is a physical and/or software change that may occur when a product change takes place at a location other than the manufacturer's facility.

Figure 8 Cable-Figure 8 fiber cable has a messenger cable located on the top of the cable to provide the load support for cable. The supporting messenger cable may be constructed of steel or a strong type of dielectric material.

File Container-A file container is a collection of data or media segments in one data file. A file container may hold the raw data files (e.g. digital audio and digital video) along with descriptive information (meta tags).

File Directory-File directories are portions of a computer storage area that are dedicated to the listing of file names and location pointers (starting addresses and file size) for the files contained on the computer storage system.

File Downloading-File downloading is the transfer of a program or of data from computer server to another computer. File download commonly refers to retrieving files from a web site server to another computer.

File Format-File formats are the sequencing and grouping of information elements (e.g. digital bits) within a block of data (file) or as organized on a sequence (stream) of information. File formats can range from simple linear (time progressive) sequences to indexed multiple file formatted blocks of data (containers).

File Integrity-File integrity is the amount of data or information that remains unchanged as compared to its original form.

File Server-A file server is a computer that stores data centrally for network users and manages access to that data. File servers may be dedicated so that no processes other than network management can be executed while the network is available, or non-dedicated so that standard user applications can run while the network is available.

File Sharing-File sharing is the providing of access rights and the ability to identify, access and potentially modify files between computing equipment which are usually connected to a network.

File Transfer Access, And Management (FTAM)-The protocol of open systems interconnection (0SI) that enable users to transfer files between computers, regardless of manufacturer. The FTAM standards were developed by the International Standards Organization (ISO).

File Transfer Protocol (FTP)-A protocol that is used to manage the transfer of data files between computers and networks. Because FTP is a standard protocol, it permits the transfer of any type of data file between different types of computers or networks.

File Type-File types are the formats of information contained within a block of data (a file). File types may be determined by the file extension (end of the file name), through identifying information

F

within the file (typically in the beginning header area of the file) or through a combination of both (such as a jpg image file that uses specific types of data encoding and compression).

File Upload-File upload is the transfer of a program or of data from one computer to a computer server. File upload commonly refers to sending files from a computer to a web site server.

Fill-(1-telecommunications) The ratio of the number of working or assigned pairs to the total number of pairs available in a cable or allocation area. A high fill number means that a cable or group of cables is almost out of spares and needs relief. The term also applies to channels in carrier systems. (2-video) In video keying, the fill is the video signal that is inserted into the hole cut in the background video by a key signal. (See also: key.)

Filler-A material used to fill in gaps or voids, such as spaces between wires in a multipair cable.

Film-Film is a material that is used to store images.

Film Transfer Kit-A film transfer kit is an assembly that allows film media to be converted to another format such as video tape or DVD.

Filter-(1-general) A network that passes desired frequencies but greatly attenuates other frequencies. (2-active) An RC filter that uses solid-state amplifiers to produce a desired frequency shaping characteristic. (3-band elimination) A band-stop or band-rejection filter that passes, with negligible loss, all signals except those in a specified band. (4-band-stop) A band elimination filter. (5-bandpass) A filter that greatly attenuates signals of all frequencies above and below those in a specified band. (6-capacitor-input) A common type of power-supply smoothing filter. The output from a rectifier is shunted by a large capacitor as the first element in the smoothing circuit. (7-cavity) A filter with precise characteristics for separating microwave frequencies using cavity resonance. (8-choke input) A low-pass power-supply smoothing filter with an inductance as its first element. (9-comb) A filter with several sharp band-stop sections for different frequencies. (10-composite) An m-derived filter made up of several filter sections, calculated to give the required impedance and sharp frequency changeover characteristics. (11-constant-k) A filter in which the product of the impedance of shunt components and of series components is a constant, independent of frequency. (12-crystal) A filter with sharp cutoff or changeover characteristics,

obtained through the use of quartz crystal components in resonant circuits. (13-high-pass) A filter that attenuates signals below a specified frequency but passes with minimal attenuation all signals above that frequency. (14-LC) A filter with inductance (L) and capacitance (C) circuit elements. (15-longitudinal suppression) A filter designed to suppress unwanted noise signals flowing in the same direction on the two wires of a pair. (16-low-pass) A filter that greatly attenuates signals higher than a specified frequency, but passes with minimal attenuation all signals lower in frequency. (17-notch) A bandpass filter in which the upper cutoff frequency is twice the lower cutoff frequency. (18-power interference) A filter in series with the utility power input to a rectifier that passes the fundamental frequency of the power supply but greatly attenuates higher interfering frequencies. (19-software) A software routine that separates computer data according to specified criteria.

This figure shows typical audio signal processing for a communications transmitter. In this example, the audio signal is processed through a filter to remove very high and very low frequency parts (audio band-pass filter). These unwanted frequency parts are possibly noise and other out of audio frequency signals that could distort the desired signal. The high frequencies can be seen as rapid changes in the audio signal. After an audio signal is processed by the audio band-pass filter, the sharp edges of the audio signal (high frequency components) are removed.

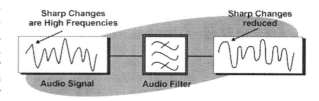

Audio Signal Filtering Operation

Filter Artifacts-Defects in the video picture caused by filtering. The most common artifacts appear as tinting and loss of resolution.

Filter Codec-A device or circuit that contains both filter and codec processing capabilities.

Filtered Noise-Noise signals that have been passed through a filter. This affects the power spectral density of noise signal that will have the same shape as the transfer function of the filter.

Filtering-Analog signal filtering is a process that changes the shape of the analog signal by restricting portions of the frequency bandwidth that the signal occupies. Because analog signals (electrical or optical) are actually constructed of many signals that have different frequencies and levels, the filtering of specific frequencies can alter the shape of the analog signal.

Filters may remove (band-reject) or allow (band-pass) portions of analog (possibly audio signals) that contain a range of high and low frequencies that are not necessary to transmit.

In some cases, additional signals (at different frequencies) may be combined with audio or other carrier signals prior to their transmission. These signals may be multiple channels (frequency multiplexing) or may be signals that are used for control purposes. If the signal that is added is used for control purposes (e.g. a supervisory tone that is used to confirm a connection exists), the control signal is usually removed from the receiver by a filter.

Filtering Database-A data structure within a bridge that provides the mapping from Destination Address to bridge port (in a D-compliant bridge), or from the combination of Destination Address and VLAN to bridge port (in a Q-compliant bridge).

Final Cut-A final cut is the edited version of a media program that is accepted by its producer or owner.

Final Draft International Standard (FDIS)-A final draft international standard is a document or industry specification that is used or referenced to obtain final changes from members, companies or individuals who are authorized or allowed to contribute final changes to a standards document that is to be issued.

Financial Posture -The strategic positioning established by upper management that defines the way that it wants to be perceived by investors, regulators and the public in financial terms. The indication of a strong financial posture would communicate to outsiders that the company was highly successful and cash rich and a weak posture would communicate the opposite. Financial posture objectives can be defined and controlled by the company through major marketing and public relations activities and they may be different than the actual financial performance of the company.

Find Me Service-A process of forwarding calls to a sequence of numbers so that the call will find the recipient. Follow me service could be password privileged to screen unwanted callers from being forward to private numbers such as home telephones or mobile phones.

Fine Granularity Scalability (FGS)-Fine granularity profile is a set of processes and protocols in the MPEG-4 standard that provides the capability of varying the resolution levels and frame rates.

Finger-Finger is a software program you use to find out either a list of users currently logged on to a system, or information about a particular user on a system. When you finger someone's Internet account you should see the person's full name, most recent log-in, and the contents of their Plan file if they have one. Most systems, however, don't let you finger from outside their network.

Fingerprint-A fingerprint is a unique identifier that is associated with a specific device, user, media or a combination of these items. Fingerprints can be used to identify or assist in the identification of devices or users.

FIR-Finite Impulse Response Filter

Fire Rating-A fire rating is the classification of the temperature ranges and exposure times that cause emissions from a substance or cable.

Firefighting-A process of rapidly addressing many problems that are occurring in a system. Solving a specific problem is referred to as putting out a fire.

Firefighting Plan-A plan that addresses how to rapidly resolve critical problems within a network when normal procedures cannot be followed. A firefighting plan usually prioritizes which systems receive priority for corrective action (such as voice communication) and who will be responsible for the non-standard procedures to correct the problems.

Firewall-A firewall is a data filtering device that is installed between a computer server or data communication device and a public network (e.g. the Internet). A firewall continuously looks for data patterns that indicate unauthorized use or unwanted communications to the server. Firewalls vary in the amount of buffering and filtering they are capable of providing. An ideal (perfect) firewall is called a "brick wall firewall."

This figure shows how a firewall works. This diagram shows that a user with address 201 is communicating through a firewall with address 301 to an external computer that is connected to the Internet with address 401. When user 201 sends a packet to the Internet requesting a communications session with computer 401, the packet first passes through the firewall and the firewall notes that computer 201 has requested a communication session, what the port number is, and sequence number of the packet. When packets are received back from computer 401, they are actually addressed to the firewall 301. Firewall 301 analyzes the address and other information in the data packet and determines that it is an expected response to the session computer 201 has initiated. Other packets that are received by the firewall that do not contain the correct session and sequence number will be rejected.

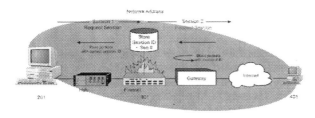

Firewall Operation

FireWire-Firewire is a short range serial bus IEEE specification that transfers data at 400 Mbps. The Firewire specification can communicate with up to 63 devices simultaneously. Firewire is described by IEEE 1394.

Firmware-Firmware is software program instructions that are stored in a hardware device that performs data manipulation (e.g. device operation) and signal processing (e.g. signal modulation and filtering) functions. Firmware is stored in memory chips that may or may not be changeable after the product is manufactured. In some cases, firmware may be upgraded after the product is produced to allow performance improvements or to fix operational difficulties.

First Application Region-For new services, the first regional company to use a new network-based service. A network-based service is one that requires the interaction of two or more types of network elements, possibly from different suppliers.

First Attachment-A hardware item comprised of porcelain, plastic, or steel fixtures that is affixed to a building to attach, buried or aerial drop wire.

First Attempt Load-An offered traffic load that excludes any load resulting from retrials.

First Draft Screenplay-The original completed draft made referring to a script including a full dialogue, usually in continuous form.

First Generation (1G)-A term commonly used to describe the first technology used in a new application. In cellular telecommunications, the first generation used analog (usually FM) radio technology. For first generation cordless telephones, the first generation of products used single channel (using AM) radios.

First Mile-The signal path between a program origination site and its entry point to the communication network or a private satellite uplink. The first mile is usually a terrestrial RF link or a local telco loop.

First Negotiation-In reference to starting agreements prior to any third party is involved, to discuss rights or employment.

First Sale Docterine-First sale doctrine is the assignment of exclusive usage rights to a particular form of intellectual property (e.g. a book) for a particular user that has purchased that copy of the work.

First-Dollar Gross-In reference to all income shares starting at the first dollar, usually acquired by the distributor or by the studio.

Fish Wire-A fish wire is a stiff rod or lead that is used to probe and connect (fish) to other cables in a wall, channel or other area that is not directly accessible.

Fishing-Fishing is the process of pushing a stiff wire or tape through the hollows of a wall or conduit. Once the fish wire has reached its destination (e.g. an outlet hole in the wall), the cable or wires are attached to one end and they are pulled through the wall or conduit until the cable or wire is pulled through the hole.

FISU-Fill In Signal Unit

FIT-Failure In Time

Fit to Width (FTW)-Fit to width is the adjusting of in image or video display size so it fits into a specific width.

Fixed Allocation-Fixed allocation is the assignment of a resource for a predetermined or scheduled amount of time.

Fixed Charges-Fixed charges are recurring and non-recurring charges that have a cost that is fixed and independent of usage quantities.

Fixed Compensation-An unconditional set rate sum of money insured to be issued to one in trade for ones services.

Fixed Count-The permanent connection of a group of cable pairs to the binding posts of a connecting block or terminal block at a terminal location.

Fixed Count Terminal-An outside plant cabling distribution terminal where only a portion of the entire cable count is readily accessible. The remainder of the cable count is behind the terminal posts and not readily visible or available for accessing facilities. Fixed count terminals are generally protected with carbon or other fusible station protection.

Fixed Mobile Convergence (FMC)-Fixed mobile convergence is the process of using or providing the same services to fixed and mobile users, regardless of location, access or terminals using existing core infrastructure and back office systems.

Fixed Reference Modulation-A type of modulation in which the choice of the significant condition for any signal element is based on a fixed reference.

Fixed Satellite Service (FSS)-A category of service devoted to point-to-point satellite communications.

Fixed Terminals-A terminal interface device that is fixed in location.

Fixed Wireless-Fixed wireless is the use of wireless technology to provide voice, data, or video service to fixed locations. Fixed wireless services include wireless local loop (WLL), point-to-point microwave, wireless broadband, and free-space optical communication. Fixed wireless systems may replace or bypass wired telephone service, high-speed telephone communication links, and cable television systems.

Fixed Wireless Access (FWA)-Fixed wireless access is the process of using a radio link to provide communication services to fixed locations such as homes or businesses.

Flags-(1-computer) An indicator used to signal a condition or to mark information for further attention. The flag may be "raised" or lowered" through the results of computation, or specifically controlled through software. (2-ISDN, PPSN) In the Integrated Services Digital Network (ISDN) and the Public Packet-Switched Network (PPSN), a unique digital pattern that is used by a link layer to delimit frames. Each link layer frame starts with an opening flag and ends with a closing flag. (See also: Call Reference FIag). (3- SS7) In the Signaling System 7 protocol, a unique pattern on the signaling data link used to delimit a signal unit. The binary flag sequence used in ISDN, PPSN, SS7 and also X.25 and Frame Relay packet systems is the HDLC binary flag 01111110. To prevent false premature end of packet indication, the contents of every packet are processed before transmission to insert a binary 0 following any natural occurrence of 5 consecutive binary 1 bits. At the receiving end, a binary zero inside a packet is deleted if it follows 5 consecutive binary 1 bits, thus accurately restoring the original packet data.

Flame-A flame is an email that is sent by a recipient of an unwanted email back to the sender of the email. The flame message usually contains a negative message for the original sender of the unwanted email.

Flaming-The process of sending one or many flame messages to the sender of unwanted emails. Flaming typically has the intent of intimidating or disabling the sender of unwanted emails.

Flapping-The continual changing of a network connection path that results from a intermittent circuit connection (or a similar condition) that indicates to the current connection path (the routers) into thinking there is a loss in connection or that a better connection path exists. This causes the continuous re-advertising of new available or unavailable packet routes within the network.

Flash-A system special service request feature that is used to indicate that a subscriber has a desire to recall a service function or to activate a custom calling feature (such as a call transfer request).

A flash feature service request can be created when the user initiates a short on-hook interval or through the sending of a special service request message. The short on-hook interval is created by a momentary operation of the telephone switch hook, in the midst of a prolonged off-hook period. This momentarily turns off the loop current in the subscriber loop. The duration of the current-off state must be typically more than 1 second but less than

F

Flash

www.IPTVMagazine.com   223

2 seconds in most systems. The special service request message can be sent by a button on a telephone (such as a PBX telephone) sometimes labeled with such words as FLASH or BREAK or SERVICE, or by pressing the SEND or TALK key on a mobile telephone. In a mobile telephone or an ISDN telephone, a digital message is sent when the subscriber presses the button, instead of momentarily turning off the loop current.

This diagram shows how the flash signal (special service request) can be sent to the telephone system to activate additional call processing features. This example shows a flash feature is sent on an analog line by momentarily opening the current loop connection. When the loop current sensing circuit senses a brief open (no current flow) period, it creates a flash message that is sent to the call processing section of the telephone system. For digital telephones, the flash message is sent via a signaling message on the digital channel. This diagram shows that on an ISDN line, the flash message is sent on the D (signaling) channel.

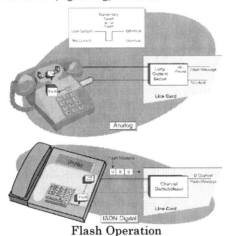

Flash Operation

Flash Memory-Flash memory is a type of memory storage that has the ability to be reprogrammed multiple times and can retain information without power.

Flash Request-A request to initiate a special processing function. Dual Mode cellular allows flash requests in both directions while analog systems only allow flash requests from the mobile phone to the cell site.

Flash SMS-Flash Short Message Service

Flashover-An arc or spark between two conductors.

Flat Face Tube-The design of CRT tube with almost a flat face, providing improved legibility of text and reduced reflection of ambient light.

Flat Noise-A noise whose power per unit of frequency is essentially independent of frequency over a specified frequency range.

Flat Response-A performance parameter of a system in which the output signal amplitude of the system is a faithful reproduction of the input amplitude over some range of specified input frequencies.

Flat Room-A room that has been acoustically designed and/or treated to reproduce equally all sound in the audible spectrum. For a room that is not acoustically flat, the audio monitoring system may be conditioned through equalization to make the room seem flat.

Flat Weighting-An amplitude and frequency characteristic that is flat (no variation over a specified frequency band). Flat weighting is used to measure noise on broadcast circuits. Flat noise power is expressed in dBrn or dBm. ionized layers, Fl and F2. Both layers exist during daylight hours but combine at night to form one layer from 200 to 400 km above the surface of the earth. These are the highest of the layers that refract radio waves.

Flex Antenna-An antenna often used with portable telephones that consists of a length of stiff wire, usually with a fiberglass core, covered with an insulating material.

Flexible Key Assignment-A telephone system feature for customizing a station to the needs of a user by assigning different features to different keys on each set.

Flexible Multiplexer (FlexMux)-A FlexMux is a set of tools that are used by a multimedia system (such as MPEG) that allows for the combining of multiple media sources (such as video and audio) so that the media streams are combined and resynchronized back into its original composite form.

FlexMux-Flexible Multiplexer

Flicker-A fluctuation in the brightness of movie images that occurs below 24 frames per second.

F-Link-Fully Associated Link

FLO-Forward Link Only

Floating-A circuit or device that is not connected to any source of potential or is not grounded.

Flocking-Assembling adHoc teams of expert resources in response to high priority issues.

Flooding-The action of forwarding a frame onto all ports of a switch except the port on which it arrived. Normally used for frames with multicast or unknown unicast Destination Addresses.

Floor-Sum of payment minimum.

Flow Chart-A graphic portrayal that shows the sequence in which functions are performed, from the beginning of a job to the end.

Flow Control-A hardware or software mechanism or protocol, that manages data transmissions when the receiving device cannot accept data at the same rate the sender is transmitting. Flow control is used when one of the devices communication cannot receive the information at the same rate as it is being sent; this usually occurs when extensive processing is required by the receiver and the receive buffers are running low. Examples are flow control algorithms are IEEE 802.3x used in Ethernet networks, Forward Explicit Congestion Notification and Backward Explicit Congestion Notification used in ATM networks and XON/XOFF used is RS-232 serial communication.

Fluorescence-The characteristic of a material to produce light when excited by an external energy source. No heat, or only a minimal amount, results from the process.

Flying Spot Scanning-Flying spot scanning is the use of a highly focused light beams to scan images (such as film). The spot beam can be aimed and adjusted to scan coarsely or finely as required to produce the desired resolution.

FM-Frequency Modulation

FMC-Fixed Mobile Convergence

FMEA-Failure Mode And Effects Analysis

FO cable-Fiber Optic Cable

FOCC-Forward Analog Control Channel

FOCIS-Fiber Cladding

Focused Overload-A condition of unusually high calling rates from many points to one point, for example, after a natural disaster or when a radio or TV station encourages mass calling.

Focusing Elements-Components used to control and confine an electron beam within an electronic device. In a device incorporating an electron gun structure, several elements are used to control the diameter of the beam. A focusing anode (plate) and several focusing coils may be used. In some devices, permanent magnets also are used to assist in maintaining correct beam diameter. In devices used to amplify microwave signals, such as klystrons, the focus system is critical because even a momentary change in the system can produce disastrous results and near instantaneous device failure.

FOD-Fax On Demand

FoIP-Fax over Internet Protocol

Follow-Me Phone Service-A service that allows calls to be routed to a customer's choice of forwarding phone numbers. Follow-me service may be automatic (e.g. when a cellular telephone automatically registers with a visited systems) or manually set (e.g. when a customer calls in with a hotel phone number where they will be temporarily located).

This diagram shows how follow-me service can allow calls to be automatically routed to one of three telephone numbers that a customer has provided to the follow-me system. In this example, the follow-me service is automatic. When the incoming call is received by the system, the system first tries the customer at the office number. Because there is no answer at the office, the system will then call the home number. When the user answers the call at home, the system will automatically move the home telephone number to the top of the follow-me calling list.

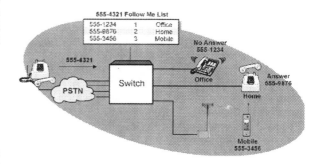

Follow-Me Phone Service Operation

Follow-Me Roaming-Follow-me calls are automatically forwarded to the area outside their home area where the mobile subscriber has registered. A cellular call generated to a user when the mobile subscriber has informed the local system that the user is roaming in another area.

FOMS-Fiber Optic Microscope

Footprint-(1-satellite) The minimum radio coverage signal strength boundaries from a satellite over in a geographic area. A single satellite may have several antennas that have different footprints. (2-equipment) The floor area occupied by a given piece of equipment. (3-software) The amount of memory required to run an operating system or software application.

Force Majeure-Unforseen or unanticipated abrupt occurance putting productions to a stop.

Forced Call-In reference to contact taken by ones form of employment, requiring one to return to work prior to time initially arranged by party's.

Forecasted Load-The predicted load (resource utilization) for a specified future period.

Foreign Exchange Office (FXO)-Foreign Exchange Office (FXO) interface or channel unit that allows an analog connection (foreign exchange circuit) to be directed at the PSTN's central office or to a station interface on a PBX. The FXO sits on the switch end of the connection. It plugs directly into the line side of the switch so the switch thinks the FXO interface is a telephone. (See also: foreign exchange station.)

Foreign Exchange Station (FXS)-A type of channel unit used at the subscriber station end of a foreign exchange circuit. A foreign exchange station (FXS) interface connects directly to a standard telephone, fax machine, or similar device and supplies ring, voltage, and dial tone. (See also: foreign exchange office.)

Forking Proxy Server-A proxy server that forwards a communication session request to more than one device on behalf of the communication connection request.

Forklift Upgrades-An upgrade that requires the removal of old equipment (possibly by a forklift) and installation of new equipment.

Form Factor-(1-product) The physical shape and size of a product. (2-mathematic) The ratio of the root-mean-squire value of a periodic function to the average absolute value, averaged over a full period of the function.

Forms-Forms are structured displays of screen information that usually allows a user to enter and edit information in predefined fields on specific areas of the form.

FORTRAN-Formula Translator

Forum-A forum is a group of people or companies that help to develop, promote, or assist with various aspects of a technology, product, or service.

Forward Acting Code-A logically constructed code with enough redundancy to correct many errors without requiring retransmission.

Forward Channel-Forward channels are the radio channels (radio links) that are used by a base station to transmit to a mobile device (also called a downlink).

Forward Congestion Notification-Forward congestion notification indicates to upstream switching devices in a data communication network that data that is being transmitted through congested switches and it is likely that some of the remaining data or packets may be discarded. The upstream switch can then change the discard priority accordingly.

Forward Error Correction (FEC)-Forward error correction is a mathematical algorithm that is used to produce extra bits of data that are sent each packet. The sending of FEC bits allows for the correction of information bits that were lost or changed due to noise or other physical effects. The extra bits increase the data rate required for a given transmission typically by 5% to 50% depending on the level of correction required so they are only used where error rates are high and retransmission is uneconomical.

Forward Explicit Congestion Notification (FECN)-Forward explicit congestion notification is the sending of a control bit within the overhead control part of a data packet in a frame-relay network that indicates network congestion exists in the forward (destination) direction of its flow. This control bit allows higher-level protocols within the data communications equipment (DCE) and data terminal equipment (DTE) to take appropriate bandwidth allocation action if necessary.

Forward Link Only (FLO)-Forward link only is a system or broadcasting process that provides information in one direction, from the broadcast station to receivers.

Forward Prediction-Forward prediction is the process of estimating the likely changes or occurrences that may occur within future media, images or media components in a sequence of media (such as audio packets or video frames).

Forward Sync Channel (F-SYNC)-A communications channel that provides information that allows receivers to acquire or determine the initial time synchronization with the system or channel.

Forward Traffic Channel (FTC)-The combination of voice and data signals existing within a forward communication channel (usually from a base transmitter to a mobile receiver).

Forwarding-The process of taking a frame received on an input port of a switch and transmitting it on one or more output ports.

Forwarding Delay-1. The amount of time that is used between the receiving of a packet or block of data to the time of its transmission through a data network (such as the Internet). 2. A measurement parameter used by the Spanning Tree Protocol that defines the delay that occurs between the transitions from a listening state to the learning state and from the learning state to the forwarding state.

Forwarding State-A stable state in the Spanning Tree Protocol state machine in which a bridge port will transmit frames received from other ports as determined by the bridge forwarding algorithm.

FOTS-Fiber Optic Transmission System

Fourier Analysis-A mathematical process for transforming values between the frequency domain and the time domain. This term also refers to the decomposition of a time domain signal into its frequency components.

Fourier Transform-A Fourier transform is a mathematical analysis that converts the time domain signal into its frequency domain components.

Fourth Generation (4G)-Fourth Generation wireless networks with bandwidth reaching 100 Mbps that allow for voice and data applications that will run 50 times faster than 3G. This capacity will enable three dimensional (3D) renderings and other virtual experiences on the mobile device.

FP LASER-Fabry-Perot

FPC-Forward Power Control

F-PICH-Forward Pilot Channel

FPLF-Field Programmable Logic Family

FPLMTS-Future Public Land Mobile Telephone System

fps-Frames Per Second

FR-Full Rate

Fractal-Fractal is the process of dividing or breaking an item or data into components or shapes that have some similarity.

Fractal Compression-Fractal compression is a data compression process that analyzes data in media (graphics) files and converts the data into mathematical equations that represents the data.

Fractional T-1 (FT-1)-A digital transmission service that provides a customer with multiple 64

kbps channels but less than the full 24 channels offered by a T-1 channel.

FRAD-Frame Relay Access Device

Fragmentation-Fragmentation is a technique that divides a data packet into smaller data packets so that they can be sent through a network that can only transfer small data packets. Fragmentation occurs during network transmission. When these packets are received at their destination, they are reassembled to their original data packet size.

Frame-(1-general) A basic repeated bit pattern in time division multiplexing systems, and/or the time duration of this pattern. (2-frame relay) A packet of data in Frame Relay systems. (3-video) In video processing, a frame is a single still image within the sequence of images that comprise the video. Note that, in an interlaced scanning video system, a frame comprises two fields. Each field contains half of the video scan lines that make up the picture, the first field typically containing the odd numbered scan lines and the second field typically containing the even numbered scan lines. (4-equipment) An electronic rack that is used to interconnect and hold electronics assemblies. (4-audio) In audio processing, a frame is a group of audio samples that are processed together (to determine an instantaneous frequency spectrum for a codec like MP3, for example), though two adjacent frames might contain common samples in an overlap region.

Frame Check Sequence (FCS)-A frame check sequence is a calculated code that is used to determine (check) if the bits within a frame have been received correctly during transmission.

Frame Dropping-Frame dropping is the process of discarding or not using all the video frames in a sequence of frames. Frame dropping is commonly performed to reduce the data transmission speed or to reduce the video and image processing requirements.

Frame Error Rate (FER)-FER is calculated by dividing the number of frames received in error by the total number of frames transmitted. It is generally used to denote the quality of a digital transmission channel.

Frame Lock-A frame lock is the synchronization of a video signal with an audio signal.

Frame Loss-The ratio of the number of frames that have been lost in transmission compared to the total number of frames that have been transmitted.

Frame Multiplexing-In a digital link the information bits used for different channels or different processes are grouped in a consecutive sequence of bits called a frame. In time division multiplexing systems, a frame has fixed time duration and contains a fixed number of bits. For example, in a DS-1 (T-1) digital multiplexing system a frame has 125 microseconds time duration and contains 193 bits. In some systems such as Frame Relay, the word frame refers to a packet of data for a single user or process, having non-fixed length and separated from preceding and following frames by special short sequences of bits used as separators, such as the HDLC binary flag sequence 01111110.

Frame Rate-(1-telecommunications) The number of frames of data that are transferred over a period of time. (2-movies) Frame rate is the number of images (frames or fields) that are displayed to a movie viewer over a period of time. Frame rate is typically indicated in frames per second (fps).

This figure shows the different types of frame rates and how lower frame rates can cause flicker in the viewing of moving pictures. This example shows that frame rates are the number of images that are sent over time (1 second). At the frame rate is reduced below approximately 24 frames per second (fps), the images appear to flicker. Increasing the frame rate results in increased bandwidth requirements. The common frame rate formats are 24 fps for film, 25 fps for European video, 30 fps for North American video, 50 fps for European television and 60 fps for North American television.

Frame Relay-Frame relay is a packet-switching technology that provides dynamic bandwidth assignments. Frame relay systems are a simple bearer (transport only) technology and do not offer advanced error protection or retransmission. Frame relay were developed in the 1980s as a result of improved digital network transmission quality that reduced the need for error protection. Frame relay systems offer dynamic data transmission rates through the use of varying frame sizes.

This figure shows a frame relay system. This diagram shows a local area network (LAN) in San Francisco is connected to a LAN in New York. A virtual path is created through the frame relay network so data can rapidly pass through each frame relay switch as its path is previously established. When data is to be transferred through the LAN (e.g. a large image file), the data file passes through a FRAD that is the gateway to the frame relay network. The FRAD divides the data file from the LAN into variable length data frames. The FRAD sends and receives control commands to the frame relay network that allows the FRAD to know when and if additional data frames can be sent.

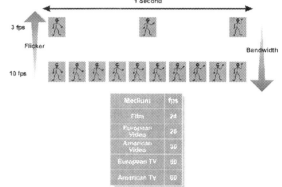

Moving Picture Frame Rates

Frame Relay Access Device (FRAD)-A frame relay access device is a communications access device that converts data from a user's network into the format that is required by a frame relay network.

Frame Relay Network Device (FRND)-The FRND is a packet switch that also operates as a gateway to the frame relay network. The FRND passes frames it receives from the frame relay access device (FRAD) to other frame relay switch that forward packets toward their destination network. Frame relay switches have buffer memory that allows them to hold, prioritize packets before they are retransmitted. Packet switches can selectively discard packets if network congestion occurs. The FRAD and FRND provide information about

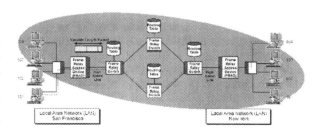

Frame Relay System

the priority of the frames (e.g. non-essential discard eligibility) and status of the system (e.g. network congestion notification).

Frame Structure-Frame structure is the division of defined length of digital information into different fields (information) parts. Frame structure fields typically include a preamble for synchronization, control header (e.g. address information), user data, and error detection. A frame may be divided into multiple time slots.

Frame Switch-A device that forwards data frames based on layer 2 addresses contained within the received data frame. Frame switches may directly switch (cut-through) or operate as store and forward switches.

FRAMES-Future Radio Wideband Multiple Access System

Frames Per Second (fps)-The number of images or video frames that are displayed over one second. Twenty four frames per second (24 fps) is considered the slowest frame rate that is suitable for movie film. Because television uses interlacing (frames of interspaced lines), the frame rate for television signals is typically 50 fps for Europe and 60 fps for the Americas.

FRAND-Fair Reasonable and Non-Discriminatory

Fraud-Fraud is the process of obtaining or using products or services without authorization.

Fraud Based Churn-A common form of involuntary churn. Fraudsters are customers who steal services without payment and who are terminated by the telephone company (Telco).

Fraud Management-Fraud management is the processes and steps taken to identify, minimize and correct the unauthorized use of products or services.

Free Agent AP-Free Agent Access Point

Free Space-An unbound medium for transmission of acoustic, electromagnetic, or photonic signal energy.

Free Space Communications-Free space communication is any form of communication that doesn't use a conductor (copper wire, waveguide, or glass fiber). Free space communication can use electromagnetic (radio) or optical signals.

Free Space Transmission-The theoretical transmission of an electromagnetic (radio) or optical communication signal through a vacuum with no absorption, no reflection, and no diffraction.

This figure shows two types of free space transmission systems: radio and optical. The microwave

transmission system shows that some of the electromagnetic energy is absorbed by the water particles in the air. The optical transmission system uses a laser and photo-detector. The optical transmission system shows that some of the optical energy is scattered in other directions as it passes through smog and water particles.

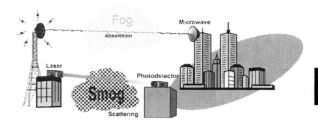

Free Space Transmission System

Free Weeks-Free weeks are additional weeks not included in arranged terms of service to an individual, allowing that individual's weekly "quota" to remain high.

Freephone-A service that allows callers to dial and telephone number without being charged for the call. The toll free call is billed to the receiver of the call. In Europe and other parts of the world, toll free calls are called Freephone and typically begin with 0-800. In the United States, freephone calls are called toll free calls and they are preceded by a 1-800, 1-888 or 1-877 exchange.

FreeWare-Freeware is software that can be transferred (e.g. downloaded from the Internet) for use without paying a fee for its acquisition. Freeware may be copyrighted and there may be restrictions on the use of the program as part of the freeware distribution process. Freeware was trademarked by Andrew Fluegelman who won an inventor of freeware and shareware award.

Frequency-(1-radio) The frequency of an electrical or optical wave is the number of complete cycles or wavelengths that the wave has in a given unit of time (second). The standard measurement for this is number of cycles per second, also known as Hertz (the scientist, not the car company), abbreviated Hz. (2-marketing) The number of times a message or advertisement appears in communication media (such as a magazine or web page) over a period of time.

Frequency Adjusting

This diagram displays how frequency is measured. In this example, there three cycles of a wave that are transmitted over a 1 second period. This equals a frequency of 3 Hertz.

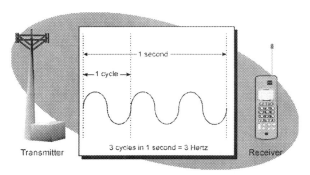

Frequency Signals

Frequency Adjusting-Frequency adjusting is a process of changing the settings or components of a device or assembly to modify its transmitter and/or receiver frequency or frequencies.

Frequency Agile-Frequency agile systems allow devices to search and establish a communication link on a variety of different frequencies or channels. Frequency agile devices first search for a beacon signal on a pre-determined list of communication channels (set of frequencies) to determine which frequencies are available for communication. The frequency agile device may then attempt to attach itself to the network by registering with the system. This allows the system to know which channel (frequency) the device is listening to (camping on) in the event the system desires to send a message (page) to the communication device.

Frequency Bands-Frequency bands are the range of frequencies that are used or allocated for radio services.

Frequency Capping-Frequency capping is the limiting of the number of advertising messages that is presented to a viewer or groups of viewers. It has been discovered that advertisements don't have to be shown too many times before the messages have been relayed. If ads are not frequency capped and are allowed to play more than around 30 times over a relatively short period of time, it can cause audience alienation. 'Frequency capping'

is the mechanism by which an ad is shown the optimal number of times to each viewer. Frequency capping is especially useful when narrative ads are used as it enables episodes to be screened in sequence, at a pace appropriate to each viewer.

Frequency Combiner-A frequency combiner is a device or assembly (such as a diplexer) that can combine two or more frequency signals to a common output line or circuit.

Frequency Coordination-Frequency coordination is the process of analyzing and/or planning for the use of frequencies in geographic areas to achieve acceptable performance for services (e.g. within interference level limits).

Frequency Division Duplex (FDD)-Frequency division duplex is the process of allowing the transmission of information in both directions (not necessarily at the same time) via separate bands (frequency division). When using FDD, each device transmits on one frequency and listens on a different frequency.

This figure shows a frequency division duplex (FDD) system. In this system, the transceivers contain a transmitter and receiver that are operating on two different frequencies. One frequency is used to send signals in one direction and the other frequency is used to send signals in the opposite direction. FDD allows for the simultaneous audio communication between users.

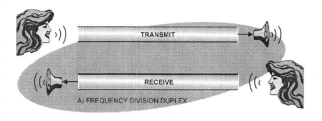

FDD Duplex System Operation

Frequency Division Multiple Access (FDMA)-Frequency division multiple access is a process of allowing mobile radios to share radio frequency allocation by dividing up that allocation into separate radio channels where each radio device can communicate on a single radio channel during communication.

Frequency Division Multiplexing (FDM)-The multiplexing of two or more signals into one output by assigning each signal its own bandwidth within a broad range of frequencies. Frequency division multiplexing is used to divide a frequency bandwidth into several smaller bandwidth frequency channels. Each of these smaller channels is used for one communications channel.

FM radio, broadcast television and some cellular telephone systems use frequency division multiplexing.

This figure shows how a frequency band can be divided into several communication channels using frequency division multiplexing (FDM). When a device is communicating on a FDM system using a frequency carrier signal, it's carrier channel is completely occupied by the transmission of the device. For some FDM systems, after it has stopped transmitting, other transceivers may be assigned to that carrier channel frequency. When this process of assigning channels is organized, it is called frequency division multiple access (FDMA). Transceivers in an FDM system typically have the ability to tune to several different carrier channel frequencies.

code. The receiver of the message or voice information must also receive on the same frequencies using the same frequency hopping sequence. Frequency hopping was first used for military electronic countermeasures. Because radio communication occurs only for brief periods on a radio channel and the frequency hop locations are only known to authorized receivers of the information, frequency hopping signals are difficult to detect or monitor. This diagram shows a simplified diagram of how a frequency hopping system transfers information (data) from a transmitter to a receiver using many communication channels. This diagram shows a transmitter that has a preprogrammed frequency tuning sequence and this frequency sequence occurs by hopping from channel frequency to channel frequency. To receive information from the transmitter, the receiver uses the exact same hopping sequence is used. When the transmitter and receiver frequency hopping sequences occur exactly at the same time, information can transfer from the transmitter to the receiver. This diagram shows that after the transmitter hops to a new frequency, it transmits a burst of information (packet of data). Because the receiver hops to the same frequency, it can receive the packet of data each time.

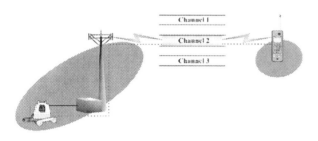

Frequency Division Multiplexing (FDM)
Operation

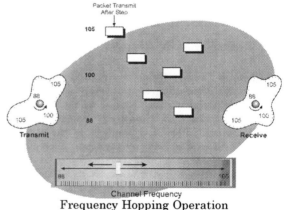

Frequency Hopping Operation

Frequency Guard Band-A bandwidth of frequency that is unused between two channels or bands of frequencies to provide a margin of protection against signal interference between energy that is transmitted outside the assigned frequency bands of channels operating near the guard frequency.

Frequency Hopping-A radio transmission process where a message or voice communications are sent on a radio channel that regularly changes frequency (hops) according to a predetermined

Frequency Hopping Spread Spectrum (FHSS)-A communication process that uses multiple frequency channels that constantly change to transfer information. FHSS systems convert information (analog or digital data) into small portions of data (e.g. data packets) and each portion of information is transmitted over assigned and constantly changing frequency channels.

Frequency Masking-Frequency masking is the process of blocking, removing or ignoring specific frequency components of a signal.

Frequency Modulation (FM)-Frequency modulation is the process of transferring an information signal onto a radio carrier wave by varying the instantaneous frequency of the radio carrier signal. In 1936, the inventor Armstrong demonstrated an FM transmission system that was much less susceptible to noise signals than AM modulation systems.

This figure shows how frequency modulation (FM) uses a modulation signal (audio wave) to change the frequency of the radio carrier signal as the voltage of the audio signal increases and decreases.

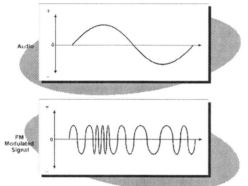

Frequency Modulation (FM) Operation

Frequency Plan-A frequency plan is the assignment of radio frequencies to radio transmission sites (cell sites) that are located within a defined geographic area. The frequency plan may use ratios that are different dependent on the number of transmitting sites to the number of antennas (sectors) on each site.

Frequency Planning-Frequency planning is the assignment of radio channel frequencies in a cellular system that allows the frequency reuse of radio channel frequencies at nearby cell sites. The frequency planning ensures that combined interference levels from nearby cell sites that are operating on the same frequency do not exceed a certain level compared to the desired signal. For the AMPS cellular system, this interference level must not exceed approximately 2% (17 dB) of the desired signal.

Frequency Response-Frequency response is a measure of system linearity, or performance, in reproducing signals across a specified bandwidth. Frequency response is expressed as a frequency range with a specified amplitude tolerance in decibels. Frequency response in digital audio systems is limited to one half the sampling frequency (Nyquist limit).

Frequency Reuse-Frequency reuse is the process of using the same radio frequencies on radio transmitter sites within a geographic area that are separated by sufficient distance to cause minimal interference with each other. Frequency reuse allows for a dramatic increase in the number of customers that can be served (capacity) within a geographic area on a limited amount of radio spectrum (limited number of radio channels).

This diagram shows that radio channels (frequencies) in a mobile communication system can be reused in towers that have enough distance between them. This example shows that radio channel signal strength decreases exponentially with distance. As a result, mobile radios that are far enough apart can use the same radio channel frequency with minimal interference.

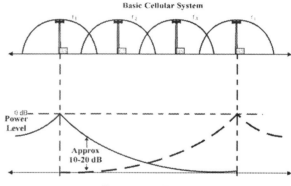

Frequency Reuse

Frequency Shift Keying (FSK)-Frequency shift keying (FSK) is a form of frequency modulation in which the modulating signal shifts the output frequency between predetermined values to represent a digital signal. Typically, one frequency shift is used to represent a digital one (sometimes called a mark) and the other frequency shift represents a digital zero (sometimes called a space).

This figure shows a sample of frequency shift keying (FSK). In this diagram, each pulse from the digital signal (on top) creates a change in carrier signal frequency (on bottom). As the digital signal voltage is increased, the frequency of the radio signal changes above the center (unmodulated) carrier frequency. When the voltage of the digital signal is decreased, the frequency changes again so the frequency of the transmitted signal is below the center (unmodulated) carrier frequency.

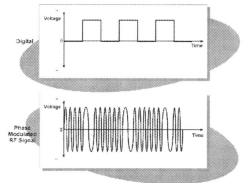

Frequency Shift Keying (FSK) Modulation Operation

Frequency Spectrum-(1-general) The distribution of electromagnetic or acoustic energy as a function of frequency. (2-Fourier analysis) The amplitude and phase of each of the sine waves that compose a signal.

Frequency Synthesizer-A frequency synthesizer is a device or electronic circuit that is capable of producing a range of frequencies based on the settings (programming) of the synthesizer. Frequency synthesizers are usually capable of producing very accurate frequencies by comparing the programmed frequency to a precise reference frequency (usually controlled by a low frequency crystal). The synthesizer uses the programming (frequency setting) to control a dividing circuit (a counter) to samples the output frequency to the frequency of a reference crystal (the precise frequency of a crystal is controlled by its physical size). If the frequency changes above or below the reference frequency, it creates an adjustment signal (voltage) that is used to correct the output frequency.

Frequency Tolerance-Frequency tolerance is the maximum allowable offset from a center frequency of the frequency band occupied by an emission from the assigned frequency.

Frequency Transform-Frequency transform is the conversion (transformation) of an image or data into frequency components that represent the original information or image.

Frequency-Agile-The capability of a transmission/reception system to operate across a broad range of frequencies.

Frequently Asked Questions (FAQs)-A list of questions and answers that are frequently asked about a particular topic or product. A list of FAQS is commonly provided on web sites to assist the visitor in finding information quickly without the need to call a customer service representative.

Fresnel Reflection-A Fresnel reflection is a loss mechanism for optical signals traveling through a fiber optic network. When a lightwave reaches the end of a fiber or is connected to another component, it typically encounters a new index of refraction at the interface between the two media. This change in index causes reflections. The reflections can be reduced by adding antireflection coatings or index-matching gels between the two media.

Fringe Area-The area outside the designated boundary of the communications system, such as a zone for radio, television, or mobile communication service may be provided.

FRND-Frame Relay Network Device

Frontload-Frontload, referring to a modified cost agenda, is an amount of money paid in percentages prior to aggreement or at the time services are rendered.

Frozen Standard-A frozen standard is an industry specification that has been accepted by a group (typically the standards body) that members or users agree that no additional changes will occur for a period of time or indefinitely.

FSAN-Full Service Access Network
FSK-Frequency Shift Keying
FSN-Full Service Network
FSO-Free Space Optics
FSS-Fixed Satellite Service
F-SYNC-Forward Sync Channel
FT-1-Fractional T-1
FTA-Fault Tree Analysis
FTAM-File Transfer Access, And Management
FTC-Forward Traffic Channel
FTP-File Transfer Protocol
FTR-Frequency Translating Repeater
FTTC-Fiber To The Curb
FTTD-Fiber to the Desktop
FTTH-Fiber To The Home
FTTN-Fiber To The Neighborhood
FTTP-Fiber To The Premise
FTTx-Fiber To The X Location
FTU-Fiber Termination Unit

FTW-Fit to Width

FUD-Fear, Uncertainty, Doubt

Fuel Cell-A chemical cell that produces electric energy from a chemical reaction.

Fulfillment-Fulfillment is the process of gathering the products and materials to complete an order and shipping the products or initiating the services that were ordered.

Fulfillment, Assurance, and Billing (FAB)-The three major functions performed by communications companies. Fulfillment is the process of preparing the network for customer access, taking orders for service, and initiating that service. Assurance is the process of making sure that customers receive continuous and high quality service. Billing is the process of receiving call detail and service usage information, grouping this information for specific accounts, producing invoices, and recording payments made for those invoices.

Full Access-An arrangement in which all traffic offered to a group of servers (trunks) has access to all the servers in the group.

Full Duplex-Full Duplex communication is the process of transferring of voice or data signals in both directions at the same time. Full duplex operation normally assigns the transmitter and receiver to different communication channels. When the communications system uses two different frequencies for simultaneous communication, it is called frequency division duplex (FDD). One frequency is used to communicate in one direction and the other frequency is required to communicate in the opposite direction.

The definition of full duplex becomes confusing when it is applied to the end result of simultaneous voice and data communication. This is because it is possible to provide information at the input and output of a communication system while not actually sending the information simultaneously in a communication system. When a communication system provides for simultaneous two way communication by time sharing, it is called time division duplex (TDD).

Full Period Service-In reference to the tariff of services, a service, circuit, facility, or piece of equipment that is continuously available for use by a customer.

Full Rate (FR)-(1- telephone) A customer data speed of 56 kbps. For the Integrated Services Digital Network (ISDN), 64 kbps is offered. (2- radio channel) The normal allocation of time slots or data rates for a mobile radio.

Full Screen-Full screen is the presenting of an image or video so it fills up the entire screen size area. During full screen presentation, frames and controls outside the digital media image are removed.

Full Service Access Network (FSAN)-A full service access network is the infrastructure and support systems that are capable of providing access to all of the communication services required by the end user.

Full Service Network (FSN)-A full service network is the infrastructure and support systems that are capable of providing all of the communication services required by the end user.

Full Services Access Network Forum-A forum that was established in 1995 to help identify technologies and network architectures that can cost effectively provide narrowband and broadband telecommunications services.

Function-A process that accepts one or more forms of information inputs or arguments and produces a single output or value that is determined by the combination of the inputs and the formal specification of the function.

Functional Block-In defined process or service that contains one or more processes that interact with other defined functional blocks.

Functional Component-An established subroutine that constructs service logic programs in an intelligent network and can pass queries, responses, and instructions between network elements that are executing service logic programs.

Functional Group Addressing-A technique used for multicasting in Token Ring LANs in which specific bits within a 48 bit MAC Destination Address are associated with predefined functional entities. A frame so addressed will be received by all devices that have implemented the functions corresponding to the bits set within the address.

Functional Specification-A functional specification is a document that provides an operational description and requirements of a device, equipment or system.

Functional Unit-An entity of hardware and/or software capable of accomplishing a given purpose.

Fundamental Frequency-The lowest frequency of a composite: signal; higher components often are harmonics or multiples of the fundamental frequency.

Fundamental Mode-The lowest order mode of an optical waveguide, the only mode capable of propagation in a singlemode (monomode) fiber.

Fuse Wire-A thin-gauge wire made of an alloy that overheats and melts at the relatively low temperatures produced when the wire carries overload currents. When used in a fuse, the wire is called a fuse link.

Fused Splice-A fiber splice that melts (fuses) the cable ends to each other. Also called a fusion splice.

Fusion Splice-A fusion splice is the connection of optical fibers that is performed by the application of heat that is sufficient to melt (fuse) the two ends, forming a solid continuous fiber. In this way, no reflection or refraction (no signal loss) can occur at the splice interface.

This figure shows the basic process of fusion splicing. This diagram shows that each optical fiber strand is inserted into the V-guide of the fusion splice machine. The fiber strands positions are viewed by a microscope so their exact alignment can be set by the positioning controls. The position of each fiber is shown on the microscope viewer or display. This diagram shows that the fusion splice machine allows the user to manually enter the arc current, arc time and overrun settings. When the fiber ends are positioned and aligned correctly, the FUSE button is pressed. This causes an electrical arc to occur across the electrodes that heats the end of the fiber. The fibers are pressed together during the fusion process so the end result produces a uniform splice.

Fusion Splicing-Fusion splicing is the joining together of two fibers media by cutting them, butting them together, and heating the connection to form a uniform interface between them. Fusion splicing can provide the lowest loss connection provided the fibers are aligned correctly and the fusing is performed correctly.

Future Proofing-Future proofing is the process of designing a system to allow for future improvements in technology or capabilities. An example of a future proof design is a video compression device that can download new compression decoding or encoding programs.

Fuzzy Logic-A branch of mathematics, used in some artificial intelligence computers, in which decisions are based on ideas and approximations rather than on mathematically rigid calculations.

FVC-Forward Analog Voice Channel

FWA-Fixed Wireless Access

FWM-Four Wave Mixing

FxM-Fax Mail

FXO-Foreign Exchange Office

FXS-Foreign Exchange Station

F

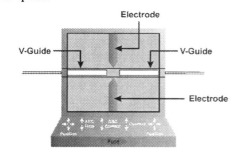

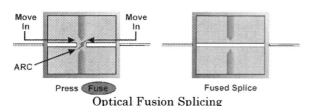

Optical Fusion Splicing

G

G.711-G.711 is a standard analog to digital coding system (coded) that converts analog audio signals into pulse code modulated (PCM) 64 kbps digital signals. The G.711 is an International Telecommunications Union (ITU) standard for audio codecs. The G.711 standard allows for different weighting processes of digital bits using mu-law and A-law coding. The G.711 standard was approved in 1965.

G.721-A standard analog to digital coding system (coded) that converts analog audio signals into ADPCM 32 kbps digital signals. The G.711 is an International Telecommunications Union (ITU) standard. The G.721 standard has been superseded by G.726.

G.723-An International Telecommunication Union (ITU) standard for audio codecs that provides for compressed digital audio over standard analog telephone lines.

G.729-G.729 is a low bit rate speech coder that was developed in 1995. It has low delay due to a small frame size of 10 msec and look ahead of 5 msec. It has a relatively high voice quality level for the low 8 kbps data transmission rate. There are two versions of G.729: G.729 and G.729 A.

G.dmt-G.dmt is an ITU standard for asynchronous digital subscriber line (ADSL). G.dmt permits data transmission rates of up to 8 Mbps downstream and 1.54 Mbps upstream.

G.Lite-The limited version of asynchronous digital subscriber line (ADSL) technology that eliminates or reduces the need to install a splitter at the end customers location. The standard allows up to 1.5 Mbps downstream and 384 Kbps upstream.

G/T-Gain-Over Noise And Temperature

ga-Gauge

GA-Grand Alliance

GaAs-Gallium Arsenide

Gaffs-Gaffs are hooks or climbing gear devices that are strapped and worn on the legs and underfoot and used in conjunction with belts, lanyards, and harnesses for accessing an aerial telephone company (telco) plant.

Gain Frequency Characteristic-The gain-vs-frequency characteristic of a channel over the bandwidth provided. Gain frequency is also referred to as frequency response.

Galactic Radio Noise-A radio noise reaching the earth from outer space, in particular from stars in our own galaxy.

Gallium Arsenide (GaAs)-Gallium Arsenide is a semiconductor material that is commonly used in high frequency communication devices.

Galvanic-Pertaining to a device that produces direct current by chemical action.

Gaming-Gaming is an experience or actions of a person that are taken on a skill testing or entertainment application with the objective of winning or achieving a measurable level of success.

Gamut-A range of voltages or signal levels allowed for a system to operate correctly. Signal voltages outside the range (that is, exceeding the gamut) may lead to clipping, crosstalk, or other distortions.

Gang-To mechanically connect two or more circuit devices so that they all can be adjusted simultaneously. An example is the ganged capacitor in a superheterodyne radio receiver, which adjusts the input and oscillator stages at the same time.

GAP-Generic Access Profile

Gas Filled-A tube or glass envelope that contains an apparatus and is filled with gas, either to improve the functioning of the apparatus or as an essential feature of the circuit itself.

Gatekeeper (GK)-A gatekeeper is a server that coordinates access to other servers. The gatekeeper receives requests from clients, determines the destination server that it needs to communicate with, and coordinates access with that server. For packet voice systems, the gatekeeper translates user names or telephone numbers into physical address for H.323 conferencing.

This figure shows how a gatekeeper sets up connections between Internet telephones and telephone gateways. The gatekeeper receives registration messages from an Internet telephone when it is first connected to the Internet. This registration message indicates the current Internet address (IP address) of the Internet telephone. When the Internet telephone desires to make a call, it sends a message to the ITSP that includes the destination telephone number it wants to talk to. The ITSP reviews the destination telephone number with a list of authorized gateways. This list identifies to the ITSP one or more gateways that are located near the destination number and that can deliver the call. The ITSP sends a setup message to the gateway that includes the destination telephone number, the parameters of the call (bandwidth and type of speech compression), along with the current Internet address of the calling Internet telephone. The gatekeeper then sends the address of the destination gateway to the calling Internet telephone. The Internet telephone then can send packets directly to the gateway and the gateway initiates a local call to the destination telephone. If the destination telephone answers, two audio paths between the gateway and the Internet telephone are created. One for each direction and the call operates as a telephone call.

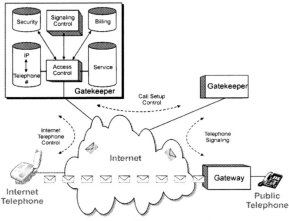

Gatekeeper Basic Operation

Gatekeeper Cluster-A group of gatekeepers that are linked together (possibly using GUP) to increase the reliability of a system.

Gatekeeper Update Protocol (GUP)-A proprietary protocol developed by Cisco to provide gatekeeper redundancy and load sharing. GUP can provide information about a gatekeeper's memory, CPU usage, number of endpoints that are registered, and available bandwidth. GUP is based on TCP.

Gateway-A gateway is a communications device or assembly that transforms data that is received from one network into a format that can be used by a different network. A gateway usually has more intelligence (processing function) than a bridge as it can adjust the protocols and timing between two dissimilar computer systems or data networks. A gateway can also be a router when its key function is to switch data between network points.

This figure shows how a gateway can convert large packets from a FDDI into very small packets in an ATM network. Not only does the gateway have to divide the packets, it must also convert the addresses and control messages into formats that can be understood on both networks.

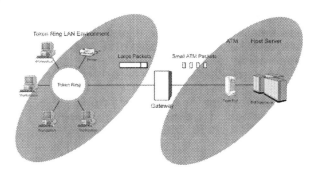

Gateway Operation

Gateway Location Protocol (GLP)-A protocol that was initially developed by a working group within the IETF to allow the gateway selection process in telephone and multimedia networks. This protocol is now called telephony routing over Internet protocol (TRIP).

Gauge (ga)-(1-wire) A measure of wire diameter. Under the American Wire Gauge (AWG) system, higher numbers indicate thinner wire. The AWG number indicates the number of times the wire has been "drawn" or pulled through successively thinner forming dies to make it thinner. The term Browne & Sharpe (B&S) gauge is a synonym for AWG. Browne & Sharpe is a manufacturer of measuring instruments. Sheet metal gauge, a similar number system indicating the number of times that the sheet metal has been rolled between rollers to make it thinner, has a different dimension for the same gauge number and for different materials. That is, 12 ga sheet copper is not the same thickness as 12 ga copper wire, nor is it the same thickness as 12 ga sheet iron. Note that outside the United States, copper wire and most other wires and sheet metals are expressed by their actual thickness in millimeters (mm) and the term gauge is not used. (2-instrument) A device that measures a value, such as pressure or temperature.

Gaussian Frequency Shift Keying (GFSK)- Gaussian frequency shift keying is a form of frequency modulation in which the modulating signal shifts the output frequency between predetermined values to represent a digital signal and that information signal (data) is passed through a Gaussian filter prior to modulation to minimize the rapid changes to the carrier signal. Typically, one frequency shift is used to represent a digital one (sometimes called a mark) and the other frequency shift represents a digital zero (sometimes called a space).

Gaussian Minimum Shift Keying (GMSK)-A form of frequency modulation in which the modulating signal shifts the output frequency between predetermined values. A form of MSK that uses gaussian low pass filtering of the binary data to reduce sideband energy.

GAZPACHO-Generation, Alignment, Zero [suppression], Polar, Alarm, Clock, Hunt, Office

GB-Gigabyte

GBN-GPRS Backbone Network

GCF-Gatekeeper Confirm

GCIDs-Global Call Identifiers

GCR-Group Call Register

GDMF-Generic Data Message Format

GDOI-Group Domain of Interpretation

GE-Gigabit Ethernet

GED-Global Engineering Documents

Geek-A geek is a person who is focused on technology, typically computers who does not tend to conform to mainstream habits such as dressing for success and/or regular bathing.

Gel Filling Compound-A gel filling compound is a substance that is used to fill space within a cable or assembly so that other substances such as water cannot enter and area.

General Parameters-In specification description language, the basic operating parameters of a unit or system.

General Purpose Interface (GPI)-(1-parallel) A parallel interconnection scheme that allows remote control access to certain functions of a device. One wire is dedicated to each function. (2-computer) An interface for data processing equipment. The term usually refers to a serial connection (RS232 or RS422 format) between computer devices.

General Register-An internal addressable register in a central processing unit that can be used for temporary storage, as an accumulator, or for any other general-purpose function.

General Switched Telephone Network (GSTN)-A name used by the International Telecommunications Union (ITU) to describe telephone networks (such as public telephone networks) that may connect to packet telephony systems.

General Telemetry Processor (GTP)-A device that is used to receive and process telecommunications equipment alarming protocols.

Generalized Mark-up Language (GML)-Generalized mark-up language is a precursor to SGML.

Generation Loss-Loss caused by the copying of scenes from one videotape to another.

Generic Requirement (GR)-A specification document controlled by Telcordia (formerly Bellcore) that defines requirements of systems that connect within a telephone network.

Genlock Module-A module that can phase lock to another source of video or sync.

GEO-Geosynchronous Earth Orbit

Geographic Number Portability (GNP)-Geographic number portability involves the transfer of telephone numbers for telephone devices or services that are used outside the normal geographic boundaries of the service provider's original system or area. Geographic number portability allows a customer to keep their same area code when they move to new cities or other distant geographic regions.

Geographic Spectral Efficiency-A measurement characterizing a particular modulation and coding method that describes how much information can be transferred in a given bandwidth within a mobile communication (cellular) system having reuse of the same radio carrier frequency at the closest permitted reuse cells. This is often given as bits per second per Hertz per square km of service area, or alternatively as bits per second per Hertz per cell. Modulation and coding methods that have high spectral efficiency often typically have very low geographic spectral efficiency and are not suitable for a cellular radio system. (See also Economic Geographic Spectral Efficiency or Spectral Efficiency)

G

Geostationary Satellite-A satellite that orbits the Earth at the same relative speed that the earth rotates, resulting in the satellites fixed relative position to the surface of the Earth. Geostationary satellites have an altitude of approximately 22,300 miles (35,680 km) above the Earth when they are in Geostationary position.

This diagram shows that a geostationary satellite is launched from Earth to achieve an orbit of approximately 22,300 miles above the Earth. At this position, it appears to have the same relative position above the surface of the Earth. This diagram shows that the combined effects of speed and gravity actually keep the satellite in the same relative position above the Earth.

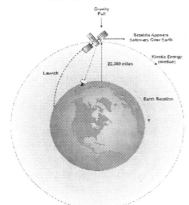

Geostationary Satellite Operation

Geosynchronous-Geosynchronous is the process of a satellite orbiting a body that has a speed that appears to be synchronized with the body it is orbiting. For Earth satellite systems, geosynchronous satellites maintain a fixed orbit approximately 24,000 miles above the Earth.

Geosynchronous Earth Orbit (GEO)-A satellite system where the satellites are located approximately 23,500 miles above the Earth. GEO systems are unique because at the specific height of 23,500 miles, the rotation of the Earth matches the rotation of the satellite resulting in the appearance of a fixed location of the satellite relative to the surface of the Earth.

Geotargeting-Geographic Targeting

GET-GET is the simplest of the SNMP operations. The GET operation is a request for an agent to return one or more objects or specific pieces of information. For each object requested (MIB object), the agent will return that object's value.

Get-Bulk-The get-bulk SNMP operation was added by SNMPv2 and improves performance of retrieving large amounts of SNMP data. This operation allows the agent to pack as many values in each get-response packet as will fit without exceeding the maximum SNMP packet size (The packet size is pre-defined or user defined on the agent).

Get-Next-The get-next SNMP operation allows a network management system (NMS) to request the "next" object or objects in the MIB supported by the agent.

Get-Response-The get-response SNMP operation is the packet type that SNMP agents send in response to receiving a get, get-next, get-bulk, or set-request packet.

Getter-A metal used in vaporized form to remove residual gas from inside an electron tube during manufacture.

GFC-Generic Flow Control Field

GFP-Generic Framing Procedure

GFSK-Gaussian Frequency Shift Keying

GGSN-Gateway GPRS Support Node

Ghost-A form of distortion, in a TV picture, in which a duplicate image is displayed, offset from the primary picture image. In VHF and UHF reception, ghosts result when the receiver antenna detects both the direct line of sight signal from the transmitter and one or more weaker signals reflected by nearby buildings, mountains, or other obstructions.

Ghost Image-The displaced image of a received TV picture, resulting from a secondary signal that is received slightly after the primary signal. Secondary signals can be caused by multipath propagation, such as a reflection from a nearby building. (See also: multipath propagation.)

GIF-Graphics Interchange Format

Giga-A prefix that represents one billion units.

Gigabit-One billion bits of data.

Gigabit Ethernet (GE)-A data communications system primarily used for computer networks based on the Ethernet IEEE standard 802.3 that transmits at 1000 Mbps (1 Gbps).

Gigabit Ethernet Alliance-An alliance between several companies that was formed in mid 1990s that is assisting in the creation of a Gigabit Ethernet system.

Gigabyte (GB)-A gigabyte is one billion bytes of data. When gigabyte is used to identify the amount of data storage space (such as computer memory or a hard disk), a gigabyte commonly refers to 1,073,741,824 bytes of information.

GIP-DECT/GSM Interworking Profile

GK-Gatekeeper

Glare Backout-The process of releasing a communication trunk that is seizure by one end office switch so that a call from another end office switch can be completed. This is also called glare release.

Glare Control-Glare is the conflict that occurs in the assignment of a communication trunk line (interconnection line) in a communication system when the line is accessed at both ends simultaneously. Glare is overcome through the use of glare hold and glare release processes.

Glare Release-A process of releasing a communication trunk that is simultaneously seized by two switches at the same time. Glare release results when one switch is given a lower priority and release the line. This is also called glare blackout.

Global-Applies to all users or devices connected to a network or system.

Global Call Identifiers (GCIDs)-A unique identifier (a tag) that is assigned to a telephone call to allow all events associated with that call to be easily identified and grouped. GCIDs were initially developed for computer telephony interface (CTI) systems.

Global Engineering Documents (GED)-The company that manages and sells TIA and EIA industry standard specifications.

Global Navigation Satellite System (Glonass)-A Russian global positioning navigation system, similar to the US Global Positioning System (GPS).

Global Positioning System (GPS)-The Global Positioning System (GPS) is a network of 24 Navstar satellites that are orbiting the Earth at 11,000 feet above the surface that provide signals that allow the calculation of position information. The transmit frequency for the GPS system is 1575.42 MHz. A GPS receiver compares the signals from multiple GPS satellites (4 satellite signals are usually used) to calculate the geographic position. In March of 1996, the military requirements for limited accuracy transmission were lifted. This allows the GPS system to provide very precise vehicle or device locations.

This figure shows a global positioning satellite (GPS) system. This diagram shows how a GPS receiver receives and compares the signals from orbiting GPS satellites to determine its geographic position. Using the precise timing signal based on a very accurate clock, the GPS receiver compares these signals from 3 or 4 satellites. Each satellite transmits its exact location along with a timed reference signal. The GPS receiver can use these signals to determine its distance from each of the satellites. Once the position and distance of each satellite is known, the GPS receiver can calculate the position where all these distances cross at the same point. This is the location. This information can be displayed in latitude and longitude form or a computer device can use this information to display the position on a map on a computer display.

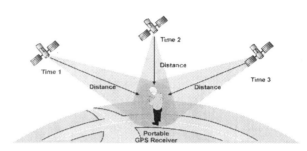

Geosynchronous Earth Orbit (GEO) Operation

Global System For Mobile Communications (GSM)-Global system for mobile communication (GSM) is a wide area wireless communications system that uses digital radio transmission to provide voice, data, and multimedia communication services. A GSM system coordinates the communication between a mobile telephones (mobile stations), base stations (cell sites), and switching systems. Each GSM radio channel is 200 kHz wide channels that are further divided into frames that hold 8 time slots. GSM was originally named Groupe Special Mobile. The GSM system includes mobile telephones (mobile stations), radio towers (base stations), and interconnecting switching systems.

This figure shows an overview of a GSM radio system. This diagram shows that the GSM system includes mobile communication devices that communicate through base stations (BS) and a mobile switching center (MSC) to connect to other mobile telephones, public telephones, or to the Internet.

Global Title (GT)

This diagram shows that the MSC connects to databases of customers. This example shows that the GSM system mobile devices can include mobile telephones or data communication devices such as laptop computers.

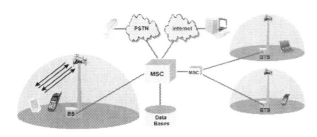

Global System for Mobile Communications (GSM) System

Global Title (GT)-A routing name (such as customer-dialed digits) that does not contain explicit information to enable routing in a communication network. The global title is usually converted to an address that allows the network to setup or route information to its ultimate destination point.

Global Title Translation (GTT)-A process used in a common-channel signaling system (such as SS7) that uses a routing table to convert an address (usually a telephone number) into the actual destination address (forwarding telephone number) or into the address of a service control point (database) that contains the customer data needed to process a call.

Glonass-Global Navigation Satellite System

GLP-Gateway Location Protocol

GMDSS-Global Maritime Distress And Safety System

GML-Generalized Mark-up Language

GMP-Group Membership Protocol

GMRS-General Mobile Radio Service

GMSC-Gateway Mobile Switching Center

GMSK-Gaussian Minimum Shift Keying

GMT-Greenwich Mean Time

GNP-Geographic Number Portability

GNSS-Global Navigation Satellite Systems

GOEP-Generic Object Exchange Profile

GOP-Group of Pictures

GOS-Grade Of Service

GP-Guard Period

GPI-General Purpose Interface

GPON-Gigabit Passive Optical Network

GPRS-General Packet Radio Service

GPS-Global Positioning System

GR-Generic Requirement

GR-303-A set of technical specifications that help define the next generation of digital loop carrier (DLC) interconnection.

Grade Of Service-(1-general) The customer satisfaction level (estimated) associated with a particular aspect of service (such as voice quality). (2 - noncompleted calls) A measure of the proportion of calls that cannot be completed because of limitations in the network capability to handle many simultaneous calls. (3 - subscriber features) A feature level granted to customers (typically in a cellular or PCS system) that allow access to basic or advanced services.

Grade Of Service (GOS)-Grade of service is the characteristics that define the capabilities of service. Some of these characteristics may include error rate, packet loss rate, service availability, and blocking rate.

Grand Alliance (GA)-The Grand Alliance is a group of companies that was formed in 1993 to develop and promote competing digital standards for HDTV. The final digital standard for HDTV was known as digital television (DTV) and was accepted by the FCC in December of 1996.

Grant Management-Grant management is the allocation (granting) of resources (such as transmission time or bandwidth) to a device or system.

Granularity-(1-network operation) The steps of network resource allocation. This is typically bandwidth allocation that is adjustable per communication session. (2-digital conversion) The increments (step size) of conversion from an analog signal to a digital signal.

Graphic Elements-Graphic elements are component parts of media images. A graphic element can be the smallest common denominator of an image component. A graphic element is considered a unique specific element such as a shape, texture and size.

Graphic User Interface (GUI)-The use of graphics (typically on a computer monitor display) to interface the output or requested input of a software application with a user. The use of buttons, icons and dynamically changing windows are typical examples of a GUI. Sometimes pronounced "gooey interface."

Graphics Interchange Format (GIF)-A graphics file data compression format that produces relatively small graphic files. A GIF image may contain up to 256 colors and uses a lossless data compression method. A version of the GIF format GIF89a permits the addition of animation features, image interleaving, and the use of transparent backgrounds.

Graticule-A fixed pattern of reference marking used with oscilloscope CRTs to simplify measurements. The graticule may be etched on a transparent plate covering the front of the CRT; for greater accuracy in readings, it may be electrically generated within the CRT itself.

Grazing Path-A radio path that does not have a clean line of sight between the transmitting and receiving antennas; a path that grazes ground level.

Great Understanding Relatively Useless (Guru)-A consultant or expert with a reputation for being helpful to other less knowledgeable users.

Green Application-A service application or software program that is designed to efficiently use system resources.

Green Light-Green light is an agreement or contract allowing a studio to commence with project development.

Greenwich Mean Time (GMT)-Greenwich mean time is the mean solar time at Greenwich England. This time is in accordance with 0' latitude that passes through Greenwich and it is commonly referred to as UTC (coordinated universal time).

GRIC-Global Roaming [For Reach] Internet

Grid-(1-computing) A distributed computing infrastructure aimed at advanced scientific research. The Grid provides access control, computational resources, and resource discovery over large sets of data. This allows for the formation of a virtual organizations centered around a specific area of research where sharing data, computing resources and applications are important for collaboration. (2-tube)A mesh electrode within an electron tube that controls the flow of electrons between the cathode and plate of the tube. The potential applied to a grid in an electron tube to control its center operating point. (3-control)The grid in an electron tube to which the input signal usually is applied. (4-screen)The grid in an electron tube, typically held at a steady potential, that screens the control grid from changes in anode potential. (5 - suppressor)The grid in an electron tube near the anode (plate) that suppresses the emission of secondary electrons from the plate.

GRIN-Graded Index Fiber

Grooming-Grooming is the process of selecting or managing the resources (such as bandwidth) within a network to allow the facilities or communication circuits to be used as effectively as possible.

Gross Add-Refers to the total addition of new subscribers to a communication system.

Gross Participation-Gross participation is the sharing of gross revenue that is generated by the sale of a program (e.g. movie), product or service.

Gross Revenues-Gross revenues are monies (or equivalent values) that are earned through the sale or providing of products and services without and reductions for costs associated with developing, producing or selling the products or services.

Ground-(1-earth) An electrical connection to earth or to a common conductor usually connected to earth. (2-reference) A reference point or reference voltage in a circuit.

Ground Loop-A ground loop is an undesirable circulating ground current in a circuit grounded via grounding connections or at multiple points where the ground connections can form another path around the cable or conductors.

Ground Plane-(1-circuit) A conducting material at ground potential, physically close to other equipment, so that connections may be made readily to ground the equipment at the required points. (2-radio)A conducting surface used to provide uniform reflection of an impinging electromagnetic wave. An example is the arrangement of buried wires at the base of an antenna tower.

Ground Rod-A metal rod driven into the earth and connected into a mesh of interconnected rods to provide a low-resistance link to ground.

Ground Segment-The communications gateway that connects a satellite to the telephone network (or other type of network such as the Internet).

Ground Wave-A radio wave that is propagated over the earth and ordinarily is affected by the presence of the ground. Ground waves include all components of waves over the earth except ionospheric and tropospheric waves, and are affected somewhat by changes in the dielectric constant of

G

the lower atmosphere. A ground wave also is known as a surface wave.

Ground Wire-A conductor used to extend a low-resistance earth ground to protective devices in a facility.

Grounded-The connection of a piece of equipment to earth via a low-resistance path.

Grounding-Grounding is the connecting of a device or circuit to an electrical ground point or to a conductor that is grounded.

Group Address-A group address is a code that is used to identify a set of stations or a number of devices as the destination for transmitted data.

Group Call-A call from one mobile radio, telephone or dispatcher to a predefined group of receivers in a group that are capable of receiving and decoding the messages.

This figure shows how voice group call service may operate in a GSM system. In this diagram, a single voice message is transmitted on GSM radio channels in a pre-defined geographic area. Several mobile radios are operating within the radio coverage limits (group 5 in this example) of the cells broadcasting the group message. In this example, a user is communicating to a group. Each user in this group (including the dispatcher) listens and decodes the message for group 5. Other handsets in the area are not able to receive and decode the group 5 message.

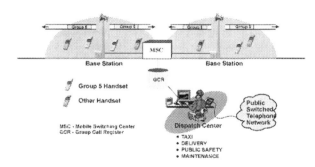

GSM Dispatch Operation

Group Delay-(1-general) A condition in which the various frequency elements of a given signal suffer differing propagation delays through a circuit or a system. The delay at a lower frequency is different than the delay at a higher frequency, resulting in a time related distortion of the signal at the receiving point. (2-optical) The transit time required for optical power, traveling at the group velocity of a given mode, to traverse a specified distance. The measured group delay of a signal through an optical fiber has a dependence on specific wavelengths because of the each wavelength reacts differently to dispersion mechanisms in the fiber.

This figure shows how group delay can cause pulsed signals, such as in digital transmission systems, can cause signal distortion. This diagram shows that a digital pulse signal is actually composed of many low, medium, and high frequency components. As the pulse is transmitted through the transmission line, some of the frequency components are delayed more than others. This results in a distorted pulse at the receiving end of the transmission line.

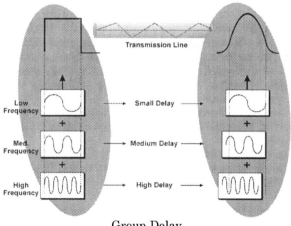

Group Delay

Group Identity (GroupID)-A groupID or telephone number that indicates the group for which a group call or broadcast message is intended.

Group Index-The ratio of the speed of light in a vacuum to the group velocity of pulsed light in a transmission medium.

Group Management-Group management is the process of defining groups of users or devices and adding and removing members (people and/or devices) to the groups.

Group Membership Protocol (GMP)-Group membership protocols are the commands, processes, and procedures that are used to send control messages and coordinate the multicasting (simultaneous distribution) of data through communication systems. Group membership protocols include Internet group management protocol (IGMP) and multicast listener discovery (MLD) protocol.

Group of Pictures (GOP)-A group of pictures is an encoding of a sequence of frames that contain all the information that can be completely decoded. For all frames within a GOP that reference other frames (such as B-frames and P-frames), the frames so referenced (I-frames and P-frames) are also included within that same GOP.

Group Paging-A telephone call-handling feature that provides a quick way to contact a specific group (list) of users.

GroupID-Group Identity

GRQ-Gatekeeper Request

GRX-GPRS Roaming Exchange

GSC-Golay Sequential Coding

GSI-Grid Security Infrastructure

GSM-Global System For Mobile Communications

GSM-Special Mobile Group

GSM PLMN-GSM Public Land Mobile Network

GSMA-GSM Association

GSN-GPRS Support Node

GSTN-General Switched Telephone Network

GT-Global Title

GTP-General Telemetry Processor

GTP-GPRS Tunneling Protocol

GTP-U-GPRS Tunneling Protocol User Plane

GTT-Global Title Translation

Guaranteed Service-Guaranteed service provides a specific bandwidth with a set maximum end-to-end transmission delay time.

Guaranteed Step-A writing process taking account assured payment, subjective only to the performance of that writer of mandatory services.

Guard Band-A guard band is a portion of a resource (frequency or time) that is dedicated to the protection of a communication channel from interference due to radio signal energy or time overlap of signals. While guard bands protect a desired communication channel from interference, the guard band also uses part of the valuable resource (frequency bandwidth or time period) for this protection.

Guard Period (GP)-A time period that is a portion of a burst period where no radio transmission can occur. The guard period is used to protect adjacent burst from transmission overlap due to propagation time from the mobile radio to the base station.

Guard Time-A time allocated within a single time slot period in a communication system to help ensure variable amounts of transit times (e.g. from close and distant transmitters) do not cause overlap (collisions) between adjacent time slots. Transmission of information does not occur within the guard period.

GUI-Graphic User Interface

Guiding Billing Records-Guiding is a process of matching call detail records to a specific customer account. Guiding uses the call detail record identification information such as the calling telephone number to match to a specific customer account.

Guild-A guild is an organization, society or association. A form of unity among members as to form a Union.

Guild Signatory-Guild signatory is referring to an individual who is to have signed a contract or treaty devising unity of an organization, society or association.

GUP-Gatekeeper Update Protocol

Guru-Great Understanding Relatively Useless

Guy Anchor-An anchor point (mount) where an antenna guy wire is attached.

246

H

H-henry
h-Plank's Constant
H Phase-Horizontal Phase
H Rate-Horizontal Rate
H&V Lock Time-Horizontal and Vertical

H.223-H.223 is a multiplexer protocol developed by the ITU that is used to coordinate the combining and separation of multiple (multiplexed) communication channels over a data transmission link.

H.225-H.225 call signaling is used to set up connections between H.323 endpoints (terminals and gateways), over which the real-time data can be transported. Call signaling involves the exchange of H.225 protocol messages over a reliable call-signaling channel. For example, H.225 protocol messages are carried over TCP in an IP based H.323 network.

H.245-H.245 is a signaling control protocol that contains a library of transmission control messages for use in packet based multimedia communication systems. H.245 control signaling consists of the exchange of end-to-end H.245 messages between communicating H.323 and H.324 endpoints. The H.245 control messages are carried over H.245 control channels. The H.245 control channel is the logical channel 0 and is permanently open, unlike the media channels. The messages carried include messages to exchange capabilities of terminals and to open and close logical channels.

H.248-H.248 is a control protocol specified by the ITU that uses text or binary format messages to setup, manage, and terminate multimedia communication sessions in a centralized communications system. This differs from other multimedia control protocol systems (such as H.323 or SIP) that allow the end points in the network to control the communication session. H.248 is also known as media gateway control protocol (MGCP) and it is the basis of the PacketCable NCS protocol.

H.261 Video Coding and Decoding (CODEC)-H.261 is a video codec standard that was intended for video conferencing over circuit switched data connections. H.261 was designed to operate with bandwidths that are multiples of 64 kbps.

H.263 Video Coding for Low Bit Rate Communication-H.263 is an ITU standard for low bit rate video encoding. The H.263 video coding

specification improves on the H.261 version that was created in (1990). There have been several enhancements to the orginal H.263 including H.263+ released in 1997 and H.263++ released in 2000.

H.264 Video CODEC For High Quality Video Streaming-The H.264 video CODEC for high quality video streaming standard is an ITU industry specification that was released in May 2003 in cooperation with the IEC joint video committee (JVC). This video codec that can be used in packet based video communication systems such as the MPEG-4 standard. This video coding (compression) technology provides standard definition (SD) quality at approximately 2 Mbps. The ISO/IEC version of this specification is the H.264 specification and it is called advanced video coding (AVC). The H.264 system offers much higher compression than is possible with its MPEG-2 predecessor.

H.26L-H.26L is the name of the technology that was created for the H.264 advanced video coding specification.

H.320-A videoconferencing standard developed by the ITU-T for transmitting video and audio over circuit-switched digital networks. The H.320 system supports the use of video compression protocols H.261 and H.263.

H.323-H.323 is an umbrella recommendation from the International Telecommunications Union (ITU) that sets standards for multimedia communications over Local Area Networks (LANs) that may not provide a guaranteed Quality of Service (QoS). H.323 specifies techniques for compressing and transmitting real-time voice, video, and data between a pair of videoconferencing workstations. It also describes signaling protocols for managing audio and video streams, as well as procedures for breaking data into packets and synchronizing transmissions across communications channels.

H.323 Gateway-The H.323 gateway is a communications device or assembly that transforms audio that is received from a telephone device or telecommunications system (e.g. PBX) into a format that can be used by a data network. A voice gateway usually has more intelligence (processing function) than a bridge as it can select the voice compression coder and adjust the protocols and timing between two dissimilar computer systems or voice over data networks.

H.323 Packet-Based Media Communications Systems-An ITU industry standard for multimedia communications that combines and coordinates multiple data compression and communication standards to allow audio, picture, and video transmission between users on packet switched networks. The H.323 system has four key components: terminals, gateways, gatekeepers, and multipoint control units (MCUs). The original name for H.323 was Visual Telephone Systems and Equipment for Local Area Networks.

This table shows that the H.323 protocol suite is composed of several different types of protocols. Some of these protocols are dedicated for audio processing, data compression, and video transmission. Other protocols are used for control and special feature processing

Audio	Data	Video	Transport
			H.225
			H.235
G.711	T.122		H.245
G.722	T.124		H.450.1
G.723.1	T.125	H.261	H.450.2
G.728	T.126	H.263	H.450.3
G.729	T.127		RTP
			X.224.0

H.323 System

This figure shows the basic structure of an H.323 system. This diagram shows that the H.323 system can interconnect standard telephones and data communication devices (multimedia computers) through the use of gateways and gatekeepers. Gateways convert the audio and multimedia information into formats that can be transmitted through the packet data network. Gatekeepers coordinate, authorize, and bill (if billing is required) access through the gateways. This diagram shows that when calls are initiated through the H.323 network, the gateway requests access from the gatekeeper. The gatekeeper reviews its database to determine if the request is authorized and may perform translation of dialed digits to a data (IP) address. The destination gatekeeper is then contacted and if it authorizes service, its associated gateway will be setup to translate the call to the communication device (e.g. telephone) that is receiving the call.

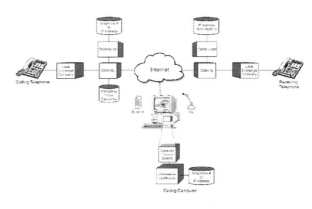

H.323 System

H.323v1-The first version of H.323 Visual Telephone Systems and Equipment for Local Area Networks created by study group 16 in the ITU in 1995.

H.323v2-The second version of H.323 that changed the video conferencing focus of H.323 to multimedia communications. H.323v2 was released in 1998.

H.450 Supplementary Services-Supplementary Services for H.323, namely call transfer and call diversion, have been defined by the H.450 series. H.450.1 defines the signaling protocol between H.323 endpoints for the control of supplementary services. H.450.2 defines call transfer and H.450.3 call diversion. Call transfer allows a call established between endpoint A and endpoint B to be transformed into a new call between endpoint B and a third endpoint, endpoint C. Call diversion provides the supplementary services call forwarding unconditional, call forwarding busy, call forwarding no reply and call deflection. The "hooks" for supplementary services are specified in H.323 Version 2.

HA-Home Agent

HAAT-Antenna Height Above Average Terrain

HAAT-Height Above Average Terrain

Hacker-A hacker is a person or a machine that attempts to gain access into networks and/or computing devices. Hackers may perform their actions for enjoyment (satisfaction), malicious reasons (revenge), or to obtain a profitable gain (theft).

Hacking-Hacking is the process that is used when attempt to gain unauthorized access into networks and/or computing devices. The term hacking has also been used by programmers to solve their programming problems. They would continually change or hack the program until it operated the way that they desire it to operate.

HACN-Home Agent Control Node

Half Duplex-Half duplex communication is the process of transferring voice or data information in either direction between communications devices but not at the same time. The information may be transmitted on the same frequency or divided into different channels. When divided into different channels, one channel of frequency is used for transmitting and the other channel or frequency is used for receiving.

The use of different frequencies is common in half duplex radio transmission because the transmitter and receiver are commonly connected to the same antenna. If the same transmitter and receiver frequency were used, the high transmitter power would probably destroy the receiver circuitry.

Half-Power Point - 3 dB Point-The point(s) on a frequency spectrum, or on the radiation pattern of an antenna, at which the power of a signal is equal to one half of its maximum power. It often is called the 3dB point because the signal there is approximately 3 dB less than maximum. The bandwidth of a filter (in kHz) or the beam width of an antenna radiation pattern (in degrees) is often measured between two 3 dB (half power) points, which is relatively easy to measure, but not always a meaningful bandwidth or beam width.

Hamming Distance-The number of bits that are different between a pair of binary values.

Hand Hole (Handhole)-A handhole is a plastic, steel, or tile enclosure used in buried cable distribution systems that includes a cover that is used as a splice or pull box. Handholes are smaller versions of manholes and do not enable outside plant personnel to enter for splice work. The splice closure or cable sheath is pulled out of the handhole, spliced or repaired above ground and then laid back in the enclosure. Handholes are common in residential and commercial subdivisions served by 200 pair or smaller cables between buildings.

Hand Portable-A cellular telephone or computing device that is self contained in a lightweight, hand carried unit.

Hand Portable Unit (HPU)-A mobile radio or computing device that is able to be carried and operated by a user.

Handheld-A wireless telephone or computing device that is self-contained in a lightweight, handheld unit.

Handheld Device Markup Language (HDML)-HDML is an internet web browsing language that evolved from hypertext markup language (HTML). HDML is optimized for low bandwidth operation using limited screen size and limited user input keys. HDML allows soft keys to be dynamically defined to allow a phone and return the key presses to the network application.

Handhole-Hand Hole

HANDO-Handover

Handoff-A process where a mobile radio operating on a particular channel is reassigned to a new channel. The process is often used to allow subscribers to travel throughout the large radio system coverage area by switching the calls (handoff) from cell-to-cell (and different channels) with better coverage for that particular area when poor quality conversation is detected. Handoff (also called handover) is necessary for two reasons. First, where the mobile unit moves out of range of one cell site and is within range of another cell site. Second, where the mobile has requested a cellular channel with different capabilities. This might mean assignment from a digital channel to an analog channel or assignment from an analog channel to a digital channel.

Handset Churn -Type of voluntary churn unique to, and common in the wireless industry. Handset churn occurs when a customer changes carriers in order to get a newer and better (and often free) handset.

Hands-Free-An audio interface to a communication device (such as a mobile telephone) that permits the user to listen and talk without holding the handset to their ear. Hands-free operation is required in some countries and states for operation as a safety feature.

Handshake-A handshake is a process of exchanging information between communication devices to ensure a connection can be established between both devices.

Handshake Protocol-A series of messages exchanged between communication devices operating on a system to establish communications links for further transmissions of voice or data.

Hang Up-In ground-start supervision, an information signal indicating that a carrier is finished with a connection but is not ready to accommodate a new request for service.

Hanover Bars-An undesirable artifact of interlaced video scanning that shows scrolling bars.

Hard Launch-Hard launch is the process of making a product available for purchase and distribution with significant promotional efforts. Companies may perform a hard launch strategy after a soft launch or it may use a hard launch campaign to maximize the effects of new product publicity.

Hard Stop-A hard stop is the ending of a software program or the termination of a specific function of that program (such as ending a media player session).

Hardware Decoding-Hardware decoding is the process of using a device or equipment to decode and/or decompress a coded signal.

Hardware Fingerprint-A hardware fingerprint is a group of characteristics that uniquely identifies a specific device or equipment. Examples of hardware fingerprint characteristics include stored specific voltages, signal processing times, firmware revision numbers and electronic serial numbers.

Harmonic Analysis-The process of breaking down a complex wave into the sum of a fundamental and various harmonic frequency components.

Harmonic Analyzer-A test set capable of identifying the frequencies of the individual signals that makeup a complex wave.

Harmonic Emission-A spurious emission at frequencies that are whole multiples of those contained in the frequency band occupied by an emission.

Hashing-Hashing is a computational process that converting a data or a message into a fixed length data message. Hashing can be used to convert text based passwords into fixed length digital password codes.

HAVi-Home Audio Video

HBS-Home Base Station

HCI-Host Controller Interface

HCS-Hard Clad Silica Fiber

HCS-Hierarchical Cell Structure

HD-High Definition TV Format

HD Radio-High Definition Radio

HDB3-High Density Binary or Bipolar 3

HDLC-High-Level Data Link Control

HDMI-High Definition Multimedia Interface

HDML-Handheld Device Markup Language

HDR-High Data Rate

HDSL-High Bit Rate Digital Subscriber Line

HDSL2-High Bit Rate Digital Subscriber Line 2

HDTV-High Definition Television

HDWDM-High Density Wave Division Multiplexing

Head-A device that erases, records, or reads information from a hard disk or other media that is passed under it.

Head Mounted Displays (HMDs)-Head mounted displays are graphics rendering devices that are mounted on the head that allow the person to view images or video.

Headend-The part of a cable television system that selects and processes video signals for distribution into a cable television distribution network. A variety of equipment is used at the headend, including antennas and satellite dishes to receive signals, preamplifiers, frequency converters, demodulators and modulators, processors, and scrambling and de-scrambling equipment.

With the advent of the cable modem, the headend also houses sophisticated digital data communications equipment that is able to transmit computer communications over a standard television signal's spectrum (6 MHz in the U.S. or 8 MHz in other parts of the world).

Headend Driver Amplifier-An amplifier that is used in a cable television network to increase the level of the RF signal for supply to the distribution network.

Header-(1-communications) The part of a data packet that contains address, routing, and origination information. (2- record)A record that contains administrative, physical, and electrical data describing a cable count, carrier facility, or type of equipment. (3-style element)A pre-defined style format that has a hierarchical structure that allows for the multi-level organization and presentation of documents through the use of header element types. Headers can have different font sizes, position structure (left, center, or right), and font types.

Header Compression-A process that removes redundant header (usually repetitive addressing and control data) information that is transmitted in the header of data packets.

Headroom-(1-signal level) The difference, in decibels, between the typical operating signal level and a peak overload level. (2-bandwidth) The amount of available data transmission bandwidth that exists for applications and services.

Headset-A headset is a combination of a microphone and earpiece that is used to enable better audio conversations. Headsets may be connected to various types of devices by wires including telephone sets, wireless devices (cellular telephone or cordless telephone), or they may integrate wireless technology (such as Bluetooth) to allow them to directly connect to a wireless receiver.

Hearing Loss-When expressed as decibels, the difference between the measured threshold of audibility in an individual ear, and the normal or standard threshold at the same frequency. When expressed as a percentage, the value is equal to 100 times the hearing loss in decibels divided by the decibel difference between the normal or standard threshold of audibility and the normal or standard threshold of vibratory sensation.

Heat Loss-The loss of useful electric energy resulting from conversion into unwanted heat.

Heat Shrink-Heat shrink is a material that has its size contract with the application of heat.

Heat Shrink Tubing-Heat shrink tubing is used to cover and protect connections. When the heat is applied, the tubing contracts and surrounds and snugly fits over the wire and connection.

Heat Sink-A device that conducts heat away from a heat producing component so that it stays within a safe working temperature range.

Hecto-A prefix meaning 100.

Height Gain-For a given propagation mode of an electro-magnetic wave, the ratio of the field strength at a specified height to that of the surface of the earth.

Help Desk-A help desk is an accessible location where questions about the operation of a product or service can be answered. A help desk may be reached by a combination of voice, instant messaging or email.

henry (H)-Unit of electrical inductance, equal to one volt second per ampere. In a 1 henry inductor, a voltage of 1 volt occurs when the electric current changes at the rate of 1 ampere per second. Named after the 19th century American scientist Joseph Henry (1797-1878).

Hertz (Hz)-A measurement unit for frequency that is equal to the number of cycles per second. This measurement unit was named after the German physicist Herrich R. Hertz (1857-1894).

This figure shows how to measure a signal wave in cycles per second (Hertz). This diagram shows a signal wave that has three cycles that moves past a point in 1 second. This equals a frequency of 3 Hertz. Radio waves typically have several million cycles per second that is called a Megahertz (MHz).

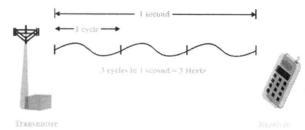

Measuring a Wave in Hertz

Hertzian Wave-Hertzian waves are electomagnetic waves that was discovered by Heinrich Hertz in 1887.

Heterodyne Frequency-The sum of, or the difference between, two frequencies, produced by combining the two signals in a modulator or similar device.

HF-High Frequency

HFC-Hybrid Fiber Coax

H-FDD-Half Frequency Division Duplex

HFS-Hierarchical File System

HI8 Video-Hi8 is a video cassette package of 8mm reel-to-reel magnetic tape that is used for high resolution (higher luminance) analog video signals and may be played or rewound on demand.

Hiatus-A hiatus is a break given to staff or employees between working hours.

HID-Human Interface Devices

Hierarchical Computer Network-A computer network in which processing and control functions are performed at several levels by computers specially suited for the functions performed.

Hierarchical File System (HFS)-A file system that is structured like trees where files are associated with higher level files or directories (roots).

Hierarchical Numbering-Multiple level numbering. An example is the telephone number made up of levels such as "Country Code," "Area Code," "Exchange Number" and "Line Number."

Hierarchy Structure-A hierarchy structure is a set of transmission speeds or components and/or protocols arranged to multiplex successively higher numbers of circuits.

High Bit Rate Digital Subscriber Line (HDSL)-High bit rate digital subscriber line is an all digital transmission technology that is used on 2 or 3 pairs of copper wires that can deliver T1 or E1 data transmission speeds. HDSL is a symmetrical service.

This figure shows an HDSL system uses two or more pairs of copper wire. This example shows that each pair of HDSL wires carries 784 kbps full duplex (simultaneous send and receive) data transmission and that the head end has an HTU-C and the remote end has an HTU-R. Because a T1 line requires 1.544 Mbps, two pairs of lines are used.

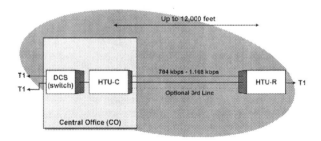

High Bit Rate Digital Subscribe Line (HDSL)
System

High Bit Rate Digital Subscriber Line 2 (HDSL2)-High bit rate digital subscriber line is a second generation of HDSL that offers several enhancements to HDSL data transmission. These improvements include the ability to transfer T1 or E1 data transmission rates over a single twisted-pair local loop instead of the two or three pairs of copper wire required for standard HDSL.

High Definition Multimedia Interface (HDMI)-HDMI is a specification for the transmission and control of uncompressed digital data and video for computers and high-speed data transmission systems.

High Definition Television (HDTV)-High definition television (HDTV) is a TV broadcast system that proves higher picture resolution (detail and fidelity) than is provided by conventional NTSC and PAL television signals. HDTV signals can be in analog or digital form.

High Definition TV Format (HD)-High definition (HD) television is the resolutions of enhanced analog television and digital television. The resolutions of HD range from 480/60p - 480 pixels (vertical) by 728 pixels (horizontal) with 60 progressive fields (60p) per second to 1080/60p - 1080 pixels (vertical) by 1920 pixels (horizontal) with 60 progressive fields per second.

High Density Wave Division Multiplexing (HDWDM)-High density wave division multiplexing (HDWDM) is a form of wave division multiplexing that increases the number of optical network units (ONUs) per passive optical networks. In 2001, HDWDM increased PONs from 32 to 64 ONUs.

High Level Control-In data transmission, the conceptual level of control or processing logic existing in the hierarchical structure of a primary or secondary station that is above the link level, and upon which the performance of link level functions are dependent or are controlled.

High Pass Filter (HPF)-A device or circuit that passes signals of higher than a specified frequency but attenuates signals of all lower frequencies.

This diagram shows a highpass filter can be used to block low and mid frequency component parts of an input signal and allow high frequency components (signals) to pass through. In this example, both the

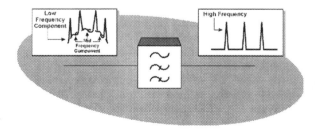

Highpass Filter Operation

low and mid frequency noise signals are highly attenuated by the bandpass filter while the desired high frequency is allowed to pass through the filter with minimal attenuation.

High Speed Circuit Switched Data (HSCSD)- An enhancement to the GSM mobile communications system that combines up to four 14.4 Kbps channels (up to 4 slots per frame) to be combined to provide 57.6 Kbps data transfer.

This figure shows how GSM can provide HCSD service. This diagram shows that the HSCSD system combines multiple time slots to provide data services that are at higher data transfer rates than is possible using single slot operation. This example shows how data from the HSCSD system can be routed to a separate data network directly from a base station or a base station controller. The multiple HSCSD system also permits asynchronous operation where there can be a higher data transfer rate in one direction (typically in the downlink when the user is downloading files or images from the Internet) than the other direction.

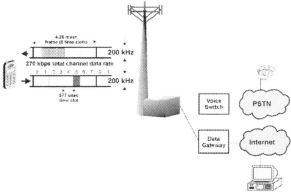

High Speed Circuit Switched Data (HSCSD)
System

High Speed Serial Interface (HSSI)-A standard serial data communications interface that can transfer data rates up to 52 Mbps. HSSI is sometimes used to connect DS3 (T3) lines to asynchronous transfer mode (ATM) connections.

High-Level Data Link Control (HDLC)-An ISO communication protocol that is located in the data link layer in that delineates the beginning and end of a data frames. HDLC is used in X.25 packet switching communication networks to transfer bit-oriented, synchronous protocol that provides error correction at the data-link layer. HDLC systems allow messages to be transmitted in variable-length frames.

High-Speed Multimedia-High-speed multimedia usually refers to image based media such as pictures, animation or video clips. High-speed multimedia usually requires peak data transfer rates of 1 Mbps or more.

High-Tier-A wireless system that serves high-speed vehicular traffic and may have higher power levels.

Hijacking-A process of gaining security access by the capture of a communication link in mid-session after the session has already been unauthorized.

This figure shows how hijacking may be used to obtain access to an authorized media session to gain access to protected media. This example shows that an unauthorized user has obtained information about a media session request between a media provider (such as an online music store) and a user (music listener). After the media begins streaming to the validated user, the hijacker modifies a routing table distribution system to redirect the media streaming session to a different computer.

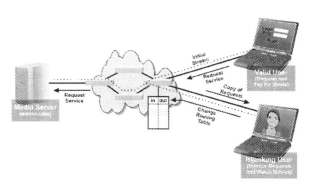

Hijacking Operation

Hinting-Hinting is the process of providing information about the contents of a file that assist the receiver of the information to gather, organize or decode the file information.

HiperMAN-HiperMAN is a fixed wireless access system developed by European Telecommunications Standards Institute (ETSI) that operates at frequencies between 2 GHz and 11 GHz. The HiperMAN system was designed to provide point to multipoint (PMP) services and it is compatible with the physical and data link layers of the WiMAX system.

Historical Data File-A set of data that is altered infrequently and supplies basic data for processing operations, such as forecasting.

Hit-(1-Internet) A request for a file transfer on a web site (web page access). (2-telephone signaling)A short off-hook or on-hook supervisory signal on a telephone line. Telephone applications may process or ignore hits by identifying signals that are shorter than a minimum specified time period, typically 10 ms to 400 ms. (3-data)A short disruption in the transmission of a data stream. (4-lightning)A lightning strike.

Hits-Hits are a measure of how many files have been requested on a Web site. Because hits are a measure of files that are requested/opened, the number of actual visitors is usually much less than the number of hits. If a web page has multiple graphic file links, the request for each graphic file may count as a hit. For example, the viewing of a web page with 10 images may count as 11 hits.

HLR-Home Location Register

HMDs-Head Mounted Displays

HMI-Human Machine Interface

HNPA-Home Numbering Plan Area

Hold-(1-general) A temporary mode of operation that is typically entered into by a device when there is no need to send voice or data information for a relatively long time. The hold mode allows the device audio to be muted or the transceiver to be turned off in order to save power. (2-Bluetooth)The Bluetooth hold mode is used to release devices from actively communicating with the master. This allows the devices to sleep for extended periods and allows the master control device to discover or be discovered by other Bluetooth devices that want to join other Piconets. With the hold mode, Piconet capacity can be freed up to do other things like scanning, paging, inquiring, or attending another Piconet sessions.

Holdback Period-Iholdback period is a timed in which one is forbidden to use some or all liberty's of an individual.

Holding Time-The amount of time that a line connection or other shared resource is in use for a call or call attempt.

Hologram-A hologram is a three-dimensional image. Holograms may be produced by lasers or through materials that bend, focus or redirect light.

Holographic Data Storage-Holographic data storage is the process of storing information using 3 dimensional optical elements.

Holographic Memory-The nonvolatile storage of information throughout the whole of a storage medium rather than in a specific location.

Home-The geographic location where communication service is based.

Home Agent (HA)-A router used to manage the current location and routing information for mobile users. A foreign agent exchanges information with the home agent to allow data packets to be forwarded to network where the user is currently operating.

Home Agent Control Node (HACN)-A signaling control node used that manages several home agent nodes.

Home Area-A home area is the radio coverage area of the system where a mobile device is registered for service.

Home Audio Video (HAVi)-A specification for home networks comprised of consumer electronics devices such as CD players, televisions, VCR's, digital cameras, and set-top boxes. The network configuration is automatically updated as devices are plugged in or removed. The IEEE 1394 protocol, also known as FireWall, is used to connect devices on the wired HAVi network at up to 400 Mbps.

Home Carrier-A home carrier is the communication service provider (such as a mobile telephone company) that a user has registered for service use.

Home Location Register (HLR)-The part of a wireless network (typically cellular or PCS) that holds the subscription and other information about each subscriber authorized to use the wireless network.

Home Numbering Plan Area (HNPA)-The numbering plan area within which a calling line appears at a local switching office.

Home Page (Homepage)-A home page is the first web page that is accessed on a web site. The default home page file name is typically index.htm.

Home Phoneline Networking Alliance (HomePNA)-The HomePNA is a non-profit association that works to help develop and promote unified information about a phoneline technologies, products and services.

Home Phoneline Networking Alliance Specification (HPNA)-The HPNA specification defines the signals and operation for data and entertainment services that can be provided through telephone lines that are installed in homes and businesses. The HPNA specification is designed to co-exist with other communication systems including POTS, ISDN and ADSL. More information about HomePNA can be found at www.HomePNA.org.

Home System-The home system is a communication network where a customer has registered for service.

Homepage-Home Page

Homeplug Power Alliance-Homeplug power alliance is an association that assists in the development of power line networking standards that allow for data to be distributed to devices that are connected through power line wiring in a home or business.

Homeplug Specification-The Homeplug specification defines the signals and operation for data and entertainment services that can be provided through electric power lines that are installed in homes and businesses. More information about Homeplug can be found at www.Homeplug.org.

HomePNA-Home Phoneline Networking Alliance

HomeRF-An industry working group that is assisting in the development of a local area RF communications that operates in the 2.4 GHz frequency band. HomeRF systems allow consumer devices such as computers, printers and fax machines to communicate with each other with data transmission rates up to 10 Mbps.

HomeRF Working Group (HRFWG)-An organization that was created to assist in the development and commercial introduction of interoperable wireless consumer devices. The HRFWG established an industry specification that allows for digital communication between computers, accessories, and other consumer electronic devices in small geographic areas (such as the home).

Homochronous-Signals whose corresponding significant instants have a constant but uncontrolled phase relation-ship with each other.

Hook Flash (Hookflash)-A special feature service request signaling process that allows the user to momentary disconnect circuit disconnection (temporary off-hook state) as a means if indicating a special service request (e.g. call forwarding request).

Hookflash-Hook Flash

Hop Count-The hop count is a measure of the number of routers through which a packet has passed from when it was initially sent into a network.

Hopoff-The process of transitioning from one network to another network. An example of hopoff is the transition from an H.323 network to a PSTN network.

Hopping Sequence Number (HSN)-The hopping sequence number identifies the frequency hopping pattern that a mobile radio should use when communicating with the system.

Horizon Program-The Horizon program developed by Motorola that helps independent software vendors (ISV) to develop television and Internet applications such as gaming, interactive TV, and video on demand (VOD).

Horizontal (Hum) Bars-A group of relatively broad horizontal bars, alternately black and white, that extends over the entire picture on a video display. The bars may be stationary or they may move up or down. Hum bars are caused by interference from the power line (utility) frequency or one of its harmonics.

Horizontal Application-A horizontal application is a program or software that is used accross multiple industries or by different types of users.

Horizontal Resolution-Horizontal resolution is the chrominance (color) and luminance (intensity) resolution (detail) expressed horizontally across a display. This parameter usually is expressed as a number of pixels or lines that can be differentiated.

Horizontal Scan-A single horizontal scan line of a camera or electron beam on a picture tube. Multiple horizontal line scans are grouped together to form a frame of video. There are 525 interlaced lines per frame in NTSC and 625 lines per frame in PAL.

Each horizontal scan signal contains a horizontal synchronizing pulse that identifies the start of each line of video. For color systems, the horizontal synchronization pulse is followed (during a short time interval called the "back porch") by a color burst (3.58 MHz in NTSC color systems, 4.43 MHz for PAL) that provides a phase reference signal for the phase sensitive decoder used to extract the three primary color signals from the composite phase modulated video waveform.

This diagram shows the electrical video signal that represents a single line (horizontal sweep) of video (there are 525 to 625 lines of video for most television systems). In this example, the horizontal video scan signal starts with a horizontal synchronizing pulse and this synchronization pulse also includes a color burst tone. Following the horizontal pulse, a composite video represents the intensity (brightness) of the image as the sweep moves across the picture tube. Because this is a color video signal, the composite signal represents the intensity of 3 colors. This diagram also shows that a horizontal blanking period occurs during the sync pulse to allow the picture tube beam to sweep back to the other side of the picture tube without being seen (blanked)

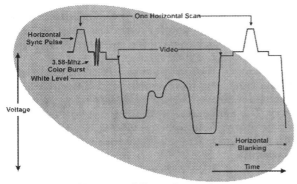

Horizontal Scan Operation

Horizontal:Vertical-Horizontal:vertical is the ratio of the number of items (such as pixels on a screen) as compared to their height and width. Horizontal:vertical (also called the aspect ratio) determines the frame shape of an image.

Host-A host is a computer or other type of data information processing device that is connected to a network that processes request and provides information services to remote users.

Host Controller Interface (HCI)-The host controller interface (HCI) is a standard communication protocol that is used to control a communication link to a data processing device. HCIs are used in many types of communication systems include universal serial bus (USB) and Bluetooth.

Host Name-The name that has been assigned to the computer responsible for one or several IP addresses.

Host Number-The part of an Internet address that identifies the node on the network or subnetwork that is being addressed.

Host Switching System-A switching system with centralized control over most of the functions of one or more remote switching units. The host usually provides trunk access to an exchange carrier's intraLATA network and access tandems.

Hosted PBX-Hosted PBX systems are telephone systems that are operated at a remote location (hosted) to provide private telephone services that are similar to private branch exchange systems (PBX). Hosted PBX systems can be operated by independent companies and systems or they can be operated by telephone companies by sharing central exchange (Centrex) telephone switch functions.

Hosted Telephony-Hosted telephony is a managed IP Telephony communication service (also known as IP Centrex). Hosted telephony is a communications service for users who prefer to use a communication system that is managed by another company. The term "hosted" or "managed" is used instead of "Centrex" because an IP-based feature set is generally broader than what existing Centrex offerings provide. Furthermore, hosted or managed solutions do not just cater to the traditional Centrex market - typically campus-based, such as universities or government. The flexibility of IP allows for very compelling offerings to other markets, such as SOHO and greenfields.

Hosting-The providing of application services (such as virtual telephone service or the providing of web service) for the benefit of a client or customer.

Hot Patch-A method of patching a failed digital line, such as a T1 line, onto a spare facility. This technique does not use bridging repeaters. Service is interrupted briefly when the patches are removed. This type of patching is used only when bridging repeaters are not available for full-service patching.

Hot Spots-(1-Mobile communication) Hot spots are geographic regions or service access points that have a higher than average amount of usage. Examples of hot spots include wireless LAN (WLAN) access points and traffic jam areas on mobile telephone (cellular) systems. (2-Web page) A screen display area on a web page image (image map) that can be selected ("clickable") by a mouse.

Hot Standby Routing Protocol (HSRP)-HSRP is a proprietary protocol developed by Cisco Systems. It is used to provide network redundancy for IP networks, especially for network edge devices or hosts. It allows user traffic to seamlessly recover from first hop (default gateway) router failures. Multiple routers on LAN segments are configured to communicate via HSRP status messages to provide this level of redundancy, thus a "virtual router" to end host devices. HSRP allows two or more routers to share the same IP address and MAC (Layer 2) address. The ACTIVE router in the HSRP group is the primary route out of the locally attached network. All other routers in the HSRP group act as STANDBY routers until the ACTIVE router goes away.

Hotline-A restricted calling class that forces a telephone (usually a wireless telephone) to be connected to an operator regardless of the digits actually dialed. Hotline is typically used when a telephone is first sold or activated to allow activation after the customer has provided the information to register for service or when the customer has not paid their bill.

Housing-An enclosure for carrier equipment and other electronic systems deployed in an outside plant. A housing may be above or below ground and may range in size from a small shelter to a large building containing many carrier systems.

HPC-Handheld Personal Computer

HPF-High Pass Filter

HPNA-Home Phoneline Networking Alliance Specification

HPU-Hand Portable Unit

HR-Half Rate

HRFWG-HomeRF Working Group

HSCSD-High Speed Circuit Switched Data

HSDPA-High Speed Downlink Packet Access

HS-DSCH-High Speed Downlink Shared Channel

HSN-Hopping Sequence Number

HSRP-Hot Standby Routing Protocol

HSSI-High Speed Serial Interface

HSTB-Hybrid Set Top Box

HTML-Hypertext Markup Language

HTTP-Hypertext Transfer Protocol

HTTP Pseudostreaming-A technique for streaming audio or video over the Internet using the HTTP protocol. When compared to streaming using other protocols, its advantages include being able to operate through any firewall that allows web traffic, and being based on a ubiquitous standard protocol. However in poor network conditions it is not suitable for real-time delivery since it is based on TCP. Many internet streaming systems can fall back to HTTP pseudostreaming as a last resort if connections using preferred protocols fail due to a firewall or other issues. Some simple Internet streaming systems support HTTP pseudostreaming exclusively. HTTP pseudostreaming cannot be used for true multicast streaming since Internet multicast is not TCP-based.

HTTP Streaming-HTTP streaming is the process of initiating and managing the sequential transferring of media files through an IP data network (such as the Internet) using HTTP commands.

Hub-A hub is a communication device that distributes communication to several devices in a network through the re-broadcasting of data that it has received from one (or more) of the devices connected to it. A hub generally is a simple device that re-distributes data messages to multiple receivers. However, hubs can include switching functional and multi-point routing connection and other advanced system control functions. Hubs can be passive or active. Passive hubs simply re-direct (re-broadcast) data it receives. Active hubs both receive and regenerate the data it receives.

This figure shows how a hub distributes communication to several devices in a network through the re-broadcasting of data that it has received from one (or more) of the devices connected to it. A hub generally is a simple device that re-distributes data messages to multiple receivers. However, hubs can include switching functional and multi-point routing connection and other advanced system control functions. Hubs can be passive or active. Passive hubs simply re-directs (re-broadcasts) data it receives. Active hubs both receive and regenerate the data it receives.

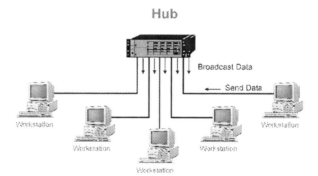

Hub Operation

Huffman Coding-A data compression technique that takes advantage of different probabilities of each data symbol occurring. More probable symbols use shorter codes while less probable symbols use longer codes, resulting in an overall average code length which is within one bit of the optimum for any compression technique. All codes are "prefix-free" that means that no valid codes can start with another valid code, which would be ambiguous when decoding.

Hum-Undesirable coupling of the 50 Hz or 60 Hz power sine wave into other electrical signals and/or circuits.

Hum Bars-Horizontal

Hum Rejection-The capability of a circuit to cancel interference in a video or audio signal, usually at the 50 Hz or 60 Hz power line frequency.

Human Factors Engineering-Engineering that applies the study of ergonomics to the design of equipment and software to create safe, easy-to-use systems.

Human Interface Devices (HID)-Human interface devices (HID) convert signals and information into forms that are accessible to humans.

Human Machine Interface (HMI)-The human machine interface is the method of how a user will enter (input) and receive information from a device or system. For the GSM/GPRS system, the HMI defines the requirements for a WAP browser (micro browser) on a mobile device.

Hunt Group-A hunt group is a list of telephone numbers that are candidates for use in the delivery of an incoming call. When any of the numbers of the hunt group are called, the telephone network sequentially searches through the hunt group list to find an inactive (idle) line. When the system finds an idle line, the line will be alerted (ringing) of the incoming call. Hunt lines are sometimes called rollover lines.

Hunting-A telephone call-handling feature that causes a transferred call to "hunt" through a predetermined group of telephones numbers until it finds an available ("non busy") line.

This diagram shows the process of hunting (also called "roll-over") for an available telephone. This diagram shows that an incoming call enters into the telephone switching system and is attempted to be delivered to the main telephone line extension (or dialed telephone number port). The switching system is programmed with a hunt list that allows the switching system to determine where to redirect a call if it is unable to deliver a call. This example shows that the system first tries 1001 that is off-hook (unavailable). The hunt group table shows

that the call should be routed to 1002. Because 1002 is also unavailable, the hunt group list instructs the switch to try extension 1003. This telephone (or port) is available and the call can be delivered (telephone rings).

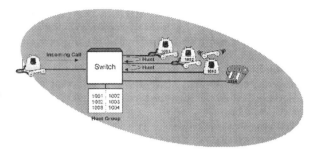

Hunting Operation

Hybrid-(1-Device)A device that combines transmit and audio signals from two-pairs of lines to one pair of lines. (2-Network)The combination of two different types of network technologies (such as fiber and coax) to form a combined (Hybrid) network.

This figure shows how a typical analog telephone transmission line operates. In this diagram, audio from customer #1 is converted to electrical energy by microphone #1. This signal is applied to the telephone line via the hybrid adapter #1. A portion of this signal is applied to the handset speaker to produce sidetone (so the customer can faintly hear what they are saying). This audio signal travels down the telephone line to hybrid #2. Hybrid #2 applies this signal to speaker #2 so customer #2 can hear the audio from customer #1. When customer #2 begins to speak, microphone #2 converts the audio to an electrical signal. This signal travels down the line to hybrid #1. Hybrid #1 subtracts the energy from microphone #1 (the combination of both signals are actually on the line) and applies the different (audio from customer #2) to the speaker #1.

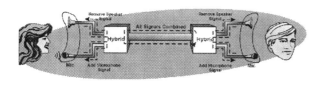

Hybrid Telephone Operation

Hybrid Backbone-A high-speed interconnection network that uses two (or more) types of transmission systems (such as SONET and Ethernet).

Hybrid Balance-The degree to which a hybrid termination prevents incoming energy on a four-wire circuit from being reflected in the opposite direction. Hybrid balance is measured in decibels of return loss, which is the difference between the levels of an input signal and its reflection.

Hybrid Cable-(1-cable television) A cable television communication system that uses a combination of optical fiber and electric conductors. (2-otpical cable) An optical cable that has multiple types of optical communication lines such as combining single mode and multiple mode optical fibers in the same cable assembly.

This figure shows a typical cable distribution system that uses a combination of fiberoptic cable and coaxial cable for the local connection. This diagram shows that the multiple video signals from the head-end of the cable television system is converted into digital form to allow distribution through high-speed fiber cable. The fiber cable is connected

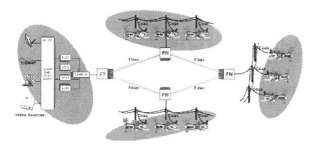

Hybrid Cable Operation

in a loop around the cable television service area so that if a break in the cable occurs, the signal will automatically be available from the other part of the loop. The loop is connected (tapped) at regular points by a fiber node. The fiber node converts the fiber signals into RF television signals that are distributed on the local coaxial cable network. The coax network distributes the RF signals to homes in the cable television network.

Hybrid Codec-A hybrid codec is a device or software that uses or combines two or more coding processes (such as waveform coding and voice coding) to compress (code) or expand (decode) information to a fewer number of bits for more efficient transmission and storage.

Hybrid Coil-A coil that is used as a bridge to connect a 2-wire circuit (combined transmit and receive) to a 4-wire circuit (separate transmit and receive pairs).

Hybrid Connector-A connector that contains both electrical conductors and optical fiber connections.

Hybrid Fiber Coax (HFC)-The hybrid fiber coax (HFC) system is an advanced CATV transmission system that uses fiber optic cable for the head end and feeder distribution system and coax for the customers end connection. HFC are the 2nd generation of CATV systems. They offer high-speed backbone data interconnection lines (the fiber portion) to interconnect end user video and data equipment. Many cable system operators anticipating deregulation and in preparation for competition began to upgrade their systems to Hybrid Fiber Coax (HFC) systems in the early 1990's.

Hybrid Integrated Circuit-A single integrated circuit that contains different types of components such as discrete (independent) components and integrated components on the same IC substrate (base).

Hybrid ITSP-An Internet telephone service provider that combines packet voice with switched voice services.

Hybrid Local Network-A local communication network that consists of more than type of local network. For example, a hybrid local network may consist of Ethernet and Token Ring.

Hybrid Loss-The transmission signal loss that occurs when a signal passes through a communication hybrid (such as a 2 line to 4 line converter)

Hybrid Set Top Box (HSTB)-A hybrid set top box is an electronic device that adapts multiple types of communications mediums to a format that is accessible by the end user. Hybrid set top boxes are commonly located in a customer's home to allow the reception of video signals on a television or computer. The use of HSTBs allows a viewer to get direct access to broadcast content from Terrestrial or Satellite systems in addition to accessing other types of systems such as interactive IPTV via a broadband network.

Hybrid System-A hybrid system is the combining of two or more systems that can accommodate multiple types of signals, physical channels, and services. Examples of hybrid systems include hybrid fiber coax (HFC) and combined analog and digital signal transmission.

Hybrid Transmission-Hybrid transmission is the merging of two transmission technologies onto one transmission channel. An example of hybrid transmission is the combining of color on black and white television onto the same television transmission system.

Hyperband Channel-The CATV channels from 37 through 59. These channels are normally down-converted to frequencies above the VHF band to alleviate the excessive signal loss that would result if UHF channels were transmitted on the cable. In order for viewers to uses these channels, they must use a converter or have a cable ready television receiver.

HyperLAN-HyperLAN is a wireless local area network (WLAN) specification overseen by the European Telecommunications Standards Institute (ETSI). It operates in the 5.7 GHz industrial, scientific, and medical (ISM) frequency band. HyperLAN provides for data transmission rates of up to 54 Mbps. HyperLAN uses the 802.3 Ethernet standard for it's fundamental structure.

Hyperlink-Hypertext Link

Hypertext-Hypertext is a syntax (structure) of text characters that allows the access of additional information within a document or display area. Hypertext is typically in the form of links on a web page that allows the reader to access additional or related information about the document.

Hypertext Link (Hyperlink)-Hyperlinks are tags, icons, or images that contain a crossed reference address that allows the link to redirect the source of information to another document or file. These documents or files may be located anywhere the link address can be connected to.

Hypertext Markup Language (HTML)-A text based communications language that allows formatting and item selection features to be transferred independent of the type of computer system. HTML is primarily used for Internet communication.

Hypertext Transfer Protocol (HTTP)-A protocol that is used to transmit hypertext documents through the Internet. It controls and manages communications between a Web browser and a Web server.

Hypervideo-Hypervideo is a video program delivery system that allows the embedding of links (hotspots) inside a streaming video signal. This allows the customer (or receiving device) to dynamically alter the presentation of streaming information. Examples of hypervideo could be pre-selection of preferred advertising types or interactive game shows.

Hyphenate-Hyphenate is when one is carrying out two or more roles in the production process.

Hysteresis-Hysteresis is a process in a system that can change when a point at which the change can occur has different start and stop change thresholds (inelastic behavior).

Hz-Hertz

I

I Frame-Intra Frame

I2R Loss-The amount of power that is lost as a result of the effect of current passing through resistance. The energy lost is transformed into heat.

IAB-Internet Architecture Board

IAD-Integrated Access Device

IAM-Initial Address Message

IANA-Internet Assigned Numbering Authority

IAP-Initial Alignment Procedure

IAP-Intercept Access Point

IAP-Internet Access Provider

IAPP-Inter-Access Point Protocol

IBA-Independent Broadcasting Authority

IBAC-In Band Adjacent Channel

IBG-Interblock Gap

IBOC-In Band On Channel

IBSS-Independent Basic Service Set

iBTS-Internet Base Transceiver Station

IC-Integrated Circuit

ICANN-Internet Corporation for Assigned Names and Numbers

I-Carrier-A system operating at one of the standard levels in Japan's digital hierarchy. Each digital signaling level supports several 64 kbps (DS0) channels. J-carrier was initially used in Japan and now is used throughout several parts of the world. The different digital signaling levels include;
- DS-1, 1.544 Mbps with 24 channels
- DS-1C, 3.152 Mbps with 48 channels
- DS-2, 6.132 Mbps with 96 channels
- DS-3, 32.064 Mbps with 480 channels
- DS-4, 397.200 Mbps with 5760 channels

ICE-Information and Content Exchange

Ice Loading-Ice loading is the pressure placed upon an antenna structure or cable by the formation of ice.

Ice Tray-A metal or plastic shield mounted over switch gear at the base of a wireless tower for it's protection from ice and other falling debris.

ICEA-Insulated Cable Engineers Association

IceCast-IceCast is a media streaming protocol that was developed by Nullsoft to stream Internet radio stations (MP3 files).

Icecast Protocol (ICY)-Icecast is an open source protocol that is used to manage streaming media.

ICMP-Internet Control Message Protocol

ICO-ICO Global Communications

I-Commerce-Internet Commerce

Icon-Icons are a small graphic or symbol that is linked to a software program or function. The purpose of the icon is to allow the user to quickly identify (find) their programs. Icons usually contain logos or unique product identification characteristics.

Iconoscope-An Iconoscope is an electronic camera tube that was created by Vladimir Zworykin in 1932. The Iconoscope converts images into electrical signals.

ICP-Intelligent Call Processing

ICQ-I See You

ICS-Intercompany Settlements

ICY-Icecast Protocol

iDEN-Integrated Dispatch Enhanced Network

Identity-An identity is a name, symbol or information element that uniquely identifies a person, device or service that can be recognized by a process or system.

IDF-Intermediate Distribution Frame

IDLC-Integrated Digital Loop Carrier

Idle Mode-A period of time that a mobile radio is not required to transmit or receive data. During the idle period, a mobile radio may measure the channel quality of other radio channels.

Idle Noise-Noise that occurs on an idle communication channel.

Idle URL-A URL address that is used by a device (such as an IP telephone) should the device come into an idle state. Using an idle URL allows for the display of customized status messages on device displays ("Lunch Special").

IDN-Integrated Digital Network

IDS-Intrusion Detection System

IDSB-Integrated Services Digital Broadcasting

IDT-Integrated Digital Terminal

IDT-Interdigitated Transducer

IDTV-Improved Definition Television

iDTV-Interactive Digital Television

IDU-Indoor Data Unit

IEC-Interexchange Carrier

IEEE-Institute Of Electrical And Electronics Engineers

IEEE 1394-IEEE 1394 is a personal area network (PAN) data specification that allows for high-speed data transmission (400 Mbps). The specification allows for up to 63 nodes per bus and up to 1023 busses. The system can be setup in a tree structure, daisy chain structure or combinations of the two. IEEE 1394 also is known as Firewire or I.Link.

IEEE 802.1-An IEEE specification that defines the interconnection of IEEE 802 LANs at the Data-Link Layer (layer 2 of the OSI reference model). The standard specifies the operations of bridge devices, including MAC address learning and aging, packet forwarding services and the Spanning Tree Protocol. Spanning Tree Protocol is used to prevent loops in bridged networks, so that packets are not continually and repeatedly forwarded between sides of the bridge.

IEEE 802.11-The IEEE standard that specifies the MAC protocol for commonly available wireless LANs that use radio frequency for transmission.

IEEE 802.14-An IEEE standard that specifies a MAC protocol two-way data communication over cable TV (CATV) systems. The standard has generally been supplanted by another industry standard, CableLabs' Data Over Cable Service Interface Specification (DOCSIS).

IEEE 802.17-An IEEE standard that specifies a MAC protocol for a metropolitan area network which utilizes fiber rings and transports packets. Commonly known as a resilient packet ring, the protocol is intended to provide better metropolitan area networking for enterprises using other IEEE 802 protocols, such as Ethernet.

IEEE 802.2-This standard applies to most IEEE 802 LANs, regardless of their topology. Commonly known as logical link control (LLC), it standardizes the commands and command formats for connectionless (type 1) or connection-oriented (type 2) service. These services define the sequence of communication between LAN nodes. LLC services were intended to provide assured delivery at layer 2 of the OSI reference model; however, in recent years and with the proliferation of TCP/IP, this service is often provided via TCP at layer 4.

IEEE 802.3-The IEEE standard that specifies the MAC protocol for a collision-based network commonly known as Ethernet. The 802.3 standard allows 10 Mbps or 100 Mbps data transmission rates. Because Ethernet can use CDMA/CD, users can share a network cable, however only one user can transmit data at a specific time.

IEEE 802.4-The IEEE standard that specifies the MAC protocol for a token passing network protocol which utilizes a bus topology. This protocol is popular in products used in manufacturing environments such as factories.

IEEE 802.5-The IEEE standard that specifies the MAC protocol for a token passing network protocol which utilizes a ring topology. This protocol was widely used by IBM products in the 1990's.

IEEE 802.6-The IEEE standard that specifies the MAC protocol for a metropolitan network based on a Distributed Queue Dual Bus (DQDB) protocol. With line rates of 1.5 to 155 Mbps, the protocol has rarely been used; more commonly ring topologies, such as FDDI are used instead.

IEEE standards-The Institute of Electrical and Electronics Engineers (IEEE) provides industry guidance and consensus building enabling various vendors to work together to produce interoperating products. The results are specifications, or standards, which corporations agree to abide by to produce interoperating equipment. One of the most influential bodies in the networking industry, the IEEE 802 committee is responsible for LAN and MAN standards, such as 802.3 Ethernet and 802.11 wireless LANs.

IETF-Internet Engineering Task Force

IF-Intermediate Frequency

IFF-Interchange File Format

I-Frame-Intra Frame

IGFs-International Gateway Facilities

IGMP-Internet Group Management Protocol

IGMP Snooping-Internet Group Management Protocol Snooping

IGP-Interior Gateway Protocol

IGRP-Interior Gateway Routing Protocol

IJ-Insulating Joint

IKP-Internet Keyed Payments Protocol

ILD-Injection Laser Diode

ILE-ISDN Line Emulator

ILEC-Incumbent Local Exchange Carrier

Illuminance-The amount of light that is received on a specific surface area.

Illuminated-An surface or geographic area to which a radio communications signal is directed and can be received.

ILMI-Interim Link Management Interface

ILS-Instrument Landing System

IM-Instant Messaging

IM-Intensity Modulation

IM-Intermodulation

IMA-Inverse Multiplexing Over ATM

Image-(1-radio) One of the sidebands that is produced by amplitude modulation. Sidebands produced in the modulation process are usually images of each other. One sideband is located above the carrier (center) frequency and the other sideband is located below it. (2-demodulation)An undesirable mixing product that is created in heterodyne mixing conversion in radio receivers. (3-graphic)Images are data files that organize their digital information in a format that can be used to recreate a graphic image when the image file is received and decoded by an appropriate graphics application.

Image Frequency-A frequency on which a carrier signal, when heterodyned with the local oscillator in a superheterodyne receiver, will cause a sum or difference frequency that is the same as the intermediate frequency of the receiver. Thus a signal on an image frequency will be demodulated along with the desired signal and will interfere with it.

Image Map-An image map is one or more predefined display areas on a web page that can be selected ("clickable") by a mouse. Image maps are commonly graphic images that are used to help users navigate through a web site.

Image Quality-Image quality is the ability of a display or image transfer system to recreate the key characteristics of an original image. Traditional analog image quality impairment measurements include blurriness and edge noise. Digital image and transmission system impairments include tiling, error blocks, edge business and object retention.

Imagery-Collectively, the representations of objects reproduced electronically or by optical means on film, electronic display devices, or other media.

Imaging-A process that may include converting an image into a different form (such as digital format), transferring, processing, storing, displaying, and reproducing (printing).

IMAP-Internet Message Access Protocol

IMD-Intermodulation Distortion

IMEI-International Mobile Equipment Identifier

Immunity-In telecommunications, a characteristic that permits equipment to operate normally when the equipment or any external lead or circuit is subjected to electromagnetic voltages, currents, or fields.

iMode-Internet Mode

i-Mode-Internet Mode

Impairment-(1-circuit) The loss of quality of service provided by an individual circuit when its transmission units are exceeded or signaling functions (such as seizure, disconnect, and automatic number identification) are experiencing intermittent failures. (2-signal transmission)Any distortion that affects a transmitted waveform or is perceivable by a video or audio user.

Impedance-(1-general) The total passive opposition offered to the flow of an alternating current. Impedance consists of a combination of resistance, inductive reactance, and capacitive reactance. It is the vector sum of resistance and reactance (R + jX) or the vector of magnitude Z at an angle. (2-characteristic) The impedance of a transmission line when it is infinitely long. (3-driving point) The input impedance of a transmission line. (4-input) The impedance looking into the input terminals of a device. (5-line) The impedance measured looking into a trans-mission line. (6-loaded) The impedance measured at the input of a device when the output is connected to its normal load. (7-mutual) The impedance between the primary and secondary windings of a transformer. Numerically, mutual impedance is equal to the secondary voltage divided by the primary current. (8- negative) The condition for an inductive circuit when an increase in applied voltage results in a decrease in current, and vice versa. (9-nonlinear) An impedance that varies with the applied voltage or with series current. (10-open circuit) The input impedance when the far-end terminals of a 4-terminal network are open. (11-output) The impedance looking into the output terminals of a device. (12-reflected) The impedance that, if added to the primary of a transformer, would change the primary current by the same amount as an identical impedance connected as the load on the transformer secondary (13-short circuit) The impedance at the input of a transmission line or network when the output is short-circuited. (14-surge) A quantity equal to the characteristic impedance, provided the resistance and leakage of the transmission line are negligible, compared with line inductance and capacitance. (15-terminal) The impedance looking into the terminals of a device.

Impedance Characteristic-A graph of the impedance of a circuit showing how it varies with frequency.

Impedance Matching-The adjustment of the impedance of adjoining circuit components to a common value so as to minimize reflected energy from the junction and to maximize energy transfer across it.

I

Implicit Rights-Implicit rights are actions or procedures that are authorized to be performed based on the medium, format or type of use of media or a product.

IMPP-Instant Messaging and Presence Protocol

Improved Definition Television (IDTV)-Improved definition television is a TV service as compared to existing analog television systems (e.g. NTSC) with improved picture quality resulting from better signal encoding, filtering and ghost cancellation.

Impulse-An impulse is a signal or energy surge that occurs in a short period of time.

Impulse Noise-A short burst of noise having random amplitude and bandwidth.

Impulse Pay Per View (IPPV)-Impulse pay per view is a service that allows a viewer to select a movie for immediate viewing without making previous arrangements to order the movie or service. IPPV fees are usually charged to the viewers account or credit card at the time the service is ordered.

IMS-IP Multimedia System

IMSI-International Mobile Subscriber Identity

IMT-InterMachine Trunk

IMT-2000-International Mobile Telephony 2000

IMTA-International Mobile Telecommunications Association

IMTC-International Multimedia Teleconferencing Consortium

IMTS-Improved Mobile Telephone Service

IMUX-Inverse Multiplexer

IN-Intelligent Network

In Band Adjacent Channel (IBAC)-In-band adjacent channel is addition of a new type of signal (typically digital) within the bandwidth of an adjacent radio channel (out-of-band) that is used to channel enhance or replace the existing channel when the adjacent channel signal is available. Use of IBAC allows existing radios to continue to receive program content and allows new radios to receive higher quality audio (digital audio) and additional program related content (such as artist title, song name, graphic icon, and other information).

In Band On Channel (IBOC)-In-band on channel is combination of multiple types of signals (typically analog and digital) within the bandwidth of an existing radio channel (in-band). There are two ways to convert a radio broadcast signal to digital transmission for IBOC. The first method adds the digital data to the existing analog signal before it modulates the radio signal and the second method adds a separate radio signal in an unused portion of the FM broadcast channel. Both of these system types allow the existing analog broadcast service to continue and are completely compatible with conventional AM or FM broadcasts. The advantage of using an IBOC system is that existing equipment and services can continue while new services and equipment can be introduced.

In Band Signaling-Signaling that occurs within the audio signal bandwidth. In band signaling while the conversation is in progress requires the users voice or data information to be momentarily interrupted or altered while signaling messages are being transferred. In band signaling is sometimes called blank and burst signaling.

Historically, the term "in-band" is related to the use of different frequency bands, but the term is also applied to signals in digital multiplexing that occur in the same time slot, better named "in-slot." Two examples of true in-band signaling are DTMF (touch tone) and MF. During the period of in-band signaling, the voice or data communication should be temporarily inhibited (muted) to allow the transfer of control messages without interference.

This figure shows how the basic process of in band signaling is used in analog cellular radio to deliver control messages sharing the same communication channel for voice and control signals. In this diagram, a radio base station desires to send a message to the mobile radio. The base initially sends a dotting sequence that indicates a synchronization word and message will follow. The mobile radio detects the dotting sequence. As a result, the mobile radio mutes the audio and begins to look for a synchronization word. The synchronization word is used to determine the exact start of the message. The mobile radio receives the message and upon completion of the message, the mobile radio will then un-mute the audio and conversation continues. Because the sending of the message can be less than ¼ second, the user may not even notice a message has been received.

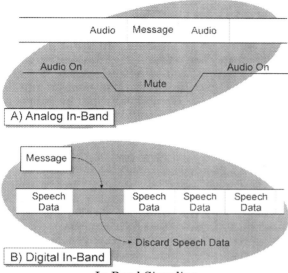

In Band Signaling

In Building Radiation Systems-In building radiation systems are supplementary systems that contain low power transmitters, receivers, indoor antennas and/or leaky coaxial cable radiators that are designed to provide radio communication services inside buildings.

In Building Service-A classification of communication services that may be provided in buildings. For the 3rd generation wireless systems, in buildings, service has the capability of providing data transmission rates of up to 2 Mbps.

In Home Accounts-In home accounts are records that contain identification codes, authorizations for each user id and possibly usage history for each user. Home media distribution systems may access and use in home accounts to determine the rights authorized for use and distribution of media along with any financial charges that may be required.

In Point-In point is an SMPTE time code that identifies a specific frame when a media clip begins.

In the Black-In the black is a saying used to portray ones beneficial income, financial success, or profits .

In the Red-In the red is a saying used to portray an individuals financial disadvantage, debts, or losses.

INA-Integrated Network Access

Inbound Call Center-A call center (group of customer service agents) that receives telephone calls from customers. Inbound call centers (teleservice centers) are often used in response to advertisements and direct marketing campaigns. The call routing to inbound call centers can be fixed (established telephone lines) or dynamically controlled (based on activity or skills based routing.)

Inbound Telemarketing-The communication of product or service information (marketing) to customers who have initiated a call to a call center. The call may be initiated from an advertisement, direct mailing, or other outbound marketing program.

Incarnation Number-A unique name or number sent within a data unit to avoid duplicate data unit acceptance.

Incident Report-An incident report lists the current unresolved problems (incidents) that have occurred within a communication system.

Incident Tracking-The process of recording and updating information about problems that occur in a communication system (incidents) and the steps taken to resolve the problem.

Incidental Churn -Incidental churn is a type of voluntary churn where the customer changes carriers because of changes in their situation or lifestyle, and where the change of carrier is only a secondary issue.

InCollects-Incollects are call detail records (CDRs) that are received by service provider A from service provider B for services provided by B to A's customers. An example of an Incollect is a roaming record.

Incoming Call Restriction-In telephone call-processing feature that disables a telephone from receiving incoming calls.

Incumbent-An existing system or service provider.

Incumbent Local Exchange Carrier (ILEC)-A telephone carrier (service provider) that was operating a local telephone system prior to the divestiture of the AT&T bell system.

InDate-Installation Date

Independent Broadcasting Authority (IBA)-A nonprofit public corporation in Britain that was established to set up and supervise commercial radio and TV operations. After the IBA builds and operates transmitter facilities, the IBS rents these services and facilities to those it authorizes.

Independent Software Vendor (ISV)-A vendor (company) that develops software that is independent of the computer hardware that the software operates on.

I

Index File-(1-general) An index file is a list of location pointers or reference values that are used to redirect software programs to the location of information for specific items such as media files or web pages. (2-database) A file that is used for indexing (reorganizing) the data in a database. Index files create a relationship between the record order of the database and the organization of values (sorted) desired by the user.

Indicator-Words, letters or numerals appended to and separated from the call sign during a radio station identification.

Indirect Modulation-Indirect modulation is the changing of a carrier signal through the use of a device or assembly that is external to the carrier signal source (such as an oscillator, LED or laser).

Indoor Data Unit (IDU)-An indoor data unit is part of a communication system that is located outside environmentally controlled areas. IDUs are typically constructed of lighter and low cost materials than outdoor units.

Indoor Splice Enclosure-An indoor splice enclosure is a plastic or metal container that is used to cover and protect wires or cables that are located inside a protective building or enclosed space. Indoor splice enclosures may contain multiple container shells. An outer shell may be used to provide mechanical and environmental protection and the inner shell may be used to hold the cables or a splice tray.

Induce-To produce an electrical or magnetic effect in one conductor by changing the condition or position of another conductor.

Induced Charge-An electrostatic charge produced on one body when it is brought near another charged body.

Inducement Agreement-An inducement agreement is a condition of agreement binding an individual to terms of his or her loan-out company.

Inductance-(1-general) The property of an inductor that opposes any change in a current that flows through it. The standard unit of inductance is the henry. (2-distributed) An inductance spread uniformly along a circuit or network. (3-mutual) The inductance between two circuits, which determines the electromotive force induced in one circuit by changes of current in the other circuit. (4-self) The property of an electric circuit analogous to the resistance to change. Self-inductance produces an electromotive force (erof) in a conductor that tends to oppose a changing current in the same conductor; the emf is proportional to the rate of change. (5-variable) A coil whose inductance can be varied.

Induction Coil-(1-electrical components in general) A single coil of insulated wire wound around a core volume. The core volume is typically in the form of a rod or a toroid (donut, bagel). The core material is typically air or a ferromagnetic material such as iron or ferrite. Also called an inductor. (2-in a telephone set)A multi-winding transformer used as part of a directional coupler device in a telephone set as a directional coupler to separate the microphone and earphone electric waveforms that both flow together in opposite directions on the subscriber loop. The so-called "hybrid coils" used at the central office directional coupler for 2-wire to 4-wire conversion are similar, but traditionally described by that distinct name.

Industry Standards-Industry standards are operational descriptions, procedures or tests that are part of an industry standard document or series of documents that is recognized by people or companies as having validity or acceptance in a particular industry. Industry standards are commonly created through the participation of multiple companies that are part of a professional association, government agency or private group.

Inert-An inactive unit, or a unit that has no power requirements.

Infinite Line-A transmission line that appears to be of infinite length. There are no transmitted signal reflections back to the source from the far end because it is terminated in its characteristic impedance.

INFO-Information

Information Access Service-A service offering that gives many telephone callers simultaneous access to selected prerecorded messages or databases furnished by private entrepreneurs known as information providers. This service also is called mass announcement network service.

Information and Content Exchange (ICE)-Information content exchange is an XML protocol that is used to describer and define content that is exchanged between networks and affiliated companies. ICE is used for automated delivery, content asset management and media backup. The ICE specification is governed by the international digital enterprise alliance (IDEAlliance). More information about ICE can be found at www.icestandard.org.

Information Bulletin-A message directed only to amateur operations consisting solely of subject matter of direct interest to the amateur service.

Information Content Provider-Any person or entity that is responsible, in whole or in part, for the subject matter of direct interest to the amateur radio service.

Information Facility-A facility whereby a data service user, by sending a predetermined address from the terminal installation, may gain access to general information regarding data communication services.

Information Highway-A term that refers to a common communication path (highway) for the transport of information. The world wide web (Web) is sometimes called the information highway.

Information Networking-The processing, delivery, and management of any type of information through public or private telecommunications networks.

Information Service-Information services involve the processing of information that is transferred through a communications system. Information services add value to information by generating, acquiring, storing, transforming, processing, retrieving, utilizing, or making available information via telecommunications. Examples of information services include fax store and forward, electronic publishing, text to voice conversion, and news services.

Information System-Information systems store, transfer, and process information for specific purposes. Information systems consist of hardware (usually computers) and software (data and applications) that add value to information by; generating, acquiring, storing, transforming, processing, retrieving, utilizing, or making available information via data and telecommunications connections. Examples of information systems include information storage, financial applications, order processing, web e-commerce, and engineering design.

Information Technology (IT)-Information technology is the processes or the study of processes used for information and data processing.

Information URL-A URL address that is used by a device (such as an IP telephone) should the user select an information button or icon. Using an information URL allows for the display of appropriate data should an information button be pressed or selected.

Information-Rich-Information Rich Products

Informs-Informs are synonymous with SNMP traps (from SNMPv1) or notifications (from SNMPv3). Events on network devices or SNMP agents send informs to the network management system when something of interest occurs on the agent. Informs were introduced in SNMPv2.

Infotainment-Infotainment is a media program type that provides information in a way that is entertaining to the viewer.

Infrared (IR)-Communication that uses non-visible infrared light (approximately 850-nanometers) for data transmission between communication devices. The data transmission rate can exceed 16 Mbps using IR transmission. Infrared signals must have an unblocked path between devices. Some wide area systems include reflection points that allow IR signals to reflect to reach their destination. The distance between many IR communication devices is commonly less than 3 feet.

Infrared Spectrum-Electromagnetic radiation in the wavelength range of 700 to 2000 nanometers, spectrally adjacent to the red portion of the visible spectrum at one end, and to the millimeter wavelength radio spectrum at the other. Sensed by a human being as heat, and not visible to the human eye. Used for optical fiber communication and by certain electronic equipment for short range cordless connections.

Infrared Transmission-Transmission of information through the use of a band of electromagnetic energy that is located between above microwave frequencies and below the low end of the visible part of the spectrum. Infrared communication may occur in free space or through an optical fiber. Data transmission rates for infrared transmission ranges from low speed data (e.g. for television remote controls) to high speed data (above 10 Mbps used for local area networks).

Infrasonic-Sound waves at frequencies too low to be heard by a human ear (below about 15 Hz).

Infrastructure-All parts of the communication systems, excluding the subscriber. This includes switches, radio carrier equipment, databases, and other network parts that enable telecommunications networks. In a traditional telephone network, the equipment includes the switches, multiplexers and other equipment used to manage the network and the copper cables that connect them together and to the users. In a cellular system, the equipment includes base stations and microwave links

I

as well as wireline equipment. In an optical network, infrastructure includes optical analogues to the traditional switches and multiplexers, as well as the optical fiber among them and connections to the traditional and cellular networks.

Infrastructure Addressing-Infrastructure addressing is the assigning of device addresses or names to network elements (e.g. routers) in a communication system.

Ingesting Content-Ingesting content is a process for which content is acquired, usually from a satellite downlink and loaded onto initial video servers (ingest servers). Once content is ingested it can be edited to add commercials, migrated to a playout server or played directly into the transmission chain.

Ingress-(1-signal) A process where a strong signal outside a communication system enters into a transmission line or system. (2-network) The process of data or signals entering into a communication network.

Ingress Filter-Ingress filtering is the process of restricting the types of connections or packets that are transferred in to a network.

Ingress Firewall-An ingress firewall is a device or software that is installed between a computer server or data communication device and a public network (e.g. the Internet) to restrict the flow of packets into a network.

Inhibit-A process or control signal that prevents a device, circuit, or system from operating.

Initial Alignment-A procedure that prepares a signaling link to carry signaling traffic either for the first time or after a failure.

Initial Alignment Procedure (IAP)-A procedure that is used when a signaling link is activated for the first time or when a link connection is restored after a communication failure.

Initial Period-(1-billing) The unit of time for billing at the beginning of each message, as defined by a tariff. (2-default) The initial (default) time required before normal processing can occur.

Initial Time Delay Gap (ITOG)-The elapsed time between a direct sound arrival at a listener and the first reflection of significant loudness.

Initialization-A process of setting the initial values and settings of the software and electrical components within a circuit or equipment is when it is first activated.

Injection-The application of a signal to an electronic device or assembly.

Injection Laser Diode (ILD)-A semiconductor diode in which lasing (emission of coherent light) takes place within the PN junction.

Inline Power Patch Panel-An inline power patch panel is used to insert power on the non-data lines in a data communications system (such as an Ethernet data network). Inline power patch panels connect to a data network. It allows for data transmission to transfer directly through it unaltered. The inline power patch panel is used to supply power to IP telephones and other data communication devices that can use these power pins. Only devices that are connected to the inline power patch panel receive power.

INMARSAT-International Maritime Satellite Organization

Inner Jacket-An inner jacket is an optional material used inside a cable to hold or separate cables or wires from each other within the cable. The inner jacket may also be used to protect or isolate cables from a strength member.

Innerduct-Inner Duct

Input-(1-signal) The waveform fed into a circuit, or the terminals that receive the input waveform. (2-data) Any data being sent to a computer from a user, another computer, or other equipment.

Input Impedance-The impedance presented at the input terminals of a circuit, device, or channel.

Input Looping-A circuit arrangement that permits the input of a device to be connected to another system downstream. The looping connector may or may not be buffered from the input signal.

Input Return Loss-The amount of signal that is returned from the circuit or device that is being supplied (return loss).

Input Selector-A device or routing switcher that is used to select a number of signals to the input of another device.

Input Transformer-A transformer at the input of a device or circuit to match the impedance of the device to that of the preceding stage, or to isolate the preceding stage from the device.

Inquiry-A process that requests specific information from a computer or communication device to determine its access code or availability for a communication session. Inquiry is a process in a Bluetooth system that is used to determine the address of other Bluetooth devices that are operating in the same area.

Inquiry Procedure-The inquiry procedure enables a device to discover other devices that are in range and determine the addresses and clocks for the device. After the inquiry procedure has been completed, a connection can be established using the paging procedure.

Inquiry Scan-A process that allows a communication device to listen (scan) for other devices that desire to discover its availability.

Insertion Gain-Insertion gain is the ration of the gain of a system (such as an optical system) with a component inserted as compared to the optical system without the optical component inserted.

An example of insertion gain is the signal level increase that results from the insertion of an amplifier in a transmission system, expressed as the ratio of the power delivered to that part of the system following the amplifier to the power delivered to that same part before insertion. If more than one component is involved in the input or output, the particular component is specified. This ratio usually is expressed in decibels. If the resulting number is negative, an insertion loss is indicated.

Insertion Loss-(1-general) The amount of signal loss that is caused by the insertion of a component or device, usually expressed in decibels (dB), as the ratio of input power to output power. (2-optical fiber coupler) The loss associated with that portion of the light that does not emerge from the nominally operational ports of an optical coupler device.

This figure shows that the basic process that is used to perform optical insertion loss testing. This diagram shows that the test path is normalized to determine a reference value. The cables that will be used for the testing are connected to the power meter and the Reference Level button is pressed. The device under test (DUT)is then inserted in the transmission line to determine. The new power level reading is the relative power level that represents the insertion loss of the device under test.

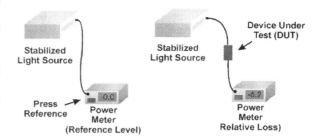

Optical Insertion Loss Testing

In-Slot-See: in-band

Inspection Lot-A collection of units of product from which a sample is drawn and inspected to determine conformance with acceptability criteria.

Installation Date (InDate)-The date on which a telecommunications link or equipment is installed, sometimes referred to as an InDate.

Installed First Cost-The installed cost of an outside plant item when it is placed in service.

Instant Messaging (IM)-A process that provides for direct connection between computers that are connected to a data communications network. Instant messaging (IM) service usually includes client software that is located on the communicating computers and an instant messaging server that tracks and maintains a list of alias names and their communication status. The IM server usually registers each client and links an address (usually an internet protocol address) so the clients can directly communicate with each other. The client software controls the presentation of information as it is sent directly between each computer.

This diagram shows the basic process used by instant messaging (IM) systems to allow IM users to directly communicate with each other. This diagram shows how IM server is primarily used as an address book for IM clients that want to directly communicate with each other. This diagram shows that IM clients sign on (register) and sign off (deregister) with the IM server each time they want to be able to send and receive messages to other IM members. The first step in this example shows that IM client "Buddy Steve" signs on with the IM server. The IM server captures the current Internet

I

address from "Buddy Steve" and sends back a list of current IP addresses for other buddies identified by "Buddy Steve." Next, IM client "Ready to Play" signs on with the IM server and receives the IP address of "Buddy Steve." Because "Buddy Steve" and "Ready to Play" now have stored the current IP addresses of each other, they can directly communicate with each other without having to go through the IM server. This communication can involve the sending of messages or the continual sending of data (such as packet voice.)

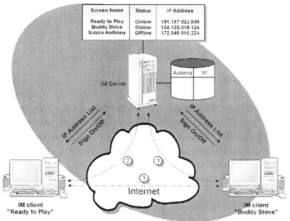

Instant Messaging (IM) Operation

Instant On-Instant on is the process of having a system or service starting the operation immediately (or perceived as immediate) after the system or service is requested. Examples of instant on is the ability of a computer to begin operating when it is turned on or the presentation of a media image or video as soon as it is requested.

Instantaneous Value-The value of a varying waveform at a given instant of time. The value may be expressed in volts, amperes, or phase angle.

Institute Of Electrical And Electronics Engineers (IEEE)-An organization formed in 1963 that represents electrical and electronics scientists and engineers. The IEEE resulted from the merger of the institute of Radio Engineers (IIRE) and the American Institute of Electrical Engineers (AIEE). Its has various societies that focus on key industry technical specialties (such as communications and robotics).

Instruction-A set of identifying characters designed to cause a processor to perform certain operations.

Instructional Television Fixed Station (ITFS)-A fixed station operated by an educational organization and used primarily for the transmission of visual and aural instructional, cultural, and other types of educational material to one or more fixed receiving locations.

Instrument Multiplier-A measuring device that enables a high voltage to be measured using a meter with only a low-voltage range.

Instrument, Scientific and Medical Band (ISM)-The industrial, scientific, and medical (ISM) band are frequency bands that are authorized for the use of instrument, scientific and medical radio devices. These bands of the electromagnetic spectrum include frequency ranges at 902-928 MHz, 2.4-2.484 GHz, and 5.725-5.825 GHz frequency bands which do not require an operator's license.

Insulate-The process of separating one conducting body from another conductor.

Insulating Joint (IJ)-Insulating joints (IJs) are openings in cable sheaths that are grounded on each side and installed in central office vaults to minimize electrolysis and to added protection from outside power surges. IJ's are located as close as possible to the second upright from the conduit entrance. The intent is to maximize the distance from the insulating joint to the first vertical bend in the cable.

Insulation-Insulation is material that holds and/or insulates a conductor (such as electrical wire) from other wires or components.

Insulator-(1-general) A material or device used to separate one conducting body from another. (2-guy) A double hole or loop shape. Insulator used in the guy-wire line of a tower or power pole. (3-stand off) A device, typically made of porcelain, used to support high-voltage conductors and components. (4-strain)A guy insulator used to provide separation between upper and lower sections of a guy-wire line.

Integrated Access Device (IAD)-A device that converts multiple types of input signals into a common communications format. IADs are commonly used in PBX systems to integrate different types of telephone devices (e.g. analog phone, digital phone and fax) onto a common digital medium (e.g. T1 or E1 line).

This figure shows an integrated access device (IAD) combines multiple types of media (voice, data, and video) onto one common data communications system. This diagram shows that three types of communication devices (telephone, television, and computer) can share one data line (e.g. DSL or Cable Modem) through an IAD. The IAD coordinates the logical channel assignment for device and provides the necessary conversion (interface) between the data signal and the device. In this example, the telephone interface provides a dialtone signal and converts the dialed digits into messages that can be sent on the data channel. The video interface buffers and converts digital video into the necessary video format for the television or set top box. The data interface converts the line data signal into Ethernet (or other format) that can be used to communicate with the computer. This diagram also shows that the IAD must coordinate the bandwidth allocation so real time signals (such as voice) are transmitted in a precise scheduled format (isochronous). The digital television signal uses a varying amount of bandwidth as rapidly changing images require additional bandwidth. The IAD also allocates data transmission to the computer as the data transmission bandwidth becomes available (whats left after the voice and video applications use their bandwidth).

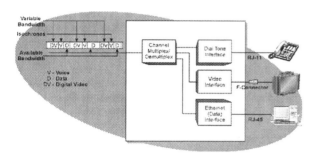

Integrated Access Device (IAD) Operation

Integrated Circuit (IC)-(1-general) An integrated circuit is an array of electronic components that may be active and/or passive which are integrated into a semiconductor substrate that are used to perform signal processing functions. (2-film) An integrated circuit whose elements are made of film materials. (3-hybrid) A hybrid integrated circuit is a signal processing device that is made up of several types or mixtures of components and/or integrat-

ed circuit types. (4-monolithic) A monolithic integrated circuit is a signal processing device that is composed of a single semiconductor substrate. (5-multichip) A multichip integrated circuit is a device or assembly that is constructed of two or more semiconductor chips that are attached to a single substrate or material.

Integrated Digital Loop Carrier (IDLC)-IDLC systems are the integration of the integrated digital terminal (IDT) and remote digital terminal (RDT). The IDT is part of the local digital switch (LDS) and it acts like a concentrator to put more channels on a digital communications line. The IDLC system moves some of the switching services from the local switches into RDTs to increase the efficiency of communication lines between customers and the central office.

BellCore's (now Telcordia Technologies) GR-303 specification defines the interconnection of the LDS and RDT.

This diagram shows how an integrated digital loop carrier (IDLC) system can be installed in a local telephone distribution network to allow a 24 channel T1 line to provide service up to 96 telephone lines. This diagram shows that a switching system has been upgraded to include an IDT and a remote digital terminal (RDT) has been located close to a residential neighborhood. The IDT dynamically connects access lines (actually digital time slots) in the switching system to time slots on the communications line between the IDT and RDT. The RDT is a local switch that can connect up to 96 residential telephone lines. When a call is to be originated, the RDT connects (locally switches) the residential line to one of the available channels on the DS1 interconnection line. The IDT communicates with the RDT using the GR-303 standard.

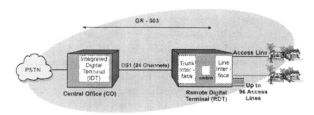

Integrated Digital Loop Carrier (IDLC) System

Integrated Digital Network (IDN)-A network in which connections carrying digital signals are established by digital switching systems. (See also: Integrated Services Digital Network.)

Integrated Digital Terminal (IDT)-An electronic assembly that is part of a local digital switch (LDS) that coordinates communication with a remote digital terminal (RDT). The IDT concentrates some of the communication channels onto high-speed digital lines that are routed to RDTs. The RDTs demultiplex the digital line and assign channels to individual access lines.

Integrated Market Planning -The process of combining market research, advertising, direct marketing, sales, and public relations activities into one, cohesive united operational plan.

Integrated Network Access (INA)-A network architecture for special services that extends digital carrier into a distribution loop to eliminate the need for an analog interface between distribution and interoffice facilities. The carrier terminates at either a remote terminal site or a customer's premises. Integrated Network Access eliminates the need for tandem digital-to-analog-to-digital conversions in a network.

Integrated Optics-A class of optical devices that perform two or more functions integrated on a single substrate.

Integrated Receiver and Decoder (IRD)-An integrated receiver and decoder is a device that can receive broadcast signals (such as from a satellite system) and has the ability to decode some or all of the encrypted or coded signals.

Integrated Service Unit (ISU)-The combination of a digital service unit (DSU) and channel service unit (CSU) into one device. The ISU interfaces a customer's digital telephone equipment to the formats used by a telephone network by adapting digital protocols, electrical levels and physical connections.

Integrated Services (Intserv)-A network in which multiple types of traffic can flow with different quality of service (QoS) requirements. The two types of services defined in Intserv are guaranteed service and controlled local service. Guaranteed service provides a specific bandwidth with a set maximum end-to-end transmission delay time. Controlled-load service provides a variable bandwidth for each communication session that varies based on factors including the amount of network activity (e.g. heavy traffic) and quality of service

requirements (e.g. real-time compared to non-real time communication application).

Integrated Services Digital Broadcasting (IDSB)-Integrated digital services broadcasting is the sending of television signals over digital transmission channels. ISDB transmission can be over different types of systems including broadcast radio, satellite systems, cable television systems and mobile communications. ISDB industry standards were created in Japan.

Integrated Services Digital Network (ISDN)-A structured all digital telephone network system that was developed to replace (upgrade) existing analog telephone networks. The ISDN network supports for advanced telecommunications services and defined universal standard interfaces that are used in wireless and wired communications systems.

ISDN provides several communication channels to customers via local loop lines through a standardized digital transmission line. ISDN is provided in two interface formats: a basic rate (primarily for consumers) and high-speed rate (primarily for businesses). The basic rate interface (BRI) is 144 kbps and is divided into three digital channels called 2B + D. The primary rate interface (PRI) is 1.54 Mbps and is divided into 23B + D for North America and 2.048 Mbps and is divided into 30B + 2D for the rest of the world. The digital channels for the BRI are carried over a single, unshielded, twisted pair, copper wire and the PRI is normally carried on (2) twisted pairs of copper wire.

This diagram shows the different interfaces that are available in the integrated services digital network (ISDN). The two interfaces shown are BRI and PRI. These are all digital interfaces from the PSTN to the end customers network termination. Network termination 1 (NT1) equipment devices can directly connect to the NT1 connection. Devices that require other standards (such as POTS or data modems) require a terminal adapter (TA). This example shows that the NT2 interface works with the NT1 interface to allow the application layers (terminal intelligence) to communicate with the ISDN termination equipment.

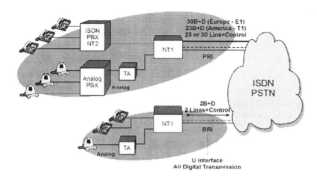

Integrated Services Digital Network (ISDN)
System

Integration-(1-component)The production of complete and complex circuits on a single chip, usually of silicon. (2-services)The combining of different services (such as voice, data, and video) onto a common communication system. (3-multiple networks)The process of interconnecting different types of networks through the use of portals or gateways.

Intellectual Property (IP)-Intellectual property is intellect that is produced that has some form of value. Intellectual property may be represented in a variety of forms and the copying, transfer and use of the intellect may be protected or restricted.

Intellectual Property Rights (IPR)-Legal rights that protect the innovations or creative content of proprietary to an individual, group or company. IPR is commonly associated with patent rights although it also applies to copyright and trademarks.

Intelligent Building-A building, often part of a commercial complex, in which local area networks, alarm circuits, and similar communications facilities are designed around a common communications strategy. Intelligent buildings are sometimes called smart buildings.

Intelligent Caching-Intelligent caching is a process by which information is moved to a temporary storage area to assist in the processing or future transfer of information to other parts of a processor or system and the decision to store the information is based on knowledge of the information type or expected future need for the information.

Intelligent Call Processing (ICP)-The dynamic processing of calls based on information obtained from or selected by the caller. This information can be used by automatic call distribution (ACD) systems to route calls to the next available agent, agents with specific qualifications, or other call centers.

Intelligent Modem-A modem that performs certain functions under computer control. As an example, an intelligent modem automatically dials a selected telephone number and hangs up when it detects a busy signal and may re-dial the same number. It is also called a smart modem.

Intelligent Network (IN)-A telecommunications network architecture that has the ability to process call control and related functions via distributed network transfer points and control centers as opposed to concentrating in switching system. (See. Advanced Intelligent Network.)

Intelligent Overlay System-An intelligent overlay system is the combination of a new system on top of an existing system where the new system has the ability to sense and interoperate with an existing communication system. The common reason for deploying an intelligent overlay system is to allow the introduction of new services without interfering with the operation of an existing system.

Intelligent Peripheral (IP)-A type of hardware that can be programmed to perform a new intelligent network capability in an SS7 network. IP's perform processing services such as interactive voice response (IVR), selected digit capture, and feature selection and account management for pre-paid services.

Intelligent Streaming-Intelligent streaming is the providing of a continuous stream of information such as audio and video content with the ability to dynamically change the characteristics of the streaming media to compensate for changes in the signal source, transmission or media application playing capabilities.

Intenret on Television (Internet on TV)-Internet on TV is the ability of a user to access the Internet through the use of television equipment.

Intensity-(1-general) Intensity is the strength or quantity of a given signal under specified conditions. (2-optics) The square of the electric field amplitude of a lightwave signal. Intensity is proportional to irradiance and may be used in place of the term irradiance when only relative values are important. (3-radio field strength) The radio field intensity is the strength of an electromagnetic wave radio signal at a particular point in space. (4-acoustic) The acoustic intensity is the sound energy per unit area at right angles to the propagation direction, per unit of time.

Intensity Modulation (IM)-A method of modulation used in fiber optic systems as a method of transmission in which an analog signal directly modulates the light source.

Inter Symbol Interference (ISI)-The appearance of part of a pulse symbol in a waveform appearing at an undesired time overlap with the following symbol(s). This is a result of undesired dispersion or time broadening of the pulses. ISI may be caused by multipath transmission where a single radio transmitted signal is reflected by objects (such as buildings) and part of the radio energy travels a path of different distance compared to another part of the signal (e.g. direct line of sight compared to reflected off a building or mountain). The interference results when the combined effect of multiple signals changes the decision points used to convert the radio signal back into its original digital form. ISI can be compensated by means of an equalizer. See: Dispersion, Equalizer.

Inter-Access Point Protocol (IAPP)-A protocol that is used to setup and manage connections between access points in a data communication network.

Interaction-The process by which a system accepts input, processes input requests, and (if necessary) returns appropriate response data to the originating terminal.

Interactive Advertising-Interactive advertising is the process of allowing a user to select or interact with an advertising message.

Interactive Communication-Interactive is the process of communication that involves exchange of information in two-direction to determine the content or information that is transferred between users and/or devices. Interactive communication usually requires real-time or near real-time communication to control the content that is being transferred.

Interactive Content-Interactive content is the information contained within a message, call, or web site display that is created after the user or visitor has selected or interacted with the system or web site.

Interactive Digital Television (iDTV)-Interactive digital television is media that can allow the user to interact with the media source in real time or near real time. An example of an iDTV service is television gaming.

Interactive Entertainment-Interactive entertainment is a process that provides a diversion or an amusement experience where information is exchanged in both directions.

Interactive On Demand Services-A service providing video programming to subscribers over switched networks on an on-demand, point-to-point basis, but does not include services providing video programming prescheduled by the programming provider.

Interactive Services-Interactive services use two-way communication to successfully satisfy the information requirements of the end customer. Examples of interactive services include conversational service and online gaming (real time), messaging services and video on demand (near-real time) and retrieval services (non-real time).

Interactive Television (ITV)-Interactive television has three basic types: "pay-per-view" involving programs that are independently billed, "near video-on-demand" (NVOD) with groupings of a single film starting at staggered times, and "video-on-demand" (VOD), enabling request for a particular film to start at the exact time of choice. Interactive television offers interactive advertising, home shopping, home banking, e-mail, Internet access, and games.

Interactive Video Services (IVS)-Interactive video services are processes that can automatically interact with a video viewer through the providing of audio and/or visual prompts to request information. The responses can be in the form of key presses, voice responses, or some other activity that can be sensed by the video delivery system. Interactive video responses are typically converted to digital form for processing the interactive video service platform.

Interactive Voice Response (IVR)-IVR is a process of automatically interacting with a caller through providing audio prompts to request information and store responses from the caller. The responses can be in the form of touch-tone(tm) key presses or voice responses. Voice responses are converted to digital information by voice recognition signal processing. IVR systems are commonly used for automatic call distribution or service activation or changes.

This figure shows a sample IVR system that is used to route an incoming call. When this call is received by the PBX system, an initial voice prompt informs the user of the system along with initial menu options. The user selects an option. This results in the playing of another prompt indicating new menu options. The user enters the data for the option and the IVR system retrieves data and creates a new verbal response.

Interblock Gap (IBG)-A signal that is inserted in

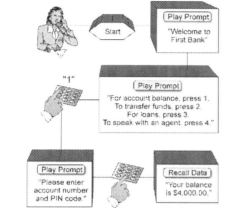

Interactive Voice Response (IVR) Operation

the helical track in the DAT format for the purposes of preventing adjacent track interference. This process also ensures that new data is written over the exact position of the previously recorded track during editing or re-recording.

Intercarrier Sound System-The method of transmitting the TV aural carrier at a fixed offset above the visual carrier, forming a composite signal. For NTSC, the separation of the two carriers is 4.5 MHz.

Intercept Access Point (IAP)-A place in a communication network where information and/or content is intercepted for the purpose of passing it to a law enforcement agency.

Interchangeable Numbering Plan Area (NPA) Codes-Codes in the format NNX previously used as central office codes but now available to supplement the supply of traditional numbering plan area (NPA) codes (NO/iX). (See also: Interchangeable Central Office Codes.)

Intercom Call-Intercom calls allows calls to directly communicate from one terminal toward another terminal. In a Bluetooth system, an intercom call between two terminals can be rapidly set up with gateway support if the two terminals are members of the same wireless user group (WUG).

Intercom Group-A group of telephone devices (the intercom group) that can be simultaneously accessed using an intercom feature. The use of the intercom group feature allow divisions or functional groups within a company to more effectively communicate with all the members of the group.

Intercompany Settlements (ICS)-Financial settlements made between carriers for usage of each other networks (such as for collect calls or roaming charges).

Interconnect-The term for manufacturers who provide telephone equipment for and on consumer premises.

Interconnect Cable-A short-distance cable (generally less than 3 m) intended for use between pieces of equipment.

Interconnection-Interconnection commonly refers to the connection of telephone equipment or communications systems to another network such as the public switched telephone network. Government agencies such as the FCC or department of communications usually regulate the interconnection of systems to the public switched telephone network.

This figure illustrates some of the different types of private to public telephone system interconnection. This diagram shows some groups of phone lines (e.g., dial line, Type 1) that provide limited signaling information (line-side) that primarily interconnect the PSTN with private telephone systems. Another group of lines (Type 2 series) are used to interconnect switching systems or to connect to advanced services (such as operator services or public safety services). The interconnection lines (trunk-side) provide more signaling information. Also shown is the type S connection that is used exclusively for sending control signaling messages between switching system and the signaling system 7 (SS7) telephone control network.

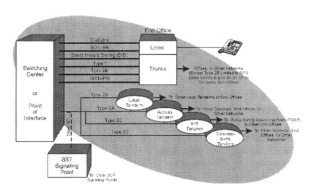

Private to Public Telephone System
Interconnection System

Interconnection System-Any system of interconnection facilities used for distribution of media or programs to telecommunication systems and/or companies.

Interdiction-Interdiction is the process of sending a signal that distorts or blocks the reception of another signal. An example of interdiction is the sending of a blocking signal through cable television networks that disables the reception of television channels (such as premium movie channels) that have not been ordered and paid for by the viewer.

Interexchange-Services and functions relating to telecommunication originating in a LATA and terminating else-where. In common usage, it is synonymous with the term interLATA.

Interexchange Carrier (IEC)-A telephone service company that provides long distance (interLATA, interstate, and/or international telecommunications service.

Interexchange Carrier (IXC)-Inter-exchange carriers (IXCs) interconnect local systems with each other. IXCs are also known as long distance carriers. In the US, from 1984 until 1997, IXC and LEC operating companies were legally required to refrain from engaging in directly competitive business operations with each other. Since 1997, one business entity can engage in both IXC and LEC business if it satisfies certain competitive legal rules. In Europe and throughout the rest of the world, the same PTT operators also usually provide inter-exchange service within their country. In any case, governments regulate how networks are allowed to interconnect to local and long distance networks.

For inter-exchange connection, networks as a rule connect to long distance networks through a separate toll center (tandem switch). In the United States, this toll center is called a point of presence (POP) connection.

Interface Equipment-The conversion equipment that enables circuits designed to one set of characteristics to communicate efficiently with circuits designed to meet different protocols and specifications.

Interference-(1-undesired signal) Radio interference is an undesired signal that interfere with the normal operation of other radio communication devices. In theory, interference sources can be turned off because it originates from a device that can, in principle, be turned off. An example is an undesired radio signal coming from another radio transmitter, or an undesired radio signal eliminating from a faulty neon sign or fluorescent light. Noise, in contrast, originates from fundamental physical mechanisms that cannot be turned off, such as the random thermal motion of electrons. Some authors lump the two types of undesired signals together under the name "noise," sometimes inappropriately. (2- combination of sine waves having the same frequency)This second meaning of the word "interference" typically occurs in the context of such phrases as "constructive interference," "destructive interference," or "interference and diffraction." These terms occur mostly, but not exclusively, in the context of radio waves and light. It describes the situation where (typically) two sinusoidal waveforms, having the same amplitude and frequency and a fixed phase relationship, are added together physically. Two sine waves that have a fixed phase relationship are described as mutually "coherent." If they exactly are in phase (zero degrees phase difference), the resulting signal has a higher power level than either sine wave alone (more precisely, double the amplitude or four times the power level of one sine wave alone). If they are not in phase, the resulting signal waveform has a lower power level, and in the special case where they are exactly out of phase (180 degree phase difference) the composite waveform has zero power. In a situation where the two sine waves have different phase difference at different places in space (because each sine wave has traveled a different distance from its own source location) the two sine waves will cancel in some locations and produce higher power (higher brightness for visible light) in other locations.

Interference Avoidance-Interference avoidance is a process that adapts the access channel sharing method so that the transmission does not occur on specific frequency bandwidths. By using interference avoidance, devices that operate within the same frequency band and within the same physical area can detect the presence of each other and adjust their communication system to reduce the amount of overlap (interference) caused by each other. This reduced level of interference increases the amount of successful transmissions therefore increasing the overall efficiency and increased overall data transmission rate.

This figure shows a PDA that uses WLAN FHSS can change its hopping pattern to avoid a video camera that is using multiple frequency channels. After detecting the presence of a continuous signal being transmitted by the video camera, the WLAN changes its frequency hopping pattern to avoid transmitting on the frequency band that is used by the video camera signal transmission. This results in more packets being successfully sent by the WLAN system and reduced interference from the WLAN to the transmitted video signal.

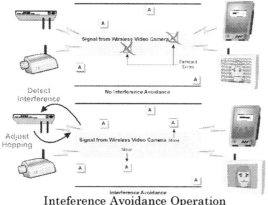

Inteference Avoidance Operation

Interference Canceling-Interference canceling is the process of extracting interference signals from a received signal and inverting and adding the interference signals to the received signals to reduce or cancel the effects of the interfering signals.

Interfering Source-An emission, radiation, or induction that is determined to be a cause of interference in a radio communications system.

Interframe Coding-Interframe coding is a video compression technique that tracks the differences between frames of video. Interframe coding results in more compression over a range of frames than intraframe coding.

Interframe Gap-The spacing between time-sequential frames.

Interim Standard (IS)-A designation of the American National Standards Institute—-usually followed by a number—that refers to an accepted industry protocol; e.g., IS-95, IS-136, IS-54.

Interim Standard 124 Data Message Handler (IS-124)-A standard billing communication protocol that allows for the real time transmission of billing records between different systems. IS-124 messaging is independent of underlying technology and can be sent on X.25 or SS7 signaling links. The development of the standard is primarily led by CiberNet, a division of the cellular telecommunications industry association (CTIA).

Interior Gateway Protocol (IGP)-IGP is an Internet protocol that is used to exchange routing information between switches and routers within the same domain network.

Interior Gateway Routing Protocol (EIGRP)-Enhanced Interior Gateway Routing Protocol (EIGRP) is an enhanced version of IGRP. The same distance vector technology found in IGRP is also used in EIGRP, and the underlying distance information remains unchanged. The convergence properties and the operating efficiency of this protocol have improved significantly. This allows for an improved architecture while retaining existing investment in IGRP

Interior Gateway Routing Protocol (IGRP)-IGRP is Cisco's interior gateway routing protocol used in TCP/IP and OSI internets. It is regarded as an interior gateway protocol (IGP) but has also been used extensively as an exterior gateway protocol for inter-domain routing. IGRP uses distance vector routing technology. The concept is that each router need not know all the router/link relationships for the entire network. Each router advertises destinations with a corresponding distance. Each router receiving the information adjusts the distance and propagates it to neighboring routers. The distance data in IGRP is represented as a composite of available bandwidth, delay, load utilization, and link reliability. This allows tuning of link characteristics to achieve optimal paths.

Interlaced-An image display that uses alternating graphic lines (e.g. odd and even) such as the lines of a television picture display during each picture scan.

This diagram shows how the lines displayed on each frame are interlaced by alternating the selected lines between each image frame. In frame one, every odd line (e.g. 1,3,5, etc) is displayed. In frame two, every even line (e.g. 2,4,6, etc) is displayed. In frame 3, the odd lines are displayed. This process alternates very quickly so the viewer does not notice the interlacing operation.

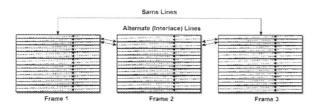

Interlaced Operation

Interlaced GIF-An interlaced GIF file contains multiple low resolution images that are combined by overlapping images to improve in resolution of the final image. Interlaced GIFs are used to allow a user to see a low resolution image while they are waiting for the entire high resolution image to finish downloading. This is particularly helpful when the Internet connection is limited (such as a dial-up connection).

Interlaced Image-An interlaced image contains multiple low resolution images that are combined by overlapping images to improve in resolution of the final image.

Interlaced Scanning-Interlaced scanning is the process of converting lines of images into other forms of information (such as converting a scene into line of video) where the scanning of lines are alternated between the scanning of an image (e.g. odd lines are sent first followed by even line scans). Interlaced scanning results in sending frames of information (images) at twice the rate using the same amount of transmission bandwidth. The use

of interlaced scanning in a video system helps to reduce the flicker effect.

Interlacing-(1-television) The process of mixing the scanning lines between successive images in a video system to reduce the flicker effect. (2-data transmission) The process of alternating the data bits of multiple communication channels.

interLATA-Telecommunication services that cross from a local access and transport area (LATA) into another LATA.

Interleaved File-An interleaved media file is the alternating (interleaving) of data (usually video and audio) over a sequence of data. The use of interleaved files simplifies the gathering and combining of multiple forms of synchronized media so it can be presented in combination to the viewer.

Interleaving-Interleaving is the reordering of data that is to be transmitted so that consecutive bytes of data are distributed over a larger sequence of data to reduce the effect of burst errors. The use of interleaving greatly increases the ability of error protection codes to correct for burst errors. Many of the error protection coding processes can correct for small numbers of errors, but cannot correct for errors that occur in groups.

This diagram shows that a block of data information may be distributed over multiple time slots or frames in a carrier line to distribute the effect of burst errors on the information signal. In this example, a block of digital audio is being transmitted through a radio channel. The digital audio is divided into blocks of 4 bits and the bits for each block are distributed (interleaved) over a communication channel. During the transmission, a lightning bolt creates a burst of electrical noise that disrupts 3 bits of data transmission. Because these

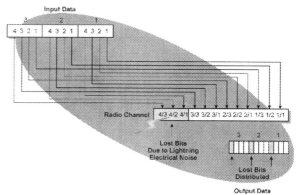

Interleaving Operation

bits are interleaved, the received data has burst errors that are distributed. This allows the audio to be continuously heard with a marginal amount of distortion instead of completely losing the audio during the burst errors.

Interlock-(1-circuit) A protection device or system designed to remove all dangerous voltages from a machine or piece of equipment when access doors or panels are opened or removed. (2-tape machines) The state of synchronous operation of two separate audio or video playback machines, achieved by a synchronizer that electronically compares the time code from the two machines and adjusts the speed of one or both until they are locked together.

Intermediate Distribution Frame (IDF)-An intermediate cross connect that is used at telephone company (telco) or riser closets to interface horizontals, laterals, risers, and/or feeders to other distribution cabling. IDF's can be 110, Krone, 66, or RJ45 patch panels.

Intermediate Frame-An intermediate frame is an image within a sequence of video images that is partially or completely created from related or adjacent frames. The creation of intermediate frames depend on receiving information from other frames.

Intermediate Station-Synonymous with internetworking device.

Intermittent-A non-non-continuous recurring event, often used to denote a problem that is difficult to find because of its unpredictable nature.

Intermittent Service Area-The area receiving service from the groundwave of a broadcast station but beyond the primary service area and subject to some interference and fading.

Intermodulation (IM)-The generation of unwanted radio or optical signal frequencies in a receiver or in a device such as a diode, or an optical medium such as glass, as a result of non-linear "mixing" of input radio or optical signals. This is typically caused when one or more input signal power level(s) is/are high enough so that the non-linear effects cause a mathematical multiplication of the two signals. This can be described mathematically by considering a very simple non-linear input-output relationship. Assume that the output of some non-linear device is proportional to the square of the input, $v = x$ squared. If the input x consists of two voltages (u+w), then when we square this we get three terms: u squared, 2 times the product of

u and w, and w squared. That middle term shows the mathematical product result, which produces IM.

Intermodulation is the same technological process as modulation, but it produces undesired output frequency components instead of desired frequency components. One scientist has compared this difference in terminology in the use of "modulation" and "intermodulation" to the two names "soil" and "dirt." If you are a botanist, and you prepare soil to grow plants in pots, that is a favorable name. But if the soil spills on an expensive carpet, you call it "dirt," an unfavorable name!

Intermodulation Distortion (IMD)-The distortion that results when mixing two input signals in a nonlinear system. The resulting output contains new frequencies that represent the sum and difference of the input signals and the sums and differences of their harmonics. IMD also is called intermodulation noise.

Intermodulation Noise-In a transmission path or device, the noise signal that is contingent upon modulation and demodulation, resulting from nonlinear characteristics in the path or device.

Internal Resistance-The actual resistance of a source of electric power. The total electromotive force produced by a power source is not available for external use; some of the energy is used in driving current through the source itself.

International Callback-International callback is a call processing service that reverses the connection of calls. International callback service is popular in countries that have high tariffs (fees) for outgoing (originating) international calls and have low tariffs for incoming (received) international calls. This process is divided into the call setup (dial-in) and callback stages. The international caller dials a number that provides access to the international callback service. This number may be local in the visited country or be an international number. The international callback gateway receives the call and prompts the caller to say or enter (e.g. by touch tone) the international number they desire to be connected to and the number they want the callback service to connect to. The international callback center then originate calls to both numbers and connects the two individuals to each other.

International Gateway Facilities (IGFs)-Systems or equipment that provide access between telephone systems in different countries. International gateways may convert SS7 and other

I

signaling formats between different signaling formats. These include ANSI standards, ITU standards, national variants of SS7 signaling standards, MF signaling, and R2 signaling. International gateways may also provide for transcoding services between mu-LAW PCM and A-LAW PCM speech coding.

This figure shows how two national SS7 systems interconnect using an international gateway between an ANSI based end office SSP in North America and an ITU based end office SSP in Asia. This example shows that an ANSI based SS7 system require address translation and circuit identifier code format changes as the messages are passed between the systems. The ANSI 24 bit destination point code (DPC) and origination point code (OPC) addressing must be translated to 14 bit DPC and OPC codes for the ITU system. It also shows that the 14 bit ANSI CIC code used in ISUP messages must be translated to 12 bit CIC codes used by the ITU system.

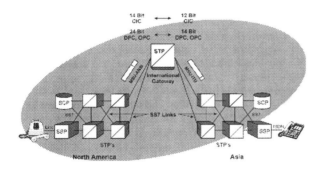

International Gateway Facilities

International Maritime Satellite Organization (INMARSAT)-An organization that was established by an international treaty to provide satellite services for wide area mobile communications. INMARSAT is based in London, England.

International Mobile Telephony 2000 (IMT-2000)-The name given to third generation wireless systems by the international telecommunications union (ITU).

International Radio Consultative Committee (CCIR)-Standards organization established by UNESCO originally having the French language name "Comité Consultatif International des Radiocomunications." As a result of a 1993 reorganization, it was supplanted by the International Telecommunication Union - Radio Sector (ITU-R or simply ITU). It is responsible for the study of technical and operating questions relating to radio communications. Its findings, if adopted as treaty agreements, have the effect of international law. (See also: CCITT, Consultative Committee for International Telephony And Telegraphy)

International Record Carrier (IRC)-A communication carrier that transfers data between international locations.

International Signaling Point-A signaling point which belongs to the international signaling network.

International Telecommunication Union (ITU)-A specialized agency of the United Nations established to maintain and extend international cooperation for the maintenance, development, and efficient use of telecommunications. The union does this through standards and recommended regulations, and through technical and telecommunications studies. Based in Geneva, Switzerland, the ITLI is composed of two consultative committees: the International Radio Consultative Committee (CCIIR) and the Consultative Committee for International Telephony And Telegraphy (CCITT).

International Telecommunications Union - T (ITU-T)-A sector of the International Telecommunications Union (ITU) that focuses on telecommunications.

International Telecommunications Union-Radiocommunication Sector (ITU-R)-The International Telecommunication Union - Radio sector (ITU-R) is responsible for studying and recommending technical and operating guidelines for radio communication. The ITU-R was formerly the International Radio Consultative Committee (CCIR).

International Toll Free Service (ITFS)-International toll free is a service that allows callers to dial and telephone number from other countries without being charged for the call. The toll free call is billed to the receiver of the call.

Internet (Net)-A public data network that interconnects private and government computers together. The Internet transfers data from point to point by packets that use Internet protocol (IP). Each transmitted packet in the Internet finds its way through the network switching through nodes (computers). Each node in the Internet forwards received packets to another location (another node) that is closer to its destination. Each node contains routing tables that provide packet forwarding information. The Internet evolved from ARPANET and was designed to allow continuous data communication in the event some parts of the network were disabled.

Internet Access-Internet access is the ability for a user or device to connect to the Internet. An example of Internet access is the requesting and connection to the Internet via a dial-up or broadband connection through an Internet service provider (ISP).

Internet Access Provider (IAP)-A company that provides an end user with data communication service that allows them to connect to the Internet. Internet access providers are also called Internet service providers (IAPs.)

Internet Address-An Internet address is a unique binary digital number that identifies a specific connection point within the Internet. An internet address is 32 bit for version 4 and 128 bits for version 6.

Internet Appliance-An Internet appliance is a computer or a device that is specially designed for communicating through Internet. When an Internet appliance is specifically designed as a user interface to the Internet, it is called an Internet terminal.

Internet Architecture Board (IAB)-A technical advisory group that is part of the Internet Society (ISOC) that manages Request for Comments (RFCs) publication standards and documents. The IAB also serves as an board to hear appeals and provides other services to the ISOC.

Internet Assigned Numbering Authority (IANA)-The Internet assigned numbering authority is a group that is responsible for the assignment and coordination of Internet addresses and key parameters such as protocol variables and domain names.

Internet Backbone-The main "pipes" that connect Internet service providers (ISPs) to one another. These interconnecting links typically operate at high-speed lines of 45 Mbps or above.

Internet Base Transceiver Station (iBTS)-The Internet base transceiver station (iBTS) is the radio access part of a wireless Internet network (typically for a cellular or wireless broadband system.) The iBTS includes the transmitters and receivers, antennas and towers that are used to communicate with mobile data devices. A BTS is connected to a packet control unit and radio network controller (RNC).

Internet Broadcasting-Internet broadcasting is a process that sends digital voice, data, or video signals simultaneously to a group of people or companies in a specific geographic area or who are connected to the Internet. Internet broadcasting is typically associated with online radio channels that send the same digital audio signal to many receivers who connect to a common server or group of servers.

Internet Codec-Internet coder/decoders (codecs) are information compression devices or processes that are used to convert information into a compressed (more efficient form) that can be sent through the Internet. Internet codecs may be designed to adapt to the varying data transmission rates and effects of packet losses that commonly occur to data that is transferred through the Internet.

Internet Control Message Protocol (ICMP)-A protocol that is used to report errors that occur during transmission of IP datagram packets. Routers send ICMP to the sender of the datagram to indicate when packets are undeliverable.

Internet Corporation for Assigned Names and Numbers (ICANN)-Internet corporation for assigned names and numbers is an international organization that is responsible for the issuing and managing of Internet addresses. The ICANN has assigned the responsibility of issuing and managing IP addresses and domain names to Internet registries.

Internet Engineering Task Force (IETF)-An organization that assists in the development and coordinates protocol standards that are used on the Internet.

Internet Explorer-Internet explorer is a software program created by Microsoft that is used to convert information that is available on the Web portion of the Internet into forms usable by a person (text, graphics and sound). Also called a web browser.

Internet Fax-The process of sending faxes through the Internet. Sending faxes through the internet can be performed unaltered (Internet is a bearer service) or it can interact using Internet fax protocol (IFP) to increase the reliability of complete fax information transfer.

Internet Group Management Protocol (IGMP)-Internet group management protocol is the commands, processes, and procedures that are used to send control messages and coordinate the multicasting (simultaneous distribution) of data through an Internet protocol network. IGMP is used to establish membership into a multicast group that is operating within a network. Using IGMP, users can inform routers within the network that they would like to receive media and control messages from a specific multicast group. IGMP is defined in RFC 1112 and 2236.

Internet Group Management Protocol Snooping (IGMP Snooping)-Internet group management protocol snooping is the process of looking inside packets for IGMP messages so that the router can update its multicast routing tables as group members are added or removed from the multicast distribution tree. Internet Group Management Protocol (IGMP) Snooping enables DSLAMs, PON Optical Line Terminals (OLTs), and routers to passively monitor subscriber traffic in order to identify and properly assign multicast group membership. Access platforms incorporating this feature check IGMP packets passing through, pick out the group registration information, and configure multicasting accordingly. Via IGMP snooping, multicast group traffic is only forwarded to ports servicing members identified as belonging

to that particular multicast group. IGMP snooping generates no additional network traffic, allowing carriers to reduce network congestion.

This figure shows how routers to perform IGMP snooping. This example shows that a router has IGMP snooping software that allows it to decode the contents of each packet as it passes through the router. The router stores the packet, decodes the packet header and determines if the packet contains an IGMP message. If so, the router can use the information in the IGMP message to update its multicast routing table.

Internet Keyed Payments Protocol (IKP)-A protocol that is used to process secure payments over the Internet. iKP uses the RSA public-key encryption system to process transactions of financial nature. iKP allows a buyer and a seller interact with a third party (such as a merchant processing company) to securely process transactions

Internet Message Access Protocol (IMAP)-A protocol that defines the access and storage procedures for electronic mail (e-mail) messages. IMAP is used with simple mail transfer protocol (SMTP) to move e-mail messages between email servers and mailboxes. IMAP is defined in RFC 2060.

The IMAP protocol defines message headers to allow the recipient to better search, select, and download specific messages or parts of messages. IMAP also includes security authentication procedures.

Internet Number-The dotted-quad address used to specify a certain system within the Internet.

Internet on TV-Internet on Television

Internet Paradigm-The sending of information over a common communication system (Internet protocol) as compared to sending information over separate dedicated networks (such as telephone, data network, and cable television systems).

Internet Phone (IPhone)-One of the first commercial Internet telephones that was introduced in February 1995. The IPhone product was developed by VocalTec Inc.

Internet Protocol (IP)-A low-level network protocol that is used for the addressing and routing of packets through data networks. IP is the common language of the Internet. The IP protocol only has routing information and no data confirmation rules. To ensure reliable data transfer using IP protocols, higher level protocols such as TCP are used. IP protocol is specified in RFC-791.

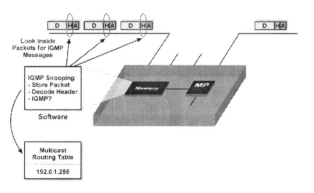

IGMP Snooping

This protocol defines the packet datagram that hold packet delivery addressing, type of service specification, dividing and re-assembly of long data files and data security. IP protocol structure is usually combined with high-level transmission control protocols such as transaction control protocol (TCP/IP) or user datagram protocol (UDP/IP).

Internet Protocol Address (IP ADDRESS)-The address portion of an Internet Protocol (IP) packet. For IP version 4, this is a 32-bit address and for IP version 6, this is a 128 bit address. To help simplify the presentation of IPv4 addresses, it is common to group each 8 bit part of the IP address is a decimal number separated from other parts by a dot(.), such as: 207.169.222.45. For IPv6 it is customary to represent the address as eight, four digit hexadecimal numbers separated by colons, such as 1234:5678:9000:0D0D:0000:5678:9ABC:8777.

This diagram shows how different types of data network addressing systems. This diagram shows an end-to-end data connection may transfer through many different networks and each network may use a different addressing system. This example shows that an end-user uses an Internet address to connect to a remote data device that has its own Internet address. The Internet address and its data is carried through the entire end-to-end communication in the data parts of each network packet. The first path connects the user to a company Ethernet network. The computers network interface card (NIC) that has a 48 bit address unique to the Ethernet. Each packet that travels in the company's Ethernet network has it's own

Ethernet address. Each packet of data from the end user includes the 32 bit Internet address. This packet (datagram) is encapsulated (stored) as part of the data message after the Ethernet address. The company's network is connected to an ISP by a high-speed frame relay connection. The frame relay access device (FRAD) has a unique identifier to the ISP. The ISP connects the data connection via asynchronous transfer mode (ATM) to the ASP.

Internet Protocol Audio (IP Audio)-Internet Protocol audio (IP Audio) is the representation of audio information in digital (discrete level) formats that are transferred using IP data packets (datagrams). The use of IP audio allows for more simple storage, processing, and transmission of audio signals through data networks.

Internet Protocol Broadcast (IP Broadcast)-A data packet that uses a frame address mask of 255.255.255.255 to identify it is intended for broadcast distribution. This allows devices within the network to identify broadcast messages and inhibits routers from constantly circulating packets through the network. The use of the address mask inhibits the normal transmission of the data packet through routers because routing protocols use the zeros at the end of the subnet mask number to identify the subnet. Because of the subnet mask (11111111.11111111.11111111.11111111 equals 255.255.255.255), the end of the address does not contain any zeros.

Internet Protocol Centrex (IP Centrex)-IP Centrex is the providing of Centrex services to customers via Internet protocol (IP) connections. IP Centrex allows customer to have and use features that are typically associated with a private branch exchange (PBX) without the purchase of PBX switching systems. These features include 3 or 4 digit dialing, intercom features, distinctive line ringing for inside and outside lines, voice mail waiting indication and others.

Internet Protocol Configuration (IPConfig)-An Internet Protocol application is used to program Internet Protocol configuration information such as the host IP address, gateway IP address, and subnet mask.

Internet Protocol Datagram (IP Datagram)-The packet of data that contains addressing header (IP header), its associated control header (such as a TCP header), and the data associated with the packet. The IP header contains addressing information (source and destination address) and basic

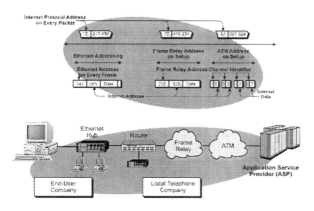

Internet and Network Numbering System

control information (such as time to live). Additional headers (such as a TCP header) contain control or additional routing information (such as a port number) related to a specific application for which the packet is related to. The data portion of the datagram has a variable length.

Internet Protocol Detail Record (IPDR)-A data record containing information related to an IP-based communication session. This information usually contains identification information of the users of the service, types of services used, quantity measurement unit type (e.g. kilobytes or time), quantities of services used, Quality of Service parameters, and the date/time (usually relative to GMT) the services were used.

Internet Protocol Header (IP Header)-The addressing portion of an IP packet that contains the routing and control information for the packet. The IP header is used by routers to determine the route the packet must be forwarded to help it reach its destination. The IP header contains the version number of the protocol (e.g. IPV4 or IPV6), source and destination address information, and other control (e.g. type of service) information.

This diagram shows the IP header field structure for version 4 IP. The IP header is located at the beginning of each IP datagram packet. The IP packet header starts with a version number that indicates which IP version is being used. The version indicates the field structure that is used for this packet (e.g. 32 bit addressing compared to 128 bit addressing). The Hlen field identifies the length

of the packet header. A service type field identifies that type of service that the IP packet is being used for. This type of service field allows for differentiated service (DiffServ) handling of packets that require real time (e.g. voice) or reliable (e.g. data) packet transfer. The total length field identifies the total length of the packet (with data). A time to live protocol field is used to allow the network to discard packets that travel through too many switching points preventing the possibility of infinite loops. Each packet contains the source IP address and the destination IP address.

Internet Protocol Next Generation (IPng)-A common term used for the next generation of Internet Protocol.

Internet Protocol Phone (IP Phone)-An Internet protocol phone (IP phone) is a device (a telephone set) that converts audio signals and telephony control signals into Internet protocol packets. These stand alone devices plug into (connect to) data networks (such as the Ethernet) and operate like traditional telephone sets. Some IP Telephones create a dialtone that allows the user to know that IP telephone service is available.

Internet Protocol Private Branch Exchange (IPBX) or (IP PBX)-A private local telephone system that uses Internet protocol (IP) to provide telephone service within a building or group of buildings in a small geographic area. IPBX systems are often local area network (LAN) systems that interconnect IP telephones. IPBX systems use a IP telephone server to provide for call processing functions and to control gateways access that allows the IPBX to communicate with the public switched telephone network and other IPBX's that are part of its network. IPBX systems can provide advanced call processing features such as speed dialing, call transfer, and voice mail along with integrating computer telephony applications. Some of the IPBX standards include H.323, MGCP, MEGACO, and SIP.

IP PBX represents the evolution of enterprise telephony from circuit to packet. Traditional PBX systems are voice-based, whereas their successor is designed for converged applications. IP PBX supports both voice and data, and potentially a richer feature set. Current IP PBX offerings vary in their range of features and network configurations, but offer clear advantages over TDM-based PBX, mainly in terms of reduce Opex (operating expenses).

IP Header	Data		

Vers	Hlen	Service type	Total length	
Identification			Flags	Fragment offset
Time-to-Live Protocol			Header	Checksum
Source IP address				
Destination IP address				
Options + padding				

IP Packet Header Version 4 Structure

Internet Protocol Security (IPSec)-A part of the Internet Protocol that helps to ensure the privacy of user data. IPSec is part of the next generation internet, IPv6. IPSec is defined in RFC 1827.

Internet Protocol Set Top Box (IP Set Top Box)-An IP Set Top box is an electronic device that adapts IP television data into a format that is accessible by the end user. The output of an IP set top box can be a television RF channel (e.g. channel 3), video and audio signals or digital video signals. IP set top boxes are commonly located in a customer's home to allow the reception of IP video signals on a television or computer.

Internet Protocol Suite-A combination of network protocols that have designed to interoperate with each other to provide a common data communication language that is used on the Internet. The layers of protocol suite include physical layer, network (or routing) layer, transport (or session) layer, and application layer.

This protocol suite is overseen by the Internet Engineering Task Force (IETF). Key protocols included in the Internet Protocol Suite include Internet Protocol (IP), Transaction Capabilities Protocol (TCP), and User Datagram Protocol (UDP). There are many other protocols that are part of the Internet Protocol suite.

Internet Protocol Telephony (IP Telephony)-IP telephone systems provide voice or multimedia communication services through the use Internet protocol (IP) networks. These IP networks initiate, process, and receive voice or multimedia communications using IP protocol. These IP systems may be public IP systems (e.g. the Internet), private data systems (e.g. LAN based), or a hybrid of public and private systems.

Internet Protocol Telephony (IPTEL)-A process of sending voice telephone signals over a data network (such as the Internet) using Internet protocol.

Internet Protocol Television (IPTV)-Internet protocol television (IPTV) is the process of providing television (video and/or audio) services through the use Internet protocol (IP) networks. These IP networks initiate, process, and receive voice or multimedia communications using IP protocol. These IP systems may be public IP systems (e.g. the Internet), private data systems (e.g. LAN based), or a hybrid of public and private systems.

Internet Protocol Version 4 (IPv4)-A revision of Internet protocol that uses 32 bit addressing.

Internet Protocol Version 6 (IPv6)-A network packet routing protocol that uses 128 bit address. IPV6 is an enhanced version of Internet protocol version (4 IPv4) that was developed primarily to correct shortcomings of IPv4 such as the 32 to bit address that limited the maximum number of devices that could be addresses and to extend the capabilities of IP to meet the demands of the future such as improved quality of service (QoS) capabilities. IPv6 addresses are denoted as 8 hexadecimal numbers, separated by colons. A typical address will look like this:

0800:5008:0000:0000:0000:1005:AABC:AD46

A short-hand notation that replaces one set of consecutive zeros with colons (::) may also be used. The above address can also be denoted by:

0800:5008::1005:AABC:AD46

IPv6 utilizes a hierarchical address, called an Aggregatable Global Unicast Address Format (AGUAF). In this format, each IP address is built by concatenating a Top-Level Aggregation (TLA) ID, a Next-Level Aggregation ID (NLA) and a Site-Level Aggregation ID (SLA) and an interface ID. This enables the IPv6 address space to be assigned in a logical manner by multiple address assignment authorities while still guaranteeing that all hosts have unique IP addresses and the addresses can be used to easily route packets without requiring switches to maintain enormous routing tables. The IPv6 header has been simplified with the introduction of extension headers. The basic IPv6 header contains just seven fields such as hop count and destination IP address. Also included is a Next Header field, which points to the next header in the packet. This greatly simplifies the logic required to parse the packet size in hosts and routers. IPv6 natively supports functions to discover neighbors, assign IP addresses dynamically and to identify multicast participants.

The most useful application of IPv6 is in next generation wireless phone networks. Many companies are moving to IP based networks to replace existing cellular technology. Over the next few years the need for IPv4 and IPv6 to co-exist will be an important factor in the deployment of IPv6 networks.

Internet Protocol Video (IP Video)-IP video is the transfer of video information in IP packet data format.

Internet Protocol Virtual Private Network (IP-VPN)-An Internet protocol virtual private network is a system that provides secure private communication path(s) through one or more public and/or private data networks that is dedicated between two or more points. VPN connections allow data to safely and privately pass over public networks (such as the Internet). The data traveling between two points is encrypted for privacy.

Internet Radio-Internet radio is the sending or broadcasting of digital audio signals through IP data networks (such as the Internet). Internet radios can be software programs that operate on multimedia computers or they can be dedicated devices (an audio player with a data plug instead of a radio antenna).

Internet Relay Chat (IRC)-Internet relay chat is an Internet protocol that was developed to allow for direct instant messaging (IM) between members of chat group.

Internet Reliability-The ability of the Internet to consistently provide data transmission between points that are connected to the Internet.

Internet SCSI (iSCSI)-A protocol used to build storage networks based on TCP/IP for transport. The protocol scales the SCSI command set used by direct attached storage so that it may be used on traditional networking technologies such as Ethernet.

iSCSI allows the creation of a storage network using familiar networking technology, while gaining the benefits of SANs.

Internet Service Provider (ISP)-An Internet service provider (ISP) is a company that receives and converts (formats) information to and from Internet connections to Internet end users. An ISP purchases a high-speed link to the Internet and divides up the data transmission to allow many more users to connect to the Internet.

Internet Signaling Transport Protocol (ISTP)-A signaling protocol that is used by PacketCable networks to provide SS7 type signaling capabilities.

Internet Society (ISOC)-An international organization assists the develop of Internet technologies and applications. The Internet society is composed of many companies that share an interest in the development of the Internet and services for the Internet. The ISOC also oversees and coordinates the activities of various Internet working groups. This includes the Internet Research Task Force (IRTF), Internet Architecture Board (IAB), the Internet Engineering Task Force (IETF), and the Internet Assigned Numbers Authority (IANA).

Internet Streaming Server-An Internet streaming server is a computer or a device that efficiently and effectively performs continuous transmission (streaming) of digital media through the Internet.

Internet Telephone (IP Telephone)-An IP telephone is a device that is specifically designed to communicate telephone signals through the Internet without the need for a voice gateway. Internet telephones contain embedded software that allows them to initiate and receive calls through the Internet using standard protocols such as H.323 or SIP.

Internet Telephony-Telephone systems and services that use the Internet to initiate, process and receive voice communications.

This figure shows how calls can be made between company telephones through the Internet to standard telephones anywhere in the world. In this example, an existing company PBX telephone system in Paris is connected to the Internet through a voice gateway. Each PBX telephone is registered with a public Internet telephone service provider (ITSP) that is located in New York. The ITSP is able to provide connections to gateways located throughout the world. In this example, when the caller in Paris dials a telephone number in Cairo, the dialed digits are first routed to the ITSP in New York. The ITSP server searches for the telephone number in its address list. If it finds that it has access to a voice gateway (the ITSP may not actually own the voice gateway) that is connected to the Internet near the destination telephone number in Cairo, it informs the destination voice gateway that an incoming call is to be received. The ITSP then provides the PBX gateway with the data network address of the destination voice gateway in Cairo. The call can then proceed from the PBX telephone, through the PBX voice gateway, through the Internet, through the destination voice gateway (in Cairo), to the dialed telephone.

Internet Telephony Service Provider (ITSP)-Internet Telephony Service Providers (ITSPs) are companies that provide telephone service using the Internet. ITSPs setup and manage calls between Internet telephones and other telephone type devices.

An ITSP coordinates Internet telephone devices so they can use the Internet as a connection path between other telephones. ITSPs are commonly used to connect Internet telephones or PC telephones to telephones that are connected to the public telephone network. This is accomplished by using gateways. Gateways convert packets of audio data from the Internet into standard telephone signals.

This figure shows how an ITSP sets up connections between Internet telephones and telephone gateways. The ITSP usually receives registration messages from an Internet telephone when it is first connected to the Internet. This registration message indicates the current Internet address (IP address) of the Internet telephone. When the Internet telephone desires to make a call, it sends a message to the ITSP that includes the destination telephone number it wants to talk to. The ITSP reviews the destination telephone number with a list of authorized gateways. This list identifies to the ITSP one or more gateways that are located near the destination number and that can deliver the call. The ITSP sends a setup message to the gateway that includes the destination telephone number, the parameters of the call (bandwidth and type of speech compression), along with the current Internet address of the calling Internet telephone. The ITSP then sends the address of the destination gateway to the calling Internet telephone. The Internet telephone then can send packets directly to the gateway and the gateway initiates a local call to the destination telephone. If the destination telephone answers, two audio paths between the gateway and the Internet telephone are created. One for each direction and the call operates as a telephone call.

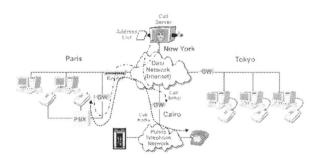

Calling Through the Internet

Internet Television-Internet television (also called broadband television) is the delivery of digital television services over broadband Internet connections. They may be able to control and guarantee the quality of television services if the underlying broadband connections have enough bandwidth. Internet service providers or media management companies usually provide unmanaged IPTV systems through broadband Internet connections.

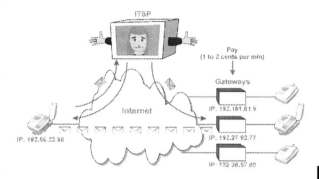

Internet Telephony Service Provider (ITSP) Operation

Internet Television Service Provider (ITVSP)-Internet Television Service Providers (ITVSPs) are companies that provide television or video services that connect through the Internet or other types of data networks. ITVSPs setup and manage television services between multimedia computers, televisions with adapters, or integrated IP television devices and media sources.

An ITVSP coordinates Internet television devices so they can use the Internet as a connection path between television media sources. ITVSPs are commonly used to connect end users to television content providers that use media gateways. Media gateways convert packets of audio data from the television source into packets that can be routed through data networks to end users.

Internet Terminal-An Internet terminal is a communication access device that is specifically designed to allow users to browse and communicate through the Internet.

Internet TV (iTV)-Internet TV is a Television service that is provided through the Internet.

Internet Video-Internet video is the transfer of video information through the Internet in IP packet data format.

Internet2-A second generation of the Internet that uses a high-speed backbone communications network. The Internet system is a result of the next generation Internet (NGI) initiative that is sponsored by the United States government. Internet2 is seen as the way to deliver multimedia content (e.g. video on demand) through the Internet.

Internetwork Operating System (IOS)-An operating system that is part of the CiscoFusion architecture that can provide centralized integrated, automated installation and management of Internet and intranet networks.

Internetwork Packet Exchange (IPX)-Internetwork packet exchange is a data communication protocol that is used to transfer data between the server and workstations within a Novell network. IPX is part of Novell's NetWare's protocol stack and it is a trademark of Novell. IPX data packets are encapsulated (both address and data are stored in the payload) and carried by the packets used in local area networks (such as Ethernet or Token Ring). IPX packet header contains 30 bytes of information that includes the address information (network, node, and socket addresses) of the source and the destination. It is followed by the data area that can vary in length from 0 bytes (only the header) to 65,535 bytes. Many networks limit the maximum packet size to approximately 1500 bytes.

Internetwork Protocol-The protocol used to move frames from originating source stations to their ultimate target destinations (through routers, if necessary) across an internetwork. IP, IPX, and DDP are all examples of internetwork protocols.

Internetworking Device-A device used to relay frames or packets among a set of networks (e.g., a bridge or router).

Interoffice Channel (IOC)-A communications link between two end office switching centers or between two points of presence (POPs) for interexchange carriers (IXCs).

Interoffice Facilities (IOF)-Communication lines or trunk (multi-channel lines) that interconnect switching systems. The facilities include the channel multiplexing and de-multiplexing equipment along with physical copper, wireless or fiberoptic lines.

Interoperability-The condition achieved among communications and electronics systems or equipment when information or services can be exchanged directly between them, between their users, or both.

Interoperability Testing-Interoperability testing is the performing of measurements or observations of a device, system or service to determine if the device will operate with other devices of a similar type or with devices that have been designed and tested to specifications (e.g. industry standards).

Interoperable/Interoperability-Interoperability is the capability of two or more devices or technologies to work (interoperate) with each other.

Interoperator Settlement-Interoperator settlement is the financial clearing process that is used to settle the usage charges for services they provide to visiting customers and the fees that other carriers have charged for services provided to their customer that have visited other systems.

InterPBX-The direct connection of calls between PBX switching systems through the use of dedicated connection lines (tie lines).

Interrupt-An event signal that is used to inform a processing system that suspension of the operation is required so another sequence of instructions can be processed. Interrupts are usually created by device under a processors control, such as an accessory device, that interrupts normal processing to gain rapid processing response. Interrupts usually cause software processing routines to branch from their current processing steps to an interrupt service routine (ISR).

Interrupts are sometimes classified to different types. These include internal hardware, external hardware, and software interrupts. Because there can be several types of interrupts in a system, interrupts can be prioritized. Interrupts used by Intel products allows up to 256 prioritized interrupts. Of these, the first 64 interrupts are reserved for use by the system or the operating system.

Interstitial Ad-An interstitial ad is a browser window that is opened for the display of an ad message "in between" the requesting and opening of new web page. Interstitial ads are also known as pop-ups (when opening a web page) and pop-downs (when leaving a web page).

Intersymbol Interference-Intersymbol interference is the effect of the overlapping of information symbols, usually due to the effects of dispersion.

Signal dispersion causes pulses to spread out so that they may overlap into adjacent time slots.

Intersystem Signaling-Control signaling that occurs between systems.

Interworking-The process of adapting the communications between two different types of networks. This may include circuit switched, packet switched or messaging services.

Interworking Function (IWF)-Interworking functions are systems and/or processes that attach to communications network that are used to process and adapt information between dissimilar types of network systems.

Interworking Profile-An interworking profile is the processes, protocol adaptations and call processing requirements that are used to adapt the operation of communication systems between two different types of networks.

Intra Frame (I Frame)-An intra-frame is an image in a motion video sequence that is independently coded and has no reference to any other frames. In MPEG other video codec such as H.263 and H.264, I-frames are encoded in a way similar to JPEG for still images.

Intra Frame (I-Frame)-Intra frames (I-Frames) are complete images (pictures) within a sequence of images (such as in a video sequence). I-frames are used as a reference to compare to other images.

Intraframe-An intraframe is an image in a motion video sequence that is independently coded and has no reference to any other frames.

Intraframe Coding-Intraframe coding is a video compression technique that is used to identify and compress image information within each individual frame.

Intranet-A private network that is used within a company to provide company information to employees. Intranets may be connected to vendors and customers through private data connections or via public Internet connections. When Intranets are connected to the Internet, they are commonly connected through firewalls to protect the company's internal data.

Intrastate Service-A broad category that includes all telecommunications services and/or activities that are covered by an intrastate or state tariff. Such services normally include state toll, local exchange, extended area service, multiple message unit, most optional calling service plans, and access services under state jurisdiction.

Intrinsic Loss-Intrinsic loss is the minimum loss of signal that is lost within a device (such as an optical coupler) that is caused by the internal properties of the device that connect and transfer the signals through the device.

Intro-Introduction

Introduction (Intro)-An intro is a video or media segment that is shown at the beginning of another video or media segment. An Intro is the opposite of an Outtro

Intrusion Detection System (IDS)-Intrusion detection system is the equipment and/or processes that are used to detect specific actions of an intruder who is attempting to gain access to a network or modify the normal operation of equipment. IDS may inform a management entity (such as a network operator) or it may disable specific operations of the equipment (self protection).

Intserv-Integrated Services

Invention-Date of Conception

Invention Disclosure-Most companies require an inventor to submit an "invention disclosure" describing an invention and providing supplementary information such as the project to which the invention is connected.

An invention disclosure is not a formal requirement in order to obtain a patent, but it does form part of the chain of evidence which can be used in the United States to prove date of invention.

Inventive Step-According to article 56 of the European Patent Convention, "an invention shall be considered as involving an inventive step if, having regard to the state of the art, it is not obvious to a person skilled in the art."

Inventive step is analogous to the US requirement that inventions be Non-Obvious.

Inventor's Notebook-An inventor's notebook is a bound, numbered, book containing information as to the conception of ideas and their subsequent development up to and including reduction to practice. A properly kept inventor's notebook can be useful in proving the date of invention in the United States.

Inventory Available Date-The date on which equipment and facilities are disconnected and made available for reuse.

Inverse Multiplexer (IMUX)-A device that divides a single telephone or data communication channel into two or more channels to be transported over multiple communication links. Inverse

I

multiplexing may be in the form of frequency division (e.g. multiple radio channels on a coax line), time division (e.g. slots on a T1 or E1 line) , or code division (coded channels that share the same frequency band) or combinations of these.

Inverse Multiplexing-The combining of information signals received on multiple communications channels to form a higher speed communication channel than is possible on a single independent communication channel. Inverse multiplexing has been used on wireless communication systems to allow high-speed digital video signals to be sent over cellular radio channels that have a limited maximum data transmission rate.

Inverse Multiplexing Over ATM (IMA)-A standard process of dividing a high speed data channel into multiple lower speed data channels (such as inverse multiplexing high-speed ATM channel over two or more T1 circuits).

Inverse Telecine-The process of transferring a video image to a film image.

Inversion-(1-signal) The change in the polarity of a signal or pulse, such as from positive to negative. (2-scrambling) A form of speech scrambling used to ensure the privacy of a transmission. A voice signal is mixed with a higher-frequency audio signal, and only the difference frequency is transmitted.

Invite Message-A message that is used to invite a person or device to participate in a communication session. The invite message is defined in session initiation protocol (SIP).

Invoice Record-A telecommunications data record that contains control counts or totals that describe an accompanying data set. Excluded are indexes and header and trailer information.

Invoice Statement-A financial statement that identifies the outstanding (unpaid) invoices. This figure shows a sample invoice. This invoice statement provides the customer account information (name, address, and account number), invoice charge totals, along with detailed billing information. This example shows that the customer pays recurring charges (monthly fees) plus additional charges such as taxes and communication costs that are not part of their rate plan. The detailed charges identify the category of the charge (rate), the amount of usage (time), and any additional charges (surcharges) that may apply.)

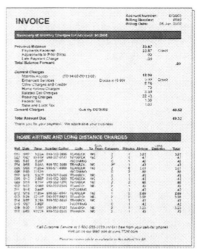
Invoice Statement

Invoke Component-A request message that is part of the transaction capabilities application part (TCAP) in the SS7 system that requests that an operation be performed at a receiving node.

Involuntary Churn-A disconnection of service that occurs when a carrier decides to disconnect service from an existing customer regardless if the customer wants to disconnect service or not.

IOC-Integrated Optical Circuit

IOC-Interoffice Channel

IOF-Interoffice Facilities

IOS-Internetwork Operating System

IOT-Interoperator Tariff

IP-Intellectual Property

IP-Intelligent Peripheral

IP-Internet Protocol

IP ADDRESS-Internet Protocol Address

IP Audio-Internet Protocol Audio

IP Backbone-IP backbones are the core infrastructure of a network that connects several major IP network components together.

IP Billing-The recording and processing of Internet protocol events for billing purposes.

IP Broadcast-Internet Protocol Broadcast

IP Centrex-Internet Protocol Centrex

IP Centrex System-A system that provides Centrex services to customers using Internet protocol (IP) connections. IP Centrex allows customers to have and use features that are typically associated with a private branch exchange (PBX) without the purchase of PBX switching systems. This figure shows a basic IP Centrex system that allows a local exchange company (LEC) in New

York City to provide Centrex services to a company in Los Angeles. In this diagram, the LEC in New York City uses a class 5 switch to provide for plain old telephone (POTS) and Centrex services to their local customers. The Centrex software is installed in the switch and existing Centrex customers in the local area continue to connect their telephone stations directly to the Class 5 switch. To provide Centrex services to new customers located outside the geographic area, the LEC has installed a network gateway in New York that can communicate with the customer gateway in Los Angeles. Because the network gateway converts all the necessary signaling commands to control and communicate with the customer gateway, the class 5 switch does not care if the customer gateway is in Los Angeles or Tokyo. It simply provides the Centrex services as the users request.

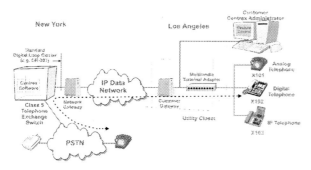

IP Centrex System

IP Centric-IP centric systems use Internet Protocol (IP) for the central part or core focus of their system.

IP Connectivity-The ability to setup communication sessions using Internet protocol (IP).

IP Convergence-IP convergence is the process of adapting one or more transmission mediums (such as radio packet or circuit data transmission) into Internet protocol data transmission formats.

IP Core-An IP core network is a central network portion of a communication system that uses Internet protocol as the basis of its operation. The core network primarily provides interconnection and transfer between edge networks.

IP Datagram-A packet of data that contains the ultimate network destination along with some control information and travels within (encapsulated) in the data portion of other network packets that are used to transport the IP datagram towards its ultimate destination.

IP Datagram-Internet Protocol Datagram

IP Demultiplexer-An IP demultiplexer is a device or system that used to separate two or more signals from one source (such as a satellite receiver) into multiple channels that use Internet protocol.

IP Enabled PBX-A non-IP PBX (such as a traditional TDM PBX) which allows for the support of IP phones or IP interconnect through the use of a special line or trunk card which converts TDM signals to IP signals. This is a common approach in migrating users of TDM PBXes to IP telephony technologies. This term often implies that IP to IP telephony audio may pass through a TDM switch.

IP Handset-Internet Protocol Handset

IP Headend-IP Television Headend

IP Header-Internet Protocol Header

IP Loopback Address-An Internet protocol address that is used to loopback data packets from another IP device.

IP Multicast-An Internet protocol that is used to broadcast the same message to multiple recipients. An IP multicast message is transferred to all the members within pre-defined group.

IP Multimedia System (IMS)-IP multimedia system (IMS) is an enhancement to the basic services for 3G mobile wireless systems that uses Internet protocol (IP) based systems to provide enhanced multimedia services.

IP Phone-Internet Protocol Phone

IP Precedence-IP Precedence utilizes the 3 precedence bits in the Type-of-Service (TOS) field in the IP header to specify class of service assignment for each IP packet. IP Precedence provides considerable flexibility for precedence assignment including customer assignment (e.g. by application) and network assignment based on IP or MAC address, physical port, or application. IP Precedence enables the network to act either in passive mode (accepting precedence assigned by the customer) or in active mode utilizing defined policies to either set or override the precedence assignment. IP Precedence can be mapped into adjacent technologies (e.g. Frame Relay or ATM) to deliver end-to-end QOS policies in a heterogeneous network environment. Thus, IP Precedence enables service

I

classes to be established with no changes to existing applications and with no complicated network signaling requirements.

IP Scrambling-IP scrambling is a process of altering or changing an IP signal to prevent interpretation of the signals by users that can receive the signal but are unauthorized to receive the signal. IP scrambling involves the changing of an IP signal according to a known process so that the received signal can reverse the process to decode the signal back into its original (or close to original) form.

IP Set Top Box-Internet Protocol Set Top Box

IP Telephone-Internet Telephone

IP Telephony-Internet Protocol Telephony

IP Television Headend (IP Headend)-An IP television headend is the part of an IP television system that selects and processes video signals for distribution into a IP distribution network. A variety of equipment may be included at the headend including satellite dishes to receive signals, content decoders and encoders, media servers and media gateways.

IP Television Service (IPTV)-IP television is the transmission of digital video and audio through data networks, usually through the Internet. IP television services may be on a subscription basis (paid for by the recipient) or may be funded by commercials or government agencies. IP television broadcasters transmit multimedia data signals to end users or to distribution points that redirect the digital television signals to end users.

IP Television Set-An IP television set s a viewing device that is specifically designed to video digital television signals through the IP data networks (such as the Internet) without the need for a signal conversion set top box. IP televisions contain embedded software that allows them to initiate and receive television through the data networks using standard protocols such as IGMP and SIP.

IP Tunnel-A logical path (a tunnel) between connection points through a data network using Internet protocol. This logical path allows data to freely flow between the connection points (entry and exit point) regardless of the underlying type of data or physical types of networks that are used to complete the tunnel connection (e.g. Ethernet LAN, ATM).

IP Video-Internet Protocol Video

IPBX or IP PBX-Internet Protocol Private Branch Exchange

iPBX System-Internet protocol private branch exchange (IPBX) systems use Internet protocols to provide voice communications for companies. IPBX systems can be separate from data network or they may share the data network systems. When the iPBX system shares the local area network (LAN), it may be called LAN Telephony or TeLANophy.

IPC-ISDN To POTS Converter

IPConfig-Internet Protocol Configuration

IPDR-Internet Protocol Detail Record

IPhone-Internet Phone

IPng-Internet Protocol Next Generation

IPNS-International Private Network Service

IPNS-ISDN PBX Network Specification

IPPV-Impulse Pay Per View

IPR-Intellectual Property Rights

IPSec-Internet Protocol Security

IPTEL-Internet Protocol Telephony

IPTV-Internet Protocol Television

IPTV-IP Television Service

IPTV Distribution Equipment-IPTV distribution equipment is the hardware (physical products) that route, switch or temporarily store and forward IPTV signals between the head end and the customer's IPTV equipment (such as a set top box). IPTV distribution includes IPTV DSLAM (with IGMP and IP Multistream capability), video distribution servers, IPTV edge routers and video switches.

IPTV Ecosystem-The IPTV ecosystem is interrelationship of systems, services and business processes that influence the consumer needs and the ability to deliver voice, data and video services over IPTV networks. The IPTV industry is characterized by a high degree of flexibility to the operator in terms of such components of the end-to-end system from video servers and encoders, through network delivery, middleware and conditional access systems. In addition to the network distribution, the IPTV provider has multiple home distribution choices including coax, cat-5, power line, phone line or wireless. Each ecosystem can be tailored for the requirements of the region or operator to enable the level of functionality, performance and security that is appropriate.

IPv4-Internet Protocol Version 4

IPv6-Internet Protocol Version 6

IP-VPN-Internet Protocol Virtual Private Network

IPX-Internetwork Packet Exchange

IR-Infrared

IR Loss-The conversion of electric power to heat caused by the flow of electric current through a resistance.

IRC-International Record Carrier
IRC-Internet Relay Chat
IRD-Integrated Receiver and Decoder
IREG-International Roaming Experts Group
IS-Interim Standard
IS-124-Interim Standard 124 Data Message Handler
IS-41-Intersystem Signaling 41
IS-826-Interim Standard 826
ISBN-International Standard Book Number
ISCP-Integrated Services Control Point
iSCSI-Internet SCSI
ISDN-Integrated Services Digital Network
ISDN S Interface-An ISDN S interface is a 4 wire digital interface between end user terminal equipment (e.g. telephones and fax machines) and the network termination point in the building. Two wires are used for a transmit signal and 2 wires are used for receiving signals. The S interface allows up to 7 devices to share the S interface communication line. The S interface is the multi-device version of the T (single device) interface.
ISDN To POTS Converter (IPC)-A device that converts an ISDN basic rate interface (BRI) into a POTS analog telephone interface.
ISDN U Interface-An ISDN U Interface is a 2 wire digital subscriber loop that transports a duplex ISDN 160 kbps digital signal between the ISDN central office and the termination point.
ISI-Inter Symbol Interference
ISM-Instrument, Scientific and Medical Band
ISO-International Standards Organization
ISO 9000-Quality management standards that are defined by the international standards organization (ISO) to help companies define and improve their production and service processes. These standards require companies to define their own product and service processes and demonstrate to independent auditors that their processes provide the ability to track the quality of the products or services they provide.
ISOC-Internet Society
ISOC-Isochronous
Isochronous (ISOC)-A communication process that sends data between communication devices in continuous form (equal transmission time for all data). Isochronous signals are used in systems that require continuous data to be sent at specific time intervals (such as digital audio communication systems).

Isochronous Devices-Devices that transmit at a regular interval, typified by time-division voice systems.
Isochronous Ethernet (IsoEnet)-A system that allows Ethernet systems to co-exist with isochronous systems (such as T1 or E1 lines). The IsoEnet consortium was formed in 1992 and is specified in IEEE 802.9a.
Isochronous User Channel-The channel used for time-bounded information like compressed audio.
IsoEnet-Isochronous Ethernet
Isolated Cable-A cable that is energized or fed with "live" Central Office facilities via drop wire. The cable's grounding integrity from the CO is interrupted and isolated and requires a special bonding and grounding process that utilizes either conventional ground rod or counter-poising methods.
Isolated Ground Plane-A set of connected circuit frames that are grounded through a single connection to a ground reference point. That point and all parts of the frames are insulated from any other ground system in a building.
Isolator-An RF device that allows radio signals to pass through in one direction but it does not allow signals to pass through in the opposite direction. An isolator is used to protect a transmitter assembly from reflected signals.
This diagram shows how an isolator allows a signal to pass through in one direction and restricts the signal flow in the opposite direction. When a signal is reflected from the antenna back towards the transmitter, the reflected signal is absorbed by the isolator.
Isophasing Amplifier-A timing device that corrects for small timing errors.
Isotropic-A quantity exhibiting the same properties in all planes and directions.
ISP-Internet Service Provider
ISSN-Internatinoal Standard Serial Number
Issuing Carrier-A service provider (carrier) that files its own tariff with a regulatory agency for a given service offering.
ISTP-Integrated Signal Transfer Point
ISTP-Internet Signaling Transport Protocol
ISU-Integrated Service Unit
ISUP-ISDN User Part
ISV-Independent Software Vendor
ISW-Inbound Service Word
IT-Information Technology

ITAD-Internet Telephony Administrative Domains

Iterative Loop-A repeated group of instructions in a software routine.

ITFS-Instructional Television Fixed Station

ITFS-International Toll Free Service

ITOG-Initial Time Delay Gap

ITSP-Internet Telephony Service Provider

ITSP System-Internet Telephony Service Providers (ITSPs) are companies that provide telephone service using the Internet. ITSPs setup and manage calls between Internet telephones and other telephone type devices.

An ITSP coordinates Internet telephone devices so they can use the Internet as a connection path between other telephones. ITSPs are commonly used to connect Internet telephones or PC telephones to telephones that are connected to the public telephone network. This is accomplished by using gateways. Gateways convert packets of audio data from the Internet into standard telephone signals.

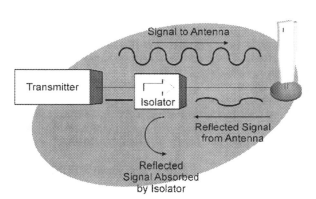

Isolator Operation

ITU-International Telecommunication Union

ITU-R-International Telecommunications Union-Radiocommunication Sector

ITU-T-International Telecommunications Union - T

ITV-Interactive Television

iTV-Internet TV

ITVSP-Internet Television Service Provider

IUA-ISDN User Adaptation Layer

IUM-Impacted User Minutes

IVHS-Intelligent Vehicle Highway System

IVR-Interactive Voice Response

IVS-Interactive Video Services

IWF-Interworking Function

IXC-Interexchange Carrier

J

J-joule
J2ME-Java 2 Micro Edition
J2ME-Java Version 2 Mobile Edition
Jack Field-The location of jacks or jack strips on a piece of equipment, or on a panel or panels in a central office.
Jacket-A jacket is a layer of material surrounding wires or fibers inside a cable. A jacket may be used to insulate the conductors and/or to hold conductors and materials together inside a cable assembly.
Jacket Slitting-Jacket slitting is the process of cutting a cable along its side to allow access to its inner conductors. The jacket slitting device includes a blade that may have an adjustable depth to allow the cutting of the outer jacket without cutting any of the inner cables or conductors. After the outer jacket has been slit (cut) down the side, the pull rope or inner conductors are pulled out of the slit. The outer jacket is then removed by cutting around the outer jacket where the slit ends.
JAE-Java Application Environment
Jamming-Jamming is the creation of interference signals to prevent the reception of channels or programs. Jamming on television or radio channels may be performed by countries who desire to stop or interfere with signals that are being broadcasted by foreign countries.
Java-An object-oriented programming language that works with a wide variety of computers. Created by Sun Microsystems, Java adds animation and interactivity to Web pages, granted you to have a Java-enabled browser. Java technology is a portable, object-oriented language that is well suited for web-based (Internet) and platform-independent applications. Java is a high level language and is architecturally neutral as it can operate on almost any underlying operating system.
Java 2 Micro Edition (J2ME)-Java 2 Micro Edition suns Java platform for developing applications for various consumer devices, such as set-top boxes, Web terminals, embedded systems, mobile phones and pagers.
Java Application Environment (JAE)-The development of applications using the Java programming language and its development kits.

Java Script-Java Script is a scripting language that is embedded in the HTML code. It is used in web pages designed to be viewed by PC-based browsers.
Java Telephony Application Programming Interface (JTAPI)-Java telephony application programming interface is an industry standard that defines the application interface between computers and telecommunications devices based on the Java programming language. The JTAPI standard allows computers to control private telephone systems (such as PBX systems).
Java Version 2 Mobile Edition (J2ME)-A compact version of Sun's Java technology targeted for embedded consumer electronics.
Java Virtual Machine (JVM)-Java virtual machine (JVM) is a software program that operates on a computer that allows the computer to use java instructions to request, transfer and process information.
JavaScript (Jscript)-Javascript is a scripting language designed to allow the development of active online content on Web severs (web sites). Javascript was created by Netscape Communications and Sun Microsystems that allows developers to add a specific information processing capabilities Web pages.
JavaTV-JavaTV is an industry middleware standard used in television systems to allow the user to access additional interactive services such as Internet browsing and electronic programming guides. JavaTV is developed by Sun Microsystems and more information about JavaTV can be found at www.java.sun.com/products/javatv/.
JCL-Job Control Language
JDC-Japanese Digital Cellular
Jeopardy-A condition resulting from any schedule change that is likely to cause a service request to be completed later than a committed due date. Failure to update the status of an order on or before a critical report date can result in a jeopardy condition. The term applies to both installation and repair jobs.
Jerkiness-Jerkiness is holding or skipping of video image frames or fields in a digital video.
JF-Junction Frequency

Jini-A connection technology developed by Sun Microsystems that provides simple mechanisms which enable devices to plug together to form an impromptu community. -The Jini system allows each device within a community system to provide services that other devices in the community may use without any required planning, installation, or human interaction. These devices provide their own interfaces, which ensures reliability and compatibility. Jini works at high (application and session layer) protocol layers.

JIT-Just In Time

Jitter-(1-general)Jitter is a small, rapid variation in arrival time of a substantially periodic pulse waveform resulting typically from fluctuations in the wave speed (or delay time) in the transmission medium such as wire, cable or optical fiber. When the received pulse waveform is displayed on an oscilloscope screen, individual pulses appear to jitter or jump back and forth along the time axis. (2-packet)The short-term variation of transmission delay time for data packets that usually results from varying time delays in transmission due to different paths or routing processes used in a packet communication network. (3-IP Telephony)The variance of interpacket arrival times.

Jitter Buffer-The jitter buffer receives and adds small amounts of delay to packets so that all the packets appear to have been received without varying delays. Jitter buffers allow for the smoothing out of digital audio signals that experience variable transmission delay across a network (such as the Internet).

This diagram shows how a jitter filter can remove the variable transmission delay for packets that experience variable transmission time through a packet switched network. This diagram shows that packets are delayed variable amounts (delay 1-3). The jitter filter receives the packets and stores the packets in memory until a specific start time. The jitter filter has a clock that provides a specific start times for the transmission of the pulse. This fixes the amount of delay to an anticipated maximum amount.

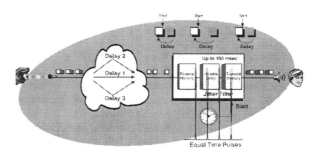

Jitter Filter Operation

Job Control Language (JCL)-A computer language that links an operating system and applications programs in order to define processing jobs and executable programs, and to provide for a job-to-job transition.

Jog-Jog is a movement forward or backward of a media segment or film clip.

Jog-Jogging

Jogging (Jog)-The process of moving a videotape forward or backward one field or frame at a time.

Join Message-A join message is a connection request from a user or device to join the service (such as a multicast session) or to access other types of shared or group resources. A join message is passed along router connections between the intended recipient and the source to determine if they should connect the member to the multicast service at their location or whether they should forward the join message to other routers.

Joint Access Costs-The costs associated with network access facilities that are used for services from two (or more) access systems (such as local and long-distance connections.) Included are costs for basic termination, installation labor, inside wiring, drop wire, a subscriber loop, and all non-traffic-sensitive equipment in a local central office. Joint access costs include those that are incurred regardless of whether a customer makes a call. Also included are those expenses that vary with the number of customers rather than the amount of use, such as commercial, directory monthly billing, and testing.

Joint European Standards Institute-The Committee for European Standardization/Committee for European Electrotechnical Standardization

Joint Photographic Experts Group (JPEG)-
JPEG is a working committee under the auspices of the International Standards Organization (ISO) with the goal of defining a standard for digital compression and decompression of still images for use in computer systems. The JPEG committee has produced an image compression standard format that is able to reduce the bit per pixel ratio to approximately 0.25 bits per pixel for fair quality to 2.5 bits per pixel for high quality.

JPEG uses lossy compression methods that result in some loss of the original data. When you decompress the original image, you don't get exactly the same image that you started with, although JPEG was specifically designed to discard information not easily detected by the human eye.

Joint Photographic Experts Group Compresion (JPEG Compression)-JPEG compression is a group of compression methods that are used to provide for high-quality images at varying levels of compression up to approximately 50:1. The JPEG compression system can use compression that is fully reversible (no loss of information) or that is lossy (reversible with some loss of quality). JPEG compression is defined by the Joint Photographic Experts Group (JPEG) committee.

This figure shows the basic process that can be used for JPEG image compression. This diagram shows that JPEG compression takes a portion (block) of a digital image (lines and column sample points) and analyzes the block of digital information into a new block sequence of frequency components (DCT). The sum of these DCT coefficient components can be processed and added together to reproduce the original block. Optionally, the coefficient levels can be changed a small amount (lossy compression) without significant image differences (thresholding). The new block of coefficients is converted to a sequence of data (serial format) by a zig-zag process. The data is then further compressed using run length coding (RLC) to reduce repetitive bit patterns and then using variable length coding (VLC) to convert and reduce highly repetitive data sequences.

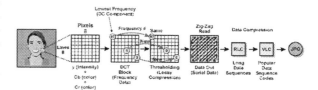

JPEG Compression

Joint Stereo (JS)-Joint stereo is a digital processing technique that is used to enhance a stereo signal.

Joint Technical Committee (JTC)-A joint technical committee is a group that is composed of members from multiple organizations or affiliations with the purpose of analyzing, recommending, solving technical issues or creating specifications.

Joint Technology Committee (JTC)-A working committee under the auspices of the International Standards Organization (ISO) with the goal of defining a standard for digital compression and decompression of still images for use in computer system.

Joint Use-A mutual agreement among two or more telephone, power, cable television, or other utility companies to use common poles, trenches, and similar facilities.

Joint Video Committee-The MPEG joint video committee is a group that is composed of members from the IETC and ITU for purpose of analyzing, recommending, solving technical issues to create an advanced video compression specification.

joule (J)-The standard unit of work (unit of energy) that is equal to the work done by one Newton of force when the point at which the force is applied is displaced a distance of one meter in the direction of the force. One joule is also equal to the energy conveyed by a power level of one watt acting for one second, so a joule is equivalent to a "watt second." 3,600,000 joules is a kilowatt hour (kWh) of energy.

The joule is named for the19th century Irish physicist James Prescott Joule (1811-1889).

Journalization-The process of booking billing charges & credits to the appropriate financial accounts. A journal is the interface between the billing system and the general ledger.

Joystick-An electromechanical control level, similar to the control stick on an aircraft, used for hand positioning graphic images on a video or computer monitor.

JPEG-Joint Photographic Experts Group

JPEG Compression-Joint Photographic Experts Group Compression

JS-Joint Stereo

Jscript-JavaScript

J-TACS-Japan Total Access Communication System

JTAPI-Java Telephony Application Programming Interface

JTC-Joint Technical Committee

JTC-Joint Technology Committee

Judder-Judder is the varying of position or fluctuating motion changes of images as they are played on a movie or video display.

Jumbo Frame-A frame longer than the maximum frame length allowed by a standard. Specifically used to describe the dubious practice of sending 9-Kbyte frames on Ethernet LANs.

Jump-A class of instruction that causes a software program to move forward or backward to a specific location.

Jumper Cable-A jumper is a pair of wires, crossbar, or optical fiber that is used to establish a connection of circuits or equipment. Jumpers may be used to create connections that allow the selection of a software or hardware configuration.

Jurisdiction-A geographical area or identifying characteristic that is assigned to a regulatory authority to determine the boundaries of their authority.

Just In Time (JIT)-Just in time is a process that provides materials, services or products to a person or process at or close to the time that the materials, services or products are needed for the next step in a process.

JVM-Java Virtual Machine

K

K-1024

k-kilo

Kazaa-Kazaa is a software program that allows users to copy content.

kbps-kilobits Per Second

kbyte-Thousand bytes.

Kc-Cipher Key

Keep It Simple Stupid (KISS)-A philosophy that states that simple things or processes are better than complex methods. KISS is commonly applied to business practices.

Kernel-A kernel is a software part of an operating system that is fundamental to its overall operation. The kernel software continually manages specific processes such as system memory, the file system, and disk operations. A kernel may also run processes such as the communication between applications, coordinating the input and output of information within an application. Once loaded, the kernel software usually remains stored and operates from computer memory and is typically not seen by the users of the system.

Kevlar®-Kevlar® is a strong fiber material (yellow) used in cable strength members and the name is a trademark of the Dupont Company. Kevlar is also used in the construction of bulletproof vests.

Kevlar® Scissors-Kevlar® scissors are specially designed cutting tools that can cut Kevlar rope or fiber strands that are commonly used as strength members and cable fillers.

Key-(1-connector) A short pin or other projection that slides into a mating slot or groove to guide two parts being assembled. The key is used to prevent a connector interface from rotating. (2-database) An attribute that uniquely identifies the sets of ordered elements in a database relationship. (3-encryption) A word, algorithm, or program used to encrypt and decrypt a message. (4-video) A signal, also called key source or key cut, that can be used to electronically cut a hole in a video picture to allow for insertion of other elements, such as text or another video image. The key signal is a switching or gating signal for controlling a video mixer, which switches between or mixes the background video and the inserted element. Key also may refer to the composite effect created by cutting a hole in one image and inserting another image into the hole. (5-radio) To initiate radio transmission.

Key Exchange-Key exchange is the process used to exchange the key value between to or more devices or users. Exchanging key information does not necessary involve the physical transfer of the key. It may only involve the transfer of information parameters that are used to create and/or validate the key.

Key Frame-A key frame is a reference video frame image that is part of a series or group of related frames that contains all the information needed to create its image. The creation of a key frame image does not depend on information from any other frames.

Key Frame Effect-A video effect consisting of a series of effects "snapshots" called key frames. When the overall effect is replayed, the video processing system automatically and gradually dissolves from one key frame to the next. A process called interleaving defines what happens between key frames. The result is an animation effect.

Key Generator (Keygen)-A key generator is a program that creates usable "keys" to allow access to content or programs. The keys generated may not be the actual serial number of the program.

Key Length-Key length is the number of digits or information elements (such as digital bits) that are used in an encryption (data privacy protection) process. Generally, the longer the key length, the stronger the encryption protection.

Key Management-Key management is the creation, storage, delivery/transfer and use of unique information (keys) by recipients or holders of information (data or media) to allow the information to be converted (modified) into a usable form.

Key Performance Indicator (KPI)-Key performance indicators are the metrics (measurements) established by upper management to determine the accomplishment of objectives set for organizational units. Examples of KPIs include net customer additions to a system, net customer loss rate (net churn) and total revenue.

Key Revocation-Key revocation is the process of deleting or modifying a key so a user or device is no longer able to decode content. Key revocation may be performed by sending a command from a system to a device.

Key Server-A key server is a computer that can create, manage, and assign key values for an encryption system.

Key Service Unit (KSU)-The central operating unit of a key telephone system (KTS) or non-PBX/ACD telephone system (small customer premises telephone switch).

Key Set-A telephone set that is part of a key telephone system (KTS). A key set usually has several buttons that allow the user to select call hold, line connection, intercom and other private telephone system features.

Key Tag-The parameter defining one of several encryption codes or methods.

Key Telephone System (KTS)-Key telephone systems are (usually small) multi-line private telephone network that allows each key telephone station to select one of several telephone lines, place a line on hold, and call via an intercom circuit between key telephones. Key systems contain a central key service unit (KSU) that coordinates status lights and lines to key telephones ("Key Sets"). Early KTS system technology was based on electro-mechanical relay hardware. They required all the outside telephone lines to be connected to all of the key telephone sets in the installation. In addition, two additional pairs of wire were used in conjunction with each telephone line, one pair for the A/A1 connection indicating if that line is off hook at that particular key telephone set, and another pair to operate a small light to indicate the status of that line. Consequently, each key telephone set was connected to the central KSU via a thick cable containing 50 wires (25 pairs). Newer KTS systems typically use only 4 wires to connect the electronic KSU to each electronic key telephone set, and are often called "skinny wire" key systems. Modern electronic key systems are small microprocessor controlled switching systems and have some of the same advanced call processing features such as call hold, busy status, multi-line conference, abbreviated dialing, and station-to-station intercom that are available in a larger PBX.

This figure shows a typical key telephone system. This diagram shows telephones wired to a key service unit (KSU) that is connected to the PSTN. The KSU allows the telephones to have access to the outside lines to the PSTN. The KSU controls lights on the telephone sets, intercom access, and call hold.

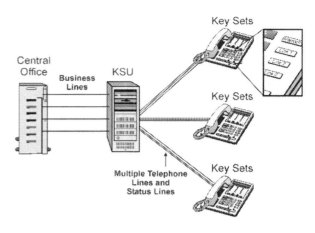

Key Telephone System (KTS) Operation

Key Video-The video key fill signal, key source, or both.

Keyboard-A physical device that allows a user to enter data to a computer or other electronic device. A keyboard usually consists of individual key switches that are assigned one (or several) alphanumeric character and/or function.

Keyer-An electronic circuit that creates a signal to control a video multiplier based on selective information contained in a video signal.

Keygen-Key Generator

Keying-(1-data) The process of converting data into a machine readable form. (2-video) The process of replacing part of one TV image with video from another image. The point along a horizontal scan line at which switching from one image to the other occurs is based on a signal level (luminance keying) or a specified color (chroma keying). (3-connector) Connector keying is the use of physical characteristics (such as a slot and grove) to maintain/force the alignment of connectors with each other.

Keypad-A physical interface device (a group of keys) that allows for a user to input data to a computer or other electronic device.

Keyword Advertising-Keyword advertising is a marketing process that uses key words that potential customers enter into search engines to find product or service information. Keyword advertising is usually paid for by a fixed fee or bidding process. To Keyword advertise, a list of keywords is selected and associated with a URL and a short message to accompany the listing. When the search term(s) matches the keyword, the URL and the descriptive text are displayed. These listings are called sponsored listings.

This diagram shows the basic keyword advertising process. In this example, four companies have submitted ads to a search engines that match a keyword. When an Internet user enters a search word into the search engine, the search engine provides the user with a list of URLs found along with a list of sponsored ads. The sponsored ad presentation (impressions) is organized with the highest bid on top and ads with lower bids positioned lower on the screen.

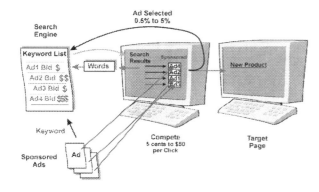

Internet Marketing Adwords

kg-kilogram
kHz-kilohertz
Killer App-Killer Application
Killer Application (Killer App)-A software application of such great importance to the end customer that it alone motivates the customer to buy the entire system. One example is a computer game having devoted users.
kilo (k)-Prefix indicating 1000 units in the metric system (Decimal 1000). Note the use of a lower case k.
kilobit-A quantity equal to one thousand bits.

kilobits Per Second (kbps)-A measure of data transmission equal to one thousand bits per second.
kilogram (kg)-The unit of mass. It is equal to the mass of the international prototype of the kilogram kept at the International Bureau of Weights and Measures laboratory at Sévres (a suburb of Paris) France. The prototype is a cylinder of iridium-platinum alloy 39 mm in diameter and 39 mm in height. Approximately equal to the mass of a liter of water, a cube of water having each edge of 100 mm length. See also mass.
kilohertz (kHz)-A unit of measure of frequency equal to one thousand hertz.
kilovar-A unit of measure equal to one thousand volt-amperes.
kilovolt (kV)-A unit of measure of electric voltage equal to one thousand volts.
kilowatt (kW)-A unit of measure equal to one thousand watts.
Kilowatt Hour (kWh)-kWh is a widely used unit of energy, equal to 1000 watts of power per hour. See also joule.
Kinking-Kinking is the collapsing or distortion of a cable or conduit in a localized area due to the excessive bending force.
Kiosk Billing System-A uniform billing system used by France Telecom in its deployment of mass-market videotex services. The system takes its name from the kiosk, or newsstand, where information is purchased on an as needed basis.
Kirchoff's Laws-(1-current) In a circuit node there is as much current flowing into the node as there is flowing away from it. (2-voltage) In a closed electric circuit, the algebraic sum of the applied voltages and the voltage drops is equal to zero.
KISS-Keep It Simple Stupid
Kit-Any number of electronic parts, usually provided parts, often provided with a schematic diagram or printed circuit board, which, when assembled in accordance with instructions, results in a device subject to the regulations of this part, even if additional parts of any type are required to complete assembly.
Klystrode-An amplifier device for UHF-TV signals that combines aspects of a tetrode (grid modulation) with a klystron (velocity modulation of an electron beam). The result is a more efficient, less expensive device for many applications. Klystrode is a trademark of EIMAC, a division of Varian Associates.

Knee-In a response curve, the region of maximum curvature.

Knee Voltage-A voltage transient on a battery during a charge cycle that indicates the battery if fully charged.

Knowledge Neighborhood -Term used to define a group of related departments and disciplines. Groups within the
telco who share a common vocabulary, view of the world and information resources. For example, all groups involved in customer relationship management form a CRM Knowledge Neighborhood.

KPI-Key Performance Indicator

KSU-Key Service Unit

KTS-Key Telephone System

kV-kilovolt

kW-kilowatt

kWh-Kilowatt Hour

L

L2CAP-Logical Link Control And Adaptation Protocol

L2TP-Layer 2 Tunneling Protocol

Laboratory Testing-Laboratory testing is a process of testing a device, assembly, or system at a location that typically involves its design, prototyping or performance certification.

LAC-Location Area Code

Lag-The difference in phase between a current and the voltage that produced it, expressed in electrical degrees.

Lambda-The Greek symbol for "L" that is used to represent wavelength, for instance in optical and electrical signals. Wavelength is the distance one cycle of a wave travels. Wavelength can be calculated by dividing the speed of the wave by its frequency. For example, in free space (in a vacuum) the wavelength of an electromagnetic wave is equal to: (300,000,000 meters/sec)/ frequency (Hz). The wavelength of a radio or lightwave traveling in vacuum at 300 MHz is 1 meter.

LAN-Local Area Network

LAN Emulation (LANE)-Local area network emulation (LAN emulation) is the adaptation information on a communication network so that it can transparently operate as a local area network (LAN) system.

The use of LANE allows the connection of standard local area networks (LANs) such as Ethernet and token through other high speed networks such as asynchronous transfer mode (ATM) and broadband wireless systems. LANE translates the services between the two types of networks and is transparent to higher-level applications that use LAN networks.

The major part of LANE include broadcast and unknown server (BUS) to manage broadcast and multicast addresses, LAN emulation client (LEC) to adapt (map) between the 48 bit Ethernet media access control (MAC) addresses and other types of addresses such as ATM. A LAN emulation configuration server (LECS) can be used to manage the LAN emulation clients, and a LAN emulation server (LES) that controls and coordinates the entire LANE system.

LAN Emulation Client (LEC)-A process used for LAN Emulation (LANE) in an asynchronous transfer mode (ATM) system to handle the adaptation of Ethernet data frames into ATM packets. The LEC is used by the each node that connects the LAN (e.g. Ethernet) system to the ATM network.

LAN Segmentation-The practice of dividing a single LAN into a set of multiple LANs interconnected by bridges.

LAN Switch-A switch that allows for segmentation of a LAN passing frames between segments only when necessary. Ethernet switches are the most common and are used to reduce congestion and collisions.

LAN Telephony (TeLANophy)-Local access network (LAN) telephony (sometimes called TeLANophy) use LAN systems to transport voice communications.

Land Mobile Service-Land mobile service is a type of wireless communication service where mobile radio telephones connect people to the public switched telephone system (PSTN) or to other mobile telephones. Mobile telephone service includes cellular, PCS, specialized and enhanced mobile radio, air-to-ground, marine, and railroad telephone services. MTS was a name used for the first public mobile telephone system used in the United States.

Landline-A conventional domestic or business telephone circuit. The term landline applies to telephone lines that are either buried or carried just over the ground.

Landline Network-The communications infrastructure that generally is associated with the public switched telephone network. (See also: landline.)

LANE-LAN Emulation

Language-(1-general) A set of symbols, characters, conventions, and rules used for conveying information. (2-programming) A set of commands and associated parameters that can be decoded to represent specific sequences of computer processing instructions.

LAPB-Balanced link access procedure

LAPB-Link Access Protocol Balanced

LAPD-Link Access Protocol on D Channel

Laptop Computer-A laptop computer is a portable computer that is small enough to fit and be used on the lap of a person.

Large Scale Integration (LSI)-Large scale integration is the combining of hundreds of transistors on an integrated circuit.

LASER-Light Amplification By Stimulated Emission of Radiation

Laser Diode (LD)-A laser diode is a junction diode that emits electromagnetic radiation or light when injected electrons under forward bias recombine with holes in the vicinity of the junction. The laser diode is used to transmit light signals over fiber optic cables. See also Injection Laser Diode (ILD).

Laser Safety Officer (LSO)-A laser safety officer (LSO) is a person who is in charge of overall safety from exposure to laser light at a company or facility.

Lashing-Lashing is the attachment of a cable to a support strand by using helically wrapping materials such as dielectric filament or steel wire to hold the new line to another line.

This figure shows a basic method of installing aerial optical cable. This diagram shows that the optical cable is being attached to a messenger wire between telephone poles. This diagram shows that the optical cable is sent through a cable guide and a cable lashing machine that is attached to the messenger cable. The optical cable is supplied from a cable reel trailer that is attached to a truck. The end of the lashing cable is attached to the messenger wire near the pole via a lashing clamp. An installer pulls the cable lasher via a pulling rope. As the cable lasher is pulled, the lashing wire inside is spun around the cable and the messenger wire holding them (lashing) together. The truck and installer travel at approximately the same speed.

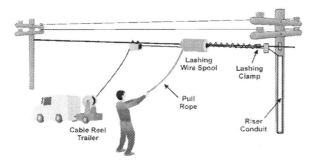

Aerial Cable Installation

Last Call Return-A telephony service that allows a telephone user to automatically call back the phone number of the last received incoming call. Last call return is normally accomplished by the customer entering the service code (e.g. "*69").

Last Mile-The last portion of the telephone access line that is installed between a local telephone company switching facility and the customer's premises.

Last Number Redial-The ability for a telephone to remember and dial the last dialed telephone number.

LATA-Local Access And Transport Area

Latch-An electronic circuit that holds a digital signal after it has been selected. To latch a signal means to hold it.

Late Entry-Late entry is the ability of a communication device to be added to a communication session after the communication session has been established.

Latency-Latency is the amount of time delay between the initiation of a service request for data transmission or when data is initially received for retransmission to the time when the data transmission service request is granted or when the retransmission of data begins.

Lateral Cables-Sections of cable between feeder and distribution networks. The feeder cable can come from underground conduit systems, buried cable systems, or aerial cable systems and extend to a distribution network through serving area interfaces, cross connect terminals, or (rarely now) control, access, or splice points.

Latitude-An angular measurement of a point on the earth above or below the equator. The equator represents 0, the north pole +90 and the south pole -90.

Launch-(1-satellite) The process of lifting a satellite from earth into orbit. (2-electromagnetic wave)The process of transferring energy from a feeder or waveguide to an antenna.

Launch Fiber-A short length of optical fiber that couples the light from an optical source into the optical fiber of a communications system.

Layer-(1- LAN) A collection of related network processing functions that constitutes one level of a hierarchy of functions. (2-video)A single video image that is processed so that it can be inserted into a final composite image. There may be other layers in the image, which can be prioritized as to location. (3-computer)An overlay which can be used to place information, so that CAD/CAM drawings can be logically subdivided, for viewing or hardcopy purposes. (4-boundaries)A group of one or more entities contained within an upper and lower logical boundary. Layer (N) has boundaries to the layer (N + 1) and to the layer (N - 1).

Layer 2 Switch-A switching device that operates at OSI link layer 2. Synonymous with bridge.

Layer 2 Tunneling Protocol (L2TP)-Layer 2 tunneling protocol is used to allow a secure communication path, a virtual private network link, between computers. It is an evolution of earlier point-to-point tunneling protocol (PTPP) as it offers more reliable operation and enhanced security. L2TP enables private communication lines through a public network. L2TP was developed via the Internet engineering task for (IETF).

Layer 3 Switch-A switching device that operates at OSI network layer 3. Synonymous with router.

Layer 4 Switch-A switching device that operates at OSI transport layer 4. A router that can make routing policy decisions based on transport layer Information (e.g., TCP port identifiers) encapsulated within packets.

Layer Interface-The boundary between two adjacent layers of the protocol model.

Layered Coding-Layered coding is a process that converts media into several component parts where each layer can be combined with other layers to produce a higher quality or improved version of the media.

Layered Network Architecture-A network structure that divides the network communication functions into layers that perform specific functions. The precise definition of each layer allows products to be developed for specific functions by different companies. Different types of networks may have different layer types. Examples of layered architecture include the 7 layer open systems interconnect (OSI) model and the 4 layer Internet model.

Layered Protocols-Protocols that are designed to communicate with higher or lower level protocols in a communication network. Each layered protocol performs a specific function and each layered protocol has specific ways to pass information to protocols that in layers directly above or below it.

Layout-(1-location) A proposed or actual arrangement or allocation of equipment or physical position of components on a circuit board. (2-

process)The design the physical layout of an electronic circuit board.

LBS-Location Based Services

LBT-Listen Before Talk

LC Circuit-An electric circuit with both inductance (L) and capacitance (C) that is resonant at a particular frequency.

LCA-Local Calling Area

LCC-Life Cycle Cost

LCM-Licensed Compliant Module

LCP-Liquid Crystal Polymer

LD-Laser Diode

LD-Long Distance

LDAP-Lightweight Directory Access Protocol

LDP-Label Distribution Protocol

LDS-Local Digital Switch

Lead-An electrical wire, usually insulated.

Leader Stroke-In lightning, the first stroke, which usually determines the path to be followed by the return stroke, where most of the energy is carried.

Leakage Resistance-The resistance of a path from a circuit to ground through which leakage current flows.

Leaky Cable-A cable that is designed to deliberately leak RF energy. A leaky cable often is used to provide radio coverage in a shielded area, such as a tunnel or basement

Leaky Coax-Leaky coax is a coaxial cable that is used to allow some of the signal it is carrying to leak out and communicate with nearby radio devices. Leaky coax cable is used to provide radio signal coverage over long (such as train tunnels) or unusually shaped radio coverage areas. Leaky coax cable is typically constructed by adding a series of holes on the side of the coaxial cable that allows a small amount of radio signal to enter and leave the coaxial cable.

Lean Back Television-Lean back television is the process of a person watching television or video programs without a significant amount of interaction allowing the viewer to relax and enjoy the media.

Lean Protocol-A lean protocol is a set of commands and processes that are used to perform a specific function and the number of commands and/or level of detail or processes is limited to the specific set of functions.

Leap Second-A time step of one second, used to adjust coordinated universal time to ensure approximate agreement with international universal time. An inserted second is called a positive leap second, and an omitted second is called a negative leap second.

Learning Process-The process whereby a bridge builds its filtering database by gleaning address-to-port mappings from received frames.

Leased Line-Leased lines are telecommunication lines or links that have part or all of their transmission capacity dedicated (reserved) for the exclusive use of a single customer or company. Leased lines often come with a guaranteed level of performance for connections between two points.

Leased Network-A data network using circuits or channels leased from a telephone company (telco) or other telecommunications carrier and dedicated to use solely by the lessee.

Leased Service-The exclusive use of any channel or combination of channels designated to a subscriber.

Leave Message-A leave message is a command that instructs a device or user to leave a service, session or to exit from a multicast group.

LEC-LAN Emulation Client

LEC-Light Energy Converter

LEC-Linear Echo Canceller

LEC-Local Exchange Carrier

LECS-Local Area Network Emulation Configuration Server

LED-Light Emitting Diode

Legacy-Legacy is established or well known systems, technology or products.

Legacy Application-Legacy refers to a system or established technology that has been used in the past.

Legacy System-A legacy system is a communication system or network that satisfies specific business needs using established technology or equipment. Legacy systems may become obsolete or is incompatible with new industry standards. To extend the life of existing investments in legacy systems, new technologies or systems are often designed to communicate with legacy systems.

Legacy Wiring-Legacy technology is commonly used to describe a previously used technology that is undergoing a change or is being replaced with a new technology to reduce costs or to satisfy new functional requirements.

Legal Rights-Legal rights are actions that are authorized to be performed by individuals or companies that are specified by governments or agencies of governments.

Lempel-Ziv-Welch (LZW)-An algebraic digital data compression algorithm originally published by Abraham Lempel and Jacob Ziv, and implemented in a practical software program by the late Terry A. Welch. Used in the .GIF file format (developed by CompuServe), and a similar scheme is used by the V.42bis modem data compression standard, in PKZip (developed by the late Philip Klein) and other file compression programs.

Length-The number of bits or bytes in a packet, data block, information field, record, or other variable length block of information.

Length Indicator-In common-channel signaling, a 6-bit field that differentiates between message, link status, and fill-in signal units. When the binary value of an indicator is less than 63, it indicates the length of a signal unit.

Lens Grade Tissue-Lens grade tissue is a soft, porous dust free material that is commonly used for dry or wet (e.g. with alcohol) cleaning of optical or coated metal materials.

LEO-Low Earth Orbit

LES-LAN Emulation Server

Letter of Credit-A letter of credit is a financial payment instrument that authorizes the transfer of funds or other valuable assets to the recipient of the letter of credit provided the terms of the letter of credit are fulfilled (such as the receipt of products at a specified location).

Letterbox-Letterbox is the method of displaying wide screen images on a standard TV receiver where the wide screen aspect ratio is much larger than the standard television or computer monitor typical aspect ratio of 4:3. This causes the display to appear within borders at the top and bottom of the image producing a horizontal box (the letterbox).

This figure shows how the use of a letterbox allows an entire video image to be displayed on a screen that has an aspect ratio lower than the video image requires. This example shows that part of an image (of a boat) is lost on the left and right parts of the display. Using a letterbox, the image size is reduced so its width can fit within the length of the screen area. The result is part of the top and bottom area of the screen area are blanked out (black) resulting in the formation of a box (a letterbox).

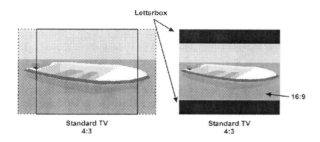

Video Display Letterbox

Level-(1-general) The strength or intensity of a given signal (2-crosstalk) The power of the crosstalk signal compared with a reference signal. (3-peak) The maximum applied sound or signal amplitude. (4-speech) The energy of speech measured in volume units (VU), and typically displayed on a VU meter. (5-speech power) The acoustic power in human speech. (6-transmission) The ratio of the power of a test signal at one point to the test signal power applied at another point in the system used as a reference. (7-video routing) An independently controllable spectrum of signals within a routing switcher. Typically, a routing switcher has a video level and one or more audio levels.

Level Setting-Adjustment of video or audio signal levels.

Libel-In reference to any printed or written publication which may be slanderous.

License-A license is a contract that grants specific rights to use of intellectual property.

License Fee-License fees are an amount charged or assigned to an account for the authorization to use a product, service, or asset. License fees can be a fixed fee, percentage of sales, or a combination of the two.

License Server-A license server is a computer system that maintains a list of license holders and their associated permissions to access licensed content. The main function of a license server is to confirm or provide the necessary codes or information elements to users or systems with the ability to provide access to licensed content.

License Terms-License terms are the specific requirements and processes that must be followed as part of a licensing term agreement.

Licensed Bands-Licensed frequency bands give the licensee (service provider or service user) the authority to use the radio spectrum within their licensed frequency band according to the requirements of the license. These requirements may include a type of service (such as paging or mobile telephone service), channel types (single or multiple channels), and power levels within a specific geographic area (amount of signal strength allowed).

Licensee-A licensee is the holder of license that permits the user to operate a product or use a service. In telecommunications, a licensee is usually the company or person who has been given permission to provide or use a specific type of communications service within a geographic area.

Licensing-Licensing is the defining, authorizing and compensating for the rights to develop, use or sell products and services.

Licensing Collective-A licensing collective is a group or organization that represents a several or many rightsholders which has the authority to negotiating and administer licenses agreements. ASCAP is an example of a licensing collectives.

Licensing Rules-Licensing rules are the processes and/or restrictions that are to be followed as part of a licensing agreement. Licensing rules may be entered into a digital rights management (DRM) system to allow for the automatic provisioning (enabling) of services and transfers of content.

Licensor-A licensor is a company or person who authorizes specific uses or rights for the use of technology, products or services.

LIDB-Line Information Database

Life Cycle-Life cycle is the time period that a particular device, assembly or a class of equipment is usable under normal working conditions.

Life Cycle Cost (LCC)-Life cycle cost is the combination (addition) of the initial acquisition cost, operational and maintenance cost, the cost of disposition (removal for upgrade or end of service) less the recovered cost from the sale of salvaged equipment (if any).

Life Safety System-A system designed to protect life and property such as emergency lighting, fire alarms, smoke exhaust and ventilating fans, and site security.

Life Test-A test in which random samples of a product are checked to see how long they can continue to perform their functions satisfactorily. A form of stress testing is used, inducting temperature, current, voltage, and/or vibration effects, cycled at many times the rate that would apply in normal usage.

Life Time-(1-general) The estimated time over which a product, assembly, or a communication cable can be used before it must be replaced. (2-cable) The end of life for a cable reached when all spare pairs or a specified maximum percentage of the total number of pairs are in use. The lifetime of pair groups within a cable can be estimated by dividing the number of spare pairs by the forecasted growth rate.

Lifeline Service-A communication service that is considered a "Lifeline" in case of emergency. Communication service that assures a person can call for assistance or be contacted.

Lifetime Revenue per Subscriber (LRS)-Lifetime revenue per subscriber is the sum of the revenues for each subscriber (customer) in a system. This value is calculated by multiplying the annual average revenue per user (ARPU) by the average time period (lifetime) that a subscriber purchases products and/or services.

Lifetime Value-Lifetime value is a technique utilized to define how valuable a customer will be to the company, over their lifetime. For telephone companies (telcos), lifetime value calculations are usually not valid or useful because of the vagaries in pricing and product structures over the average lifetime of an individual. As an alternate, many telco's use customer value assessments.

Light-Electromagnetic radiation visible to the human eye. The visible wavelength of light ranges from approximately at 400 nm to 700nm. Commonly, the term is applied to electromagnetic radiation in most fiber optic communication systems. An electromagnetic radiation with wavelengths from 400 nm (violet) to 740 nm (red), propagated at a velocity of roughly 300,000 km/s (186,000 miles/s), and detected by the human eye as a visual signal in the optical communication field, the term also includes the much broader portion of the electromagnetic spectrum that can be handled-died by the basic optical techniques used for the visible spectrum. This extends the definition of light from the near-ultraviolet region of approximately 300 nm through the visible region, and into the mid-infrared region of 3.0 to 30 nm.

Light Amplification By Stimulated Emission of Radiation (LASER)-A Laser is a device that emits coherent light of essentially one wavelength in a narrow beam. Lasers can be made using gaseous, liquid or solid state. In optical networks, solid state or semiconductor lasers are used as high performance light sources. Photons are generated in the semiconductor by application of a voltage. Photons with the right wavelength, phase, and direction of travel are selected by an optical cavity in the laser. See also Optical Cavity, LED. Laser light is monochromatic (it has only one frequency or "color," although infrared or ultraviolet lasers produce optical frequencies that are not visible to the human eye), in contrast to white light that has, at the other extreme, many frequency components, and the phase of the sine wave electromagnetic waveform maintains a fixed (coherent) timing relationship over a long interval of time. The peak amplitudes of each cycle of a coherent wave form occurs at absolutely uniform time intervals. Lasers create monochromatic coherent light by combining the light radiation due to the oscillation of individual electrons in individual atoms as these electrons change their electric charge configuration from a higher energy level to a lower energy level. First these electrons are "pumped" up to a higher energy level by a primary source of power. We call these the "excited" electrons. One way to excite electrons is to accelerate the atoms in a gas by applying a high voltage to electrodes at two ends of a container of the gas, so that some atoms collide with each other and transfer their kinetic (motion related) energy to some of the electrons, or alternatively by shining a non-coherent light source of higher frequency (shorter wavelength) than the desired output light frequency on a solid material. Some excited electrons in an atom subsequently "fall" to a lower energy level, and when doing this they emit light at a frequency proportional to the difference in energy between the high and low energy levels. The energy difference $(E2-E1)$ is related to the frequency of the light, f, by the formula $E2-E1= hf$, where h is Planck's constant. Excited electrons "fall" naturally, but at unpredictable times, from the higher to the lower energy level for no apparent reason. One way to make even more electrons fall from the high energy level to the low energy level is to shine a light on these atoms, that light having the same frequency f as the expected output light frequency. This latter process is called stimulated emission. Because the gas or solid is made up of atoms having the same energy level structure, all the atoms then emit light at the same frequency. By placing two parallel reflecting mirrors at two ends of the material, and precisely locating the distance between these mirrors so that their distance is an integral number of wavelengths of the light in question, a standing wave of multiply reflected light is set up in the gas or solid. This is what makes the light emission from all the different electrons coherent. By making one of the two mirrors either partially reflecting and partly transparent, or by having a small transparent spot on one otherwise fully reflective mirror, some of the light is able to escape in a straight, monochromatic, coherent light beam. This is the laser beam. It can be guided into the core of an optical fiber for communication purposes. The light output of the laser can be turned on and off electrically to produce light pulses to convey digital information.

This figure shows the basic operation of a semiconductor light amplification stimulated emission of radiation (LASER) device. This example shows that a semiconductor LASER is constructed of a specific type of p-type and n-type semiconductor material. When a forward current is applied to the device, photons are produced within the optical cavity. As photons travel down the cavity, the produce other photons along their same path. This diagram shows that the optical cavity has a fully reflective mirror on one end that reflects all photons back into the cavity. At the other end, the mir-

ror is partially reflective allowing some of the photons to exist from the LASER. This diagram shows that photons exit from the LASER in the same direction (coherent light).

Light Emitting Diode (LED)-Light emitting diodes produce light as a result of their forward electric current that can be transferred from the component package. Electrons decrease their energy as they pass from the N to the P side of the diode, and this energy difference is equal to the energy radiated. While some diodes are made with an opaque material surrounding the junction, LEDs are made specifically with transparent material around the junction to allow the light to be visible from the outside. The color (frequency or wavelength) of the light is dependent upon the difference in electron energy level, which in turn is dependent on the amount of added material dopants used in the two parts of the diode. A low energy change corresponds to infrared light (not visible to the human eye), a medium energy change corresponds to red or yellow or green color, and a high energy change corresponds to blue or ultraviolet "color." Ultraviolet light is not visible to the human eye.

This figure shows the basic operation of a light emitting diode (LED). This example shows that a LED is constructed of a specific type of p-type and n-type semiconductor material. When a forward current is applied to the device, photons are produced at the emitting junction. This example shows that LEDs typically produce light in several different directions.

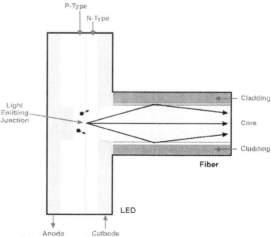

Light Amplification By Stimulated Emission Of Radiation (LASER) Operation

Light Emitting Diode (LED) Light Source

Light Energy Converter (LEC)-A photovoltaic semiconductor device that converts light energy into electrical energy.

Light Propagation-Light propagation is the process of transferring a light signal (electromagnetic signal) from one point to another point. The speed of light is the velocity that lightwaves travel. In vacuum (similar to air), the wave speed of a lightwave is 300 million meters per second (186,281.6 miles per second). In other materials, the speed that lightwaves travel is lower.

Light Ray-The path of a given point on a wavefront. The direction of a light ray is generally normal to the wavefront.

Light Receiver-A photodiode or other transducer that is used for receiving optical signals.

Light Source-A generic term that includes lasers and LEDs, even though these may operate outside the visible light band.

Lightning-A flow of current between a charged cloud and the ground resulting from an electric discharge due to large potential differences between cloud charge and ground (or the lightning strike point).

Lightning Flash-An electrostatic atmospheric discharge. The typical duration of a lightning flash is approximately 0.5 seconds. A single flash is made up of various discharge components, usually including three or four high-current pulses called strokes.

Lightning Protector-(1-general)A device that limits impulse voltages from lightning to prevent damage to people and electronic equipment. Basic spark-gap ("carbon block" or gas tube) protectors installed in buildings are inadequate to protect modern electronic equipment. Supplementary lightning protectors (surge protectors) are frequently installed at the equipment to protect against excess voltages on both signal and AC connections. (2-rod) A lightning rod system that routes lighting voltages to ground.

Lightweight Directory Access Protocol (LDAP)-A standard protocol that allows users to find other devices and services in a communication network. It provides directory services for LAN and the Internet. LDAP is a subset of the X.500 protocol that operates over TCP/IP.

Lightweight Protocol-A protocol that is a simplified version of another protocol. Lightweight protocols often simplify the information access, control, and transfer processes at the expense of extended capability or reliability (a more limited number of commands and status messages).

Likeness Parity-In accordance to requirements issued to any given producer, usually regarding matched characters, to integrate images of one actor with images of another actor(s) in any advertisement in correlation to the film in which they co-act.

Limited Access-An arrangement in which only some traffic offered to a group of servers has access to all the servers in the group.

Limiter Circuit-A circuit of nonlinear elements that restricts the electrical excursion of a variable in accordance with some specified criteria.

Line Card-A line card is a plug-in electronic circuit card that connects telephone switching environment to telephone lines. The line card adapts signal levels, senses and inserts control commands and tones and performs other functions that allow the line card to communicate with specific types of telephone lines.

Line Coding-The process of modulating and formatting data for transmission on a communications line.

Line Frequency-Line frequency is the number of horizontal scans per second in a video system. For the NTSC analog video system, the line frequency is 5,734.26 times per second (HZ).

Line Of Sight (LOS)-Line of sight (LOS) is a direct path in a wireless communication system that does not have any significant obstructions. LOS systems can use optical or radio signals for transmission.

Line Resistance-Copper cable has resistance (impedance) that is dependent on the size (diameter) of the cable. The resistance of the copper wire increases as the diameter decreases (gauge number increases). The higher the line resistance, the more of the signal energy is dissipated by the line and less energy is transferred to the receiving device.

This figure shows how line resistance attenuation and the wire size decreases. This diagram shows that cables with larger diameter copper wires are typically used to in the distribution system. As the distribution system nears its destination, the size of the wire often decreases.

Line Resistance Attenuation

Line Side Connection-Line side connections are an interconnection line between the customer's equipment and the last switch (end office) in the telephone network. The line side connection isolates the customer's equipment from network signaling requirements. Line side connections and are usually low capacity (one channel) lines.

Line Signal-A signal sent over a line; included are call progress, supervisory, control, address, and alerting signals.

Line Source-(1-spectral) An optical source that emits one or more spectrally narrow lines, as opposed to a continuous spectrum. (2-geometric) An optical source whose active (emitting) area forms a spatially narrow line.

Line Station Transfer (LST)-A process of clearing communication line pairs for new subscribers in telecom distribution areas with marginal or depleted facilities. After a review by outside plant technicians and discussions with engineering, an existing customer may be switched to a new shorter length cable pair. This frees up the higher capacity pair as an available spare to serve the newly-signed customer at the far end.

Line Tapping-Line tapping is the connection and monitoring of a communication line.

Line Terminal (LT)-A line terminal is a device that is an end point in a communication system (a communication line) that is used to convert information from that network into a form that can be used by a user or another type of communication device.

Line Terminating Equipment (LTE)-A device or system that terminates a line in an optical system. LTE assemblies are used in a SONET network to originate and/or terminate Optical Carrier (OCn) signals. LTE equipment contains optical transmitter/receiver assemblies and an LTE assembly can decode, modify, creates the overhead control messages used in the optical network.

Line Testing-Line testing is the measurements of the characteristics of a communication line (such as the dialtone level on a telephone line) to determine that the line is in service (operational) or is operating within expected performance levels (such as error or distortion levels).

Line Trunk-A line trunk is a transmission channel but is not limited to, transmission media such as radio, satellite, wire, cable and fiber optic cable means of transmission.

Linear-(1-general) A circuit, device, or channel whose output is directly proportional to its input. (2-frequency) A circuit, device, or channel whose response is constant over a specified frequency range. (3-video effects) A straight-line motion path for objects being manipulated by a digital effects device.

Linear Amplifier-An amplifier in which the output signal is linearly proportional to the input.

Linear Combiner-A diversity combiner that adds two or more receiver outputs.

Linear Power Amplifier (LPA)-A linear power amplifier produces an output signal that is linearly proportional to its input (with minimal distortion).

Linear Predictive Coding (LPC)-Linear predictive coding is an analog-to-digital conversion technique that employs a level or multilevel sampling system in which the value of the signal at each sample time is predicted to be a particular linear function of the past values of the quantized signal.

Linear Programming-(1-televisoin) Linear programming is the time sequencing of media programs in a progressive order. (2-software) A software process that is used to find an optimum solution to a linear function, typically through the use of multiple equations.

Linear Receiver-A radio receiver that operates in such a manner that the signal-to-noise ratio at the output is proportional to the signal level at the input, and/or to the degree of modulation.

Linear Television (Linear TV)-Linear television is the providing of television programs in a time sequence.

Linear TV-Linear Television

Line-Rate-(1-Transmission) The raw speed, in bits-per-second, of a transmission protocol on a particular media. (2-Device) A term used to specify a

L

device, such as a switch, is able to operate with all ports at maximum speed and traffic levels. For example, an Gigabit Ethernet switch is said to be able to handle line rate if it can process packets on all ports simultaneously at a gigabit per second AND all the packets are 64-bytes (minimum packet size) in length.

Link-(1-telecommunications) A transmission facility in a telecommunications network. (2-common channel signaling) A communications path between two adjacent common channel signaling nodes. (3-computer program) The part of a computer program, in some cases a single instruction or address, that identifies the location (link) to another program or module. (4-web) Web links (Hyperlinks) are tags, icons, or images that contain a crossed reference address that allows the link to redirect the source of information to another document or file. These documents or files may be located anywhere the link address can be connected to.

Link Access Protocol Balanced (LAPB)-In the integrated services digital network (ISDN) and the public packet-switched network (PPSN), the layer 2 data-link-layer procedures of the X.25 packet-switching protocol. LAPB is similar to the High-level Data Link Control (HDLC) asynchronous balanced mode. This allows a terminal to start a transmission without receiving permission from a system control unit.

Link Access Protocol on D Channel (LAPD)-In the Integrated Services Digital Network (ISDN), the primary data link-layer protocol on the D channel. LAPD is similar to LAPB except it uses a different framing sequence.

Link Budget-Link budget is the maximum amount of signal losses that may occur between a transmitter and receiver to achieve an adequate signal quality level. The link budget typically includes cable losses, antenna conversion efficiency, propagation path loss, and fade margin.

Link Establishment-(1-general)The process of establishing a communication link. This may involve the creation of a physical link and/or the creation of a logical channel on the physical link. (2-Bluetooth)A procedure is used to setup a physical link-specifically, an Asynchronous Connectionless (ACL) link- between two Bluetooth devices using procedures from the Bluetooth IrDA Interoperability Specification and Generic Object Exchange Profile.

Link Layer-The link layer facilitates the detection of and recovery from transmission errors on a specific link connection.

Link Layer Discovery Protocol (LLDP)-A draft standard within the IEEE 802.1 working group which intends to provide a common way for devices connected to each other (say via ethernet) to identify a directly connected device. For example, an IP telephone using LLDP to a connected Layer 2 switch could share information about power requirements, VLAN configurations, and location information for E911 purposes.

Link Manager (LM)-The functional assembly that creates link setup, authentication, link configuration, quality of service (QOS) capabilities, and other management functions. The link manager in a Bluetooth system coordinates the different modes of operation (park, hold, sniff, and active).

Link Margin-Link margin is the amount of signal loss, usually expressed in decibels, that a signal in a communication path can provide an expected quality level of service.

Link Set-A set of signaling links in an SS7 network that connects a pair of adjacent nodes.

Link State Table-A link state table is information that is stored within a router that is used to determine the forwarding path (route) for incoming packets for multicast connections.

Link Supervision-(1-general)The process of monitoring the link status and managing physical and logical changes to the link. (2-Bluetooth)Each Bluetooth link has a timer used to link supervision. This timer is used to detect link loss caused by devices moving out of range, a device's power-down, or failure cases. The scheme for link supervision is described in Bluetooth's Baseband Specification.

LiON-Lithium Ion

Lip Sync-Lip Synchronization

Lip Synchronization (Lip Sync)-Lip synchronization is the process of adjusting the relative timing of audio information so that the playing of audio is aligned with the facial characteristics of a presentation (such as a video display).

Liquid Crystal-A material of low viscosity that matches the shape of the vessel in which it is contained (like a liquid) but has different refractive indices for light, depending on the path direction of the light through the material (like a solid crystal). Under the influence of an electric field, molecules align themselves in specific directions, changing the polarization plane and enabling characters to be made visible in a display panel.

Lissajous Pattern-The looping patterns generated by a CRT spot when the horizontal (x) and vertical (y) deflection signals are sinusoids. The Lissajous pattern is useful for evaluating the delay or phase of two sinusoids of different frequencies. Named for the French physicist Jules Antoine Lissajous.

List Management-(1-marketing) List management is the acquiring, sending and updating of lists of people or companies that share common interests (2-video) Video editing list management is the process that allows the system operator or manager to change the edit the lists of programs.

Listen Before Talk (LBT)-A process of listening to an access channel to determine if the channel is busy before attempting access to the channel or system.

Listen Interval-The time period a communication device will remain in low power. For the 802.11 system, the listen interval is indicated by a 16 bit number.

Listing Services System-An interactive software system that manages customer listing information, including name, address, and telephone number. The data can be used to support customer and network services as well as to compose listings for use in white pages directories and other specialized products.

Literary Options-A literary option is an uncompromisable right to a potential buyer, at any stipulated time, to buy a literary property or the rights there to.

Lithium Ion (LiON)-Lithium ion is a type rechargeable battery technology that is commonly used in portable electronic devices. Lithium ion batteries typically store more energy in the same weight when compared to Nickel Cadmium (NiCd) and Nickel Metal Hydride (NiMH) batteries.

Litz Wire-Litzendraht Wire

Litzendraht Wire (Litz Wire)-A braided wire, with individually insulated fine strands, that gives low resistance at high radio frequencies.

Live-(1-electric circuit) A device or system connected to a source of electrical potential. (2-acoustical) An area in which sound is not greatly absorbed by the walls and timings; the room, therefore, reverberates. (3-media) A media source that is transmitted when the conversion of media first occurs or

within a perceived live time period (up to several seconds of delay).

Live Asset-A live asset is a media source that is being provided in real time.

Live Content-Live content is the real time or near real-time transfer of information from a non-stored content source (such as a news camera) to viewers of that information.

Live Streaming-Live streaming is the process of transferring audio or video streaming for which the clients may not control the playback time of the media. That is, the clients may not control when the stream starts, pause the stream, skip to a different time within the presentation, and so on. Live streaming is often used for broadcast of an event happening in real time.

Live Television-Live television broadcasting is the transmission of video and audio to a geographic area or distribution network in real time or near-real time (delayed up to a few seconds).

Live Video-Live video is image media that is viewed immediately (or within a short delayed period such as a few seconds) when the program media is created (such as at a sports event).

LLC-Logical Link Control

LLDP-Link Layer Discovery Protocol

LM-Link Manager

lm-Lumen

LMDS-Local Multichannel Distribution Service

LME-Layer Management Entity

LMI-Layer Management Interface

LMR-Land Mobile Radio

LNA-Launch Numerical Aperture

LNA-Low Noise Amplifier

LNB Converter-Low Noise Block Converter

LNC-Line Not Cutting

LNC-Low Noise Converter

LNP-Local Number Portability

Load-(1-general) The work required of an electrical or mechanical system. (2-data) The process of inputting programs or data to a computer for storage or manipulation. (3-device output) A circuit or device that receives the output of an amplifier or transmission line. (4-generator) The amount of electric power taken from a generator. (5-magnetic tape) The process of placing a magnetic tape reel on a drive in preparation for recording or playback. (6-telecommunications) A volume of traffic that equals the sum of the holding times for a number of calls or call attempts. Such loads are expressed in either hundred call seconds or erlangs.

Load Balancing-Load balancing is a process of equalizing or redistributing the usage load of line concentrators in a switching system. Depending on the switching system, loads can be balanced simply by controlling the number of subscribers in each class of service assigned to each concentrator. In the case of a severe load, working lines can be physically rearranged to achieve load balance.

Load Box-A box or circuit that simulates a load (power sink) that is used to test the ability of a system to supply energy to a system that uses or absorbs energy.

Load Factor-The ratio of the average load over a designated period of time to the peak load occurring during the same period.

Load Gauge-A load gauge is a meter on a cable winch (cable pulling device) that indicates the load stress (tension) on the cable as it is pulled through a channel or conduit.

Load Monitoring-Load monitoring is the process of observing the stresses such as tensile load or current load during the installation of a cable or operation of a system.

Load Monitoring Cable Puller-A load monitoring cable puller is a pulling device that is used to pull wire or cables through conduits or other cable channel guides that has a tensile load meter to monitor the tensile load on the cable during its installation.

Load Set-In traffic engineering, the matrix of loads that results from the statement of a load from each specified pair of points in a network. A load set is further defined as a time consistent load set.

Loaded Cable-A cable with uniformly spaced loading coils to improve transmission quality.

Loaded Loop-A cable pair with loading coils placed at intervals along its length.

Loading-(1-circuit) The addition of electrical inductance to a metallic transmission line to improve the frequency characteristics of the line. Loading a line increases the distance over which a quality signal can be sent. (2-antenna) The addition of an inductance to enable an antenna to be tuned to a frequency lower than its natural frequency. (3-multichannel communications) The insertion of white noise or equivalent dummy traffic at a specified level to simulate system traffic performance. (4-system) The total signal power of a multichannel system, expressed as the total of the average power on all channels, or as the per channel load that may be carried by all channels. (5-cable) (6-wind) The total ice and wind pressure allowed for in the design of a tower, pole, or line.

Loading Coil-A loading coil is an inductive device (temporary storage of energy in a magnetic field) that is installed in a telephone line to help enhance the frequency response of the line at specific audio frequencies. Unfortunately, loading coils significantly add distortion to high-speed data signals on those lines (such as DSL signals).

This diagram shows that there may be several installed audio loading coils on a single local loop line. Although these loading coils improve the audio frequency response, they must be removed to allow for high-frequency transmission for systems such as DSL.

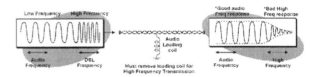

Loading Coils Operation

Loan-Out-Loan-out is an agency that an actor, or talent, hires to settle contracts with the employing studio or producer providing certain revenue advantages.

Lobe-(1-general) A representation of the transmission directional efficiency of a radio antenna; the larger the major lobe, compared with minor lobes, the more directive the system. (2-back) A lobe in an antenna radiation directivity pat-tern pointing directly away from (at 180' to) the intended direction. (3-front) The lobe in the required direction of an antenna radiation directivity pattern. The front lobe is the main or major lobe. (4-minor) Any of the lobes in the radiation directivity pattern of an antenna, except the major or front lobe. Also called side lobe.

Local Access And Transport Area (LATA)-A geographic region in the United States where a local exchange carrier (LEC) is permitted to provide interconnected telephone service. LATAs were created as a result of the division of the company

AT&T by the designated by the Modification of Final Judgment (MFJ). A LATA contains one or more local exchange areas, usually with common social, economic, or other interests.

Local Area Network (LAN)-Local area networks (LANs) are private data communication networks that use high-speed digital communications channels for the interconnection of computers and related equipment in a limited geographic area. LANs can use fiber optic, coaxial, twisted-pair cables, or radio transceivers to transmit and receive data signals. LAN's are networks of computers, normally personal computers, connected together in close proximity (office setting) to each other in order to share information and resources. The two predominant LAN architectures are token ring and Ethernet. Other LAN technologies are ArcNet, AppleTalk, and fiber distributed data interface (FDDI).

This figure shows several of the most popular LAN topologies and their configurations. Some data networks are setup as bus networks (all computers share the same bus), as start networks (computers connect to a central data distribution node), or as a ring (data circles around the ring). This diagram shows for popular types of LAN networks: Thinnet, Thicknet, token ring networks, and Ethernet star network.

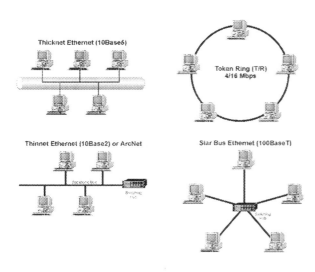

Local Area Network (LAN) Systems

Local Area Network [LAN] Telephone-A telephone that provides telephone services through the use of a local area network (LAN) system.

Local Calling Area (LCA)-Applies only to originating minutes of use and foreign carrier (OHX) account and second dialtone (OHY) accounts. The file contains the subscriber line counts by interexchange carrier (IXC) for each end office in the local access and transport area (LATA). The line counts are used to calculate ratios (factors) that are then multiplied by the IXC's OHY actual or assumed originating minutes of use (MOU) in that LATA to assign MOU to end offices for reclassification. In billing, LCA usually refers to an area within which a customer (typically residential) is not charged for usage.

Local Control-A function of the mobile unit which has been designated to provide special features in addition to those specified by the cellular standard.

Local Digital Switch (LDS)-A digital switch that is the final switching point between the end customer and the public switched telephone network.

Local Exchange-Another term for a end office (EO) telephone switching system. The local telephone company is sometimes called the local exchange.

Local Exchange Carrier (LEC)-Local exchange carriers (LECs) or post and telegraph and telecommunications (PTT) companies provide telephone services directly to residential and business customers located within a localized geographic area. Typically, these telephone companies provide services via copper lines that extend from a local carrier's switching facilities to the end customer's premises equipment (CPE). This is referred to local loop.

Until the early 1990's, most countries had a single company that provided local telephone services. This company was either owned or highly regulated by the government. To increase competition and reduce telephone service prices to consumers, some governments have begun to allow other companies to provide basic (local) telephone service. These competitive local exchange company (CLEC) or competitive access providers (CAPs) provide alternative connections to the public switched telephone networks (PSTN). The established telephone companies are now called the incumbent local exchange carriers (ILECs),

Local Headend-A local headend is part of a broadcast system that selects and processes video signals into local broadcast distribution system.

Local Insert-Local Insertion

Local Insertion (Local Insert)-Local insertion is the process of directing or redirecting media or content from a local source into a broadcast distribution system (such as a television system).

Local Loop-The local loop is the connection (wired or wireless) between a customer's telephone or data equipment and a LEC or other telephone service provider. Traditionally, the local loop (also called "outside plant" or the "last mile") has been composed of copper wires that extend from the local central office, also known as the end office (EO). The EO got its name since it is part of the public switched telephone network (PSTN) that is at the edge, providing physical connections and dial tone to customers.

This diagram depicts a traditional local loop distribution system. This diagram shows a central office (CO) building that contains an EO switch. The EO switch is connected to the MDF splice box. The MDF connects the switch to bundles of cables in the "outside plant" distribution network. These bundles of cables periodically are connected to local distribution frames (LDFs). The LDFs allow connection of the final cable (called the "drop") that connects to the house or building. A NT block isolates the inside wiring from the telephone system. Twisted pair wiring is usually looped through the home or building to provide several telephone connection points, or jacks, so telephones can connect to the telephone system.

Local Loop Unbundling-A requirement that requires incumbent local exchange carriers (ILECs) to provide access on a cost-based rate structure to companies. These companies, such as competitive local exchange carriers (CLECs), desire to provide local access services and require cost effective access into a customer premise.

Local Measured Service-A method of charging customers based on actual usage. Factored into local measured service are the number of local messages, the duration of those messages, the time of day, and the distance within a local exchange area.

Local Multichannel Distribution Service (LMDS)-Local multipoint distribution service is wireless broadband distribution system that operates in the 28 GHz to 31 GHz frequency band. In the United States, LMDS entered into the FCC auction process in 1997.

LMDS uses approximately 1.3 GHz wide spectrum band at around 28 GHz. This provides a typical data rate for each LMDS channel of 1 Gbps. Because of the extremely high frequencies used, the transmitter must be located within 3 to 5 miles of the receiver. The limitation of short distance is that LMDS signals from one antenna will not interfere with other antennas placed 10 or more miles apart. This allows the radio bandwidth to be reused (frequency reused) in a cellular like fashion.

This figure shows a LMDS system. This diagram shows that the major component of a wireless cable system is the head-end equipment. The head-end equipment is equivalent to a telephone central office. The head-end building has a satellite connection for cable channels and video players for video on demand. The head-end is linked to base stations (BS) which transmits radio frequency signals for reception. An antenna and receiver in the home converts the microwave radio signals into the standard television channels for use in the home. As in traditional cable systems, a set-top box decodes the signal for input to the television. Low frequency wireless cable systems such as MMDS wireless cable systems (approx 2.5 GHz) can reach up to approximately 70 miles. High frequency LMDS systems (approx 28 GHz) can only reach approximately 5 miles.

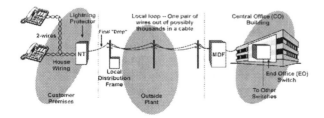

Telephone System Local Loop Operation

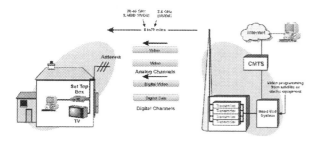

Local Multichannel Distribution Service (LMDS) System

Local Number Portability (LNP)-LNP is the process that allows a subscriber to keep their telephone number when they change service provider in their same geographic area. Local number portability requires that carriers release their control of one of their assigned telephone numbers so customers can transfer to a competitive provider without having to change their telephone number. LNP also involves providing access to databases of telephone numbers to competing companies that allow them to determine the destination of telephone calls delivered to a local service area.

This figure shows an example of the typical operation of local number portability (LNP). In this diagram, a caller in Los Angeles is calling someone in Chicago who has kept (ported) their old phone number when they connected their service to a competitive local exchange carrier (CLEC). This required the incumbent local exchange carrier (ILEC) to move (port) the telephone number to a LNP database. The line connected to the customer from the CLEC actually has a new telephone number (which the customer is not likely to be aware of). The LNP database associates the new number with the old number. This example shows how the call can be routed from an LEC in Los Angeles to the new telephone line in Chicago using the old telephone number. The call is routed from Los Angeles, through a long distance provider (IXC) who knows by the dialed area code that it needs to connect the call into a local telephone company in Chicago. Because there are several local telephone service providers in Chicago, the IXC must look first into a LNP database to see if the number has

been ported to a different service provider. This LNP database (ported telephone number list) must be available to the next to last switch (called "N-1") before the call reaches the end office switch. This LNP database search instructs the last switch to the actual number used for the final connection. The call is then routed to the correct local switching office (new line) so the call can be completed.

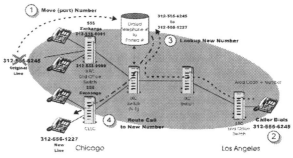

Local Number Portability (LNP) Operation

Local Programming-Local programming is the selection of shows and programs that are offered by a local television network provider. An example of a local program is a news program that is created and broadcasted by a local broadcaster.

Local Routing Number (LRN)-A local routing number is a 10 digit telephone number used for local number portability.

Local Service Request (LSR)-A form used by a competitive local exchange carrier (CLEC) to request local service form an Incumbent LEC (ILEC).

Local Switching System-A switching system that connects lines to lines, and lines to trunks in an end office. The system may be located entirely in a wire center or it may be geographically disposed, as in host remote configurations

Localization-(1-sound) The perception of sound as originating from a particular direction or distance. (2-troublshooting) The process of localizing equipment failures or below tolerance equipment.

Localized Ad Insertion-Localized ad insertion is the process of inserting an advertising message into a media stream such as a television program at a location near (local) to the receiving device. For broadcasting systems, localized ad insertions are performed at the head end near the viewing audience. For IPTV systems, localized ad insertion can be performed at locations that have access to media source and the address of the IP set top box.

Locating Receiver-In a cellular system, a locating receiver is a radio receiver that is located in a base station that can tune to any frequency in an allocated band to find a transmitting mobile radio. The locating receiver can determine the approximate energy level of the transmitting radio to determine if mobile radio requires a handoff to a new cell site that. A locating receiver is also called a scanning receiver.

Location Based Advertising-Location based advertising is the communication of a message or media content to one or more potential customers where the advertising message can vary based on the location of the recipient.

Location Based Services (LBS)-Location based services are information or advertising services that vary based on the location of the user. Mobile radio system may permit the use of different types of location information sources including the system itself or through the use of global positioning system (GPS).

Location Routing Number (LRN)-A telephone number (e.g. 10 digit number) that is used to route calls to and end office switch that allows for the processing of portable (assignable) telephone numbers.

Location Server (LS)-Location servers provide information regarding the location of resources that are located within a network (such as the Internet or within a SIP system). Location servers are typically databases that maintain a binding (mapping) for each registered user. This binding maps the address of the user to one or more addresses at which the user can be currently reached. The Location Service supports user mobility within a communication system. In a SIP system, the Location Service database is updated as a result of SIP User Agents performing a registration.

Location Updates-Location updates are a process where a communication device informs a system as to its physical or logical location within a network. Location updates may be performed periodically to identify a specific physical location or logical address that a device is operating at so that a system can alter the routing or transfer of information

so it can reach the communication device as it changes location.

LocDev-Local Device

Lock-To time synchronize two or more signals, lock to each other.

Lock Code-Wireless unit's built-in functionality which prevents unauthorized use by entering in the user-controlled lock code. It may lock out the keypad or prevent the unit from powering up altogether.

Log File-A file that contains a list of events that have occurred for a particular application or service. The log file is continually updated (added to) as new events occur. Log files are used to analyze problems that have or may occur with a particular application or service.

Log Normalization-Normalizing logs on Network Management Stations (NMS) is a method to organize log files and events, like syslog, snmp traps, RMON statistics, or polled MIBS. Normalizing the data saves on drive space and increases performance on the NMS.

Log Time-(1-general) The time at which a service or program was initiated or terminated. (2- video) new video source is placed on the program bus, usually recorded in the station log for FCC accounting and customer billing purposes.

Logarithmic Scale-(1-meter) A meter scale with displacement proportional to the logarithm of the quantity represented. (2-graph paper) A printed graph paper with one or both of the grids on a logarithmic, rather than an arithmetic, scale.

Logging-Logging is the recording of data about events that occur in a time sequence. Logging can be used with syslog to monitor network events. Logging also applies to monitoring event, application, and system logs on Windows PC based systems. Logging is very useful in troubleshooting and correlating network and system environments and events.

Logging In (Login)-Logging in is a process that allows user to gain access to the system by the identification of the user account (login ID) and the password associated with the account.

Logic Analyzer-A test instrument that is used for monitoring computer system logical operating states and state sequences.

Logic Element-A device that performs a logical function, also known as a logical element or gate.

Logic Gate-The basic decision-making circuit used in digital equipment. A logic gate usually has two or more binary inputs and one binary output.

Simple functions can be implemented by single gates, but several gates of different types, together with various forms of memory, often are combined to form complex decision-making networks.

Logical Channels-Logical channels are a portion of a physical communications channel that is used to for a particular (logical) communications purpose. The physical channel may be divided in time, frequency or digital coding to provide for these logical channels.

Logical Link Control (LLC)-A logical link control layer is used to manage the link transmission across a link. This layer provides access between the communications stack and the transmission medium. It can be used to provide data reformatting and repackaging functions to allow communication between different network types.

Logical One-A logical one is one of two possible states in a binary system. One is normally considered to be the presence of a signal, such as a voltage pulse, in contrast with a zero, or no pulse.

Logical Topology-Logical topology of a network is its' logical interconnection layout. The physical and logical topology does not have to be the same. The Logical topology is the data communication paths that messages take to move between locations on the network.

Logical Zero-A logical zero is one of two possible states in a binary system. Zero is normally considered to be the absence of a signal, such as a 0 Volts.

Login-Logging In

Login ID-A login ID is a name or another form of identification given in order for a user to access a computer, site or network

LOM-Learning Object Metadata

Long Distance-Services charged at a toll rate, or services offered by interexchange companies for traffic that crosses LATAs (InterLATA). (See also: long-haul communications, toll.)

Long Distance (LD)-The connection of calls outside the local service calling area.

Long Haul System-(1-general) A communication system which includes a number of drop/add points, repeaters locations, over long distances that extend outside the local service area. (2-microwave) A microwave system that the longest radio circuit of tandem radio paths exceeds 402 km (250 miles). This diagram shows a terrestrial microwave system-connecting IXC switches in Philadelphia and New York City. The microwave signals are moved between the two switching offices through a series of relay microwave systems located approximately

30 miles apart. Microwave is a line-of-sight technology that must take the earth's curvature into consideration. Also note that microwave towers are not limited to only facing one or two directions. A single tower can be associated with several other towers by positioning and aiming additional transceiver antennas at other microwave antennas on other towers.

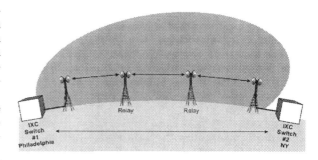

Long Haul Microwave

Long Persistence-A type of phosphor in a cathode ray tube that continues to glow after the original election beam has ceased to create light by producing the usual fluorescence effect.

Longitude-The angular measurement of a point on the surface of the earth in relation to the meridian of Greenwich (England). The earth is divided into 360, of longitude, beginning at the Greenwich mean. As one travels west around the globe, the longitude increases.

Longitudinal Time Code (LTC)-Time code information encoded as an audio like (FM) signal and recorded on audio channels of a videotape or audiotape recorder. LTC is readable at standard tape speed, and greater/slower than standard play speeds, but becomes unusable in still-frame mode.

Look Ahead Preview-The output of a video switcher that permits the operator to observe an effect before it is aired.

Loop Analyzer-A device that analyzes the performance characteristics of a local loop line.

Loop Assignment Center-An operations center that assigns customer loop facilities, telephone numbers, and central-office lines and equipment.

Loop Back Testing-Loop back testing is a process of configuring and sending test signals or information that is relayed back (looped back) to the sender for analysis. The successful reception of information indicates that system parts that the signals or information passed through are working correctly. Loop back testing commonly uses successively larger test loops (e.g. local, mid-distance, remote system) that validate which sections of a communication system are operating correctly.

Loop Battery-A direct current voltage source applied between the conductors of a line and used for loop start, ground start, and loop reverse battery supervision. For loop start supervision, loop battery is used to detect a request for service. For ground start supervision, loop battery indicates that a request for service has been recognized. For loop reverse battery supervision, the loop battery polarity indicates the supervision state of the equipment (on-hook or off-hook) connected to one end of the loop.

Loop Gain-The total gain of all the active devices in a closed loop minus the losses of all the passive devices in the loop.

Loop Plant Improvement Evaluator (LPIE)-A system that analyzes the economics of proposed changes to facilities, such as serving area interface redesign and cable replacement.

Loop Pulsing-Signaling accomplished by the repeated opening and closing of a loop at the originating end of a circuit. Rotary telephone dials are loop pulsing devices.

Loop Reach-Loop reach is the maximum distance that a communication line can provide service to end users.

Loop Signaling-Signaling protocols and processes used in a distribution network, or loop.

Loop Signaling System-Methods for sending signaling information over a communication loop. Loop signals can be transmitted by opening and closing the loop path, reversing the voltage polarity, or varying the line resistance.

Loop Start-A form of line supervision in which a service request is indicated to a network when a terminal enables loop current to flow.

Loop Wire-A wire that links several terminals or adjacent components.

Loopback IP Address-The loopback address 127.0.0.1 is IP address that is used to test a communication link and the communications capability of an IP device by sending back information that it receives. The range of IP addresses 127.0.0.0 through 127.255.255.255 are reserved for loopback testing purposes.

Loopback Testing-Loopback testing is a process of testing the transmission capability and functioning of equipment within a system in which a signal is transmitted through a loop that returns the signal to the source. The test verifies the capability of the source to transmit and receive signals.

This figure shows how loopback testing can be used in an IP Telephony system to progressively test, confirm, and identify failed equipments or portions of a network. In this example, the test signals is created by a test device that is connected to a local area data network. This example shows that the first test involves programming the media gateway to loopback mode so the received test signal from the test device can be returned to the test device. The test device can report if the signal was received and what the quality of the signal is (how many errors). The second test involves programming a remote gateway to loopback mode. This test confirms that the local data network, local media gateway, and wide area network are functioning correctly. The third test in this example sets a remote test device to loopback mode. This test confirms that the local data network, local media gateway, wide area network, remote media gateway, remote data network, and remote test device are working correctly. Failure of one or more of these tests can be used to isolate and help diagnose problems with the system.

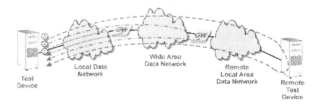

SIP System LoopBack Testing

Looped Clock-An option on digital terminals that enables a digroup transmit clock to be locked to a receive clock. The receive clock always is derived from an incoming DS1 bit stream.

Looping-(1-software) A programming technique by which a portion of a program is repeated until a certain result is obtained. (2-post production) The replacement of dialogue in post production. The term looping is derived from earlier film processing techniques that used loops of film and magnetic film stock to facilitate dialogue replacement.

Loose Construction-A type of fiber optic cable construction in which the fibers are permitted to float freely to relieve stresses and minimize bending losses.

LORAN-Long Range Navigation

LOS-Line Of Sight

Loss Deviation-The change of actual loss in a circuit or system from a designed value.

Loss Variation-The change in actual measured loss over time.

Lossy Coding-Lossy coding is a process of changing information into another for where the new coded representation of the signal may not have the exact characteristics of the original signal. Lossy coding is typically used for data reduction for images or video where the lost or approximated information has limited changes in the perception of information with significant reductions in the data storage or transmission requirements.

Lossy Compression-Lossy compression is a process of reducing an amount of information (usually in digital form) by converting it into another format that represents the initial form of information. However, lossy compression does not have the ability to guarantee the exact recreation of the original signal when it is expanded back from its compressed form.

Lot Size-A specific quantity of similar material or a collection of similar units from a common source; inspection work, the quantity offered for inspection and acceptance at any one time. The lot size may be a collection of raw material, parts, subassemblies inspected during production, or a consignment of finished products to be sent out for service.

Loudspeaker-A transducer (converter) that transforms audio electrical signal into sound waves (audible signals).

Loudspeaker Baffle-An assembly that is mounted on a loudspeaker to help focus the sound waves in a particular direction (such as to the front or side of the speaker).

Loudspeaker Paging-A feature on a communication system (such as a PBX) that permits a user to transmit their voice over one or several loudspeaker systems. Loudspeaker paging systems were commonly used to alert people in a geographic area

(such as on a retail sales area) that they are receiving a call or they are needed at a specific location.

Louver-The slots or holes on the front of a loudspeaker that permit sound to pass, but provide mechanical projection to the device.

Low Earth Orbit (LEO)-A satellite system where the satellites are located approximately 500-1,000 miles above the Earth. LEO systems typically provide mobile satellite services (MSS) to handheld or mobile satellite telephones.

This figure shows an LEO satellite system. In this diagram, a portable satellite telephone is communicating with a landline telephone. The satellite telephone communicates with the closest LEO satellite. Because LEO satellites fly very close to the surface of the earth, they go across the visible horizon in approximately 10 minutes in reference to a mobile satellite customer's location. When the first satellite moves out to the horizon, another LEO satellite becomes available to continue the call. However, robust network communications need to be in place to maintain calls (especially data transmission) within this period. Some systems will use satellite diversity to allow talking through more than one satellite at a time, avoiding call "dropouts" from signal blockage.

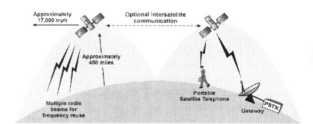

Low Earth Orbit (LEO) Operation

Low Level-(1-MPEG) Low level media formats is a low complexity, low bit rate version of the media. (2-Programming) Low level programming is the creation of programs using commands or instructions that are at or near the level of instructions that are used by the machine or microprocessor (e.g. assembly language).

Low Level Language-A programming language that reflects the structure of a computer or that of a given class of computers. A low level language consists of instructions that are converted directly into machine code.

Low Noise Amplifier (LNA)-A sensitive pre-amplifier used at a focal point (the feedhorn) of a satellite antenna to strengthen the weak satellite signal. The most important parameter of the LNA is its noise temperature, as described in degrees Kelvin. In general, the lower the noise temperature, the better the signal quality. There is a generally a tradeoff between noise temperature of the LNA and the size of the satellite receive antenna. A higher noise temperature rating for an LNA requires a larger diameter antenna to maintain the same level of performance.

Low Noise Block Converter (LNB Converter)-A device that shifts a band of received frequencies to a different (usually lower) frequency band with a small amount of added (unwanted) signal noise. A common application of a LNB converter is the conversion of extremely high-frequency satellite receiver signals (such as the KU frequency band) to a lower frequency (e.g. C frequency band). The LNB converter is often located on or near the satellite receiver antenna to allow the transfer of lower frequency received signals (instead of extremely high frequency signals) for transfer from the satellite antenna (satellite dish) to a nearby head-end building using coax cable or other types of transmission line.

Low Pass Filter (LPF)-A filter that passes frequencies below a frequency cutoff point. Lowpass filters are often used in telephone networks to pass audio frequencies below 4 kHz and block (attenuate) high frequencies.

Low Power Television (LPTV)-Low power television is the ability of broadcasters (e.g. small communities) to offer television services (origination or subscription) via low powered television transmitters. LPTV transmitter output is limited to a 1000 Watts for a UHF stations and 10 watts for a VHF station.

Lower Sideband (LSB)-The sideband of an amplitude-modulated signal containing all frequencies below the carrier frequency.

Low-Tech-Low Tech Products

Low-Tier-A wireless system which uses low-power levels intended for pedestrians and other slow moving traffic.

LPA-Linear Power Amplifier

LPC-Linear Predictive Coding

LPDE-Local Position Determining Entity

LPF-Low Pass Filter

LPFM-Low Power FM

LPIE-Loop Plant Improvement Evaluator

LPTV-Low Power Television

LRN-Local Routing Number

LRN-Location Routing Number

LRS-Lifetime Revenue per Subscriber

LS-Location Server

LSB-Lower Sideband

LSC-Link State Control

LSE-Laser Safety Eyewear

LSI-Large Scale Integration

LSMS-Local Service Management System

LSO-Laser Safety Officer

LSR-Local Service Request

LSSU-Link Status Signal Unit

LST-Line Station Transfer

LT-Line Terminal

LTC-Longitudinal Time Code

LTE-Line Terminating Equipment

LTR-Logic Trunked Radio

Lubricant Sleeve-Lubricant sleeve helps to guide the lubricant into the conduit or cable duct.

Luma-Luminance

Lumen (lm)-A unit of total visible light power output in all directions from a luminous object. Light used for this comparison has a visible spectrum distribution of power corresponding to the spectral radiation from a piece of "black" surface platinum at its standard (normal atmospheric pressure) melting/solidification temperature.

Luminance (Luma)-Luminance is the amount of visible optical energy (intensity), measured in Lumens.

Luminance Border-A non-color, luminance-only fill video for key banners and drop shadows.

Luminance Key-A key effect in which the portions of a key source that are greater in luminance than the clip level cut a hole in the background video.

Luminance Nonlinearity-A video distortion in which the luminance gain of the TV system changes as a function of luminance amplitude. The resulting TV picture will display poor resolution between brightness levels in the nonlinear range.

Luminance Signal (Y)-Luminance (Y) is the part of a video signal that describes the amount of light in each pixel. Luminance is equivalent to the signal provided by a monochrome camera. It may be generated as a weighted sum of the RGB signals in accordance with the formula: $Y = 0.3R + 0.5G. + 0.11B$. Luminance is differentiated from brightness in that the latter is non-measurable and sensory. The color video picture in-formation contains two components: luminance (bright-ness and contrast) and chrominance (hue and saturation). Luminance is the photometric quantity of light radiation.

Luminous Flux-The amount of visible light intensity per square meter (or other area unit).

Lumped Constant-A resistance, inductance, or capacitance connected at a point, and not distributed uniformly throughout the length of a route or circuit.

LW-Long Wave

LZW-Lempel-Ziv-Welch

M

M-Mega

m-Milli

M Format-A component video format for use in videotape recorders. The signal set consists of separate Y, I, and Q signals. The terminology M refers to the way in which the tape is routed through the recording mechanism. M-format is a registered trademark of Panasonic.

M Regions-The areas of the surface of the sun that appear to be responsible for many of the electromagnetic disturbances experienced on earth.

M, m-(1-Metric prefix) Capital or upper case M represents "Mega" or one million (1,000,000). Small or lower case m represents "milli" or 1/1000 or 0.001. (2-Roman Numeral) The Roman Numeral M represents one thousand or 1000. (3- metric unit) Lower case m represents the length unit "meter."

M3UA-MTP3 User Adaptation Layer

MAC-Medium Access Control

MAC-Moves, Adds, And Changes

MAC-Multiplexed Analog Components

MAC Address-Media Access Control

MAC Address-Medium Access Control Address

MAC Algorithm-A set of procedures used by communication devices to coordinate access to a shared communications medium or channel. Examples of MAC algorithms include CSMA/CD and Token Passing.

MAC Channel-Medium Access Control Channel

MAC Layer-The MAC layer is composed of one or more logical communication channels that are used to coordinate the access of communication devices to a shared communications medium or channel (copper, radio, or optical). MAC channels typically communicate the availability and access priority schedules for devices that may want to gain access to a communication system.

MAC Layer-Medium Access Control

MAC Protocol-Medium Access Control Protocol

MACA-Mobile Assisted Channel Allocation

Machine Binding-Machine binding is the process of linking media or programs to unique information that is located within a computer or machine so the media or programs can only be used by that machine.

Machine Code-The instruction code designed into the hard-ware of a microprocessor. Machine code is the direct representation of the computer instruction in memory.

Machine Language-A low-level, native programming language to a specific type of computer or processor, whose instructions consist only of computer instructions. Machine language is a program of binary coded instructions stored in memory.

Macro-(1-computers) An abbreviation for macroinstruction, an instruction that generates a larger sequence of instructions for a computer. (2-video) A special function of some zoom lenses that permits an object to be in focus at closer than usual distances to the objective element. The function usually offers magnification of the object. (3-application) A set of stored keystroke sequences or processes that are grouped to allow the user to perform repetitive control or editing sequences of application commands.

Macro Virus-A macro command that attaches itself to application documents that are capable of running macro (multiple keystroke) commands. Macros are commonly used in Word processing or spreadsheet applications to execute repetitive commands and to open, edit, and delete files on a computer.

Macrobend-A macrobend is a relatively large fiber optic cable bend that can cause attenuation of the optical signal as it is redirected around the bend. Macrobends with a radius of 10 m or more often cause negligible signal loss.

Macroblock-A macroblock is a region of a picture in a digital picture sequence (motion pictures) that may be used to determine motion compensation from a reference frame to other pictures in a sequence of images. Typically a frame is divided into 16 by 16 pixel sized macroblocks, that is, groupings of four 8 by 8 pixel blocks.

Macrocell-A cell site providing coverage over a relatively large geographical area (radius 1-5 miles).

Macrovision-Macrovision is a type of copy protection that is used on commercially produced VHS tapes that creates distortion if the video is copied on a standard VHS tape player.

MAE-Metropolitan Area Exchange

Mag Stripe-Magnetic Stripe

Magenta-A subtractive primary color, also known as "process red.

Magnet-A device that produces a magnetic field and can attract objects of iron, cobalt or nickel. The magnetic field developed around a magnet can attract or repel the fields of other magnets.

Magnetic Card-A card with a magnetizable layer in which data can be stored.

Magnetic Disk-A memory device employing magnetic material coated on a circular base. Data is stored by changing the direction of magnetization of small localized areas or domains along concentric tracks on the surface of the disk. A read/write head moves radially to access any of the tracks.

Magnetic Field-An energy field that exists around magnetic materials and current-carrying conductors. Magnetic fields combine with electric fields in lightwaves and radio waves.

Magnetic Field Strength-The strength of a magnetic field at a point following the direction of the lines of force at that point.

Magnetic Flux-The field produced in the area surrounding a magnet or electric current The standard unit of flux is the Weber.

Magnetic Recording-A method of storing information in magnetic material, such as tape or magnetic disks. Metal particles are magnetically oriented in relation to the frequency and amplitude of the recorded signal. Analog recording stores information as varying frequency and amplitude changes that relate to the orientation of the particles. In digital recording, transitions with changing polarity at a fixed amplitude are recorded at saturation.

Magnetic Stripe (Mag Stripe)-A strip of magnetic material affixed to a badge, credit card, or other item on which data can be recorded and read.

Magnetic Stripe Card-A card that stores information on a magnetic strip.

Magnetism-A property of iron and some other materials, including conductors carrying an electric current, by which external magnetic fields are maintained, other magnets being thereby attracted or repelled.

Magnetization-The exposure of a magnetic material to a magnetizing current, field, or force.

Magnetron-A high-power, ultra-high-frequency electron tube oscillator that employs the interaction of a strong electric field between an anode and cathode with the field of a strong permanent magnet to cause oscillatory electron flow through multiple internal cavity resonators. The magnetron may operate in a continuous or pulsed mode.

MAH-Mobile Access Hunting

MAHO-Mobile Assisted Handoff

MAHO-Mobile Assisted Handover

Mail Robot (Mailbot)-A mailbot is a software program or function that allows the automatic processing and routing of email messages. Mailbots may be autoresponders that automatically reply to messages or mailbots may be used to automatically sort and forward messages to specific recipients.

Mail Server-A host, with its associated network software, that offers electronic mail reception and (optionally) email forwarding service. Users may send messages to, and receive messages from, any other user in the system.

Mailbot-Mail Robot

Mailbox (MBX)-A system for storage and transmission of electronic text messages. Mailboxes are often storage areas on computer hard disks that are managed by mail server computers that interconnect to data networks such as the Internet. Mailbox systems often provide notification of an incoming message and confirmation of delivery.

Main Distribution Frame (MDF)-The wire connection point (wire rack) that is located at or near the central switching that is the point where all local access loops are terminated. The MDF connects cable pairs to the line and trunk equipment terminals of a switching system. The frame also serves as a test point between individual telephone lines and central office equipment. The vertical side carries the outside lines and protective devices. All connections to central office equipment are made on the horizontal side. The main distributing frame also is referred to as a mainframe.

Main Lobe-The main portion of a radiation pattern from an antenna.

Main Profile-Main profile is a common set of protocols and processes that are used to provide standard services. The main profile used in the MPEG system allows for the use of Intra frames (I-Frames), predicted frames (P-Frames) and bidirectional frames (B-Frames). The MPEG main profile also allows for the incorporation of background sprites, interlacing and object shapes with transparency.

Main Titles-Main titles is the scrolling text showed prior to the first scenes of a series or film.

Mainframe-Computer systems that are used for handling large quantities of central data processing and information storage applications. Mainframe computers are used for applications including invoice creation, account reconciliation, and management information reporting.

Maintainability-The probability that a failure will be repaired within a specified time after it occurs.

Maintenance-Any activity intended to keep a functional unit in satisfactory working condition. The term includes the tests, measurements, replacements, adjustments, and repairs necessary to keep a device or system operating properly.

Maintenance Center-An operations center that administers all upkeep and repair work in an outside plant network.

Maintenance Fees-Periodic fees which must be paid over the life of a patent in order to keep the patent in force. Most countries require the payment of maintenance fees. Failure to pay maintenance fees can result in premature expiration of a patent.

Maintenance Measurements-Counts of events and their duration that provide information about the maintenance condition of a network element, especially a switching system. Maintenance measurements can include a subset of traffic measurements.

Maintenance Records-Maintenance records are the history of services and test measurements that are performed on networks, systems and transmission lines. Maintenance records help technicians to troubleshoot communication lines and systems as they provide locations and expected performance results (such as optical communication line losses) at the time the systems were installed and setup.

Maintenance Terminating Unit-The equipment located at a network interface that isolates a terminal from a network for testing purposes.

MAIO-Mobile Allocation Index Offset

Major Trading Area (MTA)-A geographic region within the United States where most of the area's distribution, banking, wholesaling is performed. The United States has been divided into 51 MTAs and personal communications services (PCS) licenses were granted based on MTA.

Make Busy-(1-general) The setting of a line, trunk, or switched equipment unit to make it unavailable for service. To anyone seeking a connection, the circuit appears to be busy. (2-automat-

ic call delivery) The marking of a customer service representative line as busy ("busy out") so the system will not transfer calls to that phone.

Make Busy Leads-Terminal equipment leads at the network interface designated MB and MB1. The MB lead is connected by the terminal equipment to the MB1 lead when the corresponding telephone line is to be placed in an unavailable or artificially busy condition.

Make Interval-In dial pulse signaling, that portion of the pulse cycle during which the dial contacts are closed.

Malfunction-An equipment failure or a fault.

Malfunction Timer-A timer that runs separate from all other functions within a communication device. It continuously counts down and needs to be reset. If the mobile is operating correctly (without failure) this timer will be reset continuously and will not expire. A malfunction timer is used in mobile radios to turn off the transmitter in event of a failure in critical parts of the transceiver.

Malicious Call Trace-A process that allows the identification of the location of an undesired caller. Malicious call trace is activated after the recipient has informed the telephone company. Malicious call trace will work even if the unwanted caller's telephone number is blocked. For privacy purposes, the telephone company may only provide the unwanted caller's telephone number to the public safety authorities (such as the police) rather than directly to the recipient of the unwanted call.

MAN-Metropolitan Area Network

Man Hole (Manhole)-An access hole that allows entry of service personnel into a system or facility.

Man Machine Interface (MMI)-The man machine interface is the definition of how a user will enter (input) and receive information from a device or system.

Man Machine Language-A language designed to facilitate direct user control of a computer. A man-machine language contains inputs (commands), outputs, control actions, and procedures sufficient to ensure the performance of all functions relevant to the operation, maintenance, and installation testing of stored-program control systems.

Managed IPTV-Managed IPTV is the delivery of IP television services over a managed (controlled) broadband access network. They can control and guarantee the quality of television services. Managed IPTV systems are traditionally provided by telephone (telco) or cable service providers.

Managed Object-An atomic element of an SNMP MIB with a precisely defined syntax and meaning, representing a characteristic of a managed device.

Management Information Base (MIB)-Management information bases (MIBs) are a collection of definitions, which define the properties of the managed object within the device to be managed. Every managed device keeps a database of values for each of the definitions written in the MIB. MIBs are used in conjunction with the simple network management protocol (SNMP) as well as RMON to manage networks. MIBs (referred to now as MIB-i) were originally defined in RFC1066.

Management Information Base Browser (MIB Browser)-A management information base (MIB) browser is a graphic user interface (GUI) that allows an administrator to review and change the stored configuration and operational parameters of equipments.

Management Information Base II (MIB-II)-MIB-II obsoletes the original MIB (MIB-I) definition defined in RFC1066. MIB-II is widely used in SNMP and RMON managed networks and was originally defined in RFC1156, but was made obsolete by the more well defined and utilized standard, RFC1213.

Management Information System (MIS)-A system that gathers, organizes, and processes information for a department or a company. MIS systems are developed and used by companies to manage its information needs.

Manchester Encoding-A digital encoding technique that divides each bit period into half periods. A negative to positive transition represents a binary 1, and a positive to negative transition represents a binary 0. The use of Manchester encoding allows for clock recovery as transitions occur on every bit transmission.

This diagram shows how Manchester encoding transfers digital information in the form of positive or negative transitions. This example shows that a logical 1 is indicated by a positive transition and a logical 0 is indicated by a negative transition. This example also shows that Manchester coding forces continual transitions during each bit period and these transitions can be used to synchronize the clock timing signal.

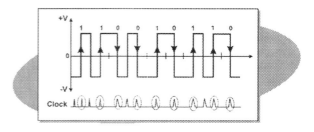

Manchester Encoding Operation

Mandatory Fixed Part-The part of a signaling message that contains those parameters that are mandatory and of fixed length.

Manhole-Man Hole

Manipulation-(1-general) The modification or reformatting of information or data. (2-video image) In a video effects system, the various processes used to alter a video image, such as transformations and programmed effects.

Manometer-A metering device for measuring gas pressure.

Manual Ingestion-Manual ingestion is a process of selecting, adapting and storing media that requires human control.

MAP-The relationship of a logical channel to a specific position in a transmission channel. The process of assigning logical channels to physical transmission channels is called mapping.

MAP-Manufacturing Automation Protocol

MAP-Mobile Application Part

MAPI-Messaging Application Programming Interface

Mapping-A process of assigning information to specific time, frame or code locations on communication channels or circuits. When the information is received, the mapping process can be used to extract the channels or information from the time, frame, or code positions as needed.

Margin-(1-performance) The difference between the value of an operating parameter and the value that would result in unsatisfactory operation. Typical parameters include signal level, signal to noise ratio (SNR) , distortion, crosstalk coupling, and/or undesired emission level. (2-receiver) The signal power available to a receiver in excess of its design limit

Mark-Mark represents a logical value of 1. Mark was defined from the closed circuit condition in a teletypewriter system that actuates a printer function. Mark is the opposite of a space.

Mark In-The point at which an edit on video tape begins, that is, the first frame that will be recorded.

Mark Out-The point at which an edit will end, that is, the first frame that will not be recorded.

Mark Signal-A sequence of marks (logical ones) that is sent before the start of a message or data block.

Market Awareness -The first objective of a brand management or marketing campaign in the Costa Model that states customers must first be aware that a company or product exists and that it offers services they are interested in.

Market Convergence Management -A formal approach used by the managers of different telephone company (telco) product groups to combine their separate smaller customer populations into a much larger, shared pool, allowing the maximizing of revenues through cross selling, brand extension and churn proofing.

Market Demand Curve -Economic model, which defines the level of service (number of ERLANGS) that the market will demand at a given price. The market demand curve is used to help set prices and define market strategies.

Market Development Funds (MDF)-The allocation of funds or sales credit allowances that are given by manufacturers as incentives to retailers to promote their products.

Market Familiarity-A second objective of a marketing activity in the Costa Model that makes the customer familiar with a company name or services and its availability.

Market Granularity -Market granularity defines the way that a company approaches its customers as a group. A telephone company with low granularity will typically divide customers into a small number of segments (business, small business and consumer for example), and then set prices, strategies and treatments for each. A company with a high granularity view will create many more segments, of varying sizes and values, and focusing more precisely on the needs of these smaller, better defined groups.

Market Preference -The third phase of any marketing activity in the Costa Model that motivates the customer to prefer a product or service from one company as compared to a product or service a different company.

Market Saturation-Market saturation is a percentage of market penetration (usage or purchase) of a product or service where the sale of additional products or services to new customers becomes difficult or the marketing cost of promoting the product or service to new customers is beyond the profit generated from the sale of the product or service. A market that experiences market saturation (e.g. above 80% market penetration) may have substantial sales of products for replacement or upgrade of existing customers.

Marketing-Marketing is the process of promoting and selling products or services. Marketing is commonly divided into product (item or service), price (retail, wholesale), promotion (communication), and place (distribution) categories.

Marketing Campaign-Any of a broad range of marketing activities designed to send messages to customers about products, services, and options that are or will be available. Marketing campaigns can be executed via advertising, direct marketing, public relations, place, or other media.

Marketing Channels-Departments, divisions, and external business partners that participate in the process of determining customer wants and needs and communicating how the company can provide these to the customer.

Marketing Program-A series of related marketing campaigns assembled to accomplish a single objective (i.e. a customer retention program could be a series of advertising, sales, and direct marketing campaigns with a related set of messages, concepts, and icons).

Markup Language-Markup languages use text based communications messages to describe formatting and item selection features to be transferred independent of the type of computer system.

MARS-Multicast Address Resolution Server

MAS-Multiple Address System

Mask-(1-semiconductor) A device used in the production of thin-film circuits and other components as a means of restricting patterns or deposits. (2-video) A video key model that allows use of a wipe pattern, box shape, or external mask signal to prevent some undesirable portions of the key source from cutting a hole in the background. The key occurs only in the area covered by the mask pattern. Areas not covered by the mask pattern consist entirely of background video (no key). (3-binary) A code or binary sequence that is used to allow, bock, or modify specific bits that are being transferred through a system or assembly. Binary masks may use AND, OR, NOT, XOR logical operators to block, allow to pass, or modify the binary number. (4-emission)

Mask Invert-A video keyer mode similar to mask except that the sense of the mask is inverted so that the key appears in the area not covered by the mask pattern. The area covered by the mask pattern will consist entirely of back-ground video (no key).

Masker-A sound that reduces the subjective audibility of another sound. An example of a masker is the noise induced into open-plan offices to reduce worker distraction caused by speech intrusion from other work areas.

Masking-(1-semiconductor) The process of covering protected areas of a semiconductor prior to depositing materials on its surface or etching them away. (2-OTDR) A process by which the detector circuit In an optical time-domain reflectometer is shielded from high-power return pulses. (3-sound) The reduction in subjective audibility of one sound by another interfering sound.

Masking Level-The subjective raising of the audibility threshold, in decibels, for a given sound by another sound.

Mass-The ratio of the force applied to an object in ratio to its resulting acceleration. Measured in kg. Proportional to but not equal to weight, although the two terms are loosely used as synonyms in everyday speech. See also weight.

Mass Announcement Network Service-A service that enables many telephone callers to access a selected, prerecorded message simultaneously. Also called Information Access Service.

Mast-A guyed structure meant to support one or more antennas.

Mast Head (Masthead)-(1-publishing) A section in a publication (such as a magazine) that contains identifying information about the publisher of the content. (2-maritime) A masthead is the top of a mast (pole) found on a ship.

Master-(1-media) An original recording data file, videotape, or audiotape. (2-system) A device within a system that is used to coordinate other devices. It is possible for devices within the network to change roles and become a master. An example of this is the Bluetooth system where the master coordinates the other devices within the Piconet.

Master Clock-An accurate timing device that generates a synchronous signal to control other clocks or equipment.

Master Control-A master control is a device, assembly or console that has overriding authority over other controls or consoles in a system.

Master Oscillator-A stable oscillator that provides a standard frequency signal for other hardware and/or systems.

Master Slave Relationship-A relationship within a communication session that assigns the control coordination to the device that assumes the role of master. The master slave relationship can be permanently or dynamically assigned. The dynamic assignment of a master slave relationship is necessary in communication systems where each device can provide similar functions as the other.

This diagram shows two personal digital assistants (PDA) that establish a master slave relationship to allow communication. In this example, PDA 1 requests to send an electronic business card to PDA 2. In this example, PDA 1 attempts to establish a communication session with PDA 2 where PDA 1 is the master. PDA 2 is in the listening mode (typical when it is idle) and hears the request from PDA 1 to establish a communication session. PDA 2 accepts the slave role and follows the commands provided by PDA 1 to allow the transfer and acknowledgement of the business card data transfer. After the transfer is complete, PDA 2 requests to send an electronic business card to PDA 1. This time, PDA 2 attempts to establish a communication session with PDA 1 where PDA 2 is the master. PDA 1 is in the listening mode and hears the request from PDA 2 to establish a session. PDA 1 accepts the slave role and follows the commands provided by PDA 2 as the business card information is transferred.

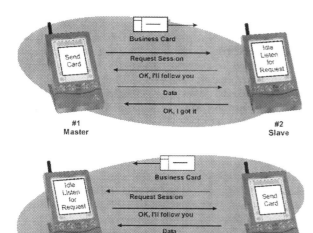

Master Slave Relationship

Master Station-A station in a multiple address radio system that controls, activates, or interrogates four or more remote stations. Master stations performing such functions may also receive transmission from remote stations.

Master/Slave-(1-video editing) A system in which one or more video tape recorders (VTR) slaves are controlled by another VTR master. (2-sync generator) A system in which several video sync generators (slaves) are controlled by one main sync generator (master). (3-Bluetooth) A relationship between devices where one device coordinates communication (the master) and the other device (the slave) follows the commands of the master.

Masthead-Mast Head

Matching-The connection of channels, circuits, or devices in a manner that results in minimal reflected energy.

Material Safety Data Sheet (MSDS)-A material safety data sheet (MSDS) is a document that provides information on the chemicals or other potentially dangerous substances that are in a workplace and potential health effects of exposure they may cause. The MSDS usually contains hazard evaluations on the storage, handling and use of chemicals and the emergency procedures to be taken in response to exposure to specific chemicals.

The MSDS usually contains more information about specific chemicals than is available on the label. MSDS information can be provided by the chemical supplier. MSDS instructions explain the potential hazards for the chemical, how to store and use the product safely and what can happen if the recommended procedures are not followed. It includes information on the symptoms that may result on exposure or overexposure and what actions should be taken if these situations occur.

Matrix-(1-general) A logical network configured in a rectangular array of intersections of input/output signals. (2-disk manufacture) Nickel electroplated onto a lacquer master, forming a negative image of it, and from which the metal is produced. (3-electronics) A routing or switching array with multiple inputs and outputs. (4-mathematics) An arrangement of numbers representing the coefficients in simultaneous linear equations. (5-microphone technique) A circuit combining a unidirectional and a directional microphone into M-S stereo. (6-optical recording) A method of recording for playback channels onto two discreet optical tracks, also referred to as a 4-2A matrix. (7-TV receiver) A circuit that combines the luminance and color signals and transforms them into individual red, green, and blue signals. In a TV set, these signals then are applied to the picture tube grids.

Matrix Switch-A switching and control system that automatically shifts the flow of data from failed lines or vices into functioning equipment. A matrix switch connects to both the front end processor of a computer network and the transmission lines that connect with remote sites.

Matte-A solid color video signal that can be adjusted for chroma, hue, and luminance to till of keys and borders.

Maximal Ratio Combining (MRC)-Maximal ratio combining is the process of combining the signals from two or more antenna elements to increase the level and quality of a received signal.

Maximum Busy Hour-The busiest hour of the busiest day of a normal week, excluding holidays, weekends, and special event days.

Maximum Installation Load-The maximum installation load is the amount of force in pounds or kg that a cable, device or assembly is designed to accept during installation without causing damage or changes in its desired properties.

MB-Megabyte

MBGP-Multicast Border Gateway Protocol

MBONE-Multicast Backbone

Mbps-Mega Bits Per Second

MBps-Mega Bytes Per Second

MBR-Multi-Bit Rate

MBR-Multiple Bit Rate

m-Business-Mobile Business
MBX-Mailbox
MC-Message Center
MC-Multicarrier Mode
MC-Multichannel Carrier
MCC-Mobile Country Code
Mcommerce-Mobile Commerce
MCU-Multipoint Control Unit
MCVD-Modified Chemical Vapor Deposition
MDF-Main Distribution Frame
MDF-Market Development Funds
MDMF-Multiple Data Message Format
MDN-Mobile Data Network
MDN-Mobile Directory Number
MDS-Multipoint Distribution Service
MDT-Mobile Data Terminal
MDU-Multiple Dwelling Unit
ME-Mobile Equipment
Meal Penalties-Meal penalties are fees that are paid to an actor by a producer for not providing the actor or talent with scheduled meal time breaks. These fees are charged when the producers violate the SAG guidelines requiring them to alot a specific amount of time to an actor for meal breaks.
Mean-In statistics, an arithmetic average in which values are added and the sum divided by the number of such values.
Mean Opinion Score (MOS)-Mean opinion score (MOS) is a measurement of the level of audio quality. The MOS is number that is determined by a panel of listeners who subjectively rate the quality of audio on various samples. The rating level varies from 1 (bad) to 5 (excellent). Good quality telephone service (called "toll quality") has a MOS level of 4.0.
Mean Output Power-The calorimetric power measured during the active part of transmission.
Mean Time Between Failures (MTBF)-For a particular time period (typically rated in hours) , the total functioning lifetime of an assembly or item divided by the total number of failures for that item within the measurement time interval.
Measured Load-The load that is indicated by the average number of busy servers in a group over a given time interval.
Measured Rate Service-A usage sensitive telephone service for which a customer pays a reduced monthly charge in exchange for a set amount of service. Usage beyond the set amount is billed at a specified rate.

Measurement-(1-general) A procedure for determining the amount of a quantity. (2-data) The output of a data collection system that indicates the load carried or service provided by a group of telecommunications servers.
Measurement Accuracy-Measurement accuracy is the maximum deviation from a level or value that a device is expected to have.
Measurement Variability-Measurement variability is the amount of change in measurement values that results from repeated measurements. Measurement variability may result from the test settings (such as instantaneous values) and variances in test components and configurations (such as the vibration of optical connectors).
Mechanical Protection-The use of outside cable plant hardware to protect optical or copper cable sheath from manmade, animal or weather-related damage. Examples of this are tree guards, u-guards, and air vents.
Mechanical Rights-Mechanical rights are the authorizations to produce content (mechanically create) in specific physical forms (such as in record or DVD formats).
Mechanical Splice-Mechanical splicing of a fiber is accomplished by attaching the fiber to external holding devices (fixtures) or materials than can align the fiber ends with each other. An index matching gel material may be applied between the two fiber ends to assist in the optical coupling of signals and prevent the entry of foreign materials (dust) in a mechanical splice.
This figure shows the basic process of making a mechanical splice. This diagram shows that the ends of a fiber are inserted into mechanical splice. In this example, the fibers are dipped into an indexing gel before they are inserted and the fiber is rotated as it is inserted. The fiber is inserted until it reaches the mechanical stop in the center of the mechanical splice. After the fiber is inserted, the clamping ring is moved into the center of the mechanical splice to hold the fiber in place.

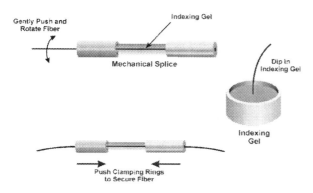

Optical Mechanical Splice

Mechanized Loop Testing-An automated testing system that verifies the condition of a loop and identifies any problems. The analysis of measured values is made available automatically for repair service personnel.

Media-Media is a term that is used to describe a type of information such as voice, data or video.

Media Access Control (MAC) Address-A physical layer address of a network element that is used to identify devices or assemblies within a network. For an Ethernet system, the MAC address is 48 bits (6 bytes) long. A MAC address is commonly called a hardware or physical address.

Media Binding-Creating a relationship of one or more media signals to a communication session. An example of media binding is synchronizing digital audio with a digital video signal.

Media Capture-Media capturing is the process of receiving, converting and storing media. Media capture typically refers to capture of audio and/or video images into digital form.

Media Distribution-Media distribution is the process of transferring information between content providers and content users. The types of media distribution include direct distribution, multilevel (superdistribution) and peer to peer distribution.

Media Elements-Media elements are component parts of media images or content programs. A media element can be the smallest common denominator of an image or media program component. A media element is considered a unique specific element such as a shape, texture and size.

Media Encoder-A media encoder is a device or circuit that converts a signal into a media format that is suitable for transmission over a communication channel.

Media Format-A method of containing audio, video, and/or other digital media within a file structure. Media formats are usually associated with specific standards like MPEG video format, or software vendors like Quicktime MOV format or Windows Media WMA format. In a few cases like MP3 files, the media "format" is little more than a single codec bitstream in a file. However in most cases the media format is not to be confused with the codecs used for any compressed bitstreams within specific files of that format.

Media Gateway (MG)-A media gateway network component which converts one media stream to another. In IP telephony this most commonly refers to a device which converts IP streams (such as audio) to the TDM or analog equivalent. A media gateway may interact with call controllers, proxies, and softswitches via proprietary or standard protocols such as MGCP, Megaco (H.248), and SIP.

There are two main types: Access gateways provide regular analog or primary rate (PRI) interfaces to a voice-over-packet (VoP) network. The inverse function is also available in VoB (voice over broadband) applications: calls are encoded digitally before entering the access network and are routed via conventional telephony once inside. Trunking gateways interface directly between the telephone network and a voice over packet (VoP) network in the core. Such gateways typically manage large numbers of digital virtual circuits.

This diagram shows the functional structure of a media gateway (MG) device. This diagram shows that this gateway interfaces between a public telephone network line side analog connection to a Internet packet (IP) data network connection. The overall operation of the voice gateway is controlled by a media gateway controller (MGC.) The MGC section receives and inserts signaling control messages from the input (telephone line) and output (data port). The MGC section may use separate communication channels (out-of-band) to coordinate call setup and disconnection.

Signals from the public telephone network pass through a line card to adapt the information for use within the media gateway. This line card separates (extracts) and combines (inserts) control signals from the input line from the audio signal. Because this audio signal is in analog form (another option could be an ISDN digital line side connection,) the media gateway converts the audio signal to digital form using an analog to digital converter. The digital audio signal is then passed through a data compression (speech coding) device so the data rate is reduced for more efficient communication. This diagram shows that there are several speech coder options to select from. The selection of the speech coder is negotiated on call setup based on preferences and communication capability of this media gateway and the media gateway it is communicating with. After the speech signal is compressed, the digital signal is formatted for the protocol that is used for data communication (IP packet.) This diagram shows that the call processing section of the media gateway is not part of the gateway. It is a separate controller that commands the gateway to insert messages in the media stream (in-band signaling) or it may communicate with the other gateway through another media gateway controller (MGC.)

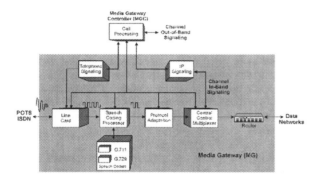

Media Gateway Operation

Media Gateway Control (MEGACO)-Media Gateway Control (MEGACO) is an IP telephony protocol that is a combination of the MGCP and IPDC protocols. MEGACO is specified in H.248.

Media Gateway Control Protocol (MGCP)-MGCP is a control protocol that uses text or binary format messages to setup, manage, and terminate multimedia communication sessions in a centralized communications system. This differs from other multimedia control protocol systems (such as H.323 or SIP) that allow the end points in the network to control the communication session. MGCP

is specified in RFC 2705 and it was first drafted in 1998. MGCP forms the basis of the PacketCable NCS protocol.

Media Gateway Controller (MGC)-The media gateway controller is the portion of a PSTN gateway that acts as a surrogate call management system (CMS). The MGC controls the signaling gateway and the media gateway (MG). The protocols between the MGC and MG include media gateway control protocol (MGCP), MEGACO, and H.323. The MGC acts as a call agent coordinating sessions between devices. Signaling between MGCs (agents) may use SIP or H.323 protocol.

Media Hub-A communication device that distributes or adapts multiple types of communication media to one or several devices in a network through the re-broadcasting of data that it has received from one (or more) of the devices connected to it.

Media Ingestion-Media ingestion is the process of transferring media into a storage or content management system.

Media Library-A media library is a list or group of media files that are associated and/or managed for or by users of the media.

Media Management-Media management is the system, software or processes that are used to store, transfer, categorize and distribute media. IPTV media management systems may allow for the modification of program content descriptive tags (meta tags) along with scheduling and transferring analog (e.g. video tape) and digital media files.

Media Player-A media player is a software application and/or device that can convert media such as video, audio or images into a form that can be experienced by humans. Media players may contain support for service different media formats, compression (codec) formats as well as being able to communicate using multiple network streaming protocols.

Media Processing-The processing of types of media such as playback of voice messages, recording of video, fax generation from computer screen, and speech recognition.

Media Relay-Media relays are servers that provide distributed points of service for media that is rich with content (large quantities of data). They are placed in strategic locations, such as on the "edge" of networks, in the "middle mile" between the media source and destination (often consumer) devices. They are also used to address network address translation (NAT) and firewall traversal issues associated with VoIP traffic.

Media Selector-A media selector is a device or circuit that is used to select a media source from a choice of several media sources.

Media Server (MS)-(1-digtial media) Media servers (sometimes called streaming servers) are computers that receive requests for media, setup a communication session to the requesting media client and provides the downloading or continuous transmission (streaming) of digital media. (2-telephone) Media servers provide common telephony features and/or specialized telephony capabilities to communication systems. The media servers' many functions are to process call connections and manage media access to media resources. Media servers can be hardware based or software-based. Hardware based media servers are specifically designed to efficiently and effectively perform call processing and media management. Software based media servers use software that operates on common computing equipment. Examples of media servers include announcement servers, conference servers, voicemail servers and CALEA server.

Media Skin-A media skin is a graphic and its associated objects that surround a media object. A media skin usually contains a graphic image and buttons that control the playing of media within the media skin.

Media Stream-A media stream is a flow of information that represents a media signal (such as digital audio or digital video). A media stream may be continuous (circuit based) or bursty (packetized).

Media Synchronization-Media synchronization is the process of adjusting the relative timing of media information (such as time aligning audio and video media). Media synchronization typically involves sending timing references in each media stream that can be used to align and adjust the relative timing of multiple media signals.

Media Value Chain-A media value chain is the operational model that describes the core functions that are required to deliver media products or services to the end customer. The blocks in a typical media value chain include content producers, content aggregators, media servers, distribution systems and media players.

Mediation Device-A network device in a telecommunications network that receives, processes, reformats and sends information to other formats between network elements. Mediation devices are commonly used for billing and customer care systems as these devices can take non-standard proprietary information (such as proprietary digital call detail records) from switches and other network equipment and reformat them into messages billing systems can understand.

This figure shows a mediation system that takes call detail records from several different switches and reformats them into standard call detail records that are sent to the billing system. This diagram shows the mediation device is capable of receiving and decoding proprietary data formats from three different switch manufacturers. The mediation device converts these formats into a standard call detail record (CDR) format that can be used by the billing system.

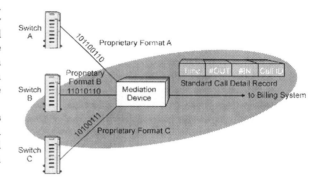

Mediation System Operation

Medium-(1-general) An electronic pathway or mechanism for passing information from one point to another. (2- data storage) A material or device on which data can be stored, such as magnetic tape. (3-transmission system) The structure or path along which a signal propagates, such as a wire pair, coaxial cable, waveguide, optical fiber, or radio path.

Medium Access Control (MAC)-A process used by communication devices to gain access to a shared communications medium or channel. Examples of MAC systems CSMA/CD and Token Passing. A MAC protocol is used to control access to a shared communications media (transmission medium) which attaches multiple devices. The MAC is part of the OSI Data-Link Layer. Each networking technology, for example Ethernet, Token Ring or FDDI, have drastically different protocols which are used by devices to gain access to the network, while still providing an interface that upper layer protocols, such as TCP/IP may use without regard for the details of the technology. In short, the MAC provides an abstract service layer that allows network layer protocols to be indifferent to the underlying details of how network transmission and reception operate.

The MAC protocol also defines the frame format, bit ordering and other characteristics to maintain a reliable network.

This diagram shows the key ways networks can control data transmission access: non-contention based and contention based. This diagram shows that non-contention based regularly poll or schedule data transmission access attempts before computers can begin to transmit data. This diagram shows that a token is passed between each computer in the network and computers can only transmit when they have the token. Because there is no potential for collisions, computers do not need to confirm the data was successfully transmitted through the network. This diagram also shows contention based access control systems allow data communication devices to randomly access the system through the sensing and coordination of busy status and detected collisions. These devices first listen to see if the system is not busy and then randomly transmit their data. Computers in the contention-based systems must confirm that data was successfully transmitted through the network, because there is the potential for collisions.

Medium Access Control (MAC) Layer-The layer of protocols that are used to coordinate access to the medium of transmission (such as radio, light, or electrical signals).

Medium Access Control [MAC] Address-The medium access control (MAC) address used to distinguish between units participating in a data network. MAC addresses are low-level address and are only associated with the MAC layer of the system that the data device is operating in.

Medium Access Control Address (MAC Address)-The physical address of the device on the medium. An example of a MAC address is the 48 bit address of a device on the Ethernet.

Medium Access Control Channel (MAC Channel)-A MAC channel is one or more logical communication channels that are used to coordinate the access of communication devices to a shared communications medium or channel. MAC channels typically communicate the availability and access priority schedules for devices that may want to gain access to a communication system.

Medium Access Control Protocol (MAC Protocol)-MAC protocol is the commands, processes, and procedures that perform functions used to send control messages and coordinate requests and transfers of information through the transmission channel.

Medium Earth Orbit (MEO)-A mobile satellite service (MSS) system where the satellite(s) has orbit heights that range from about 1,000 to 6,500 miles above the Earth.

This figure shows a MEO satellite system. In this diagram, several satellites circle the earth at several thousand miles per hour. In this example, a landline telephone is communicating with a portable satellite telephone. The telephone call is routed through the gateway to the satellite. The satellite transponder converts the frequency and retransmits the signal back to earth. The portable satellite telephone receives the radio signal and converts it back to the original audio signal.

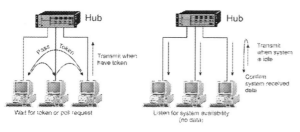

Non-Contention Based   Contention Based

Medium Access Control (MAC) Operation

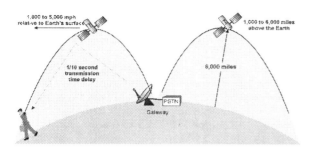

Medium Earth Orbit (MEO) Operation

Medium Frequency (MF)-The frequency band between 300 kHz and 3000 kHz (3 MHz). The wavelength of the MF band ranges from 10m 100m. The MF band also is known as medium wave.

Meet Me Conference-A telephone conference call arrangement, usually on a private PBX system that enables callers to use an access code to connect to a specific conference call.

Meet Point-A point, designated by two exchange carriers, at which one carrier's billing responsibility for service begins and the other's ends. There can be one or more meet points on a circuit.

Meet Point Billing (MPB)-Billing systems that must meet when unbundled network elements (UNE) or access services are provided by two or more providers, or by one provider in more than one state.

Mega (M)-A prefix meaning one million.

Mega Bits Per Second (Mbps)-A measurement of digital bandwidth where 1 Mbps =1 million bits per second (1,000,000 bits per second). The word "mega" is sometimes used to describe the nearest integral power of 2, namely 1,048,567.

Mega Bytes Per Second (MBps)-A measurement of the amount of information being transferred on a communications link in one second where 1 MBps =1 million bytes per second (1,000,000 8 bit bytes per second).

Mega Flops-An acronym for millions of floating point operations per second, a figure of merit for the processing speed of a computer system.

Megabyte (MB)-A megabyte is one million bytes of data. When megabyte is used to identify the amount of data storage space (such as computer memory or a hard disk) , a megabyte commonly refers to 1,048,576 bytes of information.

MEGACO-Media Gateway Control

Megahertz (MHz)-One million hertz, or cycles per second.

Megapixel-A Megapixel is one million image elements (pixels).

Mel-The subjective unit of pitch; 1000 mels is the apparent pitch of a 1 kHz tone 40 dB above the threshold of audibility.

Member Address-A member address is an identification code that is used to identify a specific device that is part of a group for the transmission and reception of data.

Member Body-A member body is an entity (such as a company) that belongs to a group (such as a standards organization).

Memorandum Of Understanding (MOU)-(1-general) A statement or agreement stating the objectives and scope of a project or plan. (2-GSM) A legal agreement between the GSM committee members to create the GSM network and its revisions. The first MOU was signed in 1987.

Memory Capacity-The total number of bits or bytes that can be stored in a device or assembly within a device.

Memory Effect-A condition in a battery where it memorizes its charge cycles. The memory effect is primarily caused when one or more of the charge cycles are not complete. Memory effect is present when the battery is provided with a full charge time interval and the battery will only supply energy for a lesser time due to the charge cycle. Memory effect was a significant problem for NiCd batteries until the early 1990's when the construction of NiCd batteries was changed that virtually eliminated the memory effect.

Memory IC-An integrated circuit incorporating from several hundred to several million logic cells for storing digital information.

Memory Management-The process used in a computing system that manages the assignment and use of memory storage. Memory management systems may involve the allocation of electronic memory (e.g. RAM and ROM) , hard disk memory, and other memory storage systems (e.g. removable disk or tape).

Memory Stick-A memory stick is a portable memory storage device. A memory stick usually can connect to a standard data connection such as a USB port.

MEMs-Micro Electrical Mechanical System

Menu Screen-A computer monitor or display screen format that lists a set of options from which users make a choice.

MEO-Medium Earth Orbit

MES-Mobile Earth Station

Mesh Network-A mesh network is a communication system where each communication device (typically computers) is interconnected to multiple nodes (connection points) in the network where data packets can travel through alternate paths to reach their destination.

Mesh Topology-Mesh topology is the physical and logical relationships between nodes in a network that allows nodes within a network to relay information through other nodes in the network to enable the data to reach its destination.

Message-(1-general) Any idea expressed briefly in a plain or secret language and prepared in a form suitable for transmission by any means of communication. (2-data communications) A set of information, typically digital and in a specific code (such as binary or ASCII), carried from a source to a destination. (3- ISDN) In the integrated services digital network (ISDN), a set of layer 3 information that is passed between customer premises equipment and a stored-program control switching system for signaling. (4-telephone communications) A communication session or a successful call attempt.

Message Body-The data information or message words that are contained in a communication message.

Message Center (MC)-The message center is a node or network function within a communications network which accommodates messages sent and received via short messaging service (SMS).

Message Center Time Stamp-In SMS, a feature that informs the recipient of the local time and date of when the message was accepted by the message center.

Message Discrimination-The process which decides for each SS7 incoming message, whether the signaling point is a destination point or if it should act as a signal transfer point (STP) for that message.

Message Distribution-The process of determining which user part the signaling message is to be delivered to in an SS7 system.

Message Format-The rules for the placement of such portions of a message as its heading, address, text, and end.

Message Investigation-A generic term used to describe the processes and group(s) responsible for investigating call detail records (CDRs) that are rejected by the rating engine portion (rate selection and cost allocation) of a billing system.

Message Processing System (MPS)-The function within a billing system that processes the events recorded in the network. MPS is sometimes referred to as the rating engine. Typically the rating engine receives events from the network, reformats each event into an internal standard, identifies the customer to be billed (see "Guiding"), and assigns a rate (see "Rating") based on parameters such as: date, day-of-week, rate period, call type, jurisdiction, an others.

Message Recording-A message coding system used to distinguish customer dialed messages from operator completed messages.

Message Retrieval-The process of locating a message that has been entered in a telecommunications system.

Message Signal Unit (MSU)-A signal unit (data packet) that carries the signaling information (messages) that are transmitted through an SS7 network. This MSU packet contains control flags (fields) that indicates the protocol that is being transmitted (e.g. mobile application part or ISDN user part) along with a variable length information (message content) field.

This diagram shows that the MSU in the SS7 system is a variable length SIF field that allows the MSU to carry many types of signaling packets. These include SCCP, ISDN-User Part, and OMAP messages.

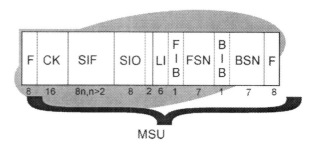

Message Signal Unit (MSU) Structure

Message Switching-A transmission method in which messages are sent to an intermediate point, stored temporarily, and transmitted later to a final destination. The destination of the message usually is indicated in an internal address field (or header) in the message itself.

Message Transfer Part (MTP)-The functional part of a common channel signaling system which transfers signaling messages as required by all the users. The message transfer part also contains, for example, error control and signaling security.

Message Type-A message that is defined according to how it is billed and the way in which it is paid. Message types include: sent paid noncom, third number, credit card, collect, special collect, coin sent paid, and coin collect. Message type also is called message class.

Message URL-An address that is preprogrammed into or used by devices (such as an IP Telephone) where information is kept regarding messages. The message URL may be associated with a message button or an icon on the display of an IP telephone.

Message Waiting Indicator (MWI)-A feature that informs a user that they have messages waiting in email, voice mail, or video mail. Optionally it may indicate how many mail messages are waiting without the user having to call their voice mailbox. MWI may use unique tone (rapidly changing dial time) or an indication on the telephone device (such as a light) as an indication of message waiting. MWI should not impact a subscriber's ability to originate calls or to receive calls. If the dial tone is altered to indicate a message is waiting, it will typically reset to a standard dialtone after a time period or after the user has re-established the connection. As a result, MWI may affect auto-dialers (such as modems) that sense for a dialtone signal before sending the dialed digits. Message waiting indication is also called message waiting notification (MWN)

Messaging-A telephone system feature that alerts station users, via a lighted lamp or other visual display, or by an interrupted dial tone, that messages are waiting.

Messaging Application Programming Interface (MAPI)-A standard program interface that was developed by Microsoft to allow the transfer of messages between software applications.

Messaging Services-Messaging services are the transfer of short information messages between two or more users in a communication system.

Mobile messaging services are typically limited to a few hundred characters per message.

Messenger-A wire that runs in parallel to the conductors which acts as a supportive strand for the attached multi-pair cable in self-supporting cable or dropwires. It also can serve as a bonding or grounding facility on either end.

Meta Tag Generator-A metatag generator is a software program tool that creates metatag codes from information that is usually provided by the web site developer along with the online marketing manager.

Meta Tag Management-Meta tag management is the process of acquiring, maintaining, distributing and the deletion of meta tags that describe program content.

Metadata-Metadata is information (data) that describes the attributes of other data. Metadata or meta tags are commonly used in databases where the attributes of a particular column of data are defined. For example: the phone number column of the employee data table contains groups of numbers 3 or 4 characters in length separated by a hyphen "-".

Metadata Management-Metadata management is the process of identifying, describing and applying rules to content assets.

Metafile-A metafile contains both data (content) and control (formatting) information. An example of a metafile is an HTML web page that contains text, images, and formatting information.

Metascheme-Meta Scheme

Metatag-Meta Tag

Meteor-A metallic or stone body that enters the Earth's atmosphere, burning up during entry. Meteors produce ionization in the atmosphere that can affect long distance radio communications.

Meter-(1-SI unit) The meter of length is now precisely defined as the length of the path traveled by light in vacuum during the time interval 1/ 299 792 458 of a second. It can be equivalently defined as 1 650 763.73 wavelengths of the orange-red light from the isotope krypton-86, measured in a vacuum. Historically, the meter was 1/40 000 000 of a meridian of longitude passing through the poles, as determined in 1797 by a land survey of a partial meridian distance between Barcelona, Spain and Dunquerque, France. (2-ampere-hour) A device that integrates current and time to indicate the number of ampere-hours of power consumed by a load. (3-field strength) A combination radio receiver and meter (calibrated for use with a particular antenna) designed to give a direct reading of the strength of a radio signal at a given point. (4-instrument) A device for measuring the value of some quantity. (5-running time) A totaling clock that runs whenever a device is in operation. Such meters are used with various types of equipment so that maintenance work can be carried out at appropriate times.

Metric System-A decimal system of measurement based on the meter, the kilogram, and the second.

Metropolitan Area Network (MAN)-A MAN is a data communications network or interconnected groups of data networks that have geographic boundaries of a metropolitan area. The network is totally or partially segregated from other networks, and typically links local area networks (LANs) together.

This diagram shows a five node MAN connecting that connects several LAN systems via a FDDI system. This diagram shows that each LAN may be connected within the MAN using different technology such as T1/E1 copper access lines, coax, or fiber connections. In each case, a router provides a connection from each LAN to connect to the MAN.

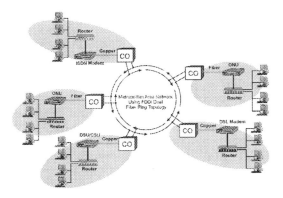

Metropolitan Area Network (MAN) System

Metropolitan Fiber Ring-A fiber optic network that provides high-speed local network capabilities for the connection of businesses and residences to long-distance carrier networks. The ring topology is used in the metropolitan area as it provides a protection switching capability that allows traffic to be quickly rerouted in the event of a fiber cut. SONET

is the most widely deployed metropolitan ring architecture, but newer packet based technologies, such as IEEE 802.17 and other resilient packet rings are being deployed.

Metropolitan Service Area (MSA)-Metropolitan service area or metropolitan statistical area. An area designated by the FCC for service to be provided for by cellular carriers. There are two service providers for each of the over 300 MSA's in the United States.

MExE-Mobile Execution Environment

MF-Medium Frequency

MF-Multifrequency Signaling

MFC-Multifrequency Compelled

MFD-Mode Field Diameter

MFJ-Modification Of Final Judgment

MFN-Multifrequency Networks

MFSK-Multiple Frequency Shift Keying

MG-Media Gateway

MGC-Media Gateway Controller

MGCP-Media Gateway Control Protocol

MH-Modified Huffman Code

MHEG-Multimedia/Hypermedia Expert Group

MHP-Multimedia Home Platform

MHz-Megahertz

MIB-Management Information Base

MIB Browser-Management Information Base Browser

MIB Instance-A MIB instance is a suffix identifier associated with a particular MIB object. Usually the instance has a value of "0" when the MIB object is to return 1 value. The instance suffix identifier can increment as well like in gathering interface statistics, where there can be more than 1 interface and many values for 1 object.

For example: In the case of the Octets MIB object it would look like: 1.3.6.1.2.1.2.2.1.10.<instance number> Where the instance number is the interface number for which you want the received octets for, so if you wanted the first interface then the MIB object definition would look like:

1.3.6.1.2.1.2.2.1.10.1

MIB-II-Management Information Base II

MIC-Media Interface Connector

MICR-Magnetic Ink Character Recognition

Micro Browser (Micro-Browser)-Micro-browser is a software application used to display web page documents in wireless devices (usually using WML formats). Various micro-browsers are available for different types of wireless devices.

Micro Channel-A personal computer bus architecture introduced by IBM in some of its PS/2 series microcomputers. Micro Channel is incompatible with original PC/AT ISA architecture. Micro Channel is a registered trademark of IBM.

Microbend-A microbend is a small bend in an optical fiber due to damage inflicted during manufacturing, installation or other handling of the fiber or to changes in its environment. If the bend is severe enough, light traveling through the core may impact the cladding at an angle beyond the critical angle, resulting in loss of light.

Microbending-In an optical waveguide, a sharp curving involving local axial displacement of a few micrometers, and spatial wavelength of a few millimeters. Such bends may result from waveguide coating, cabling, packaging, or installation.

Microbrowser-An Internet Browser designed for small display screens on smart phones and other handheld wireless devices.

Micro-Browser-Micro Browser

Microcell-A radio coverage area that has a radius of between 200 feet and 1,000 feet.

Microchip-A common term for an integrated circuit component.

Microcode-A set of control functions that are performed by the instruction decoding and execution logic of a computer and define the instruction repertoire of that computer.

Microcomputer-A small-scale program or machine that processes information; it generally has a single chip as its central unit and includes storage and input/output facilities in the basic unit.

Microelectronics-A technology used to build integrated circuits and other small electronic devices and components, sometimes referred to as micro-miniaturization.

Microfilter-A microfilter is a small filter device that attaches to a communications jack (such as a telephone jack) that blocks unwanted signals from entering into the communication device (such as a telephone). A DSL microfilter is a blocking filter device that attaches to a telephone jack that blocks unwanted high-speed data signals from entering into the telephone.

Microinstruction-An instruction of a micro-program.

Micromedia-Micromedia is media or programming that is divided into smaller segments (micro chunks) which are combined (aggregated) with other micro chunks into other format such as blogs or album tracks.

Micrometer-One-millionth of a meter. Usually abbreviated as um.

Micron-A unit of length equal to one millionth part of a meter. Also called a micro-meter, written μm (not to be confused with a measuring device called a micrometer).

Microphone-A transducer that converts sound waves into electrical signals.

Microphonics-Undesirable noise introduced into an audio or video system by mechanical vibration of electric components.

Microportable-A lightweight handheld cellular or PCS mobile telephone.

Micropositioner-A device used to hold and align small parts, such as integrated circuits or optical fibers.

Microprocessor-A single package (normally a single chip) electronic logic unit capable of executing from external memory a series of general-purpose instructions contained in the external memory. The unit does not contain integral user memory although memory on the chip may be present for internal use by the device in performing its logic functions.

Microsecond (usec)-One millionth of a second.

Microvolts Per Meter (MPM)-A measure of the field intensity of a radio signal.

Microwatt (mW)-One millionth of a Watt.

Microwave-(1-radio) The portion of the electromagnetic spectrum between approximately 1 GHz and 100 GHz. (2-heating device) A radio oven that operates at approximately 2.4 GHz..

Microwave Dish-An antenna system that uses a parabolic-shaped reflector to reflect received signals to a specific focal point (signal feed element). Because of the high signal gain and the requirement for precise positioning, dish antennas are commonly used for transmission and reception from point-to-point microwave stations and fixed position GEO communications satellites.

Microwave Frequency-Any of the frequencies suitable for microwave communication. The most commonly used microwave frequencies in use range from approximately 1-10 GHz.

Microwave Link-A microwave link uses microwave frequencies (above 1 GHz) for line of sight radio communications (20 to 30 miles) between two directional antennas. Each microwave link transceiver usually offers a standard connection to communication networks such as a T1/E1 or DS3 connection line. This use of microwave links avoids the need to install cables between communication equipment. Microwave links may be licensed (filed and protected by government agencies) or may be unlicensed (through the use of low power within unlicensed regulatory limits).

Microwave Monolithic Integrated Circuit (MMIC)-Microwave monolithic integrated circuits combine multiple types of active components such as diodes, transistors with passive components such as resistors, capacitors, and inductors on a single semiconductor substrate, typically Gallium Arsenide (GaAs).

Microwave Radio Relay System-A point-to-point radio transmission system in which microwave radio signals are received, amplified, and retransmitted by one or more transmission radios or systems. These systems may use Plesiochronous Digital Hierarchy (PDH), Synchronous Digital Hierarchy (SDH), or other forms of communication channels.

Microwave Semiconductor-Microwave semiconductors are components used for radio frequency signal processing (amplification and filtering) that are constructed from semiconductor materials.

Mid Air Meet-The point midway between separately owned radio transmission facilities at which responsibility for the radio system changes from one company to another.

Mid Section-In a lumped cable loading system, the middle of a load section between loading coils.

Mid Span Meet-Carrier spans of wire or optical fiber whose ownership changes at a demarcation point generally located at a terminal site, but sometimes located at a repeater or splice point. The spans are jointly maintained by its two owners.

Midamble-A midamble is a sequence of bits that the receiving device can recognize and lock onto to help decode the bits surrounding the midamble.

Middleware-Middleware is the software programs that operate between the core application layer of a system and a lower layer of the network. An example of middleware is electronic programming guide (EPGs) that reside in cable converter boxes that allow a customer to select from a list of available video programs.

This diagram shows how middleware is used on an IP Television system to link together the end user's equipment (IP Set Top box) to the media management and delivery systems. This diagram shows that the users set top box has an operating system (e.g. Linux or Windows) which controls the compo-

nents of the set top box (such as infrared remote control and digital video creation). Middleware is the software that communicates with the user and the operating system to send and receive commands and media from the IP television service provider. The IP television service provider has middleware on their media management system that communicates with end customers and manages the selection and delivery of media to the end user. The middleware system also provides information to the billing system to allow the IP television service provider to track usage and create customer billing records.

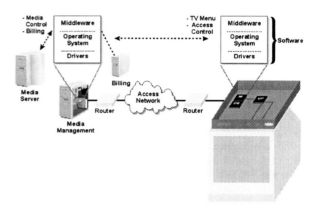

IP Television Middleware

MIDI-Musical Instrument Digital Interface
MIDP-Mobile Information Device Profile
Midrange-Frequencies in the range spanned by the human voice, from approximately 200 to 2000 cycles per second.

Mid-Side Encoding-In stereo audio processing, an alternate representation of the stereo signal by taking the sum of right and left channels (the mid channel) and the difference of left and right channels (the side signal). Especially for audio compression, it is advantageous to encode the mid and side channels separately due to pychoacoustic effects (for example the complete loss of the side channel merely causes the signal to become monaural). This is analogous to the use of YUV color representation in image coding.

Midspan-In aerial telephone company (telco) outside cable plant, the midpoint between two telephone poles. Also in relation to digital loop carrier copper plant, any point on either transmit or receive pairs between repeater locations.

Migration-Migration is the process of converting customers, software applications, or equipment from one technology or system to another technology or system.

Mileage Band-A group of individual mileage steps, such as from 0 to 50 miles, or 50 to 100 miles, measured in airline miles and used to determine billing rates for telecommunications services. This billing rating process is also called "banding."

Milli (m)-A prefix used in a unit of measure meaning one thousandth.

Milliammeter-A measurement instrument that is used to quantify the amount of electric current flowing in a circuit.

Millihenry-A unit of measure equal to one thousandth of a Henry.

Million Instructions Per Second (MIPS)-A unit of measure for the millions of processing steps that can be accomplished by a computer processing device in one second.

Millisecond (msec)-One-thousandth of a second (0.001 S).

Millivolt (mV)-One-thousandth of a volt (0.001 V).

MIME-Multipurpose Internet Mail Extensions

MIMO-Multiple In Multiple Out

MIN-Mobile Identification Number

Mini BNC Connector-A mini-BNC connector is relatively low cost non-contact and non-keyed fiber optic connector. The mini-BNC is commonly used in IBM Token Ring applications.

Mini Digital Video Cassette (MiniDV)-A mini digital video cassette is a small self-contained package of reel-to-reel magnetic tape that is used for video signals and may be played or rewound on demand.

MiniDV-Mini Digital Video Cassette

Minimum Commitment Size-In outside plant administration, the smallest number of pairs that can be committed or recommitted to a permanently connected pair group.

Minimum Cost Routing-In automated facility planning, a circuit-routing scheme that determines a path through the network for each point-to-point demand for each year so that, when point-to-point demands are provided on these paths and the resulting capacity expansion problem is solved, the total cost of transmission facilities is minimized. Minimum cost routing is not related to least cost routing.

Minimum Shift Keying (MSK)-A form of frequency modulation in which the modulating signal shifts the output frequency between predetermined values. Sometimes called fast frequency shift keying.

Minimum Usable Field Strength-The minimum value of the transmitted field strength necessary to permit a desired reception quality, under specified receiving conditions, in the presence of natural and man-made noise, but in the absence of interference from other transmitters.

Minimums-Minimums are the lowest wages (pay) assigned by any union or guild.

Minutes Of Use (MOU)-A measurement (usually billing related) of the number of minutes, actual or assumed, in traffic-sensitive (usage) equipment and facilities.

MIPS-Million Instructions Per Second

Mirror Site-A duplicate data or Internet Web site. Mirror sites are used to process communication traffic to local or regional areas as each mirror sites contain the same information as the other mirror site. Mirror sites are also used for mission-critical applications that allows a company or user to continue processing information in the event of system failure or network loss due to natural disasters.

Mirroring-A fault prevention architecture in which a backup data storage device maintains data identical to that on the primary device, and can replace the primary if it fails.

MIRS-Motorola Integrated Radio System

MIS-Management Information System

Misframe-An error condition in which a line signal entering a digital terminal contains a data bit pattern that simulates an actual framing pattern. Such a simulated pattern can cause a terminal that has previously lost framing to lock falsely onto the undesired pattern.

Mismatch Loss-Mismatch loss is the amount of signal, usually expressed in decibels (dB) , that is lost between incident signal on a transmission line (forward signal) and the reflected signal (return or reflected signal) due to the impedance discontinuity between the transmission source (e.g. transmission line) and receiver (e.g. receiving device.)

Mission Critical-Mission critical are applications or services (missions) that are essential to the operation, performance or safety of individuals or companies.

Mix-(1-video) A transition between two video signals in which one signal is faded down as the other is faded up. (2-audio) The result of a audio mixing session, wherein various inputs, often from a multi-track recording are combined into fewer tracks. This may also refer to a particular set of level of the mixing equipment.

Mix Down-In audio recording, the combining of multiple sources or tracks into a lesser number.

Mixed Media-Mixed media is the combining of media of different types. An example of mixed media is the combining of video and text graphics on a video or television monitor.

Mixer-(1-general) A circuit used to combine two or more signals to produce a third signal that is a function of the input waveforms. (2-audio) An audio console used to switch and combine various audio sources to produce a finished output (3- broadcast) The studio control console or other unit used to combine or "mix" the various program elements into a final program that is sent to the transmitter. (4- receiver) The stage in a superheterodyne radio re receiver at which the incoming signal is modulated with the signal from the local oscillator to produce an intermediate frequency signal. (5 - video) A European term for production switcher. The complete term is vision mixer.

This diagram shows how a mixer combines two signals to produce a sum or difference frequency. This diagram shows this mixer contains a diode (non-linear device) that allows the two-incoming signals to interact with each other to produce the difference (subtractive) frequency and sum (additive) frequencies. The output of this mixer circuit contains a tuned circuit (resonant circuit) that only allows the difference frequency to transfer out of the mixer.

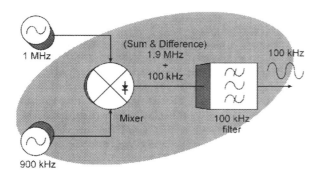

Mixer Operation

MJPEG-Motion JPEG

MKS System of Units, see: Systém International (SI)-The Systém International or meter-kilogram-second version of the metric system.

MLD-Multicast Listener Discover

MLS-Microwave Landing System

MM-Mobility Management

MMC-Multimedia Card

MMDS-Multichannel Multipoint Distribution Service

MMI-Man Machine Interface

MMIC-Microwave Monolithic Integrated Circuit

MMS-Multimedia Messaging Services

MNC-Mobile Network Code

Mnemonic-A memory aid in which an abbreviation or arrangement of symbols has an easily remembered relationship to the subject.

Mnemonic Address-A simple address code with some easily remembered relationship to the actual name of the destination, often using initials or other letters from the name to make up a pronounceable word.

MNP-Mobile Number Portability

Mobile Access Gateway-Mobile access gateways are communications devices that transforms and control data that is received from one network into a format that can be used by a mobile communication devices. A mobile access gateway usually has more intelligence (processing functions) that include authentication, secure encoding of media, and adapting of media into formats suitable that match the capabilities of mobile devices.

Mobile Application Part (MAP)-A set of call processing messages, originally defined for use with GSM, for setup and control of wireless calls via the public switched telephone network. It is normally implemented in conjunction with SS7 call processing messages. The North American standard IS-41 is similar in principle but different in details.

This illustration is a functional block diagram of a wireless network and how it uses MAP protocols between the equipments. This diagram shows that the different versions of SS7 MAP are used between network elements in a wireless network. It also shows that MAP is not used in the radio link. Instead, the relative parts of MAP are transformed into commands that can be sent on the radio links.

SS7 MAP Network

Mobile Carriers-(1-service provider) Companies that provide mobile communication (e.g. cellular or PCS) services. (2-electrons) Free electrons and holes that move through a conductor or semiconductor.

Mobile Commerce (Mcommerce)-A shopping medium that allows wireless devices in a telecommunications network to present products to the customer and process orders.

Mobile Computing-Mobile computing is the process of using computers or information processing devices at more than one location. Mobile computing typically involves the ability of computing devices to access communication systems at different locations.

Mobile Coverage Area-A geographical area that has a sufficient level of radio signal strength to allow two-way communication with mobile radios.

Mobile Data-Mobile data is the transmission of digital information through a wireless network. The term mobile data is typically applied to the combination of radio transmission devices and computing devices (e.g. computers electronic assemblies) that can transmit data through a mobile communication systems (such as a wireless data system or cellular network).

Mobile Directory Number (MDN)-A mobile directory number (MDN) is a number that can be dialed through the public telephone network. A full mobile directory number contains the country code, national designation code, and subscriber number and it can be up to 15 digits.

Mobile Gaming-Mobile gaming is an experience or the actions of a person that are taken on a skill testing or entertainment application with the objective of winning or achieving a measurable level of success on a mobile device (such as a mobile telephone).

Mobile Internet Protocol (Mobile IP)-Mobile IP is a protocol that allows IP communication devices to use the same IP address as it moves between locations and even different types of networks (e.g. Cellular to Ethernet).

Mobile Internet Television-Mobile Internet television is the ability for a user or device to connect to digital television provided through the Internet via mobile connections. An example of mobile Internet television is the requesting and connection to a television gateway or media server via a data connection on a mobile telephone.

Mobile IP-Mobile Internet Protocol

Mobile Number Portability (MNP)-Mobile number portability is the process that allows a subscriber to keep their telephone number when they change telephone service providers.

Mobile Portal-A mobile portal is an Internet web site that acts as an interface between a mobile telephone user and an information service.

Mobile Relay-A mobile transceiver (transmitter and receiver) that is used to repeat (retransmit) a signal received from a mobile radio. When setup as a mobile relay, the mobile radio receiver automatically activates the transmitter when a received signal of sufficient level or code is received.

Mobile Reported Interference (MRI)-The reporting of interference signals from a mobile to its base station.

Mobile Service-A service of radio communication between mobile and land stations, or between mobile stations.

Mobile Station (MS)-A mobile radio telephone operating within a wireless system (typically cellular or PCS). This includes hand held units as well as transceivers installed in vehicles.

Mobile Streaming-Mobile streaming is a method that provides a continuous stream of information that is commonly used for the delivery of audio and video content with minimal delay (e.g. real-time) to mobile devices. Mobile streaming signals are usually compressed and error protected to allow the receiver to buffer, decompress, and time sequence information before it is displayed in its original format.

Mobile Switching Center (MSC)-Switching system that are used for mobile communication networks (cellular, PCS, and 3G.) The MSC was formerly called the mobile telephone switching office (MTSO).

The MSC consists of controllers, switching assemblies, communications links, operator terminals, subscriber databases, and backup energy sources. The controllers, each of which are powerful computers, are the brains of the entire cellular system, guiding the MSC through the creation and interpretation of commands to and from the base stations. In addition to the main controller, secondary controllers devoted specifically to control of the cell sites (base stations) and to handling of the signaling messages between the MSC and the PTSN are also provided. A switching assembly routes voice connections from the cell sites to each other or to the public telephone network. Communications links between cell sites and the MSC may be copper wire, microwave, or fiber optic. An operator terminal allows operations, administration and maintenance of the system. A subscriber database contains features the customer has requested along with billing records. Backup energy sources provide power when primary power is interrupted. As with the base station, the MSC has many standby duplicate circuits and backup power sources to allow system operation to be maintained when a failure occurs.

Mobile Telephone-A mobile telephone is a wireless telephone that operates within a wireless communication system (typically cellular or PCS). This includes hand held units as well as transceivers installed in vehicles.

Mobile Television-Mobile television is the delivery of digital television services to portable devices over wireless connections. Mobile television may be provided by mobile telephone radio channels, satellite channels or other types of wireless channels.

Mobile Video-Mobile video is the transferring of signals that carry moving picture information to mobile devices. Mobile video is commonly associated with supplying video signals to mobile telephones.

This diagram shows the basic operation of a mobile video system. This example shows that a movie is transferred to a mobile system through a video gateway (GW). The video gateway converts the high-speed digital video signal into a low-speed digital video signal that can be sent to the mobile telephone over the mobile radio channel. The mobile telephone contains software that controls the operation of the mobile telephone (operating software) and application software (such as a media player) that allows the mobile telephone to decode and display the digital video signals.

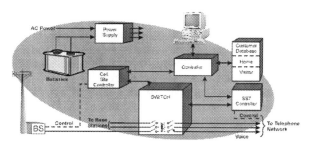

Cellular Mobile Switching Center (MSC)

Mobile Video System

Mobile Virtual Network Operator (MVNO)-A Mobile Virtual Network Operator (MVNO) is a mobile communications service provider that resells the communication services of other wireless communication network operators. MVNO providers purchase airtime (minutes of use) in quantity and resell the airtime to customers they obtain and manage.

MVNO's may provide value added services such as information services, brand labeling, special sales support, and support of unique distribution channels. MVNO's attempt to position their services so customers do not recognize that the operator does not own a network. To provide advanced services, some MVNO operators may own network equipment that interfaces with wireless networks. This may allow MVNO's to have more control over customer databases and SIM cards.

Mobility-Mobility is the capability of a device to operate at different locations. Mobility commonly refers to the ability of communication devices to operate in different geographic areas.

Mobility Management (MM)-Mobility management is the processes of continually tracking the location of mobile telephones or devices that are connected to a communication system. Mobility management typically involves regularly registering telephones or communication access devices. Mobile telephones typically automatically register when they are first turned on (attach) and when they are turned off (detach).

Mobility Services-Mobility services use communication to successfully satisfy the information requirements of the end customer that may be moving throughout a geographic area. While mobility services typically involve the use of wireless communication, mobility services may be applied to other technologies such as Internet access and IP Telephony that can be used at many geographic connection points. Mobility services involve interaction between the mobile device and the communication system to facility registration (location updates), authorization of services (authentication) and delivery (distribution) of information.

MOD-Music On Demand

Mode-(1-general) An electromagnetic field distribution that satisfies theoretical requirements for propagation in a waveguide or oscillation in a cavity. (2-circuit) The method of operation of a device or circuit, such as full duplex or half duplex. (3-dominant waveguide) The simplest mode for propagation of radio waves through a particular waveguide. The longest possible wavelength can be transmitted when using the dominant mode. (4-TE, transverse electric) In a homogeneous isotropic medium, an electromagnetic wave in which the

electric field vector is perpendicular to the direction of propagation. (5-TE, transverse electromagnetic) In a homogeneous isotropic medium, a wave in which both the electric and magnetic field vectors are perpendicular to the direction of propagation. (6-TM, transverse magnetic) In a homogeneous 1" tropic medium, an electromagnetic wave in which the magnetic field vector is perpendicular to the direction of propagation. (7-transmission line) One of the permitted electro-magnetic field distributions. The field pattern of a given mode depends on the wavelength, refractive index, and waveguide geometry.

Mode Of Failure-The physical description of the manner in which a failure occurs and the operating condition of the equipment or part at the time of the failure.

MoDem-Modulator/Demodulator

Modem Pool-A grouping of modems that are shared by users and other network devices to allow different types of communication devices (such as mobile telephones and PSTN telephone lines) to communicate with each other.

Modification Of Final Judgment (MFJ)-A 1982 settlement reached between AT&T and the Department of Justice (DOJ) that required AT&T to divest itself of exchange and exchange access telecommunications services, as well as Yellow Pages directory functions. As a result, the Bell operating companies were separated from AT&T Long Lines, Western Electric, and Bell Telephone Laboratories. The settlement was reached on Jan. 8, 1982, and was approved by a federal district court in Washington, D.C., on Aug.24, 1982. The MFJ replaced the 1956 Consent Decree and settled the 1974 antitrust case of the United States vs. AT&T.

Modified Huffman Code (MH)-A one-dimensional data compression technique that compresses data in an horizontal direction only and does not allow transmission of redundant data. Huffman encoding is a lossless data compression algorithm that replaces frequently occurring data strings with shorter codes. Often used in image compression.

Modular-(1- general) A unit or software program that is independent or may be defined as a separate part of a larger system. (2- telephone cord connector) Trade name for small molded plastic electrical connectors used on handset and mounting (wall) cords. The RJ11 connector is one example of a modular connector. Called Teledapt in Canada.

Modular Design-The process of designing systems or networks as groups of modules. The use of well defined modules (such as a printer module) allow for more simplified upgrading and troubleshooting.

Modular Engineering-An engineering process that provides communication lines in multiples of specific data rate multiples or quantities.

Modulate-The process of varying the characteristics of a carrier waveform to convey information supplied by a modulating signal.

Modulation-(1-general) The process of changing the amplitude, frequency, or phase of a radio frequency carrier signal (a carrier) to change with the information signal (such as voice or data). See also: AM modulation, FM modulation and phase modulation. (2-digital) A modulation process where a code represents the original analog signal. (3) A controlled variation of any property of a carrier wave for the purpose of transferring information.

Modulation Capability-The maximum percentage of modulation that can be successfully carried by a transmitter without introducing distortion which is deemed unacceptable.

Modulation Depth-The modulation factor expressed as a percentage.

Modulation Efficiency-Modulation efficiency is a measure of how much information can be transferred onto a carrier signal. In general, more efficient modulation processes require smaller changes in the characteristics of a carrier signal (amplitude, frequency, or phase) to represent the information signal.

Modulation Index-The ratio of the frequency deviation of the modulated signal to the frequency of the modulating signal.

Modulation Monitor-A test device that is used to measure characteristics of a modulated radio signal. This is typically the percentage of modulation.

Modulation Noise-Intermodulation distortion in an analog amplifier or modulator whose level varies as the input power changes or from other distortions.

Modulator-A device that transfers the intelligence in an information signal onto another signal that is used to better able to transport the information. A modulator modifies a carrier wave by amplitude, phase, and/or frequency as a function of a control signal that carries the intelligence.

Modulator/Demodulator (MoDem)-Modems are devices that convert signals between analog and digital formats for transfer to other lines. Data modems are used to transfer data signals over conventional analog telephone lines. The term modem also may refer to a device or circuit that converts analog signals from one frequency band to another. This figure shows how a data modem converts digital information into analog signals that can be transmitted on an analog communications network. In this example, the data signal comes from a computer (called the data terminal equipment (DTE)), via an RS-232 serial data interface. The RS-232 data interface uses pre-defined signaling commands and data transmission rates to communicate with the data modem. The modem performs a digital-to-analog conversion and from the line to the DTE an analog-to-digital conversion.

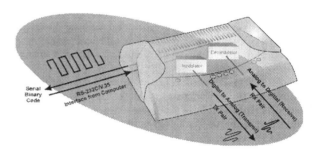

Data Modem Functional Operation

Module-(1-circuit) An assembly replaceable as an entity, often as an interchangeable plug-in item. A module is not normally capable of being disassembled. (2-software) A program unit that is discrete and identifiable with respect to compiling, combining with other modules, and loading.

Module Extender-An assembly that is used to extend the position of circuit board or assembly to permit servicing on that board. A module extender is usually a circuit board with connectors on each end.

MOH-Music On Hold

Moire-A video distortion in which a wavy pattern appears in the picture, caused by two similar high-frequency signals in the image that mix to create a visible low-frequency feat pattern.

Moisture Resistance-Moisture resistance is the ability of a device, assembly or transmission line to resist the entrance or effects of water. In addition to the potential for corrosion from moisture and changes in the electrical performance (shunting or absorbing of signal energy), water that has entered into a device, assembly or transmission line can expand and contract during freezing which may cause physical damage.

Monitoring-The process of listening to or viewing a communication service for the purpose of determining its quality or whether it is free from trouble or interference.

Mono-Mono is one channel of sound.

Monochromatic-A single color or wavelength of light. Monochromatic is an idealized concept. In communication systems, optical radiation has a narrow band of wavelengths.

Monochrome-Monitors and other devices with low-pixel resolution presented in a single color, such as green or amber. Occasionally referred to as green screen.

Monochrome Signal-A single color video signal, usually a black-and-white signal. The term monochrome signal also may refer to the luminance portion of a composite or component color signal.

Monochrome Transmission-The transmission of television signals which can be reproduced in gradations of a single color only.

Montage Effect-In a digital picture manipulator, a recursive effect that develops over time; a composite picture made of several key frame pictures.

Monthly Recurring Cost (MRC)-A cost for service or equipment usage that is continuously charged on a monthly basis.

Monthly Service Charge-A recurring fee that is charged to a customer for the monthly maintenance of their communication service.

Moore's Law-Named after Gordon Moore who in 1965 predicted that the number of electronic devices that can be made upon a circuit chip would double every year.

MOS-Mean Opinion Score

MOS-Metal Oxide Semiconductor

Mosaic Effect-In a digital picture manipulator, an effect in which the picture seems to be made up of a number of small squares or tiles.

MOSMS-Mobile Originated Short Message Service

MOSPF-Multicast Extensions to Open Shortest Path First

Mosquito Noise-Mosquito noise is a blurring effect that occurs around the edges of an image that has a high contrast ratio. Mosquito noise can be created through the use of lossy compression when it is applied to objects that have sharp edges (such as text).

This figure shows an example of mosquito noise artifacts. This diagram shows that the use of lossy compression on images that have sharp edges (such as text) can generate blurry images.

MOSQUITO NOISE

Mosquito Noise Artifacts

Most Significant Bit (MSB)-In a digital word, the first bit in the word sequence defining the largest increment of resolution. In some cases, the MSB is a sign bit signifying the polarity of the word value.

MOT-Multimedia Object Transfer Standard

Motherboard-A circuit board that accommodates plug-in cards or daughter boards and provides for interconnections between them. A motherboard also may provide input/output connections.

Motion Artifacts-Defects in a video picture that are evident during motion.

Motion Compensation-In video compression, a macroblock within a frame can be described as a difference from a macroblock-sized region in another reference frame. The spatial difference between the two macroblocks is called the motion vector, and presumably it implies probable motion of an object or camera perspective between the two frames. If indeed there is motion, encoding only the difference between the two macroblocks (hopefully requiring fewer bits) compensates for that motion.

Motion Decay-A digital picture manipulator effect in which objects in motion are blurred.

Motion Estimation-Motion estimation is the process of searching a fixed region of a previous frame of video to find a matching block of pixels of the same size under consideration in the current frame. The process involves an exhaustive search for many blocks surrounding the current block from the previous frame. Motion estimation is a computer-intensive process that is used to achieve high compression ratios.

This figure shows how a digital video system can use motion estimation to identify objects and how their positions change in a series of pictures. This diagram shows that a bird in a picture is flying across the picture. In each picture frame, the motion estimation system looks for blocks that approximate other blocks in previous pictures. Over time, the digital video motion estimation system finds matches and determines the paths (motion vectors) that these objects take.

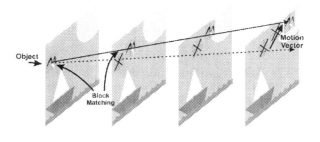

Digitial Video Motion Estimation

Motion JPEG-A video codec that uses JPEG image compression on each frame of video. Since it does not take advantage of the high correlation between frames in a typical video sequence, it requires many times the number of bits for the same quality when compared to a codec with P-frames and B-frames.

Motion JPEG (MJPEG)-A motion JPEG is a digital video format that is only composed of key frames.

Motion Path Animation-A method of choreographing a scene by specifying a "path" for an object and the number of frames over which the motion is to occur: Camera movement can be specified using this technique.

Motion Picture Asociation of America (MPAA)-In reference to an established association responsible for representation of producers and/or distributors of supreme or better quality motion pictures.

Motion Picture Experts Group (MPEG)-Moving picture experts group is a working committee that defines and develops industry standards for digital video systems. These standards specify the data compression and decompression processes and how they are delivered on digital broadcast systems. MPEG is part of International Standards Organization (ISO).

Motion Picture Experts Group 1 and 2, Layer 3 (MP3)-Interlaced scanning is the process of converting lines of images into other forms of information (such as converting a scene into line of video) where the scanning of lines are alternated between the scanning of an image (e.g. odd lines are sent first followed by even line scans). Interlaced scanning results in sending frames of information (images) at twice the rate using the same amount of transmission bandwidth. The use of interlaced scanning in a video system helps to reduce the flicker effect.

The MP3 codec was standardized by the ISO/IEC Moving Picture Experts Group (MPEG) committee in 1992. MP3 is intended for high-quality audio (like music) and expert listeners have found some MP3-encoded audio to be indistinguishable from the original audio at bit rates around 192 kbps. The design of the Layer 3 (MP3) codec was constrained by backward compatibility with the Layer 1 and Layer 2 codecs of the same family. In 1997 the MPEG committee subsequently standardized Advanced Audio Codec (AAC), an improved but non-backward-compatible alternative to MP3.

Motion Picture Experts Group Level 3 (MP3)-An audio signal compression system that is an extension of the Moving Picture Experts Group standard for audio compression.

Motion Sequences-Motion sequences are the creation of moving images through the display of pictures that are displayed in time sequence.

Motion Vector-A motion vector is a line that describes the direction (angle) and how much (length) that an object moves within a digital video sequence.

Motor-(1-electric) A machine that converts electric energy into mechanical energy. (2-series wound) An electric motor with its armature and field windings connected m series. (3-shunt-wound) An electric motor with its armature and field windings connected in parallel. (4-stepper) A type of rotary motor that converts pulses of direct current into rotary steps, one step per pulse. (5-synchronous) An ac motor that operates at a speed controlled by the frequency of the ac supply.

Motor Effect-The repulsion force exerted between adjacent conductors carrying currents in opposite directions.

MOU-Memorandum Of Understanding

MOU-Minutes Of Use

Mount Point-A mount point is the logical location of where a resource is located on a media server, network or storage device. For example, a mount point can be a sub-directory of a computer hard disk on a media server.

Mounting-A support, such as a mounting plate, that holds and integrates a plug-in unit to other circuitry.

Mouse-A handheld device that translates movement or click-button instructions into corresponding movement of a cursor on a video display. Mouse movement, measured in mickeys, relates the on-screen distance a cursor moves to the distance the mouse moves across the desk.

MOV-Movie Format

Movement Estimator-A movement estimator is a process or value that is determined through the analysis of changes in image components (objects). The movement estimator is used to determine where an object (image component) will be repositioned to in one frame (picture) as referenced to another frame in a sequence of moving pictures.

Moves, Adds, And Changes (MAC)-These define a system administration activity responsible for reconfiguring telephone sets and computer workstations on an existing switch or host system. This also includes the reconfiguration of local area network (LAN) devices.

Movie Format (MOV)-Movie format (MOV) is Apple's Quicktime multimedia digital video format that interleaves digital audio and digital video frames into a common file. MOV files have the ability to synchronize digital audio and digital video along with managing other forms of media.

Movies-Moving Pictures

Moving Pictures (Movies)-A movie is a set of images that are played in rapid sequence to create the illusion of movement.

MP-Multifrequency Pulsing

MP-Multipoint Processor

MP3-Motion Picture Experts Group 1 and 2, Layer 3

MP3-Motion Picture Experts Group Level 3

MP4-MPEG-4

MPAA-Motion Picture Association of America

MPB-Meet Point Billing

MPC-Mobile Position Center

MPEG-Motion Picture Experts Group

MPEG (.MPG)-MPEG is a digital media container file format that identifies the digital audio and digital video components into a common MPEG stream or file format.

MPEG Compression-The compression of video signals as the conform to the motion picture experts group (MPEG). There are various levels of MPEG compression; MPEG-1 and MPEG-2. MPEG-1 compresses by approximately 52 to 1. MPEG-2 compresses up to 200 to 1. MPEG-2 typically provides digital video quality that is similar to VHS tapes with a data rate of approximately 1 Mbps. MPEG-2 compression can be used for HDTV channels, however this requires higher data rates. This figure shows how MPEG transmission can be used to combine video, audio, and data onto one packet data communication channel. This example shows that multiple types of signals are digitized and converted into a format suitable for the MPEG packetizers. This example shows a MPEG channel that includes video, audio, and user data for a television message. This example shows that each media source is packetized and sent to a multiplexer that combines the channels into a single transport stream. The multiplexer also combines program specific information that describes the con-

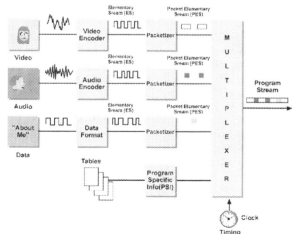

Motion Picture Expert Group (MPEG) Channel Multiplexing

tent and format of the media channels. The multiplexer uses a clock to time stamp the MPEG information to allow it to be separated and recreated in the correct time sequence.

MPEG Level-MPEG levels are the amount of capability that a MPEG profile can offer. MPEG levels can range from low level (low resolution) to high level (high resolution).

MPEG Profile-MPEG profiles are a particular implementation or set of required protocols and actions that enables the providing of features and services for particular MPEG applications. These applications range from providing standard television services over a broadcast system to providing video services on a mobile wireless network. The use of profiles allows an MPEG device or service to only use or include the necessary capabilities (such as codec types) that are required to deliver media to the applications.

MPEG Stream-An MPEG stream is a continuous sequence of digital video information that has a frame structure that conforms to the MPEG standard. MPEG streams are typically composed of video and audio streaming signals that are compressed and error protected to allow the receiver to buffer, decompress, and time sequence information before it is displayed in its original format.

MPEG Tables-MPEG tables are groups of structured information that describe programs, program components or other information that is related to the delivery and decoding of programs. There are many types of MPEG program tables and the more common tables contain listings of programs in a transport channel (PAT), program components (video and audio streams) and conditional access information (to enable decryption and decoding).

MPEG-1-MPEG-1 is a multimedia transmission system that allows the combining and synchronizing of multiple media types (e.g. digital audio and digital video). MPEG-1 was primarily developed for CDROM multimedia applications.

MPEG-2-MPEG-2 is a frame oriented multimedia transmission system that allows the combining and synchronizing of multiple media types. MPEG-2 is the current choice of video compression for digital television broadcasters as it can provide digital video quality that is similar to NTSC with a data rate of approximately 3.8 Mbps.

MPEG-21-MPEG-21 is a multimedia specification that adds rights management capability to MPEG systems.

MPEG-3-MPEG-3 is a discontinued digital video encoding process that was supposed to compress high-definition television (HDTV) media. The capability to compress HDTV was included in MPEG-2 so a separate specification was not required.

MPEG-4 (MP4)-MPEG-4 is a digital multimedia transmission standard that was designed to allow for interactive digital television and it can have more efficient compression capability than MPEG-2 (more than 200:1 compression).

MPEG-7-MPEG-7 is an object oriented (as opposed to the frame oriented MPEG-2) digital video compression and transmission process.

MPLS-MultiProtocol Label Switching
MPLS-Multiprotocol Lambda Switching
MPM-Microvolts Per Meter
MPO Connector-Multifiber Push On Connector
MPOE-Minimum Point Of Entry
MPP-Mobile Party Pays
MPS-Message Processing System
MPT 1327-Ministry of Posts and Telegraph 1327
MPTS-Mulitprogram Transport Stream
MR-Mobile Radio
MRC-Maximal Ratio Combining
MRC-Maximum Ratio Combining
MRC-Monthly Recurring Cost
MRI-Mobile Reported Interference
MRouter-Multicast Router
MRVT-Management Routing Verification Test
MS-Media Server
MS-Mobile Station
MSA-Metropolitan Service Area
MSA-Metropolitan Statistical Area
MSA-Mobile Service Area
MSAN-Multiservice Access Network
MSB-Most Significant Bit
MSC-Mobile Station Class
MSC-Mobile Switching Center
MSC-Mobile-service Switching Center
MSCM-Mobile Station Class Mark
MSDP-Multicast Source Discovery Protocol
MSDS-Material Safety Data Sheet
MSDU-MAC Service Data Unit
msec-Millisecond
M-Services-Mobile Services
MSIC-Mobile System Identification Code
MSID-Mobile Station Identity
MSISDN-Mobile Subscriber ISDN
MSK-Minimum Shift Keying
MSO-Multiple System Operator
MSS-Mobile Satellite Service

MSU-Message Signal Unit

MTA-Major Trading Area

MTA-Multimedia Terminal Adapter

MTBF-Mean Time Between Failures

MTP-Message Transfer Part

MTP level 3-SS7 Message Transfer Part

MTS-Mobile Telephone Service

MTS-Multichannel Television Sound

MTSMS-Mobile Terminated Short Message Service

MTSO-Mobile Telephone Switching Office

Mu (μ) - Law-The type of non-linear digital voice coding (digital signal companding) that is commonly used in the Americas and other parts of the world. The U Law (pronounced Mu Law) coding process is used to compress the 13 bit sampling of a digitized audio signal into the equivalent of an 8 bit sample. It does this by assigning a non-binary (non-linear) value to each of the binary bits. Another non-linear voice coding system is the A Law coding system that is used in Europe and other parts of the world.

MUF-Maximum Usable Frequency

Multiprogram Transport Stream (MPTS)-A multiprogram transport streams is the combining (multiplexing) of multiple program channels (typically digital video channels) onto a signal communication channel (such as a satellite transponder channel). These channels are statistically combined in such a way that the bursty transmission (high video activity) of one channel is merged with the low-speed data transmission (low video activity) with other channels so more program channels can share the same limited bandwidth communication channel.

Multicast Channels-Multicast channels are used to transfer information from one device (or point) to multiple devices (multiple receiving points).

Multi Party Gaming-Multi party gaming is an experience or actions that are taken on a interactive skill testing or entertainment application with the objective of winning or achieving a measurable level of success against or with other game participants.

Multi User Software-An application designed for simultaneous access by two or more network nodes, typically employing file and/or record locking.

Multi Vendor Integration Protocol (MVIP)-A standard protocol used for telephony and data switching that allows the design of advanced telephony applications such as voice mail, PBX,

faxback service and others that are interoperable with other hardware equipment and software programs.

Multibeam Satellite-A satellite that uses multiple directional antennas to allow the same frequency to be reused in different geographic locations.

This figure shows a satellite system that uses spatial division multiple access (SDMA) technology. In this example, a single satellite contains several directional antennas. Some of these antennas use the same frequency. This allows a single satellite to simultaneously communicate to two different satellite receivers that operate on the same frequency. Usually beams that are separated by more than two or three half-power beamwidths can use the same frequencies, as shown in the figure.

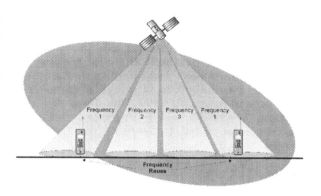

Multibeam Satellite Operation

Multi-Bit Rate (MBR)-Multi-bit rate is the capability of a communications service or transmission lines to operate at different data transmission rates. Multi-bit rate services usually require some real-time interactivity with bursts of data transmission. An example of a MBR application is a video server that can provide digital video signals at different transmission rates.

Multi-Brand Ad-A multi-brand ad is an advertising message that contains brands or logos from two or more products or companies.

Multicast-Multicast service is a one-to-many media delivery process that sends a single message or information transmission that contains a multicast address (code) that is shared by several devices (nodes) in a network. Each device that is part of a multicast group needs to connect to a router (node) in the network that is part of the multicast distribution tree. This means that the multicast media (such as an IPTV channel) is only sent to the users (viewers) who have requested it. The benefit of multicasting is the network infrastructure near the user (e.g. a home) only needs to provide one or two channels at once, drastically reducing the bandwidth requirements.

This figure shows examples of how multicast services can be implemented. The first method uses encoded video broadcast transmission and encoded messaging to allow only a select group to view the received information. While all the television broadcast receivers all receive the same radio signal, only the receivers with the correct code will be able to descramble (decode) the television signal. The second method uses multicast routing in the Internet to store and forward data to an authorized group of recipients that are connected to its router. When a router in the Internet that is capable of multicast service receives a multicast message, it will store the message for forwarding. It then uses the multicast address to lookup a list of authorized recipients in its routing table. The stored message is then forwarded to the authorized receiving device or next router that is part of the multicast service.

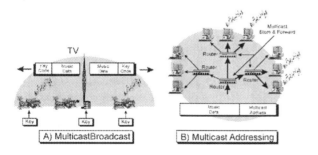

Multicast Operation

Multicast Address-A multicast address is a unique identification code that is used by a group of routers that receive and forward packets for a multicast group. The multicast address is stored in a multicast routing table.

Multicast Address Notation-Multicast address notation is the format of addresses that indicate the multicast group and its source of information. For source trees, the notation is source and group pair (S,G) and for shared trees the notation is any source and group pair (*,G).

Multicast Address Resolution Server (MARS)-Multicast address resolution server is a computer or device that provides IP multicast addresses for a group of communication nodes (known as a cluster). Each node in the MARS cluster is configured with the ATM address of the MARS. A MARS system supports multicasting by overlaying point-to-multipoint connections or through the use of multicast servers.

Multicast Backbone (MBONE)-Multicast backbone is a high-speed data communications system that interconnects the Internet that allows multicast services. The MBONE network is composed of interconnected multicast LANs.

Multicast Border Gateway Protocol (MBGP)-Multicast border gateway protocol are the commands and processes used by routers that are located between different networks to evaluate each of the possible multicast routes for the best one before choosing the routing path for multicast packets it receives and forwards.

Multicast Extensions to Open Shortest Path First (MOSPF)-Multicast open shortest path first is a dense mode multicast protocol that uses a routing process that chooses the next unused shortest path when building a multicast tree.

Multicast Forwarding-Multicast forwarding is the process of receiving data packets, checking multicast routing tables to determine if the packet should be forwarded and copying and sending data packets to paths or destinations that are identified in the multicast routing tables.

Multicast Group-A multicast group is a list of devices or communication points that have been identified as belonging to a specific group. A multicast group can be centrally managed (e.g. setup by an IP broadcaster) or it can be dynamically changed by users who request to be added to the multicast group.

Multicast Listener Discover (MLD)-Multicast listener discover, is the commands, processes, and procedures that are used to send control messages and coordinate the multicasting (simultaneous distribution) of data through an Internet protocol network version 6. MLD is used to establish membership into a multicast group that is operating within a network. Using MLD, users can inform routers within the network that they would like to receive media and control messages from a specific multicast group.

Multicast Polls-Multicast polls are requests for data transmission or responses to commands that are sent from a polling device to several receiving devices.

Multicast Protocol-Multicast protocols are the languages, processes, and procedures that perform functions used to send control messages and coordinate the setup of simultaneous or distributed content to multiple recipients or users. Examples of multicast protocol include IGMP and MSDP.

Multicast Router (MRouter)-A multicast router is a device that receives data packets, determines if the data packets are in its list of multicast addresses (from a multicast routing table) and copies, processes (possibly changing) and forwards (routes) the packets toward members of the multicast group. Multicast routers may use several different protocols to add, manage and remove addresses from its multicast routing table.

Multicast Routing-Multicast routing is the process of reviewing the destination address against a group routing table and forwarding the packet to all the paths that are listed in its group routing table.

Multicast Routing Table-A multicast routing table is a list (a database table) that is located within a router capable of multicasting that is used to determine which multicast packets will be copied and what addresses they should be forwarding to.

Multicast Source Discovery Protocol (MSDP)-Multicast source discovery protocol is a set of commands and process that are used to allow rendezvous points (RPs) to transfer information about media sources between RPs in other domains (other networks).

Multicast Stream-A multicast stream is a flow of data or information that is one-to-many media delivery process that sends a single message or information transmission that contains a multicast address (code) that is shared by several devices (nodes) in a network sent to devices that are part of a multicast group. Each part of the multicast group connects to a router (node) in the network that is part of the multicast distribution tree. A stream may be continuous (circuit based) or bursty (packetized).

Multicast Streaming-Streaming audio or video over the internet based on the IP multicast standard, analogous to a broadcast of the media. If implemented and deployed properly, multicast streaming can enable a single server to stream media to an unlimited number of clients. Since the number of clients receiving a multicast stream may be enormous, multicast streaming systems must be designed to avoid overwhelming the original source server with information from each client, such as requests to retransmit lost packets. Examples of techniques to overcome these problems include multicasting of redundant information, such as the use of forward error-correction codes.

Multicasting-Multicasting is the process of transmitting media channels to a number of users through the use of distributed channels (copying media channels) as they progress through a network. Using multicast medium access control (MAC) or Internet protocol (IP) addresses, multiple end users may "tune" to the same stream of data as it is transmitted over the network. This is in contrast to a unicast transmission whereby multiple copies of the stream, each individually addressed to an end user, are transmitted over the network. Multicasting provides much more efficient use of the network resources, however, individual users are not able to use functions such as pause and fast-forward. Multicasting is very useful for non-interactive applications, such as monitoring many stock prices in real-time.

Multichannel Carrier (MC)-Multichannel carrier (MC) is a communication system that combines or binds together two or more communication carrier signals (carrier channels) to produce a single communication channel. This single communication channel has capabilities beyond any of the individual carriers that have been combined.

This figure shows how a wireless communication system can combine multiple radio communication channels to provide a communication channel that has higher data transmission rates. This diagram shows a wireless communication system that contains multiple radio channels. This example shows that an incoming 1 Mbps data signal has a higher data transmission rate than a single radio channel can provide. To transfer the high-speed data signal, is it split into two (or more) data channels with a lower data transmission rates. Each lower-speed data channel is then sent to an RF channel transmitter. To coordinate the overall flow of data, an additional control function is needed that coordinates the flow of data on the RF channel and to insert multicarrier control messages that allow the receiver of the multiple channels to know how the channels are combined. This diagram shows that the mobile device is able to simultaneously receive and combine each of the radio channels to produce the original high-speed digital signal.

antenna) to allow the same signal to be shared by multiple devices.

This diagram shows a 16 port multicoupler device that allows an antenna signal to be connected to multiple receivers. This multicoupler is composed of one main 4 port (channel) coupler that supplies four additional 4 port (channel) couplers. This example shows that each time a signal passes through a coupler, the signal energy drops, usually in proportion to the number of ports (signal splits) for the multicoupler.

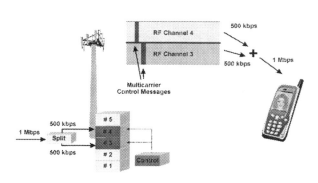

Multichannel Carrier

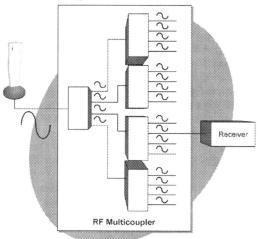

Multicoupler Operation

Multichannel Multipoint Distribution Service (MMDS)-Multichannel multipoint distribution service is the providing of television services through the use of 2.5 GHz microwave frequencies. MMDS is commonly called "wireless cable."

Multichannel Television Sound (MTS)-Any system of aural transmission that utilizes aural baseband operation between 15 kHz and 120 kHz to convey information or that encodes digital information in the video portion of the television signal that is intended to be decoded as audio information.

Multichannel Video Programming Distributor (MVPD)-A person such as, but not limited to, a cable operator, a multichannel multi point distribution service, a direct broadcast satellite service, or a television receive-only satellite program distributor, who makes available for purchase, by subscribers or customers, multiple channels of video programming.

Multicoupler-A device that splits (couples) two or more channels from a signal source (such as an

Multicrypt-Multicrypt is the ability of one device (such as a cable set top box) to have or decode multiple encryption codes.

Multifiber Cable-An optical cable that contains two or more optical waveguides, each of which provides a separate information channel.

Multifiber Connector-An optical connector designed to mate two multifiber cables, providing simultaneous optical alignment of all individual waveguides. It is also called a multi-filter joint.

Multiformat-Multiformat is the capability of a piece of equipment, service or program to access (as inputs) and/or provide (as outputs) multiple signal types, such as digital, analog component or analog composite.

Multiformat Router-A multiformat router is a packet switching device that has the capability to access (as inputs) and/or provide (as outputs) multiple signal types, such as ATM, IP or Ethernet.

Multiframe-Multiframes are groups of frames.

Multiframes are used to allow for the definition of information or logical channels that occur less often than the typical frame period.

Multiframe-Multiple Timeframe

This figure shows the different types of GSM frame and Multiframe structures. This diagram shows that a single GSM frame is composed of 8 time slots. When a radio channel is used to provide a control channel, time slot 0 and the other time slots are used for traffic channels. Fifty one frames are grouped together to form control multiframes (for the control channel). Twenty six frames are grouped together to form traffic Multiframes (for the traffic channels). Superframes are the composition of 26 control multiframes or 51 traffic Multiframes to provide a common time period of 6.12 seconds. Two thousand forty eight Superframes are grouped together to form a Hyperframe. A Hyperframe has the longest time period in the GSM system of 3 hours, 28 minutes, and 53 seconds.

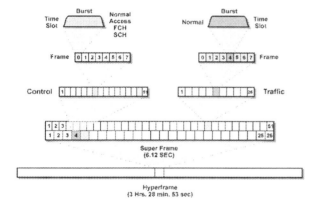

GSM Multiframe Structure

Multifrequency Networks (MFN)-A multifrequency network is a radio system that operates one or several radio transmitters on each radio tower that have difference frequencies to provide one or several communication channels. Because radio frequency signals that operate on one radio channel would interfere with nearby radio towers that could use the same frequency, multifrequency networks can coordinate the frequencies on radio towers within their system so they do not overlap.

Multi-Function Subscriber Identity Module (SIM) Cards-SIM cards that are capable of performing more functions than the providing the identity of a wireless telephone subscriber. These functions may include electronic cash, prepaid calling cards, security access card, medial record storage, electronic airline travel ticket and many other functions.

Multihop-(1-passive) A long distance radio communication service that operates via more than one reflection from the ionosphere. (2-active) A microwave system with several repeaters between the two terminal stations.

Multilateral Peering-Multilateral peering is the process of inquiring or exchanging information among network elements (such as routers) between multiple companies.

Multilayer-A type of printed circuit board that has several layers of circuitry interconnected by electroplated holes from one plane to another.

Multilevel Distribution-Multilevel distribution is the transferring of products, services or content through multiple types or levels of distribution. An example of multilevel distribution is the selling of books from a publisher to a wholesaler who sells books to retail stores who sells to book readers.

Multimedia-Multimedia is a term that is used to describe the delivery of different types of information such as voice, data or video. Because Internet service is often used with broadband (high-speed) data services, it is possible to send multiple types of information at the same time.

Multimedia Card (MMC)-A memory storage card that is used to save and provide multiple types of information media.

Multimedia Computing-A term referring to the delivery of multimedia information via computer.

Multimedia Home Platform (MHP)-Multimedia home platform is an industry middleware standard used in the digital video broadcasting (DVB) system that defines the communication of the DVB system with the integrated receiver device (IRD) and other set top box receivers that have MHP capability. MHP is designed to allow the user to access additional interactive services such as Internet browsing and electronic programming guides.

Multimedia Messaging Services (MMS)-A system that allows pager and short messaging service (SMS) messaging to include graphics, audio or video components.

Multimedia Object Transfer Standard (MOT)- Multimedia object protocol is an object oriented protocol that is used for transferring objects (as opposed to transferring complete frames) during multimedia communication sessions.

Multimedia Telephone-A multimedia telephone is a device or interface that allows for a user to communicate using multiple types of media such as voice, data and video.

Multimedia Terminal-An electronic instrument that is connected to a network that allows a user to communicate by voice, data or video communication. Multimedia terminals vary from personal computers with telephone software to dedicated Ethernet telephones.

Multimedia Terminal Adapter (MTA)-A customer premises device that connects the subscriber's telephone to a managed broadband IP network (HFC cable, ADSL, fiber, wireless) and call control elements in the network to deliver high-quality telephony services. Multimedia Terminal Adapters (MTAs) provide the codecs and all signaling and encapsulation functions required for media transport and call signaling.

Multimedia/Hypermedia Expert Group (MHEG)-The multimedia/Hypermedia expert group oversees the creation of industry standards for multimedia hypermedia information objects that can be exchanged between different services and applications. Information can be found about MHEG at http://www.MHEG.org.

Multimeter-A test instrument fitted with several ranges for measuring voltage, resistance, and current and equipped with an analog meter or digital display readout. The multimeter also is known as a volt-ohm-millimeter (VOM.)

Multimode-(1-mobile telephones) The ability of a mobile telephone to operate on multiple types of radio system. For example, as a cellular telephone and a cordless telephone. (2-Fiber Optics) A term used to describe an optical waveguide that permits the propagation of more than one mode.

Multimode Fiber-An optical fiber that can support many modes of a given wavelength of light. In general, multiple modes are undesirable because they are susceptible to dispersion effects. However, multimode fiber is less expensive to manufacture and less fragile than single mode fiber, so it is often used in less demanding applications.

This diagram shows how multimode fiber transmission uses a relatively wide (50-125 micron) fiber strand to allow several wavelengths of light to pass through the fiber. This example shows that relatively wide optical channel allows optical signals with different wavelengths to make it through the fiber. Some of these signals primarily travel through the center of the fiber while other signals travel from side to side (by refraction) through the fiber strand. Because multiple wavelengths can occupy the same fiber strand, this allows in a higher total bandwidth than single mode fibers. However, this also results in larger group delay as some signals must travel farther (zig-zag as opposed to direct) than others. As a result, this increases signal dispersion as compared to single mode fibers. Signal dispersion can be observed as pulse distortion (edge rounding) increases as the distanced of the fiber transmission increases.

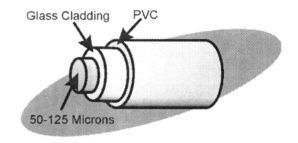

Multimode Fiber Operation

Multimode Laser-A laser that produces simultaneous emission at two or more discrete wavelengths and/or in two or more transverse modes.

Multimode Optical Fiber-An optical fiber that has a large core diameter relative to the wavelength of the light it carries and thus transmits light pulses over many paths. Multimode fiber has greater loss than singlemode fiber and can carry less information, but it is easier to handle and splice.

Multipath-Multipath is the propagation of a radio signal through multiple paths for which part of the signal energy is received before another part of the signal is received that is delayed in time. The delay is due to the extra travel time for the other part of the radio signal that may have been reflected from a building or mountain.

Multipath Fading-The fading (dynamic reduction of signal level) of a radio communications signal at specific locations due to the combining of incoming signals that travel more alternate (multiple) paths. Multipath fading occurs because the path lengths differ and the incoming multipath signals cancel each other as specific points where the signal levels are inverted (opposite).

Multipath Propagation-Multipath propagation the transmission of a radio signal over two or more paths from a transmitter to a receiver. An example of how multipath transmission can effect the reception of a signal is the production of audio distortion in a radio receiver or ghost images (slightly shifted images) on a television display.

Multiple Access-The capability of a communications system to allow more than one user to access to one ore more channels in the system.

Multiple Address System (MAS)-A multiple address radio system is a point-to-multipoint communications system, either one-way or two-way, utilizing frequencies and serving a minimum of four remote stations. If a master station is part of the multiple address system, the remote stations must be scattered over the service area in such a way that two or more point-to-point systems would be needed to serve those remotes.

Multiple Bit Rate (MBR)-Multiple bit rate is the storing and/or streaming of media using different bit rates to represent the media (such as low or high resolution digital video formats).

Multiple Dwelling Unit (MDU)-Multiple dwelling units are facilities or buildings that have multiple living areas. An example of a MDU is an apartment building.

Multiple Frequency Shift Keying (MFSK)-A form of frequency shift keying in which multiple frequency codes are used in the transmission of digital signals.

Multiple In Multiple Out (MIMO)-Multiple input multiple output is the combining or use of two or more radio or telecom transport channels for a communication channel. The ability to use and combine alternate transport links provides for higher data transmission rates (inverse multiplexing) and increased reliability (interference control).

Multiple System Operator (MSO)-A company that owns more than one telecommunications system that provides communications services. In the United States, MSO is the term that is commonly used to describe a company that owns and operates more than one cable television system.

Multiple Timeframe (Multiframe)-The combining of frames or portions of frames (such as a time slot) in a system that continuously sends frames of information to compose a one or more new channels of information.

Multiplex-(1-general) The use of a common channel to make two or more channels. This is accomplished either by splitting of the common-channel frequency band into narrower bands, each of which is used to constitute a distinct channel (frequency-division multiplex), by allotting this common channel to multiple users in turn, to constitute different intermittent channels (time-division multiplex), or by allowing the simultaneous transmission of channels using unique identification codes (code-division multiplex). (2-frequency division) A multiplexing system in which different frequency bands are used by different channels, enabling many different channels to be carried by a single frequency bearer channel. (3-time division) A multiplexing system in which the original analog signals are converted into digital form. The digital signals (for each of many channels) are transmitted sequentially at different time instants. (3-code division) A multiplexing system in which the original signals are converted into digital form and multiplied by a unique identification code. The digital signals (for each of many channels) are transmitted in parallel using different code identifiers.

Multiplex Transmission-The simultaneous transmission of two or more signals within a single channel. Multiplex transmission as applied to FM broadcast stations means the transmission of facsimile or other signals in addition to the regular broadcast signals.

Multiplexed Analog Components (MAC)-Multiplexed analog components is a video system that was developed in Europe during the 1980s that was projected to become a common standard for broadcast transmission systems. One of the systems that evolved from this effort was the D2-MAC hybrid digital audio and analog video system.

Multiplexer (MuX)-A device that conveys two or more telephone or data conversations or connections on a single channel or link. Multiplexing may be in the form of frequency division (e.g. multiple radio channels on a coax line), time division (e.g. slots on a T1 or E1 line), code division (coded channels that share the same frequency band) or combinations of these.

Multiplexer Protocol-A multiplexer protocol is a language, process, or procedure that performs functions used to send control messages that coordinate the combining and separation of multiple (multiplexed) communication channels over a transmission link.

Multiplexing-Multiplexing is a process that divides a single transmission path to parts that carry multiple communication (voice and/or data) channels. Multiplexing may be time division (dividing into time slots), frequency division (dividing into frequency bands) or code division (dividing into coded data that randomly overlap).

This diagram shows how multiplexing can combine two or more low speed channels into one higher speed communication channel. In this diagram, there are eight 8 kbps communication channels that are supplied to a multiplexer. The multiplexer stores and sends 8 bits of each slow speed communication channel during each 125 usec time slot on the 64 kbps channel.

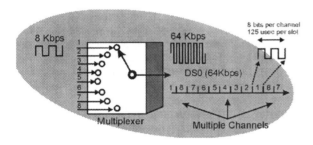

Multiplexing

Multiplier-A circuit in which one or more input signals are mixed under the control of one or more control signals. The resulting output is a composite of the input signals, the characteristics of which are determined by the scaling (multiplication) specified by the circuit.

Multiplier - Billing-The value used in determining a billed partial charge. For instance, if an average calculation is used, every month has an average of 30.417 days (365/12); therefore a customer who subscribes to a service on the 8th of the month will be charged (Y/30.417) x8 (where Y is the total charge for the month).

Multipoint-Multipoint systems can transfer information from one device (or point) to multiple devices (multiple receiving points).

Multipoint Control Unit (MCU)-A control system that allows the coordination and interaction between multiple communication devices. Usually, these devices are part of a conference call or multicast communication session.

Multipoint Distribution Service (MDS)-A one-way domestic public radio service rendered on microwave frequencies from a fixed station transmitting (usually in an omni-directional-directional pattern) to multiple receiving facilities located at fixed points.

Multipoint Video Distribution Systems (MVDS)-Multipoint video distribution system is wireless broadband distribution system that operates in the 40.5 GHz to 43.5 GHz frequency band. The MVDS system was defined by the European radiocommunications committee (ERC) in 1999.

Multiprocessor-A processing method in which program tasks are logically and/or functionally divided among a number of independent central processing units, with the programming tasks being simultaneously executed.

Multiprogramming-Programming that enables a single central processing unit or computer to execute two or more interleaved programs concurrently.

MultiProtocol Label Switching (MPLS)-A network routing protocol that is based on switching through the use of tag labels. The MPLS standard is being developed by the IETF.

Multipurpose Internet Mail Extensions (MIME)-A data communication format this allows information blocks (such as binary images and multimedia data) to be sent with email messages that may be developed primarily for text (7-bit) characters. MIME is defined in RFC 1521.

Multirate-Multirate is a process that offers multiple coding or data transmission rates in a communication system.

Multiroom Digital Video Recorder (Multiroom DVR)-A multiroom digital video recorder is a device that stores video images in digital format that allows multiple devices in a home to access the stored video programs. Multiroom

STBs commonly use Internet protocols to control and distribute media. By keeping TV channels as IP single program streams, it makes the task of allowing content to be available in any room in the home over an internal IP network much easier. Secondary TV's in bedrooms and dens, with their own STB can access live content, or can share the hard disk of a Digital Video Recorder attached to the main TV, to give access to Pause-live-TV or EPG recorded programs throughout the home, transparently.

Multiroom DVR-Multiroom Digital Video Recorder

Multiservice Access Network (MSAN)-A multi-service access network is a portion of a communication network (such as wired or wireless networks) that allows individual subscribers or devices to connect to the core network and receive multiple types of services (such as IPTV, IP Telephony and Internet web access).

Multiservice Switch-A switch that provides multiple channel connections that can each have varying bandwidths and different levels of quality of service (QoS).

Multi-Site Enterprise-A business or corporation that has multiple locations that are part of its overall business.

Multitasking-The process of switching from one task to another on a computer without losing track of either. Multitasking usually is accomplished by time slicing shared resources.

MUSA-Multiple Unit Steerable Antenna

Music On Demand (MOD)-Music on demand is a service that provides end users with the ability to interactively request and receive music or audio content. These audio services are from previously stored media (alblums) or have a live connection (music events in real time).

Music On Hold (MOH)-A feature that connects a source of music to a telephone line that is on hold.

Musical Instrument Digital Interface (MIDI)-Musical instrument digital information is an industry standard connection format for computer control of musical instruments and devices. MIDI contains voice triggers to initiate the creation of new sounds that represent musical notes.

MUSICAM-MUSICAM is an audio coding standard that uses a relatively medium-complexity audio analysis system to characterize and compress audio signals. The MUSICAM system achieves medium compression ratios dividing the audio signal into sub bands, coding these sub bands and multiplexing them together. The MUSICAM system is used in the (DAB) digital audio broadcasting system.

Mute-A control function that turns the audio input or output of a device to silent (off) mode.

Muting-Muting is the process of inhibiting audio (squelching). Muting can be automatic (such as when interference is detected) or can be manually enabled (by the user).

MuX-Multiplexer

mV-Millivolt

MVDS-Multipoint Video Distribution Systems

MVIP-Multi Vendor Integration Protocol

MVNO-Mobile Virtual Network Operator

MVPD-Multichannel Video Programming Distributor

MW-Medium Wave

mW-Microwatt

MWI-Message Waiting Indicator

MWNE-Managed Wireless Network Entity

N

n-Nano

NA-Numerical Aperture

NA Mismatch-Numeric Aperture Mismatch

NAB-National Association Of Broadcasters

NACD-Network Automatic Call Distribution

NACK-Negative Acknowledgement

Nack Message-Negative Acknowledgement Message

NACN-North American Cellular Network

Nailed Up Connection-The assignment of a long term (permanent) dedicated path that created a network.

Naked DSL-Naked DSL is digital subscriber line service that does not include dialtone service.

NAM-Number Assignment Module

Name Discovery-(1-general) A procedure that provides the initiating device with the device name of other connectable devices. (2-Bluetooth) The process of identifying other devices located nearby in a Bluetooth system. These Bluetooth devices within range that will usually respond to paging and requests for name identification information using service discovery protocol (SDP).

Name Resolution-A process of translating a text based name into a numeric address, such as Internet Protocol (IP) address. See domain name server (DNS).

Name Server-The data processing device that translates text names to address information (such as an Internet addresses).

NAMPS-IS-88

NAMPS-Narrowband Advanced Mobile Phone Service

Nano (n)-A metric preface representing 0.000 000 001.

Nanosecond (nsec)-A measurement of time using 0.000 000 001 of a second.

NANP-North American Numbering Plan

Narrative Ads-Narrative ads or information segments that provide narrative audio that provides additional information about the images in the information segment. The use of narrative ads allows advertisers to engage with their audiences and can become especially powerful when used in combination with frequency capping.

Narrowband Channel-A transmission channel whose bandwidth can be wholly contained within the bandwidth of the information being transmitted (e.g. bandwidth a 4 kHz voice signal is carried in a radio channel with a frequency bandwidth of 4 kHz.

Narrowcasting-The transmission of information to an audiences that have specific characteristics such as automobile owners of a specific type of car.

NAS-Network Access Server

NAS-Network Attached Storage

NAT-Network Address Translation

National Association Of Broadcasters (NAB)-An association representing radio and television stations as a lobbying group in interacting with the FCC.

National Electrical Code (NEC)-The national electrical code is a document that provides rules for the installation of electric wiring and equipment in public and private buildings, published by the National Fine Protection Association. The NEC has been adopted as law by many states and municipalities in the United States.

National Institute Of Standards And Technology (NIST)-A non-regulatory agency of the Department of Commerce that serves as a national reference and measurement lab oratory for the physical and engineering sciences. Formerly called the National Bureau of Standards, the agency was renamed in 1988 and given the additional responsibility of aiding U.S. companies in adopting new technologies to increase their international competitiveness.

National Music Publishers Association (NMPA)-National Music Publishers Association is an organization that assists and represents people and companies that are involved in the creation, licensing and production (mechanical licensing) of music content.

National Number-The telephone number identifying a calling subscriber station within an area designated by a country code.

National Security Agency (NSA)-The United States agency responsible for the development of cryptographic and other security measures.

National Telecommunications and Information Administration (NTIA)-A policy unit of the Department of Commerce which assigns frequencies in the spectrum used by the federal government. The NTAI also advises the President and Congress on telecommunications issues.

National Television Standards Committee (NTSC) Signal-The NTSC signal uses analog modulation where a sync burst precedes the video information. The NTSC system uses 525 lines of resolution (42 are blanking lines) and has a pixel resolution of approximately 148k to 150k pixels.

National Television System Committee (NTSC)-The industry group that established the standard for TV transmission currently in use in the United States, Canada, Japan, and other countries. The abbreviation NTSC is often used to describe the analog television standard that transmits 60 fields/seconds, 30 frames or pictures/second, and a picture composed of 525 horizontal scan lines, regardless of whether or not a color image is involved. The NTSC created its first analog television standard in 1953.

National Transcommunications Limited (NTL)-The authority that owns, operates, and maintains the terrestrial transmission services for all independent TV stations in the United Kingdom.

NAUN-Nearest Active Upstream Neighbor

Navigation Application-A navigation application is a software program and/or service that can be used to provide routing information to a user or system. This may include specific traveling directions or the location of items or facilities within a geographic region.

NB-Normal Burst

NC-Normally Closed

NCH-Notification channel

NCOS-Network Class of Service

NCP-Network Control Point

NCS-Network-Based Call Signaling

NDD-National Direct Dialing

NDIS-Network Driver Interface Specification

NDM-U-Network Data Management - Usage

NE-Network Element

Near End Cross Talk (NEXT)-Near end crosstalk (NEXT) is the leakage of signal that is coupled to a nearby cable or electronics circuit (called crosstalk) where the unwanted signal is received on the originating end (opposite direction) of the cable. NEXT is usually more troublesome than far end crosstalk, as the crosstalk signal levels of NEXT are higher.

This diagram shows how near end crosstalk (NEXT) can cause interference at the sending end of a transmission line. NEXT occurs when some of the transmitted signal energy leaks from one twisted pair and is coupled back to a communications line that is transferring a signal in the opposite direction at the sending end. Generally, NEXT has a higher level of signal interference as the crosstalk levels at the transmitter end are higher than crosstalk that may occur at the receiving (far) end.

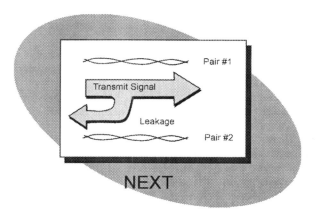

Near End Crosstalk (NEXT)

Near Line of Sight (NLOS)-Near line of sight (NLOS) is a wireless communication system that does not require a direct path (can have significant obstructions) between the transmitter and receiver. NLOS systems can use optical or radio signals for transmission.

This figure shows how near line of sight radio propagation can allow a radio signal to reach its destination in congested areas. This example shows that a radio tower is transmitting through an urban area which does not allow a radio signal to travel a direct path from the tower to the receiver. This example shows that multiple alternate paths are reflected off a building to reach its destination. This diagram shows that a main signal (shortest reflected signal) and another signal (delayed signal) become part of the received signal.

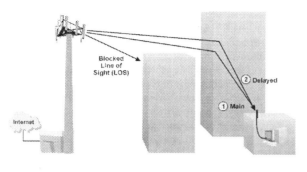

Near Line of Sight Radio Propagation

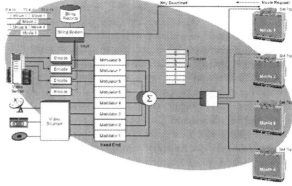

Near Video on Demand (NVOD) Operation

Near Real Time-Actions that occur within a short time period that is perceived or used (such as within a few minutes) to perform or record events when they are required or used.

Near Video On Demand (NVOD)-A video delivery service that allows a customer to select from a limited number of broadcast video channels when they are broadcast. NVOD channels have pre-designated schedule times and are used for pay-per-view services.

This diagram shows a near video on demand (NVOD) system. This NVOD system allows a customer to select from a limited number of broadcast video channels. These video channels are typically movie channels that have pre-designated schedule times. This system allows the user to unblock an encoded channel during pre-scheduled play times.

NEBS-Network Equipment Building System

NEC-National Electrical Code

Negative Acknowledgement (NACK)-Negative acknowledgement (NACK) is a process or control code that is used to indicate that a message has not been received or a process has not been started or completed.

Negative Cost-Total sum of money necessary to produce a film, before it has completed the final process of becoming a motion picture.

Negative Feedback-The return (feedback) of an output signal that subtracts (adds 180 degrees out of phase.) from the output signal. Negative feedback decreases the output signal amplitude and usually stabilizes the amplifier. This may result in reduced distortion and noise.

Negative Impedance-An impedance characterized by a decrease in voltage drop across a device as the current through the device is increased, or a decrease in current through the device as the voltage across it is increased.

Negotiation -The fourth phase of the marketing process in the Costa Model that motivates the customer to engage in negotiation with a company for the sale of products or services. Negotiation can be explicit (talking with the sales rep about options) or implicit (the customer chooses between options without direct interaction).

NEP-Noise Equivalent Power

Nested Program-A software program that is included as a component of another larger program to fulfill a specific task.

Net-Internet

Net Gain-The overall gain of a transmission circuit. Net gain is measured by applying a test signal of some convenient power at the local termination of a given communications circuit, measuring the power delivered at the other end of the circuit, and taking the ratio of the powers as expressed in decibels.

Net Loss-The overall loss of a transmission circuit. Net loss is measured by applying a test signal of some convenient power at the local termination of a given communications circuit, measuring the power delivered at the other end of the circuit, and taking the ratio of the powers as expressed in decibels.

Net Proceeds-Term often used in contract communicating direction stated amounts will take to be paid.

NetBEUI-NetBIOS Extended User Interface

NetBIOS-Network Basic Input/Output System

Netbits-Network Bits

Netcasting-A process of sending information directly to the desktop of a recipient list, usually through the Internet.

Netiquette-A set of unwritten rules (etiquette) that define the normal (socially acceptable) use of electronic mail (e-mail) and other network communication services.

Netizen-Net Citizen

Netpreneur-A netpreneur is an entrepreneur who focuses on developing online Internet business.

Network-A series of points that are interconnected by communications channels, often on a switched basis. Networks are either common to all users or privately leased by a customer for some specific application.

Network Access-Electronic circuitry that determines which, when, and how a communication device may transmit and communicate with the system. This circuitry may be centrally located or may be located in each of the network interface controllers.

Network Access Charge-A fee paid by an operator of a system for access to other network systems.

Network Access Revenue-Service revenue that results from charge for service access between carriers, such as when inter-exchange (IXC) carriers

pay to connect through the local telephone access network infrastructure.

Network Access Server (NAS)-A server that coordinates access to network systems. NASs are used in MGCP systems.

Network Adapter-A device or assembly that converts the format of information that is transferred between one network or device to the format used by another network or device. An example of a network adapter is an Ethernet network adapter card that converts information in an Ethernet packet into a format that can be transferred into a computer's internal communication bus.

Network Address-(1-general) A unique number associated with a network host that identifies it to other hosts and devices during network communication. (2-SS7) The Signaling System 7 signaling code that contains a network identification, a network duster, and network cluster member fields.

Network Address Translation (NAT)-Network address translation (NAT) is a process that converts network addresses between two different networks. NAT is typically used to convert public network addresses (such as IP addresses) into private local network addresses that are not recognized on the Internet. NAT provides added security as computers connected through public networks cannot access local computers with private network addresses.

This figure shows the basic operation of a network address translation (NAT) system. In this diagram, the NAT receives a message with a desired pubic IP address (209.67.22.59) originating from a local computer with a private IP address of 10.01.01.01. The NAT translates this originating address to a public IP originating address. The NAT then initiates a session with the Internet server (web site) using the network's public IP address 118.54.23.11 as the originating address. The Internet server receives the request for information and responds with data messages address to the NAT's public IP address. When the NAT receives these data messages for that particular communications session, they are translated to the local (private) IP address 10.01.01.01 and forwarded to the originating computer. If messages are received to the NAT's public IP address that are not part of a communications session that it knows about, the NAT will not route the messages to computers connected to the LAN.

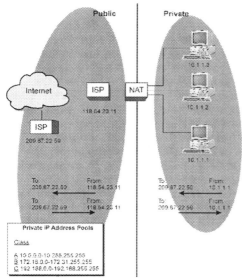

Network Address Translation (NAT) Operation

Network Administration Center-An operations center with administrative responsibility for local and tandem switching systems.

Network Administrator-(1-coordinator) A person responsible for managing the day-to-day operations of a network. (2-humor) The person blamed for all computing problems, whether they are related to the network or not.

Network Analyzer-A test instrument that receives, decodes, and analyzes data transmitted through a network. A network analyzer may be an integrated software program or separate hardware device.

Network Architecture-The design, physical structure, functional organization, data formats, operational procedures, components, and configuration of a network. Network architectures usually divide network functions into layers of software and hardware. Each network layer serves a specific purpose. There are often specific relationships between network layers that allows different manufacturers or equipment that operate at different layers to interoperate with each other (e.g. a router interfacing with a network hub.)

Network Attached Storage (NAS)-(1-product) A collection of mass-storage devices contained in a single chassis with a built-in operating system. Typically connected to a local area network, these devices usually support Network File System (NFS) and common Internet file system (CIFS) as a

means to share data in a departmental or enterprise environment. NAS products are marketed to small and medium businesses as self-contained, plug-and-play, easy to operate storage expansions. (2-architecture) Network Attached Storage is an architecture in which traditionally LAN-oriented technologies, such as Ethernet and TCP/IP are used to connect storage. NAS utilizes LAN technology in place of traditional storage protocols such as Fibre Channel and parallel-SCSI to produce large arrays of disk drives with a virtualized interface. One NAS disk array may by configured to appear as a single disk drive or as multiple volumes of varying sizes.

Network Automatic Call Distribution (NACD)-NACD is a is a call processing system that routes (distributes) incoming telephone calls to specific telephone sets or stations calls based on the characteristics of the call or network settings. These characteristics can include an routing on network congestion, time of day routing, and other criteria.

Network Busy Hour-The hour in a given 24-hour period during which the total load carried by all trunks in a network is greater than the total load carried during any other hour.

Network Call Center-A network call center is a system that processes calls between a company and a customer. Network call centers typically assist customers with requests for new products and service along with providing information about product and service features. A network call center usually has many stations for call center agents that communicate with customers.

Network Centric-Network centric is the primary use of a network system or function to provide features or services.

Network Class of Service (NCOS)-The types of access or services that users are authorized to receive from a communication system.

Network Control-Network control is the transmission of signals or messages that perform call control, equipment configuration, or information management functions. Network control can be centralized or distributed. The control of public telecommunications networks is a centralized system as call processing is coordinated through a common channel signaling (CCS) network. The Internet uses distributed control as the switching information dynamically changes in packet switching centers (routers) throughout the Internet network.

Network Control Point (NCP)-A special applications processor that provides network access to a variety of centralized database services. The corresponding term in common channel signaling is the service control point.

Network Cost-Network cost is the charges and fees associated with the setup and operation of networks.

Network Data Collection Center-An operations center that administers network data collection and supervises the operation and maintenance of the Engineering and Administrative Data Acquisition System.

Network Data Management - Usage (NDM-U)- The network data management - usage (NDM-U) is a standard messaging format that allows the recording of usage in a communication network, primarily in Internet networks. The NMD-U defines an Internet Protocol detail record (IPDR) as the standard measurement record.

The IPDR record structure is very flexible and new billing attributes (fields) are being added because Internet services are now offered in almost all communications systems. The NMD-U standard is managed by IPDR organization at www.IPDR.org.

Network Driver Interface Specification (NDIS)-An interface specification that was developed by Microsoft to provide a common set of rules for network adapters to interface with operating systems. NDIS is independent of hardware and types of network interface cards and it allows multiple protocol stacks to co-exist in the same computer.

Network Element (NE)-A facility or the equipment used in the provision of a communications service. The term includes subscriber numbers, databases, signaling systems, and information sufficient for billing and collection or used in the transmission, routing, or other provision of a communications service.

Network Equipment-The telecommunications equipment and facilities owned, installed, and maintained by a telephone company or service provider and that are part of a telecommunications network.

Network Failure-Network failure is a cessation of capability to perform a specified function or functions within a network.

Network Gateway (NGW)-A media and signaling adapter (gateway) used in a network to interface between different types of networks. A network gateway can convert both the media and signaling control messages between the systems.

Network Harms-Adverse effects on carrier provider systems (such as telephone company), employees or customers. The four basic harms are excessive signal power, hazardous voltage, improper network control signaling, and line imbalance.

Network Identity-A network identity is a unique identification code that can be recognized by or through a network.

Network Integration-The joint provision of telecommunications services and the joint assumption of risk through a partnership arrangement among telephone companies. The expression often is used to describe both technical and economic integration.

Network Interface (NI)-(1-interface) The point of connection between customer premises equipment and a public switched telephone network. This is also called a standard network interface (SNI). (2-boundary) The physical and electrical boundary between two separately owned telecommunications systems.

Network Interface Card (NIC)-A network interface card (NIC) adapts data communication network protocol (such as Ethernet) to a data bus or data interface in a computer or data terminal. The NIC is installed between a computer network (such as the Ethernet) and a computer data bus (such as a PCI socket). The NIC is usually a PC expansion board connector and operating system. Software in the computer is installed and setup to recognize the NIC card.

Network Interface Device (NID)-A connection point between the end customers equipment and the telecommunications network. This is also called the demarcation point.

Network Interface Unit (NIU)-Network interface units are electronic assemblies that terminate a telecommunications line into an end user's facility. For optical networks, the NIU may terminate wireless, fiber, or copper lines and convert the signals into other forms such as analog (telephone) and digital (computer network or multimedia) signals.

Network Intrusion Detection Expert System- An expert system that provides the final connection between an end customers premises and the telecommunications network.

Network Layer-The Network layer performs the switching and routing of data through the network, controls the flow of data within the network, segments (divides) or reformats data packets if necessary between network types, and performs error control functions specific to the address decoding and routing functions. The network layer receives data for transmission from an upper layer (such as a transport or session layer) and converts it into network addressable data formats that can be transferred through a network or transmission line. An upper layer provides the network layer with the necessary addressing and network routing control requirements (e.g. priority codes) to allow the network layer to send data through the network. The location of the network layer within the protocol stack is usually above a physical layer and below a transmission or session layer. The network layer is layer 3 in the open system interconnection (OSI) protocol layer model.

Network Maintenance Center (NMC)-A facility that allows monitoring, testing and maintenance of a telecommunications network. The NMC is typically operational 24 hours a day, 7 days per week.

Network Management (NM)-Network management is the process of configuring equipment in the network, the setup (provisioning) of services, system maintenance, and repair (diagnostic) processes. Network management systems are commonly composed of a network management server computer and network management software.

Network management systems usually include a set of procedures, equipment, and operations that keep a telecommunications network operating near maximum efficiency despite unusual loads or equipment failures.

Network Management Center (NMC)-An operations center that monitors and controls traffic flow to help ensure the most efficient and economical use of available network capacity. The center plans strategies and works to minimize the effects of disasters, abnormal traffic loads, and switching system or facility failures.

Network Management Layer (NML)-A layer of communication control protocols and systems in a network that an operator to setup, control, and optimize their systems.

Network Management Protocol (NMP)-Communication protocols that were developed by AT&T to control network equipment and assemblies including modems data multiplexers.

Network Management Station (NMS)-A device that communicates with network management or SNMP agents throughout a network. Typically it comprises a workstation operated by a network administrator, equipped with network management software or other relevant applications that assist in the monitoring of the network. Some, if not all the components of the FCAPS model are usually designed into the NMS. NMS is also known as a network manager.

Network Management System (NMS)-A network system is a combination of equipment and software that is used to setup, control, monitor and manage the operation of a communication network.

Network Monitor-A software program or graphic display that monitor and identifies network-related problems. Network Monitor tracks data as it moves through the network layer. It may insert, filter packets, or perform packet analysis.

Network News Transport Protocol (NNTP)-The protocol that governs the transmission of network new, a threaded messaging system for posting messages to form newsgroup discussions.

Network Node Interface (NNI)-The defined (typically standardized) interface between functional elements (network nodes) in a system or network. NNI interfaces may add additional test functionality for connections between network elements and reduce access control functions as network physical connections are typically fixed for extended periods of time. See: user network interface (UNI)

Network Operating System (NOS)-A software program that manages communication between devices within a network. The NOS oversees resource sharing and often provides security and administrative tools.

Network Control Point (NCP)-(1-Surveillance) A center responsible for the surveillance and control of telecommunications traffic flow in a service area. (2-Service) A facility or organization responsible for maintaining, monitoring, and troubleshooting a network infrastructure. Responsible for applying the FCAPS model to the network.

Network Operator-A company that manages the network equipment parts of a communications system. A network operator does not have to be the service provider. Also see Service Provider and Reseller.

Network Personal Video Recorder (NPVR)- Network personal video recorder is the use of network media storage and media streaming technology to allow viewers to control the time, method and what video program viewing occurs. In addition to the ability to delay the viewing time for television programs, NPVR usually enables viewers to pause, rewind live TV programs and adds a fast-forward functionality up until the point at which a subscriber reaches parity with the system-wide live broadcast. NPVR offers subscribers VCR-like control. The rules governing which content can be recorded and how it may be used are controlled by the content protection system or rights management system.

Network Planning System (NPS)- An interactive computer program that assists strategic planners in the development of interoffice facilities and wire centers, and aids in the planning of traffic and distribution routes.

Network Port- A communication input/output access point to a network. The network port usually has specific network access protocols and security levels associated with it.

Network Processor- A programmable device (silicon chip) which is capable of receiving packets from one or more network interfaces and processing these packets. The processing of the packets includes address lookup, packet modification and transmission and is under software control. Network processors are special purpose microprocessors, specifically designed to meet the needs of systems that require each packet to be inspected in detail. By using network processors, system's designers and vendors are able to change the function of routers, switches and other devices in the field as new protocols are developed.

Network Program- (1-computer system) A program that operates on a server within a network. (2-distribution) Any program delivered simultaneously to more than one broadcast station regional or national, commercial or noncommercial.

Network Programming- Network programming is the selection of shows and programs that are offered by a television network provider.

Network Requirements Manager (NRM)- A network requirements manager is a person or function that identifies, determines and assigns network resources that are required to provide features or services.

Network Security- The processes used within a network to validate the identity of users (authentication), access control of services (authorization), and information privacy protection (encryption).

Network Server Interface Specification (NSIS)- A specification for network interface cards (NICs) that is independent of hardware and protocol.

Network Servers- Hosts, and sometimes personal computers, may function as specialized types of nodes called network servers. These specialized nodes serve the other nodes by storing many of their files and running much of their common software.

Network Service Center (NSC)- An operations center that performs quality control functions related to a grade of network service. These functions include the completion of a call to a desired number, the capability of hearing and being heard, accurate billing of calls, and the capability of receiving incoming calls.

Network Service Provider (NSP)- Any company that provides network services to customers or devices.

Network Subsystem (NSS)- (1-general) The network parts of a communication system. network. (2-GSM) The system parts of a GSM network this includes the mobile switching center (MSC), home location register (HLR), visitor location register (VLR) and equipment identity register (EIR).

Network Termination (NT)- A final end point in a network that is usually owned by the network service provider. After the network termination, the equipment is commonly owned by the customer (called customer premises equipment - CPE). When the network termination (NT) is an active device, it typically has standard communications parameters such as protocols, timing and voltages to allow specific types of equipment to correctly communicate with the network.

Network Termination 1 (NT1)- A standard network termination in the ISDN network that adapts the physical characteristics of the ISDN network.

Network Termination 2 (NT2)- A standard intelligent network termination in the ISDN network that contains the intelligence (application layers) for ISDN termination equipment.

Network Termination Equipment (NTE)- Network termination equipment (NTE) are the devices used by end-user to access the network. In the traditional PSTN world, network termination equipment was generally confined to the telephone, headset or conference phone. The cellular industry expanded this to include cell phones and pagers. In IP Communications network termination equipment can include all these traditional devices as well as computers and PDAs.

Network Time Protocol (NTP)-Network time protocol is a method for synchronizing clocks on physically separated machines. NTP is a complex protocol that uses multiple methods to synchronize clocks on a computer to a more accurate time source. NTP is defined in RFC 1129.

Network Topology-The physical and logical relationships between nodes in a network, typically star, bus, tree, ring, or hybrid. Network topology is the layout and structure of a network.

This figure shows several of the most popular LAN topologies and their configurations. Some data networks are setup as bus networks (all computers share the same bus), as start networks (computers connect to a central data distribution node), or as a ring (data circles around the ring). This diagram shows for popular types of LAN networks: Thinnet, Thicknet, token ring networks, and Ethernet star network.

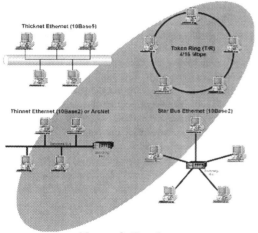

Network Topology

Network Tunneling-Network tunneling is the process that creates a secure communication path through the use of a virtual connection within a network link through the encryption of data that is transmitted between the virtual connection points.

Network Voice Protocol (NVP)-An older voice protocol that was developed to allow real-time voice transmission over the ARPANET (packet voice) system.

Networking-The connection of geographically separate computers and communication devices using transmission line facilities.

Neural Network-A computer system whose information processing is modeled on the structure of the brain and its neurons. Information processing is parallel, in that many processes take place at the same time. Processes are also asynchronous because each step in a process has no time relationship to steps in other processes. To mimic the brain, some functions are stochastic, or random.

Neutral-(1-charge) A device or object having no electric charge. (2-power system) A conductor in an electric power-distribution system that carries no current when the power load is balanced.

New Matter-When a continuation, or divisional, application is filed which contains descriptions, figures, or other information which was not part of the original pending application, the additional information is deemed "new matter" and the application called a continuation-in-part, or CIP.

When determining patentability over the prior art, new matter is awarded the filing date of the CIP, while matter which is common to the CIP and the original parent application is awarded the priority date of the parent application.

Newbie-A newbie is a person who is new to the use of the Internet.

NewsML-News Markup Language

Newton-The standard SI metric unit of force. One Newton is the force that, when applied to a body mass of 1 kg, gives it an acceleration of 1 meter per second (or one meter per second squared.)

NEXT-Near End Cross Talk

Next Generation Operations Support System (NGOSS)-Next generation operations support system is a set of commands, processes and procedures that is used by a network to allow a network operator to perform the administrative portions for business operations such as billing and customer care.

Next Hop Resolution Protocol (NHRP)-An name resolution protocol designed to allows Internet Protocol (IP) datagrams to route across multiple types of access networks (e.g. ATM, SMDS, and X.25).

NextGen or Next Generation Network (NGN)-Next generation networks (NGN) refers to the infrastructure service providers will require as they migrate from circuit switched systems to packet switching systems. Nextgen networks are

N

still evolving, but have five core components - media gateways, softswitches, media servers, application servers and session border controllers. Also known as next generation network or NGN.

NF-Noise Factor

NGN-NextGen or Next Generation Network

NGO-Non Governmental Organization

NGOSS-Next Generation Operations Support System

NGW-Network Gateway

NHRP-Next Hop Resolution Protocol

NI-Network Interface

Nibble-A 4-bit unit of data (half of a byte).

NIC-Network Interface Card

NiCd-Nickel Cadmium

Nickel Cadmium (NiCd)-Nickel Cadmium is a type of rechargeable battery technology that is commonly used in portable electronic devices.

NID-Network Interface Device

Nielsen Rating-Nielson ratings are measurements of viewer habits. Neilson ratings include the percentage of households that watch specific programs.

Night Service-The processing state of a telephone system (such as a PBX) during hours of operation when the company is closed or in a different state of business operation. Night service usually provides a different greeting messaging and call routing (transfer) capability. Night service may prompt callers to leave messages instead of being routed to an operator.

This diagram shows how a telephone system can change its basic operation for daytime and nighttime telephone service. In this example, during the day, all the incoming calls are routed to (received by) a receptionist at extension 1001. At night (between 5 pm and 8 am), the calls are automatically redirected to an automated telephony call processing system that is connected to extension 1014. When the automated attendant detects a ring signal, answers the phone (off-hook signal) and plays a pre-recorded messaging informing the caller of options they may choose to direct the call to a specific extension. In this example, the automated call attendant software decodes DTMF tones or limited list of voice commands to determine the routing of the call. The automated call attendant software then determines if the destination choice is within the option list and if the extension is available. If the extension is available, the automated attendant will send a command to the computer tele-

phony board (voice card) that can switch the call to the selected extension. If the extension is not valid or not available, the automated attendant will provide a new voice prompt with updated information and additional options.

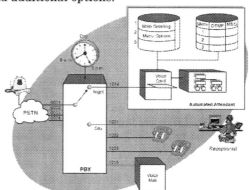

Night Service Operation

NIH-Not Invented Here

Nipkow Disk-A Nipkow disk is a device that is used to transmit pictures over low speed communication lines. The disk was composed of a sequence (spiral) of holes that could show parts of an image as the disk was rotated. Light that would transfer through the holes at the disk rotated was turned into electrical signals. On the receiving end, the electrical signal would then be converted back into light on a disk that was spinning at the same speed allowing the image to be recreated at the receiving end on a viewing screen. This form of Image transmission uses a relatively small amount of information bandwidth to transfer images. The Nipkow disk was invented by Paul Nipkow in the 1920s.

NIST-National Institute Of Standards And Technology

NIU-Network Interface Unit

NLOS-Near Line of Sight

NLOS-Not Line of Sight

nm-Nanometer

NM-Network Management

NMC-Network Maintenance Center

NMC-Network Management Center

NML-Network Management Layer

NMP-Network Management Protocol

NMPA-National Music Publishers Association

NMS-Network Management Station

NMS-Network Management System

NMT-Nordic Mobile Telephone

NNI-Network Node Interface
NNTP-Network News Transport Protocol
NO-Normally Open
No Charge Traffic-Traffic, such as 611 service request and 911 emergency calls, classified as "no-charge" in a tariff on file with an appropriate regulatory agency.
No Circuit Tone-A low tone, interrupted 120 times per minute (02 5 on and 0.3 5 off) indicating that no trunk is available. This term also is known as a reorder, all-circuits busy, or fast-busy tone.
No Return To Zero (NRZ)-A digital code in which the signal level is low for a 0 bit and high for a 1 bit and does not return to 0 between successive 1 bits.
This diagram shows that no-return to zero (NRZ) encoding uses the logical level voltage during the entire period of each logical bit. In this example, the data is transmitted at 1 kbps so each logical bit period is 1 msec. During this entire period, the logical level remains at the same voltage associated with the logical level.

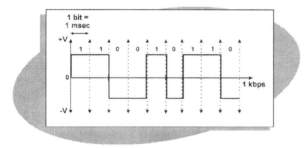

No Return to Zero (NRZ) Operation

NOC-Network Operations Center
Nodal Multiplexer-A multiplexer that has the capability of dynamically routing channels onto different communication circuits.
Node-(1-network) In network topology, a terminal of any branch of a network, or a terminal common to two or more branches of a network. (2-ascending) The point where a satellite crosses the plane of the equator when moving north. (3-current) The points at which the current is at minimum in a transmission system in which standing waves are present. (4-descending) The point where a satellite crosses

the plane of the equator when moving south. (5-network) A terminal on any branch of a network. (6-switching) The switching points in a switched communications network, including patching and control facilities. (7-transmission line) A point of interconnection on a transmission line. (8-tree structure) A point where subordinate data originates. (9-telephony) A switching office or facility junction. (10-test facility) A remote test facility. (11-voltage) The points at which the voltage is at a minimum in a transmission system in which standing waves are present.
Noise-An undesired signal generated by a physical mechanism that cannot be "turned off" in theory. The most typical example is the electric current produced by the random motion of electrons, a manifestation of temperature related kinetic energy that is zero only when the material involved is cooled to superconducting temperatures. In contrast to noise, a distinct type of undesired signal, called interference, in theory can be "turned off" because it originates from a device like a separate radio transmitter, a faulty electric device, or other artifact. In some cases, writers use the term "noise" to describe all undesired signals, both physical noise and interference together, often inappropriately.
Noise Factor (NF)-The ratio of the noise power measured at the output of a receiver to the noise power that would be present at the output if the thermal noise resulting from the resistive component of the source impedance were the only source of noise in the system.
Noise Figure-A measure of the noise in decibels generated at the input of an amplifier, compared with the noise generated by an impedance method resistor at a specified temperature.
Noise Filter-A network that attenuates noise frequencies.
Noise Floor-The power level of background noise signals.
Noise Generator-A generator of noise signals (usually a test instrument).
Noise Immunity-The capability of a device to receive and decode signals or information in the presence of noise.
Noise Level-The ratio of noise on a given circuit to a reference noise, expressed in decibels above reference noise for an electrical system or decibels sound pressure level for an acoustical system.

N

Noise Margin-An assigned minimum signal-to-noise ratio, expressed in decibels, that is required for a specific type of signal to be useful.

Noise Power Ratio (NPR)-The ratio, expressed in decibels, of signal power to intermodulation product power plus residual noise power, measured at the baseband level.

Noise Suppressor-A filter or digital signal processing circuit in a receiver or transmitter that automatically reduces or eliminates noise.

Nominal-The most common value for a component or parameter that falls between the maximum and minimum limits of a tolerance range.

Nominal Value-A specified or intended value.

Non Governmental Organization (NGO)-A non-governmental organization is a company or group that operates without specific controlling interest or significant influence from government agencies or authorities.

Non-Blocking-A characteristic of a switch fabric implying that it is capable of handling traffic at the maximum number of circuit switched channels for a circuit-switched fabric, or for a packet switch the maximum frame arrival rate, on all interfaces simultaneously, without requiring some packets to wait for resources to become available before transmission. This term is used with the assumption that no port will be presented more then the line-rate. In older switching technologies, blocking was common. This means that, due to the design of the switch, it was possible for a packet to arrive, destined for a port that currently has no traffic, but the packet was still required to be buffered. Except for the most massive switches on the market, most modern day packet switching equipment is non-blocking. Most small to medium size circuit switches are non-blocking today, but large circuit switches having 10000 lines or ports are typically configured for less than 1 percent probability of blocking rather than being non-blocking.

Non-Disclosure Agreement-A binding contract between parties not to disclose and to keep confidential information shared among the parties from being spread to other parties. Commonly referred to as an NDA, in certain circumstances such an agreement can preserve the novelty of an invention. Documents exchanged under a properly written and executed NDA may not be considered as a publication, or public dissemination of an invention and can preserve the right of the inventor to apply for patent protection.

Non-facilities Based Carrier-Refers to carriers that do not operate their own switches and networks

Nonlinear-A device or circuit whose output is not directly proportional to its input.

Nonlinear Distortion-The usually undesirable difference between a signal at the input to a system and at the output caused by the nonlinear functioning of the system.

Non-Obvious-According to 35 USC 103 a) a patent may not be obtained if the differences between the subject matter sought to be patented and the prior art are such that the subject matter as a whole would have been obvious at the time the invention was made to a person of ordinary skill in the art.

Non-Obvious is analogous to Inventive Step.

Non-Payment Churn-The most common form of involuntary churn. Non-payment churn occurs when a customer fails to pay their bills and the telco terminates their service.

Nonprofit-Any corporation, foundation, or association that is not operated to benefit any private shareholder or individual.

Non-Recurring Charge (NRC)-A cost for a facility or product that only occurs one time or is not periodically charged.

Non-Recurring Engineering Costs (NRE)-Costs associated with a product or service that are associated with the development of the product and not associated with the marketing or support of the product sales.

Nontransparent Access-Nontransparent access is the connection of a user to communication system (such as the Internet) that requires assistance and/or validation by a system operator.

Nonvolatile-A memory device or system who's stored data is unaffected by the removal of operating power.

Nonvolatile Memory-A form of computer memory that will store information for an indefinite period of time with no power applied. If the storage area cannot be rewritten with new information, the nonvolatile memory can also be called read only memory (ROM).

Normal-A line perpendicular to another line or to a surface.

Normal Business Hours-Those hours during which most similar businesses in the community are open to serve customers. In all cases, "normal business hours" must include some evening hours at least one night per week and/or some weekend hours.

Normal Routing-The routing of a given signaling traffic flow under normal conditions, that is, in the absence of failures.

Normalization-Normalization is the adjustment of frequency, time, and/or power characteristics of a communication signal to compensate for changes or distortions that have occurring during the transmission of the signal.

Normally Closed (NC)-Switch contacts that are closed in their non-operated state, or relay contacts that are closed when the relay is de-energized.

Normally Open (NO)-Switch contacts that are open in their non-operated state, or relay contacts that are open when the relay is de-energized.

North American Numbering Plan (NANP)-The NANP is an 11 digit-dialing plan that is used within North America. It contains 5 parts: international code, optional intersystem code (1 +), geographic numbering plan area (NPA), central office code (NXX), and station number (XXXX). The NPA code defines a geographic area for the serving telephone system (such as a city). The NXX defines a particular switch that is located within the telephone system. Finally, the station code identifies a particular line (station) that the switch provides service to.

NOS-Network Operating System

Not Invented Here (NIH)-Not invented here is a concept or a viewpoint that new products or services should be invented within a company rather than being developed outside the company.

Notch Filter-A circuit designed to attenuate a specific frequency band; also known as a "band-stop" or "band reject" filter. Notch filters are sometimes used to restrict access to video signals that are transmitted through a cable television distribution system.

This diagram shows a notch (band reject) filter that is used to block a specific frequency band (television channel) from a multi-channel input signal. In this example, a television system is broadcasting many television channels. This diagram shows how a notch filter can block a specific channel (such as a pay for subscription channel) from being received by a customer.

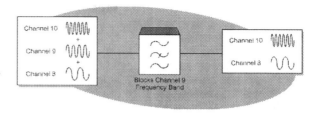

Notch Filter Operation

Notification-Notification is the process of sending a message that indicates the status of an event or action that will occur. An example of notification is the alerting of mobile telephones that a group or broadcast call is occurring.

Notification Server-A computing device (typically a computer with communications software) that provides notification to users or devices when specific events occur.

Notifications-Notifications are synonymous with SNMP traps (from SNMPv1) or informs (from SNMPv2). Events on network devices or SNMP agents send notifications to the network management system when something of interest occurs on the agent.

Notify Message-A message that is used to provide information to a person or device about an event that has occurred. The event criteria may have been set by sending a Subscribe message along with the parameters of the event (such as exceeding a maximum count or a level that has been exceeded.) The notify message is defined in session initiation protocol (SIP) toolkit.

NPA-Numbering Plan Area

NPA Codes-Interchangeable Numbering Plan Area

NPCS-Narrowband PCS

NPR-Noise Power Ratio

NPS-Network Planning System

NPVR-Network Personal Video Recorder

NRC-Non-Recurring Charge

NRE-Non-Recurring Engineering Costs

NRM-Network Requirements Manager

NRZ-No Return To Zero

NSA-National Security Agency

NSC-Network Service Center

N

nsec-Nanosecond
NSFNet-National Science Foundation Network
NSIS-Network Server Interface Specification
NSP-National Signaling Point
NSP-Network Service Part
NSP-Network Service Provider
NSS-Network Subsystem
NSV-Nullsoft Video
NT-Network Termination
NT1-Network Termination 1
NT2-Network Termination 2
NTE-Network Termination Equipment
NTIA-National Telecommunications and Information Administration
NTIZ-Network Identity And Timezone
NTL-National Transcommunications Limited
NTP-Network Time Protocol
NTS-Number Translation Service
NTSC-National Television System Committee
NTSC Signal-National Television Standards Committee
NTSC Video-The NTSC television system standard was developed in the United States and is used in many parts of the world. The NTSC system uses analog modulation where a sync burst precedes the video information. The NTSC system uses 525 lines of resolution (42 are blanking lines) and has a pixel resolution of approximately 148k to 150k pixels.

This figure demonstrates the operation of the basic NTSC analog television system. The video source is broken into 30 frames per second and converted into multiple lines per frame. Each video line transmission begins with a burst pulse (called a sync pulse) that is followed by a signal that repre-

sents color and intensity. The time relative to the starting sync is the position on the line from left to right. Each line is sent until a frame is complete and the next frame can begin. The television receiver decodes the video signal to position and control the intensity of an electronic beam that scans the phosphorus tube ("picture tube") to recreate the display.

Null-A zero or minimum amount or position. A binary character that has all the binary digits set to zero. ASCII 0 is a 7 bit character that represents the null (no) value. A null character is used in programming to indicate padding information (filler) or it may be used to indicate an end of a field or block of information (delimiter).

Null Modem Cable-A cable that is configured to cross-connect computers without the need of a modem. A null modem cable reverses the data and control lines (e.g. transmit to receive and receive to transmit).

Nullsoft Video (NSV)-Nullsoft video (NSV) is a multimedia digital video format developed by Nullsoft that can download or stream digital audio and digital video.

Number Portability-Number portability involves the ability for a telephone number to be transferred between different service providers. This allows customers to change service providers without having to change telephone numbers. Number portability involves three key elements: local number portability, service portability and geographic portability.

The first part of the telephone number (NPA-NXX) usually identifies a specific geographic area and specific switch where the customer subscribes to telephone service. If a telephone number is assigned to another system (different NXX) in the same geographic area (same NPA), the interconnecting carriers (IXCs) connecting to that system must know which local system to route the calls based on the selected local service providers. In this case, the IXC must look up the local telephone number in a database (called a database dip) prior to delivering the call to the end customer.

Numbering Plan-A numbering plan is a system that identifies communication points within a communications network through the structured use of numbers. The structure of the numbers is divided to indicate specific regions or groups of users. It is important that all users connected to a telephone network agree on a specific numbering plan to be

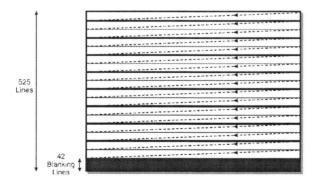

NTSC Video

able to identify and route calls from one point to another.

Telephone numbering plans throughout the world and systems vary dramatically. In some countries, it is possible to dial using 5 digits and others require 10 digits. To uniquely identify every device that is connected to public telephone networks, the Consultative Committee for International Telephony And Telegraphy (CCITT) devised a world numbering plan that provides codes for telephone access to each country. These are called country codes. Coupled with the national telephone number assigned to each subscriber in a country, the country code telephone makes that subscribers number unique worldwide. The International Telecommunications Union (ITU) administers the World Numbering Plan standard E.164 and publishes any new standards or modifications to existing standards on the Internet.

Numbering Plan Area (NPA)-A 3-digit code that designates one of the numbering plan areas in the North American Numbering Plan for direct distance dialing. Originally, the format was NO/IX, where N is any digit 2 through 9 and X is any digit. From 1995 on, the acceptable format is NXX.

Numbering Scheme-British English synonym for North American "Numbering Plan."

Numerical Aperture (NA)-Numerical aperture is a measure of the angle for light to exit (from a LED or Laser) or enter into an optical device (e.g., an optical fiber). The numerical aperture is equal to the square root of the difference between the squares of the indexes of refraction for the two media (e.g., air and fiber).

This figure shows that optical fiber has a numerical aperture area that can accept (couple) optical signals of specific wavelengths from a light source. This diagram shows that that light signals that enter the fiber core outside the NA area are not able to be redirected down the core of the fiber so they are not coupled into the fiber core.

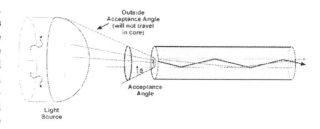

Optical Fiber Numerical Aperture (NA)

NVOD-Near Video On Demand

NVP-Network Voice Protocol

NXX-A term used for qualifying dialing digits. N stands for any number 2 through 9 and X stands for any number 0 through 9. An example of a valid NXX is 201.

Nyquist Frequency-The lowest sampling frequency that can be used for analog-to-digital conversion of a signal without resulting in significant aliasing. Normally, this frequency is twice the rate of the highest frequency contained in the signal being sampled.

Nyquist Rate-The maximum rate at which data can be transmitted over a limited-bandwidth channel without inter-symbol interference. The Nyquist rate, in bauds, is twice the channel bandwidth in Hertz. The term was named for the American physicist who determined the rate, Harry Nyquist.

O

O/E-Optical To Electronic Conversion

OA-Optical Amplifier

OA&M-Operations Administration And Maintenance

OACSU-Off Air Call Set Up

OADM-Optical Add Drop Multiplexer

OAN-Optical Access Network

OB-Outside Broadcast

OBEX-Object Exchange

Object Code Program-The representation of machine language computer programs in binary form.

Object Exchange (OBEX)-A session-layer protocol for object exchange originally developed by the Infrared Data Association (IrDA) as IrOBEX. Its purpose is to support the exchange of objects in a simple and spontaneous manner over an infrared or Bluetooth wireless link.

Object Identifier (OID)-The object identifiers or OIDs are defined in the Structure of Management Information (SMI) for SNMP (RFC1065). Object identifiers define the structure of all objects defined in a tree format. All names of objects have corresponding numbers. For example, the path to SNMP objects is represented in the following two ways as an OID: iso.org.dod.internet or 1.3.6.1. OIDs can be represented as all text, all numeric, or a combination of both.

Object Retention-Object retention is the keeping of a portion of a frame or field on a digital video display when the image has changed.

Obsolete Technology-Obsolete technology are processes, components or systems that have been replaced by better solutions or are no longer needed to perform the functions they provide.

OC-Operations Center

OC&C-Other Charges & Credits

OC&C-Other Charges and Credits

OC1-Optical Carrier 1

OC-1-The SONET optical carrier 1, operating at a data rate of 51.84 Mbps.

OC12-Optical Carrier 12

OC-12-The SONET optical carrier 12, operating at a data rate of 622.08 Mbps.

OC192-Optical Carrier 192

OC3-Optical Carrier 3

OC-3-The SONET optical carrier 3, operating at a data rate of 155.52 Mbps.

OC-48-The SONET optical carrier 48, operating at a data rate of 2488.32 Mbps.

OC768-Optical Carrier 768

OC9-Optical Carrier 9

OCA-Outside Collections Agency

OCAP-Open Cable Application Platform

OCC-Other Common Carrier

Occupancy-The fraction of the time that a circuit or equipment is in use, expressed as a decimal. Occupancy is the erlangs carried and is equal to the hundred call seconds (CCS) carried divided by 36. It includes both message time and setup time.

Occupational Safety and Health Administration (OSHA)-The occupational safety and health administration (OSHA) is a regulatory authority that publishes and enforces regulations on workplace safety.

Occupied Bandwidth-Occupied bandwidth is the portion of a frequency band that is used to transmit a signal.

OC-n-Optical Carrier Hierarchy

OCR-Optical Character Recognition

Octal-A numbering scheme with the base 8.

Octave-Any frequency band in which the highest frequency is twice the lowest frequency.

Octet-A group of eight binary digits, usually operated upon as an entity. See also Byte.

OCWR-Optical Continuous Wave Reflectometer

ODBC-Open Database Connectivity

Odd Parity-A data error detection method in which one extra bit, the parity bit, is added to the cede signal for each data character such that the total number of ones in the data, including the parity bit, is an odd number.

ODI-Open Data Link Interface

ODMA-Opportunity Driven Multiple Access

ODRL-Open Digital Rights Language

ODS-Operational Data Store

ODU-Outdoor Data Unit

OEM-Original Equipment Manufacturer

OEO-Optical to Electronic to Optical

OFC-Optical Fiber Conductive

OFCP-Optical Fiber Conductive Plenum Cable

OFCR-Optical Fiber Conductive Riser

OFDM-Optical Frequency Division Multiplexing

O

OFDM-Orthogonal Frequency Division Multiplexing

Off Line (Offline)-(1-general) A condition of devices or subsystems not connected into, not forming a part of, and not subject to the same controls as an operational system. (2-computer system) A circuit or device that is disconnected from a system, usually a remote computer, and not available for use.

Off Peak-A time period where a telecommunication system usage is lower, typically after normal business hours. Some telecommunications service providers charge a reduced rate for the use of services during off-peak hours.

Off Premises Extension (OPX)-A call processing feature that allows a call to be forwarded to a telephone at a secondary location that is located off the premises of the phone system that is transferring the telephone call.

Offer Management-Offer management is the process of assigning and tracking specific product and service offers from people and companies.

Off-Hook-An electrical signal that occurs when a customer typically removes a telephone receiver off of its cradle, thus releasing the hook switch. When the hook switch is released (off-hook), this typically causes a drop in telephone line voltage due to connecting of the local loop telephone wires together. Automatic devices such as a computer modem can also initiate an off-hook signal.

Offline-Off Line

Off-Net-Off-The-Net Calls

Offset-An offset is an intentional difference between the realized value and the nominal value. An offset is used to establish a reference point for normalized measurements.

Off-The-Shelf-Off-the-shelf products or equipment are commercially available devices or systems that can be immediately used or can be combined into other systems.

OFN-Optical Fiber Nonconductive

OFNG-Optical Fiber Nonconductive General Purpose Cable

OFNP-Optical Fiber Nonconductive Plenum Cable

Ogg-Ogg is a digital multimedia file container format that was developed by ziph.org for digital audio and digital video. The Ogg file format structure is stream oriented allowing it to be easily used for media streaming applications. The container formats allow for the ability to interleave audio and video data and it includes framing structure, error detection capability along with the insertion of timestamps that can be used to synchronize streams. Ogg is an open royalty free standard which is available for anyone to use. More information about Ogg and supporting protocols can be found at www.XIPH.org.

OHCI-Open Host Controller Interface

Ohm-The unit of electrical resistance through which one ampere of current will flow when there is a difference of one volt. The unit is named for the German physicist Georg Simon Ohm (1787-1854).

Ohmic Loss-The power dissipation in a line or circuit caused by electrical resistance.

Ohmmeter-A test instrument used for measuring resistance. Often part of a multimeter.

Ohm's Law-A law that sets forth the relationship between voltage (E), current (I), and resistance (R). The law states that E=IxR.

Ohms Per Volt-A measure of the sensitivity of a voltmeter.

OI-Operator Interrupt

OID-Object Identifier

OLAN-Onboard Local Area Network

OLC-Overload Class

OLE-Object Linking And Embedding

OLT-Optical Line Termination

OLTP-Online Transaction Processing

OMAP-Operation, Maintenance and Administration Part

OMC-Operations And Maintenance Center

Omnidirectional Antenna-An antenna that transmits its radio signal in all directions equally at a particular azimuth.

On Demand Programming-On demand programming is providing or making avaialble programs that users can interactively request and receive.

On Demand Services-On demand services interactive programming services that provide or entitle a customer to receive or gain access to specific services after they request (demand) the service. On demand services are typically provided for a single use or session with a fixed termination event (download complete) or time (viewing period). On demand services users are often billed per event for the specific service that is requested. On demand services through the television may include Internet access, videoconferencing, instant messaging and a variety of other interactive services. On demand services may be made be simplified and made more valuable through the use of sophisticated Electronic Program Guides (EPGs) that can be dynamically changed or updated.

On Line (Online)-(1-general) A device or system that is energized and operational, and ready to perform useful work. (2-computer system) A circuit or device that is connected to a system, usually a remote computer, and available for use.

On Peak-A time period where a telecommunication system usage is higher, typically during normal business hours. Some telecommunications service providers charge a premium rate for the use of services during peak hours.

On Screen Display (OSD)-On screen display is the insertion of graphics or images onto the display portion of a screen. The graphics insertion typically occurs at the graphics card or set top box assembly that creates the signals for the display assembly (such as a computer monitor or a television set).

On TV Display-On TV display is the creation of text or images from a television accessory that is displayed on a television or video image.

ONA-Open Network Architecture

On-Demand Streaming-In audio or video streaming, a stream which begins at the time that the client requests it. Usually the client may also pause the stream, skip to a different time in the presentation, or fast-forward or rewind.

One Flat Business Line (1FB)-A telephone line used by a business that is charged a single monthly fee regardless of how many calls that are originated or received during each month.

One Off Payment-A one off payment is the processing of a payment each time one order is processed.

Oneway Trunk-A trunk that can be seized at only one end.

On-Hook-An electrical signal that occurs when a customer typically replaces a telephone receiver onto its cradle, thus opening the hook switch. When the hook switch is opened (on-hook), this typically causes an increase in telephone line voltage due to removal of the connection between the telephone wires on the local loop line.

ONI-Operator Number Identification

ONIX-Online Information Exchange

Online-On Line

On-Net-On-The-Net Calls

ONT-Optical Network Termination

ONU-Optical Network Unit

OOB Channel-Out of Band Channel

OOB Transmitter-Out of Band Transmitter

OOO-Optical to Optical to Optical

OP-Outside Plant

Opacity-Opacity is the abiltiy of an image to visually display images that are behind the image.

OPB-Optical Power Budget

OPC-Originating Point Code

Open-An interruption in the flow of electric current, as caused by a broken wire or connection.

Open Architecture-A design that permits the interconnection of system elements provided by many vendors. The system elements must conform to interface standards.

Open Cable Application Platform (OCAP)-Open cable application platform is an industry middleware standard used in cable television systems to allow the user to access additional interactive services such as Internet browsing and electronic programming guides. OCAP uses parts of the multimedia home platform (MHP) capability that is used in DVB systems.

Open Database Connectivity (ODBC)-An interface for accessing data in a environment of relational and non-relational database management systems. ODBC provides a vendor-neutral way of accessing data in a variety of personal computers, minicomputer and mainframe databases.

Open Digital Rights Language (ODRL)-Open digital rights language is an XML industry specification that is used to manage digital assets and define the rights associated with those digital assets. Open digital rights language was initially designed by Renato Iannella of IPR Systems Ltd. of Australia and its use has been endorsed by several companies including Nokia. The ODRL is a relatively simple language that has been optimized for the independent definition of rights associated with many types of content and services. For more information on ODRL visit www.ODRL.net.

Open Host Controller Interface (OHCI)-The universal serial bus (USB) host interface used by Compaq, Microsoft and National Semiconductor.

Open Interface-A connection or access point between two assemblies or systems that is well defined and is readily available to manufacturers or users of the interface. Open interfaces are usually defined to encourage competition as multiple manufacturers can compete to produce products that have open interfaces.

Open Internet Access-Open Internet access is a connection to the Internet that does not include network address translation (NAT) or a proxy server between the end user and the Internet. Open

Internet access allows users to run applications that require the assignment of Internet ports (logical channels) that may not be allowed or processed correctly by NAT or proxy servers.

Open Network Architecture (ONA)-In the context of the FCC's Computer inquiry III, the overall design of a communication carrier's basic network facilities end services to permit all users of the basic network to interconnect to specific basic network functions and interfaces on an unbundled, equal-access basis. Note: the ONA concept consists of three integral components (a) basic service arrangements (BSAs), (b) basic service elements (BSEs), and (c) complementary network services.

Open Numbering Plan-A numbering plan in which local numbers comprise a different quantity of digits, even in the same city, and each area or zone code typically comprises a different quantity of digits. The national telephone numbering plans of many European countries are open plans. For example, there are both 6 digit and 7 digit telephone numbers in the same city in some countries. Some area codes for small towns have more digits than the area codes for larger towns in the same country, etc. The ITU international numbering plan is an open plan, with different national telephone systems being reached via "country codes" having different numbers of digits. The country code for North America is 1, a single digit. The country code for the United Kingdom is 44, a pair of digits. The country code for Israel is 972, comprising three digits. In each case the quantity of digits that must follow the country code is also different for each destination country, and may vary among different cities in the same country. Please note that an assembly of closed numbering plans may comprise an overall open numbering plan!

In most local numbering plans the quantity of digits in a number is tied to the leading digits of that number. This is called a "deterministic" open numbering plan. For example, in such a plan, all numbers beginning with the digits 23, 24 or 25 are 5 digits in length, while all numbers beginning with any other two digits are 6 digits in length. When an open local numbering plan is not deterministic, as in some cities in Austria, the originating telephone switch must use a "time out" method to determine when the originator has dialed the last digit. An open numbering plan has the advantage of allowing residents of small towns to dial a minimum quantity of digits actually required to distinguish the small quantity of local telephones in their local dialing area. However, it also increases the complexity of determining accurately when an originator has dialed the final digit of a non-local call. Most systems that handle open numbering plans use the "time out" method. That is, they assume that the originator has completed dialing when an interval of typically 6 seconds elapses without the originator dialing any further digits. Some systems will wait as long as 20 seconds to ensure that no further digits are dialed. This either prolongs the time to set up the call, if the waiting time is very long, or occasionally causes incorrect number connection attempts if the waiting time is too short. In some systems such as in North America, the "time out" method is used for international calls, but an originator dialing an international call can also indicate the end of the dialed digits by using the # key, but this is only possible for a originator who has a touch-tone dial. See also Closed Numbering Plan.

Open Settlement Protocol (OSP)-A standard protocol that is designed to transfer billing information to allow inter-carrier billing between voice and data communication systems. The OSP format is approved by the European Telecommunications Standards Institute (ETSI). OSP allows communication gateways to transfer call routing and accounting information to clearinghouses for account settlements between carriers (service providers).

Open Shortest Path First (OSPF)-An Internet routing protocol used that provides all the routers within a network domain to know the topology and to use this information when determining the optimal routing (shortest path) through the network. OSPF can also use network loading and bandwidth cost when determining the optical routing path through the network. Routers within an OSPF domain continuously update their stored maps of the network by swapping information with each other.

Open Source Software (OSS)-Software that includes the original source code from which the product was developed to allow other developers to make changes to the software to meet their specific application needs.

Open Standards-Open standards are operational descriptions, procedures or tests that are part of an industry standard document or series of documents that is recognized and available to all people or companies as having validity or acceptance in a particular industry. Open standards are commonly created through the participation of multiple companies that are part of a professional association, government agency or private group.

Open Switching Interval (OSI)-A time period that occurs in switching systems (primarily analog switching systems) where circuits or equipments are temporarily disconnected from a line or when other circuits are connected to the line. During OSI periods, other transmission systems (such as custom calling features) may be connected to circuits or equipments.

Open System-A system whose characteristics comply with specified standards and that therefore can be connected to other systems that comply with these same standards.

Open Systems Interconnection (OSI)-The open systems interconnection (OSI) standard layer model was developed by the international standards organization (ISO) and the CCITT. The OSI model helps to standardize the inter-connection of computers and data terminals to their applications, regardless of their type or manufacturer. The protocols specify seven layers: physical, link, network, transport, session, presentation, and application. Each layer performs specific functions for data exchange and is independent of the other layers.

This diagram shows the seven layers of the open systems interconnection (OSI) model and how they interact with each other. This example shows how an email application can use the OSI model to allow communication between an email client (user that is checking email) to an email server (computer providing the email information) independent of who controls each layer, provided the interfaces between each layer are specifically defined. This diagram shows that the application layer is the interface to the user that permits the user to request delivery of their email. The application layer presents this request to the transport layer as a data file. The data file is divided up into smaller blocks of data and presented to the session layer. The session layer determines a new session is required (communication link) between the client and the server and this session information is passed on to the transport layer that will oversee the transfer of data during the session. The transport layer sends the destination address of the email server to the network layer. The network layer sends this information to the data link layer that establishes and maintains a data link connec-

tion to the network. The data link layer sends information to the physical layer that converts to data signals to either radio, electrical, or optical formats suitable for transmission.

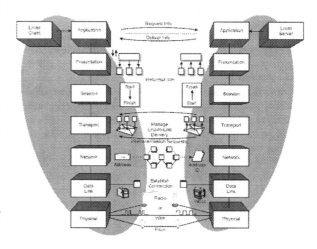

Open System Interconnection (OSI) Protocol Operation

Open Wire-A type of wire installation, now obsolete, in which the electrical conductors need no insulation or sheath for protection from the environment. Open wire is mounted on insulators.

Opening Ticket-An initial work order that is requested by outside plant personnel from the network operations center (NOC) or other maintenance center prior to opening an underground splice closure for either repair or splicing activity. The ticket is closed out at the end of the days activities and a new ticket issued upon request for subsequent work activity.

Operand-Any of the qualities arising out of or resulting from the execution of a computer instruction, a constant, a parameter, the address of any of these quantities, or the next instruction to be executed.

Operating Lifetime-The period of time during which the principal parameters of a component or system remain within a prescribed range.

Operating System (OS)-An operating system is a group of software programs and routines that directs the operation of a computer in its tasks and assists programs in performing their functions. The operating system software is responsible for coordinating and allocating system resources. This includes transferring data to and from memory, processor, and peripheral devices. Software applications use the operating system to gain access to these resources as required.

Operating System Footprint-The amount of memory required to run an operating system or a piece of equipment (such as a computer or a television set top box).

Operation, Maintenance and Administration Part (OMAP)-The application entity that is dedicated to the communications aspects of the operation, administration and maintenance of the signaling system network

Operational Expenses (OpEx)-The term OpEx is used to define the day-to-day short term expenses paid to a telephone company (telco) to support continued business operations (e.g. salaries, rents, commission fees).

Operational Testing-Operational testing is the configuring of system equipment, application of test signals (if required) and measurements or observations of signals and test responses that ensure a system is operating correctly.

Operations-The term denoting the general classifications of services rendered to the public for which separate tariffs are filed, namely exchange, state toll and interstate toll.

Operations Administration And Maintenance (OA&M)-The functions that are necessary to operate, perform administration functions and maintain a communications network.

Operations And Maintenance Center (OMC)-The OMC includes alarms and monitoring equipment to help a network operator diagnose and repair a communications network.

Operations Center (OC)-An operations center is a facility and the associated equipment that is responsible for the operations and monitoring of communication services and system operations.

Operations Cost-Operations cost is the charges and fees associated with the administration, provisioning and management of a business or system.

Operations Support System (OSS)-A system that is used to allow a network operator to perform the administrative portions of the business. These functions include customer care, inventory management and billing. Originally, OSS referred to the systems that only supported the operation of the network. Recent definition includes all systems

required to support the communications company including network systems, billing, customer care, etc.

Operations System-A general-purpose software system that supports the operations of a telecommunications company. Operations supported include planning, engineering, ordering, inventory tracking, automated designs, provisioning, assignment, installation, maintenance, and testing.

Operator-(1-general) A person who assists customers with the operation or use of a system or service (2-carrier) In telecommunications, this is the company that provides communication services.

Operator Assisted Call-A telephone call made by a customer who dials for an access code for assistance (such as 0 +) or is automatically connected to an operator for assistance in placing person-to-person, collect, coin, third-panty-billed, or credit card calls.

Operator Interrupt (OI)-An operator service whereby the operator may interrupt an ongoing conversation. Sometimes called an emergency interrupt (EI)

Operator Relay Services-A program to assist those with hearing and/or speech disabilities to communicate over telephone networks through the use of a relay operator who translates written text into speech, and spoken replies into text.

Operator Services-Operator services use an operator to assist in the handling of a processing of a call. These special handling services include collect calling (billing to a called number), third party charging (billing to another phone or calling card), identification of a person who has called (call trace services), call information services (assistance with directory number location), rate information services (call charge rates), or any other service that requires an operator for special call processing services.

OpEx-Operational Expenses

OPGW-Optical Power Ground Wire

OPS-Order Processing System

OPT-Open Packet Telephony

Optical Add Drop Multiplexer (OADM)-In an optical network carrying multiple optical signals (WDM or DWDM), an Optical Add/Drop Multiplexer adds or removes (drops) individual optical signals (also known as lightwaves or wavelengths) from the network. This action provides access to specific wavelengths for other networks or access points.

This figure shows the different types of ADM con-figurations. The direct through ADM is a non-ter-minating network node that allows the optical sig-nal to pass straight through the ADM. The drop and continue ADM allows for a connection to ter-minate part of the incoming signal and allow the remaining part of the signal to continue through the network. A multi-drop ADM extracts two or more channels from the incoming signal. A multi-drop and continue ADM extracts two or more chan-nels from an incoming signal and allows the remaining signal to go through. An add drop ADM drops one or more channels from the incoming sig-nal while inserting one or more channels as the sig-nal passes through. A hairpin ADM redirects the inbound electrical signal to the outbound electrical lines (loopback).

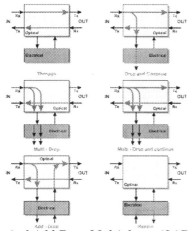

Optical Add Drop Multiplexer (OADM)

Optical Amplifier (OA)-A means of amplifying an optical signal through the sensing of energy at particular optical wavelengths and adding optical energy at the same wavelength so the resultant signal is a replica of the input signal at higher (amplified) energy level. The most common optical amplifiers are the Erbium doped fiber amplifier (EDFA), RAMAN, and semiconductor laser amplifi-er.

Optical Attenuator-A device that is used to reduce the intensity (strength) of lightwaves in a fiber optic transmission system when inserted seri-ally into an optical link. Optical attenuators are composed of semitransparent material that

absorbs a significant amount of photonic energy within the attenuator.

This figure shows how optical attenuators may be inserted in the transmission path to reduce the sig-nal strength so that the optical receiver doesn't become overloaded and is able to receive the optical signal. This diagram shows that an attenuator is installed at the receiving end of an optical commu-nication system to reduce the maximum level of the optical signal to below the maximum signal that the optical detector can process.

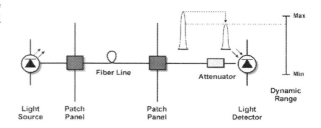

Optical Attenuator Operation

Optical Cable-A cable that contains fibers or bun-dles of fiber lines that is designed for physical and optical (e.g. optical signal loss at specific wave-lengths) specifications that allow it to be used in specific types of optical communication applica-tions (e.g. undersea or in-building.)

Optical Cards-A card similar to a credit card that stores information (such as account codes and account balances) that is stored and retrieved from the card optically.

Optical Carrier 1 (OC1)-Operates at 51.84 Mbps

Optical Carrier 12 (OC12)-Operates at 622.08 Mbps (12 X OC1)

Optical Carrier 192 (OC192)-Operates at 9.95 Gbps

Optical Carrier 3 (OC3)-Operates at 155.52 Mbps (3 X OC1). This rate is the lowest at which asynchronous transfer mode (ATM) is implement-ed.

Optical Carrier 768 (OC768)-Operates at 39.81 Gbps

Optical Carrier 9 (OC9)-Operates at 466.56 Mbps (9 X OC1)

Optical Carrier Hierarchy (OC-n)-Optical carrier (OC-n) transmission is a hierarchy of optical communication channels and lines that range from 51 Mbps to tens of Gbps (and continues to increase). The "n" is an integer (typically 1, 3, 12, 48, 192, or 768) representing the data rate. Lower level OC structures are combined to produce higher-speed communication lines. There are different structures of OC. The North American standard is called synchronous optical network (SONET) and the European (world standard) is synchronous digital hierarchy (SDH).

Optical Character Recognition (OCR)-The recognition of printed or handwritten characters by automatic systems, often laser- and photoelectric-based.

Optical Combiner-A device that combines several input optical signals (usually from fibers) into one or several output fibers.

Optical Continuity Check-An optical continuity check is the sending of an optical signal through an optical line or assembly and sensing if the light reaches the end of the optical circuit or assembly. A quick optical continuity check involves placing a white light near one end of an optical fiber or cable and viewing the other end to see if the light can be viewed at the opposite end.

This figure shows a diagram of how a visual test light can be used to verify the continuity of a fiber. This example shows that a flashlight has been modified with an adapter that allows it to focus its strong visual light source into the end of the fiber. This visual optical signal travels through the fiber where it can be seen at the end of the fiber.

Optical Data Bus-An optical fiber network used to interconnect terminals in which any terminal can communicate with any other terminal.

Optical Demultiplexing-Optical demultiplexing is the separation of an optical signal that contains multiple optical signal wavelengths (optical channels) into individual optical signals (or wavelengths or lightwaves) from each other so that they can be rerouted or processed individually. Optical demultiplexing is used in wave division multiplexing (WDM) systems.

This figure shows how optical demultiplexing can extract one or more optical channels from an incoming optical communication channel. In this diagram, there is an optical signal that is supplied to an optical demultiplexer. The optical demultiplexer allows one (or more) of optical signals with specific wavelength to pass through while blocking the other optical signals.

Optical Detector-Usually a semiconductor

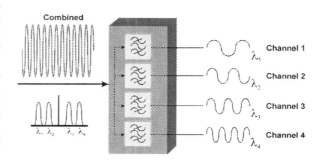

Optical Demultiplexer Operation

device, such as a PIN or avalanche photodiode, that converts light to an electrical signal in fiber optic communications systems. An optical detector also is called an optical receiver.

Optical Disk-A form of data storage using a laser to optically record the data on a disk which is read with a low-power laser pickup. The primary types of optical discs are: read only (RO), write once read many (WORM), erasable/record-able (thermo-magneto-optical TMO) and phase change (PC).

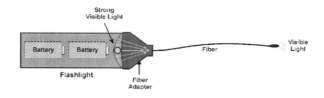

Optical Cable Continuity Testing

Optical Fiber-A thin filament of glass (usually smaller than a human hair) that is used to transmit voice, data, or video signals in the form of light energy (typically in pulses).

Optical Fiber Conductive Plenum Cable (OFCP)-Optical fiber conductive plenum cable has some form of conductive material such as armor or a central strength member as part of the cable. OFCP is used in spaces or areas within a building that have airflow through the area and have a need for added strength (cable protection) or conductors (e.g. ground wires or conductive strength member). OFCP plenum cable is constructed of materials that are flame resistance.

Optical Fiber Connector-A device whose purpose is to transfer optical power between two optical fibers or bundles, and that is designed to be connected and disconnected repeatedly.

Optical Fiber Coupler-A device whose purpose is to distribute optical power among two or more ports. The term also may be used to describe a device whose purpose is to couple power between a fiber and a source or detector.

Optical Fiber Splice-A permanent joint whose purpose is to couple optical power between two fibers.

Optical Filter-A passive or active device that selectively blocks (rejects) or transmits (passes) a range of wavelengths in a fiber optic transmission line.

Optical Frequency Division Multiplexing (OFDM)-A process of transmitting several high speed communication channels through a single fiber through the use of separate wavelengths (optical frequencies) for each communication channel. OFDM is now commonly called wave division multiplexing (WDM). However, WDM usually refers to optical channels that have very small spacing between them and OFDM refers to multiple optical channels that can have any amount of wavelength spacing between them.

Optical Isolator-A device used to allow light to travel in only one direction through an optical element. They are used to prevent light from reflecting back into lasers or fibers causing damage or signal loss. Optical isolators typically use two polarization filters and a Faraday rotator to accomplish this effect. Light entering in one direction is vertically polarized by the first filter, rotated by 45 degrees, and then passed through the second filter, which is designed to pass linearly polarized light

rotated by 45 degrees from the first filter. Light entering from the opposite direction is linearly polarized at an angle of 45 degrees from vertical by the first filter. The Faraday rotator then rotates the angle of polarization another 45 degrees so that the polarization is horizontal. The second filter blocks most of this light because it only passes light polarized in the vertical direction.

This figure shows how an optical isolator allows an optical signal to pass through in one direction and restricts the optical signal flow in the opposite direction. This example shows that a signal is reflected back toward the light source when a cable is cut. To protect the light source from the reflected signal, the optical isolator attenuates the reflected signal (isolation dB) while allowing the forward signal to pass through with minimal attenuation (insertion loss)

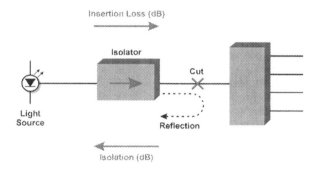

.Optical Isolator Operation

Optical Line Termination (OLT)-An optical line termination is a device in a passive optical network that typically resides at the service provider's central office or network cabinet and communicates through an optical distribution network to control and transfer optical signals to devices located at a customer's site.

This figure shows the how an optical line termination (OLT) cam be used to communicate with multiple optical network units in an optical communication system. This diagram shows that an OLT receives signals from the telephone switching system (voice channels) and from data gateways (data channels). The OLT than converts the electrical signals into optical signals for distribution to ONUs. This example shows that the OLT may co-exist with other optical signals such as cable television systems.

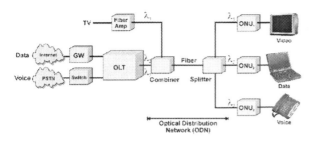

Optical Line Terminal Operation

Optical Link-Any optical transmission channel designed to connect two end terminals or to be connected in series with other links. Terminal hardware also may be considered within the bounds of this term.

Optical Multiplexer-An optical multiplexer is a device or assembly that combines two or more optical signals onto a single optical channel or optical link. Optical multiplexing may be in the form of wavelength division (e.g. multiple wavelength channels on a fiber line) or time division (e.g. slots on a optical link).

This figure shows how optical multiplexing may be used to combine optical signals with different wavelengths. This diagram shows a system that has multiple optical signal sources at different wavelengths. These signals are combined into a single fiber channel through the use of an optical multiplexer.

Optical Network-Optical networks are a series of points that are interconnected by optical communications channels or systems. Optical networks are either common to all users or privately leased by a customer for some specific application.

Optical Network Unit (ONU)-An optical networking unit is a device that can receive, multiplex and demultiplexes optical signals and converts the signals to a format suitable for distribution to other systems or to a customer's equipment such as copper lines.

This figure shows the basic operation of an optical network unit (ONU). This diagram shows that an ONU receives optical signals from OLT on multiple wavelengths. The ONU can receive and process one or more optical wavelengths and convert these optical signals into an electrical form. This diagram shows an ONU that can process electrical signals into video (television), data (Internet browsing) and digital audio (telephone) formats.

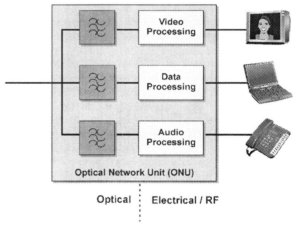

Optical Network Unit (ONU) Operation

Optical Node-An assembly within an optical network where signals are transferred between optical fibers to other transmission media such as wires or coaxial cable.

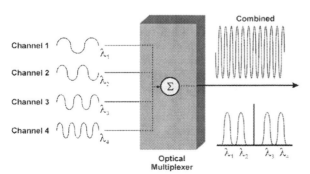

Optical Multiplexer Operation

Optical Path-The path that an optical signal travels from transmitter to receiver. The path can be within optical fiber, through various optical elements, or through free space.

Optical Receiver-An optical receiver is a device that is used to convert optical signals into electrical signals.

Optical Repeater-An optical repeater is a device or circuit that is located between transmitting and receiving devices to improve the quality the signal that is delivered between them. A repeater obtains some or all of the signal from the transmitter, amplifies and may adjust (change a wavelength) or filter the signal, and retransmits (repeats) the signal.

Optical Repeaters can be all-optical if the signal only needs to be amplified. For more sophisticated regeneration, the optical signal is converted to an electronic signal, in which form the signal-to-noise ration (SNR) can be improved among other processing.

This figure shows the basic operation of an optical repeater. This diagram shows that an optical repeater receives a pulsed (digital) optical signal that may have varying signal levels. This example shows that the repeater has two signal thresholds that allow it to recreate the original digital signal. When the repeater receives the weak and distorted signal, senses the high or low level changes and creates and transmits a new signal that is a repeated version of the received signal.

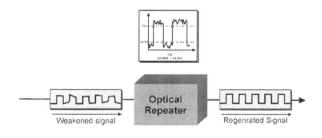

Optical Repeater Operation

Optical Source-An optical emitter, usually a semiconductor device (such as a laser or light-emitting diode), that converts an electrical signal into a light pulse for use in a fiber optic communications system. This device also is called an optical transmitter.

Optical Spectrum-Optical spectrum is the distribution of optical energy as a function of frequency or wavelength.

This diagram shows the different portions of the optical spectrum and their primary characteristics.

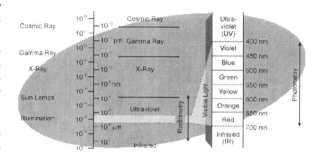

Optical Spectrum Diagram

Optical Switching-Optical switching is the process of connecting optical signals between multiple ports or time periods on optical communication lines. Optical switching systems may convert the optical signals to electrical form or they may directly connect the optical signals using optical switches.

Optical Time Domain Reflectometer (OTDR)- A test instrument that measures the time for transmission and return of optical signal pulses through a fiber transmission line to determine the distance to specific points in that transmission line (such as a cable break or damaged portion). The OTDR tester operates by sending an optical pulse of energy, timing the response, and analyzing the shape of the pulse to determine the type of change in the transmission line that caused the signal reflection (e.g. un-terminated-terminated cable break).

Optical To Electronic Conversion (O/E)

This figure shows the basic operation of an optical time domain reflectometer (OTDR). This diagram shows that an OTDR test instrument sends an optical pulse into an optical communication line (a fiber). As the light from the light burst travel down the optical transmission line, some of the light energy is reflected back due to scattering effects. The returned light signal level is shown on the OTDR display over time. The amount of time it takes between the forward traveling light signal (incident light) and the time the reflected light is received back at the OTDR determines the length of the optical signal has traveled in the optical transmission line.

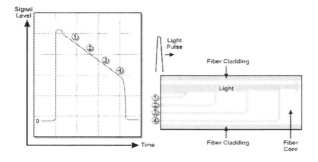

Optical Time Domain Reflectometer Operation

Optical To Electronic Conversion (O/E)-The process of changing an optical signal into an electronic one. Optical signals, or lightwaves, are used to transport information rapidly. However, many critical network processing steps can only be performed on electronic signals. For this reason, the optical signals are converted to electronic signals at the endpoints of a transmission and usually several points in between for processing. See also Electronic to Optical Conversion (E/O).

Optical to Electronic to Optical (OEO)-Optical to electronic to optical (OEO) refers to network elements that convert optical signals to electronic ones for signal processing, then convert them back to optical signals for further transport through the optical network. See also OOO.

Optical to Optical to Optical (OOO)-Optical to optical to optical (OOO) is a term used to describe all-optical network elements, such as optical switches, to differentiate them from OEO (Optical to Electronic to Optical) elements.

Optical Transmission-Optical transmission is the transfer of information through the use of optic signals. These optical signals may be transferred through optical cable (fiber) or through another medium (such as air).

Optical Transmitter-An optical transmitter is a device or assembly that converts an electrical, acoustical or other form of information signal into a transmitted optical carrier signal.

Optimize-The process of adjusting for the best output or maximum response from a circuit or system.

Optimized Network-A network in which each trunk group has been sized to operate at its specified economic or service objective when traffic is routed according to a specified plan.

Optimum Layered Pricing Strategy-A pricing plan that allocates different prices to different customers for the same service levels.

Opt-in-Option in

Option in (Opt-in)-Opt-in is the authorization of an option (option in) to a process such as to be included on a mailing list or subscription service.

Option out (Opt-out)-Opt-out is the declining of an option (option out) for a process such as to be included on a mailing list or subscription service.

Optional Calling Plan-A service offering that gives customers the choice of expanding a local calling area for an additional monthly charge, or selecting a smaller calling area and paying toll charges for all calls outside that area. The plan includes Extended Area Service (EAS).

Options Message-A message that is used to define the parameters of a communication session. An options message is defined in session initiation protocol (SIP).

Optoelectronics-The range of materials and devices that generate, amplify, detect, and control light. Each of these functions requires electric energy to operate and depends on electronic devices to sense and control this energy.

Opt-out-Option out

OPX-Off Premises Extension

Order Handling-Order Handling is the process of entering the orders gathered by the sales organization into the billing and network management systems.

Order Processing-Order processing is the defining of terms that are agreeable to your prospect for the acquisition of your product or service. Most sales people refer to the order processing step as closing the sale. Asking for the order seems to be a fearful process for most

Order Processing System (OPS)-(1-general) Order processing systems gather information related to orders, process the information into specific orders, and create actionable information that allows the fulfillment of the orders. (2-IP Telephony) IP telephone systems can be integrated with order processing systems to allow interactive control with customers to allow the capturing of order information directly from customers and to assist in fulfillment of the order.

Order Wire-A dedicated voice grade line for communications between maintenance and repair personnel. In digital carrier systems, the order wire is a communication (talk) channel that allows near and far end telephone company (telco) personnel to communicate using telephone sets.

Organizational Learning-Process whereby the entire organization is able to receive feedback from the environment (the market, competition, new technology providers) and dynamically adapt itself accordingly.

Original Equipment Manufacturer (OEM)-The original manufacturer of equipment regardless of who sells the equipment or the name marked or associated with the equipment. The term OEM is sometimes used to refer to companies that use other manufacturers to produce products for them. These companies sell, name, and/or use their distribution system for the product that was produced by the other manufacturer. When a company adds assemblies, software, or documentation to products produced by OEMs. These are referred to as a value-added reseller (VAR). See also Badget.

Original Programming-Original programming is content that is owned, developed and controlled by a network operator who provides the media to its viewers. Original programming may be in the form of news, documentaries, education and other programming that is created for the network. While the creation of original programming may reduce the cost for programming content, it may still involve the payment of fees or royalties for the use of brands, actors or other images.

Originating Number-In identifying number of a device (such as a telephone number) that originated a call or request for service. For telephone systems, the originating number is often provided as an ANI.

ORL-Optical Return Loss

Orthogonal Coding-A method of multiplexing that allows multiple communication channels to exist on the same medium through the use of coded channels. The codes for these channels are chosen so that interference between information elements (symbols) are mutually exclusive.

Orthogonal Frequency Division Multiplexing (OFDM)-A process of simultaneously transmitting several communication channels through the use of different frequencies for which each communication channel is independently managed and optimized.

This figure shows how OFDM divides a single radio channel into multiple coded sub-channels. This example shows that a high-speed digital signal is divided into multiple lower-speed sub channels that are independently from each other and can be individually controlled. The OFDM process allows bits to be sent on multiple sub channels and the channels selected can be varied based on the quality of the sub channel. In this example, a portion of a sub channel is lost due to a frequency fade. Due to the OFDM encoding process, the missing bits from one channel can transmitted on other channels.

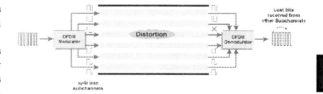

Orthogonal Frequency Division Multiplexing

Orthogonal Variable Spreading Factor (OVSF)-The use of OVSF allows the spreading factor to be changed and remain orthogonal (no interaction between codes) between different spreading codes of different lengths to be maintained.

OS-Operating System

OS-Optical Splitter

OSA-Optical Spectrum Analyzer

Oscillation-(1-general) A periodic change in a voltage, current or other quantity above and below a mean value. (2-parasitic) An oscillation, usually unwanted, that occurs in a self-resonant element of a device.

Oscillator-(1-general) An electrical device or circuit that converts direct current into a periodically varying output. (2-audio) A circuit used to produce audio frequency alternating current. (3-beat-frequency) A circuit used to generate a signal that is combined with a received radio signal to produce an audio frequency beat signal. (4- crystal-controlled) An oscillator circuit whose frequency is accurately controlled by a quartz crystal. (5-local) An oscillator in a radio receiver whose output is mixed with the received radio signal to produce an in intermediate frequency (IF) signal, which is amplified, then detected. (6-master) A stable oscillator used to control the operating frequency of other oscillators and/or systems. (7- relaxation) An oscillator whose frequency is dependent on the charging time of a capacitor.

Oscilloscope-A test instrument that uses a display, usually a cathode ray tube or liquid crystal display (LCD) to show the instantaneous values and waveforms of a signal that varies with time or some other parameter.

OSD-On Screen Display

OSHA-Occupational Safety and Health Administration

OSI-Open Switching Interval

OSI-Open Systems Interconnection

OSI Reference Model-This reference model was created in 1982 by the International Standards Organization (ISO) to standardize communication systems. The model standardized nomenclature across existing protocols and provided guidelines for new protocols using 7 layers. Each successively higher layer builds on the functions of the layers below, as follows:

Application layer 7 The highest level of the model. It defines the manner in which applications interact with the network, including database management, e-mail, and terminal-emulation programs.

Presentation layer 6 Defines the way in which data is formatted, presented, converted, and encoded.

Session layer 5 Coordinates communications and maintains the session for as long as it is needed, performing security, logging, and administrative functions.

Transport layer 4 Defines protocols for structuring messages and supervises the validity of the transmission by performing some error checking.

Network layer 3 Defines protocols for data routing to ensure that the information arrives at the correct destination node.

Data-link layer 2 Validates the integrity of the flow of data from one node to another by synchronizing blocks of data and controlling the flow of data.

Physical layer 1 Defines the mechanism for communicating with the transmission medium and interface hardware.

OSNR-Optical Signal To Noise Ratio

OSP-Open Settlement Protocol

OSP-Out Side Plant

OSPF-Open Shortest Path First

OSS-Open Source Software

OSS-Operations Support System

OSW-Outbound Service Word

OTAP-Over The Air Programming

OTDR-Optical Time Domain Reflectometer

Other Charges & Credits (OC&C)-Charges and credits which do not fall under any other billing category. An example of OC&C: One time waiver of charge.

Other Common Carrier (OCC)-A pre-divestiture term for a telecommunications common carrier, other than a former Bell operating company, authorized to provide a variety of private line services. This term has been replaced by interexchange carrier. The Federal Communications Commission (FCC) also uses the terms miscellaneous, or specialized common carrier.

Out of Band Channel (OOB Channel)-An out of band channel is a communication signal that is transferred on a communication channel (such as a radio frequency channel) that is not normally used for transferring the data or media signals.

Out Of Band Emission-Emission on a frequency or frequencies immediately outside the necessary bandwidth that results from the modulation process, but excluding spurious emissions.

Out of Band Signaling-Signaling, typically related to call processing such as setup, disconnection, (or handover in a cellular system) that is transmitted without using any part of the transmission channel capacity reserved for subscriber traffic. In older frequency division multiplexing equipment, some transmissions of this type utilized a different frequency band than the speech signal, so the name was literally accurate. In modern digital transmission systems, this term is sometimes loosely used for signals that use different digital bits than those reserved for the subscriber traffic. The term is not precisely accurate in this case, because there are not different bands of frequencies involved. The opposite term is "In band signaling"

Out of Band Transmitter (OOB Transmitter)- An out of band transmitter is a circuit, device or assembly that is used to send information or control signals on a channel that is not used for transferring the user's data or media. An example of an OOB transmitter is the sending of program request information (such as a pay per view selection) from a television set top box (STB) to the head end of a cable television systems.

Out of Slot Signaling-The transmission of signals in a time division multiplexing system in different time slots (or in general in different bit fields) than those reserved for the digital subscriber traffic. A more accurate term than "out of band signaling" when used with digital systems, but actually only precisely accurate when used for time division multiplexing systems.

Out Side Plant (OSP)-All telephone company (telco) facilities that are located from the main distribution frame (MDF) outward toward the subscriber, interoffice or toll facility. This includes all toll, trunk, exchange grade facilities whether copper, fiber, or wireless.

Outage-Any disruption of service that persists for more than a specified time period.

Outage Probability-The probability that the outage state will occur within a specified time period. In the absence of specific known causes of outages, the outage probability is the sum of all outage duration's divided by the time period of measurement.

Outage Threshold-A defined value for a supported performance parameter that establishes the minimum operational service performance level for that parameter.

Outbound Call Center-A call center (group of customer service agents) that originate telephone calls from customers. Outbound call centers (telemarketing centers) are often used by companies to solicit new business or to obtain statistical or other business related information.

OutCollects-Outcollects are charges a network operator sends to other telecommunication companies for services they provided to customers that are not registered in the local network (such as completing their calls in local networks). Outcollects are call detail records (CDRs) that are sent by service provider A to service provider B for services provided by A to B's customers. An example of an Outcollect is a roaming billing record that is sent to the home service provider of a customer that details the usage of the customer for services that were provided by the visited system.

Outdoor Data Unit (ODU)-An outdoor data unit is part of a communication system that is located outside environmentally controlled areas. ODUs are typically constructed of more rugged materials than indoor units.

Outdoor Service Loop-An outdoor service loop is a bundle (loop) of cable that is part of an outdoor communication line (such as a pole mounted cable TV or telephone lines) that provides additional cable length that may be necessary to perform a splice or cable path reconfiguration at a later time. Outdoor cable loops can store 100 to 200 feet of cable for each 1,000 feet of installed cable line.

Outdoor Splice Enclosure-An outdoor splice enclosure is a plastic or metal container that is used to cover and protect wires or cables. Outdoor splice enclosures may contain multiple container shells. An outer shell may be used to provide mechanical and environmental protection and the inner shell may be used to hold the cables or a splice tray.

Outer Jacket-An outer jacket is the outermost material used on a cable. The outer jacket typically provides protection from the environment and forces applied during cable installation and mounting.

Outgoing Call Restriction-In telephone call-processing feature that restricts telephone use to specific authorized dialing patterns (typically local phone number).

Out-of-Band-Communication-A type of communication that uses frequencies outside the range being used for data or message communications. This name is sometimes questionably applied to digital time division multiplexing systems in which a communication occurs out of the time slot used for subscriber traffic, better called "out-of-slot." Out-of-band communication is generally done for diagnostics or management purpose.

Outpulsing-The process of transmitting address information over a trunk from one switching environment to another.

Output Power-Output power is the final output stage of a transmitter (radio or light energy) as measured at the output terminal while connected to a load of the impedance recommended by the manufacturer.

Output Stage-The final driving circuit in a piece of electronic equipment.

Outside Broadcast (OB)-A radio or TV program that originates outside the studio. If the program is presented live, the signals must be sent back to the permanent control equipment by temporary links.

Outside Collections Agency (OCA)-An external organization that attempts to collect past due money from delinquent customers. OCAs generally perform collection services for a fee (commission) that is based on the success of their collection activity.

Outside Plant (OP)-The part of a telephone system that is outside of local exchange company buildings. Included are cables and supporting structures. Microwave towers, antennas, and cable system repeaters are not considered to be outside plant equipment.

Outsourcing-The use of an outside firm to produce products, assemblies or to perform specific business functions that would or could be conducted internally. When a communication carrier contracts to another company to provide teleservices or facilities management is an example of outsourcing.

Outtro-Outtro is a video or media segment that is shown at the end of another video or media segment. An Outtro is the opposite of an Intro.

OVD-Outside Vapor Deposition

Ovenized Crystal Oscillator (OXO)-A crystal oscillator that is enclosed within a temperature regulated heater (oven) to maintain a stable frequency despite external temperature variations.

Over Modulation-(1-general) Amplitude or frequency modulation that exceeds 100 percent, resulting in distortion of the transmitted signal and/or excessive bandwidth. (2-digital coder) The condition when a modulating signal level exceeds the dynamic coding range of a digital coder, causing increased quantization distortion in the output.

Over Specification-Over specification is a standard or operations procedure that contains detailed information that limits the ability of a device or system to be designed or produced with innovative and new features.

Over The Air Activation-The ability for a wireless service provider to program or activate service features for a mobile telephone or radio receiver after the product has been purchased.

Over The Air Programming (OTAP)-The ability for a service provider to directly program information stored in a mobile telephone or radio device. This allows over the air activation after the customer initially purchases a mobile telephone.

Overcoupling-A degree of coupling greater than the critical coupling between two resonant circuits. Over-coupling results in a wide bandwidth circuit with two peaks in the response curve.

Overflow Traffic-That part of an offered traffic load that is not carried.

Overgauging-The installation of a relief cable with pairs of a thicker gauge than prescribed by transmission or signaling requirements. Over-gauging is an economical alternative to using two separate cables of different gauges to serve a route that has differing minimum gauge required.

Overhead Bit-Any bit in a digital data stream other than an information bit. Also called a control bit or, simply over-head.

Overhead Information-The digital information transferred across the functional interface separating a user and a telecommunication system (or between functional entities within a telecommunication system) for the purpose of directing or controlling the transfer of user information.

Overhead Messages-System messages that are sent from a system (such as a cellular base station) to receivers (such as mobile telephones) giving the communication device the necessary parameters (such as access codes or initial transmitter power levels) to operate within that system.

Overlap Sending-A sending condition that occurs in an integrated services digital network (ISDN) when the setup message sent from a user's equipment to a stored-program control system does not contain all of the called-party number information. Certain functions at the terminal will overlap those of the switching system because the system must wait for a complete address.

Overlay-(1-data storage) The technique of repeatedly using the same areas of internal storage during different stages of the execution of a program. (2-radio system) The process of overlaying another radio system technology or capability over an existing (underlying) system.

Overlay Network-An overlay network is the combination of a new system on top of an existing system. The common reason for deploying an overlay system is to allow the introduction of new services while maintaining the operation of an existing system.

Overload-In a transmission system, a power greater than the amount the system was designed to carry. In a power system, an overload could cause excessive heating. In a communications system, distortion of a signal could result.

Overload Class (OLC)-A field within the global system messages sent to communication devices such as mobile telephones that indicates if the device is authorized to attempt access or if its overload class has been restricted from accessing the system. The use of OLC allows systems to reduce the number of call access attempts during periods of high system activity.

Overload Control-A process used by the system to control the access attempts initiated by mobiles. Overload Class (OLC) bits sent in the overhead message inhibit operation of groups of mobile telephones.

Overprovisioning-The providing more capacity than is actually required for a given application. Overprovisioning may include reserving excessive switching capacity or using a higher speed communications link. Overprovisioning is sometimes performed to reduce the transmission delay of data transmission.

Overrun-(1-computer) The loss of data that occurs when a receiving device is unable to accept data at the rate at which it is being transmitted. (2-fiber fusion splice) Overrun is the amount of distance that an optical fiber is inserted toward the splice area during the application of heat (the arc) to the splice. Overrun is necessary to ensure enough fiber material is available to fill the gap during the fusion process. Excessive overrun will result in a bulge in the fiber and low setting on the overrun will result in a narrowing (necking) of the fiber at the splice area.

Overshoot-The first maximum excursion of a pulse beyond the 100% level. Overshoot is the portion of the pulse that exceeds its defined level temporarily before settling to the correct level. Overshoot amplitude is expressed as a percentage of the defined level.

Oversubscription-A situation that occurs when a service provider sells more capacity to end customers than a communications network can provide at a specific time period. This provides a benefit of reduced network equipment and operational (reduced leased line) cost.

Over-subscription is a common practice in communications networks as customers do not continuously use the maximum capacity assigned to them and customers access the networks at different time periods. Unfortunately, over-subscription in telecommunications can cause problems when customers do attempt to access the network at the same time. For example, when customers open their presents at a holiday event (e.g. Christmas) and attempt to access the Internet at the same time.

OVSF-Orthogonal Variable Spreading Factor

OXO-Ovenized Crystal Oscillator

P

P Frame-Predicted Frame

P2P-Peer-to-Peer

PABX-Private Automatic Branch Exchange

PACCH-Packet Associated Control Channel

Packet-A small group of digital bits that are routed through a network to their destination. The bit sequence of the packet (field structure) may be arranged to include the destination address in addition to the data that is being transported and other data such as the packet originator and error protection bits.

Packet Assembler And Disassembler (PAD)-A PAD divides or converts blocks of data (such as data files) to and from small packets of information. In the disassembly process, a PAD usually assigns sequential numbers to the packets as they are created to allow the reassembly PAD to identify the correct sequence of data packets to reproduce the original data signal.

This diagram shows a packet assembler and disassembler (PAD) system operation. This diagram shows that a large file is to be sent over a packet data network. The large file is supplied to the PAD circuit that divides the data file into smaller packets. These packets are sent toward their destination through a data communications network. When they are received, the are reassembled into the original large data file by a packet assembler.

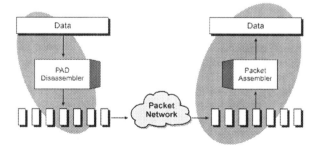

Packet Assembler and Disassembler (PAD)
Operation

Packet Billing-Packet billing involves the authorizing, gathering, rating, and posting of account information for the transmission of data packets. Packet billing may be based on the number of packets, amount of packet data transferred, the time packet data session was in progress or other measurable service information.

Packet Buffer-Memory space set aside for storing a packet awaiting transmission or for storing a received packet. The memory may be located in the network interface controller or in the computer to which the controller is connected.

This figure shows how packet buffering can be used to reduce the effects of packet delays and packet loss for streaming media systems. This diagram shows that during the transmission of packets from the media server to the viewer, some of the packet transmission time varies (jitter) and some of the packets are lost during transmission. The packet buffer temporarily stores data before providing it to the media player. This provides the time necessary to time synchronize the packets and to request and replace packets that have been lost during transmission.

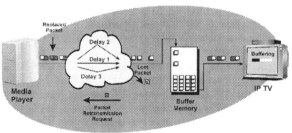

Packet Buffering

Packet Buffering-Packet buffering is the process of temporarily storing (buffering) packets during the transmission of information to create a reserve of packets that can be used during packet transmission delays or retransmission requests. While a packet buffer is commonly located in the receiving device, a packet buffer may also be used in the sending device to allow the rapid selection and retransmission of packets when they are requested by the receiving device.

Packet Consolidation-Packet consolidation is the process of the regrouping of sequencing of packets according to their communication session or quality of service.

Packet Data-The sending of data through a network in small packets (typically under 100 bytes of information). A packet data system divides large quantities of data into small packets for transmission through a switching network that uses the addresses of the packets to dynamically route these packets through a switching network to their ultimate destination. When a data block is divided, the packets are given sequence numbers so that a packet assembler/disassembler (PAD) device can recombine the packets to the original data block after they have been transmitted through the network.

Packet Data Protocol (PDP)-Packet data protocol is the language, processes, and procedures that perform functions used to send packets through a communication network. Examples of PDP include Internet protocol (IP) and X.25 protocol.

Packet Data Switched Network (PDSN)-The packet data serving node (PDSN) is used to control (route) data packets to and from the PCF functions (BS packet controllers) that communicate with access terminals. The PDSN is responsible for originating, maintaining, and terminating data interfaces between the access terminals (via the PCF and base stations) and packet data networks (such as the Internet).

Packet Data Transmission Delay-Packet data transmission delay is the accumulated delays that occur for the transmission of a packet from entry into the packet data network (e.g. from entry into a router) to its exit of the system. Common causes of packet data transmission delay include router switching time, alternate routing through other paths, priority queuing delays (other packets may have priority), and retransmission requests.

Packet Elementary Stream (PES)-A packet elementary stream is a raw information component stream (such as audio and video) that has been converted to packet form (a sequence of packets) which is part of a program stream (such as a digital television program).

Packet Encapsulation-Packet encapsulation is the process of inserting the entire contents of a data packet into the payload (data portion) of another packet.

Packet Filter-The device or circuit that decodes and searches incoming packets to determine and alter its contents or change its routing or contents based on filtering information (e.g. type of service).

Packet Format-A set of rules governing the structure of data control information in a packet. The packet format defines the size and content of the various fields that make up a packet

Packet Forwarding-The process of copying a packet to another node without looking at the destination address.

Packet Header-The front part of a packet that gives the receiving device information such as the message length, the terminal to which the packet is addressed and the packet priority.

Packet Identifier (PID)-A packet identifier is information that is contained within a packet that identifies the contents and format of the packet. A packet identifier in an MPEG system identifies the elementary stream (ES) of a program channel.

Packet Internet Groper (Ping)-A command that is used for data networks that tests for network connectivity by transmitting Internet control message protocol (ICMP) diagnostic packets to a specific node on the network. The ICMP packets for the receiving node to acknowledge that the packet reached the correct destination. If the node responds, this is referred to as a "ping" and it indicates the node is operational.

Packet Loss-Packet loss is a ratio of the number of data packets that have been lost in transmission compared to the total number of packets that have been transmitted.

Packet Mode-The switching of packets of information from different users by statistically multiplexing, or combining, them over shared transmission facilities.

Packet Radio-Packet radio is the process of sending voice communication by converting voice signals (audio) into digital signals and sending the digital signal through a radio communication network using data packet transmission.

Packet Segmentation-The dividing of a block or packet of data into several segments (pieces.) Packet segmentation is often performed to divide a large data packet into smaller data packets so that they can be sent through a network that can only transfer small data packets. When these packets are received at their destination, they are reassembled to their original data packet size. See fragmenting.

Packet Sniffer-A program or process used to monitor a data stream for a specific pattern such as an address or specific content. Packet sniffers are sometimes used inappropriately to discover passwords or credit card numbers.

Packet Switch-A device in a data transmission network that receives and forwards packets of data. The packet switch receives the packet of data, reads its address, searches in its database for its forwarding address, and sends the packet toward its next destination.

Packet switching is different than circuit switching because circuit switching makes continuous path connections based on a signal's time of arrival (TDM) port of arrival (cross-connect) or frequency of arrival. In a packet switch, each transmission is packetized and individually addressed, much like a letter in the mail. At each post office along the way to the destination, the address is inspected and the letter forwarded to the next closest post office facility. A packet switch works much the same way.

Packet Switched Data-Packet switch data is the transfer of information between two points through the division of the data into small packets. The packets are routed (switched) through the network and reconnected at the other end to recreate the original data. Each data packet contains the address of its destination. This allows each packet to take a different route through the network to reach its destination.

This figure shows data communication using packet switching technology. In this example, a laptop computer is sending a file to a company's remote computer that is connected to a packet data network. The laptop computer data communication software requests the destination address for the packets for the user to connect to the remote computer (202.196.22.45). In this example, the source computer divides the data file into three parts and adds the packet address to each of the 3 data packets. The packets are sent through routers in the packet network that independently determine the best path at the time that will help the packet reach its destination (smart switches). This diagram shows the three packets take 3 different routes to reach their destination. When the 3 packets reach their destination, the remote computer reassembles the data packets into the original data file.

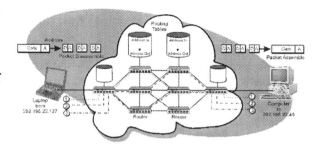

Packet Switched Data Operation

Packet Switched Public Data Network (PSPDN)-Public data networks interconnect data communication devices (e.g. computers) with each other through a network that is accessible by many users (the pubic). To allow many different users to communicate with each other, standard communication messages and processes are used. The Internet is an example of a public data network (there are other public data networks) that uses standard Internet protocol (IP) to allow anyone to transfer data from point to point by using data packets. Each transmitted packet in the Internet finds its way through the network switching through nodes (computers). Each node in the Internet forwards received packets to another location (another node) that is closer to its destination. Each node contains routing tables that provide packet-forwarding information.

Packet Switching-A mode of data transmission in which messages are broken into increments, or packets, each of which can be routed separately from a source then reassembled in the proper order at the destination.

Packet Switching Exchange (PSE)-Packet switches that are used within the X.25 network.

Packet Video-Packet video is the transfer of video information in packet data format.

Packet Voice-Packet voice is the process of sending voice communication by converting voice signals (audio) into digital signals and sending the digital signal through a data communication network (such as the Internet) using data packet transmission..

PacketCable-The name of the project to define a suite of interoperability specifications to allow for devices within a packetized telephony-over-cable network to function correctly even if provided by many vendors.

Packetization-Packetization is the process of dividing data files or blocks of data into smaller blocks (packets) of data.

Packetized Voice-Packetized voice is the process of converting audio signals into digital packet format, transferring these packets through a packet network, reassembling these packets into their original data form, and then recreating the audio signals.

Packet-Switched Connection-In the public data wide-area network, the transmission medium is shared by all devices and it is by the address in the packet header that data reaches its destination. Thus, with packet switching the circuit established during call set-up is a virtual circuit. The packet switch directs the packet to the device to which it is addressed.

Packet-Switching Network-A network that uses packet-switching technology to transfer blocks of data. (See also: packet switching and public packet switched network)

Packing-Packing is a technique of inserting more than one packet of data into the payload of a transport packet. For example, several short IP datagram packets could be packed into the payload (data portion) of a long Ethernet packet.

PACS-Personal Access Communications System

PAD-Packet Assembler And Disassembler

PAD-Program Associated Data

Padding-A process of adding data to a packet or message so that the total number of elements in the data block (usually bits) is a specific multiple of a common data size.

Padding Stream (PS)-A padding stream is a flow of data or information that adds or replaces data of other streaming media or data.

PADS-Product Acquisition and Development System

PAGCH-Packet Access Grant Channel

Page Views-Page views are the number of times a web page has been requested and displayed.

Paging-Paging is a process used to deliver a message, via a public communications system or radio signal, to a person whose exact whereabouts are unknown by the sender of the message. Paging can be a dedicated service (such as numeric pagers) or it may be a general process of alerting devices that they are receiving a call, command, or message.

For paging service, users typically carry a small paging receiver that displays a numeric or alphanumeric message displayed on an electronic readout or it could be sent and received as voice message or other data.

Paid Download-A paid download is a data file or program that requires a user to pay for the transfer (downloading) of the information. While the user may pay for a file or media download, the rights to play (render), transfer (copy) or alter may be limited.

Paid Placement-Paid placement is a marketing program where companies or people pay a fee for a specific location on a media page (such as a web page). The use of a paid placement program for keyword advertising assures the position (usually on top) of a specific listing regardless of its actual popularity ranking.

Pair-(1-general) The two wires of a transmission circuit. (2-shielded) A pair of wires wrapped with an electrostatic shield to minimize induced interference. (3-twisted) A pair of wires with the conductors twisted together to reduce the effect of inductive interference.

Pair Gain-The effective increase in the capacity of a pair of wires that results when signal concentration, or multiplexing, enables the pair to carry not one, but many, signals simultaneously. A pair gain system can include digital or analog carrier systems, and line concentrator systems. A typical example of pair gain is % subscribers on eight pairs, or a ratio of 12 to 1.

Paired Device-In a Bluetooth system, a device with which a link key has been exchanged, either before connection establishment was requested or during the connecting phase.

Pairing-For a Bluetooth system or device, an initialization procedure whereby two devices communicating for the first time create a common link key that will be used for subsequent authentication. For first-time connection, pairing requires the user to enter a Bluetooth security code or PIN.

Pairs Terminated-The feeder pairs terminated on a main distributing frame.

PAL-Phase Alternating Line Video

PAL-Programmable Array Logic

PAL M-Phase Alternating Line Video M

PAL N-Phase Alternating Line Video N

PAL+-Phase Alternating Line Video +

PAM-Pass Along Message

PAM-Pulse Amplitude Modulation

PAMR-Public Access Mobile Radio

PAN-Personal Area Network

Pan, Tilt and Zoom (PTZ)-Pan, Tile and Zoom is the ability to remotely control the angular position and zoom level of a video surveillance camera.

Panel-A plate, usually vertically mounted, on which components, keys, controls, and/or meters are mounted.

PAP-Password Authentication Protocol

Parabolic Antenna-A type of antenna that takes its name from the shape of the structure, described mathematically as a parabola. The function of the parabolic shape is to focus the weak microwave signal hitting the surface of the dish into a single focal point in front of the dish. It is at this point that the feedhorn, or a sub sector that reflects the energy back into the feedhorn, usually is located.

Paradigm-An example or pattern that is commonly associated with thought processes or business operations.

Paradigm Shift-A significant change in the thought patterns or business operations that have been previously established. An example of paradigm shift includes sending mail through the Internet instead of sending mail through the postal service.

Parallax-An apparent change in the position of an object because of a change in the viewing position of the observer.

Parallel-Parallell transmission is the simultaneous transmission of a given signal on different paths. For example, a bit signal can be sent in a serial format, meaning one bit after another on a single path, or a parallel format, meaning that each bit is sent on a separate path so that all eight are sent at the same time.

Parallel Circuit-Circuits that operate in parallel with each other. The electrical current flows through multiple paths in a parallel circuit. If one electrical load is not operating the electricity can switch paths to power the other loads.

Parallel Codes-Parallel error correction codes

Parallel Connection-A group of circuit elements connected in such a way that the same voltage appears across each. The current is divided among them in inverse proportion to impedance.

Parallel Data-The transmission of data bits in groups simultaneously along a cable or collection of circuit board traces referred to as a bus. A typical parallel bus may accommodate transmission of one 8-, 16, or 32-bit byte at a time.

Parallel error correction codes (Parallel Codes)-Parallel error correction codes are efficient data error correction codes that add information redundancy through the use of parallel coding systems where one of the parallel coding systems is delayed with respect to the other. The use of parallel correction codes offers significant reductions in data transmission overhead (additional bits for error detection/correction).

Parallel To Serial Converter-A device that converts parallel input data into a sequenced stream of signal elements (usually data bits).

Parameter-(1-component or circuit) A specific value of some variable as applied to a particular component or circuit. Examples include the resistance of a resistor or the operating frequency of a transmitter. (2-machine language) A variable that identifies and contains information needed to execute a command. (3-variable) A variable that must have a constant value for a specified application.

Parameter Negotiation (PN)-The process of requesting and agreeing on the preferred characteristics for a communication session.

This diagram shows how two data communication devices negotiate for data transmission rates and protocols selection in a data network using the preferences assigned by a user along with the options determined by equipment availability. In this example, data terminal 1 sends a connection request message to data terminal 2. This connection request indicates that the data terminal prefers to use a 56 kbps data transmission rate because it has enough bandwidth. Unfortunately, the data terminal cannot accept the request for 56 kbps because it's access bandwidth is low speed (28 kbps). The receiving data terminal sends back a request to use 28.8 kbps data transmission rate. When the originating data terminal receives this request, it accepts the request because it has that data transmission rate and protocol capability available. It then confirms the request and both devices use a data transmission rate of 28.8 kbps.

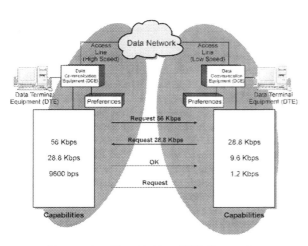

Parameter Negotiation (PN) Operation

Parasitic Oscillation-Parasitic is a self feeding oscillation that occurs due to the combination of capacitance and inductance in an electronic circuit.

Parity-A method of verifying the accuracy of transmitted or stored data. An extra bit is appended to an array of data as an accuracy check during transmission. This bit is referred to as the parity bit. Parity may be even or odd. In an odd parity system, if the number of ones in the array is even, a one is added as the parity bit to make the total odd. For even parity, if the number of ones in the array is odd, a one is added as the parity bit to make the total even. The receiving computer checks the parity bit and indicates a data error if the number of ones does not add up to the proper even or odd total.

Parity Bit-A check bit indicating that the total number of binary "one" bits in a character or word is odd or even.

Park Mode-In a Bluetooth system, park mode is a long term state that allows the slave device to become inactive so it does not need to participate on the Piconet channel. During park mode, the slave device still remains synchronized to that channel. The parked slave wakes up at regular intervals to listen to the channel in order to resynchronize and to check for broadcast messages.

Parking Customers-Parking customers is the process of assigning customers to an alternative service while a new service is being made available or optimized.

PARS-Paging And Radiotelephone Service

Particle Behavior-Particle behavior is the characterization of light as particles of light to help explain and quantify how light energy can be manipulated or converted into other forms (such as optical to electrical detection).

Partner Management-Partner management is the process of identifying, assigning terms and tracking performance of companies that have a collaborative (partnering) relationship.

Passband-A range of frequencies that can be passed through a filter or medium essentially unchanged.

Passcode-A passcode is a unique string of characters that is entered by a person or device to uniquely identify them when requesting access to services or processes.

Passive-A component, device, circuit, or network that is not powered. This term also may refer to a part of an electric circuit that is not a source of power; examples include resistors and capacitors.

Passive Filter-A filter that does not add energy in the filtering process.

Passive Optical Network (PON)-A passive optical network (PON) combines, routes, and separates optical signals through the use of passive optical filters that separate and combine channels of different optical wavelengths (different colors). The PON distributes and routes signals without the need to convert them to electrical signals for routing through switches.

PON networks are constructed of optical line termination (OLT), optical splitters and optical network units (ONUs). OLTs interface the telephone network to allow multiple channels to be combined to different optical wavelengths for distribution through the PON. Optical splitters are passive devices that redirect optical signals to different locations. ONU's terminate or sample optical signals so they can be converted to electrical signals in a format suitable for distribution to a customer's equipment. When used for residential use, a single ONU can server 128 to 500 dwellings. In 2001, most PON's use ATM cell architecture for their transport between the provider EO or POP and the ONU (in some case even to the user workstation). When ATM protocol is combined with PON system, it is called ATM passive optical network (APON).

This figure shows an ATM passive optical network (APON) system that locates optical network units (ONUs) near residential and business locations. This passive optical network routes different optical signals (different wavelengths) to different areas in the network by using optical splitters instead of switching devices. In this example, the optical distribution system uses ATM protocol to coordinate the PON. ONU interfaces are connected via fiber to an OLT located at the provider's EO or POP. Each ONU multiplexes user channels (between 12 and 40) into an optical frequency spectrum allocated to that ONU. Up 32 ONU's can share access to a single PON using the features of dense wave division multiplexing (DWDM). Some newer PON's use high-density wave division multiplexing (HDWDM). Use of HDWDM increases the number of ONU's per PON from 32 to 64. This diagram shows that a PON that uses HDWDM can support approximately 2500 residential customers.

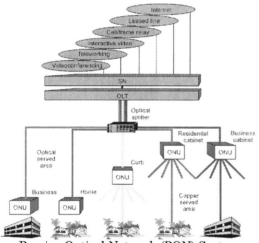

Passive Optical Network (PON) System

Passive Repeater-In a radio communication system (such as microwave links), passive repeaters are antennas that are placed at an intermediate point in a signal path to redirect the signal around terrain or other obstacles.

Passive Sensor-A measuring instrument in the earth exploration-satellite service or in the space research service by means of which information is obtained by reception of radio waves of natural origin.

Passive Splitter-A passive splitter is a coupler that has one input and several output ports where the input port is connected to all the output ports and the energy is passively (no amplification) distributed between the input and output ports.

Passive Star Coupler-A passive star is an optical or RF coupler that has several input and output ports where each input port is connected to output ports and the energy is passively (no amplification) distributed between the input and output ports.

Password-Passwords are a string of characters or an equivalent combination of characters that is associated with a specific user ID. Generally, passwords are not directly stored on a computer network. The passwords selected by users are usually processed by a mathematical formula to produce a result code (a number) that is stored. When the user enters a password or the equivalent value of a password (other character sequences may also work), the result is calculated and compared to the stored result. If the result matches for the specific user ID, access to the system or resources is usually granted.

Password Authentication Protocol (PAP)-A security protocol that prompts the user to enter a user name and password before that is transferred to the system before access to service is granted.

Password Hacking-Password hacking is the repeating of attempts to access a system. Password hacking can be a systematic process such as sequentially trying password characters in sequence (e.g. A followed by B) or it can by a random process that attempts to use additional information (e.g. birthdays, names of family members).

Password Hashing-Password hashing is a computational process that converting a password or information element into a fixed length code. Password hashing can be used to convert variable length text based passwords into fixed length digital password codes.

Password Protection-Password protection is the restriction to access or use information or media by requiring the entry of one or more identifying codes (passwords) before allowing access.

PAT-Program Association Table

Patch-(1-circuit) A temporary circuit rearrangement made by using jacks to bypass faulty circuit components or transmission facilities. (2-software) A temporary software program update that is used to correct software problems between official software release updates.

Patch Cable-A patch is a short cable that is configured to quickly connect and disconnect computers or devices with each other. The wires or optical connections in a patch cable usually connect the pins of one end of the cable to the exact same pins on the other end of the cable.

Patch Cord-A patch cord is a short cord that is used to connect assemblies to each other.

Patch Panel-An optical patch panel is an array of switches or connectors that can be used to reroute or reconfigure an optical transmission system. Optical patch panels are commonly installed to allow reconfiguration of connections such as in an office building where people are moved or reassigned without the need to install new wiring or network distribution equipment.

This figure shows how a patch panel can be used in a data communications network to interconnect computers with data network equipment. This patch panel contains an array of switches and connectors that reroute or reconfigure an electrical or optical transmission system. This patch panel is located within the telecommunications room (a common practice) and it uses drop cables to route the connections to wall jacks located near each workstation. Wall jacks are connected to network interface cards (NICs) by a patch cable.

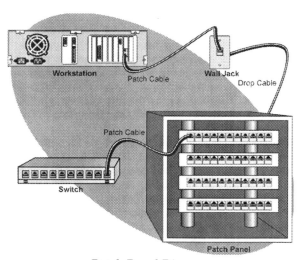

Patch Panel Diagram

Patching-Creating business units with overlapping business functionality in order to encourage competition and excellence in delivery.

Patent-A patent is a document that grants a monopoly for a limited time to an invention described and claimed in the body of the document. A patent describes the invention and defines the specific aspects (claims) of the invention that are new and unique. There are several forms of patents including mechanical patents (processes) and design patents (appearances).

Patent-Design Around

Patent-Enablement

Patent Claim-A description of a specific use of technology that is new and described in the patent. There are usually many patent claims that are part of a patent.

Patent Examiner-A patent examiner is a person having a technical degree, employed by a national or regional patent office, and having the responsibility for granting valid patents according to national, and international laws.

Patent License-A license that is given for the production, use, or sale of products that may use technologies or processes defined by the claims in a patent.

Patent Review Board-A patent review board is a group of people (typically within a company) that review invention applications to determine if they are worthy of submission to the patent office.

Patent Rights-Patent rights are intellectual property rights which give the owner, or assignee, the right to prevent others from making, using, or selling the invention described and claimed in the patent. Patent rights must be applied for in the country or region where they are desired. Most countries have a national patent office, and an increasing number of countries grant patents through a single regional patent office.

Patent Royalty-A fee paid for the use of technology that is defined in the claims of a patent.

Patent Search-The process of searching for patents.

Patent Statement-A patent statement is a disclosure by a company or person for the patents they believe relate to an industry specification or system which may include their intended licensing requirements or restrictions.

Path-(1-general) The route a signal travels through a channel, circuit, or network. (2-communications system) In radio and optical communications systems, the route that an electromagnetic wave travels through space. (3-data) A logical connection between the point where a signal is assembled into a data format and the point at which the signal is disassembled. (4-file name) The complete name location for a file or information element in a computing system.

Path Length-The time required for a signal to travel through a piece of equipment or a length of cable. This time also may be referred to as propagation delay.

Path Loss-A decrease in signal energy during transmission from one point to another. The path loss in free space transmission is 20 dB per decade. This means that for each 10 times in distance the signal travels, the energy will decrease by a factor of 100 (99% decrease in energy).

This diagram shows that the amount of signal energy decrease (path loss) in free space as compared to an obstructed path for an omnidirectional (all direction) antenna. The path loss in free space is approximately 20 dB per decade. This diagram shows that for a 10 times increase in distance, the signal energy will decrease by 99%. The path loss through objects will vary by frequency. This diagram shows that the path loss of radio signals through buildings and trees for radio signals below 2 GHz is approximately 40 dB to 60 dB per decade. The lower diagram shows that a 100 Watt signal will be reduced to less than 10 milliWatts as distance increases by a factor of 10.

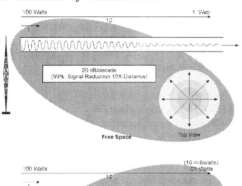
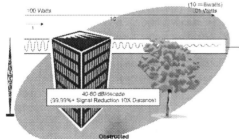

Path Loss Operation

Path Profile-A graphic representation of a propagation path, showing the surface features of the earth, such as trees, buildings, and other features that may cause obstruction or reflection in the vertical plane containing the path.

Path Survey-An assembling of the geographical and environmental data required to design a microwave communication system.

Patriot Act-The Patriot Act is a U.S. aw that was enacted on 26 October 2001 that requires companies to store and provide information to government agencies to help deter terrorism.

Pattern-(1-antenna system) A description of how much power is sent in any given direction from a directional antenna system. The pattern often is documented in the form of a graph. (2-video effects) In a production switcher, a variety of geometric shapes that are available for use as wipe transitions, key masks, and other functions.

Pay Per Listen (PPL)-Pay per listen (PPL) is the providing of audio programming such as music, news, and other education audio that customers view for a fee. PPL services may be ordered by telephone or via an Internet data radio.

Pay Per View (PPV)-Pay per view (PPV) is the providing of television programming such as sports, movies, and other entertainment video that customers view for a fee. PPV services may be ordered by telephone or via an interactive set top box.

Pay Phone-A telephone that requires coins to be inserted or a calling card swiped or inserted before a toll call can be placed. Some pay telephones allow toll free or freephone calls to be originated without the need for coins or calling cards.

Payload-As applied to a data field, a payload is a discrete package of information, also called a packet, which contains a header, user data, and a trailer. In an industry specification (such as Bluetooth), the format for a data field consists of a payload header that specifies the logical channel, controls the flow on the logical channel, and indicates the length of the payload body. The payload body consists of the user data. The "trailer" consists of a Cyclic Redundancy Check (CRC) code for ensuring the accuracy of the packet.

Payload Aware Device-Payload aware devices are network components or assemblies that can detect and adjust the operation of network equipment or services in response to the type of payload that is being carried and may also be able to modify the payload to adjust for network conditions (such as fragmenting or adding error protection).

P

Payload Pointer-In the synchronous optical network (SONET), the pointer that indicates the location of the beginning of a synchronous payload envelope.

Payload Type Identifier (PTI)-A control code that is located at the beginning of a packet payload (such as in an Asynchronous Transfer Mode [ATM] cell) that indicates the type of data in the payload. The PTI may contain length of the packet, type of information (e.g. voice or data), and additional routing and control information.

Payment Processing-Payment processing is the tasks and functions that are used to collect payments from the buyer of products and services. Payment systems may involve the use of money instruments, credit memos, coupons, or other form of compensation used to pay for one or more order invoices.

This figure shows some of the different options available for bill payment. This example shows that customers can make cash payments to a company office of authorized agent, can send checks to the company, can pay via credit cards, via a 3rd party financial processor (e.g. paypal) or through electronic funds transfer (transfer fund). This example shows that payments on account are applied to specific invoices and when the invoices are paid in full, they are marked as pay and are not available for additional payments to be applied.

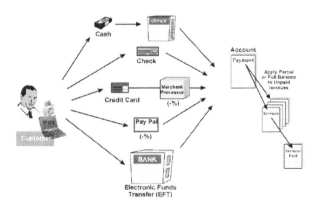

Bill Payment Options

PBCCH-Compact Packet Broadcast Control Channel
PBCCH-Packet Broadcast Control Channel
PBX-Private Branch Exchange
PBX Emulation-Private Branch Exchange Emulation
PC-Personal Computer
PC-Point Code
PC Connector-Physical Contact Connector
PC Telephone-Personal Computer Telephone
PCB-Printed Circuit Board
PCCCH-Packet Common Control Channel
PCCH-Paging Control Channel
PCCPCH-Primary Common Control Physical Channel
PCF-802.11 Point Coordination Function
PCF-Packet Control Function
PCF-Physical Control Field
PCF-Point Coordination Function
PCH-Paging Channel
PCI-Peripheral Component Interconnect
PCI Card-Peripheral Component Interconnect Card
PCM-Pulse Coded Modulation
PCMCIA-Personal Computer Memory Card International Association
PCN-Personal Communications Network
PCPCH-Physical Common Packet Channel
PCR-Program Clock Reference
PCS-Personal Communication Services
PCS-Plastic Clad Silica Fiber
PCSA-Personal Computing System Architecture
PCU-Packet Control Unit
PD-Portable Device
PDA-Personal Digital Assistant
PDCH-Packet Data Channel
PDCP-Packet Data Convergence Protocol
PDD-Post Dial Delay
PDE-Position Determining Entity
PDF-Portable Document Format
PDFA-Praseodymium Doped Fiber Amplifier
PDH-Plesiochronous Digital Hierarchy
PDN-Premises Distribution Network
PDP-Packet Data Protocol
PDP Context-Packet data protocol context is the association of the packet data protocol used in a communication network and the application that it is supporting.
PDS-Packet Driver Specification
PDSCH-Physical Downlink Shared Channel
PDSN-Packet Data Switched Network

PDTCH-Packet Data Traffic Channel

PDU-Protocol Data Unit

PE-Polyethylene

Peak-(1-signal) The maximum amplitude or value, usually of a voltage or current. In a periodic function, the peak is the instantaneous maximum value, either positive or negative. (2-category of usage) A category of usage that indicates that the request or usage of service is at its highest level. (3-billing) A rating time period that may assign a different billing rate (generally a higher rate) when customers use services during period of higher network usage.

Peak Amplitude-The maximum amplitude of a periodically repeating quality.

Peak Bandwidth-Expressed in bytes per second, this limits how fast packets may be sent back-to-back from applications. Some intermediate systems can take advantage of this information, so that more efficient resource allocation results.

Peak Hour-The one or two hours of the day where the number of calls to a call center is at its highest level.

Peak Intensity-Peak intensity is the maximum amplitude, quantity or intensity of a signal or energy that occurs over a period of time.

Peak Load-The load that results from higher-than-average traffic volume. Peak load usually is expressed as the load during a 1-hour period. It also can be expressed in terms of any of several functions of an observing interval, such as a peak hour during the day.

Peer Protocol-A formal language used by peer entities to exchange information. In the Signaling System 7 protocol, a formal language used by peer entities to exchange user data.

Peering-Peering is the process of inquiring or exchanging information with a similar (peer) company or system. Peering is commonly used to assist in the setup and management of communication connections, identification of users, and to determine billing charges. To allow the viewing or transfer of information, a peering agreement may be completed which defines when information can be gathered, how it may be acquired, and in what ways may the information be used or not used.

Peering Point-A peering point is a location or device in a network that is used to exchange information with devices in other networks.

Peer-To-Peer-A system in which two or more network nodes or processes can initiate communica-tion with each other. Peer-to-peer usually describes a network in which all nodes have the capability to share resources with other nodes so that a dedicated server can be implemented but is not required.

Peer-to-Peer (P2P)-Peer to peer is the exchange of information between devices or systems that are capable of operating as both a server (provider) of information and a client (consumer) of information.

Peer-To-Peer Network-Peer-to-peer networks or communication systems that allow users within the communication network to directly interact and exchange data with other users (peers) who are connected to other users in the network.

Peg Count-The total number of any traffic event that occurs during a given period. An example is the number of times a switching-system component functions during one hour.

PEL-Picture Cell

Penalties-Penalties are fees that are charged for actions or service usages that fall outside the agreed limits of service usage. Penalties can include early contract termination fees, charges for lost equipment or the usage of services by unapproved devices (e.g. media storage devices).

Penetration-(1-skin effect) A measure of the depth of current in a conductor. As the frequency through a conductor is increased, the current tends to travel only on the outer surface (skin) of the conductor. (2-broadcasting) The extent to which the population within a service area can receive a broadcast signal, or the extent to which receivers in a service area are available. Penetration may be measured as a percentage of all households, or of the total population. (3-cable TV) The ratio of the number of subscribers to the total number of households passed by a cable TV system.

Performance Gateway (PG)-A performance gateway is a communications device or assembly that monitors the performance of data communication that is received from one network into a format that can be used by a different network. A performance gateway usually has more intelligence (processing function) that can change system configurations to adjust the performance of a system to provide established performance objectives or levels.

Period-The time required for one complete cycle of a regular, repeating series of events.

Periodic-A recurring function or signal that repeats at regular time intervals.

Peripheral-An auxiliary device, such as a printer, mouse or graphics pad, that is connected to a computer and provides service to it but does not participate in principal computer functions.

Peripheral Bus-A serial communications bus between a master controlling device and peripheral or slave devices. The master sends out commands to remotely control the peripherals.

Peripheral Component Interconnect (PCI)-A standard data communication connection that allows accessory cards to be installed into a personal computer. The PCI specification defines both the electrical and physical (connector) requirements. PCI was introduced by Intel and it allows up to 10 PCI-compliant expansion cards in a PC. The PCI standard has replaced the previous industry standard architecture (ISA) bus.

Permanent Virtual Circuit (PVC)-A PVC is a virtual circuit that is manually created for a continuous communication connection.

After a permanent communications circuit is established, a data path (logical connection) is maintained.

This diagram shows how a permanent virtual circuit (PVC) is used to allow the transfer of data through a communications network through a preestablished logical (virtual) path. In this example, a PVC is created by programming routing tables in 4 switches before any data is sent. These routing tables assign data transfer connections between input and output channel on each switch. For example, as data from the sending computer

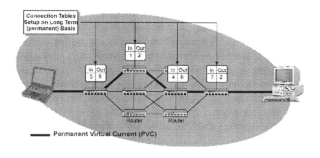

Permanent Virtual Circuit (PVC) Operation

(portable computer) is sent into input channel 3 of the first switch, it is transferred to the output channel 5. This process will repeat for any data that is sent from the sending computer to the destination computer. This example also shows that the PVC path remains active even if the portable computer is disconnected for a period of time.

Permeability-The magnetic permeability of a material is the ratio of the magnetic field intensity, usually denoted B (measured in units of volt seconds per square meter, or the equivalent unit tesla), to the magnetic induction, usually denoted H (measured in amperes per meter). For a linear magnetic material, this ratio is a constant at all levels of magnetic field intensity, and is usually denoted by the Greek letter Mu (μ). The unit of magnetic permeability is one henry per meter. The magnetic permeability of vacuum and most nonmagnetic materials, usually denoted μo, is symbolically 4 pi · 0.000 000 1 henry per meter, where pi represents 3.14159. The numerical value of μo is 0.000 001 256 henry per meter (1.256 microhenrys per meter).

Permissible Interference-Observed or predicted interference which complies with quantitative interference and sharing criteria contained in these [International Radio] Regulations or in CCIR Recommendations or in special agreements as provided for in these Regulations.

Permission Marketing-Permission marketing is the process of promoting and selling products or services only to people who have indicated they desire to receive information from a particular company or related to a specific subject.

Permissions Department-A permissions department is a group or groups within a company that manage the assignment of usage rights to content or portions of content.

Permittivity-The dielectric permittivity of a material is the ratio of the so-called displacement field, usually denoted D (measured in units of ampere seconds per square meter), to the electric field intensity, usually denoted E (measured in volts per meter). For a linear dielectric material, this ratio is a constant at all levels of electric field intensity, and is usually denoted by the Greek letter epsilon (e). The unit of dielectric permittivity is one farad per meter. The dielectric permeability of vacuum, usually denoted eo, is numerically 0.000 000 000 008 85 farad per meter (8.85 picofarads per meter)

Persistence-The characteristic of a material to continue producing light after the exciting energy source is removed.

Persistence Of Vision-A characteristic of the human eye. The sensory cells of the eye are quickly saturated and require a short period of time to recover after the light source is extinguished. It is this "persistence" that allows the phenomenon of television, which relies on the use of individual image frames that are repeated at a given rate.

Personal Advertising-Personal advertising is the creation and delivering of advertising messages that are specifically designed for the needs of a specific person.

Personal Area Network (PAN)-A network concept in which all the devices in a person's life communicate and work together, sharing each other's information and services.

Personal Communication Services (PCS)-Refers to the emerging market of wireless communications that is personalized with services selected by the individual. The wireless PCS networks use radio signals as the access point to the network; the wireless network is then tied back into the public switched network for call routing to or from the wireless subscriber to the other party.

Many PCS carriers occupy the newer PCS frequencies auctioned by the government.

Personal Communications Network (PCN)-A digital cellular system which is based on the DCS 1800 specification. It is a short-range wireless communications network whose radio frequencies carry personal communications service (PCS) signals between portable handsets and radio base stations that are connected to a public switched network. Typically, service is available from 100 to 2000 feet from a radio base. See also: Personal Communications Service (PCS).

Personal Computer (PC)-A computer that is primarily used for home, non-business applications.

Personal Computer Memory Card International Association (PCMCIA)-A standard physical and electrical interface that is used to connect memory and communication devices to computers, typically laptops. The physical card sizes are similar to the size of a credit card 2.126 inches (51.46 mm) by 3.37 inches (69.2 mm) long. There are 4 different card thickness dimensions: 3.3 (type 1), 5.0 (type 2), 10.5 (type 3), and 16 mm (type 4 - unofficial size). The number of connections (pins) is 68. The first PCMCIA standard was approved in 1991 with the concept of standardizing computer memory cards. Because the interface defined how to address, control, and transfer data in standard formats, the PCMCIA standard was quickly used for other types of communication and storage accessories such as modems, network interface cards (NICs), hard disk storage, and controller adapters.

Personal Digital Assistant (PDA)-Small computing devices that contain it's own software operating system that allows the user to run software processing applications. Personal digital assistants are often used to organize personal activities and may provide access to communication services (such as web browsing and email).

Personal Firewall-A device or software program that runs on your computer that provides protection from Internet or data network intruders. Firewalls can restrict access types and may monitor for advanced security threats by analyzing certain types of data communication activities. Although firewalls are important for protecting data that is connected to pubic networks, they can be complicated to setup, can cause problems with desired communications and generally can slow down the transfer of data communication.

Personal Handiphone System (PHS)-Personal handiphone system (PHS) is a digital cordless telephone system that allows wireless telephones to be used at home and in public places. The PHS system uses a 300 kHz radio carrier signal that is divided using TDMA.

This figure shows a PHS radio system. This diagram shows that a PHS system includes radio devices (portable station - PS), radio base stations (cell station- CS), and interconnection equipment (PHS switching center - PSC). The PHS system radio channel has a 300 kHz bandwidth with a gross data transfer rate of 384 kbps. The radio channel is divided into 5 msec frames and each 5 msec frame is divided into 2.5 msec transmit and a 2.5 msec receive frames that contain 4 time slots each. The PHS system uses time division duplex (TDD) multiplexing so that one slot in the 2.5 msec transmit group is used in the downlink direction and one slot in the 2.5 msec receive group is used in the uplink direction to provide full duplex (simultaneous) communication. This example shows that the PHS system can be used in a home environment or in an office or public cordless system.

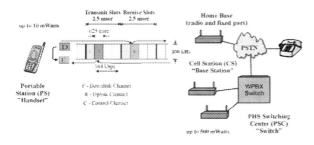

PHS System

Personal Identification Number (PIN)-A number assigned to an individual subscriber which is used to gain access to specified services, such as credit card calling or prepaid wireless services.

Personal Information Manager (PIM)-Are like computerized appointment books, and are used on many personal computers, to help a person keep track of appointments, telephone numbers, and other contact information. A telephony enabled PIM could provide point and click telephone dialing, automated conference calling, or automated access to databases and other applications.

Personal Media Channel (PMC)-A personal media channel is a communication service that allows a media user (e.g. viewer) to select and view media (typically video or music) from a variety of media sources.

Personal Navigation-Personal navigation is the use of position information to assist a person in determining traveling directions (e.g. travel map) or to find services or other locations of interest.

Personal Protective Equipment (PPE)-Personal protective equipment is any articles of clothing or devices that are used or worn by a person to protect against environmental hazards. Examples of personal protective equipment include gloves, goggles and flame retardant clothing.

Personal Video Channel-A personal video channel is a stored or live media channel that is accessible to an individual or group of users that are setup by the owner(s) of the media channel. Personal video channels allow users to create, store and share self-produced video content on networks (such as an IPTV system).

Personal Video Recorder (PVR)-A personal video recorder is a device that stores video images in digital format for personal use.

Personalized Ringing-The ability for a user to select and setup the audio sound that is used to indicate an incoming call. Personalized ringing helps individuals to identify their telephone is ringing when there are several telephones located within the same area.

Pervasive Computing-Refers to the emerging trend toward numerous, casually accessible, often invisible computing devices. The computing devices are frequently mobile or embedded in the environment, and connected to an increasingly ubiquitous network structure. The objective is to facilitate computing anywhere it is needed.

PES-Packet Elementary Stream

PFM-Pulse Frequency Modulation

PFP-Pay for Placement

P-Frame-Predicted Frame

PG-Performance Gateway

PGP-Pretty Good Privacy

PGPhone-Pretty Good Privacy Phone

Phantom Call-A phone that rings inadvertently without a calling party online.

Phantom Circuit-An electric circuit using center-tapped connections to a transformer coil for the purpose of transmitting an additional electric current signal on the transmission wires that connect to these transformer coils. There are several arrangements of this type, but typically the phantom current flows to the center-tapped connection, where it divides into two parts, each part flowing through half of the turns of the coil but in opposite directions. The current then returns to its source via the two parallel wires that are connected to the two ends of the coil. Because this center-tap current flows in equal but opposite amounts in the two parts of the transformer coil, it does not produce a net signal on the other (secondary) coil of the transformer. This current can exist without causing any effect on a second simultaneous current that flows from one end of the coil to the other, not using the center-tap wire.

Phantom circuits are used to transmit direct current for power purposes in some ISDN s-interface customer premises wiring, and also in some proprietary PBX equipment. Phantom circuits were first used in the 19th century on telegraph circuits to transmit an additional telegraph channel over existing wires.

Phase-(1-general) Any distinguishable state in a periodic phenomenon. (2-displacement) The time displacement between two currents or two voltages or between a current and a voltage measured in electrical degrees, where an electrical degree is 1/260 part of a complete cycle. (3-power) The number of separate voltage waves in a commercial alternating-current supply where each signal is a separate phase (such as single-phase or three-phase).

Phase Alternating Line Video (PAL)-Phase alternating line video is a television system that was developed in the 1980's to provide a common television standard in Europe. PAL is now used in many other parts of the world. The PAL system uses 7 or 8 MHz wide radio channels. The PAL system provides 625 lines of resolution (50 are blanking lines).

Phase Alternating Line Video + (PAL+)-Phase alternating line video + is an improved form of the PAL video format that allows for wide screen (16:9 aspect ratio) displays. PAL+ is compatible with existing PAL receivers that have an aspect ratio of 4:3.

Phase Alternating Line Video M (PAL M)-Phase alternating line video version M is an analog television transmission system that is used in Brazil. The PAL M system has 525 scan lines at 59.94 Hz with a subcarrier at 3.575611 MHz.

Phase Alternating Line Video N (PAL N)-Phase alternating line video version N is an analog television transmission system that is used in Argentina. The PAL N system has 625 scan lines at 50 Hz with a subcarrier at 3.582056 MHz.

Phase Distortion-Distortion resulting from the selective phase shifting of the frequency components of a signal as it travels through a transmission medium. (See also: dispersion, phase velocity.)

Phase Jitter-(1-general) The unwanted phase variations of a signal, expressed in degrees. (2-PCM) An abrupt variation in the phase of a pulse code modulated signal caused by impulse noise, cross-talk, or timing changes in a digital bit stream.

Phase Locked-A condition in which signals of the same frequency are in phase at all times.

Phase Locked Loop (PLL)-A circuit that compares the output of a voltage controlled oscillator with a reference signal and uses the resulting difference signal to adjust the oscillator so that it will

be locked to the reference in both phase and frequency. Phase-lock loops are used in frequency synthesizers.

This diagram shows how a phase locked loop circuit allows the precise creation of any frequency using a standard precision (crystal controlled) signal source. This example shows that a voltage-controlled oscillator (VCO) is used to provide a radio frequency (RF) signal based on an approximate DC voltage level. This RF signal is sampled and divided by a frequency divide. The frequency divider is programmed from the frequency control panel so it produces an output frequency that is the same as the crystal controlled (precision) frequency reference. This divided signal phase is compared against the reference frequency. As the phase begins to change, it produces a small DC voltage that is used to fine tune the VCO therefore phase locking it to the reference crystal frequency.

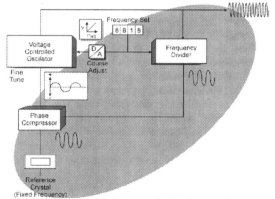

Phased Locked Loop (PLL) Operation

Phase Shift Keying (PSK)-A method of modulation used for digital transmission wherein the phase of the carrier is discretely varied in relation to a reference phase, or the phase of the previous signal element, in accordance with the data to be transmitted.

Phased Array-An array antenna whose directivity pattern is controlled largely by the relative phases of the excitation coefficients of the radiating elements.

PhCH-Physical Channel

Phone Doubler (TM)-The telephone service that provides a customer an indication that an incoming call (usually a voice call) is waiting for them while they are connected to the Internet. The service allows the customer to temporarily hold the Internet connection while they communicate on the other communications channel (answer the voice call). When the customer hangs up the phone, the Internet connection is restored and Internet service resumes.

Phone Patch-A device or circuit that is used to temporarily connect a radio system to a public telephone network.

Phoneline Network-A phoneline network is a local area data network technology that allows standard telephone wiring to be used as network cabling without the need to disconnect standard telephones. The phoneline network uses high frequency signals that are above standard telephone and DSL frequency bands. To install a phoneline network, end users install Phoneline NICs in a similar method to adding an Ethernet card. The Phoneline networking system allows computers to be connected to each other without the use of a hub (daisy chain).

Photocell-A device that produces an electric output from a visible light input.

Photodetector-A photodetector is a device used to detect light by converting photons into electrical signals. A photodetector is typically a semiconductor p-n junction diode designed to generate a voltage or current when it absorbs a photon. The voltage or current is representative of the number of photons that reach the photodetector, so it converts an optical signal to an electronic one that can be processed or measured.

Photon-A Photon is a particle of light as an electron is a particle of electrical energy. Photons are the smallest measurable part of light. Each Photon is a packet of energy that resonates at a specific optical frequency.

Photonic Device-An optical device, such as an optical switch, that processes light signals without first converting them to electrical form.

Photonic Switches-Devices that can directly permit or inhibit (switch) the optical transmission between two or more optical connections. Photonic switches eliminate the need for the conversion of optical signals to electrical form to allow switching to different communications lines or ports.

Photonics-The field of science and engineering that deals with photons of light and their utilization.

Photons-Quanta of electromagnetic radiation. Light can be viewed as either waves or as a series of Photons

PHS-Personal Handiphone System

PHY-Physical Layer

Physical Address-A grouping of numbers that identifies a particular piece of computer hardware connected to a local area network or other data communications system. Contrast with logical address.

Physical Channel (PhCH)-Physical channels are the electrical, radio, or optical transmission channels that are connected between transmitters and receivers.

Physical Layer (PHY)-The physical layer performs the conversion of data to a physical medium (such as copper, radio, or optical) transmission and coordinates the transmission and reception of these physical signals. The physical layer receives data for transmission from an upper layer, such as the Open System Interconnection (OSI) Data Link layer, and converts it into physical format suitable for transmission through a network (such as bursts, slots, frames, and superframes). An upper layer provides the physical layer with the necessary data and control (e.g. maximum packet size) to allow conversion to a format suitable for transmission on a specific network type and transmission line. The physical layer is layer 1 in the OSI protocol layer model.

Physical Link-A sequence of transmission slots on a physical channel alternating between master and slave transmission slots.

Physical Media Dependent (PMD)-The physical media dependent layer is a functional layer of a communication system that is dependent on the physical medium used by the communication system. This dependence may include timing and data transmission flow processes.

Physical Topology-Physical topology of a network is its' physical interconnection layout. The physical and logical topology does not have to be the same. The physical topology describes the connection of cables to equipment.

PIC-Plastic Insulated Cable

PIC-Preferred Interexchange Carrier

PICC-Primary Interexchange Carrier Charge

PICH-Page Indicator Channel

Pico-A prefix of a unit of measure meaning one-trillionth.

Picocell-A radio coverage area that has a radius of less than a few hundred feet (usually less than 100-200 feet).

Piconet-A small network of Bluetooth communication devices (e.g. less than 8). Bluetooth is a wireless personal area network (WPAN) communication system standard that allows for wireless data connections to be dynamically added and removed between nearby devices. Multiple Piconets can be linked to each other to form Scatternets.

This diagram shows Bluetooth devices that have created temporary connections. In this diagram, the personal digital assistant (PDA) device is synchronizing (deleting, changing, and adding) addresses with a laptop computer. The laptop computer is also connected to the Internet through a Bluetooth enabled access node. A mobile phone is also synchronizing its phone book listing with the laptop computer. However, because it is out of direct range of communicating with the laptop, it communicates through the access node. The mobile phone is also communicating with a wireless headset.

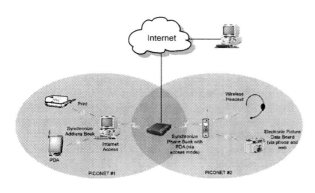

Piconet System

Picosecond (ps)-One trillionth of a second. In one picosecond, light travels about one-third of a millimeter.

Picture-A picture is an image that can be independently viewed or can be contained in a sequence of images that are used for motion pictures.

Picture Cell (PEL)-An abbreviation for picture cell (or picture element). A PEL is one sample of digital picture information, which can be an individual sample of R, G, B, luminance or chrominance or a collection of such samples if they are co-

sited and as a group produce one picture element. Also called a pixel.

Picture Element (Pixel)-(1-video) The smallest distinguishable and resolvable area in a video image. A pixel is a single point on the screen. The term pixel is derived from the words picture element.

Picture in a Picture (PIP)-Picture in picture is a video process that places one or more video images on a viewing screen at the same time.

Picture in Picture (PIP)-Picture in picture is a television feature that allows for the display of one (or more) video images within the display area of another video display. Picture in picture typically allows a viewer to watch two (or more) programs simultaneously.

Picture in Picture Box (PIP Box)-A PIP box is the window that holds the picture in picture video image.

Picture Tube-A tube that is used to display an image by the varying of intensity of a scanning electron beam onto phosphors that are deposited on the inner surface of the display portion area of the tube. The beam of electrons is produce by a electron gun. The electron beam is directed to different positions on the tube by horizontal and vertical magnetic fields (the yoke assembly). To create the image, the electron beam is swept across the picture tube line by line horizontally from top to bottom until a complete frame is displayed. The electron beam is momentarily blanked as it is repositioned to across the screen (horizontal blanking) and when it is repositioned to the top of the screen to begin the next frame image (vertical blanking).

This diagram shows how a picture tube directs an electron beam to a screen to produce a video image. This example shows that a gun produces the electron beam (stream of electrons) that is focused and positioned (directed) towards a high voltage screen. The screen has a phosphor coating that emits light when the electrons hit the screen area. The electron beam is directed from side to side (horizontal scan) with the intensity of the electron beam varying so the amount of light (brightness) varies. When the electron beam reaches the end of the sweep, the electron beam is stopped (blanked) during the changing of the focusing (directing) controls. Horizontal lines are produced from top to bottom until the entire picture is created. This diagram shows that the final lines of the video signal are video blanking lines that indicate the next frame of the picture is to begin. This allows the television system to reposition the electron beam to the top of the picture tube.

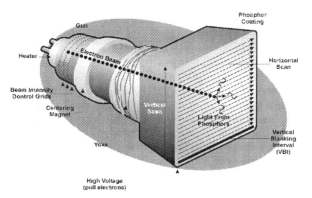

Picture Tube Operation

PID-Packet Identifier

Pigtail-(1-copper pair) A short length of copper-pair cable, typically for connections from a repeater or loading coil case. (2-coax) A short length of coaxial cable extending from a transmitter or receiver and used to make a connection to that equipment. (3-optical) A pigtail splice is a splice that adds a length of optical fiber with one end terminated with a connector and the other end attached to a light source or detector. The pigtail couples light from a source to a fiber optic cable or from a cable to a connector or device.

Pilot-(1-general) A signal, usually a single frequency, transmitted over a system for supervisory, control, synchronization, or reference purposes (2-FM) A signal at 19kHz transmitted with the FM broadcast carrier. The FM stereo subcarrier at 38kHz, modulated with a L-R difference signal, is generated from the pilot, leaving the two in-phase. In FM receivers, a pilot detector circuit serves the present of the 19kHz signal, which is used to decode the stereo information.

Pilot Burst-A pilot burst is a synchronization packet or code within is a data packet that is used to define the beginning of transmission of data or the start of a video frame and to assist in the demodulation and/or decoding of the data.

Pilot Channel-A communication channel that provides a reference (pilot) signal.

Pilot Signal-A pilot signal is a reference channel used by communication systems to allow devices to

identify and obtain timing information from the system to decode other communication channels.

Pilot Subcarrier-A subcarrier that serves as a control signal for use in the reception of FM stereophonic sound broadcasts. A subcarrier used in the reception of TV stereophonic aural or other sub-channel broadcasts.

Pilot Tone-An unmodulated tone of a specified frequency appropriate to the transmission system concerned. The pilot tone is transmitted over the system together with the information channels. The pilot performs synchronization and level control functions at the receiver.

PIM-Personal Information Manager

PIM-Protocol Independent Multicast

PIMS -Profit Impact of Market Share

PIN-Personal Identification Number

PIN Blocking-Personal Identification Number Blocking

PIN Code-Personal Identification Number

PIN Number-Personal ID Number given a subscriber to access service. Prepaid wireless services often use PIN numbers when using phones with the billing platform built into the handset to access the network.

Ping-Packet Internet Groper

Ping-Pong-A process of time compression multiplexing where transmitters and receivers constantly exchange information.

Pink Noise-A random noise whose frequency spectrum has been contoured to appear flat across the audible spectrum of human auditor perception.

Pinout-The pin configuration for a connector or system cabling.

PINT-PSTN and Interworking

PIP-Picture in a Picture

PIP-Picture in Picture

PIP Box-Picture in Picture Box

Pipe-A software interface or a hardware device that acts as an interface or buffer between network applications and devices.

Piracy-(1-transmission) The operation of unauthorized commercial stations, usually in international waters near target regions. Pirates violate flu frequency regulations, copyright laws, and national broadcasting laws. (2-programming) The duplication and sale of program materials, particularly TV programs on videotape, in violation of copyright laws. (3-signals) The unauthorized use of cable TV or satellite signals.

PIU-Percent Interstate Usage

PIU-PHS Interface Unit

Pixel-A pixel is the smallest component in an image. Pixels can range in size and shape and are composed of color (possibly only black on white paper) and intensity. The number of pixels per unit of area is called the resolution and more pixels per unit area provides more detail in the image.

Pixel-Picture Element

Pixel Density-Pixel density is the number of pixels that are located in a specific area.

Pixelized Screens-A pixelized screen is a graphic or video display that is composed of small image elements (pixels).

Pixels Per Inch (PPI)-Pixels per inch is the number of pixels that appear per inch on the horizontal axis of a display device.

PL-Power Level

PL-Preferred Language

PLA-Programmable Logic Array

Place Shifting-Place shifting is the viewing of a program or information at a location other than that at which it was originally received. Place shifting allows for the viewing of media programs (such as movies or television channels) at any location that has a multimedia computer and a broadband connection.

Plain Old Telephone Service (POTS)-Plain old telephone service is the providing of basic telephone service without any enhanced features. It is the common term for ordinary residential telephone service. The POTS system uses in-band signaling tones and currents to determine call status (e.g. call request). Because POTS allow for the transfer of audio signals below 3.3 kHz, POTS systems are also used for modems that allow data transmission (called dial up connection). Whenever a new service or feature is described, the author may refer to the previous available package of features and services as POTS, even when the previous package included several very sophisticated capabilities.

Plant-A general term applied to all of the physical property and facilities of a telephone company that contribute to the provision of communications services.

Plant Location Record-Records that show the placement of distribution terminals and the location, length, size, date, and nomenclature of cables.

Plasma-(1-arc) An ionized gas in an arc-discharge tube that provides a conducting path for the discharge. (2-solar) The ionized gas at extremely high temperature found in the sun.

Plasma Display-A type of flat visual display device, usually based on ionization of gas for light emission.

Plate-(1-electron tube) The anode of an electron tube. (2-battery) An electrode in a storage battery. (3-capacitor) One of the surfaces in a capacitor. (4-chassis) A mounting surface to which equipment may be fastened.

Platform-A platform is the combination of system hardware and software that programs or services operate.

Platform Dependent-Platform dependent refers to devices or systems that can only communicate with specific platforms that contain specific hardware types and software programs.

Play-List-Playlist

Playlist (Play-List)-A playlist is a sequence of objects or links that is used by a playing device (such as a media player) to select and play media files. A playlist may be contained within a multimedia file (the server side) or it may be kept in a separate file by a user (client side) of the media player. A playlist may be composed of the actual file names or it may contain URL links where the files can be found. A client side playlist for Windows media player may have the file extension .asx.

Playout-Playout is the process of streaming or transferring media to a user or distributor of the media.

Playout Automation-Playout automation is the process of using a system that has established rules or procedures that allows for the streaming or transferring media to a user or distributor of the media at a predetermined time, schedule or when specific criteria have been met (such as user registration and payment).

PLC-Power Line Carrier

PLCP-Physical Layer Convergence Protocol

PLD-Programmable Logic Device

Plenum Cable-Plenum cable is constructed of flame retardant material that generates little smoke when exposed to fire. Plenum cable may be installed in air ducts.

Plesiochronous-The relationship between two signals whose corresponding significant instants occur at nominally the same rate, any variation in rate being constrained within specified limits. Two signals having the same nominal digit rate, but not stemming from the same clock are usually plesiochronous.

Plesiochronous Digital Hierarchy (PDH)- Plesiochronous digital hierarchy is a grouping of digital communication formats that have levels of relationships where the communication links operate in a synchronous like manner (plesiochronous translates to almost synchronous). Plesiochronous is the original multiplexing hierarchy that is used in T1/E1 and T3/E3 systems. While multiplexing to higher rates, justification and synchronization bits had to be added to the original T1/E1 channels. These bits were discarded when demultiplexing thus creating a very inefficient and inflexible structure.

PLICF-Physical Layer Independent Convergence Function

PLL-Phase Locked Loop

PLMN Code-Public Land Mobile Network

PLU-Percent Local Usage

Plug And Play (PnP)-A compatibility system that simplifies the installation and removal of hardware devices into a personal computer. PnP includes the automatic recognition of hardware installation and removal, activation and deactivation of software drivers, and system accessory management functions.

Plug-In-A plug-in is a software program that works with another software application to enhance its capabilities. An example of a plug-in is a media player for a web browser application. The media player decodes and reformats the incoming media so it can be displayed on the web browser.

PM-Phase Modulation

PM-Portable Media

PMC-Personal Media Channel

PMD-Physical Media Dependent

PMD-Polarization Mode Dispersion

PMR-Private Mobile Radio

PMR-Public Mobile Radio

PMT-Program Map Table

PN-Parameter Negotiation

PNAP-Private Network Access Point

PNCH-Compact Packet Notification Channel

PNCH-Packet Notification Channel

PnP-Plug And Play

PNP-Private Numbering Plan

PNS-Perceptual Noise Substitution

POC-Push to Talk Over Cellular

Pocketphone-A handheld portable cellular telephone typically small enough to be carried in a purse or pocket.

POCSAG-Post Office Code Standard Advisory Group

POD-Print on Demand

POD Module-Point of Deploymnet

Podcasting-Podcasting is the process of broadcasting media or content to portable media devices (PODs).

PoE-Power over Ethernet

POF-Plastic Optical Fiber

POI-Point Of Interconnection

Point-A point is a measurement unit for print type sizes. Each point equals approximately 1/72 of an inch.

Point Applications-Point applications are programs, devices or systems that perform a specific purpose or function.

Point of Deploymnet (POD) Module-A point of deployment module is a portable, credit card size device that can store and process information that is unique to the owner or manager of the POD. PODs are used in cable and satellite television systems and are a form of smart card.

Point Of Presence (POP)-A physical location that allows an interexchange carrier (IXC) to connect to a local exchange company (LEC) within a LATA. The point of presence (POP) equipment is usually located in a building that houses switching and/or transmission equipment for the LEC.

Point Of Termination (POT)-The point within a local access and transport area (LATA) at which a local exchange carrier's responsibility for access service ends and an inter-exchange carrier's responsibility begins.

Point To Point Microwave-Point to point microwave radio service provides a fixed communication connection between fixed stations using microwave radio transmission. Point to point microwave systems focus their radio energy into a very narrow beam which avoids interference from most other users. Because microwave transmission is not well suited for mobile and low power applications, the number of users in the microwave band is relatively low. This means that there is more available frequency bandwidth in the microwave bands. Because the available bandwidth is relatively large and interference is low, point to point microwave systems usually provide for very high-speed communication connections (Mbps to Gbps).

Pointer-A pointer is a number or reference value that is used to identify the location or a file, block of data or index location in a software stack.

Pointer File-A pointer file is a list of location pointers or reference values that are used to redirect software programs to the location of information for specific items such as media files or web pages.

Point-To-Point Communication-The transmission of signals from one specific point to another, as distinguished from broadcast transmission which blankets the general public. Radio relay links are a common type of point-to-point communication.

Point-To-Point Link-A connection between two devices or a node that uses point-to-point protocol (PPP).

This diagram shows the structure of a point-to-point protocol (PPP) frame. This example shows that the field structure includes a start frame flag, address code for the PPP connection, a control flag (for PPP link), a link protocol identifier, data field (to hold the payload packet), and frame check sequence (FCS) for error detection. The PPP frame is only used to transport the data between points so the data payload includes the user's data along with the destination address of the packet

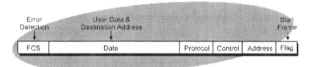

.PPP Frame Structure

Point-To-Point Protocol (PPP)-Point to point protocol (PPP) is a connection oriented protocol that is established between two communication devices that encapsulates data packets (such as Internet packets) for transfer between two communication points. PPP allows end users (end points) to setup a logical connection and transfer data between communication points regardless of the underlying physical connection (such as Ethernet, ATM, or ISDN). PPP is described in the IETF RFC 1661.

This diagram show how a point-to-point protocol (PPP) connection allows a computer connection to be established, verified, and maintained. This example shows that PPP protocol can be used on

any type of access connection. The PPP protocol allows for the separation of the link protocol on the PPP link and the network control protocol that operate above the high level data link control (HDLC) protocol that is used to coordinate the PPP link. The PPP protocol also includes security features such as password authenticated protocol (PAP) and a more secure challenge handshake authentication protocol (CHAP).

Point-To-Point Protocol Over Ethernet

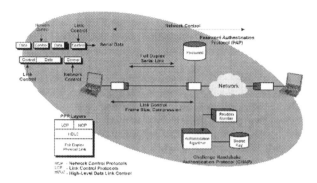

Point to Point Protocol (PPP) Operation

(PPPoE)-Point-to-point protocol (PPP) data transmission over an Ethernet connection. PPPoE works with a network interface card (NIC) to create a point-to-point communication link between the NIC and a connection point to the Internet. The use of PPPoE allows the setup, management, and disconnection of a communication link between a customer and an Internet service provider (ISP). The PPPoE manages the transport of wide area IP packets that use 32-bit or 128-bit IP addressing over a local area Ethernet system that uses 48-bit Ethernet addressing.

Point-To-Point Tunneling Protocol (PPTP)-Point-to-point tunneling protocol (PPTP) allows systems to setup virtual point to point connections in data networks (such as the Internet).

Polar-(1-general) Relating to one or more separated points, such as the north and south poles of the earth, or more figuratively two intellectual positions such as saying that, "The organization has become polarized into two groups, one that supports the proposal and another that opposes it." (2-science and technology) A distinction based on direction in space for vector quantities such as force or electric field, or based on positive vs. negative sign for scalar quantities that do not have a space direction.

Polarization-Polarization is the electro-magnetic orientation (horizontal or vertical) of a radiated field relative to a reference plane.

Polarizing Connector-In an optical network, a connector designed to join separate fibers carrying polarized light so that the direction of polarization is the same in both fibers, from input to output.

Pole Attachment-Any attachment by a line (such as a cable television system) to a pole, duct, conduit, or right-of-way owned or controlled by a utility.

Policy Server-A policy server is a communications server (computer with a software application) that coordinates the allocation of network resources based on predetermined policies on the priorities and resources required by communication services and applications within the network. A policy server is used to help manage network operation in the event of loss of resources. The pre-set policies define which communication services are critical (such as voice) and how much resources should be allocated to these critical services at the expense of other communication services (such as web browsing).

Polling-Polling is the process of sending a request message (usually periodically) for the purpose of collecting events or information from a network device. The receipt of a polling message by a device starts an information transfer operation. Polling of network devices is regularly performed to gather usage information that is processed by the billing system rating engine. Polling involves the use of a network control system that periodically invites its tributary stations to transmit in any sequence specified by the control station.

Polling Cycle-Polling cycle is the amount of time or processes that occur between the sending of polling messages (requests for data transmission or responses).

Polyethylene (PE)-Polyethylene thermoplastic synthetic polymer commonly used in cable jackets that is resistant to moisture and weather and it can be very flexible and easy to work with in warm temperatures. Without additives, it is not flame retardant and it will become stiff and harder to work with in colder temperatures.

Polyfluorinated Hydrocarbon-Polyfluorinated Hydrocarbon is a plastic material that is used in cable jackets. It has good flexibility while maintaining flame resistance and produces low smoke levels when exposed to heat.

Polyphonic Tones-Multiple tones that are used to indicate an event (such as an incoming telephone call).

Polystyrene-A plastic frequently used as an insulator in electrical products.

Polyvinyl Chloride (PVC)-Polyvinyl chloride is a thermoplastic made of polymers that is tough, resistant to moisture and weather and has flame retardant properties. PVC can be less flexible and more expensive than Polyethylene.

PON-Passive Optical Network

POP-Point Of Presence

POP-Post Office Protocol

popdown-Pop-down

Pop-down (popdown)-A window that is opened on an Internet browser when the user closes the window or attempts to leave a web address.

POPs-Persons of Population

popup-POP-Up

POP-Up (popup)-A pop-up is a browser window that is opened for the display of an ad message "in between" the requesting and opening of a new web page.

Port-(1-general) A physical or logical connection point between a computer or computer-based machine and other hardware devices. (2-network) A place of access to a device or network where energy may be supplied or withdrawn, or where the device or network variables may be measured. (3-software) The process of moving source code and executable programs from one computing system to another of a different type without substantive changes to the source code.

Port Growth-Port growth is the number of telephone stations that a telephone system can expand to. In some cases, additional switching modules or cards can be added to accommodate port growth.

Port Identifier-A value assigned to a port that uniquely identifies it within a switch. Port Identifiers are used by both the spanning tree and link aggregation control protocols.

Port Mirroring-A process whereby one switch port (the Mirror Port) is configured to reflect the traffic appearing on another one of the switch's ports (the Monitored Port).

Port Number-For internet protocols such as TCP and UDP, when a machine with a particular IP address receives an IP packet, a destination port number in the packet header can be used to determine how to handle the packet. Typically certain applications are expected to process packets sent to specific well-known port numbers for specific protocols. For example, if a machine runs a web server application, that application typically receives all TCP packets directed to port number 80.

Port Sharing-The process of allowing multiple virtual connections (logical channels) to share the same port connection in a data network.

Portability-This means that application software can be dragged across different computing platforms and operating systems. With the TAO spec, developers can write an application that will run on a PC, Alpha Server or Tandem host. SCSA's operating-system-independent APIs give developers a uniform method for supporting multiple operating systems.

Portability Rights-Portability rights are the permissions granted from an owner or distributor of content to transfer the content to other devices (such as from a set top box to a portable video player) and other formats (such as low bit rate versions).

Portal-An Internet web site that acts as an interface between a user and an information service.

Ported Number-A ported number is a telephone number that has been setup to be redirected to a new switch port (new destination number).

POS-Point of Sale. That moment or place in which the sale occurs.

Position Location System-A position location system that gathers and processes information to determine the geographic location of devices or equipment. The position location information may be gathered by the network sensing the position of the device relative to one or several antennas in a network (such as the Teletrac system) or by the vehicle reporting its location using external position locating devices (such as the Global Positioning System).

Post Dial Delay (PDD)-The time period between when a user dials the last digit of a phone number and hears the phone at the other end begin to ring.

Post Dialing-Refers to the ability of a terminal to send dialing information after the outgoing call request setup message is sent.

Post Processing-Post processing is performing operations on information or data that occurs after the event that created the information or data has occurred.

Post, Telephone And Telegraph (PTT)-A term used for a government agency in many countries that supplies and maintains the infrastructure and provides basic telecommunication services.

Posting-Posting is the process of sending a message or information to a newsgroup or bulletin board.

Postpay Service-A billing or coin telephone service that allows for payment of service after communication has started (e.g. called party answers.) Postpay, now rarely used, is the simplest form of coin service but requires additional operator attention on toll calls.

POT-Point Of Termination

POTS-Plain Old Telephone Service

Potting-The process of sealing components under a plastic cover to keep out moisture and other contaminants.

Power Class-A classification of the maximum power level or range of power level associated with mobile telephones.

Power Consumption-Power consumption is the amount of energy that a device or system uses over a period of time.

Power Control-(1-RF level) The process of regulating the power level of a transmitter. (2-cellular) The process of controlling the power level in a cellular system where the base station receiver monitors the received signal strength of a mobile telephone and control messages are transmitted from the base station to the mobile telephone commanding it to raise and lower its transmitter power level as necessary to maintain a good radio communications link.

Power Dialing-The process of dialing calls in lists, connecting the call to a customer service representative if it is answered by a person, remembering unanswered calls for a later callback, and updating records to indicate the calls that have been completed to avoid repeat calling. Power dialing systems are sometimes linked to customer information databases to allow the customer service representative to see the call recipients account information when the call is connected.

Power Efficiency-The ratio of the emitted (transmitted) power of a source to the input (source) power.

Power Factor-The ratio of the true power consumed in an alternating-current circuit to the apparent power (the power actually supplied by the source). The value of the power factor is the cosine of the phase angle between the voltage and current in the circuit.

Power Level (PL)-(1-measurement) The amount of power produced by a transmitter. (2-cellular classification) A power level setting for a transmitter as commanded by a base station.

Power Line Carrier (PLC)-A carrier wave signal that can be simultaneously transmitted on electrical power lines. A power line carrier signal is above the standard 60 Hz powerline power frequency (50 Hz in Europe).

Power Line Carrier Systems-The use of the electrical power distribution system used by an electric power utility to transmit information signals. Power line carrier systems can be used for power system equipment control and monitoring and can be used to transfer other information signals such as commercial voice or data signals.

Power Line Ethernet-A data transmission system uses power lines to transmit data in Ethernet packet format on standard (110-220 VAC) electric power lines.

Power Meter-A power meter is a test instrument that is used to measure the level of electrical energy (power).

Power over Ethernet (PoE)-Power over Ethernet (PoE) is an industry specification that supplies 48V to devices connected to an Ethernet data line. The 802.11af specification defines that the 48V can be supplied on either of the unused lines in the Ethernet cable (only 4 of the 8 lines are typically used) or it can be supplied directly on the data lines.

Power Saving-A mode of operation where electronic circuits are de-energized and energized to conserve battery life. See also sleep mode.

PPC-Pay Per Click

PPCH-Packet Paging Channel

PPDN-Public Packet Data Network

PPE-Personal Protective Equipment

PPI-Pixels Per Inch

PPL-Pay Per Listen

PPM-Pulse Position Modulation

PPP-Point-To-Point Protocol

PPP Session-The events that occur during a period of time when logical connection exists between two communication devices that are using point to point protocol (PPP). PPP is a connection oriented protocol that can be established between two communication devices that encapsulates Internet protocols such as transmission control protocol and Internet protocol (TCP/IP) into the underlying transmission system (such as dialup connections). PPP allows end users (end points) to setup a logical connection and transfer data between communication points regardless of the underlying physical connection.

PPPoA-Point-To-Point Protocol Over Asynchronous Transfer Mode

PPPoE-Point-To-Point Protocol Over Ethernet

PPS-Pay Per Sale

PPSN-Public Packet Switched Network

PPTN-Public Packet Telephone Network

PPTP-Point-To-Point Tunneling Protocol

PPV-Pay Per View

PR Agency-Public Relations Agency

PRACH-Packet Random Access Channel

PRBS-Pseudo-Random Binary Sequence

PRBS-Pseudo-Random Bit Stream

Preamble-A preamble is a sequence of bits in synchronous transmission that the receiving device must recognize and lock onto. Once the receiving device locks onto the preamble, it knows where to find the rest of the packets.

PreConditions Met (COMET)-A process that is used between communication devices that is used to determine if all the preconditions have been met to allow a communication session to begin. The use of COMET prevents the establishment of communication sessions (such as a telephone call through the Internet) that cannot be completed because the communication devices do not have all the capabilities (such as the same type of speech coding) to communicate with each other.

Predatory Acquisition Campaign-Marketing campaign developed to encourage prospects to switch from a competitor.

Predicted Frame (P Frame)-A predicted-frame is an image in a motion video sequence that is coded using difference and motion compensation information from previous frames.

Predicted Frame (P-Frame)-Predicted frames (P-Frames) are images (pictures) within a sequence of images (such as in a video sequence) that are created using information from other images (such as from Intra Frames (I-Frames). Because image components are often repeated within a sequence of images (temporal redundancy), the use of P-Frames provides substantial reduction in the number of bits that are used to represent a digital video sequence (temporal data compression).

Prediction Error-Prediction error is the difference between the projected estimated signal and an original signal. Prediction error is used to determine the best compression process in a lossy compression system (such as digital video compression). The lower the prediction error, the better match the compression process has to the original digital video signal.

Predictive Dialing-Predictive dialing is an automated method for making outbound telephone calls in which a scheduling (pacing) algorithm determines the number of calls placed in advance of actual operator availability.

Predictive Frames-Predictive frames are images whose data contents are used to calculate previous and future frames in a sequence of images.

Pre-Emphasis-Pre-emphasis is the increase in the amplitude of the high-frequency components in a transmitted signal to overcome noise and attenuation in the transmission system. The relative balance of high- and low-frequency components is restored in a receiver by de-emphasis. Frequency-modulation broadcast stations use pre-emphasis.

Preemptive Multitasking-Preemptive multitasking is the process of executing instructions of an operating system that allows the operating system to interrupt or change the processing priority (preempt) of an application based on the needs or settings of the applications. Preemptive multitasking allows time sensitive applications such as processing real time video or audio to be performed during the operation of less critical tasks such as web page browsing.

Pre-Encryption-Pre-encryption is the process of encrypting media before it is selected and distributed through a communication system (such as encrypting digital television channels before they are sent to viewers). Receivers of pre-encrypted media use a decryption key to convert the media to a form that they can use or view.

Preferred Interexchange Carrier (PIC)-The assignment or use of an inter-exchange carrier to complete calls from a customer outside their systems calling area. The PIC code is obtained when a customer dials a number that requires inter-exchange carrier (IXC) service. The PIC code is used to route the call to the IXC carriers point of presence (POP) switching center.

Preferred Language (PL)-Provides a telephone service subscriber with the ability to specify the language for network services. This service allows the subscriber to specify service in English, Spanish, French or Portuguese.

Prefix-Any digit dialed before a destination address. Prefixes indicate service options. For example, a prefix of "1" indicates a 10-digit call address in some areas, and a "0" indicates a request for the services of an operator.

Preloaded Connector-A preloaded connector is a connection device that is preloaded with an epoxy or adhesive. This ensures the correct amount of epoxy is used when assembly the connector.

Premise-Premise is the space occupied or owned by a user or customer or business within a building facility.

Premises Distribution Network (PDN)-A premises distribution network is the equipment and software that is used to transfer data and other media in a customer's facility, home or personal area. A PDN is used to connect terminals (computers) to other networks and each other and to wide area network connections. Some of the common types of PDN are Ethernet, Wireless LAN, Wireless LAN, and Phoneline Data.

This figure shows the common types of premises distribution systems that can be used for IP television systems. This diagram shows that an IP television signal arrives at the premises at a broadband modem. The broadband modem is connected to a router that can distribute the media signals to forward data packets to different devices within the home such as IP televisions. This example shows that routers may be able to forward packets through power lines, telephone lines, coaxial lines, data cables or even via wireless signals to adapters that receive the packets and recreate them into a form the IP television can use.

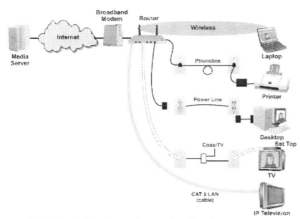

IP Television Premises Distribution Systems

Preorigination Dialing-Preorigination dialing is a process where the dialing sequence takes place before a telephone's first communication with the telephone system. Mobile telephones commonly use preorigination dialing.

Prepaid Calling Card-A card that is issued by a telecommunications service provider that contains coded identification information that permits the card holder to initiate a call or request and receive an information service. Calling cards contain a number or code contained on a magnetic stripe that uniquely identify the card and authorized services to the system.

Prepaid Service-Prepaid communication services is a process that is used by service providers (carriers) to be paid for services they provide in the future. This allows the carriers to obtain revenue for services without the risk of bad debt and it eliminates the need and cost for billing operations. Prepaid service is often associated with customers that may be credit challenged or who want more control over bills.

Prepaid Wireless-Wireless connection whereby service is prepaid before usage is accumulated. Typical users hear an announcement prior to the call noting how many minutes or dollars or units they have remaining on their account.

Prepay Service-Coin telephone service in which an initial coin deposit is required before a connection is established on chargeable calls. Prepay service is provided either by coin-first or dial-tone-first coin service.

Pre-Provisioning-Pre-provisioning is the process of assigning and/or programming information into

a system or device before it is sold or provided to a user to setup or assist in the setup in the providing service to the system or device.

Preroll-A preroll is the specific amount of time that is allowed for media to run (such as a video tape) before broadcasting or editing to allow the media device to obtain normal operating speed.

Presence-(1-communication) Presence is sensing and triggering the sending of information alert messages on specific activities or status changes that have occurred by a user or device. Presence is one of many features that can be offered as part of multimedia services bundle (such as IPTV). Based on a common feature in most of today's PC-based instant messaging applications, presence refers to the ability to view what friends and family are on the network through a "buddy list" that can be graphically displayed on-screen. Advanced presence features include the ability to alert others when you are on the phone or what program you are watching. (2-noise level) The natural noise level of a room or environment without dialogue or other produced sounds. Presence also is known as room tone. Presence is used to fill holes created during editing or footage recorded without sound.

Presence Entity-Presentity

Presence Server-The Presence Server is a server that is responsible for receiving, storing and distributing presence information received from user agents. In SIP the presence service may be co-located with other SIP servers.

Presence User Agent (PUA)-A SIP user agent which support presence subscriptions and notifications and interacts with software known as the presentity. The presentity is responsible for providing presence information to the Presence Service.

Presentation Control-Presentation control is the ability to setup and control the display and operation of media on a display such as a television along with the placement and operation windows on the users display.

Presentation Layer-The presentation layer is layer 6 of the OSI model. This layer responds to service requests from the Application Layer and issues service requests to the Session Layer. The Presentation Layer relieves the Application Layer of concern regarding syntactical differences in data representation within the end-user systems. Note: an example of a presentation service would be the conversion of an EBCDIC-coded text file to an ASCII-Coded file.

Presentation Time Stamp (PTS)-A presentation time stamp is reference timing values that are included in MPEG packet elementary streams (digital audio, video or data) that are used to control the time alignment of media within a program.

Presentity (Presence Entity)-The presentity provides presence information to a presence service (server), which can distribute that information to any user that has subscribed for it.

Pretty Good Privacy (PGP)-Pretty good privacy is an open source public-key encryption and certificate program that is used to provide enhanced security for data communication. It was originally written by Phil Zimmermann and it uses Diffie-Hellman public-key algorithms.

PRI-Primary Rate Interface

PRI Gateway-Primary Rate Interface Gateway

PRI Offload-Primary Rate Interface Offload

Price Elasticity -Economic indicator used to define how sensitive consumers will be to changes in prices. A high elasticity means that customers will change their demand drastically with small changes in price.

Pricing Models -Mathematical models developed to help finance determine what prices should be charged to what customers for different services.

Primary Channel-Primary channels are the initial or main communication channel that is setup for a communication session.

Primary Colors-Colors that combine to produce a full range of colors within the limits of a system. non-primary colors are mixtures of two or more primary colors. In television, the primary colors are specific sets of red, green, and blue.

Primary Interexchange Carrier Charge (PICC)-A recurring fee that is added to the bill of the local telephone service customer that charges them for access to inter-exchange carrier (IXC) services. This fee is charged in addition to other fees or tariffs (such as a percentage of billed long distance usage) that are paid by the IXC to the local exchange carrier (LEC). The purpose of the PICC charge is to help recover the cost of providing local loop access service in a marketplace that has declining long distance revenue charges.

Primary Rate Interface (PRI)-Primary rate interface (PRI) is a standard high-speed data communications interface. This interface provides a standard data rates for T1 1.544 Mbps and E1 2.048 Mbps. The PRI interface can be divided into several channels (channelized). These channels can be combinations of 64 kbps (B) 384 kbps (H) channels or other channels. PRI connections must include at least one 64 kbps (D) control channel.

Primary Ring-A primary ring is the preferred communication path used in a communication network ring. A primary ring is used in the fiber distributed data interface (FDDI) system.

Primitives-(1-general) Primatives are the fundamental elements of a system or service. (2-SS7) In Signaling System 7, commands and their respective responses associated with the services requested between adjacent levels. (3-graphics) Basic graphic elements, such as a points, markers, lines, circles, text character, etc., used in generating images on computer graphics systems.

Prior Art-The term prior art means different things in different countries, but generally refers to information in the public domain which affects the patentability of an invention.

For European Patent Applications, prior art includes any public display, offer for sale, or use of the invention prior to the filing date of the application. Publications, books, other patents which were published and publicly available prior to the filing date of the application are also considered as prior art.

Prioritization-Frame and packet prioritization assigns different priority codes to packet that are transmitted through a communication network. This allows some frames or packets to receive a higher transmission priority for time sensitive data communications (such as packetized voice).

Priority Call Operation-A system access control process that allows users with a higher priority level than other users to gain access to communication system resources. Priority call operation may allow high priority users to automatically disconnect lower priority users so they may gain access to the system.

Priority Date-In most cases, the priority date is equivalent to the filing date of the first application for patent in a family of patents.

Foreign applications claiming priority to a United States application for patent can only claim the filing date of the US application for patent even if the applicant is entitled to another US priority date due to an earlier Date of Invention.

PRISM-Publishing Requirements for Industry Standard Metadata

Privacy-(1-encryption) A term used with regard to encryption to indicate a level of protection that is minimal, corresponding to a moderate amount of effort on the part of the eavesdropper to understand the private communication, but not so good as the better levels designated by the words "secret" or "secure."

(2-channel separation) An electrical capability to prevent other extensions on a multi-extension analog telephone line or key telephone system from connecting during a conversation, typically via diodes or other devices actuated by the decrease in subscriber loop voltage when one set is off hook. Also the name of an equivalent capability using an electronic key system.

Privacy Policy-A privacy policy is the self proclaimed rules a receiver of information claims to follow when a customer or visitor sends or provides information. Privacy policy rules typically state how the information may be used and who the information may be distributed to.

Private-(1-general) Private means secure or not shared with others. (2-Bluetooth) A mode of operation whereby a device can only be found via Bluetooth baseband pages, it only enters into page scan.

Private Automatic Branch Exchange (PABX)-A telephone switch that is generally located on a customer premise. Often referred to as a PBX, CBX, EPABX. This provides for the transmission of calls internally as well as to and from the public network.

Private Branch Exchange (PBX)-PBX systems are private local telephone systems that are used to provide telephone service within a building or group of buildings in a small geographic area. PBX systems contain small switches and advanced call processing features such as speed dialing, call transfer, and voice mail. PBX systems connect local telephones ("stations") with each other and to the public switched telephone network (PSTN).

This diagram shows a private branch exchange (PBX) system. This diagram shows a PBX with telephone sets, voice mail system, and trunk connections to PSTN. The PBX switch calls between telephone sets and also provides them switched access to the PSTN. The voice mail depends on the PBX to switch all calls needing access to it along with the appropriate information to process the call.

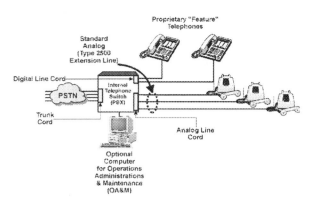

Private Branch Exchange (PBX) System

Private Carrier-An entity licensed in the private services and authorized to provide communications service to other private services on a commercial basis.

Private Internet Address-An Internet address that can be transferred by routers on a private data network.

This figure shows how private Internet address numbers are used to uniquely identify devices connected within a private network that uses Internet protocol transmission. This diagram shows that packets that are designated for the Internet must be translated to an Internet address that can be routed through the Internet.

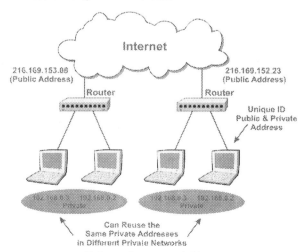

Private Internet Address

Private IP Address-Private Internet protocol addressing are ranges of IP addresses that are reserved for use by private networks (Intranets). Private IP addresses cannot be sent through the Internet. Private address include: 10.0.0.0 through 10.255.255.255 (class A), 172.16.0.0 through 172.31.255.255 (class B), and 192.168.0.0 through 192.168.255.255 (class C).

Private Key-A key used to decrypt a message encrypted with the owner's public key or to sign a message from the key's owner. Sometimes called a secret key.

Private Line-A dedicated communications circuit that is leased by a customer from a telephone service provider for voice, data, or video services. While a private line may be connected through a switching facility, the connection resources are constantly dedicated to the customer who is leasing the line.

Private Network-A network designed for the exclusive use of one customer. Often, such a network is nationwide in scope and serves large corporations or government agencies.

Private Network Access Point (PNAP)-An key private physical (layer 2) interconnection point that interconnects regional Internet systems and sub-systems.

Private Numbering Plan (PNP)-A feature that enables subscribers to call defined private-network extensions using an abbreviated dialing pattern. When a PNP subscriber dials a private-network extension, the cellular network translates the dialed extension to a number in the North American numbering plan (NANP) and routes the call.

Private Peering-Private peering is the process of inquiring or exchanging information in a private communication system (such as an Intranet).

Private Tables-Private tables are user defined data elements that are sent along with broadcast programs (such as a television show).

Private Telephone System-Private telephone systems are independent telephone systems that are owned or leased by a company or individual. Private telephone networks include key telephone systems (KTS), private branch exchange (PBX) and computer telephone integration (CTI).

This figure shows the different types of private telephone systems. This diagram shows that key telephone systems (KTS) allow each telephone station to access some or all of the lines available to the company. Private branch exchange (PBX) systems include a local telephone line switching with call processing software to allow simple stations to connect to many different telephone lines. Computer telephony (CT) systems merge computer intelligence with telecommunications devices and often link telephone systems with company information systems. Private telephone systems that use Internet protocol data networks to interconnect telephone stations are called IPBX systems. This diagram shows an IPBX system that shares a data network for both telephone calls and computer workstations.

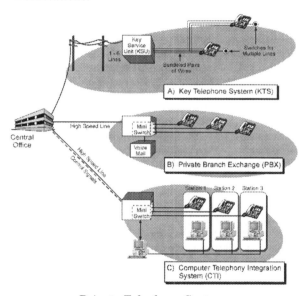

Private Telephone Systems

Privatization-The process of selling or reallocating government owned assets or resources to companies or investors. Privatization in the telecommunications industry often includes the issuance or re-allocation of licenses to provide a telecommunications services.

Probe-(1-test) A test prod used to check components for the presence of signals. (2-cavity) A wire loop inserted in a cavity for coupling energy to an external circuit. (3-communication) A process of sending a message or alert signal to another device or group of devices that discovers if there is information to be gathered.

Probe Response-The process of returning information after a probe message has been received.

Processing Delay-The time required for processing information. Processing delay for coding and decoding of voice or data information is usually measured in msec.

Processing Gain-A ratio of signal resistance to attenuation that results from the processing of the information signal with another signal (such as a code signal in a spread spectrum system). An example of a processing gain would be a 10 kbps signal that is multiplied by a 100 kbps chipping code. This provides a processing gain of 10.

Processing Power-The number of computations that a computer, microprocessor, or digital signal processor can complete in a fixed time interval. May be measured in millions of instructions per second (MIPS) or millions of floating point operations (MFlops.) Typical low-cost digital signal processing (DSP) chips provide 10-20 MFlops.

Processor-A circuit or device that systematically executes specific operations on digital data, as directed by a program stored in memory.

Product Acquisition and Development System (PADS)-A process that is used for the identification and coordination of product development activities. The PADS system is divided into 5 phases; concept, feasibility, planning, development, and introduction.

This diagram shows a basic product development process to help ensure the successful development of communication products and services. This example shows that this product development process evolves through a series of gates (called toll gates) that are used to determine if a product should continue in the normal development process or eliminated (discarded) so additional resources can be allocated to other products. In this example, the product development cycle begins with product ideas (concept), business evaluation (feasibility), resource scheduling (planning), technology design and production testing (design), and market introduction and distribution support (marketing).

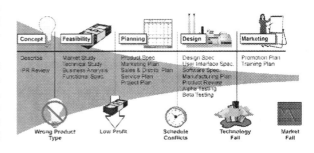

Product Acquisition and Development System (PADS) Operation

Product Development-Product development is a process identifying, defining, developing, and producing a new or improved product.

Product Distribution-Product distribution is the process of transferring products between product sellers product purchasers (buyers). Some types of product distribution includes wholesale (to other distribution companies), power retailers (high volume industry focused stores with limited product selection), mass merchants (large retail chains) and direct to consumers (direct distribution).

Product Management-Product management is the process of assigning and tracking specific tasks and functions related to ensuring the success of products or services.

Product Marking-According to 35 U.S.C. § 287(a) a patent holder is required to give notice that an item is patented. This may be accomplished by physically marking the item, or accompanying packaging, with the US patent number, or alternatively with the phrase "Patent Pending" where an application has been applied for, but not yet granted.

The purpose of marking is to provide constructive notice to the world of the patent, and is an important issue in many patent infringement suits.

Defendants may be able to avoid some damages if patentee did not properly mark the article.

Product Monitoring-Product monitoring is the process of looking for the unauthorized transfer of digital media. This information may be identified by file names, media components or embedded digital watermarks. Product monitoring may be performed by searching the web using web crawlers or by locating data sniffers at well known transfer points on the Internet. When unauthorized transfer is observed, action can be taken to inform the media host that they are violating copyright laws and they must remove the content or risk loosing their Internet access authorization (disabling the site).

Product Packaging-Product packaging is the combining of a product and its supporting elements (accessories or files) into a form or container that is used to deliver the product to the consumer or user of the product.

Product Plan-A document that defines the planned offering and development of products by a company.

Product Specification-A document that defines the key features and characteristics of a product.

Production Bonus-Sum of money due paid to contracted writer or other staff members upon the occurrence of an initiation of a production.

Production House-A facility that typically does everything to generate final video productions except shooting the original videotape. Services typically include editing raw master tapes, modifying recorded material, and creating new effects. Projects include advertising, training, promotion, music videos, and TV shows.

Production Schedule-A time schedule that identifies what products or assemblies will be produced and when they will be produced.

Production Testing-Production testing is the measurements and adjustments that are performed on products during the production process.

Production Yield-Production yield is the percentage of acceptable products that are produced compared to the total number of products that are manufactured.

Profile-A profile is a particular implementation or instantiation of a more general protocol. Many protocols are extremely general and allow one to specify a restricted set of messages and their actions for a particular purpose. Such a set is known as a profile. For example, NCS is a profile of MGCP with a few extensions.

Profiling-Profiling is the process of monitoring, analyzing and characterizing a user of a product or service as to their usage patterns or preferences.

Program-A sequence of instructions used to tell a computer how to receive, process, store, and transfer information.

Program Associated Data (PAD)-Program associated data is ancillary or related material that is received in combination with program content that enhances or supplements the original program. Examples of program associated data include music title and artist information, graphics, or news services.

Program Association Table (PAT)-A program association table contains the identification codes and system information associated with MPEG programs that are contained in a transport stream. The PAT is usually sent every 20 msec to 100 msec to allow the receiver to quickly acquire a list of available programs.

Program Clock Reference (PCR)-A program clock reference is a source of timing information that is used to coordinate the decoding and sequencing of media streams (elementary streams) that are contained in an MPEG-2 program stream.

Program Map Table (PMT)-Program map table contains information that identifies and describes the components (such as the video and audio elementary streams) that are part of a program (such as a television show).

Program Specific Information (PSI)-Program specific information is data that defines elements of characteristics of program media (such as the video, audio and meta descriptive components of a television program).

Program Stream (PS)-A program stream is a combination of elementary streams (such as video and audio) that compose a media program (such as a television program).

Programmable Read Only Memory (PROM)-A ROM device that can be programmed by the equipment manufacturer.

Programmer-A person who prepares sequences of instructions for a computer.

Programming-(1-software) Programming is the process of developing an assembly of instructions for a computer to enable it to carry out a particular job. (2-television) Television programming is the selection of shows and programs that are offered by a television service provider.

Programming Language-A language designed to be under-stood by a computer. A high-level programming language is converted into the required machine code by a program called a compiler.

Programming Language-C+ Or C++

Progressive Downloading-Progressive downloading is the transferring of a file or data in a sequential process that allows for the using of portions of the data before the transfer is complete.

Progressive Frames-Progressive frames are a motion display where each image in a sequence of a movie adds a complete image (as opposed to a partial interlaced image) progression in a moving picture.

Progressive Video-Progressive video is a movie that is composed of individual images where each image contains complete information about a new time period. Progressive frames are symbolized by adding the letter "p" to the frame rate. For example, 30p represents a image rate of 30 progressive frames.

PROM-Programmable Read Only Memory

Promiscuous Mode-A mode of operation of a network interface that allows it to attempt the reception of all data regardless of the destination address of the incoming data.

Prompt-A cue to help the operator choose the next action.

Propagation-The process of transfer of a radio signal (electromagnetic signal) or acoustic signal (sound) from one point to another point.

Propagation Constant-A measure of signal propagation through a transmission line or other medium through which a wave passes. Propagation constant consists of the attenuation constant in decibels or nepers per unit length and the phase constant in radians per unit length.

Propagation Delay-The delay in the reception of a signal caused by its travel from the transmitter to the receiver. Propagation delay can be quite lengthy when signals travel great distances, such as over satellite circuits.

Propagation Path Obstruction-A physical material or assembly that effects the travel (path) of a radio signal (loss) other than the effect of reflecting the signal. Obstructed path loss from an omnidirectional antenna (all directions) for 900 MHz mobile communications through buildings and foliage is approximately 40 dB per decade. That means for each 10 times increase in distance, the amount of energy decreases by 99.99% (10,000 times less). A signal of 1 Watt will be reduced to 0.0001 Watts (0.1 mWatts).

Propagation Time-The time required for a signal to travel between points on a transmission path.

Proprietary-Proprietary means owned or controlled by another person or company.

Proprietary Software-Proprietary software is files, programs, or applications that are privately owned and controlled by a person or company. Proprietary software is usually created or used for a specific purpose or company. Applications for specific devices or operating systems used for communications equipment (such as a mobile phone) may be considered proprietary even though many devices and applications may be available for them.

Proprietary Systems-Proprietary systems are equipment, assemblies, or networks that are unique to a specific manufacturer or company. While proprietary systems may contain components that are non-proprietary (industry standard), the interoperation of the components or systems may be unique and proprietary.

Prorating-The process of fractionalizing charges for a partial period. In order to determine the number of days for which to charge, a "multiplier" is used (See "Multiplier").

Protected Field-That portion of a packet that is to be protected by some mechanism, typically either encryption or a cyclic redundancy check.

Protector-A device or circuit that prevents damage to lines or equipment by conducting dangerously high voltages or currents to ground. Protector types include spark gaps, semiconductors, varistors, and gas tubes.

Protector Frame-Protector frames are usually part of a main distributing frame (MDF) and it serves as the termination mounting assembly for local loop cables. The protector frame includes devices that will isolate loops and/or electrically ground communication equipment to help protect the equipment when undesired voltages (e.g. high voltage spikes) are detected (such as lightning strikes.)

Protocol-Protocols are the languages, processes, and procedures that perform functions used to send control messages and coordinate the transfer of data. Protocols define the format, timing, sequence, and error checking used on a network or computing system. While there are several different protocol languages used for Internet telephone service, the underlying processes (setup and disconnection of calls) are fundamentally the same.

Protocol Adaptation-The process of adapting one protocol to another protocol. This may involve syntax changes (text format and command name changes), timing relationships, and other functional processes.

Protocol Analyzer-A test instrument that is designed to monitor a network and provide analysis of the communication taking place on the network. This allows a technician to monitor a network and provides information for problem determination and resolution.

Most modern day protocol analyzers are aware of all commonly used, industry standard protocols. More advanced protocol analyzers sit "in-line" between two devices, without the devices being aware that the analyzer is present. Other less sophisticated protocol analyzers can be created using standard PCs with network interface cards in "promiscuous mode", whereby the copy all packets that appear on the network, irregardless of destination address.

Protocol Conversion-Protocol conversion involves the translation of the protocols of one system to those of another to enable different types of equipment, such as data terminals and computers, to communicate. This is done by an inter-working function (IWF). An IWF system (such as a data bridge) adapts the communications between two different types of networks. Protocol conversion may be used to interconnect circuit switched or packet switched networks.

Protocol Converter-A device for translating the protocols of one terminal or system to those of another, enabling equipment with different formats and procedures to intercommunicate.

Protocol Data Unit (PDU)-The package of data that contains header and data protocol information that is used to communicate with a specific layer in a software stack.

Protocol Encapsulation-Protocol encapsulation is the process of inserting protocol messages into the payload of other protocol messages. Protocol encapsulation may be used to send commands that are disguised as other protocols for the purpose of tunneling through a communication device or system.

Protocol Extension-Protocol extensions are a set of commands or protocols that are used to extend the capabilities of the core protocol. Protocol extensions are commonly used to rapidly extend the capabilities of an existing application or protocol without changing the underlying application or protocol.

Protocol Independent Multicast (PIM)-Protocol independent multicasting is the setup and distribution of packets in a multicast tree structures using other protocols (such as OSPG, static routes or BGP) than multicast protocols.

Protocol Interworking-The ability of different protocols to work with each other. An example of protocol interworking is how well all of the commands and messages in VoIP protocols (such as H.323, SIP, or MEGACO) interwork with commands and messages for telephone protocols (such as SS7 and E&M).

Protocol Layers-Protocol layers are a hierarchical model of network or communication functions. The divisions of the hierarchy are referred to as layers or levels, with each layer performing a specific task. In addition, each protocol layer obtains services from the protocol layer below it and performs services to the protocol layer above it.

Protocol Multiplexing-(1-general) The identification and sharing of multiple protocols on a single logical channel. (2-Bluetooth) A function performed at the Logical Link Control and Adaptation Protocol (L2CAP) layer. L2CAP must support protocol multiplexing because the baseband protocol does not support a "type" field to identify the higher-layer protocol being multiplexed above it. L2CAP must therefore be able to distinguish between upper-layer protocols such as the Service Discovery Protocol (SDP), Radio Frequency Communication (RFCOMM), and the Telephony Control Specification (TCS).

Protocol Oppression-Protocol oppression is the blocking or restricting of the transmission of data packets that contain specific types of protocols. Examples of protocol oppression is the blocking of UDP packets to restrict access to streaming media.

Protocol Stack-A hierarchical structure of information processing functions that are logically separated into layers that theoretically only interact with higher and lower functional layers. The use of a protocol stack allows software programs and devices to be independently created (such as from different manufacturers) that only provide parts of the overall operation. Protocol stacks can be proprietary (owned exclusively by a company or group of companies) or protocol stacks can be created as an industry standard.

Protocol Suite-A hierarchical set of related protocols.

Protocol Toolkit-Protocol toolkits are a group of software programs that assist designers or developers to create and test protocols that are used in software and communication systems.

Protocols-(1-rules) Protocols are a precise set of rules, timing, and a syntax that govern the accurate transfer of information between devices or software applications. Key protocols in data transmission networks include access protocols, handshaking, line discipline, and session protocols. (2-connection) A procedure for connecting to a communications system to establish, carry out, and terminate communications.

Prototype-A prototype is a device or product that has been produced for the purposes of demonstration and/or testing but is not made commercially available. Prototype units may use temporary components and/or parts that are not fully designed or are not durable.

Provisioning-A process within a company that allows for establishment of new accounts, activation and termination of features within these accounts, and coordinating and dispatching the resources necessary to fill those service orders.

Provisioning Server-A server that coordinates the activation setup, authorization of features, and elimination of users from a communications system.

Proximity Effect-A non-uniform current distribution in a conductor, caused by current flow in a nearby conductor.

Proximity Services-Proximity services are information services that indicate the proximity of people or items to a specific location or locations.

Proxy-A proxy is a type of server (computer with specific application software) that is used to communicate with other devices or users on behalf of another user or device (a proxy).

Proxy Server-Proxy servers are computing devices (typically a server) that interface between data processing devices (e.g. computers) and other devices within a communications network. These devices may be located on the same local area network or an external network (e.g. the Internet). A proxy server usually has access to at least two communication interfaces. One interface communicates with a device requesting services (e.g. a client) and a device that is being requested for a service (the server).

PS-Padding Stream

ps-Picosecond

PS-Program Stream

PSAP-Public Safety Answering Point

PSD-Power Spectral Density

PSDN-Public Switched Digital Network

PSE-Packet Switching Exchange

Pseudo-Random Binary Sequence (PRBS)-A pseudo-random binary sequence is a succession of digital bits that appear to be completely random but have, in fact, been carefully drawn up and repeat after a significant time interval. PRBS codes may be used to modify a bit stream to make it appear more random.

Pseudo-Random Bit Stream (PRBS)-A sample data signal that is used for telecommunications testing.

PSI-Program Specific Information

PSK-Phase Shift Keying

PSPDN-Packet Switched Public Data Network

PSS1-Private Signaling System 1

PSTN-Public Switched Telephone Network

PSTN Gateway-A PSTN gateway is a communications device or that transforms data that is received from one network (such as the Internet or DSL network) into a format that can be used by the PSTN network. The PSTN gateway may be a simple device that performs simple call completion and adaptation of digital audio into a compatible signals for the PSTN. Or, it may be a more complex device that is capable of advanced services such as conference calling, call waiting, call forward and other PSTN like services. The PSTN gateway must create signaling protocols and compensate for timing differences between a end users computer and the public switched telephone network (PSTN).

PTC-Positive Temperature Coefficient

PTE-Path Terminating Equipment

PTI-Payload Type Identifier

PTN-Public Telephone Network

PTO-Public Telecommunications Operator

PTS-Presentation Time Stamp

PTT-Post, Telephone And Telegraph

PTT-Push To Talk

PTZ-Pan, Tilt and Zoom

PUA-Presence User Agent

Public-(1-general) Avaiable to everyone (2-Bluetooth) A mode of operation whereby a device can be found via Bluetooth baseband inquiries; that is, it enters into inquiry scans. A public device also enters into page scans.

Public domain-Products which may be expired or unable to be copyrighted, therefore making them accessible to the public.

Public Duct-Public ducts are owned by government agencies or groups for the good of the general public and these ducts may be shared by multiple users.

Public Exchange Points-Public exchange points or major communication networking points where Internet providers, companies, schools, government divisions and other types of communication systems meet to interconnect with each other.

Public Internet Address-An Internet address that can be recognized and transferred (routed) on the public Internet.

This figure shows that Internet address number formats are divided into large networks, smaller networks, and local networks. This example shows that an Internet address only identifies a specific device (data connection point) within the Internet. Because devices can be connected to the Internet anywhere in the world, the Internet address does not identify a specific location.

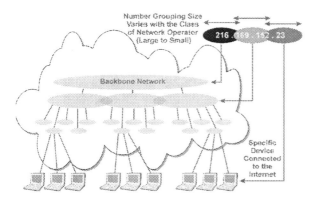

Public Internet Address

Public key-A public key is a cryptographic key that is made available to the public for the purpose of encrypting messages to add privacy (security) to messages that are sent and received.

Public Key Encryption-Public key encryption is an authentication and encryption process that uses two keys, a public key and a private key, to setup and perform encryption between communication devices. The public key and private keys can be combined to increase the key length provider and a more secure encryption system.

Public Network-A network operated by common carriers, local and long-distance, that is made available for private users and the public.

Public Notice-A message that is communicated to the pubic in general (usually through advertising or public media channels) that alerts interested parties that a change in regulation is occurring.

Public Packet Data Network (PPDN)-A packet data network that is generally available for commercial users (the public). An example of a PPDN is the Internet.

Public Packet Switched Network (PPSN)-An exchange access capability that provides packet-switched data transport for terminals, value-added networks, host computers and interexchange carriers.

Public Packet Telephone Network (PPTN)-A public telecommunications network that uses packet based communication as its core communication technology.

Public Peering-Public peering is the process of inquiring or exchanging information in a public communication system (such as the Internet).

Public Safety Answering Point (PSAP)-Public safety answering points (PSAPs) are facilities that receive and process emergency calls. The PSAP usually receives the calling number identification information that can be used to determine the location of the caller. The PSAP operator will then initiate and/or route calls to assist with the emergency situation.

Public Switched Digital Network (PSDN)-A high-speed data network that uses a public switched telephone system to link users over greater distances than can be provided by local area networks.

Public Switched Telephone Network (PSTN)- Public switched telephone networks are communication systems that are available for public to allow users to interconnect communication devices. Public telephone networks within countries and regions are standard integrated systems of transmission and switching facilities, signaling processors, and associated operations support systems that allow communication devices to communicate with each other when they operate.

This diagram shows a basic overview of the Public Switched Telephone Network (PSTN) as deployed in a typical metropolitan area. PSTN customers connect to the end-office (EO) for telecommunications services. The EO processes the customer service request locally or passes it off to the appropriate end or tandem office. As different levels of switches interconnect the parts of the PSTN system, lower-level switches are used to connect end-users (telephones) directly to other end-users in a specific geographic area. Higher-level switches are used to interconnect lower level switches.

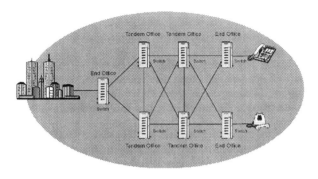

Public Switched Telephone Network (PSTN)

Public Telephone-A telephone provided by a telephone company through which an end user may originate interstate or foreign telecommunications for which he pays with coins or by credit card, collect or third party billing procedures.

Public Telephone Network (PTN)-An unrestricted dialing telephone network that is available for public use. The network is an integrated system of transmission and switching facilities, signaling processors, and associated operations support systems that is shared by customers. PTN is the common term used for public telephone networks outside North America. Inside North America, the public telephone network is called the public switched telephone network. In the United States, the PTN is referred to as the Public Switched Telephone Network (PSTN).

Public Utilities Commission (PUC)-A state regulatory body that sets rates and rules for local exchange carriers and interexchange carriers that provide long distance (interLATA) service within a state boundary. In some states, it is called a public service commission (PSC), board of public utilities (BPU), or corporation commission (CC).

Publication Date-With some exceptions, applications for patents are published and made public 18 months after the date of filing. The publication date is important in that it is the date on which pending patents can be used as prior art against non-US applications for patent. According to US law, pending US applications for patent may be considered as prior art as of the date of filing against a later US application for patent.

Publisher-A publisher is a person or company that gathers or converts intellectual property (content) into a form that it can make available to distributors and or content users.

Publisher Database-A database that contains billing and usage information about a subscriber to a service. This usually includes authorized features, billing records, and other related data that may be regularly updated (read-write). The original subscriber database that provides information to the publisher database is often read-only.

Publishing-Publishing is the process of making available intellectual property (content) to distributors and/or users of the content.

Publishing Point-A publishing point is a logical location of where a media resource is located on a computer server, network or storage device.

Publishing Requirements for Industry Standard Metadata (PRISM)-Publishing requirements for industry standard metadata is an industry standard that defines how metadata can be used to create, categorize, manage, combine and distribute content in magazines, journals books and catalogs. The PRISM standard is divided into two parts; one for the overall PRISM structure and the other defines how content will be distributed to web sites, content aggregators and syndicators. More information about the PRSIM standard can be found at www.prismstandard.org.

Publishing Rights-Publishing rights are the authorized uses that are granted to a publisher to produce, sell and distribute content. Publishing rights may range from global distribution in any media format to specific media formats and geographic areas.

PUC-Public Utilities Commission

PUK-Personal Unblocking Key

Pull Box (Pullbox)-A pullbox is a rectangular or square steel box installed inline within a buildings conduit run. Pullboxes assist with the initial installation and subsequent maintenance of indoor plant cabling.

Pull Notification-Pull Notification is when a device (such as a wireless telephone) polls the server for any new events or information.

This figure shows how pull notification works with a WAP server. This example shows a WAP push proxy gateway that receives email messages that are addressed to the WAP client. The push proxy gateway stores these messages until it receives a request from the WAP client for the delivery of messages. The WAP client will then download (pull) the messages from the push proxy gateway so the messages can be displayed on the users phone.

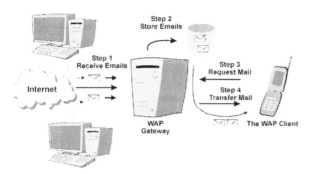

WAP Pull Notification

Pull Proof-Pull proof is the ability of a cable connector to withstand the pulling of a cable without damaging the connection of the cable to the connector.

Pull Rope-Pulling Rope

Pullbox-Pull Box

Pulldown Detection-Pulldown detection is the process of sensing that 3:2 pulldown has been used to convert 24 frame per second film to 60 frame per second video. Pulldown detection allows non-interlaced digital displays to convert the playback to the original (undistorted) 24 frames per second film format.

Pulling Eye-A pulling eye is a device that is attached to the strength member(s) at the end of a cable which allows a pull rope or line to be attached. A pulling eye is usually shaped (rounded) to allow easy pulling through conduit or ducts. Some pulling eyes may contain an enclosure to hold connectors that are already attached to the end of the cable.

Pulling Motor-A pulling motor is the powered assembly that is used on a cable winch to pull cable through a channel or conduit.

Pulling Tension-Pulling tension is the amount of pulling load (tensile load) that exists on a cable during the pulling of the cable through a duct or conduit.

Pulse-A signal of short duration characterized by a rapid change in amplitude. A pulse signal is often characterized by a rise time from 10% to 90% of the signal height, its fall time from 90% to 10%, and it's pulse duration between the beginning and end of the pulse.

This diagram shows a series of pulse signals and typical characteristics associated with pulse signals. In this diagram, these pulse signals are characterized by a rise time from 10% to 90% of the signal height, their fall time from 90% to 10%, and the pulse duration between the beginning and end of the pulse. This diagram also shows how a pulse signal may include signal ringing. Signal ringing is caused by a rapid signal level change in a system that does not have effective impedance matching.

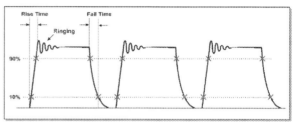

Pulse Wave Diagram

Pulse Amplitude Modulation (PAM)-A form of modulation in which the amplitude of the pulse carrier is altered in accordance with some characteristic of the modulating signal.

Pulse Coded Modulation (PCM)-Pulse coded modulation is a process that transfers information (such as an analog audio signal) into discrete form (e.g. pulses) where the information signal is sequentially sampled, quantized, and coded into a binary form for transmission over a digital link.

Pulse Interval-The time between the start of one pulse and the start of the next

Pulse Position Modulation (PPM)-A method of conveying information by transmitting a series of extremely brief pulses of equal amplitude and duration but varying time occurrence within each time window of a sequence of consecutive equal-duration time windows or time slots. To encode a telephone-quality voice signal, each time window is typically about 100 microseconds in duration. For example, the position (time relative to the beginning of the time window) of a pulse within the window is typically proportional to the sampled voltage of an analog speech waveform that is represented by the PPM signal. A PPM waveform is approximately the time derivative of a corresponding pulse width modulation (PWM) waveform.

Pulse Rate-The number of pulses that are transmitted during a time interval, usually measured in pulses per second (pps).

Pulse Repetition Period-The time interval between the repetitive pulse signals.

Pulse Train-A series of pulses having similar characteristics.

Pulsed Laser-A laser that emits light in pulses rather than continuously.

Punchdown Block-A device used to connect one group of wires to another. This is also referred to as a terminating block, a connecting block, and a cross-connect block.

Punctured Codes-Punctured Error Correcting Code

Punctured Error Correcting Code (Punctured Codes)-Punctured codes are linear error correction codes that remove a symbol from the code word. This can significantly reduce the data transmission overhead (additional bits for error detection/correction) without significantly reducing the error detection/correction performance.

Purchase Order-A purchase order is a document or record from a person or company that authorizes the sale of products or services to that person or company.

Push Notification-Push notification is when the server contacts the device (such as a wireless telephone) without any request coming from the device, and then pushes the information down to the device. One of the most common examples of push notification is when the server notifies the wireless device that a new email has arrived.

Push Technology-The ability of a communication user to push content or control to another communication user. Push technology is commonly used in web seminar (webinar) systems where an instructor can push presentations to others who are participating in a webinar session.

Push To Talk (PTT)-Push to talk (PTT) is a process of initiating transmission through the use of a push-to-talk button. The push to talk process involves the talker pressing a talk button (usually part of a handheld microphone) that must be pushed before the user can transmit. If the system is available for PTT service (other users in the group not talking), the talker will be alerted (possibly with an acknowledgement tone) and the talker can transmit their voice by holing the talk button. If the system is not available, the user will not be able to transmit/talk.

Pushbutton Dialing-The use of keys or pushbuttons instead of a rotary dial to generate the sequence of digits that establishes a circuit connection. Touch tones are the most common form of pushbutton dialing.

Push-Pull-Under the Generic Object Exchange Profile (GOEP), a client sends (pushes) data objects to the server, or retrieves (pulls) data objects from the object exchange server.

PVC-Permanent Virtual Circuit

PVC-Polyvinyl Chloride

PVR-Personal Video Recorder

Q

Q-Quality Factor

Q.931-A telecom call processing signaling protocol that is used in telephone communication systems. The Q.931 protocol defines the messages and formats are control messages that are created by the end communication device. Some of the common information contained in Q.931 messages include call setup and tear down messages, called and calling party telephone numbers, and other access control signaling messages.

This diagram shows the basic structure of a Q.931 signaling message. The examples in this diagram show that Q.931 messages are used to provide information to setup, maintain, and tear down communication connections. Although Q.931 is ISDN (digital) related, the signaling message structure is also used for some analog communication systems.

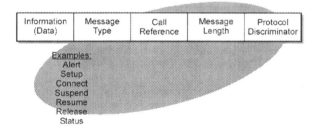

Q 931 Message Structure

QA-Quality Assurance
QAM-Quadrature Amplitude Modulation
QBE-Query By Example
QBone-Quality Of Service Backbone
QC-Quality Control
QCIF-Quarter Common Interchange Format
QCIF-Quarter Common Intermediate Format
QEF-Quasi Error Free
QoE-Quality of Experience
QoS-Quality Of Service
QoSM-Quality of Service Metrics

Quote-The set price someone would ask as wage for tasks or services rendered, as he or she has done in the past.

QPSK-Quadrature Phase Shift Keying

QSIG-QSIG is a peer-to-peer signaling standard that enables call setup between voice-enabled equipment made by different manufacturers. QSIG is also known as Private Signaling System No. 1 (PSS1) and falls under the auspices of both the European Telecommunications Standards Institute (ETSI) and the International Standards Organization (ISO). In addition to QSIG's ability to set up a basic call, supplementary QSIG services specify how calling features, such as line forwarding, call transfer, call forwarding, and many others, can work across different vendors' platforms. QSIG grew out of the Integrated Services Digital Network (ISDN) standard called Q.931.

Quadraphonic-Surround-sound reproduction involving the recording and playback of four channels of sound.

Quadrature-A state of alternating current signals separated by one-quarter of a cycle (90 degrees).

Quadrature Amplitude Modulation (QAM)-QAM is a combination of amplitude modulation (changing the amplitude or voltage of a sine wave to convey information) together with phase modulation. There are several ways to build a QAM modulator. In one process, two modulating signals are derived by special pre-processing from the information bit stream. Two replicas of the carrier frequency sine wave are generated; one is a direct replica and the other is delayed by a quarter of a cycle (90 degrees). Each of the two different derived modulating signals are then used to amplitude modulate one of the two replica carrier sinewaves, respectively. The resultant two modulated signals can be added together. The result is a sine wave having a constant unchanging frequency, but having an amplitude and phase that both vary to convey the information. At the detector or decoder the original information bit stream can be reconstructed. QAM conveys a higher information bit rate (bits per second) than a BPSK or QPSK signal of the same bandwidth, but is also more affected by interference and noise as well.

This diagram shows that amplitude and phase modulation (QAM) can be combined to form an efficient modulation system. In this example, one digital signal changes the phase and another digital signal changes the amplitude. In some commercial systems, a single digital signal is used to change both the phase and the amplitude of the RF signal. This allows a much higher data transfer rate as compared to a single modulation type.

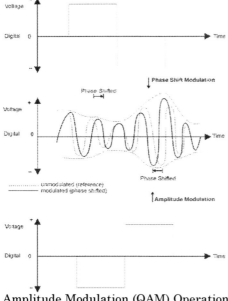

Amplitude Modulation (QAM) Operation

Quadrature Distortion-A distortion resulting from the asymmetry of sidebands used in vestigial sideband TV transmission. Quadrature distortion appears when envelope detection is used, but can be eliminated by use of a synchronous demodulator.

Quadrature Modulation-Modulation of two carrier components 90° apart in phase by separate modulating functions.

Quadrature Phase Shift Keying (QPSK)-Quadrature phase shift keying (QPSK) is a type of modulation that uses 4 different phase shifts of a radio carrier signal to represent the digital information signal. These shifts are typically +/- 45 and +/- 135 degrees.

Quake Compression-Quake compression is a vector and texture based compression process used for games and animation. Because vector and texture based images can be represented by mathematical formulas and texture codes, the amount of information that needs to be transferred to represent 2D and 3D images is relatively low.

Quality Assurance (QA)-All those activities, including surveillance, inspection, control, and documentation, aimed at ensuring that a given product will meet its performance specifications.

Quality Control (QC)-A function whereby management exercises control over the quality of raw material or intermediate products to prevent the production of defective devices or systems.

Quality Factor (Q)-(1-inductor) A figure of merit that defines how a coil comes to functioning as a pure inductor. High Q describes an inductor with little energy loss resulting from resistance. Q is determined by dividing the inductive reactance of a device by its resistance. (2-capacitor). A figure of merit that defines how close a capacitor comes to functioning as a pure capacitance. High Q describes a capacitor with little energy loss resulting from resistance. Q is determined by dividing the capacitive reactance of a device by its resistance.

Quality of Experience (QoE)-Quality of experience (QoE) is one or more measurement of the total communications experience or the entertainment satisfaction from the perspective of the end user. QoE measures may include service availability, audio and video fidelity, types of programming and the ability to use and the value of interactive services.

Quality Of Service (QoS)-Quality of service (QoS) is one or more measurement of desired performance and priorities of a communications system. QoS measures may include service availability, maximum bit error rate (BER), minimum committed bit rate (CBR) and other measurements that are used to ensure quality communications service.

Quality Of Service Backbone (QBone)-A high speed backbone network used for Internet2 that allows for different levels of Quality of Service.

Quality of Service Metrics (QoSM)-Quality of service metrics is one or more parameters that are used to define acceptable or desired performance for systems and services. Examples of QoSM include latency, jitter, peak throughput and error rate.

Quantization-Quantization is the process of representing a value with a less precise value. In analog-to-digital conversion, a continuous analog value is represented by one of a finite number of quantized values. In lossy signal compression (such as an A-law encoding) one digital value is represented by another one which is usually not precisely the same. Except in lucky cases where the quantized value is exactly the same as the original, quantization introduces error (or noise).

Quantization Error-The difference between the actual value of the analog signal at the sample time and the resulting digital word value. The greater the resolution of the analog-to-digital converter, the lower the quantization error.

Quantization Noise-Quantization noise (or distortion) is the error that results from the conversion of a continuous analog signal into a finite number of digital samples that can not accurately reflect every possible analog signal level. Quantization noise is reduced by increasing the number of samples or the number of bits that represent each sample. This term also is known as quantization distortion.

Quantizing-The process of sampling an analog waveform to convert its voltage levels into digital data.

Quantizing Distortion-The distortion resulting from the quantization process.

Quantum-A discrete package as opposed to a continuously variable process. In the electronic field, the quantum of electromagnetic radiation is the photon, but this unit is so small that, for all large scale purposes, it may be considered that energy is carried by continuously varying electromagnetic fields obeying classical laws.

Quarter Common Interchange Format (QCIF)-Quarter common interchange format (QCIF) is an image resolution format that is 180 pixels across by 144 pixels high (180x144). The QCIF standard was developed in 1990 is defined by the ITU in as H.261 and H.264 compression standards. H.261 coding includes interframe prediction (using key frames and difference frames), mathematical transform coding and motion compensation.

Quarter Common Intermediate Format (QCIF)-An ITU standard H.261 for video encoding that was created in (1990). QCIF includes interframe prediction (using key frames and difference frames), mathematical transform coding and

motion compensation.

Quartz-A crystalline mineral that, when electrically excited, vibrates with a stable period. Quartz typically is used as the frequency determining element in oscillators and filters.

Quasi Error Free (QEF)-Quasi error free is a condition where the transmission system or storage medium used to transfer a signal has a relatively low bit error rate.

Query Language-A query language is a software programming language containing commands that allow users or software applications to retrieve and/or process information from databases.

Queuing-Queuing is a process of delaying or sequencing messages. Queuing involves receiving requests for service, prioritizing these requests, storing them in appropriate order and transferring the messages when the facilities (channels) are available to send them.

Queuing systems may change the order of messages or services to be provided based on priority access. For example, communication requests from a public safety official may be given priority over a communication request from a consumer.

QuickTime-Quicktime is a computer video format (sound, graphics and movies) that was developed by Apple computer in the early 1990s, QuickTime files are designated by the .mov extension.

Quiescent-An inactive device, signal, or system.

R

R-17OA-A document prepared by the Electronics Industries Association describing recommended practices for NTSC color TV signals in the United States. The changes from RS-170 involve pulse timing parameters.

RA-Real Audio

RA-Regulatory Authority

RAB-Random Access Burst

RACE-Research In Advanced Communications In Europe

Raceway-A raceway is a covered trough or conduit channel that is used to hold internal wiring and cabling.

RACH-Random Access Channel

Rack-An equipment rack, usually measuring 19 inches (48.26 cm) wide at the front mounting rails.

RACON-Radar Beacon

Radar-Radio Detection And Ranging

Radial Splice Enclosure-A radial splice enclosure is a container that holds splices and the cables enter in one end and leave from a different direction or end.

Radiate-The process of emitting electromagnetic energy.

Radiation-(1-verb) The emission and propagation of electromagnetic energy in the form of electromagnetic waves. (2-noun) Radiation also is called radiant energy.

Radiation Pattern-(1-radio frequency) The magnitude of the relative electric field strength radiated from an antenna in a given plane as a function of the angle from a given reference direction. (2 - fiber optic) The output radiation of an optical waveguide, specified as a function of angle or distance from the waveguide axis. The near field radiation pattern is specified as a function of angle. The near field radiation pattern is specified as a function of distance from the waveguide axis. The radiation pattern is a function of the length of waveguide measured, the manner in which the waveguide is excited, and the wavelength.

Radiator-(1-radio) Any part of an antenna that radiates electromagnetic waves. (2-heat) Devices that radiate heat rather than electromagnetic waves.

Radio-(1) Radio is the transfer of information signals through the air by the means of electromagnetic waves. These electromagnetic waves repeat their cycle in a frequency range of approximately 150 kHz to 300GHz. (2) The electronic equipment used to receive or transmit radio (electromagnetic waves).

Radio Access-A method or technology used to coordinate access to radio channels or portions of radio channels.

Radio Beam-A radiation pattern from a directional antenna such that the energy of the transmitted electromagnetic wave is confined to a small angle in at least one dimension.

Radio Broadcast Data System (RBDS)-Radio broadcast data system is a broadcast communication system that simultaneously transmits encoded data alongside their regular programming. RPDS service allows radio broadcast systems to provide additional broadcast information services.

Radio Broadcast Data System (RDBS)-The radio broadcast data system is a modified version of the radio data system (RDS) that allows it to operate in the United States on FM and AM radio channels.

Radio Broadcast Service-Radio broadcast service is the radio transmission of an information signal to a specified geographic area or network. This allows the same information to be received by all customers in that geographic area that can successfully receive (demodulate) and decode the information.

Radio Broadcasting-The transmission by program material (audio or video) that can be simultaneously received by receivers that are capable of receiving and decoding the radio signal to recover the original audio or video signal.

Radio Carrier-(1-radio signal) A radio wave at a specific frequency. (2-broadcast company) A company that broadcasts information services using radio transmission.

Radio Channel-A radio channel is a communications channel that uses radio waves to transfer information from a source to a destination. A radio channel may transport one or many communication channels and communication circuits.

R

This diagram shows how a radio carrier wave is modulated by a user's information signal to produce a radio channel signal. In this example, the user data is multiplexed (time shared) with a control information signal. The combined data signal (user data and control data) is supplied to the radio frequency (RF) modulator along with the radio carrier signal. This diagram shows that the modulation process (changing) of the radio carrier signal causes the frequency of the carrier to change and the energy of the carrier wave to be distributed within a pre-defined frequency band (channel bandwidth). This example shows that some of the modulated signal energy does fall out of the prescribed channel bandwidth. This is typical for most systems and radio system specifications (and/or government transmission requirements) usually exist to ensure the levels transmitted outside the prescribed frequency band do not interfere with other systems.

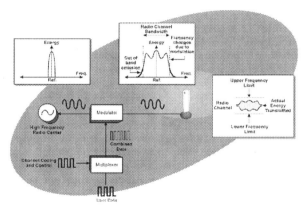

Radio Channel Operation

Radio Communications-The process of telecommunication by means of radio waves.

Radio Coverage-Radio coverage is a geographic area that receives a radio signal above a specified minimum level.

Radio Frequency (RF)-Those frequencies of the electromagnetic spectrum normally associated with radio wave propagation. RF sometimes is defined as transmission at any frequency at which coherent electromagnetic energy radiation is possible, usually above 150kHz.

Radio Frequency Interference (RFI)- Undesired signals at radio frequencies that interfere with radio reception or the proper operation of equipment. When they interfere with television reception, such signals are called television interference (TVI). Many electronic devices that are not intended to produce radio frequency power, including radio and television receivers, computers, and peripherals, can interfere with other signals in the radio-frequency range by producing incidental electromagnetic radiation. The amount of such incidental radiation is regulated by FCC Regulations Part 15 in the United States.

Radio Landline-A circuit that connects a cellular switching office to a cell site or to a public switched network. It also denotes any wireline circuit from a control station to remote transmitters or receivers.

Radio Link- A radio frequency communications channel between a fixed cellular radio site and mobile units. Also, a radio system established between two points.

Radio Link Protocol (RLP)-Radio link protocol are the commands and processes used to coordinate the delivery of data between radio transmitter and receiver links.

Radio Propagation-Radio propagation is the process of transferring a radio signal (electromagnetic signal) from one point to another point. Radio propagation may involve a direct wave (space wave) or a wave that travels along the surface (a surface wave). Radio propagation characteristics typically vary based on the medium of transmission (e.g. Air) and the frequency of radio transmission.

Radio System-A system that is composed of mobile or fixed communication devices. The types of radio systems include broadcast (one to many), multicast (one to several) or point-to-point (one-to-one). When a radio system allows communication devices to select from or be assigned to one of several radio channels, this is called trunking. The radio access control procedure allows the mobile to access alternate radio channels in the system in the event that the channel it has requested is busy.

Radio Tower-Radio towers are poles, guided towers or free standing constructed grids that raise one or more antennas to a height that increases the range of a transmitted signal. Radio towers typically vary in height from about 20 feet to more than 300 feet.

Radio Waves-Electromagnetic waves at frequencies between 9 kHz to 3000 GHz (3 THz). Above 3 THz the waves are considered optical signals.

Radiovision-A system that combined television broadcasting with radio audio broadcasting to provide for early interactive television programming.

RADIUS-Remote Access Dial In User Service

Radome-A cover that protects an antenna from the extremes of climate while allowing electromagnetic signals (radio waves) to pass through without attenuation. The Radome is usually constructed from plastic or fiberglass.

RADSL-Rate Adaptive Digital Subscriber Line

RAI-Resource Availability Indicator

RAID-Redundant Array of Inexpensive Disks

Rain Attenuation-The attenuation of a radio signal traveling through rain or moisture bearing clouds.

RAM-Random Access Memory

RAM-Real Audio Metafile

RAN-Radio Access Network

RAND-Reasonable and Non-Discriminatory

Random Access-(1-Computer storage) Random access of data is the process of reading and/or writing data to and from a storage media (such as a computer hard disk). (2-Wireless system) The process used to coordinate access (competition) to radio channels in a wireless communication system.

Random Access Memory (RAM)-A data storage device, often an integrated circuit, from which a word of data can be removed without regard for the time it was stored or its location. Internal semiconductor RAM, static or dynamic, is volatile in nature and does not retain the data stored within if power is removed from the device.

Random Noise-Electromagnetic signals that originate in transient electrical disturbances and have random time and amplitude patterns. Random noise generally is undesirable, but it is often generated for testing purposes.

Random Number-A number formed by a set of digits in which each successive digit is equally likely to be any of the digits in a specified set.

Random Separation-A method by which facilities of two or more utilities are placed with no deliberate separation of the facilities.

RAO-Revenue Accounting Office

RARP-Reverse Address Resolution Protocol

RAS-Registration, Admission, Status

RAS-Remote Access Server

Raster-A predetermined pattern of scanning the screen of a CRT. Raster also can refer to the illuminated area reduced by scanning lines on a CRT when no video is present.

Rate-The price charged a customer for a particular service, as specified and defined in a tariff.

Rate Adapter-A rate adapter is a unit or process that converts the data transmission rate from a system or communication line to the data transmission rate of another system or communication line.

Rate Adaptive Digital Subscriber Line (RADSL)-A hybrid analog and digital subscriber line technology that allows the data transmission rate to dynamically change dependent on the electrical transmission characteristics and/or the settings provided by the DSL service provider. RADSL features have been incorporated into standard ADSL technology.

Rate Band-A rate band is comprised of the combination pair of originating and terminating numbers where the "banded rating" amount (rate) is fixed within the same band. International calling is an example of banded rating (e.g. all calls from anywhere in the US to anywhere in France will carry the same rate.)

Rate Center-As defined by rate map coordinates, the point within an exchange area that is used as the primary basis for determining toll rates. Rate centers also can be used for determining selected local rates. A rate center also is called a rate point.

Rate Class-A rate classification determined by the type of message, station, or customer, and the rate in effect at the originating rate center at the time a telephone call begins.

Rate Control-Rate control is the process of setting and/or adjusting the rate of a process or transmission.

Rate Management-Rate management is the assignment, monitoring and adjustment of equipment configurations and resources to achieve or to change in the direction of desired rates. An example of rate management is the adjustment of lossy data compression (higher or lower compression) in a digital video system to maintain or limit the maximum data transmission rate so it can be sent on a transmission channel that has a fixed transmission rate.

Rate Plan-A rate plan is the structure of service fees that a user will pay to use services. Rate plans are typically divided into monthly fees and usage fees.

Rate Shapers-Rate shapers are devices or assemblies in a communication system that adapt and/or transform the transmission rate of one system to the transmission rate of another system. Rate shapers are used as digital turnaround products in a media distribution system (such as a cable television system).

Rate Table-A rate table is a group of billing codes or product identifiers and their associated billing rates and characteristics that are used to calculate the charges for billing events or billing records.

Rating-Rating is a function within the billing system that assigns a rate (cost parameter) to a usage record. Rating typically involves using the originating number or network address, terminating number or destination network address, date the service was used, amount of usage or time period, usage type and tax jurisdiction to determine the initial charge assigned to the usage record. The actual cost of the usage record may be adjusted based on volume discounts or other rate plan considerations.

Rating Engine-A rating engine is a function within a billing system that assigns the charging rates to a call event or CDR.

This figure shows the how a rating engine of a billing system can calculate the fees associated with specific usage events. This diagram shows that the rating engine receives usage records from network usage, incollects from other companies and other billing records (such as adjustments). The rating engine first identifies the account associated with the usage records and checks for duplicate records. The rate table for the customer account is selected for these records and the appropriate fee is calculated for each record. After the calculated rate is added to the usage detail record, it is sent to the bill pool to await the processing of invoices.

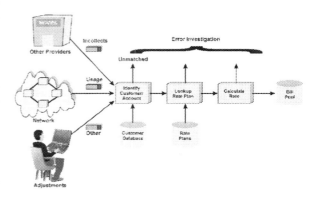

IPTV Billing Record Rating

Rating System-Ratings systems measure usage or viewer habits. Ratings systems such as Neilson ratings identify the percentages of households that watch specific programs.

Ratings Based Sales-Ratings based sales is the offering of services (such as advertising) where the value or cost of the services is determined by the popularity or subscription to the services.

Raw Media File-A raw media file is a collection of data (bits) that represents information such as voice, data or video.

Raw Stream-A raw stream is the data or streaming media that represents a particular media object (such as digital video and images).

Rayleigh Fading-The signal strength fading resulting from multipath that results from phase interference between multipath signals and is approximated by Rayleigh distribution.

Rayleigh Scattering-(1-optical fiber) The scattering of light that is transmitted through an optical fiber. Most scattering causes a phase delay in the propagated light. Light also can be re-radiated out of a fiber and lost Rayleigh scattering varies inversely with the wavelength and increases the effective attenuation of an optical fiber. (2-electromagnetic) The scattering of electromagnetic radiation by particles smaller than the radiation wavelength.

RBDS-Radio Broadcast Data System

RC-Recurring Charge

RCA Connector-A connector developed by the Radio Corporation of America that is commonly used to connect low to medium frequency electrical signals.

RCDD-Registered Communications Distribution Designer

RCF-Registration Confirmation

RCF-Remote Call Forwarding

RCST-Return Channel Satellite Terminal

RDBS-Radio Broadcast Data System

RDC-Regional Data Center

RDF-Radio Direction Finder

RDF-Resource Descriptor Framework

RD-LAP-Radio Data Link Access Protocol

RDS-Radio Data System

RDT-Remote Digital Terminal

RDTC-Reverse Digital Traffic Channel

Reach Extended Asymmetric Digital Subscriber Line (RE-ADSL)-Reach extended ADSL is an improvement to ADSL systems that optimizes the allocation of DMT transmission channels to extend the reach of ADSL systems. This increases the maximum distance that an ADSL system can operate.

Reach Sensitive Service-Reach sensitive service is the distance limitations on the ability to provide services. Reach sensitive services may have gradual distance limitations. An example of reach sensitive service is digital subscriber line (DSL) service. The data transmission rate for DSL service declines as the distance from the DSL modem increases.

Reachability-A measurement used in network management that indicates the capability of successfully pinging or successfully communicating with a device in the network from another device in the network.

Reactance-The part of the impedance of a network resulting from inductance or capacitance. The reactance of a component varies with the frequency of the applied signal. Reactance represents an opposition to the flow of alternating current.

Read Only Memory (ROM)-A memory circuit or device in which any address can be read from, but not written to, after initial programming. The ROM is an asynchronous device with an access time dictated by internal circuit time delays. Semiconductor ROM storage is nonvolatile and retains data when power is removed.

Readout-A visual display of the output of a device or system.

RE-ADSL-Reach Extended Asymmetric Digital Subscriber Line

Real Audio (RA)-Real Audio (RA) is Real's (the company) digital multimedia file format that is used to stream digital audio. The format was designed to be an efficient form of streaming and works with Real media players. Real audio files can have the file extension .ra.

Real Audio Metafile (RAM)-A real audio metafile is a text container file (index pointer) that identifies Real audio media sources.

Real Media (RM)-Real Media (RM) is Real's (the company) digital multimedia file format that is used to stream digital audio and digital video. The format was designed to be an efficient form of streaming and works with Real media players. Real (.rm) files have the ability to synchronize digital audio and digital video along with managing other forms of media.

Real Media Player-Real media player is a software application from Real that allows for the playing of multiple types of digital media formats.

Real Time-Actions that occur instantly or within a time period that is perceived or used (such as within a few seconds) to perform or record events when they are required or used.

Real Time Billing-Real time billing involves the authorizing, gathering, rating, and posting of account information either at the time of service request or within a short time afterwards (may be several minutes). Real time billing is primarily used for prepaid services such as calling cards or prepaid wireless.

This figure shows a real time prepaid billing system. In this example, the customer initiates a call to a prepaid switching gateway. The gateway gathers the account information by either prompting the user to enter information or by gathering information from the incoming call (e.g. prepaid wireless telephone number). The gateway sends the account information (dialed digits and account number) to the real time rating system. The real time rating system identifies the correct rate table (e.g. peak time or off peak time) and inquires the account determine the balance of the account. Using the rate information and balance available, the real time rating system determines the maximum available time for the call duration. This information is sent back to the gateway and the gateway completes (connects) the call. During the call progress, the gateway maintains a timer so the caller cannot exceed the maximum amount of time. After the call is complete (either caller hangs up), the gateway sends a message to the real time rating system that contains the actual amount of time that is used. The real time rating system uses the time and rate information to calculate the actual charge for the call. The system then updates the account balance (decreases by the charge for the call).

tion. The RTP system then adds a third header (the RTP control header). The RTP system uses a precise clock to add time stamp information to each packet along with other signal recreation control information. Because the packets may have different types of compression and their recreation time can dramatically vary, the RTP protocol header uses the time stamp and other information to decode the and recreate the data packet.

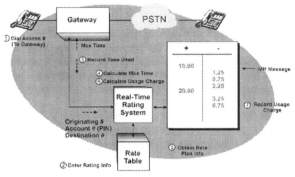

Real Time Billing Operation

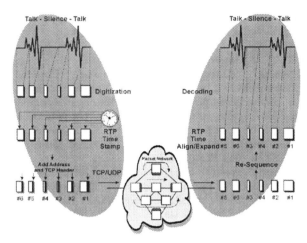

Real Time Transport Protocol (RTP) Operation

Real Time Streaming Protocol (RTSP)-Real time streaming protocol is an Internet protocol that is used for continuous (streaming) audio and video sessions. RTSP is capable of bi-directional (two-way) sessions. RTSP is defined in RFC 2326.

Real Time Transport Protocol (RTP)-RTP is a packet based communication protocol that adds timing and sequence information to each packet to allow the reassembly of packets to reproduce real time audio and video information. RTP is defined in RFC 1889. RTP is the transport used in IP audio and video environments.

This diagram shows how real time transmission control protocol (RTP) operates to send real time data through a packet network that may have variable transmission delays. This diagram shows that an RTP system requires that a real time signal (e.g. audio signal) be converted to digital form (digital audio) prior to transmission. This digital signal is divided into small packets. The RTP protocol is a high-level protocol and each packet of data each of the transmitted packets starts with an IP header that contains the destination address of the packet. An additional flow control protocol header is added (usually UDP protocol header) to identify the specific port the data will be routed to at it's destina-

RealPlayer-RealPlayer is a software application for playback of digital media developed by Real Audio. The RealPlayer media player can play Real Audio and Real Video file formats along with other industry standard audio and video formats.

Real-Time Transport Control Protocol (RTCP)-A signaling control protocol that allows for the controlling and reporting on the flow of real-time data that is transmitted in a packet network. The RTCP protocol sends messages in addition to the communication data to monitor the transmission quality within a packet network. RTCP is defined in RFC 1889.

Reasonable and Non-Discriminatory (RAND)-Reasonable and non-discriminatory are licensing terms (fees and restrictions) offered by owners of intellectual property that allows other companies to build products that incorporate their technology while providing the potential for sustainable profits.

Reassembly-The process of reconstructing a packet from its fragments.

Reassociation-A process of re-registering a wireless data device (station) with a specific access point (AP) in an 802.11 specified wireless local area network (WLAN) system.

Reboot (Reset)-To restart an electronic assembly or computer.

Rebroadcast-The broadcast of material picked up directly (off-the air) from another broadcasting station.

RECC-Reverse Control Channel

Received Signal Strength Indicator (RSSI)-A signal level indicator, usually on a mobile radio, that regularly displays the approximate level of a received signal. The RSSI indicator allows a user to determine if the radio signal strength in that area is sufficient to initiate or complete a call.

Receiver-(1-general) A device that receives a transmitted signal and makes it perceptible to a human user or converts it into some other useful form. (1- telephone handset) Older name for the earphone of a telephone handset, or used as an adjective for signals associated with that earphone. Also the entire handset (microphone and earphone, et al). (2-radio) Radio receiver, a device that responds to a modulated radio signal by producing a replica of the original modulating information. (3-tone receiver) A device in a telephone system responsive to in-band telephone tone signals such as MF, SF or DTMF signals, interpreting the significance of such tone signals by producing the corresponding digital representation in a form useable by the destination telephone switch.

Receiver Threshold Test-Receiver threshold test is the measurement of receiver characteristics when a threshold (such as minimum bit error rate) is achieved.

Receptacle-An electrical or optical socket that is designed to receive a mating plug.

Reception-The act of receiving, listening to, or watching information carrying signals.

Recharge-(1-battery) To store energy in a battery. (2-calling card) A process of adding additional time to a calling card or prepaid wireless account using either a credit card or debit card.

Reciprocal Compensation-The 1996 Telecommunications Act mandated that local telecommunications companies exchange revenue for the cost of terminating calls that originated on the wireline network. Previously, only wireless companies were obligated to pay compensation for calls originated on their networks but terminated on the wireline network.

Recombination-Combination of an electron and a hole in a semiconductor that releases energy, sometimes leading to light emission.

Recommended Standard (RS)-A set of standards documents that are managed by the electronic industries association (EIA) that specify the format of electronic signals and protocols.

Recon-Reconnect

Reconnect (Recon)-In outside plant construction, the moving of subscriber dropwires onto new cable facilities as engineered on a work order. This often relieves congestion at poles and terminals in addition to improving quality of service for customers.

Reconstructors-Reconstructors are smart aggregation companies that provide consumers with the ability to better select or manipulate the content they use.

Record-An assemblage of data elements, all of which are in some way related and handled as a unit.

Recording Industry Association of America (RIAA)-Recording industry association of America is an organization that advocates technologies and policies that will assist the recording industry.

Recovery-Recovery is a process of reconnecting a communication session, restarting a system or reusing previously discarded or unused equipment or facilities.

Recurring Charge (RC)-A predetermined charge associated with a product or service that is assessed on a regular interval (i.e. monthly, quarterly, annually).

RED-Random Early Discard

Red Button Advertising-Red button advertising allows a user to select an action or initiate a response to an advertising message using the red button on their remote control. After the viewer has pressed the red button on their remote control during the advertising message, the user access device (e.g. set top box) processes an action and/or sends a message into the network to initiate an action (such as to start an interactive exchange).

Red, Green, Blue, (RGB)-The three additive primary colors used in video processing, often referring to the three unencoded outputs of a color camera. The sequence of GBR indicates the mechanical sequence of the connectors in the SMFTE standard.

Although GBR is the preferred name, the term RGB commonly is used to denote the same signals.

Redirection Server-A redirection server assists in the establishment of communication sessions by providing alternative locations where the designated recipient can be found. The redirect server does not initiate any action. It only provides information back to the requesting device as to the potential locations of the designated recipient.

Redrawing-Redrawing is the process of recreating an image for a specific display size. Redrawing requires additional information processing to scale or reformat images. If the capability of the media device is limited (such as a low power mobile device), redrawing may result in the jittering, distortion or loss of image frames.

Redundancy-A system design that includes additional equipment for the backup of key systems or components in the event of an equipment or system failure. While redundancy improves the overall reliability of a system, it also increases the number of equipment assemblies that are contained within a network. Redundancy usually increases cost.

Redundancy Plan-A network structure plan that defines the alternate equipment configurations and routes that are used when specific failures occur.

Redundant Array of Inexpensive Disks (RAID)-Redundant array of independent disks (RAID) is a computer information storage system that uses multiple independent disk storage devices (hard disks). RAID systems were first defined in 1988 and they were originally called redundant array of inexpensive disks. RAID systems can use standard hard disk drive interfaces such as SCSI or Integrated Drive Electronics (IDE).

RAID systems can be configured in various ways to ensure data integrity and to provide the ability to remove and replace disks while the system continues to operate in the event of equipment failure ("hot-swap".) The different types of configurations are called RAID levels. There were six original levels (RAID 0 through RAID 5). Several manufacturers have combined the RAID levels to produce new unique levels above RAID 5.

For RAID 0, the data is distributed (striped) over several drives but there is no redundant storage of data. RAID 1 combines two hard disks of equal storage capacity that simultaneously store the same information (mirror). RAID 2 allows data to be corrected on one drive from another drive by interleaving (distributing) the information across multiple drives and using a data protection formula (algorithm) that relates the information. RAID 3 stores information on several disk drives where only one drive is used for parity bits that are used for error detection and correction. RAID 4 systems store data in multiple drives without the need to mirror information at the bit level. RAID 5 writes data and parity information on the same disk but to different sectors to increase the data transfer performance.

Redundant Power Supply-Redundant power supply is an energy supply that provides power supply if the primary power supply fails.

Redundant Server-The inclusion of a second communication server that allows the automatic transfer of service (reconfigure) in the event a failure occurs on part or all of the other server. In a redundant server, an equipment failure should cause no loss of data or information.

Redundant System-A redundant system is a duplicate configuration of equipment in communication systems that allows some of the equipment to automatically reconfigure in the event a failure occurs on part or all of the duplicated system. In a redundant system, a failure should cause no loss of data or information.

Reed Solomon Code (RSC)-A 2-dimensional error correction code that derives the correction information by applying two different mathematical formulas on the same data. Two correction words, P and Q, are generated from this process. Simply stated, the P word is a binary sum, and the Q word is a binary product of the individual bits in the data word. As a result of this powerful cross correlation of data, Reed Solomon codes are particularly effective in correcting large scale burst errors.

Reel Stand-A reel stand is a fixture that is designed to hold a cable spool and allow it to rotate during the installation of cable.

Refarming-A process of moving or eliminating radio users from a frequency band to allow new users to use that frequency band.

Refer Message-A message that is used to request the transfer of a communication session from one device to another device. The refer message is sent from one device (the device that desires to be transferred) to a new device that is not part of the original communication session. If the new device accepts the invitation request (it is willing to accept the transfer request), the transfer device informs the device that it is communicating with of the desire to transfer the session to a new device along with the parameters (such as the new device network address) that will be used by the existing device to connect to the new device. After the two devices have successfully established a new communication session, the transfer device is no longer part of the original communication session. The refer message is defined in session initiation protocol (SIP) toolkit.

Reference Clock-A reference clock is a timing signal that is used as the basis for other systems or timing signals that or used in a system or communication link.

Reference Model-A reference model is a representation of a product, system or process in a form that can simulate or assist in the understanding of the operation and estimation of likely results provided specific inputs or events that occur to the product, system or process.

Reference Test Leads-Reference test leads are wires or optical cables with stable characteristics that are used in testing configurations. Reference test leads include measured performance characteristics (e.g. measured insertion loss values) that can be used in testing configurations.

Reflected Multicast-A reflected multicast is the process of receiving and the simultaneous retransmission of media to many other viewers.

Reflecting Satellite-A satellite intended to reflect radiocommunication signals.

Reflection-Reflection is the sharp changing of the direction of a sound wave, radio wave, or lightwave without as it passes from a medium of one density to a medium of another density. When a traveling sound wave, radio wave or light wave encounters certain types of changes in the properties of the medium through which the wave travels, a reflected wave will be generated. A reflected wave is generated if the wave impedance is different just before and just after the region of interest. Wave impedance of a sound wave is the ratio of the sound pressure to the acoustic fluid velocity in the wave. Wave impedance of an electromagnetic wave is the ratio of the transverse electric field intensity to the transverse magnetic induction intensity in the same polarization direction. A different acoustic wave impedance is the result of a different mass density and/or a stiffer or less stiff compressibility of the acoustic medium. A different electromagnetic wave impedance is the result of a different dielectric permittivity (usually) and/or a different mag-

netic permeability of the medium. These differences may in turn be the result of a higher or lower mass density of the medium for the case of a compressible fluid. Since many of these same properties also affect the wave speed, the reflection properties can sometimes be described in terms of the change in wave speed instead of being described in terms of a change in wave impedance. When the wave impedance on both sides of the region of interest are exactly the same value, there is no reflected wave power. The fraction of the wave amplitude that is reflected can be described by a reflection coefficient, which approaches 1 in absolute magnitude (that is, 100% reflection) in the two extreme cases where the ratio of the wave impedance before and after the region of interest approaches either zero or infinity. Full 100% reflection can also theoretically occur from a perfectly conducting surface for an electromagnetic wave, or from a perfectly rigid reflecting surface for an acoustic wave.

Reflector-(1-radio) The metal elements placed behind the active element of an antenna to make it directive (2-datacom) A computer or electronic device that receives and retransmits packets to other communication devices.

Reframing-The process of recovering or relocating the framing pulse in a bit stream after a mis-frame.

Regeneration-Regeneration is the process of reception and restoration of a signal, digital pulse or lightwave signal to its original form after its amplitude, waveform, or timing have been degraded during transmission.

This figure shows the process of digital signal regeneration. This example shows how an original digital signal (a) has a noise signal (b) added to it that produces a combined digital signal with noise (c). The regeneration process detects maximum and minimum expected values (threshold points) and recreates the original digital signal (d).

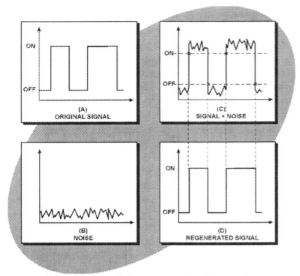

Digital Signal Regeneration Operation

Regenerative Repeater-A repeater that receives, amplifies, reshapes, and retransmits digital signals.

Regenerator-A receiver and transmitter combination used in fiber optic systems to reconstruct signals for digital transmission. The receiver converts incoming optical pulses to electrical signals, decides whether the pulses are logic 1s or logic 0s, regenerates the electrical pulses, and converts them to an optical signal for further transmission on the fiber.

Regional Data Center (RDC)-A regional data center (RDC) is a collocation facility that allows a carrier interconnection point-of-presence (POP) within a rate center to provide information services.

Regional Headend-A regional headend is part of a broadcast system that selects and processes video signals into television network locations with a regional system.

Register Message-A message that is used to inform a service provider or communication system that a user is connected to the system. The register message usually contains information about the connection of the device including the address (or temporary address) that the device is using. The register message is defined in session initiation protocol (SIP).

Registrar Server-A registrar is a server that accepts registrations from users and places these registrations, (which is essentially location infor-

mation), in a database known as a location service. The process of registration associates a user with a particular location, (IP address); this association is known as a 'binding' in SIP. When there is an incoming session for a user within a domain, the proxy server will interrogate the location server to determine the route for the signaling messages.

Registration-(1- FCC) A legally required procedure whereby vendors must submit their telephone equipment for testing and certification before it can be directly connected to a public telephone network. (2-wireless system notification) A process where a mobile radio transmits information to a wireless system that informs it that it is available and operating in the system. This allows the system to send paging alerts and command messages to the mobile radio. (3-VoIP) The process of an Internet telephone terminal registering with a gatekeeper. (4-Internet) The process of a web site visitor entering information into a web site to record their contact information (name, address, and email address).

Registry-A database of information that contains computer's hardware and software information that is used by it's operating system to manage communication between hardware and software applications. Information store in the registry is usually updated automatically as configuration information changes through the addition, removal, or modification of hardware and software controlled by the system. The registry database holds information that was commonly stored in initialization files (.INI) files.

Regular 8 mm Film-Regular 8mm film is 8 mm wide material that is used to store moving images. 8 mm film was commonly used for consumer movie cameras and projectors before videotape was available.

Regulation-(1-electrical quantity) A process of adjusting the parameters of some signal or system (such as circuit gain). (2- government) Rules established by a government agency that are designed to maintain the service public communications systems.

Regulator-(1-gas pressure) A control device that acts as an interface between gas in a high-pressure tank and the low-pressure gas in a cable pressurization system. (2-voltage) A device that maintains its output voltage at a constant level. (3-government) An agency or department of a government that establishes, changes, or enforces rules and regulations.

Regulatory Authority (RA)-The regulatory authority (RA) is a government agency that establishes and enforces laws and regulations regarding radio and wired communications services within a country. The RA must certify (radio and computer equipment before it can be sold.

Re-Invite-A message that is used to inform a person or device that the communication session parameters of the original communication session invite request have changed. The re-invite message usually contains information about the new communication parameters such as the bandwidth and media types. The re-invite message is defined in session initiation protocol (SIP).

REL-Rights Expression Language

Relay-(1-general) A device by which current flowing in one circuit causes contacts to operate that control the flow of current in another circuit. (2-general) The process of re-transmitting signals through an intermediate system. A relay point is a place in a network where a signal is "repeated" after being modified or regenerated to remove impairments or other communication degrading effects, as in a telegraph or digital repeater device, or in a telecommunications relay service (TRS) for the deaf. (3-hermetically sealed) A relay in which the contacts are sealed in an airtight glass or metal enclosure. (4-latching) A relay that latches into its operated position and is held there without the need for a holding current. Release is accomplished by energizing a release winding. (5 - mercury-wetted) A relay with contacts in sealed en-closures constantly wetted by mercury to provide a low-resistance contact. (6 - multi-contact) A relay in which a large number of spring sets are operated simultaneously. (7 - overload) A relay that operates only when current in a circuit exceeds a specified value. (8 - reed) A relay in which ferromagnetic reeds are sealed in small glass tubes and surrounded by operating coils. (9- solid-state) A device in which a current or voltage in one circuit controls the switching of another circuit, but involves no mechanical movement, armatures, moving contacts, or reeds. (10 - thermal) A heat operated relay that normally depends on the bending of a bimetallic strip for contact closure. (11 - radio system) An intermediate station on a multi-hop radio system.

Relay Server-A relay server is a computer or data server that is used to retransmit a data signal without altering the data packets in any way. Relay servers may be used to serve (relay) streaming media through firewalls.

Release-A signal that indicates that a line or circuit has been released from use.

Release Patterns-Web of distribution starting from initial release time and place(s), extended to secondary time and place(s), and so forth, in which a motion picture debuts.

Reliability-The ability of a network or equipment to perform within its normal operating parameters to provide a specific quality level of service. Reliability can be measured as a minimum performance rating over a specified interval of time. These parameters include bit error rate, minimum data transfer capacity or mean time between equipment failures (MTBF).

Remote Access Dial In User Service (RADIUS)-Radius is a network protocol that can receive identification information from a potential user of a network service, authenticates the identity of the user, validates the authorization to use the requested service and creates event information for accounting purposes. RADIUS is specified in RFC's 2138 and 2139, RADIUS is a client/server protocol that uses UDP.

Remote Access Server (RAS)-Remote access server (RAS) is software that is a part of Microsoft Windows NT Server that allows remote users or devices to connect to a server and access resources through a data network connection (such as through the Internet or through a data modem).

Remote Console-The computer access device that allows an operator to communicate with or control a network from a remote location (not directly connected to the network). A remote console commonly consists of a display monitor, keyboard, data modem or other interconnection line.

Remote Digital Terminal (RDT)-The RDT provides an interface between a high speed digital transmission line (e.g. DS1) and the customer's access line. The RDT can dynamically assign time slots from a high speed line to customer access lines. Because customer access lines are not used at the same time, an RDT that interfaces to a DS1 line usually provides service to 96 customer access lines.

The RDT is divided into three major parts; digital transmission facility interface, common system interface and line interface. The digital transmission interface terminates the high speed line and coordinates the signaling. The common system interface performs the multiplexing/de-multiplexing, signaling insertion and extraction. The line interface contains digital to analog conversions (if the access line is analog) or digital formatting (if the line is digital).

Remote Monitoring Specification (RMON)- Remote monitoring specification is a protocol that allows network management software to configure, poll and trap events on network elements.

Remote network MONitoring (RMON)-Remote network MONitoring or RMON is in essence just another SNMP MIB, but allows a network management station (NMS) to request that an agent do many types of functions. There are 9 functions or groups associated with RMON as originally defined in RFC1757: Ethernet statistics, history statistics, alarm and threshold setup, MAC-layer host statistics, busiest hosts, MAC-layer conversation pair statistics, filters, capture packets, and event and alarm notification. This is also known as RMONv1. RFC2819 now obsoletes the original RFC1757.

Remultiplexing-Remultiplexing is a process that recombines multiple communication channels into a single transmission path after it has been previously multiplexed. An example of remultiplexing is the recombining of television channels into a transport stream after a multichannel transport stream has been received, separated and decoded from a satellite or network feed.

REN-Regional Enterprise Network

REN-Ringer Equivalence Number

Render Right-Render rights are the authorizations to view, print or listen to content.

Rendering-Rendering is the process of converting media into a form that a human can view, hear or sense. An example of rendering is the conversion of a data file into an image that is displayed on a computer monitor.

Rendering Application-A rendering application is a software program that converts digital content into human-understandable form.

Rendezvous Point (RP)-A rendezvous point is a data network address of a node (such as a router) in a communications network where one or multiple data sources will be directed to so they can be retransmitted to other points in a multicast distribution tree.

Renewing Certificate-A renewing certificate is an authorization command and necessary information (such as decryption keys) that enables or continues the allowance of use of media or information.

Reorder Tone-A tone applied 120 times per minute indicating that all switching paths or trunks are busy. The reorder tone also is called a channel busy or fast busy tone.

Repeat Dialing-A service that automatically dials the last dialed number a repeated number of times or until an event occurs.

Repeatability-Repeatability is a measure of the ability of a device or connector to provide the same performance it is repeatedly operated (e.g. connected and disconnected). For example, Contact connectors offer relatively low insertion loss. However, over repeated uses, the insertion loss of the connector will rise due to the scratching and damaging of the connector surfaces.

Repeater-(1-general) A device or circuit that is located between transmitting and receiving devices to improve the quality the signal that is delivered between them. A repeater obtains some or all of the signal from the transmitter, amplifies and may adjust (change a frequency) or filter the signal, and retransmits the signal to the receiver(s). (2-LAN) In a local area network, a device which operates at layer 1 (physical layer) of the OSI reference model. This device does not inspect packets, but instead regenerates all input signals on its output(s). Repeaters were common in shared-media Ethernet based on IEEE 802.3 10-Base-2 and 10-Base-5 protocols. In recent years, the need for repeaters has been greatly diminished as new physical layer transmission technologies have provided better transmission capabilities.

Repeater Spacing-Repeater spacing is the distance between signal repeaters in a communication system. Larger the maximum repeater spacing distance capability reduces the number of repeaters that are necessary for communication lines.

Replay Attack-A replay attack is an attempt to gain unauthorized access into networks and/or computing devices by replaying captured information such as user identification codes, passwords and other forms of digital credentials.

Reports-A report is a display of information that represents specific aspects of data.

Reprint Rights-Reprint rights are the permissions granted from an owner of content to reprint the content in specific formats (such as reprinting a magazine article).

Reproduction Rights Organization (RRO)-A reproduction rights organization is an agency that is authorized to administer reproduction rights for intellectual property (content).

Republocrat-A political group or viewpoint that both individuals and companies are responsible for their actions and how their actions affect others.

Repurpose Content-Repurposing content is the process of using segments or converting content into a form that can be used for a purpose that is different than the original purpose the content was created for.

Request For Comments (RFC)-A requirement or draft standard document created by a standards body that solicits comments from manufacturers, carriers and industry experts to finalize the standard. When used by the Internet engineering task force (IETF) every major Internet Protocol is specified first by an RFC. There are many RFC documents available and they are a significant method used to define Internet protocols and technical standards.

Request For Information (RFI)-A formal statement of requirements for a technical system's acquisition that is submitted to vendors for information on the feasibility of designed concepts.

Request For Proposal (RFP)-A formal statement of requirements and technical specifications that constitute user needs when actually procuring a system and seeking bid information for the evaluation and selection of services or products.

This figure shows an outline of key elements of an RFP. This diagram shows that an RFP is typically starts with a general overview and the scope of what is being requested. This is usually followed by several pages that detail key requirements such as product types, services, and support needs. The RFP also includes the terms & conditions that define how the vendor should respond such as who to respond to (contact information), what format to respond (printed and/or electronic file formats), and key response submission dates.

Request For Quotation (RFQ)-A document that is prepared by a manufacturer, carrier, or other company that defines the needs of a company and solicits a response for quotation for equipment and services that fill the requirements needs as defined in the RFQ.

Request To Send (RTS)-A control signal on a communication line that indicates the device is requesting to send communication information.

Required Availability-Required availability is a measure of the amount of time that a service must be available provided specific conditions exist.

Reroute-Reroute is the process of transferring connections or circuits from one path (route) to an alternate path.

Rerouting-Rerouting is the process of transferring connections or circuits from one path (a route) to an alternate path (a new route). Rerouting may result from loading or unloading of communication circuits through a switching system or to establish new connections after a connection has been lost.

Resale Rights-Resale rights are the authorizations that allow a person or company to resell a product, service or content. Resale rights may specify the formats which a product may be sold (e.g. eBook) and where the product may be sold (geographic regions).

Research In Advanced Communications In Europe (RACE)-A cooperative research program started in Europe commissioned to develop the technology for Broadband Integrated Services Digital Network (B-ISDN) systems. The RACE members are reviewing a Mobile Broadband System operating in the 60 GHz bands for mobile service applications in the approximate range of 2-100 Mb/s.

Reseller-A reseller buys network services in bulk from an existing carrier for resale to the public or other customers. The reseller provides sales and support services to the customer and the customer usually pays the reseller for the communication services it receives.

Reserved Bandwidth-Reserved bandwidth is the allocation of a portion of transmission or processing capability of a device or system for specific users, purposes or applications.

Reset-The act of restoring a device to its default or original state. Reset also may refer to restoring a counter or logic device to a known state, often a zero output.

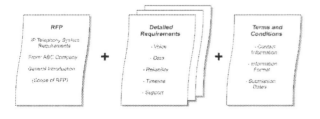

RFP Elements

Reset-Reboot

Residential Gateway-Residential gateways facilitate the circuit-to-packet conversion from analog endpoints in the home out to an IP network. Typical devices include IADs, cable modems, DSL modems, and wireless broadband units.

Resilient Packet Ring-An emerging network architecture and technology designed to meet the requirements of a packet-based metropolitan area network. Standardized by the IEEE 802.17 committee, RPR is designed to effectively manage a shared metro fiber ring by transporting packets. Unlike SONET, which uses TDM, RPR is designed to interoperate with packet-based networks such as Ethernet, while reducing the non-deterministic nature of metro Ethernet.

Resin-A resin is a material or compound that can bind together to create reinforcement fibers.

Resistance-(1-general) The opposition of a material to the flow of electric current. Resistance is equal to the voltage drop through a given material divided by the current flow through it. The standard unit of resistance is the ohm, named for the German physicist Georg Simon Ohm (1787-1854). (2-antenna) The impedance of an antenna at its operating frequency. (3-contact) The resistance measured across a pair of contacts. Mercury-wetted contacts normally will have zero contact resistance. Good, dry contacts have resistance measured in milliohms. (4-effective) At all alternating frequencies, the total resistance, including dc resistance and resistance from the skin effect, eddy currents, hysteresis, and dielectric losses. (5-insulation) The resistance offered to the flow of dc through insulation. (6-leakage) The resistance of a path, normally to

Resistive Load-A load in which the sine wave voltage is in phase with the current.

Resistor-(1-general) A device whose primary function is to introduce resistance into an electric circuit. (2-carbon) A small cylindrical resistor made of carbon. (3-current limiting) A resistor, inserted in a circuit for safety reasons, that limits the value of the current flow under a fault condition. (4-noninductive) A resistor wound in such a way as to minimize inductance.

Resolution-(1-display) The number of pixels per unit of area. A display with a finer grid contains more pixels, and therefore has a higher resolution, capable of reproducing more detail in an image. (2-digital) The number of bits into which an analog signal has been converted. The greater the number of bits, the more accurate the digital representation. (3-horizontal) The amount of resolvable detail in the horizontal direction in a picture. Horizontal resolution usually is expressed as the number of distinct vertical lines, alternately black and white, that can be seen in three quarters of the width of the picture. This information usually is derived by observation of the vertical wedge of a test pattern. A picture that is sharp, clear, and detailed has good (high) resolution. If the picture is soft, blurred, and does not show small details distinctly, it has poor (low) resolution. Horizontal resolution depends upon the high frequency amplitude and phase response of the pickup equipment, the transmission medium, and the picture monitor, as well as the size of the scanning spots. (4 - instrument) The smallest increment that can be acted upon, or read from a measuring instrument. (5- vertical) The amount of resolvable detail in the vertical direction in a picture. Vertical resolution usually is expressed as the number of distinct horizontal lines, alternately black and white, that can be seen in a test pattern. Vertical resolution is fixed primarily by the number of horizontal scanning lines per frame. Beyond this, it depends on the size and shape of the scanning spot of the pickup equipment and picture monitor.

Resonance-(1-general) A tuned condition conducive to oscillation, when the reactance resulting from capacitance in a circuit is equal in value to the reactance resulting from inductance. (2-parallel) The condition in a circuit with capacitance and inductance in parallel, when the frequency is such that the current entering the circuit from outside is in phase with the voltage across the parallel circuit. (3-series) The condition in a circuit with capacitance and inductance in series, when the frequency is such that the current through the circuit is in phase with the voltage across the circuit.

Resonator-A resonant cavity.

Resource Availability-The amount of available system capacity to provide services to additional users. Resource availability may be measured by amount of data transfer (Mbps), number of users, or number of calls in a specific time period.

Resource Availability Indicator (RAI)-A message that indicates the ability of a gateway to provide more communication services. A RAI message is defined in the H.323 packet voice system.

Resource Reservation Protocol (RSVP)-RSVP is a quality of service (QoS) protocol that is used to control the amount of bandwidth dedicated to packet flow for specific communication sessions in a packet data network. RSVP is primarily used in real-time communication sessions (such as voice over packet).

Response-(1-message) A reply to a query. (2-audio) The fidelity with which equipment reproduces an audio signal. (3-frequency) The gain or loss of a system over its specified frequency band. (4-spurious) The response of tuned equipment to an undesired signal at a frequency to which the equipment is not tuned. (5-transient) The response of an amplifier or other circuit to high-frequency signals, such as those represented by a square wave test signal.

Response Time-The elapsed time between the generation of an inquiry and the receipt of a reply in a communications system. The response time includes the transmission time, processing time, time for searching records and files to obtain relevant data, and transmission time back to the inquirer. In a data system, it is the elapsed time between the end of transmission of an inquiry message and the beginning of the receipt of a response message, measured at the inquiry originating station.

Response Time Reporter (RTR)-RTR is a feature set found on Cisco routers that allows them to send ICMP pings and other types of packets that devices may respond to. RTR can be configured with thresholds for event and alarm notifications, similar to RMON.

Responsivity-Responsivity is a measure of the sensitivity of a photosensor. Responsivity is the ratio of the output current or voltage to the input flux in watts or lumens. When responsivity is indicated at a particular wavelength (in amperes/watt), it denotes the spectral response of the device.

Restoration-The repair or returning to service of one or more telecommunication services that have experienced a service outage or are unusable for any reason, including a damaged or impaired telecommunications facility. Such repair or returning to service may be done by patching, rerouting, substitution of component parts or pathways, and other means, as determined necessary by a service vendor.

Retaining Nut-A retaining nut is a threaded adapter that is used to secure devices or receptacles to assemblies and connectors.

Retention-Retention is the process of actively or passively preventing customers from leaving and switching to other service providers or products. A measurement of retention is churn rate.

Retrace-Flyback

Retransmission-A method of network error control in which hosts receiving messages acknowledge the receipt of correct messages and either do not acknowledge, or acknowledge in the negative, the receipt of incorrect messages. The lack of acknowledgment, or receipt of negative acknowledgment, is an indication to the sending host that it should transmit the failed message again.

Retransmission Interval-A retransmission interval is the amount of time that is assigned before a communication device is allowed to attempt access again in a random access contention based communication system (such as a mobile telephone or Ethernet system) after an access transmission attempt has failed. Retransmission intervals typically increased based on the number of failed access attempts and are slightly varied (randomized) to minimize the possibility of multiple devices from reattempting to access the system at the same time.

Re-transmission Rights-Re-transmission rights are the authorization(s) that allow the retransmission of content through a distribution network. Retransmission rights typically specify compensation levels for the owner(s) of the content that will be distributed.

Retrial-Any subsequent attempt by a customer, operator, or switching system to complete a call within a measurement period.

Retrofitting-Retrofitting is the conversion of a product or system to allow it to have new capabilities.

Return Loss-The difference, usually expressed in decibels (dB) compared between the incident voltage, current or optical signal on a transmission line (forward signal) and the reflected current, voltage or optical signal as measured at a particular point (e.g. termination point or impedance discontinuity).

Return on Investment (ROI)-A financial measurement that compares the profit with the original investment. ROI is to evaluate the impact of an investment on the telephone company's profitability or operational efficiency. Return on Investment is reported in terms of dollars spent compared to benefits gained.

Return on Marketing Activity (RoMA)-A financial measurement that is utilized to evaluate the effectiveness of a marketing activity. Return on Marketing Activity is reported in terms of dollars spent / market share realized.

Return To Zero (RZ)-A data format in which the logic level for a data 1 is a 1 during the time the data clock is high, but returns to 0 during the time the data clock is low. For a data 0, the logic level is 0 for both high and low states of the data clock.

This diagram shows that return to zero (RZ) encoding uses the logical level voltage during half of the bit period of each logical bit. In this example, the data is transmitted at 1 kbps so each logical bit period is 1 msec. During this entire period, the logical level remains at the same voltage associated with the logical level for 1/2 a bit period (0.5 msec).

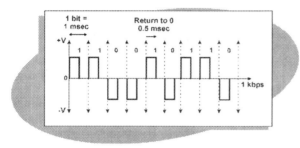

Return to Zero (RZ) Operation

Reuse-The sharing of identical radio channels in two or more cells of a cellular network. Sufficient cell separation is required to obtain adequate signal to noise ratio.

Revenue Assurance-Revenue assurance is reviewing of systems and records that are associated with revenue streams to ensure billable services are correctly recorded and collected.

Revenue Sharing-Revenue sharing is the process of transferring revenues generated by one or more companies to one or more other companies for the exchange of products or services. The products or services that are provided by the recipient(s) of revenue sharing may or may not be related to the creation of the revenues that are shared. Revenue sharing is commonly performed in the communications industry between companies that provide revenue generating services to customers and companies that provide underlying services necessary to

provide for these services. An example of revenue sharing is the access fees that are paid to local telephone companies (LECs) by long distance interexchange companies (IXCs). The long distance revenues of the IXCs are shared with the LECs in return for the LECs providing telephone access between the end-user and the IXC.

Revenue Stream-A revenue stream is a source of money or value received that is associated with the sale or providing of services or products or can be associated with a specific type of sales or marketing process.

Revenue Volume Pricing Plan-Revenue volume pricing plans apply discounts to billing charges or usage fees that are based on volume of revenue amounts.

Reverberation-The persistence of sound resulting from repeated echoes, as in a large hall, after the sound source has stopped.

Reverse Address Resolution Protocol (RARP)-An Internet protocol that controls the translation of a data link control (DLC) address to an Internet protocol (IP) address. RARP allows a computer to obtain an IP address from a server when only the hardware address is known.

Reverse Channel-Reverse channels are the radio channels (radio links) that are used by a mobile device to transmit to a base station (also called an uplink).

Reverse Engineering-Reverse engineering is the process of studying, testing, evaluating and analyzing a product to decipher how it works.

Reverse Link-The portion of a communication link used for transmission of signals from a mobile to a base station.

Reverse Path Forwarding (RPF)-Reverse path forwarding is the process of forwarding the transmission of data packets to paths that are downstream (in a forward direction) in a data communication system. Reverse path forwarding operates by reviewing the source and destination addresses of the data packet to its routing table to determine if the destination address is not located between the source and the current router address. If the address is not between the source and the current router (in the forward path), the packet can be forwarded towards its destination.

Reverse Path Forwarding Check (RPF Check)-Reverse path forwarding check is the process of comparing the source of a packet and the incoming port or path to determine if the packet has been sent from a router that is upstream (between the router and the source) or downstream (from a router that is after the router). RPF is used to only accept multicast packet requests from upstream routers to avoid the possibility of data communication loops.

This figure shows how a reverse path forwarding check is used to prevent the potential for data communication loops in multicast networks. This example shows a data communication network that has multiple media servers (M1 and M2) that are connected to the data network. Media server 1 has initiated a multicast session and routers in the network receive a multicast request packet. As the multicast request packet is received by each router, the multicast routing table is checked to determine if the packet was received from a source that is upstream (correct path) or downstream. If the packet is received on a port that is downstream, the packet is discarded.

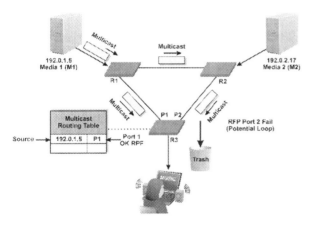

Multicast RPF Check

Rework-Rework is the process of repairing or modifying of assemblies that have already passed through a production process with failures or poor performance.

Rewrite-Action taken to revise an original product, and any other revisions made there after, as to perfect the final production

RF-Radio Frequency

RF Channel-An RF channel is a communication link that uses radio signals to transfer information between two points.

RF Energy-Radio Frequency

RFBP-Request For Budgetary Proposal

RFC-Request For Comments

RFCOMM-Radio Frequency Communication Port

RFI-Radio Frequency Interference

RFI-Request For Information

RFID-Radio Frequency Identification Tag

RFP-Request For Proposal

RFQ-Request For Quotation

RGB-Red, Green, Blue,

RGB System-The basic parallel component set (red, green, and blue) in which a signal is used for each primary color. The same signals also may be called GBR in recognition of the mechanical sequence of connections in the SMFTE interconnect standard.

RIAA-Recording Industry Association of America

RIC-Residual Interconnection Charge

Rich Media-Rich media is the use of video, graphics and/or audio media to enhance the information experience of the media viewer.

Right-A right is a legally recognized authorization or entitlement to use, transfer or display content.

Right of Way (ROW)-Right of way is the authorization provided by a government or a property owner to install, build, or maintain communication lines and/or facilities.

Rights Clauses-Rights clauses are specific terms and their requirements (such as financial consideration) that are associated with the rights (ability to use, distribute or sell) associated with a product, service or intellectual property (content).

Rights Clearance-Rights clearance is the obtaining of rights from a rightsholder. Rights clearance may be in the form of a contract, permission letter or through a financial transaction.

Rights Expression Language (REL)-Rights expression language (REL) is a protocol that is used to specify rights to content, fees or other consideration required to secure those rights, types of users qualified to obtain those rights, and other associated information necessary to enable e-commerce transactions in content rights.

Rights Holder-A rights holder is a person or company who has obtained authorization to exercise (use) the rights associated with a specific item or media content.

Rights Management-Rights management is a process of organization, access control and assignment of authorized uses (rights) of content. Rights management may involve the control of physical access to information, identity validation (authentication), service authorization, and media protection (encryption). Rights management systems are typically incorporated or integrated with other systems such as content management system, billing systems, and royalty management.

This figure shows basic rights management processes. This example shows that rights management can involve rendering, transferring and derivative rights. Rendering rights may include displaying, printing or listening to media. Transfer rights may include copying, moving or loaning media. Derivative rights can include extraction, inserting (embedding) or editing media.

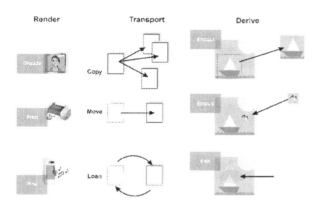

Rights Management

Rights Metadata-Rights metadata is descriptive information that defines the rights associated with specific content (e.g. a book or movie) and how the rights may be used.

Rights Model-Rights model is a representation of how the content rights are transferred and managed between companies or persons.

Rights Specification-Rights specification is the authorized usage rights for the individual aspects of intellectual property (content). Rights specification define the ability to render (display), transport (copy and send) and derive (modify or use portions) for a specific content item.

Rights Specification Language-Rights specification language is a set of commands and associated parameters that can be decoded to represent the

specific rights and authorizations of use for intellectual property (content).

Rights Transaction-Rights transaction is the transfer of rights from a content owner to a content user or distributor. A rights transaction may involve the use of a formal agreement (e.g. a publishing agreement) or it may occur through an action (e.g. a customer buying a book).

Rightsholder-A rightsholder is a company or person that has been given authorization to a set of rights for a specific amount or form of content.

Ring-(1-wire) One of the two subscriber loop/line wires. It is connected to the so-called ring conductor on a manual switchboard plug, hence its name. In North America, red insulation color is used to identify it. Corresponds to the European "B" subscriber wire. (2-audio) Audible alerting signal at subscriber set indicating incoming telephone call. May be a ringing bell or other sound.

Ring Generator-A component of virtually all phone systems, ranging from large offices to small key systems, that supplies the power to ring the bells inside phones, typically 90 volts AC at 20 Hz.

Ring Network-A network where each communication device (typically computers) are connected to a neighboring computer and the interconnected computers form a ring where the last computer in the string is connected to the first computer in the network. The data transmission process in a ring network involves computers passing all network data from its neighboring computer to the next neighboring computer. When a computer in the ring network receives data, it looks for data information designated for its address and removes (does not retransmit) the data.

This figure shows a ring network. In this diagram, each data communication device receives information from one computer in the ring and sends (and possibly forwards received information) information to another computer in the ring.

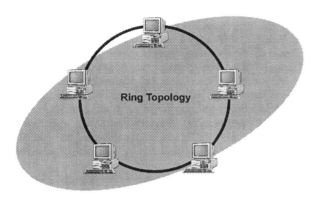

Ring Topology System

Ring Tone-The tone that is used to announce an incoming telecommunications call. There may be several different types of ring tones and some telecommunications devices allow the user to program in their own unique selection for a ring tone.

Ring Topology-A design for a local area network (LAN) formed by placing stations along a closed loop, connected by sections of a medium, such as optical fiber or copper wire. Data circulates around the ring from station to station. Some networks, such as Token Ring Networks, utilize a logical ring topology while being physically wired in a star topology. The drawback to a physically wired ring topology is the inability to quickly "wire around" a failing station. Star wiring removes this drawback. (See also: bus topology, star topology, tree topology.)

Ringback-An indication (usually audible) that the phone at the far end is ringing. Ringing may be either local or remote.

Ringer-A bell, or an electronic equivalent, that responds to an alternating current signal to produce a ringing, or audible alerting sound, from a telephone set.

Ringer Equivalence Number (REN)-A ringer equivalence number is a rating (measurement) of the amount of energy a telephone ringer or group of telephone ringers consume when the telephone device is ringing. This rating typically indicates the energy required for a mechanical ringer. A ringer equivalence of 1 indicates the device consumes the energy equivalent to one mechanical ringer. A REN of 0 means the device consumes no power (it only senses the ring signal) when it is ringing.

Ringing-(1-telephone) An alternating-current signal that rings a bell or other sounder in a telephone set to alert a subscriber to an incoming call. A ringing voltage is applied to a customer loop and, through ringers or ring-up circuitry, provides an audible and/or visual signal to the called person.

Ringing Generator-A circuit or device in a phone system that generates the ringer signal. The ringer signal voltage usually ranges from 90 to 115 VAC (typically 105) with a frequency of 30 Hertz.

RIP-Routing Information Protocol

Ripcord-Ripcord is a thread or a string that is located below the jacket of a cable that allows for the splitting of the cable jacket when it is pulled. Using the ripcord exposes the inner conductors of the cable without damaging them.

Ripping Media-Ripping media is the process of extracting (ripping it from its source) or storing media as it is streamed.

This figure shows how users may rip media from its original packaging to alter its form and potentially change or eliminate rights management attributes. This example shows that an audio CD is inserted into a personal computer that has had ripping software installed. The ripping software instructs the microprocessor in the computer to make a copy of the data that is being played to the sound card (audio chip) in the computer and to store this data on the storage disk in the computer.

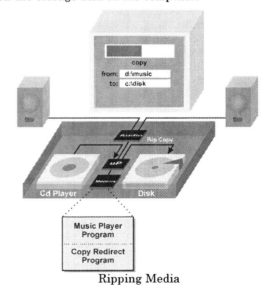

Ripping Media

Ripple-An AC voltage superimposed on the output of a DC power supply, usually resulting from imperfect filtering.

RISC-Reduced Instruction Set Computer

Rise Time-Rise time is the amount of time that it takes a signal to rise from a low level to its high level (e.g. peak) value. Rise time is usually measured when the signal increases between 10% to 90% of maximum output.

Riser Cable-Riser cable is high-strength wiring that is specifically designed for vertical installation (such as in building riser shafts). For fiber lines, riser cables use tightly buffered tubes to keep the weight of the fibers from pulling and potentially snapping the fiber cables.

RJ-11-A modular connector that has 2 to 6 conductors that is commonly used to interconnect end-user telephone equipment.

RJ-12-A six-wire or three-pair connection using the same connector as RJ-11.

RJ-14-A jack that looks and is exactly like the standard RJ-11 that you see on every single line telephone. Where as the RJ-11 defines one line — with the two center, red and green, conductors being tip and ring, the RJ-14 defines two phone lines. One of the lines is the "normal" RJ-11 line — the red and green center conductors.

RJ-45-A standard 8 wire modular connector. RJ-45 connectors are commonly used in telephone and data communication systems.

RLC-Radio Link Control Protocol

RLC-Run Length Coding

RLL-Radio Local Loop

RLP-Radio Link Protocol

RLV-Re-Usable Launch Vehicle

RM-Real Media

RMON-Remote Monitoring Specification

RMON-Remote network MONitoring

RMONv2-RMONv2 is an extension to the original Remote network MONintoring (RMON) standard as defined in RFC1757. RMONv2 introduces 9 more groups and functions, mostly concentrating on higher layer protocol statistics: protocols supported by the device, statistics per protocol, MAC to higher-layer protocol mapping, higher-layer host statistics, higher-layer conversation pairs, application-layer host statistics, application-layer conversation pairs, history statistics for user-defined criteria, and configuration of RMON features. RMONv2 is defined in RFC2021.

RNC-Radio Network Controller

Roamer-A subscriber operating in a system (such as a cellular system) other than its registered home system.

Roamer Validation-Roamer validation is the verification of a visiting user's (such as visiting user in a mobile telephone system) identity using registered subscriber information. Validation is necessary to limit fraudulent use of a service. The two types of roamer validation are "post-call" and "pre-call". Post-call validation occurs after a call is complete, and pre-call validation occurs before granting access to the system.

Roaming-Roaming is the capability to move from one carrier's system area to another carrier's service area and obtain service. While it is desirable to roam without loosing functionality of the phone or device, some communication systems offer advanced features (such as high speed data) that other systems may not have installed. This may limit the operation of advanced features.

Roaming Agreement-A roaming agreement is a service agreement between service providers that defines the services that will be provided to subscribers who are visiting in each others systems and the charges that will be charged to each carrier for providing these services.

Roaming Fees-Roaming fees are the billing charges to a customer for the usage of a product or service while operating in another service providers' communication system. Roaming fees may be composed of daily access charges and usage charges.

Robbed-Bit Signaling-Robbed-bit signaling is used to convey supervisory (idle vs. busy channel) status in some installations of the DS-1 (T-1) digital multiplexing carrier for pulse code modulation (PCM), used in North America and Japan. First a synchronized sequence of either 12 frames or 24 frames (called extended super frame [ESF]) is established by means of a specific binary sequence of 0 and 1 signals in the framing bit of each frame. Then the eighth bit of every channel of the sixth and twelfth frames (and also the 18th and 24th frames in ESF installations) is preempted, and typically replaced with a bit whose value (0 or 1) indicates the idle vs. busy (on-hook vs. off-hook)status of that channel. In some types of signaling (example, E&M tie line signaling) these bits have other meanings as well. Robbed bit signaling is not necessary when common channel signaling or associated bit signaling is used.

Robot (Bot)-A software application that works with the world wide web that automatically locates and gathers information within Web sites that meet specific criteria. Bots can be used to create databases of Web sites of specific categories.

ROI-Return on Investment

Rollout-Rollout is the introduction of a product or service into the marketplace.

Rollover Lines-Telephone lines that are placed in a hunt sequence where calls will move to the next available line when those higher in the hunt group are found busy. This also refers to the movement of inbound call traffic from one telephone switch or ACD queue to another based on current busy conditions or programmed instructions within the telephone switch.

ROM-Read Only Memory

RoMA-Return on Marketing Activity

Root-A root is the main focal point in a system or the base of a tree structure or hierarchy system.

Root Directory-A root directory is the highest level directory from which all other sub directories must branch from.

Root DNS-Root Domain Name Server

Root Domain Name Server (Root DNS)-A root domain name server is the highest level directory that identifies top level domains and their associated IP addresses.

Rotary Dial-Rotary dialing is a process of sending dialed digits through the use of a spring-loaded mechanical switch that produces pulses as it rotates through 10 positions (1 through 9, and 0). As the rotary dial turns, a switch briefly interrupts the loop current. The number of pulses per rotation is counted to determine the number dialed. A time pause between rotary dials is used to determine when additional digits are dialed or when the caller is finished dialing.

This diagram shows how a mechanical rotary switch can be used to gather dialed digit information. In this example, a spring-loaded rotary dial is turned to produces pulses that represent the numbers on the dial. As the rotary dial turns, a contact switch briefly interrupts the loop current. The line card in the telephone system counts number of pulses to determine the number dialed. This diagram shows that after a specified time pause occurs after a series of pulses, the counter resets and is ready to count another dialed digit.

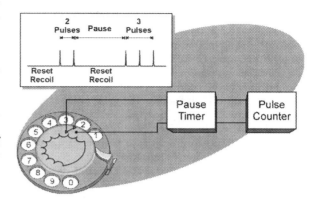

Rotary Dial Operation

Rotary Joint-Rotary joints are connection devices that allow signals to continually pass through an object that can revolve or rotate.

Rotate-Rotate is a graphics editing function that allows an image object to be turned (rotated) on one or more axes.

Rotor-The rotating part of an electric generator or motor.

Rounding Error-Rounding error is the amount of error that results from the conversion of an analog value (level) into a digital value (level) that does not exactly match the original analog level. Rounding error occurs because the limited number of discrete values of the digital signal force the level to be converted (rounded) to the closest digital level (number).

Roundtrip Delay-The time required for a wave to travel from a signal source and return from either the distant end of a circuit or an impedance mismatch within the circuit.

Route Flapping-The continual changing of a network connection path that results from a intermittent congestion or loss of circuit connection that indicates to the current router connection path that there is a loss in connection or that a better connection path exists. This causes the packet routing path to continually change. These different paths can result in significant variance in transmission delay times (excessive jitter). Router flapping is overcome by newer IPV6 protocols and reservation protocols.

Route Flutter-The variance of packet delay due to the continual changing of the routing paths in a communication network. See route flapping.

Router-A router is a device that directs (routes) content (data, voice or video) from one path to another on a network. Routers base their switching information on one or more parameters contained in the packet of content. These parameters may include availability of a transmission path or communications channel, destination address contained within a packet, maximum allowable amount of transmission delay a packet can accept, along with other key parameters.

Routers forward data packets between multiple interfaces based on the network layer. Most modern day routers support one or more of the following protocols: Internet Protocol (IP) , Novell IPX or AppleTalk. Routing occurs at layer 3 of the OSI reference model and can be used to limit the broadcast domain of a bridged network.

Routine-A group of instructions for carrying out a specific processing procedure. Routine usually refers to part of a larger program. Routine and subroutine have essentially the same meaning, but a subroutine could be interpreted as a self contained routine nested within another routine or program.

Routing-(1-telecommunications traffic) The physical path, circuit group, and switching systems through which telecommunications traffic flows. (2-communications path) The selection of a communications path. Traffic routing pertains to the path between switching offices, regardless of the physical paths of the connecting circuits. Circuit routing pertains to the alternative physical paths used in connecting switching offices. (3-packet network) The process of reviewing the destination address against a routing table and algorithm and the forwarding to a path that will help it reach its destination.

This figure shows a how a router can dynamically forward packets toward their destination. This diagram shows that a router contains a routing table (database) that dynamically changes. This diagram shows a router with address 100 is connected to two other routers with addresses 800 and 900. Each of these routers periodically exchanges information allowing them to build routing tables that allow them to forward packets they receive. This diagram shows that when router 100 receives a packet for a device number 952, it will forward the packet to router 900. Router 900 will then receive

that packet and forward it on to another router that will help that packet reach its destination.

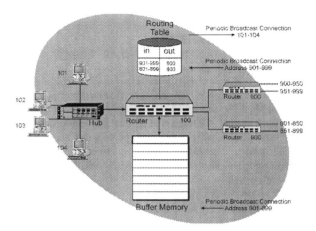

Data Network Router Operation

Routing Information Protocol (RIP)-An Internet routing protocol that uses a list of reachable networks to calculate the degree of difficulty that may be involved in reaching a specific, usually by determining the lowest hop count. RIP has been succeeded by the routing protocol Open Shortest Path First (OSPF).

Routing Label-A route identifying part of a data message that is used for message routing in the signaling network. The routing label of an SS7 message includes the destination point code (DPC), the originating point code (OPC) and the signaling link selection field.

Routing Loop-The process of passing a data packet from a router to other routers that return the packet back to the original sending router. This may occur because routing tables are dynamically changing as routers continually learn of connections to new routers and this results in new paths for forwarding packets. For packets that get caught in a routing loop, the time to live (TTL) field in the packet header will eventually be depleted (decreases each time a routing hop occurs) so the packet will eventually be discarded.

Routing Tables-A routing table is a list (a database table) that is located within a router that is used to determine the forwarding path (route) for incoming packets.

ROW-Right of Way

Royalties-Royalties are compensation for the assignment or use of intellectual property rights.

Royalty-A royalty is a fee that is paid to an author or composer for the right to use each copy of a work that is sold, performed, or produced under license of an exclusive right (such as patent rights). When patents were issued by English Queens and Kings, it was common to provide the monarch with a small (or large) payment in return for the grant to use the patent.

RP-Radio Port

RP-Rendezvous Point

RPC-Reverse Power Control

RPF-Reverse Path Forwarding

RPF Check-Reverse Path Forwarding Check

RRC-Radio Resource Control Protocol

RRI-Reverse Rate Indicator

RRO-Reproduction Rights Organization

RRQ-Registration Request

RS-Recommended Standard

RS-232-RS-232 is an electronics industry association (EIA) standard protocol that is used to transfer serial data information in asynchronous (unscheduled) form. The RS-232 specification defines optional physical (connector), electrical (signal levels), and software data formats. RS-232 serial connectors are a common connector that is used on the back of computers to connect to modems and other external devices. RS-232 communication uses universal asynchronous receiver and transmitter (UART) technology and adds communication negotiation features to the serial data communication session.

This diagram shows how an RS-232 communication system transfers information in serial form. To control the flow of data, the RS-232 system uses control lines to indicate the status of transmitting and receiving devices (e.g. ready or not ready). The control lines include a ring indicator to indicate a modem connection is requested (not used for this direct connections), carrier detect (CD) lines to sense a modem signal, data set ready (DSR) and data terminal ready (DTR) to indicate the data communication device is ready to operating. The control lines also include clear to send (CTS) and request to send (RTS) as a method of flow control. This example shows that the data can be transmitted in two directions on the transmit data (TD) and receive data (RD) lines.

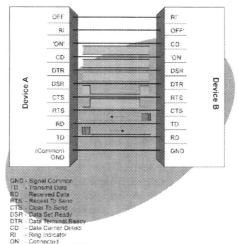

RS-232 Operation

RS-232-C-A technical specification published by the Electronic Industries Association (EIA) that establishes electrical interface requirements between computers, modems, terminals, and communications lines. Also known as EIA/TIA-232.

At the time of original issue, the EIA was known as the Radio Manufacturers Association, and its standards were then designated "Radio Standards" or RS and then the identifying number. The standard defines the specific electrical and functional characteristics used in asynchronous transmissions between a computer (data terminal equipment, or DTE) and a peripheral device (data communications equipment, or DCE). The suffix C denotes the third revision of that standard. RS-232-C is substantially compatible with the CCITT V.24 and V.28 standards, as well as ISO IS2110. (The original version of RS-232, now obsolete, used contact closures for some signals. The modern version uses different voltage levels to convey all control information.)

RS-232-C is almost always implemented using a 25-pin or 9-pin DB type connector, although the standard does not specify the mechanical shape or form of the connector. The accompanying illustration shows the pinouts used in a DB-25 male connector. It is used for serial communications between a computer and a peripheral device, such as a printer, modem, or mouse. The maximum cable limit of 15.25 meters (50 feet) can be extended by using high-quality cable, line drivers to boost the signal, or short-haul modems. A related stan-

R

dard, RS-449, uses separate balanced 2-wire circuits (instead of using a "one wire" circuit for each control pin, with a common return wire shared by several circuits) and can operate over longer cables, but is not as widely available. Peripheral device (data communications equipment, or DCE). RS is the abbreviation for recommended standard, and the C denotes the third revision of that standard. RS-232-C is compatible with the CCITT V.24 and V.28 standards, as well as ISO IS2110.

RS-232-C uses a 25-pin or 9-pin DB connector. The accompanying illustration shows the pinouts used in a DB-25 male connector. It is used for serial communications between a computer and a peripheral device, such as a printer, modem, or mouse. The maximum cable limit of 15.25 meters (50 feet) can be extended by using high-quality cable, line drivers to boost the signal, or short-haul modems.

RS-422-RS-422 is an electronics industry association (EIA) standard protocol that is used to transfer serial data information in asynchronous (unscheduled) form on a balanced line. The RS-422 specification defines optional physical (connector), electrical (signal levels), and software data formats. Because of the use of balanced (differential lines), RS-422 can communicate over a longer distance than the RS-232 serial data standard.

RSA-Rural Service Area
RSA-Rural Statistical Area
RSA Algorithm-The RSA algorithm is an encryption process that is owned by RSA Security, Inc.
RSC-Reed Solomon Code
RSH-Remote Shell
RSID-Residential System Identity
RSS-Radio Sub-System
RSSI-Received Signal Strength Indicator
RST-Running Status Table
RSVP-Resource Reservation Protocol
RTC-Reverse Traffic Channel
RTCP-Real-Time Transport Control Protocol
RTP-Real Time Transport Protocol
RTP Timestamp-An indication of the amount of time that an RTP packet holds.
RTR-Response Time Reporter
RTS-Ready to Send
RTS-Request To Send
RTSP-Real Time Streaming Protocol
RTT-Round Trip Time
RU-Route Update Protocol

Run Length Coding (RLC)-Run length encoding is a method of compressing digital information by representing repetitive data information by a notation that indicates the data that will be repeated and how many times the data will be repeated (run length).

This diagram shows how a block of repetitive data can be substantially compressed using run length coding. This example shows a long string of 1s (such as in a black image). The RLC process begins with identifying the RLC code (a unique sequence or location in a data file), the character that will be repeated and the number of times that the character (or string of characters) will be repeated.

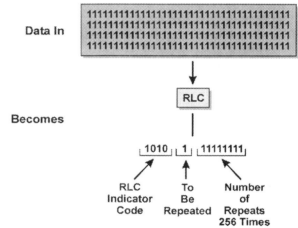

Run Length Coding (RLC) Data Compression

Run of the Picture-Term used to express a period of time from the first day to the last day of film productions.
Run-Length Encoding-Run-length encoding is a process of compressing information by substituting lengthy runs (groups) of information with a code that represents the group of information. The longer the length of the same information or pattern, the more efficient run-length encoding becomes.
Running Status Table (RST)-A running status table provides information about event changes.
Rural Service Area (RSA)-A cellular service area in rural (small population) regions.

Rural Statistical Area (RSA)-A geographic area designated by the FCC for service to be provided for by cellular carriers that falls outside the metropolitan statistical area (MSA) regions. There are over 400 RSAs in the United States.
RVC-Reverse Voice Channel
RZ-Return To Zero

www.IPTVMagazine.com   463

s-Second

S,G-Source and Group Pair

S/A-Selective Availability

SA-Source Address

Sabin-A unit of sound absorption that is the equivalent of one square foot of surface of perfectly absorptive material.

SAC-Service Access Code

SACCH-Slow Associated Control Channel

Safety Margin-Safety margin is the amount of signal loss or changes in other system parameters that can occur while providing an expected quality level of service.

Safety Procedures-Safety procedures are the learned skills, documents and supporting materials that define the actions and procedures that are used for the save installation of cables, equipment or systems.

Sag-Sag is the vertical distance a cable drops from the center of the line that connects the two end points of the cable support.

Sag Allowance-Sag allowance is the amount of cable length that is calculated or added to a cable run to compensate for changes in cable length due to environmental loading (e.g. ice) and thermal expansion and contraction.

Sales-(1-operations) The operational department responsible for identifying potential new customers, negotiating offers, and closing contracts. (2-Revenue) The amount of products or services that are sold or committed to be sold.

Sales Plan-A sales plan contains the objectives of the sales process, the responsibilities and incentives of those involved in the sales process, and the resources that will be available or used for the sales process.

The sales plan usually includes objectives (sales targets) and it may identify the salespeople or representatives and their responsibilities along with their territories and commission structures.

This diagram shows how a sales process can be managed. In this example, the sales process is divided key steps that can be defined and managed. The prospecting step is used to identify new customers or expanded needs for existing customers. The qualification step is used to determine how many of the prospects are qualified (real candidates) for the product or service. The interest assessment step is used to determine how motivated the prospect is to take action to satisfy their need for the product or service. The fact-finding stage is used to determine who are the decision makers and what steps are necessary to complete the sale. The close involves the consolidation of the previous steps (coordination of motivated decision makers) so purchases can result. The progress between each step can be tracked and optimized. Included in the chart is an example activity time sheet showing that step may require different levels of time commitment and that the allocation of resources (time for each salesperson) is usually distributed along each step.

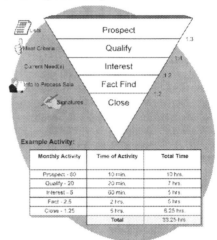

Sales Management Diagram

Sales Program-A series of related sales campaigns targeting the same objectives.

Salutation-A procedure for looking up, discovering, and accessing services, and information. This architecture defines abstractions for devices, applications, and services; a capabilities exchange protocol; a service request protocol; standardized protocols for common services; and application programming interfaces (APIs) for information access and session management. See also salutation Architecture.

Sampling-(1-signal) Sampling is the process or rate at which a signal is measured (sampled) for other processing functions such as quantization (level measurement), modulation (conversion), or coding (information manipulation). (2-marketing) The process of extracting or copying a portion of a marketing campaign to analyze the performance or results of the campaign.

Sampling Rate-The rate at which signals in an individual channel are sampled for subsequent modulation, coding, quantization, or any combination of these functions. The process of taking samples of an electronic signal at equal time intervals to typically convert the level into digital information. Frequency usually is specified as the number of samples per unit time.

SAN-Storage Area Network

SAP-Secondary Audio Program

SAP-Service Access Point

SAP-Session Announcement Protocol

SAS-Single Attached Station

SASL-Structured Audio Score Language

SAT-Supervisory Audio Tone

Satellite-(1-general) An object that revolves around another object of greater mass (such as the earth) and has a motion that is determined by the force of attraction (gravity) of the larger object. (2-communications) A space vehicle that orbits the earth which contains one or more radio transponders that receive and retransmit signals to and from the earth. (3-equipment) A piece of equipment or system that operates at a remote location from a central control system.

Satellite Availability-The probability that a satellite will be on station and available for a particular task.

Satellite Communications-The use of orbiting satellites to relay communications signals from one station to many others.

Satellite Downlink-A microwave radio link from a satellite to a ground station on earth.

Satellite Master Antenna Television (SMATV)-Satellite master antenna television is a private (closed circuit) network that provides television channels on one or more televisions or video monitors.

Satellite Relay-An active satellite repeater that relays signals between two earth terminals.

Satellite System-A network of satellites and associated ground stations that transmit telephone, audio, video, and data signals between terrestrial points.

This diagram shows the different types of satellite communication systems. The GEO satellite system is primarily used for television broadcast services, as their satellites appear stationary above the Earth. MEO and LEO systems are used for mobile communications as they are located much closer to the Earth. However, these satellites continuously move relative to the surface of the Earth.

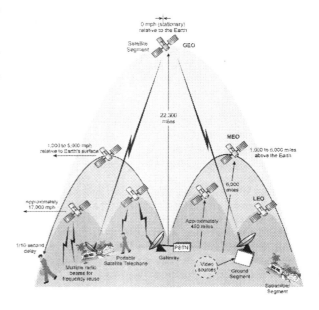

Satellite System

Satellite Telephone-A satellite telephone is a wireless telephone that uses mobile satellite service to send voice and data.

This figure shows a functional diagram of a portable satellite telephone. This diagram shows that a typical portable satellite telephone consists of a high gain antenna that has multiple antennal elements, antenna control circuitry, radio signal transceiver (transmitter and receiver) and audio signal processing circuits.

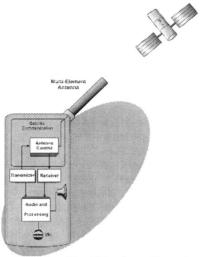

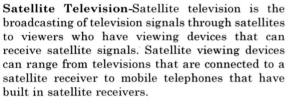

Portable Satellite Telephone Functional

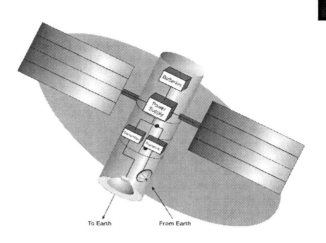

Satellite Transponder Operation

Satellite Television-Satellite television is the broadcasting of television signals through satellites to viewers who have viewing devices that can receive satellite signals. Satellite viewing devices can range from televisions that are connected to a satellite receiver to mobile telephones that have built in satellite receivers.

Satellite Transponder-A satellite transponder is a combination of a receiver and transmitter that receives a signal from an Earth station (uplink) and retransmits it to ground receiving stations (downlink).

This diagram shows how a satellite transponder operates. In this example, the transponder receives a signal from an Earth station. This signal is translated (converted) into a higher frequency, amplified, and re-transmitted back to the Earth.

Satellite Uplink-The transmission equipment used to modulate, amplify, and radiate a signal to an orbiting satellite.

Saturation-(1-chroma, color) The intensity of the colors in the active picture; the voltage levels of the colors. Saturation relates to the degree by which the eye perceives a color as departing from a gray or white scale of the same brightness. A nth percent saturated color contains no white; that is, adding white reduces saturation. In NTSC and PAL video signals, the color saturation at any particular instant in the picture is conveyed by the corresponding instantaneous amplitude of the active video subcarrier. (2-amplifier circuit) The point on the operational curve of an amplifier at which an increase in input amplitude will no longer result in an increase in amplitude at the output. (3-inductor or transformer) A condition in which the ferromagnetic core material of an inductor or transformer is subjected to strong magnetic induction so that its internal magnetic field has reached its maximum possible value of physical magnetization. At the atomic level, all of the magnetic domains in the core material are now oriented in the same space direction, and all of the elementary intrinsic electron spins in the outer electron shells of the atoms in the core are oriented in the same space direction. (4-market) Market saturation is a percentage of market penetration (usage or purchase) of a product or service where the sale of additional products or services to new customers becomes difficult or the marketing cost of promoting the product or service to new customers is beyond the profit generated from the sale of the product or service. A market that experiences market saturation (e.g. above 80% market penetration) may have substantial sales of products for replacement or upgrade of existing customers.

SAW-Surface Acoustic Wave

SAW Device-Surface Acoustic Wave

SB-Status Bit

SBC-Session Border Controller

SC-Single Carrier

SC-SMS Service Centre

SC-Subcarrier

SC Connector-Subscriber Connector

SCADA-Supervisory Control And Data Acquisition

Scalability-The ability of a system to increase the number of users or amount of services it can provide without significant changes to the hardware or technology used.

Scalable Codec-A scalable codec is capable of being changed for operating with different data rates and/or varying types of channel quality (packet and frame loss rate).

Scalar Analyzer-A test instrument that measures the characteristics of a transmission signal that includes return loss (reflected signals), voltage standing wave ratio (VSWR), and insertion loss.

Scan Lines-Scan lines are electrical signals that represent the scanned (optically converted) line areas (rows) of an image. In the United States and Japan, television signals have 480 scan lines (rows) and in Europe, television signals have 576 scan lines.

Scanner-(1-optical) An optical scanner is light intensity conversion system that is used for converting or interpreting images, documents, invoices, bar-codes or photos into other forms of information (such as converting a picture into a fax image). (2-radio) Devices that can receive or continously search through a list of radio channels.

Scanning-(1-radio) The process of searching through multiple radio channels. (2-image) The process of converting an image into information signals by sequentially converting portions of the image (e.g. lines) into signals that represent light levels at particular positions of the image.

Scanning Receiver-A receiver that is capable of searching through multiple radio channels.

SCCP-Signaling Connection Control Part

SCCP-Skinny Client Control Protocol

SCCPCH-Secondary Common Control Physical Channel

SCE-Service Creation Environment

SCF-Service Control Function

SCH-Synchronization Channel

SCH Laser-Separate Confinement Heterostructure

Schedule F-In reference to section of SAG agreements in titled, insuring exclusive rights to a producer, which would less likely be given to other staff or cast members due to inferior status.

Scheduled Conferencing-A conference bridge that allows callers to become members of a conference group during a predetermined time period.

Scheduled Transmission-A feature allowing the user to schedule a fax transmission at a specific data or time in the future. Key benefits are convenience and cost savings. Scheduling jobs at a period of low telephone rates can have immediate considerable savings.

Scheduling Algorithm-A scheduling algorithm is a program that coordinates the sequences of processes or information. The scheduling algorithm in a communication system is used to coordinate the flow of data to multiple users who have varying needs and access priority levels.

Schematic-A diagram of a circuit or flow of a system.

Scintillation-Scintillation is the random fluctuation of an optical or radio signal about its transmitted path (beam wandering and beam spreading) that results from changes in transmission characteristics. An example of a cause of scintillation is air changes due to heating ducts or smokestacks.

SCIP-Simple Conference Invitation Protocol

SCP-Service Control Point

SCP-Session Configuration Protocol

SCR-System Clock Reference

Scrambler-A device that transposes or inverts signals, or otherwise encodes a message at the transmitter to make it unintelligible at a receiver that is not equipped with a descrambling device.

Scrambling-Scrambling is a process of altering or changing an electrical signal (often the encoding or distortion of a video signal) to prevent interpretation of the signals by users that can receive the signal but are unauthorized to receive the signal. Scrambling involves the changing of a signal according to a known process so that the received signal can reverse the process to decode the signal back into its original (or close to original) form.

Scratch Pad-A read/write random access memory space used for the temporary storage of data; it is the working area of a memory unit.

Screen Based Telephony-The use of computer screens to provide a telephone user with information about a call in progress and/or the status of their telephone calls. Screen based telephony may allow the telephone user to dial, answer, and control their telephone calls.

Screen Name-An identifying name that is used for instant messaging systems to identify a instant messaging person or device. The instant messaging system provides the relationship between a screen name and an email or IP address. The use of a screen name allows a person to avoid providing an email address (for privacy) and it allows the use of dynamically changing IP addresses (for systems that dynamically assign IP addresses when the user logs onto the system).

Screen Pop-Screen pops are the display of an information screen on a computer monitor that is automatically triggered by an event such as an incoming call or customer selected feature request.

Screen Size-Screen size is the visible area of an image display (such as a computer monitor or a television screen). Screen size may be specified in resolution in pixels across (horizontal) and pixels between top and bottom (vertical). Visible screen size may also be defined by the diagonal length of visible area in inches or centimeters.

Script-A script (or a Servlet) is a small program or sequence of operations (macro) that is written in a predetermined language that can be understood by the calling program to allow automatic interaction between programs or devices. Examples of scripts include login scripts that are used to provide identification information when accessing a system or Javascripts that provide advanced features on Internet web pages.

Script doctor-Title given to a hired staff to rewrite a script as to make more entertaining then original version.

Scripting-Scripting is the creation of instructions and information to use by a software program into a file called a script. Scripts may be composed of commands from a variety of languages including Java, Perl, Microsoft Visual Basic or in a form of unique (proprietary) commands and text that only the software program can interpret.

Scripting Language-Scripting language is the specific syntax and processes that are used by a software program to execute or interpret com-

mands in a script file.

Scrubbing-Scrubbing is the process of fast forwarding or rewinding a video while pre-viewing the video.

SCSA-Signal Computing System Architecture

SCSI-Small Computer Systems Interface

SCTP-Stream Control Transport Protocol

SCTP Packet-Stream control transmission protocol (SCTP) packets contain a common header and variable length blocks (chunks) of data. The SCTP packet structure is designed to offer the benefits of connection-oriented data flow (sequential) with the variable packet size and the use of Internet protocol (IP) addressing.

This figure shows that the SCTP packet structure includes a common header format along with chunks of data. The header includes the source and destination port numbers that are associated with the specific IP addresses to uniquely identify the packet for a specific communication application. The verification tag uniquely identifies (validates) the sender of the packet. Each packet has a checksum to ensure all the data has been reliably sent. The data is sent in chunks to allow the near real-time streaming (continuous one-way delivery) of information.

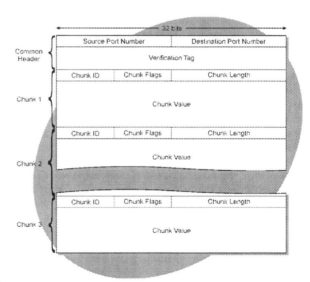

SCTP Packet Structure

SD-Standard Definition TV format

SDAP-Service Discovery Application Profile

SDARS-Satellite Digital Audio Radio Service

SDCCH-Stand-Alone Dedicated Control Channel

SDD-Space Division Duplex

SDH-Synchronous Digital Hierarchy

SDI-Serial Digital Interface

SDI Router-An SDI router is a device that directs (routes) SDI digital video content from one path to another on a system or network.

SDI Video-Serial Digital Interface Video

SDK-Software Development Kit

SDL-System [or State-machine] Description Language

SDLC-Synchronous Data Link Control

SDMA-Spatial Division Multiple Access

SDMF-Single Data Message Format

SDMI-Secure Digital Music Initiative

SDP-Service Delivery Platform

SDP-Service Discovery Protocol

SDP-Session Description Protocol

SDPng-Session Description Protocol Next Generation

SDR-Service Detail Record

SDR-Software Defined Radio

SDSL-Symmetrical Digital Subscriber Line

SDTI-Serial Data Transport Interface

SDU-Service Data Unit

SDV-Switched Digital Video

Sealing Current-Sealing current is a method that is used to decrease the effects of corrosion is to continuously run current through the copper wire pair. This "sealing current" is a small amount of direct current that is passed through a copper wire to reduce the corrosion effects of the splice points. The sealing current effectively maintains conductivity of mechanical splices that are not soldered. The direct current effectively punches holes the corrosive oxide film that forms on the mechanical splices.

Search Engine-A web portal (web site) or software that searches through web pages or data records to find matches to specific words or items. Web page search engines allow users to enter key words (search words) to find web pages that contain the key words or links that are associated with the key words.

SECAM-Sequential Couleur Avec MeMoire

Second (s)-(1-time unit) Unit of time. It is defined as 9,192,631,770 cycles of the radiation associated with the transition between the two hyperfine levels of the ground state of the cesium-133 atom. Originally historically defined as 1/ 31 557 600 of a

year. (2-order) The item in a sequence following the first item.

Second And A Half Generation (2 1/2G)-A term commonly used to describe one or more interim technologies that are used to help a specific application or industry transition from one capability to a much more advanced capability. In cellular telecommunications, 2 1/2 generation systems used improved digital radio technology to increase their data transmission rates and new packet based technology to increase the system efficiency for data users.

Second Generation (2G)-A term commonly used to describe the second technology used in a specific application or industry. In cellular telecommunications, second generation systems used digital radio technology with advanced messaging and data capabilities. For second-generation cordless telephones, second generation of products used multiple channels (using analog FM technology) radios.

Secondary-The output winding of a transformer.

Secondary Audio Program (SAP)-The second audio program channel in the BTSC stereo TV system. The SAP channel often is used for second-language or other specialized programming.

Secondary Channel-Secondary channels are transmission channels that are setup or are subject to the operation of a primary communication channel.

Secondary Ring-The secondary path used in a communication network ring. A secondary ring circulates data in the opposite direction as the primary ring. It is used in the fiber distributed data interface (FDDI) system to allow communication to continue when a communication line is cut or disabled.

Secondary Route-The circuit to be used when the primary route is congested. In manual and semi-automatic operations, a secondary route also may be used when transmission on the primary route is not of sufficiently good quality, or if traffic is to be handled outside the normal hours of service on the primary route.

Section Terminating Equipment (STE)-A device or assembly that terminates a section in an optical system. STE network equipment terminates the physical and section layers of a Synchronous Optical Network (SONET).

Section Throw-An aerial or underground engineering order involving the splicing of new cabling in parallel with live working cable. This is conduct-

S

ed by half-tapping the new cable into the existing live cable and then trimming out the old, defective section. Cable counts remain the same in the new working cable.

Sector-(1-radio) One of the radio transmission portions of a sectorized cell site. See sectorization. (2-computing) A logical divisions of storage disk's tracks. Each track is typically divided into several sectors.

Sectorization-The dividing of transmitter radio coverage areas into smaller coverage areas by using directional antennas. Most cellular systems initially use cell sites that transmit in omni-directional 360-degree coverage patterns. To increase the system capacity, the antenna system is changed so 3 or 6 focused antenna patterns are created.

This diagram shows how an omni-directional cell site radio coverage area has been divided (sectorized) into three smaller areas (sectors). This diagram shows that frequency number 2 now can be reused in an adjacent cell site due to the directional properties of the antennas (directional antennas have reduced gain in the opposite direction.)

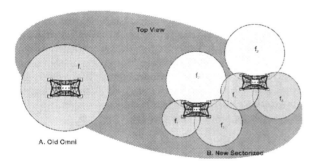

Cell Site Sectorization

Secure Digital Music Initiative (SDMI)-Secure digital music initiative (SDMI) is a digital rights management standards development project that is intended to control the distribution and ensure authorized use for digital music. Members of SDMA come from information technology, consumer electronics, security, recording companies and Internet service provider industries. The focus (charter) of SDMA is to develop open industry specifications that protect the rights of media content

owners and publishers for storing, playing and distributing of digital media while allowing users convenient access to media content through new emerging distribution systems (such as the Internet).

The initial development for SDMI (phase I) was to create a digital watermark system. Digital music access technology is a trademark for the technology used for the content protection process developed by the secure digital music initiative (SDMI).

For additional information on SDMI, visit www.SDMI.org.

Secure Electronic Transaction (SET)-A secure electronic transaction is an exchange of assets or other quantifiable information represented by data (electronic media) that is encoded in such a way to be private (not viewable by others) and unaltered (not able to be changed by others).

Secure Real Time Protocol (SRTP)-An enhanced version of real time protocol (RTP) that provides increased security (e.g. confidentiality and message authentication).

Secure Server-A secure server is a computer that can receive, process, and respond to an end user's (client's) request in a secure mode. Secure servers can use secure socket layer (SSL) protocol to establish and maintain authentication (identity) and encryption privacy.

Secure Shell (SSH)-A secure alternative to RSH, SSH allows remote command execution and remote terminal sessions. SSH creates an encrypted and authenticated channel between hosts for all communication. SSH may also be used to securely tunnel TCP traffic between two hosts' networks.

Secure Sockets Layer (SSL)-A secure socket layer (SSL) is a security protocol that is used to protect/encrypt information that is sent between end user (client) and a server so eavesdroppers (such as sniffers on a router) cannot understand (cannot decode) the data. SSL version 2 provides security by allowing applications to encrypt data that goes from a client, such as a Web browser, to a matching server (encrypting your data means converting it to a secret code) SSL version 3 allows the server to authenticate (validate the authenticity) the client.

Secure Transaction-A secure transaction is an exchange of assets or other quantifiable information that is encoded in such a way to be private (not viewable by others) and unaltered (not able to be changed by others).

Secure Video Processing (SVP)-Secure video processing is a method of ensuring the rights of video content are enforced through the use of hardware, software and rights management systems.

Secure Video Processor Licensing Authority (SVPLA)-Secure video processing licensing authority is a company that oversees the licenses that are required to implement the SVP system created by NDS Ltd.

Security-Security is the ability of a person, system or service to maintain its desired well being or operation without damage, theft or compromise of its resources from unwanted people or events.

Security Hole-A security hole is a weakness or potential access process that can be used that bypasses or disables security controls and processes.

Security Layer-A security layer is a functional process in a communication system that controls the security procedures of a device or system.

Security Management-(1-General) Security management of a network involves identity validation (authentication), service authorization, and information privacy protection. Authentication processes identifies the device or person that is requesting the use of the telecommunications device or network services. Authorization is the process of determining what services devices are customers are permitted to use. Privacy or encryption services are used to help ensure that the information transmitted or received is not available to unauthorized recipients. (2-FCAPS) Security is one of the five functions defined in the FCAPS model for network management. Security is responsible for protecting the network against hackers, unauthorized users, and physical or electronic sabotage.

Security Proxy-A proxy server that receives and filters information designated for a user it is serving. The security proxy restricts information based on rules it has received from the network operator and/or user of the information. A security proxy is also called a firewall.

Security System-Security systems are monitoring and alerting systems that are configured to provide surveillance and information recording for protection from burglary, fire, environmental hazards, and other types of losses.

Seek Time-Seek time is the amount of time required to find a record of data, generally used in reference to a disk file or request a media segment through a data network (such as requesting a new section of a digital video media file).

Seeking-Seeking is the process of jumping to other time points within a video.

SEF-Source Explicit Forwarding

Segmentation-(1-network) The dividing of a network into multiple networks. (2-data segmentation) The division of data blocks or packets into smaller segments or packets. Segmentation is also known as fragmentation as in the data and Internet community. (3-database) In database management, the process of separating filed information or records into groups (segments) that allow for more rapid information access. (4-software) The dividing of a program into functional parts (segments). These segments can usually be changed or updated independently making the software program easier to develop and change.

This diagram shows how a large data packet may be divided into several smaller segments to allow transmission through networks that have a smaller packet size than packet of data traveling through the network. In this example, a packet that is 1200 bytes enters into a newtork that can only transport packets with a maximum size of 400 bytes. If the packet header indicates it is acceptable to divide the packet, each smaller packet will include a the destination address plus additional information that indicates which part of the packet segment it represents. When the packet reaches its destination, the smaller fragments will be reassembled into the original data packet. Some time-sensitive systems (such as real-time voice systems) do not want packets to be divided.

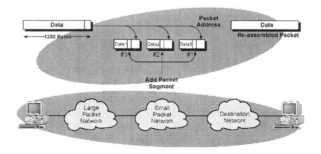

Packet Segmentation

Segmentation And Reassembly-Segmentation and reassembly is a process of dividing blocks of information into smaller size blocks (segmentation) and reassembling these smaller blocks into their original large block form when they are received.

Selective Call Acceptance-A service feature that only delivers calls to their dialed destination if they are on a previously specified selective call acceptance telephone number list. Calls that are received by other numbers are provided with a pre-recorded announcement that states the number is not accepting their call or the call may be routed to an alternate directory number.

This diagram shows how selective call acceptance can be used to only deliver calls from a specific list of callers. This diagram shows that regional managers from a service center can call from numbers that are pre-defined (their office numbers) and that when their calls are received, the will be connected to the specified telephone or extension. Call that are received from numbers that are not on the selective call acceptance list are transferred to an automated message unit that plays a pre-recorded announcement that states the number is not accepting their call.

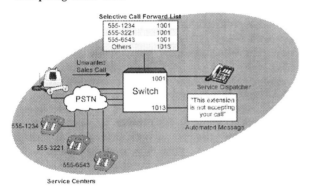

Selective Call Acceptance Operation

Selective Call Forwarding-A service feature that forwards calls to one (or multiple) telephone number dependent on the incoming call forwarding criteria. Selective call forwarding can be used to redirect calls of a specific type (such as fax calls) to a pre-designated number (such as an office fax machine.)

This diagram shows how selective call forwarding can be used to deliver calls to alternate number based on a specific criteria type. This diagram shows a selective call forwarding service that routes fax calls to different telephone number or extension after it detects the call is a fax call. After the system detects that the incoming call is a fax (by the fax tones), the switch call processing software transfers the call to the destination number that is connected to a fax machine.

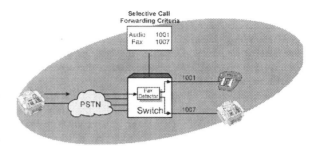

Selective Call Forwarding Operation

Selective Call Rejection-A service feature that restricts the delivery of calls to their dialed destination if they are on a previously specified call rejection telephone number list. Selective call rejection is used to block calls from undesired callers such as prank callers or harassing bill collectors. Calls that are received by numbers on the call rejection list are provided with a pre-recorded announcement that states the number is not accepting their call or the call may be routed to an alternate directory number.

Selective Combiner-A circuit or device that dynamically selects the stronger of two or more signals (usually from different antennas) so that the output signal is selected from the received signal that has the higher signal strength.

Selective Ringing-A technique for ringing the telephone of only one customer on a multiparty line.

Selectivity-The capability of equipment to select and operate upon a signal of a particular frequency despite the presence of other signals at frequencies close to that of which the equipment is tuned.

Self Certification-Self certification is the process of testing and declaring that products or services conform to specifications or other requirements. Self certification may be performed by companies or people when systems or technologies are initially released and testing and validation facilities are not yet established.

Self Correcting Protocol-A self correcting protocol is a set of commands and procedures that can dynamically change the controlling process based on unexpected results of the currently used protocol operation.

Self Destructing DVD-A self destructing DVDs is a storage disk that must be used within a period of time when the disk container is opened (such as 24 or 48 hours). After that time, a chemical reaction occurs that renders the disk useless.

Self Healing Ring-A transmission system that is designed to cover automatically from cable breaks or equipment failures, using fiber optics or radio facilities in a closed loop.

Self Install-Self Installation

Self Installation (Self Install)-Self installation is the process of allowing a user of a product or service to install and possibly configure (program optional features and parameters).

Self Oscillation-Self oscillation is the creation of a periodic signal in an electronic circuit without the need for an external signal. Self oscillation may be desired (in the case of a frequency generator) or it may be undesired (in the case of a speaker and microphone self oscillation audio squeel).

Self Rerouting-Self rerouting is the process of the automatic transferring connections or circuits from one path (route) to an alternate path when an equipment or system has detected that a connection has been lost.

Semantic Web-The next generation World Wide Web characterized by the ability to search for information based upon metadata content.

Sensitivity-(1-general) The capability of a circuit or device to respond to a minimal signal input level and still produce an acceptable level of reliable operation. (2-optoelectronics) An imprecise synonym for responsivity. (3-radio receiver) The minimum input signal required to produce a specified output waveform having a specified signal-to-noise ratio. (4-voltmeter) The current required to produce a full-scale deflection.

Sensor-A detection device that is sensitive to changes in level or state.

SEO-Search Engine Optimization

Separate Card-In reference to a credit exclusively appearing on screen at which time no other credit is shown.

Separate Video (S-Video)-S-video is a set of electrical signals that represent luminance and color information.

Sequence Layer-Sequence layer is a sequence of images or portion of a digital video that specific characteristics apply (such as a group of pictures format).

Sequenced Packet Protocol (SPP)-Sequenced packet protocol is a packet data networking protocol that uses destination identification reference numbers to identify communication connections along with sequence numbers to allow the correct routing and sequencing of packets between communication devices.

Sequential Couleur Avec MeMoire (SECAM)-SECAM is a color TV system that provides 625 lines per frame and 50 fields per second. This system was developed by France and the former U.S.S.R., and the former Eastern Block Countries, and in some Middle East Countries. In order to transmit color the information is transmitted sequentially on alternate lines as a FM signal.

Serial-An arrangement whereby one element of data is linked to the next so that progress must proceed from the first element through the next, then the next, and so on.

Serial Data Bus-A communication line or system that transfers information sequentially, bit by bit. The serial data bus (such as an RS-232 system) may use additional lines to help coordinate the transmission (e.g. ready, wait, or error).

This diagram shows the process used to transmit data in serial form. In this diagram, parallel data is converted to serial form by a parallel to serial data converter. Prior to sending any bits, a framing circuit inserts a pre-established sequence (frame indicator) to identify the start of a frame of information. This diagram shows that each frame of data contains groups of bits that represent each logical channel. Data arrives at the serial trans-

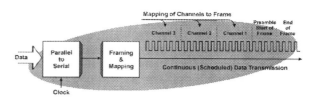

Serial Data Transmission

mitter in groups of bits or bytes where each group represents a portion of a logical channel. The relationship between the frame of data and bits assigned to logical channels is called the mapping. The framing circuit will usually append some bits to the end of the frame that are used for error checking.

Serial Data Transport Interface (SDTI)-Serial data transport interface is a flexible packetized digital media format that allows the addressing and transmission of professional digital media files between digital media devices such as camcorders, DVRs and editing systems. SDTI is compatible with and builds on the SDI interface and it is defined in SMPTE 305M.

Serial Digital Interface (SDI)-Serial digital interface is a standard digital video connection that transfers digital video at 270 Mbps. It uses a common 10 bit signal scrambling for composite and component video along with four channels of digital audio. The standard connector for SDI is a BNC connector. SDI is defined in ITU R BT 601 standard and it uses 4:2:2 sample (720 samples per line and two of each color). The HDTV version of SDI has a data transfer rate of 1.5 Gbps.

Serial Digital Interface Video (SDI Video)- Serial digital interface video is uncompressed digital video that transfers at a rate of 270 Mbps or 360 Mbps depending on the number of lines of resolution.

Serial Line Interface Protocol (SLIP)-A simple protocol that was developed to allow computers to connect to the Internet using a serial connection (such as a dialup connection). The SLIP protocol is being replaced by the point-to-point protocol (PPP).

Serial To Parallel Converter-A serial to parallel converter is a device or assembly that accepts information in time sequence and groups the information into groups of digits or elements where each group is transmitted at the same time (in parallel).

Serial Transmission-A sequential rather than simultaneous parallel transmission of data bits.

Series Regular-Secondary character, appearing in a majority of all episodes in a series, not necessarily partaking in every episode.

Series Sales Bonus-In reference to a paid sum of money paid by a studio/producer that a network uses to purchase a pilot series, thereby ordering production of a series.

Server-A computer that can receive, process, and respond to an end user's (client's) request for information or information processing.

Server Ingest-Server ingest is the process of transferring media into a computer storage system (server).

Service-(1-telecommunications) The provision of telecommunications to customers by a common carrier, administration, or private operating agency using voice, data, and/or video technologies. (2-performance) The overall quality of telephone system performance, sometimes stated in terms of blocking or delay.

Service Activation-Service activation is the processes used to enable a service to become operable.

Service Area-The region covered by a given cellular carrier. It is also a landline term which means a geographical area in which local exchange carriers provide local exchange service to end users as well as network access to interexchange carriers.

Service Bundles-Service bundles are the combining of different products and services into a "package" offer, and then offering it to a customer at a separate, combined price.

Service Bureau-A company that owns its own network or switching facilities and provides turnkey services to other companies.

Service Class-Service classes are sets of communication parameters that are used or assigned to provide transmission flows that can provide services that meet specific quality of service (QoS) requirements.

Service Delivery Network-A service delivery network is a combination of distribution hubs, switches, and transmission lines that are used to provide the transfer of specific types of services.

Service Delivery Platform (SDP)-Service delivery platform refers to hardware types and software programs that are required to deliver specific types of services.

Service Detail Record (SDR)-A service detail record holds information related to a service or group of service usage events related to a communication session. This information usually contains the origination address of the user, time of day the service or services were requested, the types and quantities of services used, and charges that may be added from supporting vendors throughout the duration of the service.

Service Discovery-The ability to discover the capability of connecting devices or hosts.

Service Discovery Application Profile (SDAP)-Service discovery application profile (SDAP) defines the processes and procedures required to allow devices to discover and allow other devices to discover services available for a Bluetooth device. The SDP allows for the dynamic addition and elimination of available services dependent on the capability and status of the Bluetooth device. The use of SDP ensures that Bluetooth devices and retrieve information pertinent to the implementation of services as they become available in other Bluetooth devices.

Service Discovery Protocol (SDP)-Service discovery protocol (SDP) is the communication messaging protocol used by a communication system (such as the Bluetooth system) to allow devices to discover the services and capabilities of other devices.

Service Flow Identifier (SFID)-A service flow identifier is a unique number that is assigned by a system (such as a 32-bit number assigned by a CMTS in a DOCSIS system) that is used to identify the flow of a communication channel that is used for a specific service type.

Service Flows-Service flows are communication channels (e.g. a stream of packets) that have particular service characteristics associated with the transfer (flow) of data. For example, a communication link might have several service flows associated with it; a real time service flow for voice communication, a high-integrity service flow (low error rate) for data file transfer and a best effort service flow for Internet web browsing.

Service Integration-Service integration is the combination of multiple services and features for a communication service and/or system. Examples of service integration include IP telephony and IP video into IP television service.

Service Level Agreement (SLA)-An agreement between a customer and a service provider that defines the services provided by the carrier and the performance requirements of the customer. The SLA usually includes fees and discounts for the services based on the actual performance level received by the customer.

Service Life-The period of time that equipment is or can be expected to be in active use.

Service Loop-A service loop is a bundle (loop) of cable that is part of a communication line (such as a pole mounted cable TV or telephone lines) that provides additional cable length that may be necessary to perform a splice or cable path reconfiguration at a later time. Outdoor cable loops can store 100 to 200 feet of cable for each 1,000 feet of installed cable line while indoor service loops typically store 10 to 20 feet of additional cable.

Service Management System (SMS)-A computer system that administers service between service developers and signal control point databases in the SS7 network. The SMS system supports the development of intelligent database services. The system contains routing instructions and other call processing information.

Service Mix-Service mix is the different types of services that an operator or service provider may provide to customers.

Service Multiplexing-The multiplexing of different types of services such as voice, data and video onto one type of communications channel. Service multiplexing allows one piece of equipment and/or communication channel to provide services that have different delay, bandwidth and other quality of service (QoS) requirements.

Service Number Portability (SNP)-Service number portability allows a customer to take their telephone number to a different type of service provider. Service number portability involves determination of the type of service provider (e.g., wireless or wired) who is responsible for completing the call using the telephone number (e.g. area code and NXX.) Service number portability may differ from local number portability as the interconnection and call processing for different types of service providers may vary.

Service Order-A record that describes a customer request to establish, change, or terminate a service. The service order contains all information required to meet a customer's needs.

Service Precedence-Service precedence is the hierarchy of the activation and operation of services in a communication system.

Service Provider-A service provider is a person or company that provides information and/or performs actions (services) to customers.

Service Provider Portability-Service provider portability is the capability of a communications customer to change their selected service provider.

Service Provisioning-Service provisioning is the process of an authorized agent or process that processes and submits the necessary information to enable the activation of a service. This typically includes: transmission, wiring, and equipment configuration.

Service Quality Churn -Voluntary churn that occurs when customers are dissatisfied with their service quality change carriers to get better quality.

Service Record-A database record in a communication system that contains a description of a service and the attributes that are necessary (e.g. protocols) to use that service.

Service Regions-Service regions are the geographic areas that provide services. Examples of service regions for optical networks include metropolitan enterprise, metropolitan access, Backbone, terrestrial long-haul, and submarine long-haul.

Service Requirements-Service requirements are the communication capabilities and processes that are necessary to provide a communication service.

Servicemark-A unique symbol, word, name, picture, or design, or combination thereof, used by firms to identify their own services and distinguish them from the services sold by others. Servicemarks are Intellectual Property Rights which give the owner the right to prevent others from using a similar mark which could create confusion among consumers as to the source of the services. Servicemark protection must be registered in the country, or region, where it is desired. In most countries and regions, servicemarks are administered by the same government agency which administers trademarks.

Services URL-An address that is preprogrammed into or used by devices (such as an IP Telephone) where information is kept regarding services that may be used by the device. The services URL may be associated with a menu button or an icon on the display of an IP telephone.

Serving Area-(1-central office) A geographic area served by a central office or exchange. (2-outside plant) In outside plant, a distribution area that connects with a feeder route through a serving area interface.

Servlet-A Servlet (or a script) is a small program or sequence of operations (macro) that is written in a predetermined language that can be understood by the calling program to allow automatic interaction between programs or devices. Examples of servlets include login scripts that are used to provide identification information when accessing a system or Javascripts that provide advanced features on Internet web pages.

Servo-An electromechanical system for relaying positional or angular information.

Session-Sessions are the time and activity between the operation of a software program or logical connection between two communications devices. In communications systems, the session involves the establishment of a logical channel with configuration transmission parameters, operation of higher level applications, and termination of the session when the application is complete. During a session, many processes or message transmissions may occur.

Session Announcement Protocol (SAP)-An Internet protocol that distributes information about multicast sessions. SAP is defined in RFC 2974.

Session Border Controller (SBC)-A session border controller (SBC) is an interface to a network firewall that facilitates the secure hand-off of voice packets from one IP network to another IP network. In an enterprise network, the SBC "controls" the communications "session" as it crosses the "border" from the LAN to IP. Conventional firewalls support the secure traversal of data streams, but for IP networks, SBCs are needed to facilitate secure, real time, multimedia communication. SBCs are a recent, but important component in today's nextgen network infrastructure.

Session Configuration Protocol (SCP)-Session configuration protocol is used to negotiate and initialize protocols that are used for communication sessions.

Session Description Protocol (SDP)-Session description protocol (SDP) is a text based protocol that is used to provide high-level definitions of connections and media streams. The SDP protocol is used with session initiated protocol (SIP). The SDP protocol is used in a variety of communication systems including 3G wireless and the PacketCable system. SDP is defined in RFC 2327.

Session Description Protocol Next Generation (SDPng)-An enhanced version of SDP that allows high-level definitions of connections and media streams independent of the media control protocols that are used (such as H.248 or MGCP). SDPng protocol includes feature descriptions that are necessary by these different types of media control systems.

Session Initiation Protocol (SIP)-SIP is an application layer protocol that uses text format messages to setup, manage, and terminate multimedia communication sessions. SIP is a simplified version of the ITU H.323 packet multimedia system. SIP is defined in RFC 2543.

Session Initiation Protocol Project Investigation (SIPPING)-An IETF working group that was formed to help the SIP working group to develop proposed standards to draft standards.

Session Invitation Protocol Version 1 (SIPv1)-The initial SIP text based protocol that was developed to provide multimedia services using Internet protocol (IP) networks. SIPv1 was first submitted to the IETF in February 1996.

Session Key-A session key is a word, algorithm or data element that is used to encrypt and decrypt messages or information during a communication session. After the communication session has ended, the session key is discarded or changed.

Session Layer-The session layer protocol coordinates the information transmission between endpoints during a communication session. The session layer receives requests for transmission from an application layer and converts it into network addressable data formats that can be transferred through a network or transmission line. The session layer usually establishes a communication session, coordinates the overall control of the session (such as handling retransmission and restart requests), and termination procedures. The location of the session layer within the protocol stacks varies dependent on the protocol. The session layer is layer 5 in the open system interconnection (OSI) protocol layer model.

Session Level-The fifth level in the ISO layered model. The session level receives the services of protocols located in the transport level and below, it performs services for the presentation and application levels above it.

Session Management-Session management is the initiation, management and termination of services on a computing device or communication system.

Session Management Protocol (SMP)-Session management protocol coordinates the initiation and termination of communication sessions.

Session Transfer-The process of transferring a communication session from one device to another device. The process used by a session transfer is the sending of an invite message from one device (the device that desires to be transferred) to a new device that is not part of the original communication session. If the new device accepts the invitation request (it is willing to accept the transfer request), the transfer device informs the device that it is communicating with of the desire to transfer the session to a new device along with the parameters (such as the new device network address) that will be used by the existing device to connect to the new device. After the two devices have successfully established a new communication session, the transfer device is no longer part of the original communication session.

SET-Secure Electronic Transaction

Set of Revsion-In reference to the rewrite chain of script, from first rewrite to last, including all those that come between.

Set Top Box (STB)-A set top box is an electronic device that adapts a communications medium to a format that is accessible by the end user. Set top boxes are commonly located in a customer's home to allow the reception of video signals on a television or computer.

Set Up Bonus-In reference to, on occurrence that a buyer makes an agreement with a studio in order to begin production of a property, the sum of additional money received by the owner in exchange for that property.

Set-Request-The set-request SNMP operation is virtually identical to the get operation, with the exception that the manager fills in the desired values for the MIB objects and the agent is requested to change the value of these objects. A read-write community string is usually required to issue a set-request operation.

Settlement-Settlements are the financial difference between revenues generated from charging for services provided by a carrier to visiting customers less the fees that other carriers have charged for services provided to their customer that have visited other systems.

Settlement Code-A code that identifies a billing message type and the preferred settlement procedures. Some sample settlement codes include interstate, intrastate, domestic, and overseas.

Settlements-Settlements are the processes or amounts of transfer of property or resources that settle the usage of products or services between companies.

Settlements Procedures-A process for distributing revenues among carriers in proportion to services provided or assets used.

Setup Routine-A software program or process that coordinates the installation of a program or system.

SF-Single Frequency Signaling

SF-Spreading Factor

SFF-Small Form Factor
SFID-Service Flow Identifier
SFN-Single Frequency Network
SG-Signaling Gateway
SG1-Service Group 1
SG2-Service Group 2
SG3-Service Group 3
SG4-Service Group 4
SGCP-Simple Gateway Control Protocol
SGCP-Skinny Gateway Control Protocol
SGML-Standardized General Markup Language
SGSN-Serving General Packet Radio Service Support Node

Shadow Market Affinity -Market strength of a company that is unaccounted for in the current financial statements because the good will, awareness, familiarity and preference have been established in the consumers mind during previous marketing cycles.

Shannon's Law-Shannon's law is a formula that defines the amount of information content that can be sent through a communications channel that has a specific bandwidth and signal to noise ratio. The formula is C (bits per second) is equal to b (bandwidth in Hertz) x log base 2 of (1 + s [signal to noise ratio]).

Shared Bandwidth-A characteristic of a communications channel in which the available capacity is shared among all of the attached stations.

Shared Bus-A type of switch fabric that uses a common communications channel (e.g., a backplane) as the mechanism for frame exchange among the switch ports.

Shared Key Authentication-An authentication process that shares a key between two or more users.

Shared Media-A physical communications medium that supports the connection of multiple devices, each of which may transmit and/or receive information across the common communications channel. Shared media systems require some method for the attached stations to arbitrate for the use of the common channel. See also Dedicated media and MAC algorithm.

Shared Secret Data (SSD)-Secret data that is stored in a mobile phone in the IS-136 and IS-95 cellular standards and in a cellular or PCS network that is used to create key codes that are transferred through the network to validate the identity of the mobile phone. The SSD information is created by an authentication A-key.

Shared Tenant Service-The sharing of centralized equipment, facilities, and telecommunication services by occupants in a building or office complex.

Shared Tree-A shared tree is a multicast structure that has the source of the multicasting located anywhere in the tree. The location of the source in a shared tree is called the rendezvous point (RP).

Shared Wireless Access Protocol (SWAP)-A common industry specification that allows wireless communication between devices in the home. The first SWAP specification was created in 1998.

Shared Writing Credit-Appearing as to recognize more then one writer, usually writers not associated to the writers team.

Shareware-Shareware is files, programs, or applications that provide a potential user to obtain the program(s) and trial the programs before deciding to keep or purchase the software. After shareware is installed, it is usually setup to operate for a predetermined period of time (e.g. 30 days) before it deactivates itself. If the user conforms to the terms of the shareware (usually through the payment of a fee), an access code is provided that activates the shareware program.

SHCCH-Shared Common Control Channel

Sheath-The outer covering of a cable, the main function of which is to provide protection for the insulated conductors that makeup up the cable.

Sheath Miles-The number of route miles installed (excluding pending installations) along a telecommunications cable path multiplied by the number of wires or fibers existing within cabling along the same path.

SHF-Super High Frequency

Shield-A metal covering used to restrict the transfer of electrostatic and electromagnetic signals (radio signals) to or from equipment or electronic circuits. Shields are used to eliminate unwanted signals that might otherwise produce interference to the normal operation of systems are circuits.

Shielded Enclosure-A screened or solid metallic housing that isolates an area from external electromagnetic fields or prevents fields from escaping such an enclosure.

Shielded Pair-A pair of wires enclosed in an electrostatic shield to prevent interference.

Shielded Twisted Pair (STP)-Wire that includes twisted pairs and a shield that surrounds the twisted pairs of wires. STP uses a metal shield (foil) that surrounds the twisted pair wires to help protect against unwanted electromagnetic interference.

Shielding-The process of enclosing a wire or circuit with a grounded metallic structure, so that electrical signals outside the structure cannot reach the wire and so that signals on the wire cannot reach beyond the structure, except to the apparatus to which the wire connects.

Shockwave Flash (SWF)-Shockwave flash (SWF) is a Macromedia multimedia digital video and animation format that is used to play animation, digital audio and digital video media on a media player that has flash plug-in.

Shockwave Video (SWV)-Shockwave flash video (SWV) is a Macromedia multimedia digital video format that is used to play digital audio and digital video media on a media player that has flash plug-in.

Shooting-Shooting is the capturing of image information onto a medium such as film or digital media files.

Shopping Cart-Shopping carts are the electronic containers that hold online store items while the user is shipping. The online shopper is typically allowed to view and change items in their shopping cart until they purchase. Once they have completed the purchase, the items are removed from their shopping cart until they start shopping again.

This diagram shows the basic operation of a shopping cart. This example shows that the shopping cart places the customers order information into a cookie (small memory area) on their own computer along with a user identification code assigned by the online store. This allows the customer's order information to remain available if they exit the store and renter at a later time.

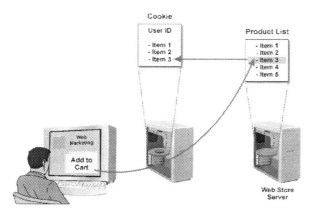

Online Store Shopping Cart

Short Circuit-A very low-impedance path between two or more conductors. The voltage output of a circuit terminated in a short circuit is zero or negligible, but the current is at a maximum, potentially capable of causing damage to the circuit. This also is called a short.

Short Message Service (SMS)-A messaging service that typically transfers small amounts of text (several hundred characters). Short messaging services can be broadcast without acknowledgement (e.g. traffic reports) or sent point-to-point (paging or email). Most digital cellular systems have SMS services. Short messaging for mobile telephones may include: numeric pages (dialed in by a caller), messages that are entered by a live operator via keyboard, an automatic message service that sends a predefined message when an event occurs (such as a fire alarm or system equipment failure), network operator announcements to customers, to and from other message capable devices in the system, from the Internet, advertisers or other information providers.

Shortened Burst-A shortened transmit burst used by the mobile when initial transmit occurs in a large diameter cell where timing information has not been established. This is required to overcome propagation delays which may cause burst collisions (overlapping received bursts).

Shortest Path Tree (SPT)-A shortest path tree is a multicast structure that has the source of the multicasting located at the base of the tree and each branch in the tree has the shortest path to other branches or nodes.

Shot Noise-The noise developed in a vacuum tube or photoconductor resulting from the random number and velocity of emitted charge carriers. The effect is noticeable as errors in a fiber optic system, in the sound output of a radio receiver, or in the picture of a TV set. Shot noise also is called Schottky noise.

Shovelware-Shovelware is a sarcastic name for a software program that has been created through the use of readily available programs or information. Shovelware is created by using (shoveling) multiple resources to create a new product that may not have significant value to its users.

Show Runner-In reference to an executive producer who takes daily responsibilities for his or her television series, and entitled superior status on set, as he or she mainly only communicates directly to networks executives to the studio.

S

SI-MKS System of Units, see: Systém International
Si-Silicon
SI-Status Indicator
SI-Système International
SID-Silence Insertion Description
SID-System Identification
SID-System Identity
Side Band-The frequency band on either the upper or lower side of the carrier frequency band where the frequencies produced by the modulation process take place. Various transmission techniques make use of one or both of these side bands when sending data.
Side Lobe-The spurious radiation of an antenna that causes signal leakage in an undesired direction. Excessive side lobes reduce antenna efficiency and can cause interference to other communication systems.
Sidetone-The hybrid coil directional coupler used in an analog telephone set is not ideal, a user hears his or her own voice appearing in his or her own earphone to some extent. This is called sidetone, and most users are accustomed to hearing a little sidetone in a correctly working telephone. Because of this, some artificial sidetone is intentionally generated in some four-wire telephone or radio systems to give the use the usual perception of a small amount of sidetone. Sidetone is very undesirable for modems and certain other systems, and echo cancellers must be used to minimize sidetone in such situations.
SIF-Signaling Information Field
SIF-Source Intermediate Format
SIFS-Short Interframe Space
SIG-Special Interest Group
Sign a Message-The process of adding a digital signature to a message. The digital signature is calculated from the contents of the message using a private key and appended or embedded within the message. The signing of a message allows a recipient to check the validity of a file or data by decoding the signature to verify the identity of the sender.
Signal-(1-electrical quantity) An electrical, optical, or other indicator, impulse, or fluctuating electric quantity that conveys information for messages, such as voice, data, or video. (2-alarm) An acoustic or visual device that attracts attention by lighting up or emitting a sound. (3-data) The information transferred over a communications system by electrical or optical means. (4-message) A message communicated by electrical, optical or other means. (See also: carrier.)(5-telecommunications network) An electrical, optical, or other indicator, impulse, or fluctuating electrical quantity used for network control, such as call routing or network management, or for the internal operation of network elements, such as timing or control of switching systems.
Signal Conditioning-The processing of a signal so as to make it compatible with a given device, including pulse shaping, pulse clipping, and other modifications.
Signal Converter-A device in which the input and output signals are formed according to the same code, but not according to the same type of electrical modulation.
Signal Distributor-A circuit in an electronic switching system that transmits control signals from a central point to other circuits.
Signal Egress-Signal egress is the emission of a portion of a transmitted signal from a transmission line. The emission is usually unwanted and may cause interference to neighboring cables or electronic circuits.
Signal Fading-A reduction of signal strength that is typically the result of combined radio signals that subtract from each other. Radio signals subtract from each other when one signal is delayed relative to the other by 1/2 signal cycle. This can be a result of the direct radio signal reception combined with a reflected signal that has to travel a longer path. See also: Multipath
Signal Generator-A test instrument that can be adjusted to provide a test signal at some desired frequency, voltage, modulation, and/or waveform.
Signal Ingress-Signal ingress is the absorption of radio signal energy from an external source into a communications circuit or communications link. Signal ingress may occur when electrical signals from sources such as radio or lightning spikes occur.
This diagram shows a source of signal ingress from a nearby radio tower that may occur in a transmission system. This diagram shows that a high power AM radio transmission tower that is located near a telephone line couples some of its energy onto the telephone line. This interference signal (radio ingress) usually reduces the data transmission capacity of a digital subscriber line (DSL).

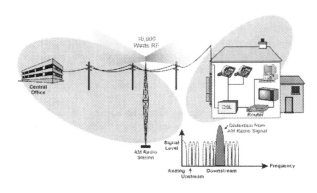

Signal Ingress Operation

Signal Path-Bus

Signal Processing-The process of modifying a signal from one form to another.

Signal Quality-Signal quality is a measure of how well a signal represents characteristics or attributes of the information it transports (voice, data or video).

Signal Regeneration-Signal regeneration is the process of reception and restoration of a digital pulse or lightwave signal to its original form after its amplitude, waveform, or timing have been degraded during transmission.

Signal Ringing-Signal ringing is an undesired oscillatory signal that is imposed on another signal as a result of a rapid signal change (such as a pulse) or signal processing element mismatches.

Signal Sampling-The process of taking samples of a particular characteristic of a signal, usually at equal time intervals.

Signal To Interference Ratio (SIR)-The ratio of a desired signal as compared to the total interference that is received with the signal.

Signal To Noise Ratio (SNR)-The ratio of information-carrying signal power to the noise power in a system. In a phone call, the signal would be your voice, while the noise could be static, clicking from relays, or any other sound that is not voice. Analogous effects occur in optical networks. SNR is often used as one measure of signal quality.

Signal Unit (SU)-(1-signaling link) Data bits that can be transferred as a group to convey information on a signaling link. (2-SS7) In the Signaling System 7 protocol, the smallest defined group of bits on the signaling channel used for the transfer of signal information.

Signaling-Signaling is the process of transferring control information such as address, call supervision, or other connection information between communication equipment and other equipment or systems.

Signaling Gateway (SG)-A signaling gateway (SG) is used to interface a signaling control system (e.g. such as SS7) and a network device (e.g. a transfer point, database, or other type of signaling system). The signaling gateway may convert message formats, translate addresses, and allow different signaling protocols to interact.

Signaling Information-(1-telecommunications) The information content of a signal that pertains to supervisory control and management of calls. (2-SS7) All information transferred over Signaling System 7 using its protocol.

Signaling Link-A communication path that carries common channel signaling messages between two adjacent signaling nodes.

Signaling Link Management Functions-Functions that control and take actions when required to preserve the integrity of locally connected signaling links.

Signaling Message-In the Signaling System 7 protocol, signaling information pertaining to a call, management transaction, or other event that is transferred as an entity.

Signaling Network-A signaling network is a system that receives, processes, and distributes messages that control a system. A signaling network may be an independent communication system or it may share some or all of the resources of the communication system it controls. An example of a signaling network is the Signaling System 7 (SS7) network. The SS7 network is a packet data communication network that is used to control public switched telephone networks. The SS7 system consists of signaling points that create or receive control commands (such as switches), signaling transfer points (for distribution of control messages), and control points that can receive and process requests for more advanced control features (such as toll free call routing information).

Signaling Protocol-A protocol that is used to coordinate the operation of devices and services within a system or network.

Signaling Route-In the SS7 protocol, a pre-determined path described by a succession of signaling points toward a specific destination.

Signaling System-A system that receives, processes and sends control information. Signaling may be in-band (replaces voice or data information) or out-of-band (is sent separately from voice or data information).

Signaling System 6 (SS6)-An internationally standardized protocol for medium-speed data communication between intelligent nodes.

Signaling System 7 (SS7)-The signaling system #7 (SS7) is an international standard network signaling protocol that allows common channel (independent) signaling for call-establishment, billing, routing, and information-exchange between nodes in the public switched telephone network (PSTN). SS7 system protocols are optimized for telephone system control connections and they are only directly accessible to telephone network operators. Common channel signaling (CCS) is a separate signaling system that separates content of telephone calls from the information used to set up the call (signaling information). When call-processing information is separated from the communication channel, it is called "out-of-band" signaling. This signaling method uses one of the channels on a multi-channel network for the control, accounting, and management of traffic on all of the channels of the network.

An SS7 network is composed of service switching points (SSPs), signaling transfer points (STPs), and service control points (SCPs). The SSP gathers the analog signaling information from the local line in the network (end point) and converts the information into an SS7 message. These messages are transferred into the SS7 network to STPs that transfer the packet closer to its destination. When special processing of the message is required (such as rerouting a call to a call forwarding number), the STP routes the message to a SCP. The SCP is a database that can use the incoming message to determine other numbers and features that are associated with this particular call.

In the SS7 protocol, an address, such as customer-dialed digits, does not contain explicit information to enable routing in a signaling network. It then will require the signaling connection control part (SCCP) translation function. This is a process in the SS7 system that uses a routing tables to convert an address (usually a telephone number) into the actual destination address (forwarding telephone number) or into the address of a service control point (database) that contains the customer

data needed to process a call.

Intelligence in the network can be distributed to databases and information processing points throughout the network because the network uses common channel signaling A set of service development tools has been developed to allow companies to offer advanced intelligent network (AIN) services

This diagram shows the basic structure of the SS7 control signaling system. This diagram shows that a customer's telephone is connected to a local switch end office (EO). The service switching point (SSP) is part of the EO and it converts dialed digits and other signaling indicators (e.g. off-hook answer) to SS7 signaling messages. The SS7 network routes the control packet to its destination using its own signal transfer point (STP) data packet switches using separate interconnection lines. In some cases, when additional services are provided, service control point (SCP) databases are used to process requests for advanced telephone services. This diagram also shows that the connections used for signaling are different than the voice connections. This diagram shows that there are multiple redundant links between switches, switching points, and network databases to help ensure the reliability of the telephone network. The links between points in the SS7 system have different functions and message structures. Access links (A-links) are used for access control between EOs and SCPs. Bridge links (B-links), cross links (C-links), and diagonal links (D-links) interconnect STPs. Extended links (E-links) are optionally used to provide backup connections from an EO to the SS7 network. Fully associated links (F-links) share (associate with) the connection between EOs.

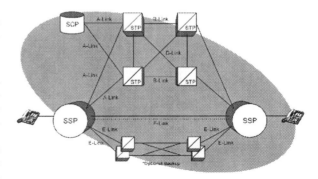

Signaling System 7 (SS7) Network

Signaling Tone-A tone that is used to indicate a status change or to transfer a signaling message on a communication system. The signaling tone is mixed with or replaces an audio signal in a communication system. Signaling tones are used on analog mobile communication system between the mobile station and the base station to indicate event changes such as handoff, the end of a call, or a special service request (e.g. hookflash).

Signaling Transfer Point (STP)-A signaling switch used in the SS7 common channel signaling network. These transfer points are used to route signaling messages (packets) to other signaling transfer points or network parts.

Signaling Transport (SIGTRAN)-A set of standards that were defined by the Internet engineering task force (IETF) that contain a set of protocols that are suitable to provide signaling control messages (such as SS7 message) over an Internet Protocol (IP) network.

This diagram shows that the Sigtran protocol stack is composed of the packet transport layer (IP), common signaling transport layer, and adaptation protocol layer. This protocol stack allows the Sigtran system to transport signaling control messages on a connectionless IP based system. The IP communication channel is managed by the SCTP connection-oriented protocol layer to allow for sequential and secure transport. The adaptation layers convert the protocols (e.g. IP addressing to Point Code addressing) between the Sigtran system to the SS7 system.

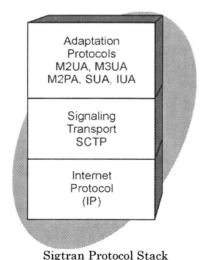

Sigtran Protocol Stack

Signature-(1-identifier) A pattern or image that uniquely identifies a person or process. (2-digital) A number calculated from the contents of a file or message using a private key and appended or embedded within the file or message. The inclusion of a digital signature allow a recipient to check the validity of the file or data by decoding the signature to verify the identity of the sender. (3-email) A short amount of information that is added at the end of each email message a person sends.

Signed Response (SRES)-(1- GSM) The calculated response value of the GSM authentication process.

SIGTRAN-Signaling Transport

Silcon Vendor-A silicon vendor is a company that manufactures the chip wafers that are used inside integrated circuits (chip manufacturer).

Silence Insertion Description (SID)-A frame of information that describes the silence period of audio. The SID information requires much less bandwidth than voice (typically 1 kbps).

Silence Suppression-A technique that is used in speech compression devices (such as speech codecs like G.729) that sends a reduced (minimal) number of bits during periods when the speaker is silent.

Silent Churn-The process where customers change their usage patterns in preparation for conversion (disconnection) from one carrier (service provider) to another. Silent churn can involve the complete or partial disconnection of service(s) by the customer. Silent churn is usually indicated by a reduction in usage of a particular service.

Silent Device-(1-general) A device that does not respond to the requests or commands of other devices (2-Bluetooth) A Bluetooth device appears as silent to a remote device if it does not respond to its inquiries. A device may be silent due to being non-discoverable, or because of baseband congestion while it is discoverable.

Silica Glass (SiO2)-Silica glass is composed using silicon dioxide SiO2 and is used in optical fibers.

Silo-A storage of information that is not accessible by one or other information systems.

SIM-Subscriber Identity Module

SIM Card-Subscriber Identity Module (SIM) card.

SIM Cards-Multi-Function Subscriber Identity Module

SIMPLE-SIP for Instant Messaging and Presence Leveraging Extensions

Simple Gateway Control Protocol (SGCP)-Simple gateway control protocol is used with SGCI for controlling Voice over IP Gateways from external call control elements.

Simple Mail Transfer Protocol (SMTP)-The protocol that provides a simple e-mail service and is responsible for moving e-mail messages from one e-mail server to another. SMTP provides a direct end-to-end mail delivery, rather than a store-and-forward protocol. The e-mail servers run either Post Office Protocol (POP) or Internet Mail Access Protocol (IMAP) to distribute e-mail messages to users. SCTP is defined in RFC 821.

Simple Message Desk Interface (SMDI)-Simplified message desk interface (SMDI) defines a way for a phone system to provide voice-mail systems with the information needed to intelligently process incoming calls. Each time the phone system routes a call, it sends an EIA/TIA-232 message to the voice-mail system that tells it the line it is using, the type of call it is forwarding, and information about the source and destination of the call.

Simple Network Management Protocol (SNMP)-Simple Network Management Protocol (SNMP) is a standard protocol used to communicate management information between the network management stations (NMS) and the agents (ex. routers, switches, network devices) in the network elements. By conforming to this protocol, equipment assemblies that are produced by different manufacturers can be managed by a single program. SNMP protocol is widely used via Internet protocol (IP) and operates over UDP well-known ports of 161 and 162. SNMP was originally defined in RFC1098 and is now obsolete and updated by RFC1157.

Simple Network Management Protocol version 1 (SNMPv1)-The Simple Network Management Protocol (SNMP) version 1 is the "original" standard protocol used to communicate management information between the network management stations (NMS) and the agents (ex. routers, switches, network devices) in the network elements. By conforming to this protocol, equipment assemblies that are produced by different manufacturers can be managed by a single program. SNMPv1 is defined in RFC1157. See also SNMP.

Simple Network Management Protocol version 2 (SNMPv2)-SNMPv2 Builds upon the SNMPv1 (Simple Network Management Protocol) protocol definition as defined in RFC1157. SNMPv2 adds the functionality of security and increased performance. It added a party-based security mechanism, the get-bulk operation, and the inform operation. SNMPv2 also introduced the use of 64-bit counters instead of the 32-bit used in SNMPv1. SNMPv2 is defined in RFCs 1441, 1445-1449, 1901, 1905, 1906, and 1908. SNMPv2 is sometimes referred to as SNMPv2p, where "p" designates support for the party-based security mechanism. SNMPv2 never really became a standard due to dissention in the standards committee about SNMPv2 security. SNMPv2c came out briefly to the same community-based security (denoted by the "c") used in SNMPv1.

Simple Network Management Protocol Version 3 (SNMPv3)-Defined in RFCs 2571 through 2575. SNMP defines a protocol designed to allow a network operator to manage the individual devices on his network. SNMPv3 specifically builds on SNMPv2 or SNMPv2c by adding in a user-based security model or USM. It supports secure authorization of the user sending the packet, as well as the privacy or encryption of the packet.

Simple Object Access Protocol (SOAP)-An XML-based protocol that can be used for simple one-way messaging and for performing Remote Procedure Call (RPC) request-response communication sessions. SOAP is not part of any particular transmission protocol group however it is commonly implemented on web based (HTTP) systems. Using the SOAP message structure, it is possible for clients and servers to use any language that can create and understand SOAP messages. Conformance to SOAP allows companies to independently develop modules for applications (building block) and applications can effectively interoperate with each other through a data network (such as the Internet).

Simplex-Simplex communication allows the transmission of information between users, but only one direction at a time on the same channel or frequency. The common use of Simplex systems is traditional television or audio broadcast radio systems that transmit a signal from a single transmitter to many receivers.

This diagram shows the communication process between two people using alternating (simplex) transmission and reception. This diagram shows that person 1 turns on (keys-on) the transmitter and begins to transmit audio. When person 1 transmits, their receiver is off or disconnected from the antenna. Person 2 hears the communication from person number one. When person 2 determines that person 1 has finished talking, person 2 turns on (keys-on) their transmitter and begins to talk on the same radio channel frequency. When person 2 transmits, their receiver is off or disconnected from the antenna. This diagram shows that simplex communications does allow two-way conversation. However, only one person can talk at any specific time.

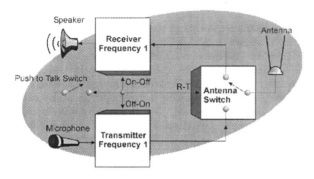

Simplex Operation

Simulation-A mathematical model that employs physical and mathematical quantities to portray a real-life situation and movement conforming natural phenomena.

Simulcast Transmission-Simulcast transmission is a process of transmitting a radio signal on the same frequency (or frequency this is very close) from multiple locations to allow the radio coverage area from adjacent radio transmitters to overlap. This overlap of radio coverage helps to ensure the radio signal more evenly covers a geographic area (preventing dead spots). Simulcast is extensively used in radio paging systems.

Because the transmission time from each of the transmitted signals may not be the same (one of the transmitter towers may be closer to the mobile radio than the other), the information sent by adjacent transmitters may be synchronized in time so that it arrives at the same approximate amplitude and phase as the other transmitter. Otherwise, this may cause radio signal distortion that could cause transmission errors. This is especially important in radio transmission systems that operate at high data rates (e.g. 9600 bps compared to 1200 bps).

This figure shows a paging system that uses simulcast transmission. In this example, the same pag-

ing message is sent to two paging transmitters. As can be seen in this example, the challenge with simulcast paging is as the pager is closer to one tower then the other, the transmit delay time can cause the signals to not directly overlap. Because radio signals travel so quickly, this delay is minor. However, it can result in some dead spots due to signal adding or subtracting. This diagram also shows that the ability to simulcast also depends on the distance and data rate of each of the transmitted signals to the radio (paging) receiver.

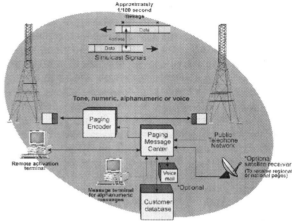

Simulcast Transmission

Simulcasting-Simulcasting is the process of transmitting multiple media channels at the same time.

Simulcrypt-Simulcrypt is the simultaneous sending of encrypted signals to multiple devices (such as a cable set top box) where each receiver is provided with dencryption codes.

Simultaneity-Simultaneity is the occurrence of a signal or pattern at the same time.

SINAD-Signal-To-Noise And Distortion

Singing-A continuous whistle or howl on a circuit, caused when repeater gain exceeds circuit losses and an imbalance exists at a 2-wire-to-4-wire conversion point.

Single Frequency Network (SFN)-A single frequency network is a radio system that operates one or several transmitters on a single frequency to provide one or several communication channels on a single frequency channel. Because single frequency signals from multiple antennas may be received a slightly different time periods (time delayed), each transmitter may be synchronized with each other or broadcast signals from each tower may include a time reference signals.

Single Mode Fiber-Single mode fiber transmission only allows optical signals to propagate in only one transmission mode. The mode of the fiber means that when photons are injected into the fiber, it has only one way of traveling in the core of the fiber

Single mode fibers are made by keeping the diameter of the fiber core small, typically less than about 10 micrometers (0.4 thousandths of an inch). Because single mode fibers limit the number of light rays that can travel down the fiber core, this reduces the distortion of optical pulses. The reduced distortion relative to multimode fiber allows singlemode fiber to support higher data rates and longer distances.

SIO-Service Information Octet

SiO2-Silica Glass

SIP-Session Initiation Protocol

SIP Extension-A set of commands or protocols that are used to extend the capabilities SIP protocol. SIP extensions are commonly used to rapidly extend the capabilities of an existing application or protocol without changing the underlying application or protocol.

Siphon-The first aerial section of cable leaving a manhole, rising vertically up a pole, and inclusive of the first splice closure. Siphon is synonymous with riser.

SIPPING-Session Initiation Protocol Project Investigation

SIP-T-A set of mechanisms that interface traditional telephone signaling with SIP. SIP-T provides protocol translation and feature transparency where the PSTN and SIP connect.

SIPv1-Session Invitation Protocol Version 1

SIR-Signal To Interference Ratio

Site Acquisition-Site acquisition is the process of acquiring the rights to locate a radio tower or communications equipment in a specific area or location.

Site Security-Site security is the ability of a web site to provide services to visitors and authorized users while protecting the information and processes it performs from unwanted and non-authorized users. Some of the site security risks include physical access, network failure, and external attacks on information and services.

Situation Comedy-A situation comedy is a television program that has a comedy focus that revolves around a specific theme or situation.

SIU-Subscriber Interface Unit

Skew-(1-digital effects) An effect in which the picture is slanted along its horizontal or vertical axis. (2-videotape machine) A curve at the top of the displayed picture resulting from improper VTR tape tension.

Skill Sets-Skill sets are groups of abilities that are developed through learning. Skill sets can be developed through training, demonstrations, practice drills and/or reviewing safety reference materials.

Skills Based Call Routing-The routing of an incoming call via an automatic call distribution (ACD) system to an extension that has a specific skill set (e.g. salesperson) based on information gathered by the caller (e.g. telephone DTMF key presses).

Skin Effect-An effective increase in the resistance of a conductor caused by alternating current flow, which tends to travel near the surface of a wire, thus reducing the total cross section of the conductor that will actually carry the current flow.

Skinny Client Control Protocol (SCCP)-Skinny protocol is used to establish, control and clear audio calls. Skinny protocol was developed by Cisco to provide for real-time calls and conferencing using the standard suite of Internet Protocols (IP). Skinny protocols permit Cisco IP phones to co-exist in an environment of other VoIP protocols such as H.323. The skinny protocol is more efficient than other VoIP protocols that allows it to operate in less memory size, using less CPU processor power, and it has lower complexity.

Skinny Gateway Control Protocol (SGCP)-A control protocol that was developed by Cisco Systems to provide communication signaling between the IP telephones and Cisco's CallManager product. It runs over TCP utilizing ports 2000-2002. MGCP, H.323, and SIP are similar protocols.

Skip-(1-radio) A phenomenon produced by ionized layers in the atmosphere that results in radio signals being received at greater distances from the transmitting point than would normally be expected. Skip varies depending on the time of day and the day of the year. (2-media) A skip is an unwanted change in the playing point of a media file. A skip may occur due to distortion or errors in the underlying media file or transmission stream.

SKU-Stock Keeping Unit

Skywave-The propagation path of a radio wave that travels through the ionosphere level of the atmosphere that reflects or refracts back to the earth.

SLA-Service Level Agreement

Slack-An extra (excess) length of wire or cable between two mounting poles or attachment points.

Slamming-Slamming is the unauthorized transfer of customer's preferred service provider to a different service provider.

Slander-Behavioral or verbal acts as to misrepresent one in a derogatory or negative way.

Slave-A component or machine in a system that does not art independently, but only under the control of another component (the master) or machine.

SLC-Subscriber Loop Carrier

Sleep Cycle-Sleep cycle is the time period when a radio receiver when a communication device determines when it should need to listen for alerting (paging) messages. The sleep cycle may be dynamically assigned in radio communication systems.

Sleep Mode-A feature where a radio receiver or other electronic circuits and a mobile radio product are deactivated or put into a low power consumption mode (such as back lighting off) to save battery energy.

Sleeve-(1-covering) A tubular covering designed to protect a splice, connection, or cable. (2-control lead) A control lead used in some patch cords, switchboards, and switching systems to represent the busy/idle status of the associated channel by means of the presence/absence of a voltage on the sleeve wire. It is connected to the sleeve conductor on a manual switchboard plug, hence its name. Corresponds to the European "C" wire. Only electromechanical switches use a sleeve wire. It is not used in electronic switches.

Slew Rate-The maximum rate of change of the output voltage of an amplifier operated within its linear region.

SLIC-Subscriber Loop Interface Circuit

Slice-A slice is a part of an image that is used in digital video and it is composed of a contiguous group of macroblocks.

Sliding Window-A sliding window is a technique that is used to provide flow and/or error control whereby the sender is allowed to transmit only that information within a specified window of frames or bytes. The window is shifted (slid) upon receipt of proper data acknowledgements from the receiver.

Slip-In a synchronous digital transmission system, the advance or delay of one or more signal elements from their expected time of arrival, resulting when transmitting and receiving terminals are not synchronized. Slip causes the loss or repetition of signal elements.

SLIP-Serial Line Interface Protocol

Slope-(1-transmission line) The rate of change, with respect to frequency, of transmission line attenuation over a given frequency spectrum. (2-telecommunications) A comparison of transmission losses at various frequencies in the voice band compared to the loss at 1004Hz. If the transmission loss is measured at 404 Hz and 2804 Hz relative to loss at 1004 Hz, it is known as three-tone slope.

Slot-(1-frequency) A narrow band of frequencies. (2-digital system) The time period for a signal element or sample of a voice channel. (3-satellite) A term used to express the longitudinal angular position geosynchronous orbit into which a communications satellite is parked. Above the United States, communications satellites typically are positioned in slots that are spaced 3' to 5' apart.

Slot Time-Slot time is a time period for a time slot. A time slot is the smallest time division dedicated to a communication device. For the GSM system, a timeslot is 577 microseconds.

Slow Frequency Hopping-Slow frequency hopping is a process of changing the radio frequencies of communications on a regular basis (pattern). The duration of transmission on a single frequency is typically much longer than the amount of time it takes to send several bits of digital information.

Slow Motion Video-Slow motion video is a method playing moving pictures to a viewer where the rate of viewing is slower than the actual event time. Slow motion video can be created by capturing or recording a program or scene at a rate that is faster than the playback rate. For example, if a scene is recorded at 60 frames per second and played back at 30 frames per second, the viewer will see the scene at ½ the speed of the scene (in slow motion).

Slow Scan Video-Slow scan video is a method of transmitting pictures over a voice grade system by scanning an image approximately 1 to 7 times per seconds rather than 25 to 60 times per second for conventional TV or video systems.

SLP-Service Location Protocol

SLP-Signaling Link Protocol

SLS-Signaling Link Selection Code

Slug Tuning-The adjustment of a resonant frequency by changing the inductance of a coil by moving a ferrite slug into or out of the coil.

SMA-Standard Microwave Adapter Connector

SMA 905 Connector-Subminiature A 905

SMA 906 Connector-Subminiature A 906 Connector

SMAE-System Management Application Entity

Small Firewire Jack-Small firewire jack is the small connector version for the Firewire personal area network system.

SMAP-System Management Application Process

Smart Aggregation-Smart aggregation is the use of detailed descriptive information about content that allows for the searching, categorizing and selecting content that helps predict its utility or value to a consumer or group of consumers. For example, smart aggregation allows for the recommendation of programs that directly relate to the desires or viewing habits of specific customers.

Smart Aggregators-Smart aggregators are companies who leverage descriptive information about content to search, categorizing and selecting content that helps predict its utility or value to a consumer or group of consumers.

Smart Antenna-An antenna technology that uses active components to allow the forming or selection of specific antenna patterns. Smart antennas may have multibeam capability that allows for the reuse of the same frequency in the same radio coverage area.

Smart Cards-A smart card is a portable credit card size device that can store and process information that is unique to the owner or manager of the smart card. When the card is inserted into a smart card socket, electrical pads on the card connect it to transfer Information between the electronic device and the card. Smart cards are used with devices such as mobile phones, television set top boxes or bank card machines. Smart cards can be used to identify and validate the user or a service. They can also be used as storage devices to hold media such as messages and pictures.

Smart Phone-An end user telephone device that combines computer functionality with telephone services. Smart phones often have enhanced display and keypad options along with advanced telephone call process feature controls. Some smart phones provide access to the Internet and they may reformat information such as telephone directories and personal schedules in a format through which the user can easily navigate and interact with.

Smart Switch-Smart switches are interconnection switching systems that can dynamically change its connection points, data rates and services that they provide. Smart switches are used to provide for voice, data, video, and advanced call processing (call feature) services.

This diagram shows how a smart switch can efficiently route information more directly to computers connected within its network. A smart switch builds a routing table based on a device's medium access control (MAC) address. As the data is addressed to a destination computer, the smart switch dynamically maintains a list of the computer addresses that are connected to each of its port's. After it has updated its table, packets are routed directly towards the port that the MAC address table has identified for that address.

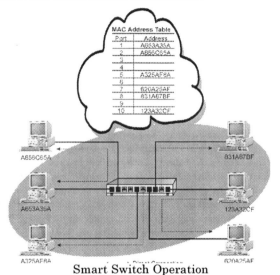

Smart Switch Operation

Smart Telephone (SmartPhone)-A SmartPhone is a telephone device that includes intelligent call processing to provide enhanced audio and display capability. Some of the advanced features offered by SmartPhones are communication list management, enhanced caller information displays, and information storage services.

Smart Terminal-An interface device (terminal) that has both independent (local) computing capability and the ability to communicate with other devices or systems.

Smart Wi-Fi-Smart Wi-Fi is a wireless LAN system that combines smart antenna array technology, multi-element MIMO diversity techniques and advanced directional beam forming to provide improved data connection performance and reliability.

SmartPhone-Smart Telephone

SMATV-Satellite Master Antenna Television

SMDI-Simple Message Desk Interface

SMDR-Station Message Detail Recording

SMDS-Switched Multimegabit Data Service

SMI-Structure of Management Information

SMIL-Synchronized Multimedia Integration Language

SMP-Session Management Protocol

SMPTE-Society of Motion Picture and Television Engineers

SMR-Specialized Mobile Radio

SMRS-Specialized Mobile Radio Service

SMS-Service Management System

SMS-Short Message Service

SMS Service Centre (SC)-The SMS service center (SC) receives, stores, delivers, and confirms receipt of short messages.

SMSC-Short Message Service Center

SMT-Station Management Task

SMT-Surface Mount Technology

SMTP-Simple Mail Transfer Protocol

SNA-System Network Architecture

Snail Mail-Snail mail is an Internet term for the process of sending messages or products through the postal service. It is called snail mail because email is near instant and the rate of delivery for the postal service seems as slow as a snail when compared to instant delivery.

Snark-A 3 headed beast!

SNDCP-Subnetwork Dependent Convergence Protocol

Sneakernet-The process of manually delivering information (such as letters) through a network of information users by walking between the distribution points.

SNI-Standard Network Interface

Sniffer-A sniffer is a device or program that receives and analyzes communication activity so that it can display the information to a person or communication system. While sniffers may be used for the analysis of communication systems, they are often associated with the capturing and displaying of information to unauthorized recipients. For example, a sniffer may be able to be setup to look for the first part of any remote login session that includes the user name, password, and host name of a person logging in to another machine. Once this information is captured and viewed by the unauthorized recipient (an intruder or hacker), he or she can log on to that system at will.

SNMP-Simple Network Management Protocol

SNMP Trap-An SNMP trap is a method for an SNMP agent, like a network device, to asynchronously deliver event notifications to a network management station (NMS) or trap receiver. SNMP traps are defined in MIB definitions. SNMP traps are sent on a best-effort basis and without any method to verify whether they were received by the trap receiver.

SNMP Views-SNMP views is a method for allowing network managers to restrict what MIBs and community-strings can be accessed by what users. SNMP views can be applied, depending on the vendor implementation, to SNMPv1 community strings, SNMPv2 party, or an SNMPv3 user. In SNMPv3 views is renamed to View-based Access Control Models (VACMs).

SNMPv1-Simple Network Management Protocol version 1

SNMPv2-Simple Network Management Protocol version 2

SNMPv3-Simple Network Management Protocol Version 3

Snooping-Snooping is the process of looking inside packets for specific message types or data elements. Snooping is typically performed to change the priorities of packets (e.g. for real time audio) or to alter or determine new locations that the packets should be sent (such as copying and forwarding packets for multicast sessions).

Snow-A form of noise picked up by a TV receiver, characterized by alternate dark and white dots randomly appearing on the screen.

SNP-Service Number Portability

SNR-Signal To Noise Ratio

Snubber-An electronic circuit that is used to suppress high-frequency noise signals.

SOAP-Simple Object Access Protocol

SOC-System Operator Code

Social/Psychological Churn -Type of voluntary churn that occurs when customers change carriers in order to enhance their image or self esteem. Often teenagers will churn for these reasons.

Society of Motion Picture and Television Engineers (SMPTE)-The SMPTE is an organization that assists with the development of standards for the motion picture and television industries.

Socket-A socket connection is the combination of a data communication device address and the logical channel (port) number.

Soft Capacity Limit-A soft capacity limit is a maximum capacity limit of a system that can be changed by altering the services provided (such as changing the type or quality of service offered) to change the maximum number of customers that can obtain service from the system. This allows a service provider to temporarily increase the system capacity in exchange for a reduction in the quality of voice. Examples of soft capacity are increasing the number of customers on a CDMA mobile communication system by reducing the audio quality through increased speech compression (lowering the average data rate per user).

This figure shows that a soft capacity limit allows for the gradual decay of voice quality in a communication system when additional users are added in a system. To provide service to more customers than is the capacity limit (over capacity) in a CDMA system, users in the system are provided with lower bit rates (higher speech compression). As a result, assigning lower bit rates to users as service demand increases trades off voice quality for increases in system capacity.

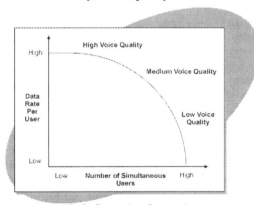

Soft Capacity Operation

Soft Client-A soft client is a software program that operates on a computing device (such as a personal computer) that can request and receive services from a network for specific applications. An example of a soft client is a software program that operates on a personal computer to operate as an Internet telephone.

Soft Console-An access point to an information system that is defined by software that allows users to monitor and/or interact with a system. The use of a soft console allows the controllable features of a console to change based on the authorization and console program software.

Soft Keys (Softkeys)-Soft keys are buttons on the keypad of an electronic device that have the ability to redefine their functions. Soft keys are typically located adjacent to a display that provides a description of the key function. This allows an electronic assembly to reduce the number of keys which is especially important for portable handheld telephones.

Soft Launch-Soft launch is the process of making a product available for purchase and distribution without significant promotional efforts. Companies may use a soft launch strategy for new products that may have limited available quantities or to test product acceptance and distribution channels prior to a hard launch.

Soft Phone-A soft phone is a software program that operates on a multimedia computing device (such as a personal computer or digital television) to operate as an Internet telephone.

Soft Switch (Softswitch)-Softswitches are call control processing devices that can receive call requests for users and assign connections directly between communication devices. Soft switches only setup the connections, they do not actually transfer the call data.

Softswitches were developed to replace existing end office (EO) switches that have limited interconnection capabilities and to transfer the communication path connections from dedicated high-capacity lines to other more efficient packet networks (such as packet data on the Internet). This allows a single softswitch to operate anywhere without the need to be connected to high-capacity trunk connections.

Softkeys-Soft Keys

Softphone-A software application that allows a computer to be used to dial numbers, make phone calls, and perform other functions that a traditional phone would be capable of doing. A softphone combines a software program that operates on a computer along with a signal processing card (a sound card) to provide telephone services. A firm-phone relies on its own circuitry for the telephone signal processing so that it does not burden the computer's operating system when processing the call. More importantly, if the operating system is busy servicing other software applications (such as a word processor), it does not degrade the quality of the telephone call.

Softswitch-Soft Switch

Software-The programs, instructions and procedures that enable a computer to perform designated tasks.

Software Application-A software application is a software program that performs specific operations to enable a user to apply the software to their specific needs or problems.

Software Decoding-Software decoding is the process of using software and a microprocessor (such as a microprocessor that is used in a personal computer) to decode and/or decompress a coded signal.

Software Defined Radio (SDR)-Software defined radios are transceiver devices that use digital signal processing to create and decode radio messages. Because they use digital signal processing for almost all functions, it is possible to change access technologies and radio transmission characteristics through software changes.

Software Development Kit (SDK)-A combination of software development tools that allow a company or person to develop software applications. The kit usually includes code editors, compilers, debuggers, utility libraries, along with technical information that instructs the developer on how to use the kit.

Software Extensions-A set of commands or protocols that are used to extend the capabilities of an existing application or protocol. Software extensions are commonly used to rapidly extend the capabilities of an existing application or protocol without changing the underlying application or protocol.

Software License-A software license is an authorization for a company and/or user to use a software program or application. Software licenses usually to contain conditions for the license that include restricting the copying of the program or the number of computers the software program can be installed on.

Software Loop-A software loop is the direction of a software program to a previous point that causes a repetition of a group of instructions in a computer software routine.

Software Patches-Software patches are program codes that are created to correct and/or update existing software.

Software Specification-A software specification is a document that describes the requirements and features of a software product or application.

Sole Writing Credit-In reference to a single writer or writing teams assigned credits.

Solenoid-An electromagnetic coil fitted with an iron core that moves when the coil is energized. Often, the core is used as an actuator when a short mechanical movement is desired, as in an electric lock.

Solid State Laser-A laser whose active medium is a semiconductor junction.

Sone-The subjective unit of loudness; one sone is the apparent loudness of a 1 kHz tone 40 dB above the threshold of audibility, or 0.0002 microbar.

SONET-Synchronous Optical Network

SOP-Standard Operating Procedure

Sorenson Codec-The Sorenson codec is the video codec (video compression) that is used in the Quicktime media player.

SOS-Silicon On Sapphire

Sound-The periodic or random oscillations in pressure, particle displacement, or particle velocity in a medium, such as air. The oscillations normally occur within an audible frequency range.

Sound Analyzer-An instrument consisting of a microphone, amplifier, and wave analyzer used to measure the amplitude and frequency of the components of a complex sound.

Sound Pressure Level (SPL)-A measure of acoustic wave force. SPL is the force a sound can exert against an object, measured in decibels.

Source Address (SA)-That address of the device that originally transmitted a message.

Source and Group Pair (S,G)-A source and group pair (S,G) is the identification information used for multicast source trees. The notation indicates the source address and the group address.

Source Code-Source code is software program instructions that are written in a language that is translated (compiled) into the operating code that a computer can understand.

Source Coding-Source coding is a process of converting an information source (such as a digital audio signal) into a coded form such as a compressed data format (such as an MP3 coded format). After a sequence of digital data bits has been produced by an analog to digital converter, the digital bits are processed to create a sequence of new bit patterns that are ready for transmission. This processing may include compression coding and the addition of control information (such as timing signals).

Source File-A tile on disk that has programming language statements for a specific assembler or compiler. From this data, the assembler or compiler creates an object file which is in machine code or language.

Source Identification-A brief message, keyed into video, that defines the originator or point of origin of the signal.

Source Intermediate Format (SIF)-Source intermediate format is a digital video format having approximately 1/2 the resolution of analog television (PAL/NTSC). SIF has a luminance resolution of 360 x 288 (625 lines) or 360 x 240 (525 lines) and a chrominance resolution of 180 x 144 (625 lines) or 180 x 120 (525 lines).

This figure the format of a SIF digital video on a display. In this example, portion of the video display 8 x 8 has been expanded to show horizontal lines and vertical sample points of luminance (intensity) and chrominance (color). This example shows that the sample frequency for luminance for SIF is 6.75 MHz and the sample frequency for color is 3.375 MHz. This example shows that the color samples occur on every 4th line and that the color samples occur for every other luminance sample.

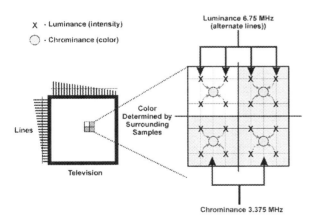

Digital Video SIF Format

Source Quench-A process of flow control used in a packet data network (such as the Internet) to indicate that a router or switching system is receiving packets faster than they can be processed. This allows the sender of the packets to slow down packet data transmission so the buffer of the receiver does not become full.

Source Tree-A source tree is a multicast structure that has the source of the multicasting located at the base of the tree.

SP-Signaling Point

Space-(1-communication) Space represents a logical value of 0. Space was defined from the open circuit condition in a teletypewriter system that actuates a printer function. Space is the opposite of a mark. (2-universe) The continuous 3-dimensional expanse outside the Earth's atmosphere.

Space Diversity-A method of transmission or reception, or both, employed to minimize the effects of fading by the simultaneous use of two or more antennas spaced a number of wavelengths apart.

Space Division Duplex (SDD)-Space division duplex (SDD) is a process of allowing two way communications between two devices by space sharing (geographic separation). When using SDD, one device transmits (device 1) on one spatial channel and the other device transmits (device 2) on a different spatial channel.

Space Shifting-Space shifting is the movement of a digital asset from one physical device or location to another. An example of space shifting is the moving of a television program from a set top box to a portable video player.

SPACH-SMS Messaging, Paging, And Access Response Channel

Spaghetti Code-Spaghetti code is the software commands or instructions that contains links to other applications, resources or programs.

Spanning Tree-A loop-free topology used to ensure that frames are neither replicated nor resequenced when bridged among stations in a catenet.

Spare Wire-A spare pair placed in a cable for use when a regular pair develops a fault.

Sparse Mode Multicasting-Sparse mode multicasting is the distribution of media to multiple users within a data network where a limited or small number of the users that are connected to the network are part of the multicast group.

Spatial Compression-Spatial compression is the analysis and compression of information or data within a single frame, image or section of information.

Spatial Interpolation-An interpolation process performed across a static video frame. This technique is used to create texturing and filtering effects, such as softening of an image. Spatial interpolation is used extensively in digital video effects systems to ensure clean anti-aliased images.

Spatial Redundancy-Spatial redundancy is the repetition of information or image elements within an image area.

Spatial Scalability-Spatial scalability is the ability of a media file or picture image to reduce or vary the number of image components or data elements representing that that picture over a given area (spatial area) without significantly changing the quality or resolution of the image.

This image shows how spatial scalability can be used to provide images in different formats through the use of progressive adding of image information. This example shows that a small portion of a data signal (a low data rate) is used to provide a low resolution mobile video (MV). Additional bits are added to the low resolution image data produce a standard definition (SD) video image. Additional bits can then be added to produce a high definition (HD) video image.

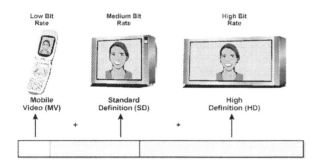

Image Spatial Scalability

SPC-Stored Program Control

Speakerphone-An audio terminal, consisting of a transmitter and loudspeaker unit, that is part of a telephone for hands-free conversations.

Special Interest Group (SIG)-A group that works to help develop and promote information about a specific technology, product, or service. A SIG is usually part of or related to an industry association.

Specification-A document intended primarily for use for procurement or operational requirements, which clearly describes the essential technical requirements for items, materials, or services, including the procedures by which it will be determined that the requirements have been met.

Spectral Efficiency-A measurement characterizing a particular modulation and coding method that describes how much information can be transferred in a given bandwidth. This is often given as bits per second per Hertz. Modulation and coding methods that have high spectral efficiency also typically are very sensitive to small amounts of noise and interference, and often have low geographic spectral efficiency. (See also Geographic Spectral Efficiency)

Spectrum-(1-general) A range or distribution of physical characteristics in a given system, such as the color spectrum and the electromagnetic spectrum. (2-frequency) A continuous band of frequencies within which waves have some common characteristics. (3-audio) The range of sound frequencies that can be detected by normal human hearing, ranging from about 20 Hz to 16 kHz - 20 kHz. Most people are not cognizant of hearing above about 17kHz, but harmonics in this area are known to be effective in music. (4-microwave) The range of radio frequencies not strictly defined but usually taken as between I GHz and 100GHz. (5-radio) The electromagnetic frequencies used for radio communication ranging from extremely low frequency (ELF) at about 200 Hz to tremendously high frequency (THF) at 3000GHz.

Spectrum Analyzer-A test instrument that presents a graphic display of signals over a selected frequency bandwidth. Amplitude in the vertical axis and frequency in the horizontal axis.

Spectrum Management-The regulation of spectrum use to achieve maximum efficiency while minimizing destructive interference from other authorized stations.

Spectrum Signature-The pattern of radio signal frequencies, amplitudes, and phases that characterize the output of a particular device, and tend to distinguish it from other devices.

Speech Coder-Speech coding (also called voice coding) is a data compression device that characterizes and compresses digital speech information. A speech coder is also called a Coder/Decoder (CoDec).

Speech Coding-Digital speech compression (speech coding) is a process of analyzing and compressing a digitized audio signal, transmitting that compressed digital signal to another point, and decoding the compressed signal to recreate the original (or approximate of the original) signal.

This diagram shows the basic digital speech compression process. The first step is to periodically sample the analog voice signal (5 - 20 msec) into pulse code modulated (PCM) digital form (usually 64 kbps). This digital signal is analyzed and characterized (e.g. volume, pitch) using a speech coder. The speech compression analysis usually removes redundancy in the digital signal (such as silence periods) and attempts to ignore patterns that are not characteristic of the human voice. In this example, this speech compression processes use pre-stored code book tables that allow the speech coder to transmit abbreviated codes that represent larger (probable) digital speech patterns. The result is a digital signal that represents the voice content, not a waveform. The end result is a compressed digital audio signal that is 8-13 kbps instead of the 64 kbps PCM digitized voice.

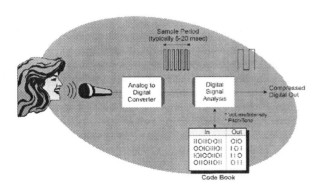

Speech Coding Process

Speech Recognition-Voice recognition is the ability of a machine to recognize your particular voice. This contrasts with speech recognition, which is different. It is the ability of a machine to understand human speech — yours and everyone else's. Voice recognition needs training. Speaker independent recognition does not require training.

Speech Scrambler-A device in which speech signals are converted into an unintelligible form before transmission and are restored to an intelligible form at reception, used for security reasons.

Speed-(1-velocity) Synonym for velocity, in some cases a synonym for the absolute value of a velocity (that is, a positive quantity regardless of whether the velocity is positive or negative). Speed is measured in meters per second (or km/s, miles/s, etc.). (2-data rate) Often loosely used when the term "bit rate" would be more precise. When a higher bit rate is used to transmit a quantity of data, the complete transmission is finished in a briefer time, thus leading to the imprecise description of this result as being "faster" or "higher speed." In fact, the wave speed of the signal through the transmission path occurs at the same speed for both low bit rates and high bit rates.

Speed of Light-The speed of light is the velocity that lightwaves travel. In vacuum (similar to air), the wave speed of a lightwave is 300 million meters per second (186,281.6 miles per second). In other materials, the speed that lightwaves travel is lower.

Spiff-The providing of marketing incentives from one company directly to sales representatives of another company (usually a retail sales company). Spiffs are often used to focus sales representatives on demonstrating and giving preference to products or services from a specific manufacturer.

Spike-A high energy, short duration pulse superimposed on an otherwise regular waveform.

SPL-Sound Pressure Level

Splice-The process of joining together two entities permanently, to provide an electrical or optical path from one wire or waveguide to another.

Splice Enclosure-A splice enclosure is a plastic or metal container that is used to cover and protect wires or cables. Splice enclosures may contain multiple container shells. An outer shell may be used to provide mechanical and environmental protection and the inner shell may be used to hold the cables or a splice tray.

Splice Loss-Splice loss is the signal loss through the cable splice (insertion loss), often expressed in decibels (dB).

Splice Tray-A splice tray is a plastic or metal pan or tray that contains ridges or guiding dividers that are used to hold and separate wires, cables or splices. A splice tray is commonly used in a splice enclosure.

Split Run Ad-A split run ad is an advertising message that has two or more different versions where the avail has been purchased by one supplier and is being used to show ads for different versions of the same brand or one product (for example different models of automobile to different age-groups). Split run ads may be used to test the effectiveness of an advertising message to specific regions to different target market segments.

Splitter-A circuit, device or component that divides a complex input signal into several outputs. A splitter may simply divide the signal energy power from a single source to 2 or more outputs or the splitter may separate frequency components to different output ports.

Spooler-A computer program that queues input for later output. For example, a print spooler can accept files at a high transfer rate, then send them to a printer at whatever rate that printer can handle.

Sporadic-An event occurring at random and infrequent intervals.

SportsML-Sports Markup Language

Spot Based Sales-Spot based sales is the offering of services (such as advertising) where the value or cost of the services is determined by the subscription or viewership of the services in a particular geographic region (spot area).

Spotting-Spotting is a step in process of production where appropriate moments of film are identified and placed with correlating music or sound effects.

SPP-Sequenced Packet Protocol

SPP-Serial Port Profile

Spread Spectrum-Spread spectrum transmission is a process of spreading information signals (typically digital signals) so the frequency bandwidth of the radio channel is much larger than the original information bandwidth.

To create a spread spectrum signal, the information signal is multiplied by a spreading code. The multiplication process produces several bits of information for each bit of information signal. This makes the transmitted signal over a wide frequency band. To receive the signal, the spread spectrum receiver must use the same spreading code to recover the original information signal.

This technique changes the frequency components of a narrowband signal so they are spread over a relatively wide frequency band. The resulting signal resembles electrical noise (called white noise).

Spread spectrum transmission is a technique that has been used to achieve very high signal security and privacy, and to enable the use of a common frequency band by many users. This is because the transmitted signal energy is relatively low compared to the natural background noise of received signals. Because the signal is spread over such as wide band, other narrowband signals (such as a high power radio jamming signal) have little effect on the overall transmission of the spread spectrum signal. This makes the spread spectrum signal very hard to find and very hard to interfere with.

Spreading Code-A spreading code is a binary number that is used to multiply an information signal to produce a wideband spread spectrum signal. Spreading codes may be unique (such as in the IS-95 CDMA system) or they may be fixed (such as in the 802.11 Wireless LAN) system.

Sprite-A sprite is a graphic object that can move and change independently of other objects and/or video on a display.

SPT-Shortest Path Tree

Spurious Emissions-Unwanted electromagnetic signal emissions from equipment or an assembly. Spurious emissions may be caused by devices such as microprocessors, oscillators, clock assemblies, electrical switching equipment, and other signal processing equipment. Unwanted spurious emissions may interfere with equipment that operate on frequencies where spurious emissions may occur.

This diagram shows a transmitter and some of the sources of spurious emissions from the equipment. This diagram shows that signal energy may leak from a local oscillator, computer clock, microprocessor, or any other signal processing or transmission device naling end point are transmitted on an "A" link. The "B" (bridge) links connect the STP to another STP. The "C" (cross) link connects STPs performing identical functions into a mated pair. "D" (diagonal) links connect the secondary (e.g., local or regional) STP pair to a primary (e.g., inter-network gateway) STP pair in a quad-link configuration. "E" (extended) links connect the SSP to an alternate STP. An "F" (fully associated) link is connected between two signaling end points (i.e., SSPs and SCPs).

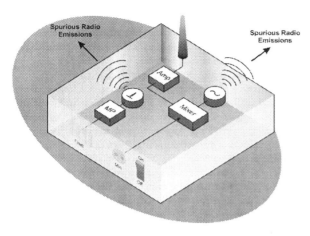

Spurious Emissions Operation

Spyware-Spyware that is software that resides on a users' computer that provides information about the user's actions on the computer to a remote location. Spyware program operations may range from simply tracking web browsing habits to the providing of private information such as user account identification and password codes. Spyware may be installed on a users' computer without the user's knowledge or consent.

SQL-Structured Query Language

Square Pixels-A square pixel is the smallest component of an image that has a square shape. Pixels on a computer monitor are typically square pixels.

SRES-Signed Response

SRTP-Secure Real Time Protocol

SS-Supplementary Service

SS6-Signaling System 6

SS7-Signaling System 7

SS7 Link-An SS7 link is a communication channel that is used in the Signaling System 7 system that transfers telephone system control messages. There are several different types of SS7 link types ("A" through "F") used in an SS7 system.

This figure shows the relationship between the link names and the link location (type). Signaling links are logically organized by link type ("A" through "F") according to their use in the SS7 signaling network. The "A" (access) links connect the signaling end points (e.g., an SCP or SSP) to the STPs. Only messages originating from or destined to the sig-

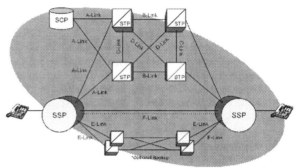

SS7 Signaling Link Types

SSB-Single Sideband

SSD-Shared Secret Data

SSH-Secure Shell

SSL-Secure Sockets Layer

SSL Certificate-SSL certificate is an authenticated document or electronic key that is signed (identified) by a trusted certificate authority (CA).

SSL Protocol-SSL protocol is the commands and procedures used by software programs to perform authentication and encryption using information or messages that are received between computing devices and trusted certificate authorities.

SSP-Service Switching Point

SSS-Switching Sub System

ST Connector-Straight Tip Connector

Stack-(1-interrupt) An area of memory needed to hold information about the status of a computer at the instant an interrupt occurs, so the computer can continue processing after the interrupt has been handled. (2-memory) A storage area for handling the accessing sequence of nested subroutines. In a UFO (last-in-first-out) stack, the last instruction pushed onto the stack is the first that may be removed or popped from the stack. Some interpreted languages permit limited manipulation of information on the stack. Others use multiple stacks for retention of "dictionaries" of special instruction features.

Stack Pointer-A counter or register in a computer system used to track the storage and retrieval of each byte stored in a system stack.

Standalone Viewer-A standalone viewer is a program and/or display device that is specifically designed to allow a user to view specific forms of media content. An example of a stand alone viewer is a portable eBook reader.

Standard-A standard is the operational descriptions, procedures or tests that are part of an industry specification document or series of documents that are recognized by people or companies as having validity or acceptance in a particular industry or technology.

Standard (STD)-A description of the processes, protocols and specific parameters that are used by a system or particular device. Standards may be developed by industry associations or government agencies.

Standard Definition TV format (SD)-Standard definition (SD) television are the resolutions of traditional analog television. Standard definition for NTSC is 480 lines with 60 interlaced fields (60i) per second. Standard definition for PAL/SECAM is 576 lines with 50 interlaced fields (50i) per second.

Standard Operating Procedure (SOP)-A procedure (or set of procedures) that are used (referred to) to ensure the same repeated step by step actions are followed each time an action is required. SOPs are often used to ensure the same quality of service or to ensure the reliable configuration and testing of equipment.

Standardization Body (Standards Body)-A standards body is a group, associated companies or people that produce or oversee the production of industry standards.

Standardized General Markup Language (SGML)-SGML is the result of generalizing and then standardizing a number of Rich Text Formats(RTF) developed for word processing software on personal computers. SGML is actually a metalanguage used to define markup languages. The definitions are called Document Type Definitions or DTDs

Standards-Documents that describe an agreed-upon way of doing things such that independent groups or companies can design and build hardware, firmware, software, or combinations there of and have them interwork with similar products designed and built by others.

Standards Body-Standardization Body

Standards Committee-(1-general) Committees that are commissioned to developed industry standards or operating procedures. Standards committees allow technical experts from several manufacturers (often from competing manufacturers) with a defined objectives and generally have a commitment to unbiased development activities. (2-humor) An excuse to travel to interesting places at nice times of the year to drink beer and discuss arcane technical issues with 400 geeks.

Standby Mode-Standby mode is an operation process that maintains a device or system in a state that is rapidly changed to a mode of preparatory or active operation.

Standby Time-The amount of time a mobile telephone or radio can listen for incoming messages between battery recharges or replacement of its disposable batteries.

Star Coupler-A star coupler is a passive signal distribution device or assembly that has several input and output ports where each input port is connected to output ports and the energy is passively (no amplification) distributed between the input and output ports.

This figure shows the functional operation of an optical star coupler. This diagram shows a star coupler that has 2 inputs and 4 outputs. This example shows that the two input signals contain 2 optical wavelengths each. The energy from each of the input signals is combined and distributed to each of the output ports.

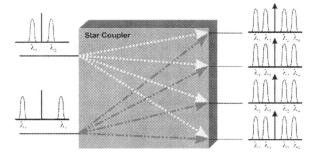

Optical Star Coupler

Star Hub-A star hub is a passive signal distribution device or assembly that has several input and output ports where each input port is connected to output ports and the energy is passively (no amplification) distributed between the input and output ports.

Star Network-A star network is a communication system where each communication device (typically computers) is interconnected to a central node in the network.

This figure shows a star topology network. Each node in this network is connected directly to a central node. Each station must communicate with or through the central node (usually a hub) to reach other stations in the network.

Star Topology (Centralized)

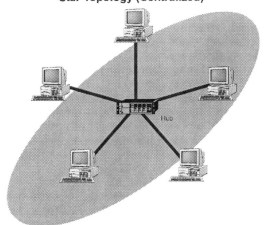

Star Topology System

Star Topology-A network connection scheme in which each station is connected directly to a central node. Some networks, such as ATM are both a logical and physical star topology. 100-Base-T Ethernet however is a star-wired, logical bus network, while Token Ring is a star-wired, logical ring network. (See also: bus topology, ring topology, tree topology.)

Start Bit-A start bit is the first element in each character that prepares the receiving device to recognize the incoming information elements in asynchronous transmission connections.

Stateful Proxy-A proxy server that remembers the call state information during a specific communication session transaction (such as a call transfer). All the call state messages associated with a specific communication state must pass through the stateful proxy. The stateful proxy is defined in the session initiation protocol (SIP) standard.

Stateless Proxy-A proxy server that does not remember any call state information during a specific communication session. Each call state messages associated with a specific communication state is completely processed by the stateless proxy and no information is stored about the event or call status. Stateless proxies are used in the center of a communication system to assist in the connection of communication sessions. The stateless proxy is defined in the session initiation protocol (SIP) standard.

Statement-(1-programming language) A single unit of program command in a high-level language. (2-billing) A list of invoices and payments for a specific user or account.

Static IP Addressing-Static IP addressing is the process of assigning a fixed Internet Protocol (IP) address to a computer or data network device. Use of a static IP address allows other computers to initiate data transmission (such as a video conference call) to a specific recipient.

Station-(1-general) A device that connects to a communication system that provides input and/or output functions. (2-telecommunications) An installed telephone, computer, or other communications instrument that communication service is used or operated.

Station Address-A grouping of numbers that uniquely identifies a station in a local area network. Data transmitted from the station will bear this grouping of numbers as a source address, and data destined for this station will bear this grouping of numbers as a destination address.

Statistical Model-Tool for testing different hypotheses through the use of statistical evaluation techniques. For example, a marketing manager would like to know how many customers will leave if the price is raised by 20%. A statistical model could be developed to predict the most likely customer reaction.

Statistical Multiplexer-A time-division multiplexing switch that dynamically assigns channel bandwidth as each connected device needs it to transmit data. Typically, a statistical multiplexer enables dozens of terminals to share a leased line and a pair of modems, thus saving line, modem, and dial-up costs.

Statistical Multiplexing-Statistical multiplexing is the process of transferring communication information on an as-needed statistical basis. For statistical multiplexing systems, connections can be initiated and maintained according to anticipated activity need. Each communication channel is dynamically allocated by time slots or codes on a main transmission facility. This allows a communication system to operate more efficiently based by transferring information only when there is activity (such as voice or video signals).

Statistical Remultiplexing-Statistical remultiplexing is a process of adjusting the compression and recombining multiple communication channels into a single transmission path after it has been previously multiplexed. An example of statistical remultiplexing is the recombining of television channels into a transport stream after a multichannel transport stream has been received, separated and decoded from a satellite or network feed.

Status-The present operational condition of a device or system.

Status Bit (SB)-A status bit is a bit of information in a transmitted signal that identifies the status of the system or information part (e.g. system busy/idle status bit).

Status Code-A code that indicates the status of a message or communication session. An example of a status code is a message that indicates that a communication session is active.

Status Indicator (SI)-A status indicator flag informs a communication device that the process of storing and transferring information has been delayed (slipped) due to the inability of transmission rate of one communication link to keep up with the supply of data or speed of another communication link.

STB-Set Top Box
STC-Space Time Coding
STC-System Time Clock
STD-Standard
STE-Section Terminating Equipment

Steady State-A condition in which circuit values remain essentially constant, occurring after all initial transients or fluctuating conditions have passed.

Steerable Antenna-An antenna that has a steerable transmission pattern. Steerable antennas may use mechanical or electrical systems to adjust the beam pattern.

Steganography-Steganography is the technique of adding (data embedding) or changing information in a file or other form of media that can be used to identify that the media is authentic or to provide other information while remaining hidden to the viewer, listener or user.

Step By Step Switching System (SXS)-A switching system that automatically selects the correct line or path for a telephone number through a switch through the use of step-by-step switches that are progressively assigned to each digit of the dialed telephone number. The SXS system operated using the dialed pulses created when a customer dialed each number using a rotary dial. The output of each switch is cascaded to the next switch in an SXS system.

Stereo-Stereo is the generation and reproduction of a 2-channel sound source. Stereo signals may be transmitting through the use of independent channels or it may be sent by sending a single (mono) channel along with difference signal(s).

Stereo Broadcasting-Stereo broadcasting is a process that sends two channels of voice, data, or video signals to a group of people or companies in a specific geographic area or who can receive signals from the broadcast network.

Sticky Customer-A sticky customer is a person or company who regularly purchases products or subscribes to a service and tends to stay with (sticks to) or use the same company or person who provided the products or services.

STK-SIM Toolkit
STL-Studio Transmitter Link
STL-IP-Studio Transmitter Link Internet Protocol
Stock Keeping Unit (SKU)-The SKU of a particular item is the package size used to store or dispense it. Each different package size has a distinct identification number, often represented by a bar

S

coded number on a label. For example, one SKU for electrical resistors is a package of 10 resistors. Another SKU is a roll comprising 1000 resistors held by disposable tapes so that the roll can be used in an automatic component inserting assembly machine.

Stop Bit-A stop bit is the last transmitted element in each character, which permits the receiver to come to an idle condition before accepting another character in an asynchronous transmission.

Stop Date-A predetermined or estimated date where an individual is retired from his or her obligated task, putting a stop to further services, can occur prior to completion of correlating project.

Stop Lock-A stop lock is a feature on a system or device that prohibits unauthorized and/or unqualified users from stopping a device from operating. A stop lock feature may be used on a video server or media converter to prevent users or operators from accidentally stopping the device from operating.

Storage-(1-non-volatile) A storage medium in which information can be retained despite the absence of power, and which becomes available again as soon as power is restored. (2-parallel) A storage medium in which all bits, characters, or words are equally available, access time not being dependent upon the order in which they were stored. (3-serial) A storage medium in which access time varies according to the order in which information was stored. Storage can be serial by word, by character or by bit. (4-volatile) A storage medium in which information is lost whenever power is removed from the store.

Storage Area Network (SAN)-A network that is used primarily in a data center or server room. The "storage area" is defined by a cluster of servers and storage devices that share common resources and users. Occasionally, the term may be used for a wide area or metro area network that is used for data center redundancy and disaster protection.

SANs typically have very low latency, high throughput and offer assured delivery of block I/O. The most common SAN implementation is Fibre Channel, however this is not the only alternative. Recently there have been efforts, such as iSCSI to implement storage networks with Ethernet and IP (Internet Protocol) infrastructure.

Storage Capacity-The quantity of information that can be retained in a memory system, usually measured in MegaBytes, Gigabytes, or TeraBytes.

Storage Medium-A storage medium is a material, device or assembly that can be modified or altered to store data or information.

Stored Program Control (SPC)-A stored program control system uses a computer program (software instructions) to control the operation of a communication system. Prior to the use of SPC, mechanical switches controlled the interconnection of communications networks.

Stovepipe-A storage of information that is not accessible by one or other information systems.

STP-Shielded Twisted Pair

STP-Signaling Transfer Point

Straight Pullbox-A straight pullbox is a rectangular or square steel box installed inline within a buildings conduit run. The use of an inline pullbox provides for interim pulling locations to minimize the length of cable that is pulled at one time. This can reduce the tensile load stress on the cable.

Straight Splice-An engineered splice method where the entire incoming cable count is spliced through. For example cable 10 pairs 1-600 is "pushed-through" or spliced straight for further distribution or feeder activity.

Strain Relief Boot-A strain relief boot is part of a connector assembly that holds the cable jacket and/or restricts the movement of the cable to minimize the stress it may cause on the connection of the cable wires or fibers inside the connector.

Stratum Clock-One of a hierarchy of digital system clocks that provide long-term accuracy, short-term stability and pull-in range to synchronized digital networks.

Stream-A stream is a flow of data or information. A stream may be continuous (circuit based) or bursty (packetized).

Stream Control Transport Protocol (SCTP)-A protocol that is used to coordinate the sending of signaling information over real time communication sessions. SCTP is defined in RFC 2960.

Stream Layer-The stream layer is used in a communication system (e.g. EVDO) to help identify and manage different types of information flows through the different layers. The stream layer is used to identify and tag packets or data blocks that are associated with specific types of information streams (such as continuous information transfers for streaming video or bursts data communications for web browsing). This allows other layers to prioritize and process packets for particular types of applications or quality of service parameters.

Stream Protocol-Stream protocol is used to identify and tag packets or data blocks that are associated with specific information streams (continuous information transfers). This allows other layers to prioritize and process packets for particular types of applications or quality of service parameters.

Stream Recorder-A stream recorder is a device and/or software that is used to capture, format and store streaming media.

Stream Thinning-Stream thinning is the process of removing some of the information in a media stream (such as removing image frames) to reduce the data transmission rate. Stream thinning may be used to reduce the quality of a media stream as an alternative to disconnecting the communication session due to bandwidth limitations.

Stream Video Format-Stream video format is the organization and flow control of digital information into a form that can be continuously streamed from a media source to a user.

Streaming-Media streaming is a method that provides a continuous stream of information that is commonly used for the delivery of audio and video content with minimal delay (e.g. real-time). Streaming signals are usually compressed and error protected to allow the receiver to buffer, decompress, and time sequence information before it is displayed in its original format.

Streaming Class-Streaming class is the delivering of audio or video signals through a network by establishing and managing of a continuous flow (a stream) of information. Upon request of streaming class of service, a server system (information source) will deliver a stream of audio and/or video (usually compressed) to a client. The client will receive the data stream and (after a short buffering delay) decode the audio and play it to a user. Internet audio streaming systems are used for delivering audio from 2 kbps (for telephone-quality speech) up to hundreds of kbps (for audiophile-quality music).

Streaming Media-Streaming media is the continuous transfer of media information that is sent through a communications network. Streaming media commonly refers to audio and/or video images that are sent through a data network (such as the Internet).

Streaming Protocol-Streaming protocols are commands, processes and procedures that are used for delivering and controlling the real-time delivery of media (such as audio and or video streaming). The true streaming protocols currently used by the major software vendors for their streaming systems are proprietary, requiring exclusive use of their software components at both server and client end. HTTP pseudostreaming, while not proprietary, is not a real-time streaming protocol strictly speaking, because it is based on TCP which is not real-time.

Streaming Server-A streaming server is a computer or a device that efficiently and effectively performs continuous transmission (streaming) of digital media.

Streaming Services-Streaming services are audio and video services delivered to customers where the content stays on the server. The delivery of information in streaming form has two main advantages: 1-The customer doesn't have to wait until the entire file downloads to view or listen to the content, 2- The content is kept on the server helps to reduce the copyright fears among content owners.

Streaming Video-Streaming video is the continuous transfer of motion picture information that is sent through a communications network. Streaming video commonly refers to audio and/or video images that are sent through a data network (such as the Internet).

Strengths, Weakness, Opportunities, and Threats (SWOT)-The analysis of strengths, weaknesses, opportunities and threats that compares the strategic position of a company in a given industry or situation.

Strip Chart Recorder-A strip chart recorder is a measurement device that continuously records information on a chart (roll or disk) as a strip line. A strip chart recorder allows for the long term (hours to months) viewing of a measured value.

Strong Password-A strong password is a user access code that is difficult to guess or obtain through the use of word lists or sequential password access attacks. Strong passwords are usually long (10 characters or more) and usually include a mix of letters, numbers and other characters such as punctuation marks.

Structure of Management Information (SMI)-SMI is used for defining how SNMP objects and the MIBs in which they are defined are structured. SMI was first introduced in SNMPv1 or in RFC1065. SMI defines that the structure of all objects is a tree, starting at or rooted at ISO. Object identifiers (OIDs) are designated as this SMI structure.

Structured Audio Score Language (SASL)- Structured audio score language is a set of commands and their associated processes that were created for MPEG-4 to control audio synthesizers. SASL is similar to MIDI and it has additional capabilities that MIDI did not have.

Structured Query Language (SQL)-A widely used computer programming language for manipulating database information.

STS-Synchronous Transport Signal

STS-1-Synchronous Transport Signal Level 1

Studio Console-A studio console is an access device that allows a studio operator or technician to manage and control content within a media distribution or production network. A studio console commonly consists of a display monitor and keyboard.

Studio Endorsed-A studio endorsement is the allowance of use of a program or service as long as the use meets with specific distribution requirements. Studios typically endorse (authorize) the use of specific content protection systems that can be used for the distribution of their programs and not all content protection systems meet their security requirements. For IPTV service providers who do not use a studio endorsed content protection system, it may not be possible to get and distribute content from that studio.

Studio Transmitter Link (STL)-A studio transmitter link is a point-to-point connection (e.g. a microwave link) that is used to deliver audio and/or video program material from a studio to the transmitter site.

Studio Transmitter Link Internet Protocol (STL-IP)-A studio transmitter link Internet protocol is an IP connection that is used to deliver audio and/or video program material from a studio to the transmitter site.

Stuffing-Stuffing is the process of inserting bits or data elements into a file or block of data to increase the size of the file or block to a specified amount (such as a packet that has a fixed length).

Stylus-(1-audio) The part of a phonograph pickup that transfers the sound vibrations, as pressed into the sides of the record groove, to the phono cartridge or transducer. In the transducer, the vibrations are converted to electrical signals. (2-comput-

er) A small pointer used like a pencil for inputting information (drawing) on a computer graphics system.

SU-Signal Unit

SUA-SCCP User Adaptation

Sub Band Signaling-Sub band signaling is a method that uses a frequency band that is located within the communication channel but outside the normal communication channel (e.g. audio band) bandwidth to transfer signaling control messages. A unique signaling feature used by some radio systems is the sub-band digital audio signaling. In many radio broadcast and mobile communication systems, an audio bandpass filter blocks the audio channel's lower range. It is possible to combine a sub-band digital signaling channel (a low speed digital signal) using the lower frequency range (below 300 Hz). This figure shows how sub-band digital and audio signals are combined with standard audio.

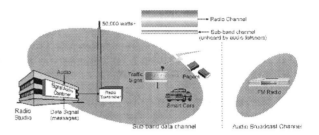

Sub-band Signaling

Sub Rate Multiplexing-A process that divides a single transmission path to sub channels that operate at lower (sub rate) data transfer rates.

This figure shows how eight compressed voice signals can be sub rate multiplexed to share a single DS0 (64 kbps) channel. In this system, an analog voice signal is converted to a PCM digital signal at 64 kbps. A speech coder analyses and compresses the voice signal to 8 kbps. This compressed digitized voice channel is time shared (time multiplexed) on a single 64 kbps DS0 channel with eight other 8 kbps compressed voice channels. Each of the 64 kbps DS0 channel is then time multiplexed onto a leased line (24 for T1 and 30 for E1 lines). The process is reversed on the receiving end where the T1 or E1 channels are split back into DS0 channels. Then each DS0 channel is further de-multi-

plexed again to produce eight 8 kbps compressed voice signals. This digital signals are then converted back into their the 64 kbps signal by a speech decoder and a digital to analog converter then changes the digital signal back into its original analog form.

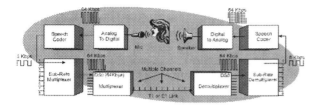

Sub Rate Multiplexing

Subassembly-A functional unit of a system.

Subcarrier (SC)-(1-general) A subcarrier is a modulation that is imposed on another carrier. (2-video) Video subcarriers for NTSC or PAL video are a continuous carrier signal that represents a portion of the color video. The NTSC subcarrier frequency is 3.579545 MHz and the PAL frequency is 4.3361875 MHz.

Subchannelization-Sub channelization is the dividing of communication channels into smaller sub-parts.

Subcommittee-A subcommittee is a group that is created by a committee to perform a specific function such as to analyze or solve a specific problem. Subcommittees are typically created on a temporary basis and are usually composed of members of a committee that created it.

Subgroup-A subgroup is a subset of a group that is created by a committee or working group to perform a specific function such as to analyze or solve a specific problem. Subgroups are typically created on a temporary basis and are usually composed of members of a committee or working group that created it.

Submarine Cable-Submarine cable is a communication line that is designed for underwater applications. There are different types of submarine cable designs that vary based on their applications. Some submarine cables are used to cross shallow waterways while others are used to cross large ocean spans. Shallow water cables have lighter armor and protective coating then deep water submarine cables.

Submarine cables may be composite cables as they can include copper lines that carry power to optical amplifiers (repeaters) that are used in long optical undersea connections. These may be called repeatered cables.

Submission Release Form-A form used by any party who desires to complete an unsolicited literary material submission to a possible buyer, exclusive to occurrence where that individual has reputable representation.

Subnet-In an Internet Protocol (IP) network, a group of devices which share a common IP address prefix, in other words, they have the same high-order bit values. The subnet mask is used to identify which bits are used for the network portion of the IP address and which bits are used for the host address portion. See also Classless Inter-Domain Routing (CIDR).

Subnet Address-The portion (sub portion) of an IP address that identifies the portion of a host, which the IP address belongs.

Subnet Mask-The assignment of a number that is used to separate sections of a network by using a specific sequence of numbers in a network address. When used with the internet, a subnet mask is 32 bits.

Subnetting-A process of configuring a network to route specific groups of network addresses to specific local domains (subnets) within the network. To create subnets, a binary subnet mask is used to mask unwanted addresses from entering into the subnet.

Subnetwork Dependent Convergence Protocol (SNDCP)-Subnetwork dependent convergence protocol layer manage services that are using the radio link. These services include connectionless service, data transfer, and user data confidentiality. The SNDCP layer coordinates information flow between the LLC layer and upper layers.

Subroutine-A program that carries out a particular function and can be called from another program. A subroutine needs to be placed only once in memory the main program can call it on multiple occasions.

Subscribe-The process of indicating to a communication server or other network service provider that the user is requesting services to be provided in the future and where those services can be delivered.

Subscribe Message-A message that is used to request services from a communication server. The process indicates to a communication server that the user is requesting services and the specific events that may occur to cause the service to be provided. The subscribe message is defined in session initiation protocol (SIP) toolkit.

A SIP message that is sent by an entity to subscribe to resource or call state information. The Subscribe message can be used for example to request presence information relating to another user, changes in the state of that user would be indicated by the Notify message.

Subscriber-A subscriber is an end user of a service. A subscriber is sometimes called a "user" or "customer."

Subscriber Database-An informational and relational database which includes subscriber information, including usage patterns, billing records, personal information and other related data. This base is often "mined" for information which helps identify potential churn candidates as well as useful marketing information.

Subscriber Fraud-Fraud perpetrated by the end-user in which false user ID information was used to obtain service.

Subscriber Identity Module (SIM)-The subscriber identity module (SIM) is a small "information" card that contains service subscription identity and personal information. This information includes a phone number, billing identification information and a small amount of user specific data (such as feature preferences and short messages). This information is stored in the card rather than programming this information into the phone itself. This intelligent card, either credit card-sized (ISO format), or the size of a postage-stamp (Plug-In format), can be inserted into any SIM ready wireless telephone.

This figure shows a block diagram of a SIM. This diagram shows that SIM cards have 8 electrical contacts. This allows for power to be applied to the electronic circuits inside the card and for data to be sent to and from the card. The card contains a microprocessor that is used to store and retrieve data. Identification information is stored in the cards protected memory that is not accessible by the customer. Additional memory is included to allow features or other information such as short messages to be stored on the card.

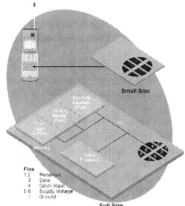

Subscriber Identity Module (SIM) Block Diagram

Subscriber Line-A telephone line that connects from a switching office to a customer's wired phone.

Subscriber Line Charge-A monthly flat-rate charge that recovers a portion of local loop costs paid by an end user. The charge is the result of a Federal Communications Commission (FCC) effort to eliminate unreasonable discrimination and undue preferences among rates for interstate service, to promote efficient use of a local network, to prevent uneconomic bypass, and preserve universal service.

Subscriber Loop Carrier (SLC)-Subscriber loop carrier works in conjunction with digital carrier (e.g. T-carrier or E-carrier) systems to increase the circuit capacity of a distribution plant without adding additional lines. SLC involves adding equipment to end offices (EOs) and remote plant locations that are connected by digital cable pairs, fiber optic cable, or digital radio media. The equipment at the end office (EO) and remote terminals allows more efficient use of cable pairs. The systems that use PAM and PCM technology can generally serve up to 96 subscribers from 10 cable pairs.

Subscriber Management-Subscriber management is a process or system that coordinates the additions, changes, and terminations of subscribers of a service.

Subscriber Number-The phone number that enables a user to reach a wireless subscriber within the same local network or numbering area. This term is a synonym for directory number.

Subscriber Penetration-Subscriber penetration is the percentage of subscribers who subscribe to one or more service offered by a service provider compared to the total number of potential customers that have the capability to subscribe (e.g. have access to or sufficient signal quality) to the services.

Subscriber Unit-A subscriber unit is a communication device or portable radio unit that can subscribe to services that are used in a communication system.

Subscriber User Interface (SUI)-Subscriber user interface (SUI) the portion of equipment or operating system that allows the equipment to interface with the user (subscriber).

Subscription Services-Subscription services are value-added services that provide or entitle a customer to receive or gain access to services. Subscription services are typically provided with no fixed termination date and subscription users are often billed periodically (e.g. monthly) for the subscription service.

Subscription Video on Demand (SVOD)-Subscription video on demand (SVOD) are on demand services that require a user to have a subscription (pre-authorization) prior to using the on demand services. SVOD service allows a customer to select and watch videos on demand from a predetermined list of video programs for an additional on demand fee. SVOD fees are usually composed of a periodic (e.g. monthly) flat rate fee and/or a usage fee (e.g. per movie).

Subsidiary Rights-Subsidiary rights are the authorizations that allow a publisher to use and/or convert content into another form of publication. For example, the republishing of portions of a book in a magazine series is a subsidiary right.

SUERM-Signaling Unit Error Rate Monitor

SUI-Subscriber User Interface

Super Channel-A super channel is a communication channel or link that has a very low bit error rate (BER).

Super Headend-A super headend is a television media reception, processing and distribution system that selects, combines and transmits the television signals to other headend distribution systems.

Super High Frequency (SHF)-The frequencies ranging from 3GHz to 30GHz; wavelengths are 10cm to 1 cm.

Super Hub-A super hub is a communication system that gathers and distributes communication to several hubs in a network through the re-broadcasting of data that it has received from one (or more) of the links or systems that are connected to it. A super hub generally is a sophisticated device that decodes media sources and repackages the signals so they can be redistributed to multiple receivers.

Super Node-A super node is used in a radio carrier or wireless distribution system that is used to provide a connection and gathering point for several smaller sites that are backhauled for interfacing the main network provider. Super nodes may be mounted on a main rooftop or tower-mounted collector site.

Super Scrambled-Super scrambling is the process of applying multiple layers of encryption to information. An example of super scrambling is the process of delivering video in a broadcast communication system where the video content may be encrypted as it is sent to individual viewers and the encrypted video program is transmitted on a channel that is also encrypted.

Super VHS (S-VHS)-Super video home system (S-VHS) is a video tape storage format that is used to store video recording images 1/2 inch magnetic tape. Super VHS offers 400 lines of resolution compared to 240 lines of resolution offered by standard VHS tape.

Superdistribution-Superdistribution is a distribution process that occurs over multiple levels. Superdistribution typically involves transferring content or objects multiple times. An example of superdistribution is the providing of a music file to radio stations and the retransmission of the music on radio channels.

Superframe-(1-GSM)A superframe is a multiframe sequence that combines the period of a 51 multiframe with 26 multiframes (6.12 seconds). The use of the superframe time period allows all mobile devices to scan all the different time frame types at least once. (2-IS-136 TDMA) For the IS-136 TDMA system, a superframe is a DCCH burst structure made up of sixteen 40ms TDMA frames equivalent to 32 consecutive TDMA blocks at full-rate, creating a sequence of 32 DCCH carrying bursts spread through 96 TDMA bursts. Each DCCH burst in the superframe is designated for either broadcast, paging, SMS messaging, or access response information. The superframe structure is continuously repeated on the DCCH channel. (3-T-1) In the North American T-1 digital multiplexing system, the historical superframe has a duration of 12 ordinary frames (with at total duration of 1.5

milliseconds) and the extended superframe has twice this duration (24 frames or 3 milliseconds). (4-E-1) In the European E-1 primary rate 2.048 Mb/s digital multiplexing system, a superframe of 16 frames duration (2 milliseconds total) is used only when channel associated bit signaling is in use (not with SS7 common channel signaling). The purpose of superframe synchronization is to allow transmission of certain predefined information types (usually call processing data) during certain pre-defined parts of the superframe time interval without further identification.

Supergroup-A bandwidth allocation in frequency division multiplexed systems that provides for 60 voice-bandwidth channels between 312 kHz and 552 kHz.

Superheterodyne Receiver-A radio receiver in which all signals are converted to a common frequency for which the intermediate stages of the receiver have been optimized for tuning and filtering. Signals are converted by mixing them with the output of a local oscillator whose output is varied in accordance with the frequency of the received signals to maintain the desired intermediate frequency.

Superstitial-Superstitial(tm) is an advertising standard that allows for non-banner ads that use interactive content.

Supervisory Signals-Signals used to indicate or control the status or operating states of circuits that establish a connection. A supervisory signal indicates that a particular state in a call has been reached and can signify the need for additional action.

Supplementary Service (SS)-Supplementary services provide a network user with capabilities beyond those of elementary call control. Supplementary services enrich the basic service functions and are not specific to a telephone or system features. Often, the subscriber (user) can specify some of the operations of supplementary services (such as call forwarding). Supplementary services may be defined or installed in systems before complete testing or industry consensus can be reached.

Supplementary Services-Supplementary services are new services that are created through the combining or modification of basic services.

Supported Rates-The data transmission rates that are available for use (hardware and software capable) in a communication device or system.

Supporting Shaft-Supporting shaft is a bar that extends through the cable roll and rests on the cable pay-off stand.

Surcharge-An additional charge for a service that is in addition to the basic charge. Examples of surcharges include additional charges for using pay telephone or toll free access lines.

Surface Mount Technology (SMT)-An electronics manufacturing process that uses components that have flat leads that are mounted direct on top of the printed circuit (PC) board. This eliminates the requirement of drilling holes in the PC board, allows the use of smaller components (allowing more parts on the PC board), simplifies the assembly process, and allow components to be mounted on both sides of the PC board.

Surfing the Web-Surfing is the process of viewing web pages on the Internet and following links from these web pages to new web pages.

Surge-A rapid rise in current or voltage, usually followed by a fall back to a normal value.

Surge Protector-A device or assembly that restricts the transfer of rapidly changing signals (a power surge) to a device or assembly.

This figure shows a surge protector that provides protection for power overage. This surge protector includes an induction coil that restricts rapid current changes and a fuse that will open in the event of a very large and rapid voltage change. Voltage transients can be caused by power surges and lightning pulses. Surge protectors may not protect against direct lightning strikes or extreme voltage changes.

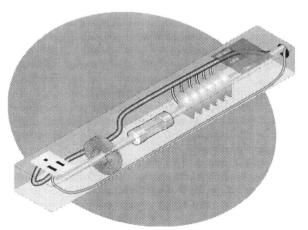

Surge Protector System

Surge Suppressor-A filter designed to protect computers and other electrical equipment from brief bursts of high voltage or surges.

Surround Sound-Sound reproduction that surrounds the listener with sound, as in quadraphonic recording and reproduction.

Surveying-Surveying is a process of determining the position of points on the Earth, measuring distances between them and using this information for a variety of purposes including the defining of boundaries of land ownership.

Survivability-The capability of a communications network to continue to provide service after major damage to any part of the system.

SV-Satellite Vehicle

SVC-Switched Virtual Circuit

S-VHS-Super VHS

S-Video-Separate Video

SVOD-Subscription Video on Demand

SVP-Secure Video Processing

SVPLA-Secure Video Processor Licensing Authority

SVS-Switched Video Service

SW-Short Wave

SWAP-Shared Wireless Access Protocol

Sweep-To vary the frequency of a signal over a specified bandwidth.

Sweetening-The process of electronically improving the quality of an audio or video signal, such as by adding sound effects, laugh tracks, and/or special effects.

SWF-Shockwave Flash

Switch-A network device (typically a computer) that is capable of connecting communication paths to other communication paths. Early switches used mechanical levers (cross-bars) to interconnect lines. Most switches use a time slot interchange memory matrix to dynamically connect different communications paths through software control. See also: Mobile Switching Center (MSC).

Switchboard-A manually operated switching system, now rarely used, that connects a limited number of telephones within a building or provides operator services.

Switched Beam-Switched beam communication systems switch radio signals transmission to one (or more) of multiple directional antennas that are located on a transmission site as a mobile radio moves through the geographic area of the switched beam system.

Switched Circuit-A circuit that may be temporarily established at the request of one or more of the connected stations.

Switched Digital Video (SDV)-Switched digital video is the process that can dynamically setup (on demand) digital video signal connections between two or more points. SDV services can range from the setup of data connections that allow digital video transfer to the organization and management of digital video content and the delivery of video programs.

Switched Multimegabit Data Service (SMDS)-A high-speed data transmission service that is often used in metropolitan areas that allows for the dynamic creation and disconnection of virtual circuits through the network. It is based on the 802.6 standard and may use T1 and T3 circuits to provide Ethernet, Token Ring, and Fiber Distributed Data Interface (FDDI) interconnection services.

Switched Video Service (SVS)-Switched video service (SVS) is the process that can dynamically setup (on demand) video signal connections between two or more points. SVS services can range from the setup of data connections that allow video transfer to the organization and management of video content and the delivery of video programs.

Switched Virtual Circuit (SVC)-A switched virtual circuit is an automatically and temporarily created virtual connection that is used for a communication session.

This diagram shows how a switched virtual circuit (SVC) uses an address provided by the user to establish a logical (virtual) path through a communications network. In this example, a SVC is created by using the destination address to determine the required programming of routing tables in switches within the data network. These routing tables assign data transfer connections between input and output channel on each switch. For example, as data from the sending computer (portable computer) is sent into input channel 3 of the first switch, it is transferred to the output channel 4. This process will repeat for any data that is sent from the sending computer to the destination computer during the switched connection.

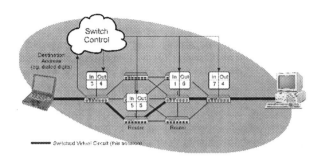

Switched Virtual Circuit (SVC) Operation

Switching-Switching is the process of connecting two (or more) points together. Switching may involve a single physical connection (such as a light switch) or it may involve the setup of multiple connections within a network through several communication devices.

Switching Hub-A data hub (broadcast device) that also has capability to directly route (switch) packets to a specific computer or device.

Switching Office-A switching center within a building (central office)

Switching Speed-Switching speed is the rate at which a device, signal or path can change from one state to another state or path.

Switching System-Switching systems are assemblies of equipment that setup, maintain, and disconnect connections between multiple communication lines. Switching systems are classified by their network types and the methods used to control the switches. The term "switch" is sometimes used as a short name for switching system. Public telephone switching systems have many switches within their network. A typical switch can handle up to 10,000 communication lines each.

Switching systems have evolved from manual switchboard systems (wires and plugs) to logical (digital) switches. The earlier types of manual switchboard systems were changed to automatic switching systems to eliminate the need for operators to setup every call. The first types of automatic switching systems used crossbar switches. Crossbar switches used mechanical arms to physically connect to wires (or busses) together. This has progressed to time slot interchange (TSI) switches. TSI switches logically interconnect communication lines through the temporary storage of data in memory time slots.

This figure shows the common types of switching systems used in telephone networks. The first diagram shows an analog crossbar switching system that makes physical (direct electrical) connections between input and output ports on the switch. The digital switching system uses a core memory section to temporarily store and distribute digital information between input lines and output lines. Switches can be connected to each other to form networks (such as the PSTN) so that connections can be made between multiple switches. The softwswitch system uses a call server to communicate data network addresses each telephone station that helps the telephone stations to send packets directly to each other through a data network. This examples shows that the routing tables forward packets towards their destination.

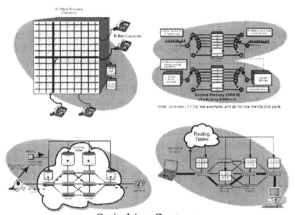

Switching Systems

SWOT-Strengths, Weakness, Opportunities, and Threats

SWR-Standing Wave Ratio

SWV-Shockwave Video

SXS-Step By Step Switching System

Symbol-A recognizable electrical state associated with a signal element. In binary transmission, a signal element is represented as one of two possible states or symbols. In ternary transmission, a signal element is represented as one of three possible states or symbols. In quaternary transmission, the signal element represents one of four possible states or symbols.

Symmetric Codec-A symmetric codec is a coder/decoder compression device or process (software program) that uses the same compression and decompression processes.

Symmetric Encryption-Symmetric encryption is the process of encoding and decoding of voice or data information so that it cannot be used by unauthorized users through the use of the same keys for the encryption and decryption process.

Symmetric Key Encryption-Symmetric key encryption is the process of using the same key for encryption and decryption.

Symmetrical-Symmetrical transmission is two-way communication that has the same data transmission rates in send (forward) and receive (reverse) directions.

Symmetrical Compression-A compression system that requires equal processing capability for compression and decompression of data. This form of compression is used in applications where both compression and decompression will be used frequently. Examples include still image database, still image transmission, color fax, video production, video mail, videophone, and videoconference. (See also: asymmetrical compression.)

Symmetrical Digital Subscriber Line (SDSL)-Symmetrical digital subscriber line is an all-digital transmission technology that is used on a single pair of copper wires that can deliver near T1 or E1 data transmission speeds. SDSL is a symmetrical service that ranges from 160 kbps to 2.3 Mbps and can reach to 18000 feet from the central switching office.

Synchronization-(1-general) The process of adjusting the corresponding significant instants of signals, such as the zero-crossings, to make them synchronous. The term synchronization often is abbreviated as sync. (2-digital) An arrangement for operating digital switching and transmission systems at a common (synchronized) clock rate to prevent the loss of portions of a bit stream during transmission. The required synchronization pulses commonly are referred to as clock pulses or clock signals. (3-video) The pulses and timing signals that lock the electron beam of the picture monitor in step, both horizontally and vertically, with the electron beam of the pickup tube. The color sync signal (in the NTSC system) is known as color burst.

Synchronization Channel (SCH)-(1-general) A logical channel that provides a mobile station with a timing reference to assist in the demodulation of the radio channel. This timing reference is a unique training sequence of digital information. (2-mobile communication) The synchronization channel (SCH) provides information that assists the mobile device to find and time-align with the system. (3-WCDMA) The WCDMA has primary and secondary synchronization channels. The primary synchronization channel uses a fixed 256 chip sequence that is used in every cell. The secondary synchronization channel provides information that allows for frame and time slot timing synchronization for each base station.

Synchronization Rights-Synchronization rights are the authorizations to use multiple forms of media (such as music and video) in a specific time relationship (e.g. synchronized audio and video).

Synchronization Word-A unique sequence of bits or information symbols that is used to provide a timing reference that is relative to other information that will follow.

Synchronized Multimedia Integration Language (SMIL)-Synchronized multimedia integration language is a protocol that is used to control the user interface with multimedia sessions. SMIL is used to setup and control the operation of media files along with the placement and operation windows on the users display.

Synchronized Television (SyncTV)-A video program delivery application that simultaneously transmits hypertext markup language (HTML) data that is synchronized with television programming. Synchronized television allows the simultaneous display of a video program along with additional information or graphics that may be provided by advertisers or other information providers.

Synchronizing Pulse Generator-A device or circuit that creates (generates) synchronization signals (pulses).

Synchronous-A system or signal that involves the transfer of information in a predefined serial time sequence.

Synchronous Data Link Control (SDLC)-A bit oriented synchronous communication protocol that organizes information into sequenced frames (groups of bits) of data. SDLC is similar to the high-level data link control (HDLC) protocol defined by the International Organization for Standardization (ISO).

Synchronous Modem-A modem that is able to transmit timing information in addition to data. The modem must be synchronized with the associated data terminal equipment by timing signals. A synchronous modem sometimes is referred to as an Isochronous modem.

Synchronous Network-A network in which the data communication lines are synchronized to each other or to a common clock signal that allows the exact determination of groups of bits (frames or fields) that are defined within the transmission of digital information.

Synchronous Operation-A process that is used to coordinate the timing of data transmission between communication devices. This allows the communication devices to know the specific timing relationship for information that is transported by the communication system.

Synchronous Optical Network (SONET)-Synchronous optical network is a digital transmission format that is used in optical (fiber) networks to transport high-speed data signals. SONET uses standard data transfer rates and defined frame structures formats in a synchronous (sequential) format.

This Figure shows a SONET system. This example shows that a SONET system may have several configurations including ring, star or point-to-point configuration. This diagram shows that the key components of a SONET system include add drop multiplexers (ADM), digital cross connects (DCS) and terminal adapter (TA). This example shows that this SONET system has been setup to have several survivable ring connections that will automatically reconfigure if one of the parts of its ring

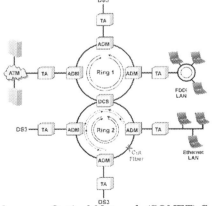

Synchronous Optical Network (SONET) System

connections becomes disabled (e.g. cut). This example shows that a fiber cable in ring #2 has been cut and that the SONET system has automatically reconfigured to continue operation to all the devices in the network.

Synchronous Orbit-An orbit in which a satellite has an orbital angular velocity synchronized with the rotational angular velocity of the earth, and thus remains directly above a fixed point on the surface. See also Geosynchronous (or Geostationary) Earth Orbit.

Synchronous Satellite-A satellite in a synchronous orbit.

Synchronous Serial Data Transmission-The sequential transmission of data, bit by bit, of that is related to a reference clock signal and a specific channel mapping format.

Synchronous Transmission-(1-general) A mode of transmission during which sending and receiving terminals operate at precisely the same frequency. Timing or synchronization is maintained by synchronizing pulses rather than by start/stop pulses, as in asynchronous transmission. (2-phase relationship) A mode of transmission in which the data transmitter and receiver are maintained in a uniform phase relationship. (See also: asynchronous transmission.)

SyncTV-Synchronized Television

Syntax-The relationships among characters or groups of characters, independent of their meanings or the manner of their interpretation and use.

Synthesizer-A device used to modify an external or internal audio signal by a preset controlled process.

Synthetic Video-Synthetic video is an image series (e.g. moving picture information) that is created through the use creating image components by non-photographic means.

Syslog-The syslog protocol was first defined as part of the UNIX operating system to log messages with the operating system (OS). Syslogs allow a computer or device to deliver messages to another computer. Syslog messages have a particular format that associates a facility, and a severity or priority with a message. Syslog servers are widely used in data networks to log information about network devices.

System-Systems are the combination of equipment, protocols and transmission lines that are used to provide communication services.

system-Autonomous Cells

System-Semi-Private Cells

System-Semi-Residential Cells

System Access-System access is the ability for a user or device to connect to a system.

System Administration-System administration are the processes and tasks performed by a system administrator to add, change, or disconnect devices, features, and equipment. System administration tasks may include monitoring the performance of the network and making adjustments to the network as necessary.

System Administrator-A person who oversees the operational functionality of a computer or related telephone equipment. Also included in this individual's responsibilities are introducing new user Ids and organizing phone numbers, commonly referred to as moves, adds, or changes (MAC).

System Capacity-System capacity is the maximum information or service carrying ability of a communications system. The unit of capacity measurement for the facility or system depends on the type of services or information content that are provided by the system.

System Clock Reference (SCR)-A system clock reference source of timing information that is used to coordinate the decoding and sequencing of media streams (elementary streams) that are contains in a MPEG-1 program stream.

System Network Architecture (SNA)-A communications system protocol developed by IBM that allows for control and data communication via different types of network communication equipment.

System Time Clock (STC)-A system time clock is the reference timing values that are inserted into MPEG program streams (combined audio, video or data) that are used to control the overall time alignment of media within a program.

Système International (SI)-The SI version of the metric system of units is compatible with practical electric units such as the volt, ampere, and watt. The basic mass unit is the kilogram, the basic length unit is the meter, and the basic time unit is the second. The basic unit of electric current is the ampere, and the basic unit of electric charge is the coulomb or ampere second (see: charge). It is also called the Georgi system, after the name of the Italian physicist who first formulated the appropriate mechanical units for its use. The older version of the metric system, sometimes called the cgs system, is based on the centimeter, gram and second.

Systems Integration-Systems integration is the process of defining, selecting, combining and configuring multiple types of systems to operate together to perform specific functions and/or services. Systems integration can range from porting (simple one-way connections) of systems to each other to full integration (two-way interactive processing) operation.

Systems Integrator-A systems integrator is a company or person who assists with the defining, selecting, combining and configuring multiple types of systems to operate together to perform specific functions and/or services.

T

T-Tera

T-tesla

T Carrier-Trunk Carrier

T Coupler-A signal coupler that has three ports.

T Coupler-Tap Coupler

T.38-A standard protocol used for the transmission of fax information over IP data networks. Created by the ITU-T, it defines real-time fax for IP-enabled fax devices and fax gateways.

T-1-T1 Carrier

TA-Transmitter Address

Table-A table is a group of structured information. The structure usually includes records (rows) and fields (columns).

Tablet PC-Tablet Personal Computer

Tablet Personal Computer (Tablet PC)-A tablet personal computer is a portable computer that is of similar size as a writing tablet. Tablet PCs commonly have a touch screen display that allows the device to be operated by a soft (display) keyboard.

TACS-Total Access Communications System

TAD-Telephone Answering Device

TADIG-Transferred Account Data Interchange Group

Tag-based Language-Tag-based Language is a language that uses start and end tags around information that specify how that information is to be displayed or treated. The markup family of languages (HDML,XML,WML, HTML, XHTML, cHTML) are tag-based languages.

Tagged Image File Format (TIFF)-TIFF is a format for storing and presenting digital images. TIFF image types include black-and-white data, halftones or dithered data, grayscale data, and color.

Take Rates-Take rates are the percentages of customers or potential customers who subscribe to a new product or service.

Take-up Reel-A take-up reel is a spool that is used to catch and hold pulling tape or rope wire during cable pulling.

Talk Group-A talk group is a predefined group of mobile radios that are capable of receiving and decoding the group messages to or from group members.

Tamperproof-Tamperproof is the ability of a product or service to have its intended use without the ability to be accessed or altered by an unauthorized user.

Tandem-(1-general) The connection of the output terminals of one network, circuit, or link directly to the input terminals of another. (2-message network) A switching system that establishes trunk-trunk connections but has no subscriber lines connected to it. Tandem types include local tandems, LATA tandems, and access tandems.

Tandem Compression-The connection of one coder/decoder (such as 32 kbps ADPCM) through another coder/decoder (such as 8 kbps G.729) when a call is routed through interconnecting tandem switches or routers. The cascading of coders (such as speech coders used in mobile or VoIP communication systems) usually results in significant degradation of voice quality.

Tandem Free Operation (TFO)-Tandem free operation involves the direct connection of switching centers in mobile communication systems without the need to decompress and re-compress (transcoding) speech information. TFO overcomes the challenges of cascading the speech coding process. Each time speech information is compressed and decompressed, some audio distortion occurs and time delay is added.

The logical function for coordinating TFO is the transcoder rate adaptation unit (TRAU). The TRAU is usually located in the MSC (it is possible to put the TRAU in the base station) and it negotiates the ability of the MSC to use TFO with another MSC. The TRAU is also responsible for disabling TFO if the call is transferred to another MSC or system that is not capable of TFO.

Tandem Office-A switching system that is used to interconnect end offices with each other.

Tandem Switch-An tandem switch is an intermediate level switch that connects to other switching exchanges.

Tandem Switching System-A switching system in a tandem office that handles trunk-trunk traffic. Local tandems switch calls from one end office to another within the same area. Access tandems switch calls to and from an interexchange carrier.

Tap-(1-circuit) A branch or intermediate circuit in a communications system. (2-optical) A device for extracting a portion of the optical signal from a fiber. (3-telephony) Short for wire tap.(4-In a local area network an electrical connection permitting signals to be transmitted onto or received from a bus.

TAP-Transferred Account Procedures

Tap Coupler (T Coupler)-A tap coupler (T coupler) is a coupler that has 3 ports where one of the ports is used as a tap (small sample) of the signal that flows through the coupler. The tap port is typically a very small portion of the signal that passes through the coupler (e.g. 20 dB) so the insertion loss of the T coupler is very low.

TAP3-Transferred Account Procedures 3

Tape Storage-A tape storage system holds data (information) on a tape.

TAPI-Telephony Application Programming Interface

TAR-Test Accuracy Ratio

Target Market-A segment of a potential market for a product or service that has common characteristics. These characteristics are often used to focus (target) groups of customers that have specific wants or needs for a product or service.

Targeted Advertising-Targeted advertising is the process of creating advertising messages that appeal to groups (target segments) and sending these messages to communication channels (such as advertising spots on television shows) that have viewers or recipients in the desired market segments.

TASI-Time Assigned Speech Interpolation

Task Group-A group of people (typically professionals with specific skill sets) that are temporarily assigned to perform a particular task or project.

TBF-Temporary Block Flow

TC-Transmission Convergence

TCAP-Transaction Capabilities Application Part

T-Carrier-A system operating at one of the standard levels in the North American digital hierarchy.
Each digital signaling level supports several 64 kbps (DS0) channels. T-carrier was initially used in North America and now is used throughout several parts of the world. The different digital signaling levels include;
- DS-1, 1.544 Mbps with 24 channels
- DS-1C, 3.152 Mbps with 48 channels
- DS-2, 6.132 Mbps with 96 channels
- DS-3, 44.37 Mbps with 672 channels
- DS-4, 274.176 Mbps with 4032 channels

TCH/F-Full Rate TCH
TCH/F-Traffic Channel Full Rate
TCH/H-Half Rate Data TCH
TCL-Tool Command Language
TCO-Total Cost Of Ownership
T-commerce-Television Commerce
TCP-Transmission Control Protocol
TCP Header-Transmission Control Protocol
TCP Ports-A logical assignment of a computer connection (such as in a TCP/IP system) that allows a communication session between data communication devices. A port is part of a socket connection (channel identifier and port number.) Port numbers are commonly assigned based on specific protocols or applications. Some of the more common port assignments for Internet based TCP communications include port 21 for file transfer protocol (FTP), port 23 for Telnet, port 25 for Simple Mail Transfer Protocol (SMTP), and port 80 for Hypertext Transfer Protocol (HTP).

TCP/IP-Transmission Control Protocol And Internet Protocol
TCS-Telephony Control Service
TDD-Telecommunications Device For The Deaf
TDD-Teleprinter Device for the Deaf
TDD-Teletypewriter Device for the Deaf
TDD-Time Division Duplex
TDM-Time Division Multiplexing
TDMA-IS-136
TDMA-IS-54
TDMA-IS-54C
TDMA-Time Division Multiple Access
T-DMB-Terrestrial Digital Multimedia Broadcasting
TDMoIP-Time Division Multiplexing over Internet Protocol
TDR-Time Domain Reflectometer
TDT-Time and Date Table
TDU-Tube Distribution Unit
TE-Terminal Equipment

Technical Working Group (TWG)-A technical working group is a team of technical experts that temporarily work togehter to complete technical projects or assignments. TWGs are commonly composed of representatives from different companies that have an interest in the technology or systems that will be developed. TWGs are commonly chartered to develop industry specifications or portions of industry specifications to meet specific technical and/or business requirements.

Technician-A skilled craftsman who is capable of diagnosing, servicing, and repairing electronic or electrical assemblies (such as a PBX system or telephone device).

Techno-Geek-A geek is a person who is focused on technology, typically computers who does not tend to conform to mainstream habits such as dressing for success and/or regular bathing.

Technological Churn-Type of voluntary, deliberate churn, where the customer changes carriers because
the new carrier has newer and better technological options. (i.e. customers switching
to digital from analog)

TEID-Tunneling End Point Identifier

TeLANophy-LAN Telephony

Telco-Telephone Company

Telcordia-Successor to Bellcore.

TelcoTV-Telephone Company Television

Telecine-An machine that is used to transfer a film image to a video image. It is also known as a film chain.

Telecommunication-The transmission and reception of audio, video, data, and other intelligence by wire, radio, light, and other electronic or electromagnetic system.

Telecommunication Closet-A room or space that is dedicated for telecommunication equipment and cable connection points.

Telecommunication Regulation-The regulation of telecommunications systems and services are developed to help or improve the ability of citizens in a country to reliably communicate with each other at reasonable cost. Telecom rules and regulations are usually imposed by a government agency. These rules are designed to maintain the quality and cost of public utility services.

In the United States, the Telecommunications Act of 1996 was created to allow competition into the telecommunications industry. It provides a national framework for the deregulation of the local exchange market, a deregulation that was already taking place on a state-by-state basis through the actions of state regulatory commissions. Its summary impact on the local exchange market is to require current LECs to remove all barriers to the competition (e.g., interconnect, white and/or yellow pages access, co-location, and wholesaling of facilities restrictions) in return for LEC access to the long distance market.

Telephone companies in the United States are regulated by the government, but not owned by the government. For most European countries and many other countries, local telephone service is provided by government owned post telephone and telegraph (PTT) operators. In some European countries, the post (mail) network has been separated from the operation of telephone and telegraph networks. In some countries, the telephone and telegraph systems have become privatized, and are no longer owned by the government.

Telecommunication Services-
Telecommunications services are the underlying communications processes that provide information for telecommunications applications. It is common to use the word services in place of applications, especially when the service is very similar to the application. Examples of communication applications include voice mail, email, and web browsing.

Telephone services include voice, data, and video transmission. Voice services can be categorized into quality of service and voice privacy. Data services use either circuit-switched (continuous connection) data or packet-switched (dynamically routed) data. Video transmission is the transport of video (multiple images) that may be accompanied by other signals (such as audio or closed-caption text).

Telecommunications-The transmission, between or among points specified by the user, of information of the user's choosing (including voice, data, image, graphics, and video), without change in the form or content of the information.

Telecommunications Act Of 1996-The U.S. Telecommunications Act of 1996 provides a national framework for the deregulation of the local exchange market, a deregulation that was already taking place on a state-by-state basis through the actions of state regulatory commissions. Its summary impact on the local exchange market is to require current LECs to remove all barriers to the competition (e.g. interconnect, white and/or yellow pages access, co-location, & wholesaling of facilities restrictions) in return for LEC access to the long distance market.

Telecommunications Device For The Deaf (TDD)-A small communications terminals with a keyboard and visual display that connects to a telephone circuit to relay writ-ten messages to and from hearing- and/or speech-impaired persons. An acoustic coupler is used to send audio tones from a TDD through the handset of a conventional telephone instrument.

Telecommunications Industry Association (TIA)-This telecom industry trade association represents the manufacturers of telecom equipment.

Telecommunications Operational Model (TOM)-An industry standard template describing the major operational components that make up a telecommunications organization.

Telecommunications Profitability Formula-A modified business profitability formula that reflects the unique nature of the telecommunications business model where Profit = (Revenue - (CapEx + OpEx))

Telecommuting-The process of an employee that is conducting business related activities at a remote location (usually at a home) through the use of telecommunications services and equipment. Telecommuting allows employees to work at home without the need to commute to the office and reduces the need for the business to maintain office space for workers.

Telecomputing-The use of remotely located computers and databases for computing and access to computer-based services, such as home shopping, banking, and electronic mail.

Teleconferencing-A process of conducting a meeting between two or more people through the use of telecommunications circuits and equipment. Teleconferencing usually involves sharing video and/or audio communications.

Telecopier-Another term for a facsimile (fax) machine.

Teledensity-Teledensity is a measure of the number of communication lines used or installed in a geographic area for a quantity of people. A common unit for teledensity is the number of lines per 100 people.

Teledesic-Teledesic is a "Big LEO" satellite system that was designed to provide high-speed data services (broadband) to consumers. The term "Big LEO" is used to indicate in excess of 200 satellites. The Teledesic system was a $9 billion project initiated by Bill Gates and Craig McCaw. Teledesic was planned to become available in the early 2000s. However, in 2002, Teledesic announced it had discontinued development and returned its frequency licenses to the FCC and ITU.

Telegraphy-A form of telecommunications, which is concerned with the process of providing transmission and reproduction at a distance of text material or fixed images. The transmission of such information may be physical transmission facilities or over the air using some form of signaling protocol.

Telemanagement-A technique involving the management of the telephone and telecommunication expenses of an organization. This also refers to third-party software packages directed at monitoring inbound and outbound calling on the PBX switch with the capability to format detailed reports by departments and per-minute charges.

Telemarketing-Telemarketing is the process of conducting marketing and sales programs using telecommunication systems. Telemarketing call centers generally combine advanced call processing systems (e.g. computer telephony automatic call distribution) with customer order processing systems.

Telematics-Telematics is the collection and distribution of measured or machine control data through a communications system.

Telemedicine-Processes that assist with health care service that employ communications services and equipment. Examples of telemedicine include delivery of medical images, remote access to medical records, remote monitoring of heath care equipment and distant monitoring of biological functions such as heart rate and blood pressure.

Telemetry-Telemetry is the transfer of measurement information to a monitoring system through the use of wire, optical fiber, or radio transmission. The term telemetry is often used with the gathering of information.

Telepayment-Telepayment is the payment of products and services by telecommunication systems.

Telephone Answering Device (TAD)-A telephone answering device is an assembly or software program in a communication device that can automatically answer telephone calls, play a prerecorded greeting message, store audio information, and allow retrieval and deletion of messages.

Telephone Answering Service-A service company that answers telephone calls and takes messages for subscribers who are away from their homes or offices.

Telephone Company-Telephone companies (also known as service providers or carriers) provide communication services to the general public. They are usually regulated by the government and in some countries, may be partly or wholly owned by the government. For most European countries and many other countries, local telephone service is provided by government owned posts, telephone and telegraph (PTT) operators. In some European countries, the post (mail) network has been separated from the operation of telephone and telegraph networks. In some countries, the telephone and telegraph systems have become privatized, and are no longer owned by the government.

Telephone Company (Telco)-A contraction of telephone company, generally signifying an operating telephone company

Telephone Company Television (TelcoTV)-Telephone company TV is the process of providing television (video and/or audio) services through telephone system broadband (e.g. DSL) transmission lines. Sending television over DSL may use proprietary digital video signals or it may use standard IP data communication protocols to initiate, process, and receive voice or multimedia communications. When television signals are sent over DSL lines using standard IP communication, it is commonly called IPTV.

Telephone Eye-A system that combined television broadcasting with telephone service to provide for early interactive television programming.

Telephone Line Simulator (TLS)-A telephone line simulator creates the signals necessary to simulate analog and/or digital telephone line service to a telephone device. A TLS may provide the dialtone, loop current, 2-wire or 4-wire pulse or DTMF signaling along with other telephone line

Telephone Numbering Plan-A numbering plan is a system that identifies communication points within a communications network through the structured use of numbers. The structure of the numbers is divided to indicate specific regions or groups of users. It is important that all users connected to a telephone network agree on a specific numbering plan to be able to identify and route calls from one point to another. See also Dialing Plan, open numbering plan, closed numbering plan.

This figure shows the complete international structure of telephone numbers and that these numbers typically identify a specific physical location of a telephone connection point. This diagram shows that a telephone number is composed of a country code, city code, switch code, and an extension code. In this example, the country 001 routes the call to the United States. The 919 city code routes the call to the city Raleigh within North Carolina. The

exchange switch code 557 directs the call to one of several switching centers located in Raleigh. The extension code 2260 directs the call to a specific extension port on that switch. This extension port is connected to the telephone by wires in the local area.

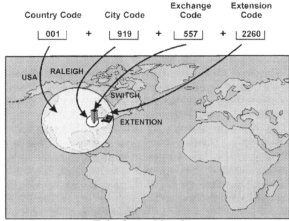

Telephone Numbers

Telephone Routing over Internet Protocol (TRIP)-A protocol that allows for the dynamic assignment of call routes through the advertising of the availability of destination devices (such as telephones) and for providing information relatives the available routes and preferences for these routes to reach the destination device(s).

Telephone Service-The providing of telephone communication through a network that allows users to interconnect to other users. Telephone systems use a numbering plan to identify communication points within a communications network through the structured use of numbers. The structure of the numbers is divided to indicate specific regions or groups of users. It is important that all users connected to a telephone network agree on a specific numbering plan to be able to identify and route calls from one point to another.

Telephone Set-The terminal equipment at a subscriber's premises used for voice telephone service. The instrument includes a microphone and earphone, called a transmitter and a receiver, as well as a switchhook, keypad or dial, ringing device, and associated circuitry. The set also is called a telephone instrument.

Telephone Station-Telephone stations are telephone instruments that are connected to a telephone network for the purpose of telephony. When these telephone stations are used as part of the private network, they are often identified by the type of system they are connected with. For example, telephone stations that are connected to key systems are called Key Telephones.

Telephone stations can vary from simple POTS telephones (sometimes called 2500 telephones) to complex Internet telephones (IP Telephones). Telephone stations usually receive their power from the telephone line (loop current) but may receive their power from an external source (such as the PBX switching system).

This diagram shows the difference between standard analog telephone stations and more advanced PBX stations. This diagram shows that analog telephones receive their power directly from the telephone line and digital PBX telephones require a control section that gets its power from the PBX system. Analog telephones also use in-band signaling to sense commands (e.g., ring signals) and to send commands (e.g., send dialed digits). Digital telephones use out-of-band signaling on separate communication lines to transfer their control information (e.g., calling number identification).

Telephone Station Functional

Telephone Trigger-A telephone trigger number is a unique telephone number that is used to trigger another service such as a callback service. An international user could use a telephone trigger number to initiate a callback service. This would reduce the fees the international traveler pays for international calls as typically it is less expensive to receive incoming calls than it is to initiate outgoing international calls.

Telephone User Part (TUP)-The User Part specified in ISDN for telephone services.

Telephony-Telephony is the use of electrical, optical, and/or radio signals to transmit sound to remote locations. Generally, the term telephony means interactive communications over a distance. Traditionally, telephony has related to the telecommunications infrastructure designed and built by private or government-operated telephone companies.

Telephony and Internet Protocols Harmonization over Networks (TIPHON)-An ETSI initiated project for multimedia communications that combines and coordinates multiple data compression and communication standards to allow audio, picture, and video transmission between users on packet switched networks. Information about TIPHON can be found at www.etsi.org/tiphon.

Telephony Application Programming Interface (TAPI)-An industry standard that defines the application interface between computers and telecommunications devices. TAPI was introduced in 1993 as the result of joint development by Microsoft and Intel. The standard supports connections by individual computers as well as LAN connections serving many computers. Within each connection type, TAPI defines standards for simple call control and for manipulating call content.

Telephony Over Internet Protocol (ToIP)-A process of providing telephony services using Internet protocol (IP).

Telescoping-Addressability of IPTV is the process of enabling separate streams to be sent to different TV sets during the same advertising time slot. Because of this, ads can be tailored according to individual demographics; location; interests; viewing habits; time of day; language and a raft of other factors.

Teleservices-Teleservices are telecommunication services that provide added processing or functionality to the transfer of information between users. Teleservices are categorized by their high level (application) characteristics, the low level attributes of the bearer service(s) that are used as part of the teleservice, and other general attributes. High-level attributes include: application type (for example voice or messaging) and operation of the application. The low-level description includes a list of the bearer services required to allow the teleservice to operate with their data transfer rate(s) and types. Other general attributes might specify a minimum quality level for the teleservice or other special condition. The categories of teleservices available include voice (speech), short messaging, facsimile, and group voice.

This diagram shows a typical teleservice in a mobile communication system. In this diagram, a telephone user wishes to send a fax to a recipient who is traveling. The designated recipient has setup a fax forwarding service where the delivery of incoming faxes can be instructed. The sender is given the recipient's fax number. When the sender dials the number, the call is routed through the telephone network to the fax forwarding service provider (step 1). When the incoming call is detected, the fax forwarding service receives the fax into a fax mailbox (step 2). Later that day, the recipient of the fax forwarding service calls in and enters a fax forwarding number (step 3). The fax forwarding service then checks the fax mailbox and automatically sends all the waiting faxes to the new number (possibly a hotel fax number) that has been updated by the recipient (step 4). Because this service involves both the transport and processing of information, it is categorized as a teleservice.

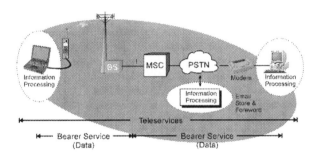

Teleservice Operation

Teletext-Teletext is a service that transfers data information along with a standard television signal to allow the simultaneous display of text and video on the television. Teletext information is usually encoded into the video blanking interval (VBI) and decoded by the receiver in the television or cable converter box.

Teletype-Trade name for teletypewriters made by the Teletype Corporation. Widely used informally as a synonym for teletypewriter.

Television-Television broadcasting is the transmission of video and audio to a geographic area that is intended for general reception by the public, funded by commercials, subscription services, or government agencies. Television broadcasters transmit at high power levels from several hundred foot high towers. A high power television broadcast station can reach over 50 miles.

This figure shows a television broadcast system. This television system consists of a television production studio, a high-power transmitter, a communications link between the studio and the transmitter, and network feeds for programming. The production studio controls and mixes the sources of information including videotapes, video studio, computer created images (such as captions), and other video sources. A high-power transmitter broadcasts a single television channel. The television studio is connected to the transmitter by a high bandwidth communications link that can pass video and control signals. This communications link may be a wired (coax) line or a microwave link. Many television stations receive their video source from a television network. This allows a single video source to be relayed to many television transmitters.

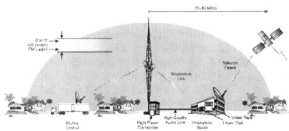

Television on Broadcast System

Television Broadcast Services-Television broadcast services are the transmission of television program material (typically video) that is typically paid for by advertising. Most television stations receive the bulk of their ad revenues from network advertising, as opposed to radio, which gets most of its revenue from local advertising.

Television Broadcaster (TV Broadcaster)-A television broadcaster is a company that transmits or provides television information to users that are connected or able to access signals on the broadcast network.

Television Channel Modulator-A television channel modulator is a device or system that converts a video and/or audio signals to a radio frequency that is used for broadcasting to television sets.

Television Commerce (T-commerce)-Television commerce is a shopping medium that uses a television network to present products and process orders.

Television Encoding-Television encoding is the signal format and data compression system that is used to deliver television signals.

Television Gateway-A television gateway is a communications device or assembly that transforms audio and video that is received from a television system (e.g. broadcast television) into a format that can be used by a different network. A television gateway usually has more intelligence (processing function) than a data bridge as it can select the audio and video compression coders and adjust the protocols and timing between two dissimilar computer systems or video over data networks.

Television over DSL (TVoDSL)-Television over DSL is the process of providing television (video and/or audio) services through DSL transmission lines. Sending television over DSL may use proprietary digital video signals or it may use standard IP data communication protocols to initiate, process, and receive voice or multimedia communications. When television signals are sent over DSL lines using standard IP communication, it is commonly called IPTV.

Television over Internet Protocol (TVoIP)-Television over Internet protocol is the process of providing television (video and/or audio) services through communication systems that can transfer IP data packets. Sending television over IP transmission networks uses standard IP data communication protocols to initiate, process, and receive

voice or multimedia communications. When television signals are sent over broadband lines using standard IP communication, it is commonly called IPTV.

Television Receive Only (TVRO)-A ground station used for the reception of TV signals transmitted from satellites. A typical station consists of a parabolic dish antenna, pre-amplifier, down-converter, tunable receiver, and video monitor.

Television Receiver-A device that receives a broadcast TV signal and converts it into a picture accompanied by its associated sound.

Television Signal-A radio signal that includes both the video component and the audio component of the transmitted program.

Television Station-A television station is a transmission system that provides video and audio (television) information on a radio communication channel. The term television station commonly is used to identify the company that operates the television transmitter.

Televoting-Televoting is a service that allows for the gathering of votes or opinions via communication systems.

Teleworker-A worker who performs their job at a remote location through the use of a telecommunications connection. An example of a teleworker is a customer service representative who answers customer service inquiry messages from their home.

Telex-An international teletypewriter exchange service for communications among its subscribers. Operation is typically at a rate of 50 bauds, via Baudot-Murray 5-bit character codes.

Telnet-Telnet is a standard network terminal control protocol that allows Internet users to login and control servers using TCP/IP protocol. The use of Telnet allows Internet users to work from any personal computer as if it were a terminal that was directly attached to another computer (e.g. data terminal). Telnet is specified in RFC-854.

Temperature Range-Temperature range is the difference between the maximum and minimum temperatures that are permitted to ensure defined performance characteristics of the device, circuit or transmission line.

Temperature Stability-Temperature stability is the ability of a device or circuit to provide its designed performance characteristics over a temperature range.

Temporal Access-Temporal access is the time that information can be accessed or is authorized for access. An example of temporal access is the

T

authorizing of a viewer to watch a television program for a 24 hour time period.

Temporal Compression-Temporal compression is the analysis and compression of information or data over a sequence of frames, images or sections of information.

Temporal Control-Temporal control is the ability of a user or device to change the relative time position of a media program or file (such as stop, play or fast forward).

Temporal Distance-Temporal distance is the time or number of images between specific types of images in a digital video.

Temporal Key Integrity Protocol (TKIP)-A new security protocol that is used in the 802.11 system that uses dynamically changing keys to replace the static security keys used in the original 802.11 system.

Temporal Masking-Temporal masking is the process of blocking, removing or ignoring specific components of a signal that occur in a specific time period or time sequence.

Temporal Redundancy-Temporal redundancy is the repetition of information or picture elements within a sequence of images.

Temporal Scalability-Temporal scalability is the ability of a streaming media program or moving picture file to reduce or vary the number of images or data elements representing that media file for a particular time period (temporal segment) without significantly changing the quality or resolution of the media over time.

Temporary Block Flow (TBF)-Temporary block flow is the sequential identifier for packets that are transferred on a GPRS data session. A TBF is established for the duration of data transfer and is used until all the radio link control (RLC) blocks have been transmitted for the TBF session.

Temporary Boot-A splice closure that is temporarily installed over a cable using rubberized coverings, cements and bindings to protect conductors during a construction or maintenance project. They are replaced with permanent plastic or lead closures at the completion of repairs or splicing and acceptance testing.

Temporary Flow Identifier (TFI)-Temporary flow identifier is a unique identifier assigned by a GPRS system to the mobile device that identifies unique TBF flows that may be concurrently sent on packet data channels (PDCHs).

Temporary Location Directory Number (TLDN)-A temporary location directory number (TLDN) is a temporary identification number that is used to route calls from a home system and a visited communication system. TLDN numbers are commonly used in mobile communication systems where mobile telephones regularly operate in other (visited) systems. The TLDN is usually assigned for each call delivery request received from the home system. When the call is received into the visited system, it is mapped (translated) to the current resources (e.g. cell site and mobile number) that are currently being used in the visited system.

Tensile Strength-Tensile strength is the maximum amount of pulling force a cable can withstand before it is deformed or damaged.

Tension Limit-Tension limit is the amount of tensile (pulling) load that a cable or an assembly can experience before distortion of damage may occur.

Tera (T)-A prefix that means 10 to the 12th power 1,000,000,000,000. Commonly referred to as one trillion in the American numbering system, and one million million in the British numbering system.

TeraByte-A quantity equal to one trillion bytes (actually 1,099,511,627,776 bytes.)

TeraHertz (THz)-A unit equal to one trillion of a Hertz (1,000,000,000,000 Hertz).

Term Plan-A product/service that offers the customer a special discount in return for committing to a certain length of time. Penalties are typically assessed for early termination.

Terminal-(1-circuit) A point at which a circuit element may be directly connected to one or more other elements. (2-communications) Any type of equipment at the end of a communications circuit. User terminals include telephone sets, teletypewriters, and computing equipment. Carrier terminals include modulation, demodulation, and multiplexing equipment used to transmit, combine, and separate communications channels in a transmission system. (3-computer) An input/output device connected to a processor or a computer in order to communicate with it and control processing. An intelligent terminal has some local computing power and an associated data store. (4-loop plant) The hardware that facilitates the connection and removal of drop or service wires to and from cable pairs. Examples include distribution terminals and cross-connection terminals. (See also: central office terminal.)(5-post) A binding past, tag, or lug to which an external circuit may be readily connected.

Terminal Emulation-A microcomputer or personal computer mimics, pretends to be (I.e. emulates) a data terminal. It does this with special printed circuit boards inserted into its motherboard and/or special software.

Terminal Equipment (TE)-Equipment that is located at the end of one or more communication lines that send or receive signals for services. Terminal equipment can be wired or wireless devices.

Terminals-Devices that typically provide the interface between the telecommunications system and the user. Terminals may be fixed (stationary) or mobile (portable).

Termination-(1-general) Termination is the end of a circuit or connection (2-circuit) A circuit termination is an impedance matching device (termination) that is connected to the end of a circuit under receive and terminate signals without causing a reflected signal. (3-service) Service termination is the ending of authorization or providing of service.

Termination Charges-Fees paid by telephone operators to other access providers for terminating (routing calls to their destination) through the other access networks. An example of this is the payment by wireless carriers to local exchange carriers (local telcos) for the termination of mobile telephone calls into the wired public telephone network.

Terminator-An electrical device that can be attached to the end of a cable to simulate the attachment of an indefinite amount of additional cable. If this device has the same impedance as the cable, signals arriving at the end of the cable and encountering this device will not experience an impedance discontinuity, and signal reflections will not be created.

Terrestrial-Terrestrial is the operation of systems or services on the ground (surface) of a planet (typically Earth). The term terrestrial as it is applied to radio or optical involves the transmission of radio signals along the surface of the Earth.

Terrestrial Digital Multimedia Broadcasting (T-DMB)-Terrestrial digital multimedia broadcasting is a radio transmission system that can broadcast media that can have multiple forms (multimedia) through surface based (terrestrial) antennas.

Terrestrial Link-A direct, overland radio circuit not relying on satellite transmission. This term also may refer to the carrier facilities, usually microwave radio or coaxial cable, between a ground station and a facility terminating office.

Terrestrial Television-Terrestrial television is the broadcasting of television signals using surface based (terrestrial) antennas.

tesla (T)-Unit of magnetic flux density, equal to one volt second per square meter, or to one Weber per square meter. Named for Serbian-American electrical engineer Nikola Tesla (1856-1943). One tesla is 10 000 gauss. (A gauss is the older cgs unit of magnetic flux density.)

Test Accuracy Ratio (TAR)-Test accuracy ratio is a comparison of the accuracy of the testing device to the accuracy that is desired in the device under test (DUT). An example of test accuracy ratio is a testing device that can measure a signal with a 0.1% accuracy and the device that it is testing has an accuracy of 0.5%. This means that the TAR is 5:1.

Test Data Form-A test data form is a record information entry sheet that is used to store test measurement information. Test data forms are often designed for specific types of measurements.

Test Equipment-Test equipment is a device or assembly that can measure or verify that a particular product or system to determine if it meets specific requirements or if it is capable of performing specific functions or actions.

Test Interface-A test interface is an electrical and/or physical adapater that allows test equipment to be attached to a unit under test (UUT).

Test Jig-A test jig is a device or assembly that holds a unit under test (UUT) so that it can be attached to testing equipment.

Test Leads-Test leads are wires or optical cables that are used for temporary connections during testing. Test leads usually have connectors (such as alligator clips) that are easy to attach or use with test equipment.

Test Line-A central office test facility, including testing equipment, circuits, and communication channels, used for the maintenance of trunks. A test line can range from a simple passive termination and tone generations to complex electronic equipment for signal and transmission testing. A test line also can be referred to as a test termination.

Test Option Deals-Contract(s) made for actors' consisting of an agreement to possibly work for 7 years on a series, yet is required to be signed prior to guaranteed role.

Test Option Period-Period of time used buy a studio to contemplate wether or not stipulated actor(s)' may proceed with production of a pilot or if he or she is to be released from his or her Test Option Pilot contract.

Test Point-A post or terminal on a circuit board to which test equipment can be connected to check a parameter of the circuit.

Test Procedures-Test procedures are a set of measurements and processes that are used to obtain the operational status or performance of a product or service.

Test Set-Any instrument designed to perform tests.

Test Signal-An electronic signal with defined standard characteristics used to check the capability of a circuit.

Test Signal Generator-A signal generator is a testing device that is used to generate test signals.

Test Specification-A test specification is a document that is primarily used for the validation of operational performance of a device or system. A test specification typically includes the device and test equipment configurations (connections) along with required signal inputs and expected signal outputs.

Testing-Testing is the performing of measurements or observations of a device, system or service to obtain the characteristics of its operation that indicate its successful operation and/or performance.

Testing Requirements-Testing requirements are the list of tests that must be performed for the acceptance of a product or service.

TETRA-Trunked Enhanced Terrestrial Radio

Text Based Protocol-Text based protocols are the languages, processes, and procedures that perform functions used to send control messages and coordinate the transfer of data using text based messages.

Text Messaging-See Short Message Service.

Text To Speech (TTS)-The conversion of text (ASCII) information into synthetic speech output. This technology is used in voice processing applications that require the production of broad, unrelated and unpredictable vocabularies, e.g., products in a catalog, names and addresses, etc.

Textual Messages-Text Messages

Texture Mapping-The ability of a digital effects system to create textured surfaces that can be applied to shapes.

TFI-Temporary Flow Identifier

TFO-Tandem Free Operation

TFTP-Trivial File Transfer Protocol

TFTP Server-A server (computer) that is commonly used to supply supplies essential startup or configuration information (boot information) to data communication devices (such as IP telephones).

TGID-Talk-Group Identifier

THD-Total Harmonic Distortion

Theatrical Release Bonus-Bonus received by or given to an individual to compensate services rendered by that person on occurrence, if it so happens, that the correlating television project be redone as a theatrical project.

Thermal Contraction-Thermal contraction is the amount of length that a material or cable shortens as a result of temperature changes.

Thermal Expansion-Thermal expansion is the amount of length that a material or cable lengthens as a result of temperature changes.

Thermal Noise-A noise resulting from the movement of electrons in conductors and semiconductors, caused by thermal effects.

THF-Tremendously High Frequency

Thicknet-A network cabling scheme using twin axial cable.

Thin Client-A thin client is a computer, hardware device or software program that is configured to request services from a network that requires a relatively small amount of memory and processing power.

Thin Ethernet-A coaxial cable (approximately 0.2-inch or RG58A/U50-OHM) that uses a smaller diameter coaxial cable than the standard thick Ethernet. Thin Ethernet systems tend to have transceivers on the network interface card rather than in external boxes. Thin Ethernet is also referred to as Thin Net and Thin Wire.

Third Generation (3G)-A term commonly used to describe the third generation of technology used in a specific application or industry. In cellular telecommunications, third generation systems used wideband digital radio technology as compared to 2nd generation narrowband digital radio. For third generation cordless telephones, products used multiple digital radio channels and new registration processes allowed some 3rd generation cordless phones to roam into other public places.

This diagram shows a 3rd generation broadband wireless system. This system uses two 5 MHz wide radio channels to provide for simultaneous (duplex) transmission between the end-user and other telecommunication networks. There are different channels used for end- user to the system (called the "uplink") and from the system to the end-user (called the "downlink"). This diagram shows that 3G networks interconnect with the public switched telephone network and the Internet. While the radio channel is divided into separate codes, different protocols are used on the radio channels to give high priority for voice information and high-integrity to the transmission of data information.

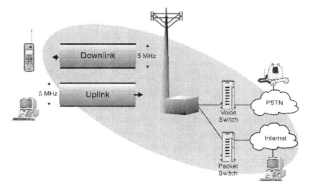

Third Generation (3G) Wireless

Third Party Billing-The billing of customers by one telephone company ("the Biller") on behalf of another Telco (the "Third Party").

Third Party Verification-Third party verification is a process of using a third entity (a person or company) to review or validate information on the products or services that are used, exchanged or transferred between two other entities. Typically, the third party is trusted by both the other parties. The third party may be used to validate the information (such as quantities of products produced under licensing agreements) through the viewing of information that may be proprietary or private.

Three Way Calling (3WC)-A service that provides a telephone service customer the capability of adding another party (third party) to an established two-party call. During a three-way call, all

three parties may communicate at the same time. However, to prevent annoying audio feedback, sophisticated audio volume control is typically offered that reduces the amplified audio signal level for parties that are not talking.

Three Way Handshake-A three way handshake is a process that exchanges information three times between the device that initiates or responds to a command to another communication device. After the initial device sends a message, the receiving device acknowledges the message and the original sending device confirms it has received the acknowledgement.

Three-Way Handshake-A series of three sets of messages that are exchanged between communication devices operating on a system to establish communications links for further transmissions of voice or data. Three-way handshake is used in the session initiation protocol (SIP) system to ensure call control messages (such as invite) can be confirmed in an unmanaged packet data transmission system (such as the Internet).

Threshold-(1-maximum) The minimum signal value that can be detected by a system. (2-FM improvement) The level at which signal peaks entering a FM receiver equal the peaks of internally generated thermal noise power. (3-noise) The radio frequency input level in a radio receiver at which the signal power equals the internally generated thermal noise power. (4-trigger) The value necessary to activate a device. (5-network Management) Threshold in network management terms applies mainly to agent-based thresholds that allow network devices to directly generate events when something interesting happens on the network. Thresholds are usually set with SNMP or the RMON MIB.

Threshold Alarm-An alarm indication that informs operators that a predetermined threshold that has been exceeded. An example of a threshold alarm is switch capacity.

Thresholding-Threshholding is the process of modifying numbers or measurements that are within a range or meet some criteria to produce a lower number or a lesser number of data elements. Thresholding is used in lossy data compression processes (such as image compression) to reduce the amount of data through the loss of accuracy of information that has little impact on the user.

Throughput-(1-telephone system) The number of telephone call attempts successfully completed each second. (2-data communication) The number of bits, characters, or blocks of data that a system working at maximum speed can process during a specified period of time. (3-ISDN/PPSN) In the Integrated Services Digital Network (ISDN) and the Public Packet-Switched Network (PPSN), the average information rate of a particular virtual circuit.

Thumbnail-Thumbnails are smaller (miniature) low resolution versions of other images. Thumbnails are commonly used as icons or images of items in online catalogs. Because the size and the resolution of thumbnails is much lower than their larger image version, the transfer time (download) is usually short, even with slow dial-up user connections.

THz-TeraHertz

TIA-Telecommunications Industry Association

Tie Wrap-A tie wrap (also called a cable tie) is a flexible strap that contains a self locking eyelet at one end that allows the other end of the strap to enter and latch so it cannot be pulled back out. The tie wrap is looped around an object that is to be fastened (such as a cable) and another object (such as a hanger or conduit pipe) and the end is pulled through the tie wrap eyelet so a snug wrap is formed.

TIFF-Tagged Image File Format

Tiling-Tiling is the changing of a digital video image into square tiles that are located in positions other than their original positions on the screen.

Time Alignment-(1-general) Time alignment is the process of adjusting the timing relationship between communication signals. (2-radio) Time alignment of transmitter and receiver timing keeps different mobile radio's transmit bursts from colliding or overlapping.

Time and Date Table (TDT)-This table provides updated time and date information (changes and corrections).

Time Delay-The amount of time for a signal to travel from one point to another point.

Time Diversity-Time diversity is the process of sending the same signal or components of a signal through a communication channel where the same signal is received at different times. The reception of two or more of the same signal with time diversity may be used to compare, recover, or add to the overall quality of the received signal.

Time Division-The separation in the time domain of a number of transmission channels between two points.

Time Division Duplex (TDD)-Time division duplex (TDD) is a process of allowing two way communications between two devices by time sharing. When using TDD, one device transmits (device 1), the other device listens (device 2) for a short period of time. After the transmission is complete, the devices reverse their role so device 1 becomes a receiver and device 2 becomes a transmitter. The process continually repeats itself so data appears to flow in both directions simultaneously.

Time Division Multiple Access (TDMA)-Time division multiple access (TDMA) is a process of sharing a single radio channel by dividing the channel into time slots that are shared between simultaneous users of the radio channel. When a mobile radio communicates with a TDMA system, it is assigned a specific time position on the radio channel. By allowing several users to use different time positions (time slots) on a single radio channel, TDMA systems increase their ability to serve multiple users with a limited number of radio channels.

Time Division Multiplexing (TDM)-Time division multiplexing is a method used to send two or more signals over a common transmission path by assigning the path sequentially to each signal, each assignment being for a discrete time interval. All channels of a time-division multiplex system use the same portion of the transmission links' frequency spectrum - but not at the same time. Each channel is sampled in a regular sequence by a multiplexer

This figure shows how a single carrier channel is time-sliced into three communication channels. Transceiver number 1 is communicating on time slot number 1 and mobile radio number 2 is communicating on time slot number 3. Each frame on this communication system has three time slots.

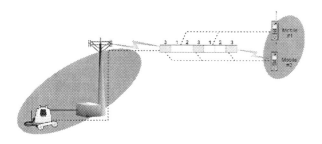

Time Division Multiplexing (TDM) Operation

Time Division Multiplexing over Internet Protocol (TDMoIP)-The providing of time based (time division) communication circuits through the Internet. TDMoIP allows for T1 and E1 communication lines to be transferred by a managed or unmanaged packet network (such as the public Internet).

Time Domain Reflectometer (TDR)-A test instrument used to identify and locate discontinuities or breaks in a transmission line. At a discontinuity, the impedance of the line is irregular and causes reflections to occur. The instrument measures the distance to a discontinuity by sending a voltage pulse through a line and measuring the time it takes for the signal to reach the point of and reflect from the discontinuity.

Time Multiplexing (TM)-Time multiplexing is the process of sending two or more signals over a common transmission path by assigning the path sequentially to each signal, each assignment being for a discrete time interval (time slot).

Time Out-A specified period of time allowed to elapse in a system before a specified event takes place, unless another specified event occurs first.

Time Shift TV-Time shift television is the use of media storage technology to allow viewers to control the time and method and that television viewing occurs. In addition to the ability to delay the viewing time for television programs, time shift TV enables viewers to pause, rewind live TV programs and adds a fast-forward functionality up until the point at which a subscriber reaches parity with the system-wide live broadcast. When coupled with network-wide VOD, time-shift TV offers subscribers VCR-like control. Time shift TV is sometimes referred to as network personal video recorder (nPVR).

Time Shifted Programming-Time shifted programming is the viewing of a program or information at a time other than that at which it was originally received or scheduled to be received.

Time Shifting-Time shifting is the viewing of a program or information at a time other than that at which it was originally broadcast.

Time Slot (Timeslot)-The smallest time division dedicated to a communication device. For the GSM system, a timeslot is 577 microseconds.

Time Slot (TS)-Time slots are the smallest division of a communication channel that are assigned to particular users in a communication system. Time slots can be combined for a single user to increase the total data transfer rate available to that user. In some systems, time slots are assigned dynamically on an as-needed basis.

Time Slot Interchange Switching (TSI)-A process of connecting incoming and outgoing digital lines together through the use of temporary memory locations. A computer controls the assignment of these temporary locations so that a portion of an incoming line can be stored in temporary memory and retrieved for insertion to an outgoing line.

Time Slot Offset-Time slot offset is the amount of delay time a transmitter must wait before transmitting a time slot signal. Time slot offset is used to compensate for the varying amount of transmission time that occurs in a mobile communication network as mobile devices move closer and further away from the radio base stations.

Time Stamp (Timestamp)-Time stamps are the insertion of information that indicates a time that an event occurred into a record. Time stamps are used to indicate the time a file was created or when an error or network failure has occurred in a system.

Time To Live (TTL)-Time to live is a field within a data packet that is used to limit the maximum number of routing or switching points a packet may pass through during transmission in a data network. The TTL counter is decreased as it progresses through each router or switching point in the network. If the TTL counter reaches 0, the packet can be discarded. The use of TTL ensures packets will not be transmitted in an infinite loop. This diagram shows a packet data network uses a time to live (TTL) control field to avoid the potential for routing packets through long travel paths or infinite loops. This diagram shows that a data packet enters into a packet network contains its destination address, time to live field, and data payload. The TTL field is initially set by the sending computer and its value may vary dependent on the type of service (e.g. real time voice compared to file transfer.) As the packets are routed through the packet network, each router (packet switching device) forwards the packet towards the destination it believes will send the packet towards its destination. Each time the packet passes through a router, the TTL field value is decreased. This diagram shows that the routers in this diagram accidentally send the packet into an infinite loop due to a broken line between routers that force packets to be rerouted through alternate paths. Eventually the routers will adjust their packet forwarding tables. However, in this case the packet travels through too many routers and the TTL field expires and the packet is discarded.

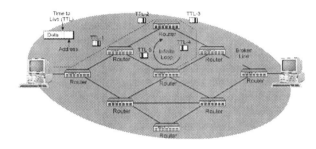

Time to Live (TTL) Operation

Timed-Release Disconnect-A disconnect process that is a result of exceeding a pre-determined maximum time interval for a process or service.
Timeslot-Time Slot
Timestamp-A timestamp is the recording of the time that an event occurs. A timestamp may be created using an internal (unsynchronized) time clock (such as a clock in a personal computer) or it may be recorded using a more accurate time through a network connection such as using network time protocol (NTP).
Timestamp-Time Stamp

Timestamp Synchronization-The process of using time information embedded in multiple sources of data (such as an audio stream and a video stream) to present them together in synchronization. An example of a streaming protocol with embedded synchronization timestamps is RTP.
Timing-Timing is the use of repetitive, typically accurate signals, to synchronize or coordinate the operations of a communication line or parts of a system.
Timing Pulse-A pulse that is used to synchronize signals.
Timing Recovery-The process of determining the appropriate sampling times for a data stream by deriving a clock signal from that bit stream.
Tinned-An element covered with a thin layer of metallic tin to inhibit corrosion and facilitate soldering.
Tip & Ring-An old fashioned way of saying "plus" and "minus," or ground and positive in electrical circuits. Tip and Ring are telephony terms. They derive their names from the operator's switchboard plug, and the ring wire was connected to the slip ring around the jack. A third conductor on some jacks was called the sleeve. That's it. Nothing more sinister. Nothing more interesting.
TIPHON-Telephony and Internet Protocols Harmonization over Networks
TIR-Total Internal Reflection
TK-Toolkit
TKIP-Temporal Key Integrity Protocol
TLD-Top Level Domain
TLDN-Temporary Location Directory Number
TLE-Telephone Line Emulator
TLS-Telephone Line Simulator
TLS-Transparent LAN Services
TM-Phone Doubler
TM-Time Multiplexing
TM Services-CLASS
TMSI-Temporary Mobile Station Identity
TND-Telephone Network For The Deaf
ToIP-Telephony Over Internet Protocol
Token-The code passed among nodes (typically computers) in a network in a particular sequence for each node. This sequence indicates which node has permission to transmit data next.
Token Passing-A method of controlling the use of a communications channel, especially a ring network. A token packet is circulated from node to node when there is no live traffic. Possession of the token gives a node access to the network for transmission of data.

Token Rate-Token rate is the rate at which traffic credits (data transfer allocations) are granted in bytes per second. An application may send data at this rate continuously. Burst data may be sent up to the token bucket size. Until that data burst has been drained, an application must limit itself to the token rate. For best-effort services, the application gets as much bandwidth as possible. For guaranteed service, the application gets the maximum bandwidth available at the time of the request. See also Token bucket.

Token Ring-Token ring is a LAN system that passes a token to each computer connected to the network. Holding of the token permits the computer to transmit data. The token ring specification is IEEE 802.5 and token ring data transmission speeds include 4 Mbps, 16 Mbps, 100 Mbps and 1 Gbps.

Token ring networks are non-contention based systems, as each device connected via the token ring network must receive and hold a token before it can transmit. This ensures only one device will transmit data at any given time. Token ring systems provide an efficient control system when many devices are interconnected with each other. This is the reason token ring systems will not see data traffic degradation when many new stations are added, compared to collision-based (non-switched) Ethernet systems which degrade exponentially as new stations are added. Passing tokens does add a small overhead (additional control messages) and can slightly increase a packet's transmission time if the transmitting station must wait for a free token to arrive in efficient control system when many computers are interconnected with each other. This is the reason token ring systems will not see data traffic degradation when many new users are added compared to Ethernet systems. However, passing tokens does add overhead (additional control messages) that reduces the overall data transmission bandwidth of the system.

Toll-A charge for telecommunications service that results when a call is routed beyond a local calling area. Due to telecommunications deregulation, toll charge boundaries have changed and in some cases, have been eliminated throughout entire country areas so all calls are charged the same rate even when calling out of their local service area.

Toll Bypass-The routing of calls or communication sessions around any other networks facilities to avoid toll charges. An example of toll bypass is the use of voice over internet protocol (VoIP) services that allows customers to bypass the public switched telephone network (PSTN) switches in order to utilize the packet network (ex. IP data network) for long-distance (also known as toll) voice calls. This technique is used in conjunction with the H.323 protocol.

Toll Fraud-The theft of long-distance service including hacking or using stolen credit cards, computers, and 800 numbers to access a switch and determine a method by which other calls can be placed.

Toll Free-A service that allows callers to dial a telephone number without being charged for the call. The toll free call is billed to the receiver of the call. In the United States, toll free calls are preceded by a 1-800, 1-888 or 1-877 exchange. In Europe and other parts of the world, toll free calls are called Freephone and typically begin with 0-800.

Toll Quality-A quality of service (QoS) that is acceptable for voice communication on a telephone system. Because the measurement of voice quality is subjective, it is measured by a mean opinion score (MOS). The MOS is a number that is determined by a panel of listeners who subjectively rate the quality of audio on various samples. The rating level varies from 1 (bad) to 5 (excellent). Good quality telephone service (called "toll quality") has a MOS level of 4.0.

TollGate-A milestone step in a product development process that is used as an evaluation point to determine if further product development steps will be taken. There are often tollgates between concept, feasibility, planning, development, and introduction phases. Tollgates may require specific types of documents such as product descriptions, marketing evaluations, and intellectual property reviews.

TOM-Telecommunications Operational Model

Tool Command Language (TCL)-A scripting language that is commonly used by communication service providers to customize their interactive voice response (IVR) systems. TCL uses relatively simple (and modifiable) syntax and can be used as a separate program or combined (embedded) in other programs. TCL is open source language so its' use is free. See http://www.tcl.tk for more information.

Toolkit-Toolkits are a group of software programs that assist designers or developers to create products or services. These software tools will allow vendors to create products such as end-points (IP telephones and messaging systems), servers (proxy, redirection, and registrar), media gateways, and application servers.

Toolkit (TK)-A graphic user extension language for tool command language (TCL). TK allows for the creation of graphic user interfaces for software applications.

Top Level Domain (TLD)-A top level domain is the uppermost group of items in a hierarchical system. For file systems it is the root directory. For the internet, it is the rightmost part of the web address (e.g. .COM, .EDU, .NET).

Top-Level Domains-The highest category of host domain name on the Internet. The domain name identifies the type of institution or the country of its registration. Common top-level domain names include:
.com - company or commercial organization
.edu - education (school, university)
.gov - U.S. government
.int - International groups and organization
.mil - U.S. military
.net - An organization that manages networks
.org - nonprofit organization
.ca - Canada
.uk - United Kingdom

Topology-The topology of a network is it's physical or logical interconnection layout. It is the connection map of a network. The physical and logical topology does not have to be the same. The physical topology describes the connection of cables to equipment. The Logical topology is the data communication paths that messages take to move between locations on the network.

Torrent-(1-file transfer) A torrent is a rapid file transfer that occurs when multiple providers of information can combine their data transfer into a single stream (a torrent) of file information to the receiving computer. (2-water) A fast moving stream or flow of water.

This diagram shows how to transfer files using the torrent process. This example shows that 4 computers contain a large information file (such as a movie DVD). Each of the computers is connected to the internet via high-speed connections that have high-speed download capability and medium-speed upload capability. To speed up the transfer speed

for the file transfer, the receiver of information can request sections of the media file to be downloaded. Because the receiver of the information has a high-speed download connection, the limited uplink data rates of the section suppliers are combined. This allows the receiver of the information to transfer the entire file much faster.

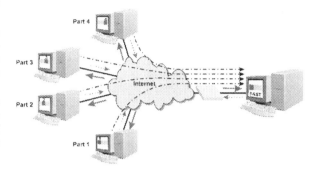

Torrent File Transfer Operation

TOS-Type Of Service

Total Cost Of Ownership (TCO)-Total cost of ownership is the direct costs of hardware and software along with the costs for operation and maintenance.

Total Harmonic Distortion (THD)-Total harmonic distortion is a ratio of the combined amplitudes of all signals related harmonically to the amplitude of a fundamental signal. THD is typically expressed as a percentage.

Touch-Tone Signaling-A signaling system that uses the combination of two tones to represent digits that are transferred in a communication system. Also known as dual-tone multi-frequency (DTMF) signaling.

Tower-A tall structure that supports antennas or antenna wires. If made of metal, the tower itself can serve as the antenna radiating element.

TPC-Transmit Power Control
TPMF-Terminating Point Master File
TPMP-Telco Product Market Planning
TPPM-Telco Product Portfolio Management
TR Switch-Transmit Receive Switch
TRA-Telecommunication Resellers Association
Trace-A trace is a displayed pattern on an test instrument (such as an oscilloscope or OTDR) that represents a measured signal.

Traceroute-A utility used on TCP/IP networks to trace the route that IP datagrams take between one network device (ex. server, router) and another system on the network. A traceroute also tells you how long each hop takes, it can be a useful tool in identifying system or network trouble spots. Traceroute utilizes the time to live (TTL) field in the IP packet and typically runs over the UDP protocol (best effort). Traceroute is sometimes indicated also by the syntax "tracer".

Track-(1-memory device) The portion of a moving-type storage medium that is accessible to a given reading station. (2-videotape or digital video) The section of a videotape or digital media file where a particular signal is recorded. In an analog recorder, there typically are separate tracks for video, audio, and time code.

Tracking-(1-services) Tracking is the monitoring or following of the flow or usage of information, transmission or services. (2-radio) The locking of tuned stages in a radio receiver so that the system changes as the receiver tuning is changed (tracking the changes).

Tracking Application-A tracking application is a software program and/or service that can be used to identify the location of a user or device within a geographic area.

Trade Secret-Information, data, documents, formulas, or anything which has commercial value and which is kept and maintained as confidential.

Trademark-A unique symbol, word, name, picture, design, or combination thereof used by firms to identify their own goods and distinguish them from the goods made or sold by others.

Trademark rights are Intellectual Property Rights which give the owner the right to prevent others from using a similar mark which could create confusion among consumers as to the source of the goods.

Trademark protection must be registered in the country, or region, where it is desired. In most countries and regions, patents and trademarks are administered by the same government agency.

Traffic-The amount of data transferred over a communications link or number of messages processed by a communication server over a specified period of time. An example of traffic measurement is centum call seconds (CCS). A single CCS equals 100 seconds of a call on the same communication circuit.

Traffic Capacity-The maximum amount of communication traffic (users or data transfer rate) per unit time that can be carried by a specified telecommunication system, sub system, or device under specified conditions.

Traffic Channel-The combination of voice and data signals existing within a communication channel.

Traffic Dimensioning-Traffic dimensioning is the configuring of a system or service to allocate resources, data transmission rates, or processing capability to particular users or systems to ensure that high volume users do not reduce or disable the services available to other users.

Traffic Load-Traffic loading is the volume of traffic that equals the sum of the holding times of several calls or attempts. Loads normally are expressed in either hundred call seconds (CCS) or Erlangs. A statement of load is inherently an average of all the instantaneous loads over a basic interval, such as an hour. This term often is called simply traffic load.

Traffic Routing-Traffic routing is the selection of communication routes, paths, or the choice of preferred communication carriers (such as long distance) based on dialed digits, cost of service, traffic congestion, line failure, or other criteria that may affect choice of routing path.

Training Sequence-A sequence of data bits that are previously known to the sender and receiver of the data bits. This allows the receiver to adjusts its reception process by using the known sequence.

Transaction-A transaction is an activity within system or domain that transfers the ownership or rights to assets or other items that have value or that can be characterized.

Transaction Processing-(1-general) Transaction processing are the steps and processes taken to complete a transaction (such as an order). (2-credit card) The processing of credit card payments by a merchant credit card processor.

Transactional Metadata-Transactional metadata is information that describes the requirements for the business requirements for the transfer or use of media. Examples of transactional metadata include the cost, usage time, authorized types of users and other usage rights restrictions.

Transactional Rights-Transactional rights are actions or procedures that are authorized to be performed by individuals or companies that granted as the result of a transaction or event. An example of a transactional right is the authorization to read and use a book after it is purchased in a bookstore.

Transceiver-A transceiver is combination of a radio transmitter and receiver into one radio device or assembly. A portable cellular phone is a transceiver.

This diagram shows a block diagram of a mobile radio transceiver. In this diagram, sound is converted to an electrical signal by a microphone. The audio signal is processed (filtered and adjusted) and is sent to a modulator. The modulator creates a modulated RF signal using the audio signal. The modulated signal is supplied to an RF amplifier that increases the level of the RF signal and supplies it to the antenna for radio transmission. This mobile radio simultaneously receives another RF signal on a different frequency to allow the listening of the other person while talking. The received RF signal is then boosted by the receiver to a level acceptable for the demodulator assembly. The demodulator extracts the audio signal and the audio signal is amplified so it can create sound from the speaker.

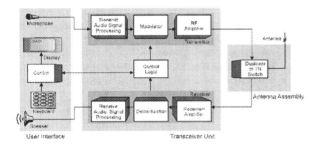

Mobile Radio Functional

Transceiver Cost-Transceiver cost is the purchase cost of a transceiver. The transceiver cost may or may not include a licensing fee that is paid for the use of technology (IPR) or software in the transceiver.

Transcoder-A device that enables differently coded transmission systems to be interconnected with little or no loss in functionality.

Transcoder and Rate Adaptation Unit (TRAU)-A trascoding and rate adaptation unit converts digital signals from one coding format to another and adjusts the data transmission rate between systems or communication lines.

Transcoding-Transcoding is the conversion of digital signals from one coding format to another. An example of transcoding is the conversion of u-Law encoded PCM to A-Law encoded PCM signals.

This diagram shows the basic transcoding process between mu-Law PCM coding and A-Law PCM coding. This diagram shows that a telephone that uses A-LAW PCM speech coding in North America system is communicating with a telephone in Europe that is using u-LAW PCM speech coding. This diagram shows that the transcoding system must identifies the type of PCM audio used by each system and the location of the transcoding gateway function. The PCM transcoder converts the A-LAW PCM signal to u-LAW PCM.

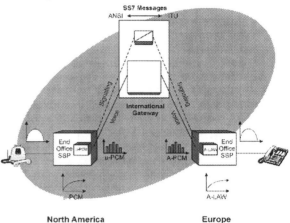

Transcoding

Transducer-Any device or substance that converts energy from one form to another. The transducer on a SAW device converts voltage to mechanical vibrations (phonons) and vice versa. A microphone converts mechanical vibrations (sound) to voltages. A photodiode converts light to electrical energy. A laser or an LED converts electrical energy to light.

Transfer Delay-Transfer delay is the time between the sending of a packet to complete delivery at the destination station.

Transferred Account Procedures (TAP)-Transferred accounting process (TAP) is a standard billing format that is primarily used for global system for mobile (GSM) cellular and personal communications systems (PCS). As of 2001, the versions of TAP, TAP II, TAP II+, NAIG TAP II, and TAP 3. Each successive version of TAP provided for enhanced features.

Due to the global nature of 3G wireless and GSM, the TAP billing standard provides solutions for multi-lingual and multiple exchange rate issues. TAP3 was released in 2000 as a significant revision of TAP2. TAP3 has changed from the fixed record size used in TAP2 to variable record size and TAP3 offers billing information for many new types of services such as billing for short messaging and other information services. The TAP standard is managed by the GSM association at www.GSMmobile.com.

Transferred Account Procedures 3 (TAP3)- Transferred accounting process 3 (TAP3) is a standard billing format that is primarily used for GSM and GPRS systems. TAP3 was released in 2000 as a significant revision of its predecessor TAP2. TAP3 uses a variable record size and it offers billing information for many new types of services such as billing for short messaging and other information services. The TAP standard is managed by the GSM association at www.GSMmobile.com.

Transformer-(1-electrical) An electrical component typically comprising two or more coils of insulated wire surrounding a common core space. The core in some transformers may be only air, but most transformers use a core of ferromagnetic material such as iron or ferrite. Transformers are used in both electric power and telecommunications applications. When an alternating current power source is connected to one coil of a transformer, a voltage having the same waveform but a different (larger or smaller) voltage amplitude is produced via electromagnetic induction at the second coil terminals. The ratio of the two voltages is typically the same as the ratio of the number of turns of wire in each coil, respectively. Thus a transformer having a primary coil (connected to the power source) comprising 100 turns and a secondary coil comprising 200 turns will produce a voltage at the terminals of the secondary coil that is double the source voltage. The current at the secondary coil in this example will be one half of the primary coil current, so the power output is not larger than the power input. Because the voltage to current ratio is four times larger at the secondary coil terminals compared to the primary coil termi-

nals, we describe this transformer as a device that changes the impedance level of the secondary circuit compared to the primary circuit by a factor of four.

Transformers with the same number of turns in the primary and secondary coils are extensively used in telecommunications to couple the alternating current component (such as speech waveform) from the primary to the secondary, but not the direct (constant) component of the current.

Certain devices that do not use coils of wire also have the capability of producing higher or lower voltages as well. For example, a resonant quarter wavelength transmission line stub, driven by a low voltage source of resonant frequency near the short circuit end, produces a higher output voltage than the input voltage at the open circuit end.

(2-toy) In non-telecom usage, there is a completely unrelated child's toy called a transformer that has pieces that can be moved about so that the toy resembles a robot in one configuration and a truck or rocket, for example, in another.

Transient-A sudden variance of current or voltage from a steady-state value. A transient normally results from changes in load or effects related to switching action.

Translation-The conversion of information from one form to another or the conversion of all or part of a telephone address destination code to routing instructions or routing digits.

Translational Bridge-A transparent bridge that interconnects LANs that use different frame formats (e.g., an Ethernet-to-FDDI bridge). A translational bridge must map the salient fields between the dissimilar frame formats used on its ports.

Transmission-(1-general) The transfer of electric power, signals, or intelligence from one location to another by wire, fiber optic, or radio means. (2-beam) The use of a directional radio antenna to concentrate radio power into a small angle. (3-parallel) The simultaneous transmission of a number of signals through one or more systems. (4-radio) The transmission of electromagnetic radiation at radio frequencies. (5-serial) The transmission of sequential signals over a given system. (6-stereophonic) The use of various methods to transmit two separate, but related, channels of audio information in order to convey the stereo effect

Transmission Channel-The assembly of circuits, components, and media necessary to transport a message produced by a source and deliver its information content to the input of a sink.

Transmission Control Protocol (TCP)- Transmission control protocol is a session layer protocol that coordinates the transmission, reception, and retransmission of packets in a data network to ensure reliable (confirmed) communication. The TCP protocol coordinates the division of data information into packets, adds sequence and flow control information to the packets, and coordinates the confirmation and retransmission of packets that are lost during a communication session. TCP utilizes Internet Protocol (IP) as the network layer protocol.

This diagram shows how transaction control protocol (TCP) operates to reliably send data through a packet network. This diagram shows that the TCP system receives the data from a specific communication port (port number). The TCP system then packetizes (divides) the sender's data into smaller packets of data (maximum 1500 bytes). Each of these packets starts with an IP header that contains the destination address of the packet. The TCP system then adds a second header (the TCP control header) that includes a sequence number along with other flow control information. The packets are sent through the system where they may be received at different time periods. The sequence numbers can be used to reorder the packets. The TCP protocol also includes a window size that indicates to the receiving device how many packets it can receive before it must acknowledge their receipt. This window defines how much data the sending device must keep in temporary memory to enable the retransmission of a packet in the event that a packet is lost in transmission. If a packet is lost, the receiving device requests the transmitting device to re-send the packet with a specific sequence number.

Transmission Control Protocol (TCP) Header-A control header that is added to the data portion of an Internet protocol datagram packet. The TCP header holds sequencing information that is used to coordinate the reliable transfer of information during a communication session.

This diagram shows the TCP header field structure for version 4 IP. The TCP header is session level protocol and is it is located after the network IP addressing portion of each IP datagram packet. The TCP control structure is used to ensure reliable communication using IP addressing. The TCP packet header starts with port numbers that indicates the logical channel where the data originated from the sending device and where it is to be routed (e.g. which software application) by the receiving communication device. The header includes sequence numbers and acknowledgement sequence numbers to track the sending and confirmation of each packet (or groups of packets) that have a TCP header. The TCP header includes a field that defines the confirmation window size. The window size indicates the maximum number of TCP packets that can be received before the receiving device must confirm their reception.

IP Header	TCP Header	Data

Source port		Destination port	
Sequence number			
Acknowledgment number			
Hlen	Reserved	Flag bits	Window
Checksum			Urgent pointer
Options + Padding			

Transmission Control Protocol Version 4 (TCP) Header Structure

Transmission Control Protocol And Internet Protocol (TCP/IP)-TCP/IP is a standard set (suite) of protocols that defines how the Internet messages are transferred reliably through a data network. The Transmission Control Protocol (TCP) portion ensures message delivery between two points and the Internet Protocol (IP) defines the routing and addressing of packets.

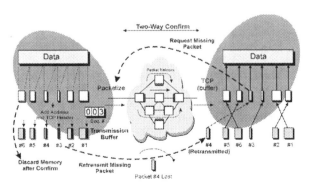

Transmission Control Protocol (TCP) Operation

Transmission Convergence (TC)-Transmission convergence is the process of adapting packet formats (such as long media packets) into a length (e.g. fragmentation) and form that can be sent on a transmission channel.

Transmission Delay-Transmission delay is the time that is required for transmission of a signal or packet of data from entry into a transmission system (e.g. transmission line or network) to its exit of the system. Common causes of transmission delay include transmission time through a transmission line (less than the speed of light), channel coding delays, switching delays, queuing delays waiting for available transmission channel time slots, and channel decoding delays.

Transmission Interval-Transmission interval is the amount of time that is allowed for the detection of authorization to transmit in a random access contention based communication system (such as a mobile telephone or Ethernet system).

Transmission Line-(1-general) Any transmission medium, including free space. (2-transmitter) A circuit that connects a transmitter to a load over a distance. (3-wave) Any circuit whose dimension is large compared with the wavelength of the signals passing through it.

Transmission Line Testing-Transmission line testing is the measurements of the characteristics of a transmission line (such as a data or fiber optic line) to determine that the transmission is operating within expected performance levels (such as error or distortion levels).

Transmission Loss-Transmission loss is the ratio, usually measured in decibels, of the power of a signal at a point along a transmission path to the power of the same signal at a more distant point along the same path. This value often is used as a measure of the quality of the transmission medium for conveying signals. Changes in power level normally are expressed in decibels of the ratio of the two powers.

Transmission Medium-Transmission medium is the physical material or free space characteristics that a signal will pass through during its transmission between two or more points. Examples are transmission medium including copper shielded twisted-pair (STP), and unshielded twisted-pair (UTP) cabling, coaxial lines, optical fiber and air.

Transmission Protocol-Transmission protocol is the commands, procedures and processes that are used to coordinate access to transmission mediums and to transfer data between two or more communication devices.

Transmission Rate-The amount of information that is transferred over a transmission medium over a specific period of time.

Transmission System-Transmission systems interconnect communication devices (end nodes) by guiding signal energy in a particular direction or directions through a transmission medium such as copper, air, or glass. A transmission system will have at least one transmitting device, a transmission medium, and a receiving device. The transmitting communication devices is capable of converting information to form electrical, electromagnetic wave (radio), or optical signals that allows the information to be transferred through the medium. The receiving communication device converts the transmitted signal into another form that can be used by the device or other devices that are connected to it. Transmission systems can be unidirectional (one direction) or they can be bi-directional (two directions).

Transmission Testing-Transmission testing is the measurement of specific performance characteristics of a transmission line that may include signal loss, frame slips or error seconds.

Transmit Delay-The time delay between when a signal is first originated to when it is first received at its destination. Also called transmission time or propagation delay for radio signals.

Transmit Diversity-Transmit diversity is the use of two or more antennas on a transmitter to improve the characteristics of the signal that is received by a device. Each antenna element of the transmit diversity system may be provided with a varied (e.g. phase shifted signal) so that the variances (fade margins) in the received signal level are reduced.

Transmit Flow Control-A means of adjusting the rate at which data may be transmitted from one terminal in a data communication system so that it is equal to the rate at which the data can be received by another terminal in the same system.

Transmit Receive Switch (TR Switch)-A switch that provides a connection from an antenna to either a transmitter or receiver. The use of a TR switch avoids the possibility of the high power transmitter being connected to a sensitive receiver. This diagram shows how a TR switch connects an antenna to either a transmitter or receiver. This TR switch is controlled by the transmitter on ("key on") switch. When the transmitter is on, the TR switch moves to connect the antenna to the transmitter. When the transmitter is off, the TR switch connects the antenna to the receiver allowing the user to hear the channel when they are not talking.

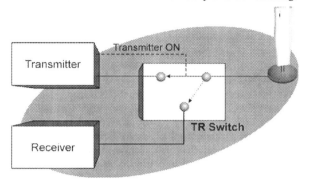

TR Switch Operation

Transmittance-A measure of the capability of a material to permit light signals to pass through.

Transmitter-(1-telephone) Older term for the microphone in a telephone handset, and used as an adjective to identify signals associated with that microphone. (2-radio) A device that converts an electrical or acoustic signal into a transmitted carrier signal.

Transmitter (XMTR)-A device or system that receives and information signal and converts it to a form that is suitable for transmission.

Transmitter Address (TA)-The transmitter address is the identifying address of the device transmitting the signal or packet.

Transmitter Power-The radio signal power level of a transmitter.

Transparency-A property of a communications medium (such as a coaxial cable or optical fiber) to carry a signal without altering or other-wise affecting the photonic or electrical characteristics of the signal.

Transparent-Transparent processes and operations are processed automatically in computer software or hardware systems but are not visible to the user or operator

Transparent Access-Transparent access is the direct connection of a user to communication systems (such as the Internet).

Transparent Bridge-A local area network bridge which relies solely on MAC address learning for all forwarding decisions. Each transparent bridge inspects each packets source MAC address to associate each MAC address with a destination port. When a packet arrives its source MAC address is inspected for forwarding table maintenance, while its destination MAC address is inspected to determine the egress port.

Contrast this to a source route bridge, which relies on explicit information in each packet to determine the path the packet will take. Transparent bridges are generally more difficult to implement than source route bridges, but allow end stations to be on bridged LAN segments without being aware of that fact. Thus they are called transparent bridges. Source route networks require each end station to maintain not only destination MAC address information, but also path route information. However, source route bridges are easier to implement and allow complex networks to be built with multiple active parallel paths for better traffic distribution and reliability.

Transponder-(1-relay) A combination receiver and transmitter, frequently part of a communications satellite, that receives a signal from an uplink station and retransmits it to ground receiving stations. (2-response system) An electronic circuit that receives a signal from an interrogating station, such as a ground-based radar unit, and transmits an appropriate response.

Transport-The process of transferring information between physical or logical connection points.

Transport Channels-Transport channels are communication channels that use one or more physical channels in a specific way (such as specific channel codes) to transfer information. Transport channels define how, when, and which physical channels are used.

Transport Connection-A transport connection is a unidirectional connection that is used to transport user data. Each transport connection has unique quality of service (QoS) parameters associated with it. Transport connections are typically assigned in pairs (uplink and downlink).

Transport Layer-The transport layer is a protocol layer that responds to transmission service requests from upper layers (such as the session layer) and creates service requests to the lower network communication layers. The primary purpose of the Transport Layer is to provide the transfer of data between end users.

Transport Overhead-In the Synchronous Optical Network (SONET), the overhead added to the Synchronous Payload Envelope of a Synchronous Transport Signal for transport purposes. Transport overhead consists of line and section overhead.

Transport Rights-Transport rights are the authorizations to move, copy or loan content.

Transport Stream (TS)-Transport Streams are the combining (multiplexing) of multiple program channels (typically digital video channels) onto a signal communication channel (such as a satellite transponder channel). These channels may be statistically combined in such a way that the bursty transmission (high video activity) of one channel is merged with the low-speed data transmission (low video activity) with other channels so more program channels can share the same limited bandwidth communication channel.

Transportable-A full 3-watt cellular mobile telephone, sometimes switchable to .3 watts, complete with battery pack and antenna, facilitating portable operation.

Transrating-Transrating is the process of converting information from one transmission rate to another transmission rate. An example of transrating is the conversion of a high-speed digital video signal that is received from a satellite into a medium-speed digital video signal that is transferred through the Internet. Transrating products are also called digital turnaround devices.

Transverter-A transverter is a pair of frequency conversion devices that change the frequency bands of transmitter and receiver into other frequency bands. The use of a transverter allows radio communications equipment to operate on different sets of frequencies without the need for equipment changes or redesign.

Trap-A trap is: (1-television) A filter that blocks (traps) a signal (such as a television channel) so it cannot pass through to a receiver (such as a television set). (2-network) A message or command (such as a SNMP message) that informs a network management system that an unusual condition has occurred.

TRAU-Transcoder and Rate Adaptation Unit

Treatment-The handling of customers or vendors for a specific purpose such as recovering overdue balances. When special treatment involves collection activities, it is called "dunning".

Treble-The highest audible sound frequencies, between approximately 2000 to 20,000 cycles per second.

Tree Topology-In a local area network, a configuration that resembles the distribution system of a tree trunk and its branches.

Trellis Coding-A method of forward error correction used in certain high-speed modems where each signal element is assigned a coded binary value representing that element's phase and amplitude. It allows the receiving modem to determine, based on the value of the preceding signal, whether or not a given signal element is received in error.

Tremendously High Frequency (THF)-The frequency band from 300GHz to 3000GHz.

Trenching-Trenching is the removal of materials from a land surface (e.g. dirt and sand) to create a narrow pit (a trench) that extends over a distance.

TRIB-Telephony Routing Information Base

Trickle Down-Trickle down is the process of sending information at a slower data transmission rate than is necessary for the instant use of the information. Trickle down transmission is commonly used to send video programming information through a limited or costly communication channel so that it can be gathered (accumulated) to allow the program to be viewed later.

Trickledown Video Delivery-Trickle down video delivery is the process of transferring media programs by streaming the program at a low bit rate to a storage device (such as a set-top box that has disk storage) so the program can be directly viewed from the storage after the transfer is complete. Trickledown video delivery can simulate a video on demand experience where the user selects to watch a movie on demand that has already been transferred to a set top box without requiring the use of significant network resources.

Trigger-The process of initiating an action as a result of a specific event occurring or a parameter associated with an event (such as exceeding a voltage level).

Trigger Number-A trigger number is a unique telephone number that is used to trigger another service such as a callback service. An international user could use a telephone trigger number to initiate a callback service. This would reduce the fees the international traveler pays for international calls as typically it is less expensive to receiving incoming calls than it is to initiate outgoing international calls.

Trigger Point-An event, incident, or development in an outside plant network that initiates a step or stimulates a reaction. An example of a trigger paint is the exhaustion of existing facilities or structures.

Triggers-Statically armed detection points.

Trim-The process of making fine adjustments to a circuit or a circuit element.

Trimmer-A small, mechanically adjustable component connected in parallel or in series with a major component so that the net value of the two can be finely adjusted for tuning purposes.

TRIP-Telephone Routing over Internet Protocol

Triple Play-Triple play refers to providing of three main services such as data, voice, and video on one network. For cable MSOs, this usually means building out the next generation network to DOCSIS 2.0 specifications, for Carriers this often means building out fibre or VDSL (very fast DSL) networks. Usually it is the larger MSOs and Telecom Carriers that roll out triple play services, and the advantage is that they can sign customers to a bundle of three services, thereby increasing revenue and customer loyalty.

Trivial File Transfer Protocol (TFTP)-Trivial file transfer protocol (TFTP) is a protocol that is used to transferring files between devices in a data communication network. TFTP is a simplified version of file transfer protocol (FTP) and it is commonly used in devices to allow for the transfer of setup and configuration information. TFTP is defined in RFC 1350.

Trojan Horse-A Trojan horse is a type of computer virus that is disguised to be a known or useful program but contains unexpected or harmful program codes that are activated when it is used. Trojan viruses are often provided in games, utilities, or other executable programs.

Trouble-A failure or fault affecting the service provided by a system.

Troubleshooting-Troubleshooting is the process of investigating, localizing and correcting (if possible) a fault or out of tolerance condition.

TRS-Telecommunications Relay Service for the Deaf

Truck Roll-A truck roll is the dispatching of an installer or technician to a customers location.

True Color-True color is represented by 24 bits or 32 bits of information.

Truncation-The removal of lower significant bits on a digital system, possibly leading to errors or artifacts.

Trunk-A communication path that connects two network elements (such as switching systems, networks or data devices). Trunks are usually shared by many users. Trunks may be classified by the type of equipment they connect. For example, a PBX trunk connects a PBX system to the public switched telephone network (PSTN). Trunks carry conversations for different subscribers at different times. The name comes from a trunk of a tree, via the prior usage for railroad trunk lines.

Trunk Group-A number of trunks that can be used interchangeably between two switching systems.

Trunk Occupancy-The percentage of some time period, usually an hour, during which a trunk is in use. Trunk occupancy also may be expressed as the carried hundred call seconds per hour per trunk.

Trunk Side Connection-Trunk side connections are used to interconnect telephone network switching systems to each other. Trunk side connections are usually high capacity lines.

Trunk Signaling-In interoffice signaling, the use of trunks carrying voice or data traffic to also transfer signals between switching systems.

Trunk Usage-The percentage of usage a trunk uses compared to its maximum capacity. Trunk usage may be measured at peak time or it may be an average measurement.

Trunking-An infrastructure dependent technique where communications resources, comprised of more than one logical channel (trunk) are shared amongst system users by means of an automatic resource allocation management technique based upon statistical queuing theory and resident in the systems fixed infrastructure. Typically usage requests follow a Poisson arrival process (statistical probability distribution) and the resource allocator assigns communications resources in response to requests from system users. As demand for service exceeds system capability at that time, service must be increasingly denied immediate access. This action is termed "blocking", with the blocked service request being queued for a later service response. The offered grade of service of the system is inversely proportional to the probability of blocking (e.g. lower probability of blocking offers a higher grade of service potential).

The dynamic resource allocation methodology of trunking results in the establishment of functional channels defining resource availability by means of dynamically allocating logical channels both to particular subscribers and for specific functions. These functional channels can be used for the conveyance of payload information, system control or a combination thereof. This results in the development of three (3) specific types of functional channels, these are: control, digital voice, and digital data.

Dedicated Control Trunking: Refers to a logical channel resource which supports only control function channel type (e.g.. resource management signaling, requests for service, etc.) between the fixed trunking system infrastructure and the subscriber and associated units.

Composite Control Trunking: Refers to a logical channel resource which supports control functional channel type as well as digital voice and/or digital data functional channel types (e.g. teleservice related payloads).

Trunking Gateway-Trunking gateways are voice gateways that provide bearer connectivity between the PSTN and the Voice-over-Packet network. Typically, T1 or T3 trunks provide the physical interface to the PSTN. The voice gateway converts voice bearer traffic from TDM voice to packetized voice (VoIP) and vice versa.

Trunk gateway is considered a "trunk-side" gateway, requiring IMTs (inter-machine trunks) for interconnection into the PSTN to carry voice traffic and "A"-link for interconnection to the SS7 network to carry signalling. In this case, a separate IP-to-SS7 signalling gateway is required to translate and mediate signals between the IP network and the SS7 network. Media Gateways also perform voice compression using CODECs; common CODECs for VoIP applications are: G.711 (64kbps), G.729a/b (8kbps), and G.723 (5 - 6 kbps).

Trunks-Groups of wires or fiber optic communication lines that are used to interconnect communications devices. Trunks usually have many physical and/or logical communication channels.

Trunks In Service-The number of trunks in a group that are in use or available to carry calls. The number of trunks in service may differ than the actual number of communication lines due to line failure or alternative assignment/uses of wire circuits.

Trunk-Side Connection-The connection of a transmission path to the trunk side of a local switching system.

Trusted Authority-A trusted authority is an information source or company that can issue or validate certificates. A trusted authority is sometimes called a "certificate authority."

Trusted Device-(1-general) A device that is previously known or suspected to only communicate information that will not alter or damage equipment of stored data. Trusted devices are usually allowed privilege levels that could allow data manipulation and or deletion. (2-Bluetooth) A device using Bluetooth wireless technology that has been previously authenticated and allowed to access another Bluetooth device based on its link-level key.

TS-Time Slot

TS-Transport Stream

TSAP-Transport Service Access Point

TSAPI-Telephony Services Application Programming Interface

TSI-Time Slot Interchange Switching

TSO-Time Sharing Option

TTA-Telecommunications Technology Association

TTL-Time To Live

TTS-Text To Speech

TTY-Teletypewriter

Tubing Cutter-A tubing cutter is a tool that contains a cutting blade that cuts around the outside circumference of a tube or cable. The tubing cutter surrounds the tube or cable and is rotated around the cable. As the tubing cutter is rotated around the cable, the pressure on the two (and blade inside) is gradually increased by more tightly gripping the tool so the cut depth is uniformly distributed around the tube or cable jacket. The depth of the cutting blade is usually adjustable to ensure the blade does not cut too deeply into the cable potentially cutting into materials or cables inside the cable.

Tune-The process of adjusting the frequency of a device or circuit, such as for resonance or for maximum response to an input signal.

Tuner-(1-radio) The radio frequency and intermediate frequency parts of a radio receiver that produce a low-level audio output signal. (2-waveguide) A device that permits adjustment of the impedance of a waveguide.

Tunneling Protocols-Tunneling protocols are commands or messages that can be embedded or disguised as other protocols that have a purpose of tunneling through a communication device or system. Tunneling protocols can be used to allow information to pass through firewalls that are designed or setup to block specific types of protocols (such as audio or video streaming protocols).

Tunneling Transport Protocols-Tunneling transport protocols is the transferring of protocol messages within data packets that are disguised as other protocols. An example of tunneling protocols is the sending of UDP packets within HTTP data packets to allow streaming media services to pass through firewalls.

TUP-Telephone User Part

Turnkey-The development of a product or system where the product or system is ready for use when it is delivered to a customer.

Turnover-(1-equipment) The point at which the installation of central office equipment is complete, and a telephone company accepts the equipment from an installation agent or supplier. (2-wiring) In a cable pair, the reversal of the tip and ring connections.

Turns Ratio-In a transformer, the ratio of the number of turns on the secondary to the number of turns on the primary. (See also: transformer.)

TV Broadcaster-Television Broadcaster

TVoDSL-Television over DSL

TVoIP-Television over Internet Protocol

TVRO-Television Receive Only

Tweaking-The process of slightly adjusting an electronic assembly circuit to optimize its performance or to bring the system within the required specifications.

Tweeter-A loudspeaker designed to efficiently handle high (3 kHz to 20 kHz) audio frequencies.

TWG-Technical Working Group

Twisted Pair-A pair of insulated copper wires used in transmission circuits to provide bi-directional communications. The wires are twisted around one another to minimize electrical coupling with other circuits. Paired cable is made up of a few, to several thousand, twisted pairs.

Two Binary, One Quaternary (2B1Q)-A line code (modulation and signaling structure) that transmits two binary bits of data at one time using a multi-level code. This code was initially used by the Integrated Services Digital Network (ISDN) standard for basic rate service (2 64 kbps + 1 16 kbps channels). 2B1Q is used by some DSL technologies including IDSL, HDSL and SDSL.

Two Wire Circuit-A circuit that carries information signals in both directions over a single pair.

TWX-Teletypewriter Exchange Service

Type 1 Connection-Type 1 connections are trunk-side connections to an end office. The end office uses a trunk-side signaling protocol in conjunction with a feature known as Trunk With Line Treatment (TWLT). This connection was originally described in technical advisory 76 published by AT&T in 1981. This interconnection was developed because dial line and DID connections did not provide enough signaling information to allow the connection of public telephone networks to other types of networks (such as wireless and PBX networks). The switch must be equipped to provide TWLT, or

its equivalent to offer Type 1 service. As a result, type 1 is not universally available. The TWLT feature allows the end office to combine some line-side and trunk-side features. For example, while trunk-side signaling protocols are used, the calls are recorded for billing purposes as if they were made by a line-side connection.

Type 1 connections are usually used as 2-way trunks. Two-way trunks are always 4-wire circuits, meaning they have separate transmit and receive paths, and almost always use MF address pulsing and supervision. The address pulsing normally uses wink-start control. One-way Type 1 connections can be provided on a 2-wire basis using E&M supervision or reverse battery like the DID connection.

Type 2A Connection-Type 2A connections are true trunk-side connections that employ trunk-side signaling protocols. Typically, they are two-way connections that are 4-wire circuits using E&M supervision with multifrequency (MF) address pulsing. The address pulsing is almost always under wink-start control. Type 2A connections allow the other public or private telephone network switching systems to connect to the PSTN and operate like any other EO.

Type 2A connections may restrict calls to specific NXX (exchange) codes and access to operator services (phone number directories, emergency calls, freephone/toll free) may not be permitted. For some interconnections, additional connections (such as a type 1) may be used to supplement the type 2A connection to allow access to other operator or network services.

Type 2B Connection-Type 2B connections are high usage trunk groups that are used between EOs within the same system. The type 2B connection can be used in conjunction with the Type 2A. When a type 2B is used, the first choice of routing is through a Type 2B with overflow through the type 2A. Because the type 2B connection is used for high usage connections, it can access only valid NXX codes of the EO providing that it is connected to. Type 2B connections are almost always 4-wire, two-way connections that use E&M supervision and multifrequency (MF) address pulsing.

Type 2C Connection-Type 2C connections were developed to allow direct connection to public safety centers (E911) via a tandem or local tandem switch. This interconnection type must provide additional such as the return phone number (complicated on mobile telephone systems) and the location of the caller. This information is passed on to a public safety answering point (PSAP).

Type 2D Connection-Type 2D interconnection lines allow direct connection from an operator services system (OSS) switch. The OSS switch is a special tandem that contains additional call processing capabilities that enables operator services special directory assistance services. The type 2D connection also forwards the automatic number identification information to allow proper billing records to be created. Type 2D connection will normally use trunks employing E&M signaling with wink start, and multifrequency (MF) address pulsing.

Type A USB Connector-A USB connector that is used to connect to the USB host (upstream) device.

Type Approval-An approval from a regulatory agency (such as the FCC) that identifies the equipment that is manufactured has passed tests certifying it meets the minimum requirements for that type of electronic or radio equipment. Most electronic devices must meet several regulatory specification requirements to receive type approval.

Type Of Service (TOS)-The field within the IP datagram header that indicates the type of packet so the system can vary the type of service (typically the priority of routing) that is performed on the IP data packet. The TOS field as specified in RFC 760 and RFC 2475 defines an application of TOS as used in DiffServ networks.

Type S-Type S connections transfer signaling messages that are associated with other interconnection types (out-of-band signaling). The type S is a data link (e.g. 56 kbps) that is used to connect the signaling interfaces between switches. Type S connections permit additional features to be supported by the network such as finding and using call forwarding telephone numbers. Because type S connections cost money, some smaller public telephone networks do not offer or use type S connections.

U

u See: μ, Mu-Occasional typographic substitute for Greek letter μ.

UA-User Agent

UAC-User Agent Client

UART-Universal Asynchronous Receiver And Transmitter

UAS-User Agent Server

UAT-User Acceptance Testing

UBR-Unspecified Bit Rate

UCD-Uplink Channel Descriptor

UCH-User Channel

UCITA-Uniform Computer Information Transactions Act

UDP-Uniform Dial Plan

UDP-User Datagram Protocol

UDP Header-User Datagram Protocol

UE-User Equipment

UGS-Unsolicited Grant Service

UHCI-Universal Host Controller Interface

UHF-Ultra High Frequency

UI-User Interface

u-Law Encoding-The type of non-linear digital voice coding (digital signal companding) that is commonly used in the Americas and other parts of the world. The U Law (pronounced Mu Law) coding process is used to compress the 13 bit sampling of a digitized audio signal into the equivalent of an 8 bit sample. It does this by assigning a non-binary (non-linear) value to each of the binary bits. Another non-linear voice coding system is the A Law coding system that is used in Europe and other parts of the world.

ULF-Ultra Low Frequency

Ultra Broadband-Ultra broadband is a term that is commonly associated with very high-speed data transfer connections. When applied to consumer access networks, ultra broadband often refers to data transmission rates of 10 Mbps or higher, and is generally associated with the delivery or ability to deliver triple play services. Ultra broadband is increasingly associated with IP-based networks and communications.

Ultra High Frequency (UHF)-The frequency range from 300 MHz to 3000 MHz.

Ultra Low Frequency (ULF)-The frequency band from 300 Hz to 3000 Hz.

Ultra Wide Band (UWB)-Ultra wideband is a method of transmission that transmits information over a much wider bandwidth (perhaps several GHz) than is required to transmit the information signal. Because the UWB signal energy is distributed over a very wide frequency range, the interference it causes to other signals operating within the UWB frequency band is extremely small. This may allow the simultaneous operation of UWB transmitters and other existing communication systems with almost undetectable interference.

This figure shows how ultra wideband (UWB) modulation technology allows the transmission of a signal over existing frequency bands with minimal interference between communication systems. This diagram shows how several high power narroband signals are operating at specific points over a relatively wide frequency range. The UWB signal is spread over a very wide frequency range and this example shows that its relative power level compared to the other signals is so low it should have essentially no effect on the quality of the narroband signals.

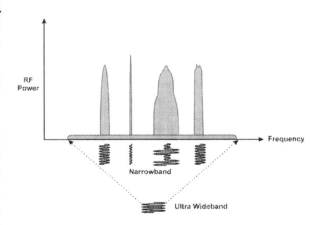

Ultra Wideband Modulation Technology

Ultrasonic-An acoustical signal of a frequency higher than can be heard by a human (above 20 kHz).

Ultraviolet Light (UV)-Ultraviolet light is electromagnetic waves that are invisible to the human eye, with wavelengths about 10-400 nm. UV light can be used for disinfecting materials and for curing compounds.

UM-Unified Messaging

Umbrella Principle-The umbrella principle is the application of the same charging rates (tariffs) to all visiting customers in a mobile communication system irregardless of which home system they are registered with.

Umbrella Statute-An umbrella statute is a law or rule created by a government or company that defines how the regulation applies to a group or category of products.

UMTS-Universal Mobile Telecommunications System

Unacknowledged Mode-Unacknowledged mode is a communication process that does not require the receiver of information to send indications back to the sender that it has successfully received the information that was sent.

Unavailability-A measure of the degree to which a system, subsystem, or piece of equipment is not operable and not in a committable state at the start of a mission when the mission is called for at a random point in time.

Unbalanced-A condition of circuits or lines in which the impedance of one side of the terminal differs from that of the other side of the same terminal. Examples include a cable pair in which the impedance of the ring and tip differ, and an amplifier in which one of the input terminals is connected to ground. (See also: single ended, balanced.)

Unbalanced Line-A transmission line in which the magnitudes of the voltages on the two conductors are not equal with respect to ground. A coaxial cable is an example of an unbalanced line.

Unbundled-A term describing services and programs that are sold separately by the manufacturers of telephone a computer-related equipment.

Unbundled Loop-The portions of a telephone system local loop that can be separated and leased or provided to other companies for their own use.

Unbundling Services-Unbundling is the process of separating portions of a telecommunication network that are owned or operated by a service provider. Unbundling is a common term used to describe the separation of standard telephone equipment and services to allow competing telephone service providers to gain fair access to parts of incumbent telephone company systems. An example of an unbundled service is for the incumbent phone company to lease access to the copper wire line that connects an end user to the local telephone company. The competing company may install high-speed data modems (such as ADSL) on the copper line to enhancing the value of the telecommunications service.

Unchannelized Carrier-An unchannelized carrier is a communication line (carrier) that allows the user to have unrestricted access to the entire data transmission capacity (after the line control overhead is removed) of the communication bearer circuit. Unchannelized carriers are sometimes called unstructured carriers.

Under Specification-Underspecification is a standard or operations procedure that does not have enough detailed information to ensure the proper operation or ability of a device or system to reliably interoperate with other devices or systems.

Underground-A type of construction in which cables are pulled through conduits buried in the ground. Access to such cables and conduits is provided by utility access holes placed at distances of every 500 to 1000 feet.

Underlying Material-In reference to written property a film or television series is based on.

UNI-User Network Interface

Unicast-Unicasting is the delivery media or data to only one client per communication session. If unicasting is used to provide broadcast services, a separate communication session must be established and managed between each user (client) and the broadcast provider (media server).

Unicast Address-An address used to route data to a specific device that is connected to a communications network. The Unicast address is also called a Physical or Hardware address.

Unicasting-Unicasting is the process of transmitting media channels to a number of users through the use of a separate channel (unicast channel) for each user.

Unicode-Universal Code

Unicom-Aeronautical Advisory Station

Unidirectional-A signal or current flowing in one direction only.

Unidirectional Carrier-A telephone company (telco) carrier system that is utilizing 2 cable runs.

Unified Messaging (UM)-Unified messaging allows you to store, manage, and transfer different forms of messages from a variety of access devices. Unified messages include audio (voice messages), electronic mail (email), data messages (such as fax or files), and video (video mail). Unified messaging provides you with access to these multiple types of messages using standard telephones, (text to audio), Internet web pages (playing back voice messages), and other devices such as fax machines and mobile telephones.

Uniform Computer Information Transactions Act (UCITA)-The uniform computer information transactions act is a regulation (statute) that covers transactions that involve digital information. UCITA is similar to the uniform commercial code (UCC).

Uniform Dial Plan (UDP)-The dialing plan (digit sequences) that are used by all standard users within a telephone system. The use of UDP in private telephone systems (such as a PBX) allows callers to dial telephone extensions (other users within the private system) using predefined dialing codes (such a 4 digit or 5 digit extension codes).

Uniform Resource Name (URN)-Uniform resource name (URN) is a naming process defined by the Internet Engineering Task Force (IETF) that is used as an identifying name for Internet resources. The URN has no relation to where the resources are located within the Internet.

Uninterruptible Power Supply (UPS)-A battery backup system designed to provide continuous power in the event of a commercial power failure or fluctuation. A UPS system is particularly important for network servers, bridges, and gateways.

Unipolar-A transistor formed from a single type of semiconductor material, n-type or p-type, as employed in a field effect transistor.

Unit Under Test (UUT)-A unit under test is a device that is attached or communicating with a test system for the purposes of testing, performance measurements and/or diagnostic purposes. A UUT may be placed in a special test mode to permit test commands to be received and processed.

Universal Code (Unicode)-A 16-bit, fixed-width character encoding standard that encompasses virtually all of the characters commonly used on computers today-this includes most written languages, plus publishing characters, mathematical and technical symbols, and punctuation marks.

Unicode support makes it easier and faster for developers to create products and localize them different languages.

Universal Connector-A connector that can be used for voice, data, or video services. Universal connectors use industry standard connection types to allow end users to connect a variety of device types to the connector.

Universal Resource Locator (URL)-A standardized addressing process used to identify resources that are connected to the Internet. The URL is a text string that defines the location of a resource (such as an address of a web site the Internet), as well as the protocol to be used to access the resource.

Universal Serial Bus (USB)-An industry standard data communication interface that is installed on personal computers. The USB was designed to replace the older UART data communications port. There are two standards for USB. Version 1.1 that permits data transmission speeds up to 12 Mbps and up to 127 devices can share a single USB port. In 2001, USB version 2.0 was released that increases the data transmission rate to 480 Mbps. This diagram shows how a universal serial bus (USB) system interconnects devices in a personal distribution network (PDN). This example shows that a USB system uses a host controller interface (HCI) to coordinate the access to all other devices that it is attached to. As each device is added, the host controller registers the device (called device enumeration) and coordinates all communication to and from the devices. This diagram also shows that there are two types of connectors used in the system to ensure that a host device is not accidentally connected to another host device. A hub is used to allow the connection of additional devices (up to 127 can be attached to one host system). The USB system allows for the supply of power through the USB cable (5 volts) or an external power supply can be used.

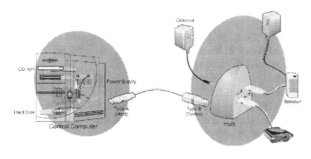

Universal Serial Bus (USB) System

Universal Service-The objective set by many state regulatory agencies and the Federal Communications Commission to keep telephone services affordable for as many customers as possible.

Universal Synchronous Asynchronous Receiver Transmitter (USART)-An integrated circuit provided in many data communications devices that converts data in parallel form from a processor into serial form for transmission.

Universal Synchronous Receiver And Transmitter (USRT or USART)-An electronic module that combines the transmitting and receiving circuitry needed for synchronous communications over a serial line.

UNIX-A computer operating system originally developed and deployed by the Bell Telephone laboratories and now an industry standard. UNIX is a registered trademark mark of UNIX System Laboratories.

Unlicensed Frequency Band-Unlicensed frequency bands are a range of frequencies that can be used by any product or person provided the transmission conforms to transmission characteristics defined by the appropriate regulatory agency.

This figure shows typical types of unlicensed radio transmission systems. This example shows that there are several different communication sessions that are simultaneously operating in the same frequency band and that the transmission of these devices are not controlled by any single operator. These devices do cause some interference with each

other and the types of interference can be continuous, short-term intermittent, or even short bursts. For the video camera (such as a wireless security system), the transmission is continuous. For the wireless headset, the transmission is on for several minutes at a time. For the microwave oven, the radio signals (undesired) occur for very short bursts only when the microwave is operating.

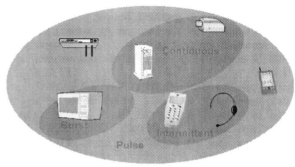

Unlicensed Radio Systems

Unmanaged Connection-A communication connection that does not provide guaranteed performance. Examples of unmanaged connections include residential DSL and cable modem lines.

Unmanaged IPTV-Unmanaged IPTV (also called Broadband television) is the delivery of digital television services over broadband data networks. They may be able to control and guarantee the quality of television services if the underlying broadband connections have enough bandwidth. Internet service providers or media management companies usually provide unmanaged IPTV systems through broadband Internet connections.

Unmodulated-A carrier signal that is not modulated by an information carrying signal.

Unordered Lists-Groups of information elements (list items) that are displayed and/or selected by users in any (unspecified) order.

UnplugFest-UnplugFests (UPF) are testing events where manufacturers or developers agree to test their Bluetooth products with other products in a secret closed environment. There are approximately 3 UnplugFests per year. The participation in UnplugFests allows manufacturers and developers to find problems with their products or areas of correction or clarification that are needed in the Bluetooth specifications.

Unshielded-Wiring not protected from electromagnetic and radio frequency interference by a conductive braid or foil.

Unshielded Twisted Pair (UTP)-The transmission line for DSL systems is typically unshielded twisted pair (UTP). UTP consists of pairs of copper wires twisted around each other and covered by plastic insulation. The twisting of the copper wire pair reduces the effects of interference as each wire receives approximately the same level of interference (balanced) thereby effectively canceling the interference. UTP is by far the most popular cabling used for local access lines and computer LANs (such as 10BaseT and 100BaseT).

Unsolicited Email-Unsolicited emails are messages that are sent to users without a request or permission granting from the recipient. Although the term unsolicited email commonly refers to unwanted email, unsolicited email can be a desirable message.

Unspecified Bit Rate (UBR)-Unspecified bit rate (UBR) is a category of telecommunications service that provide an unspecified data transmission rate of service to end user applications. Applications that use UBR services do not require real-time interactivity nor do they require a minimum data transfer rate. UBR applications may not require the pre-establishment of connections. An example of a UBR application is Internet web browsing.

Unterminated-A device or system that is not terminated.

Untrusted Device-Untrusted devices are hardware components or software applications that are unknown or not validated with a provider of data or information. An untrusted device may require authentication based on some type of user interaction before access is granted.

Unwanted Emission-A spurious or out-of-band emission.

Up Converter-A frequency translation device in which the frequency of the output signal is greater than that of the input signal. Such devices are commonly found in microwave radio and satellite systems.

Uplink-(1- Satellite) The earth-to-satellite microwave link and related components such as earth station transmitting equipment. The satellite contains an uplink receiver. Various uplink components in the earth station are involved with the processing and transmission of the signal to the satellite. (2- cellular systems) The radio link between the mobile station and the base station.

Uplink Encoder-An uplink encoder is a device that processes one or more input signals into a specified format for transmission on an uplink communication channel. An example of an uplink encoder is a satellite encoder that combines and compresses multiple media channels into a single satellite uplink radio channel.

Upload-Upload is the transfer of a program or of data from one computer to a computer server. Upload commonly refers to sending files from a computer to a web site server.

Upper Sideband (USB)-The higher of the two bands of frequencies produced by amplitude modulation. The upper sideband is equal to the sum of the carrier and the modulating signal frequencies. It can be transmitted with or without a full-level carrier and retains all information impressed during the modulation process.

UPR-User Performance Requirements

UPS-Uninterruptible Power Supply

Up-Sell Campaign-Marketing campaign with the objective of encouraging customers to purchase more of a product that they are already buying.

Upset Price-Defined by the WGA, the lowest amount of money due to be paid to the writer of a product, which allows a buyer to acquire abstracted rights to that product.

Upstream-(1-general) A device or system placed ahead of other devices or systems in a signal path. (2-network) The direction opposite the direction of distribution of network timing signals. (See also: down-stream.) (3-video keyer) A term that describes the location of keyers in a mix/effects level or in the overall switcher architecture. (4-video switcher) A term relating the priority of the video signals as they are combined through a production switcher.

Uptime-The uninterrupted period of time that network or computer resources are accessible and available to a user.

Urban Service-Urban services are communication services provided to users located in urban (developed) areas. For 3rd generation systems, rural services may obtain data transmission rates of 2 Mbps.

URI-Uniform Resource Identifier
URL-Universal Resource Locator
URN-Uniform Resource Name
US TDMA-IS-136
Usage-Usage is a measurement of a service provided by a server or system. For voice communication, this is commonly expressed in hundred call seconds (CCS) per hour or Erlangs. For data communication, this is usually expressed in bytes of information used or transferred.

Usage Accounting-Usage accounting is the tracking of data obtained from network transactions for applications in billing, traffic analysis, settlement, fraud, and customer analysis.

Usage Based Charging-Usage based charging is the rating of billing cost that is determined by the amount of data or service used regardless of the duration (start to end) time of the service.

Usage Fees-Usage fees are the billing charges to a customer for the usage of a product or service.

Usage Level-Usage level is a measure of an amount of resource that is used (such as minutes used or amount of data that is transferred) over a given time period or for a particular device or service.

Usage Management-Usage management is the analysis and application of data obtained from network transactions that are used for authorization, mediation, charging, accounting, and various types of business intelligence.

Usage Metering-Usage metering is the process of tracking a quantity of service or material over a period of time or event period.

This figure shows some of the common types of service usage metering. This diagram shows that the types of usage metrics may include the amount of time a service has been used, how much of a service may be used, the number of times a service has been used or activated, the type of use (e.g. single viewer or public viewers), quality of service (e.g. high resolution or low resolution) or the location of the service access point (e.g. home or at a visited/away location).

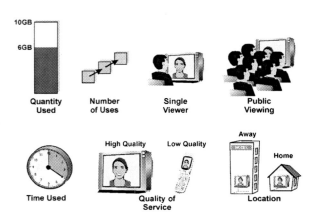

Billing Usage Metering

Usage Model-A usage model is a representation of the typical usage of a product or service. A usage model is created to help designers define the required operations and communication that is required to satisfy the needs of the usage model.

Usage Profiling-User profiling is the process of monitoring, measuring and analyzing usage characteristics for a user of a product or service.

Usage Royalties-Usage royalties are fees that are paid to an author or composer for the right to use each copy of a work that is sold, performed, or produced under license of an exclusive right (such as patent rights).

Usage Sensitive Pricing-Usage sensitive pricing is the selection of usage charging rates based on actual usage of services or resources. Usage sensitive pricing may vary based on the time of day, the duration of data transfer or call, or the type of media transferred.

Usage Statistics (Usage Stats)-Usage statistics is the mathematical analysis and representation of usage information that can be qualified and quantified. Usage statistics allows for the expression of characteristics of information or data in a form that can be used to help understand specific aspects of information (such as average usage rates for specific types of users). Usage statistics can also be used to predict or estimate the usage of a product or service based on the related factors (such as changes of price or types of customers).

Usage Stats-Usage Statistics

Usage Tracking-Usage tracking is the recording of a quantity of service or material that is transferred over a period of time or between events.

USART-Universal Synchronous Asynchronous Receiver Transmitter

USB-Universal Serial Bus

USB-Upper Sideband

USB Handset-Universal Serial Bus Handset

USB Version 2.0-A high speed version of the USB bus that was released in mid 2001. This specification increases the data transmission rate to a maximum of 480 Mbps.

USCH-Uplink Shared Channel

usec-Microsecond

User-A user is a person, company, or group that receives, processes or takes some form of action on services or products. A communication user transfers or processes voice, data, video or other information.

User Acceptance Testing (UAT)-A set of tests that are performed by or for a potential user (often a buyer) of a piece of equipment or system that is supposed to ensure the equipment or system will meet the functional requirements of the user.

User Agent (UA)-End user devices in a SIP system are called user agents (UA). The UA is a conversion device that adapts signals from a data network into a format that is suitable for users. Examples of user agents include dedicated IP telephones (hardphones), analog telephone adapters (ATAs), or software (softphones) that operate on a computer that has multimedia (audio) capabilities.

User Agent Client (UAC)-User agent clients are the requesting user part of the session initiation protocol (SIP) communication session.

User Agent Server (UAS)-User agent servers are the request processing part of the session initiation protocol (SIP) session. The user agent server receives requests from a user agent client (UAC) and generates responses.

User Authentication-A security strategy that verifies the identity of a person requesting access to a network, system, and/or computer.

User Behavior-User behavior is the trends or characteristics of user actions during the usage of services by customers in order to obtain useful data needed to allocate resources for the delivery of those and future services.

User Datagram Protocol (UDP)-UDP is a high-level communication protocol that coordinates the one-way transmission of data in a packet data network. The UDP protocol coordinates the division of files or blocks of data information into packets and adds sequence information to the packets that are transmitted during a communication session using Internet protocol (IP) addressing. This allows the receiving end to receive and re-sequence the packets to recreate the original data file or block of data that was transmitted. UDP adds a small amount of overhead (control data) to each packet relative to other high-level protocols such as TCP. However, UDP does not provide any guarantees to data delivery through the network. UDP protocol is defined in request for comments 768 (RFC 768).

This diagram shows how user datagram protocol (UDP) operates to efficiently send data through a packet network. This diagram shows that the UDP system first packetizes (divides) the sender's data into smaller packets of data (maximum 1500 bytes). Each of these packets starts with an IP header that contains the destination address of the packet. The UDP system then adds a second header (the UDP control header) that includes a destination port. The packets are sent through the system where they may be received or lost in transmission. Because the UDP protocol does not contain any guarantee of delivery, it is up to the user on how to handle lost packets of data.

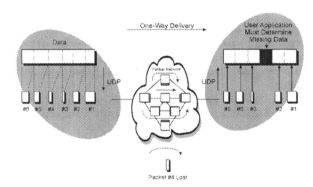

User Datagram Protocol (UDP) Operation

User Datagram Protocol (UDP) Header-A control header that is added to the data portion of an Internet protocol datagram packet. The UDP header holds a limited amount of control data and it is used to coordinate the transfer of information during a communication session that does not need to retransmit (confirm) information.

This diagram shows the UDP header field structure for version 4 IP. The UDP header is session level protocol and is it is located after the network

IP addressing portion of each IP datagram packet. The UDP header structure is simplified used to allow efficient unacknowledged communication using IP addressing. The UDP packet header contains with port numbers that indicates the logical channel where the data originated from the sending device and where it is to be routed (e.g. which software application) by the receiving communication device.

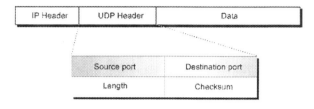

UDP Packet Version 4 Header Structure

User Equipment (UE)-A mobile radio telephone operating within a universal mobile telephone system (UMTS). This includes hand held units, transceivers installed in vehicles and fixed wireless units.

User Friendly-User friendly is a product or service that is designed to be easily understood and operated by a user.

User Group-A user group is a number of users of a specific product or software system that share information. User groups may have newsletters and chat rooms to help gather and distribute information relative to a product or service.

User ID-User Identification

User Identification (User ID)-A user identification number is a unique identifer that is assigned to a user. The User ID may be a tempoaray or permanent number.

User Interface (UI)-The portion of equipment or operating system that allows the equipment to interface with the user. Also called the man machine interface.

User Interface Developer-User interface developers are companies or people that develop software and/hardware that allows equipment (such as television set top boxes) to interface with the user.

User Interface Specification-A user interface specification describes the requirements and operation of the interaction (interface) with a device or system.

User Manual-A users manual describes the typical operation and requirements of devices or equipment.

User Network Interface (UNI)-The interface between an end user and a telecommunications network. A UNI could be a industry standard set of protocol rules and data transmission specifications or may be a proprietary protocol.

User Part-A functional part of a common channel signaling system that transfers signaling messages via the message transfer part. Different types of user parts exist (e.g. for telephone and data services), each of which is specified to a particular use of the signaling system.

User Performance Requirements (UPR)-A set of requirements that are necessary to meet the needs or desires of typical users of a system and/or service.

User Plane-The portion of a network system that is involved in the transfer of the users media (voice, data, or video).

User Rights Management-User rights management is the process of selecting, assigning, and terminating (revoking) the authorization of access and use privileges of a user of a product or service. The rights that may be assigned may include a length of time, number of users and the ability to copy or modify the product or service.

User Session-(1-web) A user session is a single visitor's access of a web site from the initial request to the termination of activity on that web site. Because a single visitor usually requests more than one web page from a web site, a session can be tracked by the IP address of the visitor or through the review of a cookie that is stored on the visitor's computer.

User Tracking-User tracking is the process of sensing and recording the activities of a user of a program, product or service.

User-Based Security Model (USM)-The security model was originally defined in RFC 2274 for version 3 of SNMP. It is obsoleted by RFC2574. It supports secure authorization of the user sending a packet, as well as the privacy or encryption of the packet.

USF-Uplink State Flag

USIM-UMTS Subscriber Identity Module

USM-User-Based Security Model

USOC-Uniform Service Ordering Code

USP-Unique Selling Point

USPTO-United States Patent and Trademark Office, a branch of the United States Department of Commerce.

USRT or USART-Universal Synchronous Receiver And Transmitter

USSD-Unstructured Supplementary Service Data

UTC-Coordinated Universal Time

Utility Rights-Utility rights are the authorizations that permit the moving of content for different uses. For example, the ability to move content from one computer to another computer.

Utilization-(1-Facilities) The use of telecommunications facilities and equipment, expressed as a percentage of working to working plus spare facilities. (2-Performance Management) Utilization as used in performance management for network management measures the use of a particular resource over time. The measure is usually expressed in the form of a percentage in which the usage of a resource is compared with its maximum operational capacity, like bandwidth utilization on a network link.

Utilization Factor-The ratio of maximum demand for service to the rated capacity of a system that must provide that service.

UTP-Unshielded Twisted Pair

UTRA-Universal Terrestrial Radio Access

UUID-Universally Unique Identifier

UUS-User-To-User Signaling Supplementary Service

UUT-Unit Under Test

UV-Ultraviolet Light

UWAG-Universal ADSL Working Group

UWB-Ultra Wide Band

V

V-Volt

VA-Variable Attenuator

VACMS-View-based Access Control Models

Vacuum Tube-An electron tube. The most common vacuum tubes include the diode, triode, tetrode, and pentode.

VAD-Vapor Axial Deposition

VAD-Voice Activity Detector

Value Added Reseller (VAR)-A company or organization that adds assemblies, software, or documentation to products produced by another manufacturer or service provider so they may be sold in their sales and distribution system. VARs may modify a standard product (such as a laptop computer) and modify for use in a specific industry (called a vertical application.)

Value Added Services (VAS)-Services that provides benefits to a customer that are not part of the standard telecommunications services associated with a basic communication service. VAS services include voice mail, information services and content delivery.

Services offered by prepaid provider (e.g., voice mail, fax store and forward, interactive voice response, and information services) in addition to calling time.

Value Added Tax (VAT)-A tax that is added on to the value of the product or service.

Value Proposition-A statement, made by a business person, which describes how the business will make use of the information delivered to improve operational efficiency, profitability, or market strength.

Vampire Tap-A type of connector that connects one cable segment to another without the need to cut and splice the cable. The vampire tap uses a needle to pierce the cable insulation so it can make a connection to the wire inside the cable. See also Penetration Tap.

This diagram shows a vampire tap connector is used to connect one cable segment to another. Within the vampire tap, a blade and needle are used to pierce the cable insulation to make a connection to the wire within. Vampire taps are typically used on a Thicknet Ethernet system and connect directly to the network backbone.

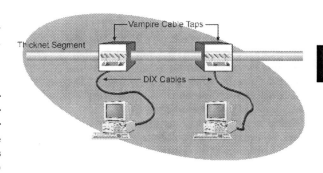

Vampire Tap Diagram

Vaporware-A sarcastic name for a product that has been announced but is not available and has not been produced. Vaporware products often do not get released.

VAR-Value Added Reseller

Variable Attenuator (VA)-A variable attenuator is a device that can be adjusted to reduce the signal energy (e.g. power) that is transferred through it. Variable attenuators are normally rated as range of attenuation in decibels (dB) as compared between the input and output signal levels.

Variable Bandwidth-A communication system that allows for a variable data transmission rate or changes in communication channel frequency bandwidth dependent on the need of the end user applications and/or the ability of the system to provided the desired data transmission or frequency bandwidth. Because variable bandwidth systems help match the system resources used to the actual data transmission needs of the end customer (e.g. reduce the bandwidth when the user has nothing to send or say), variable bandwidth systems are more efficient that constant bandwidth systems.

Variable Bit Rate (VBR)-A category of telecommunications service that provide an variable data transmission rate of service to end user applications. Applications that use VBR services usually require some real-time interactivity with bursts of data transmission. An example of a VBR application is videoconferencing.

Variable Gain Amplifier-An amplifier whose gain can be controlled by an external signal source.
Variable Graphics Array (VGA)-Variable graphics array is a graphic display standard that was developed by IBM for personal computers. VGA technology enables computer graphics cards to generate four different levels of resolution for screen displays.
Variable Length Coding (VLC)-Variable length encoding is a method of compressing digital information by representing repetitive groups of data information by code that are used to look up the data sequence along with how many times the data will be repeated (variable length).

This diagram shows how groups of repetitive data can be substantially compressed using variable length coding. This example shows a data file with groups of repetitive data blocks (such as repetitive shapes in an image file). The VLC process begins with identifying the common groups of bits in the data file and assigning unique codes to represent the groups. The compressed file identifies the group of bits by the VLC code of (unique sequence or location in a data file), the character code that is assigned for this file (used to identify the VLC process), the code for the bits that will follow (that will be found in the code table) and the number of times the sequence of bits will be repeated.

Variable Length Coding (VLC) Data Compression

Variable Rate Speech Coding-Variable rate speech coding is a speech compression process that offers multiple speech coding rates. The use of vari-

able compression rates allows a lower bit rate (higher compression rates) coding process to be used when system capacity is limited and more users need to be added to the system.

This figure shows the basic variable rate speech coding process. This diagram shows that a user is talking into a CDMA radio with a variable speech coder. In this example, the speech coder dynamically changes the data rate based on the speech activity level. During periods of high activity, the speech coder transmits at 9600 bps. At periods of low speech activity level (e.g. periods of silence), the speech coder transmits at 1200 bps.

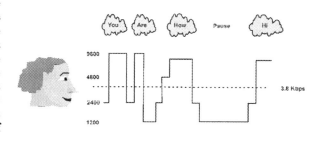

Variable Rate Speech Coding Operation

Variable Spreading Rates-Variable spreading rates are the ability of the ratio of chips to information bits (spreading rate) to change in a spread spectrum communication system. The use of variable spreading rates allows the system to assign different data transmission rates on the same radio channel without having to change the chip rate.
Variant-A variant is a product or service that is related to other products or services but has slightly different features or characteristics.
VAS-Value Added Services
VAT-Value Added Tax
VAT-Visual Audio Tool
VBI-Video Blanking Interval
VBR-Variable Bit Rate
VBS-Voice Broadcast Service
VC-Virtual Circuit

VC-1-VC-1 is the designation for Microsoft's Windows Media Player codec by the SMPTE organization.

VCC-Virtual Channel Connection

VCD-Video Compact Disk

V-Chip-Video Chip

VCI-Virtual Channel Identifier

VCO-Voltage Controlled Oscillator

VCR-Video Cassette Recorder

VCSEL-Vertical Cavity Surface Emitting Laser

VCXO-Voltage Controlled Crystal Oscillator

VDSL-Very High Bit Rate Digital Subscriber Line

VDT-Video Display Terminal

Vector-A vector is an item that is described by magnitude and direction parameters.

Vector Base Register-A vector based register is a memory storage are that used during computer processes that has its memory address identified by a vector (pointing information).

Vector Based-Vector based images are created from lines and curves. Because lines and curves can be represented by mathematical equations which can require a much lower number of bits to represent a shape, shade or graphic, vector based images typically occupy much less memory space than equivalent bit mapped (pixel) images.

Vector Sum Code Excited Linear Predictive (VSELP)-Vector sum code excited linear prediction coding is an efficient speech coding (voice compression) process.

Vector Sum Excited Linear Predictive Coding (VSELP)-A speech analysis and compression technology by that uses digital software processing which linear audio (voice) samples are collected, analyzed, and then compressed using an encoding algorithm. VSELP technology is the algorithm used to code and decode speech information in IS-54B and IS-136 TDMA cellular environments.

Vectorscope-A vectorscope is a type of oscilloscope that is used to display the color parameters of a video signal. A vectorscope converts color information into R-Y and ~Y components and displays these components on the X and Y axes of the scope. The intensity of color on a vectorscope is presented by the distance the image appears from the origin (0 reference) of the vectorscope.

Versatility-Versatility is the ability of a product, system, or service to be expanded or adapted to support or provide other services and applications.

Version Number-A number that identifies a particular software or hardware product that uses the same name as other products. These products usually undergo revisions or updates and the version number often relates to the date of release. The version number is usually assigned by the manufacturer or developer of the product. that often includes numbers before and after a decimal point; the higher the number, the more recent the release. The version number is important as features and operation of a product or software program may vary between different versions. The ability to determine the specific version number of a product may allow more reliable interaction between programs and products. Version number 1.0 often indicates an initial version of a product or software.

Vertical Application-A vertical application is a program or software that is designed or used for a specialized industry application or profession. An example of a vertical application is a software programmed to provide wireless meter reading services to utility companies.

Vertical Resolution-Vertical resolution is the amount of chrominance (color) and luminance (intensity) detail expressed vertically in the picture. Vertical resolution is limited by the number of system scan lines.

Vertical Rise Limit-Vertical rise limits is the maximum length of a cable that can be installed vertically before damage or changes in the characteristics of a cable occur.

Very High Bit Rate Digital Subscriber Line (VDSL)-Very high bit rate digital subscriber line is a communication system that transfers both analog and digital information on a copper wire pair. The analog information can be a standard POTS or ISDN signal and the typical downstream digital transmission rate (data rate to the end user) can vary from 13 Mbps to 52 Mbps downstream and the maximum upstream digital transmission rate (from the customer to the network) can be 26 Mbps. The data transmission rate varies depending on distance, line distortion and settings from the VDSL service provider. The maximum practical distance limitation for VDSL transmission is approximately 4,500 feet (~1,500 meters). However, to achieve 52 Mbps, the maximum transmission length is approximately 1,000 feet (~300 meters).

This diagram shows a VDSL system is commonly used with a fiber distribution network that reaches a neighborhood or small group of buildings. The fiber terminates in an optical network unit (ONU).

The ONU converts the optical signal into an electrical signal that can be used by the VDSL modem in the DSLAM. The DSL modem signal is supplied to a splitter that combines the analog and digital signal to copper access line. The splitter is actually attached to the last few hundred feet of the copper access line. The figure shows that the analog POTS signal from the local telephone company may still travel thousands of feet back to the central office. At the customers' premises, the VDSL signal arrives to a splitter that separates the analog signal from the high-speed digital VDSL signal. Because VDSL has a much higher data transfer rate, the CPE may include a digital video set top box that allows for digital television.

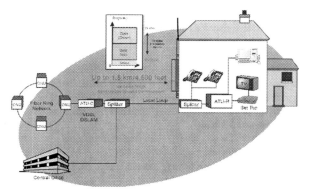

Very High Bit Rate Digital Subscriber Line (VDSL) System

Very High Frequency (VHF)-The frequency band from 30 MHz to 300 MHz (wavelengths 10 m to 1 m).

Very Low Frequency (VLF)-The radio frequency band from 3 kHz to 30 kHz.

Very Small Aperture Terminal (VSAT)-A Very Small Aperture Terminal (VSAT) consists of a small, dish-shaped antenna and associated electronics which allow satellite access to a geosynchronous, communications satellite. A VSAT system is an entire network which includes the central hub, the remote sites, and the network software to run the system. VSAT utilizes geosynchronous satellites located 22,500 miles above the equator, as the communication backbone. The satellite connects the VSAT locations to the central hub facility, which routes messages to the appropriate destination.

Vesting Schedule-A wage agenda, showing on which days one may receive their pay.

VF-45™ Connector-Volition™ Fiber Connector

VFD-Vacuum Fluorescent Display

VFL-Visible Fault Locator

VGA-Variable Graphics Array

VGCS-Voice Group Call Service

V-Groove-V Groove

VHE-Virtual Home Environment

VHF-Very High Frequency

VHS-Video Home System

Vibration Testing-A testing procedure in which subsystems are mounted on a test base that vibrates, thereby revealing any faults resulting from badly soldered joints or other poor mechanical design features.

Video-An electrical or optical signal that carries moving picture information.

This figure demonstrates the operation of the basic NTSC analog television system. The video source is broken into 30 frames per second and converted into multiple lines per frame. Each video line transmission begins with a burst pulse (called a sync pulse) that is followed by a signal that represents color and intensity. The time relative to the starting sync is the position on the line from left to right. Each line is sent until a frame is complete and the next frame can begin. The television receiver decodes the video signal to position and control the intensity of an electronic beam that scans the phosphorus tube ("picture tube") to recreate the display.

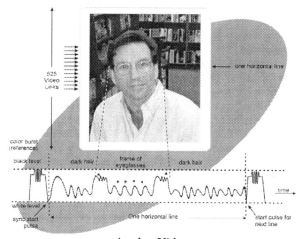

Analog Video

Video 8-Video 8 is a video cassette package of 8mm reel-to-reel magnetic tape that is used for analog video signals and may be played or rewound on demand.

Video Amplifier-An amplifier designed to operate over the band of frequencies used for TV signals.

Video Bandwidth-Video bandwidth is the difference between the upper frequency limit and lower frequency limit of allowable transmission of a video signal.

Video Blanking Interval (VBI)-Video blanking interval is a number of video lines that are transmitted in addition to the video display lines of a television signal to allow the television to blank the picture tube while the electron beam is repositioned from the bottom of the picture tube to the top of the picture tube. For NTSC signals, the VBI is 21 lines of the 525 transmitted video display lines. Because the VBI signal is not used for video display, some systems (such as Teletext or closed caption) use the VBI signals

Video Broadcasting-The process of transmitting video images to a plurality of receivers. The broadcasting medium may be via radio waves or wired systems such as CATV or the Internet.

Video Camera-A device that converts images (light signals) into electrical video (multiple frame) signals.

This diagram shows how a video camera uses a cathode ray tube (CRT) to convert light energy into a video signal. This diagram shows that the CRT includes a photosensitive plate that receives an optical signal through a lens and also receives energy from an electron beam signal. When there is light on the plate and the electron beam hits the plate in a specific spot, a small amount of current flows from the CRT tube. The video signal generator controls the horizontal and vertical position of the electron beam through the deflection coils. Because the video generator knows the exact position of the beam, it can create a composite video signal that represents the intensity (amplitude) and position (timing) of the image (light).

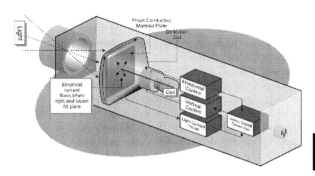

Video Camera Operation

Video Capture-Video capturing is the process of receiving and storing video images. Video capture typically refers to capture of video images into digital form.

Video Capture Card-A capture card is a printed circuit board or electronics assembly that is designed to be inserted (plugged-in) into a computer or electronic device and capture video information. A capture card may have several types of inputs including several formats of analog video, RF television channels, and digital video.

Video Card Hacking-Video card hacking is the process of modifying or using a video card to convert, capture or redirect video or graphics that is unauthorized for viewing, storage or manipulation.

Video Cassette-A video cassette is a self-contained package of reel-to-reel magnetic tape that is used for video signals and may be played or rewound on demand.

Video Cassette Recorder (VCR)-A videocassette recorder is a video media storage device that is used to record and play back audio-visual programs on magnetic cassette tapes. VCRs were developed in the 1960s and become commercially available by the 1970s.

Video Catalog-A video catalog is the presenting of items available for selecting or ordering in a video format. Video catalog formats can range from a linear progression of products (such as a television shopping channel) to an interactive video shopping cart that allows users to search and find items.

Video Chip (V-Chip)-A video chip is an integrated circuit (chip) that is designed to identify and block the viewing of certain video or television programs.

Video Clip-Video clips are short movie or video segments. Video clips may be isolated video segments (such as the launch of a space shuttle) or they may be a short portion of a larger movie or video (a video clip of a famous movie scene).

Video Coding-A coding algorithm that converts video signals into streaming data signals. Some of the common compression video compression technologies include H.261 and H.263.

Video Communication-Video communication is the transmission and reception of video (multiple images) and other signals that can be represented by the frequency band used for video signal transmission. Telecommunications systems can transfer video signals in analog or digital form.

This figure shows the basic process used for video signal transmission. In this example, a television camera converts an image and audio sounds to electrical signals. The video signal is created by a camera scanning the viewing area line by line. At the beginning of each line scan, the camera create a synchronization pulse and the image (light level) is created by varying the electrical signal level after the synchronization pulse. The audio signal is created by using a microphone. These video and audio electrical signals are combined to form a composite video electrical signal. The composite video signal (baseband) modulates the radio transmitter frequency (broadband) signal. This low level radio signal is amplified to a very high power level for transmission. A video receiver (typically a television) receives the radio signal and many others from its antenna. It's receiver selects the correct radio signal by using a variable frequency filter (television channel selector) that demodulates the incoming radio signal to create the original video and audio electrical signals. The video signal is connected to a display device (typically a picture tube) and the audio signal is connected to the speaker.

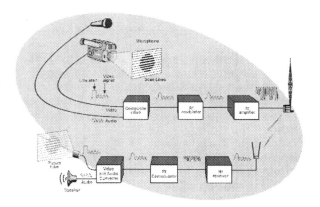

Video Transmission Operation

Video Components-Video components are separate electrical signals that represent the intensity and color of video signals.

Video Compositor-A video compositor is a device or system that can take two or more video inputs or graphic images and combine them into one composite video signal.

This diagram shows that a video compositor can take two or more video inputs or graphic images and combine them into one composite video signal. This example shows a video compositor that takes a video format (news clip), text graphics (scene cap-

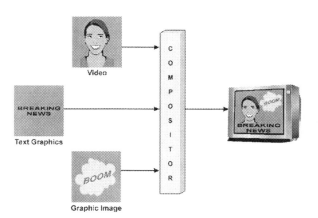

Video Compositor

tion) and a graphic image (explosion picture) and combines them (renders) onto a single video display.

Video Compression-Video compression is the process of reducing the amount of transmission bandwidth or data transmission rate by analog processing and/or digital coding techniques. When compressed, a video signal can be transmitted on circuits with relatively narrow channel bandwidth or using data rates 50 to 200 times lower than their original uncompressed form.

Video Conferencing-A process of conducting a face-to-face meeting between two or more people in different locations through the use of telecommunications circuits and equipment that allows video and audio communications. Video conferencing usually requires real-time two-way transmission of audio and video communications between two or more locations. Transmitted video images may be in the form of full TV-quality images or freeze frame still images, where the picture is repainted every few seconds.

This figure shows the basic operation of sending video over an Internet connection. This diagram shows a computer with video conferencing capability that calls a destination computer. Computer #1 initiates a video conference call to computer #2 using the address 223.45.178.90. When computer #2 receives a data message from computer #1, a message is displayed on the monitor and an audio tone (ring alert) occurs. If the user on computer #2

wants to receive the call, they select the answer option (via the mouse or keyboard) that is generated by the software. Computer #1 then initiates a data connection with computer #2. The video conferencing software and data processing software in the computers (e.g., USB data bus and sound card) convert the analog audio signal from the microphone and digital video signal into a digital form that can be transmitted via the data link between the computers.

Video Container-A video container is a file or media streaming data that contains an organization of objects (such as digital video and images).

Video Dial Tone-An access and transport service for carrying full-motion video in much the same way as a dial-up call is carried on a conventional voice network.

Video Digitization-Video digitization is the conversion of video component signals or composite signal into digital form through the use of an analog-to-digital (pronounced A to D) converter. The A/D converter periodically senses (samples) the level of the analog signal and creates a binary number or series of digital pulses that represent the level of the signal.

This figure shows the basic process used to digitize images for pictures and video. For color video, each line of video is divided into its intensity (brightness) and color components. These components are periodically sampled and converted into a digital format.

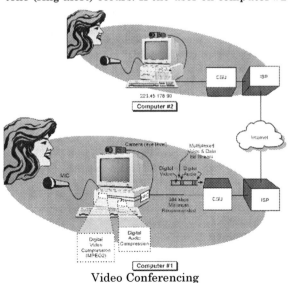

Video Conferencing

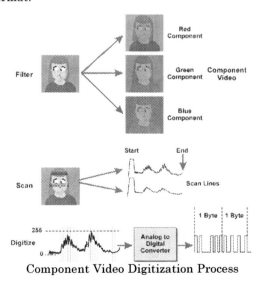

Component Video Digitization Process

Video Display-A computer output device that presents data to the user in the form of an image, including text and/or graphics.

Video Display Terminal (VDT)-A computer terminal equipped with a keyboard and an electronic readout, such as a cathode ray tube or liquid crystal display. Video display terminals often are used to connect remote locations to a distant host computer.

Video Encoder-Video encoders convert (format) video information into a form that allows it to be efficiently and reliably sent on a communication channel. Common forms of video encoding include MPEG-2 and MPEG-4.

Video Entertainment Services-Video entertainment services are leisure related multimedia content that is used for personal stimulation and satisfaction.

Video Home System (VHS)-Video home system is a video tape storage format that is used to store video recording images 1/2 inch magnetic tape. VHS offers 240 lines of resolution.

Video Interface Hacking-Video interface hacking is the process of using the interface connections to a video display (such as a monitor or television) to convert, capture or redirect video or graphics that are unauthorized for viewing, storage or manipulation.

Video Inversion-Video inversion is a process that inverts some of the components of a video signal to prevent unauthorized customers from viewing premium (paid) subscription channels. The inverted video components are corrected using a decoder circuit.

Video Mail (VMail)-A process of recording and sending short video messages (typically 1-2 minute video clips) in digital form via an electronic mail (email). Video mail messages may be sent as an attachment to standard Email addresses.

Video Monitor-A high utility TV set (without RF circuits) that accepts video baseband inputs directly from a TV camera, videotape recorder, or other TV source.

Video Network-A video network is a system that contains a series of points that are interconnected by communications channels, often on a distributed (tree structure) basis.

Video Object (VOB)-Video object is a media file format that is used for MPEG video.

Video On Demand (VOD)-Video on demand is a service that provides end users to interactively request and receive video services. These video services are from previously stored media (entertainment movies or education videos) or have a live connection (news events in real time).

This figure shows a video on demand (VOD) system. This diagram shows that multiple video players are available and these video players can be access by the end customer through the set-top box. When the customer browses through the available selection list, they can select the media to play.

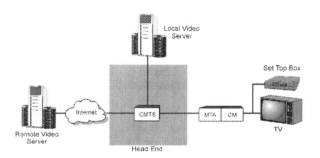

Video On Demand (VOD) Operation

Video on Demand Server (VOD Server)-The video on demand server is an application server that receives requests for video or multimedia programs and provides access to the requested media for authorized users.

Video Quality-Video quality is the ability of a display or video transfer system to recreate the key characteristics of an original video signal. Traditional video quality impairment measurements include blurriness and edge noise. Digital video and transmission system impairments include tiling, error blocks, smearing, jerkiness, edge business and object retention.

Video Segment-A video segment is a portion of a media file (such as a video or multimedia file) that contains the video media.

Video Server (VS)-The video server is an application server that provides video and/or specialized television capabilities. Video servers receive requests for video and/or media delivery, find the matching media, and deliver the video program as requested.

Video Signal-An electrical signal that includes all of the intensity and position information related to a sequence of images. The video signal includes a horizontal sync pulse that indicates the start of an image sweep across the screen. The video signal then includes a composite signal that indicates the intensity of the image at each position along sweep. The video signal also includes a vertical blanking pulse or signal to indicate the end of a frame and to allow the image sweep to be repositioned at the top of the screen.

Video Size-Video size is the resolution images in pixels across (horizontal) and pixels between top and bottom (vertical).

Video Stream-A process of delivering a continuous stream of digital video information that is commonly with minimal delay (e.g. real-time). Video streaming signals are usually compressed and error protected to allow the receiver to buffer, decompress, and time sequence information before it is displayed in its original format.

Video Streaming-Video streaming is the process of delivering video, usually along with synchronized accompanying audio in real time (no delays) or near real time (very short delays). Upon request, a video media server system will deliver a stream of video and audio (both can be compressed) to a client. The client will receive the data stream and (after a short buffering delay) decode the video and audio and play them in synchronization to a user.

Video Sync Suppression-Video sync suppression is a process that removes or reduces the synchronization portion of a video signal. The video synchronization signal is recreated using other information that is inserted into the video signal. Video sync suppression is used in analog television scrambling systems to prevent unauthorized customers from viewing premium (paid) subscription channels.

Video Synchronization-Video synchronization is a process that adjusts the timing of the video signals to match the presentation of other media (such as audio or a slide presentation).

Video Tape Recorder (VTR)-A device that permits audio and video signals to be recorded onto magnetic tape.

Video Telephony-A telephone service that allows customers to hear and see another telephone user or video source. Video telephony applications include video on demand (VOD) movies, distance learning (remote education), telemedicine, teleconferencing and other applications that can benefit from the combination of video and audio signals.

Video Track-A video track is a section of a media file (such as a video tape or multimedia file) that can be associated with an audio signal. Multiple video tracks may be used to provide video signals at different resolutions and data rates.

Video Transmission-Video transmission is the transport of video (multiple images) that may be accompanied by other signals (such as audio or closed caption text).

Video Watermarking-Video watermarking is a process of adding or changing information in an analog or digital video media tape, streaming media or other form of video media to uniquely identify the media and/or its authorized uses. Video watermarking may be performed by adding or slightly modifying the colors and/or light intensities in the video in such a way that the viewer does not notice the watermarking information.

This figure shows how digital watermarks can be added to digital video to provide identification information. The digital watermark is added as a color shift video component that is typically not perceivable to the listener of view of the media.

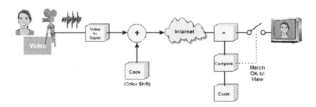

Video Watermarking

Videoconferencing-Conducting conferences via a video or multimedia telecommunications system.

Videophone-A videophone is a communication device that can capture and display video information in addition to audio information. A videophone converts multiple forms of media; audio and video into a single transmission format (such as Internet Protocol). The use of videophones with an Internet telephone service allows the video portion of the communications session to share the data connection.

Videotape-A magnetic tape used for recording video programs.

Videotex-Videotex is an interactive electronic information retrieval system that allows users at home or office to select and return information to computer centers or data banks.

Viewer-A viewer is a software program and/or display device that allows a user to view media content. An example of a viewer is a web browser that allows a user to view media files that are transferred through the Internet.

Viewer Consumption-Viewer consumption is the programs viewed or accessed from a video distribution system (such as from a television broadcast network).

Viewer Experience-Viewer experience is the interactions a user has with a video product or service.

Viewer Tracking-Viewer tracking is the recording of a program selections and possibly usage habits and options for program viewers..

Viewership-Viewership is the people who watch or access a viewable media product (such as a television show).

Viewing Platform-A viewing platform is the combination of system hardware and software that allows a user to view programs or media files.

Viewing Statistics (Viewing Stats)-Viewing statistics is the mathematical analysis and representation of media viewer (e.g. people watching television) information that can be qualified and quantified. Viewer statistics allows for the expression of characteristics of information or data in a form that can be used to help understand specific aspects of information (such as average viewing time per day). Viewer statistics can also be used to predict or estimate the viewing habits of a media product or service based on the related factors (such as changes of type of customers and availability of viewing media).

Viewing Stats-Viewing Statistics

Virtual-A facility or arrangement that gives the effect of being a dedicated facility but in fact, is relatively shared.

Virtual Channel Identifier (VCI)-(1-general) The identification of a logical channel on a virtual path. (2-ATM) In an Asynchronous Transfer Mode

(ATM) cell header, a 16-bit field used to identify virtual channels between users or between users and networks.

Virtual Circuit (VC)-A logical connection between two communication ports in one or more communication networks. Virtual circuits are used to temporarily connect data terminals to host computers. Because virtual circuits logically connect communication ports together, a single network switching system may be used to provide for many virtual circuits.

Virtual Desktop-A desktop workplace for the employee that consists primarily of computing devices. The virtual desktop devices usually include a computer, printer, and a telephone.

Virtual Extension-A communication extension that is created through the use of system programming rather than through the installation of physical equipment.

Virtual Home Environment (VHE)-A concept that a network supporting mobile users should provide them the same computing environment on the road that they have in their home or corporate computing environment.

Virtual Keypad-A software program that operates on a computer or other interactive display device to provide a user with the ability to enter keypad or keyboard information.

Virtual LAN-A number of devices (a subset) that are linked to each other within a larger network by logical channels to allow each device to communicate with other devices in the virtual network using these logical channels. Virtual networks often appear as a separate network to the users of the network. An example of a virtual network is the connection of computers in a city to computers in another city via logical channels (and encrypted channels for security) through the Internet. This allows the computers in one city to access the computers in the other city as if they were connected as a separate network.

Virtual Local Area Network (VLAN)-A local area network (LAN) that is logically (virtually) setup through one or more data networks that are independently managed. VLAN connections are setup to allow data to safely and privately pass over other types of data networks (such as the Internet).

Virtual Machine-A virtual machine is a data processing device (such as a computer or television set top box) which is designed to allow software programs to operate as if they were on processing devices of another type (a different computer type). A virtual machine allows software that was designed for another type of computing device to operate within its system without changes to the software.

Virtual Memory-A memory management operating system technique that allows programs or data to exceed the physical size of the main internal directly accessed memory. Program or data segments or pages are swapped from disk storage as needed. The swapping is invisible to the programmer.

Virtual Path Connection (VPC)-(1-general) An identifier of a physical channel and a logical channel that is used as a connection path between two points. (2-ATM) In Asynchronous Transfer Mode (ATM), a set of logical Virtual Channel Connections (VCCs) between two end stations. All channels in a specific VPC connect the same two end stations.

Virtual Path Identifier (VPI)-(1-general) An identifier of a physical channel or portion of a physical channel that is used as a connection path between two points. (2-ATM) In an asynchronous transfer mode (ATM) cell header, an 8-bit field used to identify virtual paths between users or between users and networks.

Virtual PBX (vPBX)-A virtual PBX offers business users the ability to make and receive calls through the company's PBX system using telephones that can be connected to any of the company's PBX systems at locations that have the ability to connect to a PBX access port (such as an Internet connection).

Virtual Phone-A software program that operates on a computer to provide telephone service.

Virtual Private Network (VPN)-Secure private communication path(s) through one or more data network that is dedicated between two or more points. VPN connections allow data to safely and privately pass over public networks (such as the Internet). The data traveling between two points is encrypted for privacy.

This figure shows the operation of a virtual private network (VPN). This diagram shows that the virtual private network is constructed of network access points that are under the control of the network operator. These network access points usually encrypt the data entering into the network to provide secure private communication path(s) through

the network. These secure VPN connections allow a company to safely and privately pass over public networks (such as the Internet). A VPN management system is used to program the access points (e.g. IXC switch) for key parameters (e.g. data rates and QoS.) While this diagram shows virtual paths, the connections may actually pass through one or more switches have been set so a reserved amount of bandwidth is assigned so the end user can reliably receive a Quality of Service (QoS) characteristics that allows the connections to appear as dedicated lines.

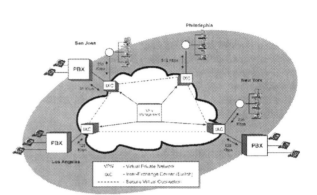

Virtual Private Network (VPN)

Virtual Reality Modeling Language (VRML)-Virtual reality modeling language is a text based language that is used to allow the creation of three-dimensional viewpoints, primarily for use with Web browsing. VRML was created by Mark Pesce and Tony Parisi in 1994 and is a subset of Silicon Graphics' Inventor File Format.

Virtual Smart Card-A virtual smart card is a software program and associated secret information on a users device (such as a TV set top box) that can store and process information from another device (a host) to provide access control and decrypt/encrypt information that is sent to and/or from the device.

Virtual Storage-An auxiliary storage mapped into real ad-dresses so that a computer user views it as an addressable memory store.

Virtual Tributary (VT)-In the synchronous optical network (SONET), a structure designed for the transport and switching of Synchronous Transport Signal Level I (STS-1) payloads, typically in units of 1.544 Mbps each.

Virus-A software program spread by automatic copying from disks or computer networks and intended to interrupt or destroy the functioning of a computer.

Visible Light-Visible light or electromagnetic signals that have wavelengths that range from 330 nm (blue) to 770 nm (red).

Visitor-(1-Web Site) A person who visits a web site. (2-Mobile Telephone) A mobile telephone that is operating in a system other than it's system of home registration.

Visitor Location Register (VLR)-A visitor location register is a database part of a wireless network (typically cellular or UMTS) that holds the subscription and other information about local or visiting subscribers that are authorized to use the wireless network.

Visual Artifacts-Visual artifacts are the unintended, unwanted visual aberrations in an image (such as blocks on a video image or speckles on a picture image around sharp edges).

Visual Voice Mail-An application displaying and controlling voice messages on a desktop computer. Usually associated with unified messaging.

Viterbi Decoder-An algorithm for maximum-likelihood decoding of a convolutionally encoded data sequence, given a limited amount of memory.

VITS -Vertical Interval Test Signal

VLAN-Virtual Local Area Network

VLC-Variable Length Coding

VLF-Very Low Frequency

VLR-Visitor Location Register

VLSI-Very Large Scale Integration

VM-Voice Mail

VM/UM-Voicemail/Unified Messaging Server

VMAC-Voice Mobile Attenuation Code

VMail-Video Mail

VMS-Voice Mail System

VOA-Variable Optical Attenuator

VoATM-Voice Over ATM

VOB-Video Object

VoB-Voice over Broadband

VoBB-Voice over Broadband

VoCable-Voice Over Cable

Vocal-Vovida Open Communications Application Library

VoCoder-Voice Coder

VOD-Video On Demand

VOD Server-Video on Demand Server

VoDSL-Voice Over DSL

VoFR-Voice Over Frame Relay

Voice Activity Detector (VAD)-The Voice activity detection (VAD) is a process or an electronic circuit that senses the activity (or absence) of voice communication signals. VAD is often used to inhibit a transmission signal during periods of voice inactivity or as a control source to allow digital speech interpolation (DSI) or time assigned speech interpolation (TASI).

Voice Band-The frequency spectrum, from approximately 300Hz to 3400Hz, that is considered adequate for speech transmission.

Voice Broadcast Service (VBS)-A voice communications service that allows a single voice conversation or message to be transmitted to a geographic coverage area to be received by subscribers that are capable of identifying and receiving the voice communications.

This figure shows the basic operation of voice broadcast service. This example shows how an urgent news message (traffic alert) can be sent to all mobile devices that are operating within the same radio coverage area.

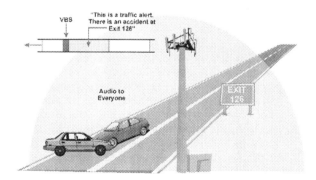

Voice Broadcast Service

Voice Card-A communication card that is inserted into a computer that can process calls.

Voice Channel-In a cellular telephone system, a channel on which voice or data communication occurs, and on which brief digital messages may be sent from a cell to a mobile unit or from a mobile unit to a cell site.

Voice Circuit-A circuit for the interchange of human speech. Normally, the standard band provided is 300 Hz to 3400 Hz, but narrower bands also provide commercially acceptable circuits in some circumstances.

Voice Coder (VoCoder)-Voice coding is a digital compression device that consists of a speech analyzer that converts analog speech into its component parts digital signals and speech synthesizer for the recreation of audio signals from the component parts. Voice coders are only capable of compressing and decompressing voice audio signals.

Voice Communication-Voice communication is the transmission and reception of audio and other signals that can be represented by the frequency band used for voice signal transmission. Telephone systems transfer voice signals in a variety of forms, by wire, radio, light, and other electronic or electromagnetic systems. These forms include analog and digital voice signals. Options for voice communications include different voice quality of service levels and voice privacy options.

Voice Compression-Refers to the process of electronically modifying a 64 Kbps PCM voice channel to obtain a channel of 32 Kbps or less for the purpose of increased efficiency in transmission.

Voice Dialing-A process that uses the callers voice to dial a call. Voice dialing involves the activation of the voice dialing feature (either by pressing a key or by saying a key word), saying words in the vocabulary of the voice dialing processor, and providing feedback to the user (usually by audio messages) of the status of the voice dialing process. Voice dialing can be a system (network provided) or device (stored in the telephone device) feature.

There are two basic forms of voice dialing; speaker independent and speaker dependent. Speaker independent voice dialing allows any user to initiate voice commands from a predefined menu of commands. Speaker dependent voice dialing requires the user to store voice commands so these voice commands can be activated by the user and others are unlikely to match the speaker dependent voice commands. Speaker dependent voice recognition

allows a user to program specific names into the telephone or network voice recognition system.

This diagram shows different types of dialing using voice commands. In this example, both the telephone set and telephone network have voice dialing control capability. When the telephone is used for voice control, the voice from the user is converted to digital form by and analog to digital converter. After the audio is converted to digital form, it is analyzed for patterns and matched to previously stored voice control digital sound patterns. This example shows that the telephone set has some speaker independent patterns (such as start and digits) that have been previously stored. It also shows that this telephone also has a speaker dependent memory storage area that allows the user to store specific names. When these specific names are spoken, the telephone set will retrieve the pre-stored telephone numbers or extensions.

This diagram shows similar voice dialing capabilities that are located in a telephone network. This network voice control system has more accurate voice processing capability than the telephone set and each voice control module can service many line cards as users only use voice control for brief periods.

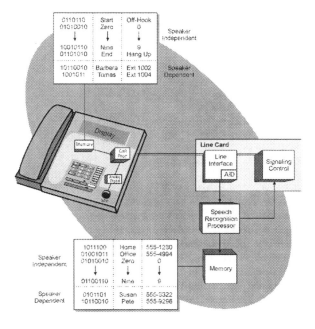

Voice Dialing Operation

Voice Digitization-This figure shows how an analog signal is converted to a digital signal. An acoustic (sound) signal is first converted to an audio electrical signal (continuously varying signal) by a microphone. This signal is sent through an audio band-pass filter that only allows frequency ranges within the desired audio band (removes unwanted noise and other non-audio frequency components). The audio signal is then sampled every 125 microseconds (8,000 times per second) and converted into 8 digital bits. The digital bits represent the amplitude of the input analog signal. This figure shows how an analog signal is converted to a digital signal. An acoustic (sound) signal is first converted to an audio electrical signal (continuously varying signal) by a microphone. This signal is sent through an audio band-pass filter that only allows frequency ranges within the desired audio band (removes unwanted noise and other non-audio frequency components). The audio signal is then sampled every 125 microseconds (8,000 times per second) and converted into 8 digital bits. The digital bits represent the amplitude of the input analog signal.

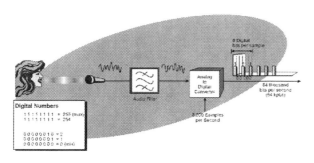

Voice Digitization

Voice Enabled Web Sites-Web sites that are capable of storing or playing voice (audio) clips.

Voice Gateway-A voice gateway is a communications device or assembly that transforms audio that is received from a telephone device or telecommunications system (e.g. PBX) into a format that can be used by a different network. A voice gateway usually has more intelligence (processing function) than a bridge as it can select the voice compression coder and adjust the protocols and timing between two dissimilar computer systems or voice over data networks.

This diagram shows the functional structure of a voice gateway device. This diagram shows that this voice gateway interfaces between a public telephone network to a packet data network. Input signals from the public telephone network pass through a line card to adapt the information for use within the voice gateway. This line card separates (extracts) and combines (inserts) control signals from the input line from the audio signal. If the audio signal is in analog form, the voice gateway converts the audio signal to digital form using an analog to digital converter. The digital audio signal is then passed through a data compression (speech coding) device so the data rate is reduced for more efficient communication. This diagram shows that there are several speech coder options to select from. The selection of the speech coder is negotiated on call setup based on preferences and communication capability of both voice gateways. After the speech signal is compressed, the digital signal is formatted for the protocol that is used for data communication (e.g. IP packet or Ethernet packet). This call processing section of the voice gateway may insert control commands (in-band signaling) to allow this gateway to directly communicate with the remote gateway. These digital signals are sent through a data access device (e.g. router shown here) so it can travel through the data communication network. The overall operation of the voice

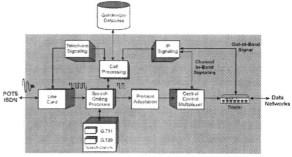

Voice Gateway Operation

gateway is controlled by the call processing section. The call processing section receives and inserts signaling control messages from the input (telephone line) and output (data port). The call processing section may use separate communication channels (out-of-band) to coordinate call setup and disconnection.

Voice Group Call Service (VGCS)-Voice group call service (VGS) is the process of transmitting a single voice conversation on a channel or group of channels so it can be simultaneously received by a predefined group of service subscribers.

Voice Mail (VM)-A service that provides a telephone customer with an electronic storage mailbox that can answer and store incoming voice messages. Voice mail systems use interactive voice response (IVR) technology to prompt callers and customers through the options available from voice mailbox systems. Voice mail systems offer advanced features not available from standard answering machines including message forwarding to other mailboxes, time of day recording and routing, special announcements and other features.

This diagram shows how a voice mail system provides electronic storage mailboxes to users within the telephone system. In this example, the voice mailbox system connects to a switching system through 2 extensions (ports) on the switching system (other voice mail systems may have many more ports). To access the voice mail system, users may select the voice mailbox system extension (usually programmed into a button on a telephone set that says voice mail). In this example, when a user dials into the telephone system to reach extension 1001, the line is busy. The system has been setup to forward calls to extension 1015 (the voice mail system) when extension 1001 is busy. To help ensure the voice mail system is accessible, if extension 1015 is busy, the call will be forwarded to extension 1016. When the call has entered the voice mail system, the interactive voice response (IVR) system will prompt the caller or user to enter information using touchtone or voice commands. This will allow callers or users to either store or retrieve messages from the digital message storage area (e.g. a computer hard disk drive).

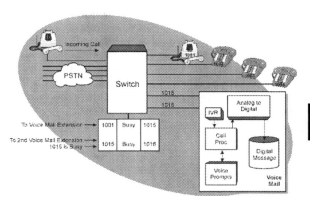

Voice Mail Operation

Voice Mail System (VMS)-The voice mail system is a telecommunications system that allows a subscriber to receive and play back messages from a remote location (such as a PBX telephone or mobile phone). The VMS consists primarily of memory storage (for messages), telephone interfaces (to connect to the communication system), and message recording, playback, and control features (typically via DTMF tones).

Voice Mailbox-A portion of memory, usually located on a computer hard disk, that stores and plays audio messages. The audio messages are often in compressed digital audio format.

Voice Messaging-A storage and retrieval system for voice messages. Commonly called "voice mail."

Voice On the Net (VON)-The process of sending voice over a data network (such as sending voice over the Internet).

Voice Over Cable (VoCable)-Voice over Cable solution is a complete Voice over Internet Protocol (VoIP) packet based broadband solution that supports DOCSIS and the PacketCable 1.0 specification.

This diagram shows how a cable television can offer telephony services. In this example, the cable television system has been modified to offer telephone service by adding voice gateways to the cable network's head-end cable modem termination system (CMTS) system and multimedia terminal adapters (MTAs) at the residence or business. The voice gateway connects and converts signals from the public telephone network into data signals that can be transported on the cable modem system. The CMTS system uses a portion of the cable modem signal (data channel) to communicate with the MTA. The MTA converts the telephony data signal to its analog audio component for connection to standard telephones. MTAs are sometimes called integrated access devices (IADs).

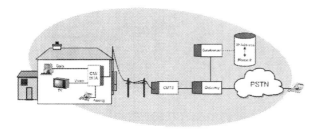

Voice over Cable Television Operation

Voice Over DSL (VoDSL)-Sending voice over a digital subscriber line system (VoDSL) is a process that sends audio band (also called "voice band") signals (e.g. voice, fax or voice band modem) via a digital channel on a digital subscriber line (DSL) system. VoDSL requires conversion from analog signals to a digital format and involves the formatting of digital audio signals into frames and time slots so they can be combined onto a digital (DSL) channel.

To communicate to other users, VoDSL requires one or more communication device that are capable of sending and receiving with the DSL network and conversion of a digital channel back into its analog voice band signal. This can be as simple as a computer with a sound card, a DSL modem and VoDSL software or as complex as a companies telephone network with an integrated access device (IAD). Optionally, some DSL systems have a PSTN gateway that can convert digital audio on a DSL system into telephone signals that can be sent through the public switched telephone network.

Voice Over Frame Relay (VoFR)-A process of sending digitized voice signals over frame relay data networks.

Voice Over Internet Protocol (VoIP)-A process of sending voice telephone signals over the Internet or other data network. If the telephone signal is in analog form (voice or fax) the signal is first converted to a digital form. Packet routing information is then added to the digital voice signal so it can be routed through the Internet or data network.

This diagram shows how an Internet network (public or private) can be used to provide telephone service. In this example, a calling telephone or multimedia capable computer dials a telephone number. This telephone number is provided to a voice gateway. The voice gateway decodes the dialed digits and determines the destination address (IP address) of the gateway that can service the dialed telephone number. The remote gateway signals the caller of an incoming call (rings the phone or alerts a multimedia computer). When the user answers the call, a message is sent between the gateways and a virtual path can be created between the gateways. This virtual path takes the audio, converts it to digital form, compresses and packetizes the information, adds the destination gateway address to each packet, routes the packets through the Internet to the destination gateway, and converts the digital audio back to its original analog form.

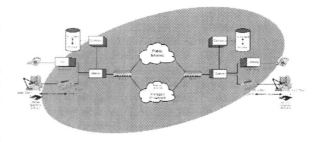

Voice over the Internet (VoIP) System

Voice Privacy-Voice privacy is a process that is used to prevent the unauthorized listening of communications by other people. Voice privacy involves coding or encrypting of the voice signal with a key so only authorized users with the correct key and decryption program can listen to the communication information.

Voice Quality-Voice quality is a measurement of the level of audio quality, often expressed in mean opinion score (MOS). The MOS is a number that is determined by a panel of listeners who subjectively rate the quality of audio on various samples. The rating level varies from 1 (bad) to 5 (excellent). Good quality telephone service (called toll quality) has a MOS level of 4.0.

Voice Recognition-A computer-based technology that analyzes audio signals (typically spoken words) converts them into digital signals for other processing (e.g. voice dialing).

Voice Response Unit (VRU)-An equipment that provides a caller with audio messages in response to their touch-tone(tm) key presses or voice commands . VRU are part of interactive voice response (IVR) systems.

Voice Service-Voice service is a type of communication service where two or more people can transfer information in the voice frequency band (not necessarily voice signals) through a communication network. Voice service involves the setup of communication sessions between two (or more) users that allows for the real time (or near real time) transfer of voice type signals between users.

Voice Synthesis-(1-audio) Computer-generated sounds that simulate a human voice. (2-voice decompression) The process of recreating voice from a compressed voice coded signal.

Voice Trigger-A voice trigger is a signal or event that begins the creation of an audio (voice) signal.

VoIP-Voice Over Internet Protocol

VoIP Testing-Voice over Internet Protocol (VoIP) testing is the process of testing a data network to determine if it meets the user requirements for sending IP Telephony and data over a data network.

This figure shows how a data network may be monitored for several days to determine the capacity and transmission delays at concentration points (routers and switches) within the network. In this example, a data network has several routers that transfer data between computers in the company and to other computers connected to the Internet. As part of the VoIP capability pre-test, each router is monitored for peak data transfer activity detection for several days. This helps to determine if the data network lines and switching points have

enough capacity to provide both the data network and VoIP system needs. In this system, an on-site company host computer stores the files for engineering and sales. An off-site company computer is used to store files from the company computer that are transferred through the Internet during late evening hours. The analysis shows that the engineering router uses bandwidth during the morning and evening hours (light lunchtime use). The sales router has capacity used from late morning through late evening (they eat lunch at the office). The analysis also shows that the company's Internet data connection has high capacity late in the evening when information backup is in process.

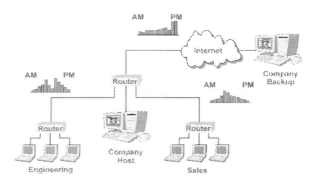

Testing Data Networks for VoIP Capability

Volatile-A term applied to information held in a memory store that depends on power being continuously available.

Volatile Memory-A type of read/write memory whose content is irretrievably lost when the operating power is removed.

Volt (V)-The standard unit of electromotive force equal to the potential difference between two points in a conductor that is carrying a constant current of one ampere when the power dissipated between the two points is equal to one watt. One volt is equivalent to the potential difference across a resistance of one ohm when one ampere is flowing through it. The volt is named for the Italian physicist Alessandro Volta. Also called voltage or "electromotive force." Called electric "tension" in some languages.)

Voltage Controlled Oscillator (VCO)-An oscillator circuit which has an output frequency that changes proportionally with an input voltage.

Voltage Drop-A decrease in electrical potential resulting from current flow through a resistance.

Voltage Regulator-A circuit used for controlling and maintaining a voltage at a constant level.

Voltmeter-A test instrument that is used to measure differences or level of electrical potential.

Volume-(1-sound) The loudness (intensity) of a sound. (2-data) A certain portion of data, together with its data carrier, that can be handled conveniently as a unit. (3-graphics) A three-dimensional array of raster data. (4-general) The amount of a cubic space measure (cubic meters, cubic inches, etc.) contained in a given three-dimensional region of space.

Voluntary Churn -A disconnection of service that occurs when a customer decides to drop a service from an existing carrier and initiate the service with another carrier.

VOM-Volt Ohm Milliammeter

VoMBN-Voice over Multiservice Broadband Network

VON-Voice On the Net

Voting Receivers-A wireless communication system that uses multiple radio receivers to receive the strongest radio signal possible from mobile radios operating in the system.

This figure shows a system that uses voting receivers to allow low power mobile radios to effectively communicate in a system that has a high power transmitter. In this system, there is one transmitter that that operates a base station that has at 100 Watts transmitter power. As the mobile radio moves throughout the transmitter site radio coverage area, the mobile radio continues to receive the signal from the same base station transmitter. However, because each mobile radio in the system can only transmit at five Watts, the system selects a receiver in the system that has the strongest received signal from the mobile radio. As the mobile radio moves from voting receiver #1 to voting receiver #2, the voting system will eventually select voting receiver #2 as the best choice to receive communication from the mobile radio.

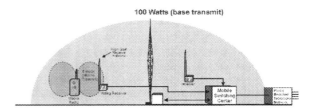

System with Voting Receivers

Voucher-A card, printed brochure, or electronic record that authorizes a customer to activate or recharge (add usage time) to their prepaid telecommunications account or device.

Vox-A voice operated relay circuit that permits the equivalent of push-to-talk operation of a transmitter by the operator.

vPBX-Virtual PBX

VPC-Virtual Path Connection

VPI-Virtual Path Identifier

VPN-Virtual Private Network

VRML-Virtual Reality Modeling Language

VRU-Voice Response Unit

VS-Video Server

VSAT-Very Small Aperture Terminal

VSELP-Vector Sum Code Excited Linear Predictive

VSELP-Vector Sum Excited Linear Predictive Coding

VSWR-Voltage Standing Wave Ratio

VT-Virtual Tributary

VTR-Video Tape Recorder

W

W-watt
W3C-World Wide Web Consortium
WAAS-Wide Area Augmentation System
WAE-Wireless Application Environment
Wallet Share Improvement-Wallet Share Improvement is the process of actively and passively doing something that result in the customer either using more services or buying more products than before.
WAN-Wide Area Network
WAP-Wireless Access Point
WAP-Wireless Access Protocol
WAP-Wireless Application Protocol
WAP-Wireless Application Provider
WAP Gateway-Wireless Application Protocol
WAP Protocol Stack-Wireless Application Protocol
WAP Server-Wireless Application Server
WAR-Wallet Share, Acquisition, and Retention
Warble-A sound that is produced during the decoding of a compressed digital audio signal that has been corrupted (has errors) during transmission. The warble sound results from the creation of different sounds than were originally sent.
WARC-World Administrative Radio Conference
Warm Start-The process of rebooting a computer system without turning the power off.
Warning Tape-Warning tape is a material that is placed above or around a cable, assembly or area to inform others of a potential hazard or of items that may be damaged by passing through the warning tape (such as buried cable).
WASP-Wireless Application Service Provider
Watchdog Process-A device or system that continually monitors specific functions of devices or systems (usually mission critical systems) to ensure they continue to operate within predetermined limits.
Watchdog Timer-A hardware timer that upon counting down to zero resets the central processing unit and therefore brings about a reset of the system. Software within the system must set the timer back to its starting point often enough that it does not reach the zero count. As long as the software functions normally it will continue to "pet" the watchdog and prevent a system restart. Software that gets stuck or that crashes will be reset once the watchdog counts down.

Water Blocking Compound-A water blocking compound is a substance that is used to fill space within a cable or assembly so that water cannot enter an area. Water that enters into a cable may cause corrosion or travel through the cable to other electrical assembles. Water that enters into cables may also freeze causing damage from expansion and contraction of the cable assembly.
Water Ingress-Water ingress is the entry of water into a cable, device or assembly.
Water Seal-A cap or water seal is a covering that is applied to the end of a cable to seal the end of the cable from the entry of water and other unwanted substances.
Watermark-An imperceptible signal hidden in another signal, such as audio or an image, which carries information. Watermarking is related to the general field of stenography, or information hiding. Ideally a watermark would not be destroyed (that is, the signal altered so that the hidden information could no longer be determined) by any imperceptible processing of the overall signal, for example high-quality lossy compression, slight equalization, or digital-to-analog-to-digital conversion. Sophisticated techniques for successfully destroying watermarks make that ideal difficult to achieve.
Watermark Extractor-A watermark extractor is a filter, software program or assembly that can separate a watermark from a media file. This watermark may be used to provide the key that is able to decode and play the media file.
Watermarking-Watermarking is a process of adding (data embedding) or changing information in a file or other form of media that can be used to identify that the media is authentic or to provide other information about the media such as its creator or authorized usage.
WATS-Wide Area Telecommunications Service
watt (W)-Unit of electric power, equal to one joule per second. One watt is produced (or consumed) when a current of 1 ampere flows through a voltage difference of 1 volt. The watt is named after the Scottish inventor James Watt (1736-1819).
WAV-Wave Audio
Wave-(1-general) A disturbance that is a function of time or space, or both, and is propagated in a medium or through space. (2-backward) A wave

with phase velocity in the verse direction to the direction of electron flow. (3-carrier) A single frequency wave that may be used to carry the information of a modulating wave. (4-damped) A wave whose amplitude is reduced with every cycle. (5-direct) A radio wave that travels directly from the transmitting antenna to the receiving antenna with no reflections or refractions. (6-electromagnetic) A wave propagated through space (and through many material substances) consisting of varying electric and magnetic fields. Some of the properties of these waves depend on frequency. Lightwaves and radio waves are two frequency band manifestations of the same electromagnetic wave mechanism. (7 - forward) A wave with group velocity in the same direction as the direction of electron flow. (8-ground reflected) A radio wave reflected from the surface of the earth. (9-horizontally polarized) An electromagnetic wave with its electric field component parallel to the surface of the earth. (10-intelligence) A wave carrying a message. (11-ionosphere.) A radio wave reflected back to the earth's surface by the ionosphere. (12-modulated) A wave whose characteristics have been varied in accordance with a modulating signal. (13-modulating) The information carrying signal that modulates a carrier so that the information may be transmitted by line, fiber, or radio more efficiently than would be possible at the original signal frequency. (14-periodic) A wave that repeats the same pattern at regular intervals. (15-plane polarized) A wave whose electrical intensity (or electric field vector) lies at all times in a plane that contains the direction of propagation. (16-radio) An electromagnetic wave with electric and magnetic field components. A radio wave is radiated by an antenna, travels at the speed of light, and may be picked up by another antenna. (17-sky) A radio wave that travels from the earth's surface toward the sky. At some frequencies and in some circumstances, the wave can be reflected or refracted down to earth again. (18-sound) A wave carried in an elastic medium by audio frequency vibrations. (19-space) The component of a radio wave that travels through the atmosphere just above the surface of the earth. (20-square) A periodic wave with the characteristic of suddenly changing from negative to positive, maintaining a steady period, then suddenly reversing from positive to negative. A square wave contains odd harmonics of the fundamental frequency. (21-surface) The part of ground-wave radiation that travels above the surface of the earth. (22-transverse electric, Th) A mode of propagation in a waveguide in which the electric field vector is in all places perpendicular to the direction of propagation. (23-transverse electromagnetic, ThM) A mode of propagation in coaxial)~ cables and in open feeders in which the electric field vector and the magnetic field vector both are perpendicular to the direction of propagation(24-transverse magnetic, TM) A mode of propagation in a waveguide in which the magnetic field vector is in all places perpendicular to the direction of propagation. (25-traveling plane) A plane wave, each of whose frequency components has an exponential variation of amplitude and a linear variation of phase in the direction of propagation. (26-undamped) A wave with constant amplitude. (27-vertically polarized) An electromagnetic wave whose electric field component is in all places perpendicular to the surface of the earth.

Wave Audio (WAV)-Wave audio is a coding form for digital audio used by Win32. Wave audio files commonly have a. WAV extension to allow programs to know it is a digital audio file in Wave coding format.

Wave Number-Wave number is the number of wavelengths of an optical signal that occurs per unit distance traveled. Wave number is the reciprocal of wavelength.

Wave Theory-Wave theory is the relationship of wavelength to the speed the wave travels is a specific medium (such as air). The formula for wavelength is the speed of the medium (such as light or radio waves) divided by the frequency of the wave. Wavelength = c/f. The speed of light is 300 million meters per second (186,281.6 miles per second) in free space. For example, the wavelength of a 300 MHz signal is approximately 1 meter when transmitted in free space.

This figure shows that light signals can be represented by electromagnetic waves that travel at 300 million meters per second (186,281.6 miles per second). The distance of one wave cycle is called the wavelength. This example shows that the frequency of the optical signal is calculated by the inverse of the wavelength.

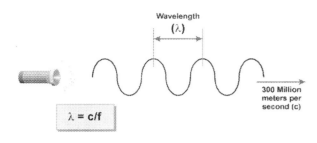

$$\lambda = c/f$$

Optical Wave Theory

Wave Trap-A tuned circuit that attenuates an undesired frequency.

Waveform-The characteristic shape of a periodic wave, determined by the frequencies present and their amplitudes and relative phases.

Waveform Coder-Waveform Coding

Waveform Coding (Waveform Coder)- Waveform coding consists of an analog to digital converter and data compression circuit that converts analog waveform signal into digital signals that represent the waveform shapes. Waveform coders are capable of compressing and decompressing voice audio and other complex signals.

Waveform Monitor-A special oscilloscope used to display and evaluate video signals. The horizontal axis of a waveform monitor is driven by a time base synchronized to the video signal. The vertical axis of the display is driven by the amplitude of the video signal.

Wavefront-A continuous surface that is a locus of points having the same phase at a given instant. A wavefront is a surface at right angles to rays that proceed from the wave source. The surface passes through those parts of the wave that are in the same phase and travel in the same direction. For parallel rays, the wavefront is planar; for rays that radiate from a point, the wavefront is spherical.

Wavelength-(1-general) The length of a wave through one full cycle, e.g., from a starting point of zero amplitude through maximum amplitude, then minimum and back to zero. In optical networks, light of about 800 nanometers (nm), 1300 nm, or 1550 nm is typically used. These wavelengths are in the infrared portion of the electromagnetic spectrum and are not visible to the naked eye. (2-light)

Each beam of light that carries information in an optical network has a specific wavelength (e.g., 1548 nm). They are sometimes referred to as wavelengths rather than beams or lightwaves.

This diagram shows that the wavelength of a signal is the distance that the signal travels over one cycle of the signal. In this example, two signals are transmitted in free space; 300 MHz and 3GHz. The wavelength in free space at 300 MHz is 1 meter and the wavelength at 3 GHz is 1/10th meter. This example also shows that the wavelength changes dependent on the transmission medium, such as coaxial cable as shown in this example. Because the signal wave travels slower in other mediums, the wavelength is shorter. This is why antennas are physically a little shorter than their ideal electrical requirements.

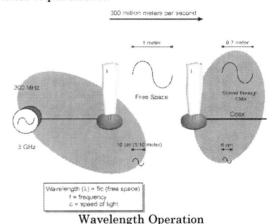

Wavelength Operation

Wavelength Division Multiplexing (WDM)- Wave division multiplexing is a process of transmitting several distinct communication channels through a single optical fiber via the use of a distinct separate infrared wavelength (optical frequency or "color") for each communication channel. Each such channel may be further subdivided into several logical channels via time division multiplexing or other methods.

This diagram shows how a wave division multiplexing over fiber operates. This diagram shows that there are several lasers operating at different optical wavelengths (different colors/frequencies). Each laser converts an electrical signal into a pulsed light signal. These optical signals (optical carriers) are combined by an optical multiplexer (lens) for transmission through the optical fiber. At the receiving end, the different optical carriers are separated by an optical demultiplexer (lens) and each optical carrier is sent to a photo-detector. The photo-detector converts the optical signal back into its original electrical form.

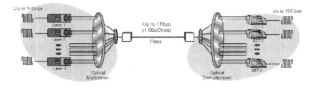

Wavelength Division Multiplexing (WDM) System

Wavelet Compression-Wavelet compression is a form of digital signal analysis that reduces the amount of data that is necessary to represent an audio, image or video signal by converting the media into wavelet components.

wb-weber

WBMP-Wireless Bitmap

WCDMA-Wideband Code Division Multiple Access

WCS-Wireless Communication Service

WDM-Wavelength Division Multiplexing

WDP-Wireless Datagram Protocol

Weak Password-A weak password is a user access code that is easy to guess or obtain through the use of word lists or sequential password access attacks. Weak passwords are usually short (9 characters or less) and usually contain well used words, names or birthdates that are related to the user.

Web-A short term used for the World Wide Web.

Web Address-The text or numeric address that can be used to access a web site on the Internet.

Web Billing-Web billing is the process of grouping service or product usage information for specific accounts or customers, producing and sending invoices, and recording (posting) payments made to customer accounts and providing access or control of this information through the Internet.

Web Browser-Software that is used to graphically view information on Web servers. Web browsers request, receive, and reformat information receives from web servers.

Web Camera (webcam)-A camera converts images into a digital form that can be sent through the Internet.

Web Conferencing-Web conferencing is the conducting of meetings or functions that use and/or provide media (such as audio and video) via the Internet.

Web Hosting-The providing of web program application services (html or file transfer), allocation of information storage space, and interconnection to the Internet for customers.

Web Interactive Toy (WIT)-A toy that connects to the Internet so that it can interact with other toys and/or computing devices that are also connected to the Internet.

Web Master (Webmaster)-The person who is responsible for the operation and maintenance of a web site.

Web Page-A file located on a computer that is connected to the Internet that has a format that allows the file to display (format) information on a user's display (web browser).

Web Phone-An Internet telephone that only requires multimedia capability to allow users to call through the Internet. One of the first web phones was NetSpeak.

Web Radio (WebRadio)-Web radio is the sending or broadcasting of digital audio signals through IP data networks (such as the Web). Web radios are typically software programs that operate on multimedia computers.

Web Seminar (Webinar)-A web seminar (webinar) is an online instruction session that uses the Internet Web as a real time presentation format along with audio channels (via web or telephone) that allow participants to listen and possibly interact with the session. Webinars allow people to participate in information or training sessions from anywhere that has Internet and audio access.

Web Server-Web servers are computer systems that are used provide access to data that is stored and retrieved by commands in Hypertext Transfer Protocol (HTTP). HTTP is a protocol that is used to request and coordinate that transfer of documents between a web server and a web client (user of information). The typical use of web servers is to allow web browsers (graphic interfaces for users) to request and process information through the Internet.

Web Site-A file or group of files located on a computer that is connected to the Internet. These files are generally accessible by other users that are connected to the Internet through the use of Internet protocols.

Web Television (WebTV)-A set-top box (cable converter) that provides the user with the ability to use and display Internet services on a standard television.

Web Video-Web video is the providing of video or television service through the Internet.

WebCam-A webcam is a PC video camera that captures and posts live images to a website. These images are refreshed every few seconds. Webcams can be used for video email and video conferencing and video instant messaging.

webcam-Web Camera

Webcast-The live presentation of information in a continuous (streaming) format delivered through the Internet web. A webcast might be associated with other web pages or other web-browser-based content in addition to the live stream.

weber (wb)-Unit of magnetic flux, equal to one volt per second. Named for 19th century German physicist Wilhelm E. Weber (1804-1891).

Webinar-Web Seminar

WebMail-A webmail reader is a media viewing application that allows a user to view media (such as emails) that is stored in a data network (such as the Internet). A webmail reader allows the user to view the media through the use of a standard web browser program without requiring email or other software programs.

Webmaster-A webmaster is the person who is responsible for maintaining and administering a Web site.

Webmaster-Web Master

WebRadio-Web Radio

WebTV-Web Television

Webzine-Web Magazine

WECA-Wireless Ethernet Compatibility Alliance

Weight-The force exerted on an object due to gravitational attraction to the planet it is on or near. In the SI metric system, this force is measured in newtons (N). In contrast to weight, which depends on the properties of the planet you are on or near, the mass of an object is the same regardless of which planet you are on or near. For example, a 1 kg object has a mass of 1 kg everywhere in space. But it has a weight of 9.8 N on the surface of the earth (more precisely, the weight is slightly small-er at the equator than at the poles, due to the effect of the earth's rotation), and the same 1 kg object has a weight of only 4.8 N on the moon's surface, and 25.9 N on Jupiter's surface. Despite this difference, the words weight and mass are loosely used as synonyms in everyday speech.

WEP-Wired Equivalent Privacy

WER-Word Error Rate

Wet Circuit-An analog trunked cable pair that carries speech or data signals, plus a direct current. The DC component is used for signaling and supervision, for talk battery feed, or for sealing current.

Wetting Current-A continuous current ("sealing current") that is applied to a communication line to minimize the effects of oxidation on splices and junctions that could cause poor communication.

WFOM-Wait For Overhead Message

WFQ-Weighted Fair Queuing

What you See is What You Get (WYSIWYG)-An expression that is used for a computer system that displays information in the same style in which it will be printed. It is pronounced "wizzy-wig."

White Hat-A which hat is a hacker who attempts to gain unauthorized access to systems or information sources but does not maliciously use the information or damage the data.

White Noise-A mathematical idealization of random noise. White noise has an equal amount of energy over a wide frequency spectrum.

Whiteboard-A device that can capture images or hand drawn text so they can be transferred to a video conferencing system. Whiteboards allow video conferencing users to place share documents, images and/or hand written diagrams with one (or more) video conference attendees.

This figure shows how a whiteboard can be used during an Internet telephone call to transfer hand drawn images. In this example, an instructor is drawing a diagram on a white pad. While the instructor is drawing, the image is being displayed on both the instructors monitor and the students monitor.

Internet Telephone Whiteboard Operation

Wide Area Network (WAN)-A communications network serving geographically separate areas. A WAN can be established by linking together two or more metropolitan area networks, which enables data terminals in one city to access data resources in another city or country.

This figure shows that a WAN is usually composed of several different data networks. Different types of communication lines such as leased lines, packet data systems, or fiber transmission lines can interconnect these networks.

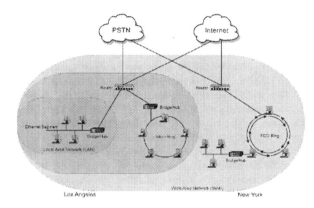

Wide Area Network (WAN) Systems

Wide Screen-Wide screen is the display of a video or motion picture on a screen that has a wider aspect ratio than standard 4:3 (width to height ratio). For example, a motion picture has a 16:9 aspect ratio.

Wideband-(1-communication channel) Wideband is a communication channel or signal that is much wider than is required to transfer the information content. An example of a wideband channel is the IS-95 CDMA cellular system that uses a radio channel which is 1.23 MHz wide to transfer user data that is 14.4 kbps or less. (2-data transmission rate) A wideband data channel typically has data transmission rates above 1 Mbps.

Wi-Fi-Wireless Fidelity

Wi-Fi Cordless-Wi-Fi cordless is short range wireless telephone systems that use wireless local area network (Wi-Fi) systems to communicate. Wi-Fi cordless telephones typically use radio transmitters that have a maximum power level below 100 milliwatts (0.10 Watts). This limits their usable range of 300 meters or less.

Wi-Fi Diversity-Wi-Fi diversity is the use of multiple antennas or multiple signals for the purpose of increasing the probability of a high quality signal path between the sender and the receiver. Diversity can be implemented at the transmit end, the receive end or at both ends of the wireless link.

Wi-Fi IP Telephony-Wi-Fi IP telephony is short range wireless telephone systems that use wireless local area network (Wi-Fi) systems to communicate voice signals using IP telephony technology. Wi-Fi cordless telephones typically use radio transmitters that have a maximum power level below 100 milliwatts (0.10 Watts). This limits their usable range of 300 meters or less.

Wi-Fi Protected Access (WPA)-Wi-Fi protected access is an encryption (privacy) process that is used in the 802.11 Wireless LAN system to prevent unauthorized receivers of information to be able to decode and use transmitted information. The WPA encryption uses an encryption key that changes for each communication session. WPA is an enhancement to the wired equivalent privacy (WEP) fixed key system that was used in the original 802.11 WLAN systems.

Wi-Fi Television (Wi-Fi TV)-Wi-Fi television is short range wireless telephone systems that use wireless local area network (Wi-Fi) systems to distribute television signals. Wi-Fi televisions typically use radio transmitters that have a maximum power level below 100 milliwatts (0.10 Watts). This limits their usable range of 300 meters or less.

Wi-Fi TV-Wi-Fi Television

Wiggle Proof-Wiggle proof is the ability of a cable connector to withstand the wiggling (side to side movement) of a cable without damaging the connection of the cable to the connector.

WiMax-WiMax is a name for the 802.16 series of industry specifications for wide area broadband wireless systems.

This figure shows the key components of a WiMax system. This diagram shows that the major component of a WiMax system include subscriber station (SS), a base station (BS) and interconnection gateways to datacom (e.g. Internet), telecom (e.g. PSTN) and television (e.g. IPTV).. An antenna and receiver (subscriber station) in the home or business converts the microwave radio signals into broadband data signals for distribution in the home. In this example, a WiMax system is being used to provide television and broadband data communication services. When used for television services, the WiMax system converts broadcast signals to a data format (such as IPTV) for distribution to IP set top boxes. When WiMax is used for broadband data, the WiMax system also connects the Internet through a gateway to the Internet. This example also shows that the WiMax system can reach distances of up to 50 km.

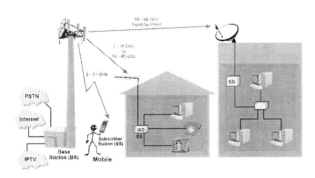

WiMax Radio System

Winch-A winch is a pulling device that is used to pull wire or cables through conduits or other cable channel guides.

Winch Line-A winch line is a rope or cable that is connected from a pulling winch to a cable that will be pulled through a conduit, overhead line or other type of cable channel.

Wind Loading-Wind loading is the pressure placed upon an antenna structure or cable by the wind.

Window-(1-transmission buffer) An indication of the amount of time or data that should not be exceeded waiting for successful reception of information. If this amount is exceeded, retransmission may occur. (2-video) Video containing information or allowing information entry from a keyboard, time code generator, or other device. A window dub is a copy of a videotape with time code numbers keyed into the picture. (3-display) A window is an area of a screen display that is used to display information associated with a specific program, application or function.

Window Size-In the X.25 and X.75 packet-switching protocols, the number of unconfirmed frames or packets that are transmitted across a connection before additional frames or packets can be sent. Window size affects the control of data transmission and reception on the user and network sides of an X.25/X.75 connection.

Windows CE-Windows CE is an operating system for consumer electronics devices that commonly have a limited amount of memory and processing power. The Windows CE system is designed to require less memory (decreased footprint) allowing it to be used in devices such as mobile telephones, personal digital assistants (PDAs), and television set top boxes.

Windows Media (WM)-Windows media .WM is a container file format that holds multiple types of media formats. The .WM file contains a header (beginning portion) that describes the types of media, their location and their characteristics that are contained within the media file.

Windows Media Audio (WMA)-Windows media audio is an audio codec format that was created by Microsoft that works with later versions of Windows media player and integrates advanced media features such as digital rights media (DRM) control capability.

Windows Media Player (WMP)-Windows media player is a software application from Microsoft that allows for the playing of multiple types of digital media formats.

Windows Media Video (WMV)-Windows media video is a video codec format that was created by Microsoft that works with later versions of Windows media player and integrates advanced media features such as digital rights media (DRM) control capability.

Windows XP Embedded-Windows XP Embedded is a version of the Windows XP operating system that can be embedded into consumer electronics devices such as point of sale systems and television set top boxes.

Wink-(1-control signal) A telephone line signal that is a single supervisory pulse. When caused by the passive end of a subscriber loop, it is usually transmitted as an off-hook signal followed by an on-hook signal where the off-hook signal is of a very short specified duration compared to the on-hook signal. When caused by the powered end of a subscriber loop, it is usually transmitted as a short removal of central office battery voltage from the loop wires. When used with robbed bit or associated bit signaling on a digital multiplexed channel, the robbed bit or associated bit changes binary value for a short time interval. The duration of any form of a wink signal is typically 200 milliseconds. (2-indicator light) A rapid flashing cadence of an indicator light, typically used to indicate which line is ringing with an incoming call on a multi-line telephone set. Winking cadence is typically 5 winks per second, faster than flashing cadence.

WIPO-World Intellectual Property Organization, a specialized agency of the United Nations

Wire Elongation-Wire elongation is the process of extending the wire length as a result of pulling on the wire (such as during cable installation).

Wired Equivalent Privacy (WEP)-An encryption (privacy) process that is used in the 802.11 Wireless LAN system to prevent unauthorized receivers of information to be able to decode and use transmitted information. The WEP encryption can use a 64 or 128-bit encryption key.

Wireless-Communication without the use of cables or devices that transmit over wireless networks rather than over telephone lines. Historically, at various times during the 20th century, this had specialized meanings that have come and gone. For many years, British English used the word "wireless" while North American English used "Radio" instead. In the past, the word "wireless" was occasionally used to describe the transmission via radio of Morse code, but not voice. Today the term wireless is used primarily for cellular systems, and secondarily also for other short range radio systems used directly by end users, such as for example 802.11b short range data transmission.

Wireless Access Point (WAP)-A wireless access point contains radio transceivers that convert digital information to and from radio signals that can be exchanged with other wireless communication devices. The most basic forms of wireless access points simply for wireless connections. A wireless access point that includes the ability of DHCP and network address translation (NAT) is typically called a wireless gateway.

Wireless Access Protocol (WAP)-Wireless access protocol is a standard protocol specification that allows advanced messaging and information services to be delivered to wireless devices independent of which wireless technology they use.

Wireless Application-Wireless applications are systems and services that are designed and perform operations using commands or information that are transferred between devices without physical connections.

Wireless Application Protocol (WAP)-WAP is a collection of protocols and standards that enable communication and information applications to run efficiently on mobile devices. WAP is to wireless devices what hypertext transfer protocol (HTTP) is to Web browsers.

Wireless Billing-The recording and processing of wireless transmission events for billing purposes.

Wireless Broadband-Wireless broadband is the transfer of high-speed data communications via a wireless connection. Wireless broadband often refers to data transmission rates of 1 Mbps or higher.

Wireless Cable-"Wireless Cable" is a term given to land based (terrestrial) wireless distribution systems that utilize microwave frequencies to deliver video, data and/or voice signals to end-users. There are two basic types of wireless cable systems, multichannel multipoint distribution service (MMDS) and local multichannel distribution service (LMDS).

This figure shows a overview of a wireless broadband communication system. This system uses radio towers (usually called "base stations") that are located within a few miles of the customer to transmit relatively low power RF signals directly to a customer's radio receiver. The radio receiver uses a directional high-gain antenna to capture and focus radio signals for transmission between the radio tower and the customer's house. The radio transceivers (transmitter and receiver pair) in the base station transmit on radio channels up to several MHz wide each. The radio receiver converts these radio channels back to its original digital form so it can be provided to the customer's computer. Many MMDS and LMDS systems provide a data transmission rate of approximately 10 Mbps from the radio tower to the customer's receiver and approximately 1 Mbps from the customer's receiver back to the radio tower.

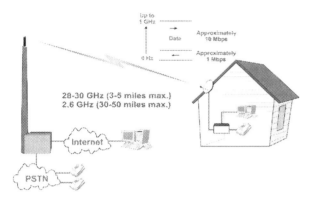

Wireless Cable Overview

Wireless Communication-Wireless communications is the transmission of information without wires. Wireless communication may use radio or optical transmission for communication.

Wireless Communication Service (WCS)-Wireless communication service is a frequency band 2305 MHz to 2320 MHz that is used for wireless multimedia communication.

Wireless Data-A system or the transmission of digital information through a wireless network such as wireless packet data systems or cellular mobile communications. Wireless data systems are specifically designed to reliably transfer information (data) between a sender and receiver. The term wireless data can apply to mobile or fixed devices and the transmission may be in the form of radio or optical (e.g. infrared systems) communication systems. Wireless data transmission can be sent over dedicated wireless data communication systems (such as Reflexion, Ardis, or Mobitex) or

the transmission may share a common channel for voice and data (such as on GSM or 3G cellular systems).

This figure shows the three key types of wireless data networks. This diagram shows a wireless LAN system that has multiple access nodes. These access nodes operate as gateways between the data communication devices (e.g., mobile computer) and the data network hub. Building 1 uses an older 801.11 wireless LAN system that operates from 902-928 MHz at 2 Mbps. Building 2 uses a newer 802.11 wireless LAN system that operates at 2.4 GHz providing up to 11 Mbps data transfer rate. This diagram also shows a microwave data link that provides a 45 Mbps interconnection between campus buildings. Finally, a user who is operating in a remote area outside the core campus is using the wide area mobile system to transfer data files (at a data transfer rate below 28 kbps).

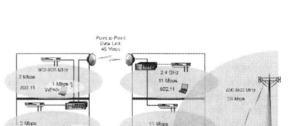

Wireless Data Networks

Wireless Datagram Protocol (WDP)-An efficient form of packet transmission that is used in wireless communication systems. WDP is similar to UDP.

Wireless DSL-Wireless digital subscriber line is the providing of broadband digital suscriber line service using radio technology.

Wireless Email-E-mail that can be downloaded through a wireless network via a wireless modem.

Wireless Ethernet-A wireless version of packet based Ethernet system. Wireless Ethernet systems typically use the ISM frequency band in the 2.4 GHz or 5.7 GHz range. Because bandwidth is limited, wireless Ethernet systems are limited to 1 Mbps to 54 Mbps compared to the 10 Mbps to 10 Gbps for wired Ethernet systems.

Wireless Ethernet Compatibility Alliance (WECA)-An industry trade group that was established to help standardize and ensure interoperability of wireless 802.11 devices.

Wireless Fidelity (Wi-Fi)-Another name for the 802.11 wireless LAN system.

Wireless Headset-A wireless headset is a combination of a earpiece and radio transmission system that is used to allow users to extend the audio portions of their communication devices to portable and flexible external speakers. Some wireless headsets include an external microphone.

Wireless Hot Spots-Wireless hot spots are geographic regions or service access points that have a higher than average amount of usage. Examples of hot spots include wireless LAN (WLAN) access points and traffic jam areas on mobile telephone (cellular) systems.

Wireless Internet Access-Wireless Internet access is the ability for a user or device to connect to the Internet via wireless connections. An example of wireless Internet access is the requesting and connection to the Internet via a wireless LAN or via a data connection on a mobile telephone.

Wireless Internet Protocol (Wireless IP)-Wireless Internet protocol is the sending of IP datagrams over wireless connections. Sending IP data over wireless connections has unique challenges that traditional wired connections do not have, such as changing (mobile) destination points. This is being address by mobile IP protocol that can maintain end-to-end connections as a data communication device moves through a wireless system or network.

Wireless Internet Service Provider (WISP)-A wireless Internet service provider (is a company that receives and converts (formats)) information to and from wireless Internet connections to Internet end users. A WISP purchases a high-speed link to the Internet and divides up the data transmission to allow many users to connect to the Internet via wireless connections.

Wireless Internet Television-Wireless Internet television is the ability for a user or device to connect to digital television provided through the Internet via wireless connections. An example of wireless Internet television is the requesting and connection to a television gateway or media server via a wireless LAN or via a data connection on a mobile telephone.

Wireless IP-Wireless Internet Protocol

Wireless Local Area Network (WLAN)-A wireless local area network (WLAN) allows computers and workstations to communicate with each other using radio propagation as the transmission medium. The wireless LAN can be connected to an existing wired LAN as an extension, or can form the basis of a new network. While adaptable to both indoor and outdoor environments, wireless LANs are especially suited to indoor locations such as office buildings, manufacturing floors, hospitals and universities.

Wireless Local Loop (WLL)-Wireless local loop (WLL) is the providing of local telephone service via radio transmission. Wireless local loop systems often use a radio conversion device located at the home or business to allow the use of standard telephones. Although WLL systems may provide for traditional dialtone service, WLL systems commonly provide for multiple types of services such as telephone service, Internet access, and video programming.

This diagram shows a wireless local loop system. In this diagram, a central office switch is connected via a fiberoptic cable to radio transmitters located in residential neighborhoods. Each house that desires to have dial tone service from the WLL service provider has a radio receiver mounted outside with a dial tone converter box. The dial tone converter box changes the radio signal into the dial tone that can be used in standard telephone devices such as answering machines and fax machines. It is also possible for the customer to have one or more wireless (cordless) telephones to use in the house and to use around the residential area where the WLL transmitters are located.

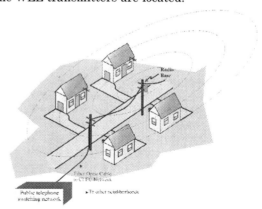

Wireless Local Loop (WLL) System

Wireless Markup Language (WML)-Wireless markup language is an efficient version of hypertext markup language (HTML) that is used between wireless communication devices (such as a PDA) and a server to allow world wide web (WWW) to efficiently operate over low speed data connections. Wireless markup language (WML) is part of wireless access protocol (WAP).

JJJ Wireless Markup Language" is a tag-based language used for describing the structure of documents to be delivered to wireless devices. It is to wireless devices what HTML (Hyper Text Markup Language) is to Web browsers. WML is used to layout "pages" (or cards as they are called in WML) to be viewed in wireless devices. WAP is the protocol for WML. WML is less forgiving and stricter on syntax than HTML. It is more like a programming language this way than HTML is.

Wireless Mesh Network-A wireless mesh network is a radio communication system where each communication device (typically radio access points) can be interconnected to multiple nodes so data packets can travel through alternate paths to reach their destination. For some wireless mesh networks, the relaying of packets through multiple access points forms the backbone connection (interconnection) of the network.

Wireless Messaging-Wireless messaging is the process of sending and/or receiving messages using a wireless connection (radio and/or optical).

Wireless Metropolitan-Area Network (WMAN)-WMANs are usually private wireless packet radio networks often that cover an urban or city geographic area. They are commonly used for law-enforcement, utility, or public safety applications.

Wireless Network-Wireless networks are primarily designed to transfer voice and or data from one point to one or more other points, (multipoint). Many networks make use of some wireless technologies as a transport medium even though we do not consider them to be wireless networks. Examples of wireless networks include cellular, personal communication service, (PCS), paging, wireless data, satellite, and broadcast radio and television. Wireless network is a term commonly used for wireless local area network (WLAN).

This figure shows the basic types of wireless networks and that these networks vary from broadcast (one-way) systems to complex switching two-way systems. This diagram shows a private land mobile

radio system, television broadcast system, paging system, mobile telephone system, and satellite communication system. Although all wireless networks can transmit information from one point to another, different types of networks better suited to provide specific types of services (e.g., paging compared to television broadcasting).

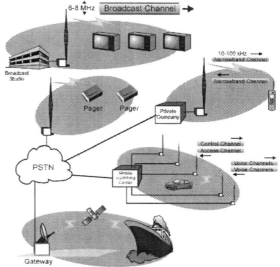

Wireless Networks

Wireless Packet Data-Wireless data transmission technology that transmits data in small packets (up to approximately 100 characters each).

Wireless Personal Area Network (WPAN)-Wireless personal area networks (WPANs) are temporary (ad-hoc) short-range wireless communication systems that typically connect personal accessories such as headsets, keyboards, and portable devices to communications equipment and networks.

Wireless Private Branch Exchange (WPBX)-A WPBX offers business users the ability to make and receive calls through the company's PBX system using cordless telephones anywhere on a company's premises that has a radio port (wireless access node).

Wireless Private Branch Exchange (WPBX) 1iii-This diagram shows a sample WPBX radio system. A WPBX system typically has a switching system that is located at the company. The WPBX switch interfaces a PSTN communication line and multiple radio base stations. Radio base stations communicate with wireless office telephones that can move throughout the system. A control terminal is used to configure and update the WPBX with information about the wireless office telephones and how they can be connected to the PSTN.

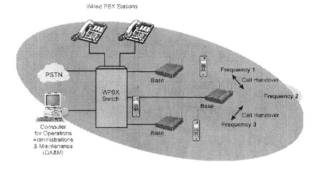

Wireless Private Branch Exchange (WPBX) System

Wireless Router-A wireless router is a device that directs (routes) content (data, voice or video) from one point to another point on a wireless network. Wireless routers base their switching information on one or more parameters contained in the packet of content. These parameters may include availability of a transmission path or communications channel, destination address contained within a packet, maximum allowable amount of transmission delay a packet can accept, along with other key parameters.

Wireless Security-Wireless security is the ability of a wireless system or service to maintain its desired operation without damage, theft or compromise of its resources from unwanted people or events. Wireless security may use access security, authentication and encryption systems to maintain the security of the system.

WirelessMAN-WirelessMAN is a licensed version of 802.16 WiMax system.

Wireline-Telecommunications services provided by wireline common carriers, such as telephone companies. This term also refers to the use of copper wire for transmission of signals rather than radio links.

Wireline Carriers-Cellular service providers that are also engaged in the business of landline tele-

phone service. Band B is allocated for these service providers. Some wireline carriers have been authorized to provide service in band A in other regions.

WISP-Wireless Internet Service Provider

WIT-Web Interactive Toy

WLAN-Wireless Local Area Network

WLAN Roaming-WLAN roaming is the capability to move from one WLAN system area to another WLAN service area and obtain service.

WLIF-Wireless LAN Interoperability Forum

WLL-Wireless Local Loop

WM-Windows Media

WMA-Windows Media Audio

WMAN-Wireless Metropolitan-Area Network

WML-Wireless Markup Language

WMLscript-A scripting language for use with wireless access protocol (WAP) devices similar to JavaScipt.

This figure shows how WMLScript can be used to respond to WML requests for information. This example shows that a user has requested a weather forecast document from his preferred weather information company. This request is sent through the cellular tower to the cellular system that forwards the request to the Internet address selected by the user. The Internet routes the request to a WAP server. The WAP server has a WML deck residing on it. Because this users request involves inputting data (the requested city), a WMLScript function is called from the WML deck to process the inserted data. The WMLScript function then sends the processed data (the temperature and forecast) back to the device via the Internet, through the cellular system, to the display of the wireless device.

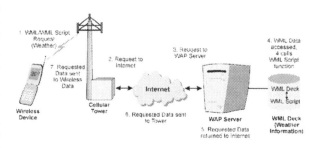

WML Script Operation

WMP-Windows Media Player

WMV-Windows Media Video

WNIC-Wireless Network Interface Card

Wobulator-A device that creates a sweeping frequency signal through the use of a variable capacitor in a tuned frequency generator circuit where one or more plates in the capacitor vibrates (wobble) resulting in a change of frequency.

Word-In data communications, a character string, binary element string, or bit string that is considered as an entity.

Word Error Rate (WER)-The ratio of words received in error with respect to the total number of words sent.

WordDAB-World DAB Forum

Work Order-A detailed drawing or print that indicates the addition, removal, or rearrangement of outside plant, also called a work print.

Workflow Software-Workflow software is the programs or applications that are used to setup, manage and complete projects.

Working Group-A working group is a team of experts that are responsible for the development of documents or standards that relate to a particular technology or industry.

Workstation-A computer that is attached to the network. A workstation has the capability of processing information in addition to requesting and sending information through a network.

World Administrative Radio Conference (WARC)-An international meeting coordinated by the ITU at which countries determine which frequencies will be allocated for what services. Each radio transmitter is assigned an identifying alphanumeric identifier ("call letters") by the licensing authority of the national government. Handsets and portable radio units of cellular and trunked radio systems are deemed to be sub-users of the license identification of their home base service provider. The WARC assigns specific initial call letters or numbers for use by each country, and has done so since 1912. Several examples are obviously based on the name of the country: A for America (USA) although originally assigned to Germany (from its Latin name Allemania), C for Canada, D for Germany (due to its German-language name Deutschland), F for France, G for Great Britain, I for Italy, J for Japan, R for Russia, and so forth. Nations with large populations also have additional initial letters chosen for various interesting historical reasons: The USA also uses K, N and W,

which were chosen to allow certain pre-WARC arbitrarily-chosen station names to continue to be used by the US military, and still fall within the WARC allocations; namely N for NAVY — the official fleet radio station of the US Navy, and W for WAR— the original name of the official US Army radio station (then part of the US War Department, now the US Department of Defense). Originally the letters W and K were used as the initial letters for all continental US broadcast stations regardless of geography, but since 1923 the initial letter K was assigned exclusively for new broadcast stations west of the Mississippi river, and W for those to the east. Legend has it that the letter V was used by the then British Empire in memory of Queen Victoria, and was assigned in combination with various other succeeding letters to various outposts of the then British Empire. The result is the continued use of V today in the independent nations of the former British Commonwealth such as: Australia (VZA), Canada (VE), India (VTA) and so forth.

World DAB Forum (WordDAB)-The World DAB Forum is a private organization who's key objective is to assist in the development and commercialization of digital audio broadcasting (DAB) radio services, primarily based on the Eureka 147 specifications. You can obtain more information about World DAB at www.WorldDAB.org.

World Numbering Plan-The numbering plan that assigns each telephone customer in the world a unique telephone. This number that consists of a country code followed by a national number.

World Wide Web (WWW)-A service that resides on computers that are connected to the Internet that allows end users to access data that is stored on the computers using standard interface software (browsers). The WWW (commonly called the "web") is associated with customers that use web browsers (graphic display software) to public users to find, acquire and transfer information.

World Wide Web Consortium (W3C)-The world wide web consortium (W3C), formed in 1994, is a group of companies in the wireless market whose goal is to provide a common markup language for wireless devices and other small devices with limited memory.

Worm Virus-A Worm is a type of computer virus that replicates itself onto other programs. Worms are commonly used with email messages allowing it to move from file to file and computer to computer.

WOTS-Wireless Office Telephone System

WPA-Wi-Fi Protected Access

WPAN-Wireless Personal Area Network

WPBX-Wireless Private Branch Exchange

WRA-Wireless Resellers Association

Wrapper-A wrapper is a playlist that is used to coordinate the playing of media or video clips such as introduction clips, station identification messages and ad bumpers.

WRED-Weighted Random Early Discard

Write-Off-A financial transaction which records as a loss to a company billed services or fees that cannot be collected, either due to customer non-payment, or due to an inability to bill for service or usage.

WSL-Wireless Session Layer

WSP-Wireless Session Protocol

WTAI-Wireless Telephony Application Interface

WTLS-Wireless Transport Layer Security

WTN-Working Telephone Number

WTP-Wireless Transfer Protocol

WUG-Wireless User Group

wVoIP-Wireless Voice over Internet Protocol

WWAN-Wireless Wide-Area Networks

WWW-World Wide Web

WYSIWYG-What you See is What You Get

WZ1-World Zone 1

X

X.25-An international standard for communications with a packet data switching network. The X.25 standard specifies the protocol between the data device (such as a computer) and the network (such as a public packet data network).

This diagram shows a X.25 packet data system that is used to transfer banking (cash flow machine) information through several different X.25 systems. This diagram shows a bank teller machine in Rome is connected to a bank processing system in London through a virtual path. This path is created through each packet switching exchange (PSE) the X.25 networks before any data is sent. This example shows that each switching point in the X.25 networks validate the transfer of each packet to the next node or switching point. This ensures that data reliably passes through each packet node to reach its previously established destination.

XBar-Crossbar Switches
XDR-External Data Representation
xDSL-A set of large-scale high bandwidth data technologies that can use standard twisted-pair copper wire to deliver high speed digital services (up to 52 Mbps).
xHTML-Extensible Hypertext Markup Language
XMCL-eXtensible Media Commerce Language
XML-Extensible Markup Language
XMTR-Transmitter
XrML-Extensible Rights Markup Language

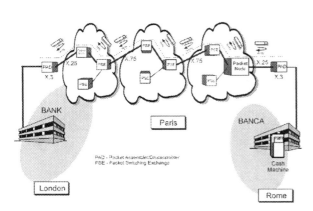

X.25 System

X.75-An international standard for communications between X.25 packet data switches. The X.75 standard specifies the protocol that is used to transfer packets between the network switches. The X.75 standard was developed by CCITT and it is sometimes referred to as the X.25 gateway.

Y-Luminance Signal

YIG-Yttrium-Iron Garnet

Yoke-A material that is used to interconnect and focus magnetic cores. A yoke is commonly used in picture tubes to control magnetic fields that adjust the direction of an electron beam.

YUV-A color space with components Y (luminance), U and V (the color difference components). A color value represented in RGB space can be converted to YUV by a simple linear formula. Image compression (such as JPEG) and video compression (such as MPEG) work on pixel colors in YUV values, often subsampling the U and V components to a lower spatial resolution because the eye is less sensitive to errors in U and V than in Y. See also RGB and CMYK.

Z

ZBTSI-Zero Byte Time Slot Interchange

Zigbee-Zigbee is a wireless technology that is used for short range network monitoring and control applications. The industry standard for Zigbee is IEEE 802.15.4 and information on Zigbee can be found at www.Zigbee.org.

ZigBee was designed for the hostile RF environments that routinely exist in mainstream commercial and industrial applications. Utilizing DSSS with features including collision avoidance, receiver energy detection, link quality indication, clear channel assessment, acknowledgement, security, support for guaranteed time slots and packet freshness; ZigBee-compliant networks offer OEMs and vendors a highly reliable, standards-based solution.

ZigZag Scan-A zigzag scan is the process of diagonally scanning a matrix of measurements (such as a digital image block) to convert the information into a serial data format.

Zip Drive-A removable storage device storing 100 MB to 250 MB of data on 3.5-inch ZIP disks.

Zombie-A process that is not used (dead) and has not yet been deleted from the process table in a Unix or other operating systems.

Zoom Ratio-Zoom ratio is a comparison between the wide image size or focal length to a more narrow image or ending focal length for a display (such as a computer monitor) or an optical device (such as a movie camera). An example of a zoom ratio is 20x where the magnified image is 20x larger than the original image.

Associations

ADSL Forum
39355 California St. Suite 200
Fremont, CA, 94538
510-608-5905
510-608-5917(F)
www.adsl.com

Advanced Television Systems Committee (ATSC)
1750 K Street, N.W., Suite 1200
Washington, DC, 20006
202-872-9160
202-872-9161(F)
www.atsc.org

Alliance for Telecommunications Industry Solutions (ATIS)
1200 G St. NW, Suite 500
Washington, DC, 20005
202.628.6380
202-393-5453(F)
www.atis.org

American Mobile Telecommunications Association (AMTA)
1150 18th St. NW, Suite 250
Washington, DC, 20036
202-331-7773
202-331-9062(F)
www.amtausa.org

American National Standards Institute (ANSI)
1819 L St. NW
Washington, DC, 20036
202-293-8020
202-293-9287(F)
www.ansi.org

American Registry for Internet Numbers (ARIN)
3635 Concorde Pkwy., ste. 200
Chantilly, VA, 20151-1130
703.227.9840
703.227.0676(F)
www.arin.net

American Teleservices Association (ATA)
3815 River Crossing Parkway, Suite 20
Indianapolis, IN, 46240
317.816.9336
www.ataconnect.org

Association for Local Telecommunications Services (ALTS)
888 17th Street, NW, 12th Floor
Washington, DC, 20006
202.969.ALTS
202.969.ALT1(F)
www.alts.org

Association of Communications Enterprises (ASCENT)
1401 K Street, NW, Suite 600
Washington, DC, 20005
(202) 835-9898
(202) 835-9893(F)
www.ascent.org

Association of TeleServices International (ATSI)
12 Academy Avenue
Atkinson, NH, 03811
(603) 362-9489
(603) 362-9486(F)
www.atsi.org

ATM Forum
Presidio of San Francisco, PO Box 29920
San Francisco, CA, 94129
415-561-6275
415-561-6120(F)
www.atmforum.com

Bluetooth Special Interest Group
www.bluetooth.com

British Standards Institution
British Standards House 389 Chiswick High Road
London, , W4 4AL
England UK
44 (0)20 8996-9000
44 (0)20 8996-7400(F)
www.bsi-global.com

Building Industry Consulting Service International (BICSI)
8610 Hidden River Parkway
Tampa, FL, 33637
813.979.1991
813.971.4311(F)
www.bicsi.org

Business Technology Association
12411 Wornall Road, Suite 200
Kansas City, MO, 64145
816.941.3100
816.941.4838(F)
www.bta.org

Cable Television Laboratories, Inc (CableLabs)
400 Centennial Pkwy.
Louisville, CO, 80027-1266
303-661-9100
303-661-9199(F)
www.cablelabs.com

California Broadband Users' Group
P.O. Box 27901-391
San Francisco, CA, 94127
415.241.9943
415.753.6942(F)
www.ciug.org or www.isdnworld.com

California Cable & Telecommunications Association (CCTA)
4341 Piedmont Ave. (P.O. Box 11080)
Oakland, CA, 94611
(510) 428-2225
510-428-0151(F)
www.calcable.org

Canadian Standards Association (CSA)
5060 Spectrum Way, Ste 100.
Mississauga, Ontario, 14W 5N6
CANADA
416-747-4044
416-747-2510(F)
www.csa.ca

Canadian Wireless Telecommunications Association (CWTA)
130 Albert Street, Suite 1110
Ottawa, ON, K1P 5G4
Canada
613-233-4888
613-233-2032(F)
www.cwta.ca

CDMA Development Group (CDG)
575 Anton Blvd., Ste. 560
Costa Mesa, CA, 92626
1-888-800-CDMA or 1-714-545-5211
714-545-4601(F)
www.cdg.org

Cellular Telecommunications Internet Association (CTIA)
1250 Connecticut Ave NW, Suite 800
Washington, DC, 20036
202-785-0081
202-785-0721(F)
www.wow-com.com

CommerceNet
10050 N. Wolfe Rd. Ste. SW2-255
Cupertino, CA, 95014
408-446-1260
408-446-1268(F)
www.commercenet.net

Communications Fraud Control Association (CFCA)
3030 North Central Avenue, Suite 707
Phoenix, AZ, 85012
602.265.2322 (CFCA)
602.265.1015(F)
www.cfca.org

Competitive Telecommunications Association (CompTel)
1900 M Street, N.W., Suite 800
Washington, DC, 20036
202.296.6650
202.296.7585(F)
www.comptel.org

Competitive Telephone Carriers of New York, Inc.
1 Columbia Place
Albany, NY, 14
518-434-8112
518-434-3232(F)

Computer and Communications Industry Association (CCIA)
666 11th St. NW
Washington, DC, 20001
202-783-0070
202-783-0534(F)
www.ccianet.org

Defense Advanced Research Projects Agency (DARPA)
3701 Fairfax Drive
Arlington, VA, 22203-1714
(703) 526-6630
(703) 528-1943(F)
www.darpa.mil

Electronic Industries Association (EIA)
2500 Wilson Blvd.
Arlington, VA, 22201
703-907-7500
703-907-7501(F)
www.eia.org

European Telecommunications Standards Institute (ETSI)
650, route des Lucioles
Sophia-Antipolis, Cedex, 06921
France
33 (0)492944311
33 (0)492385299(F)
www.etsi.fr

Federal Communications Commision (FCC)
445 12 St. SW
Washington, DC, 20554
888-CALL-FCC
202-418-0232(F)
www.fcc.gov

Home Phoneline Networking Alliance (HomePNA)
Bishop Ranch 2, 2694 Bishop Drive, Suite 105
San Ramon, CA, 94583
925-277-8110
925-277-8111(F)
www.homepna.org

Indiana Telecommunications Association (ITA)
54 Monument Circle, Suite 200
Indianapolis, IN, 46204
317-635-1272
317-635-0285(F)
www.itainfo.org

Industrial Telecommunications Association (ITA)
1110 North Glebe Rd. Suite 500
Arlington, VA, 22201-5720
703-528-5115
703-524-1074(F)
www.ita-relay.com

Infared Data Association (IrDA)
P.O. Box 3883
Walnut Creek, CA, 94598
www.irda.org

Information Technology Association of America (ITAA)
1401 Wilson Blvd. Ste. 1100
Arlington, VA, 22209
703-522-5055
703-525-2279(F)
www.itaa.org

Information Technology Industry Council
1250 I Street NW, Suite 200
Washington, DC, 20005
202.737.8888
202.638.4922(F)
www.itic.org

Institute of Electrical and Electronics Engineers, Inc. (IEEE)
1828 L Street, N.W., Suite 1202
Washington, DC, 20036-5104
202-785-0017
202-785-0835(F)
www.ieee.org

Insulated Cable Engineers Association (ICEA)
P.O. Box 1568
Carrolton, GA, 30117
508-394-4424
www.icea.net

International Multimedia Teleconferencing Consortium, Inc.
Bishop Ranch 2, 2694 Bishop Drive, Suite 275
San Ramon, CA, 94583
925-275-6600
925-275-6691(F)
www.imtc.org

International Municipal Signal Association (IMSA)
165 East Union Street (PO Box 539)
Newark, NY, 14513-0539
315-331-2182 1-800-723-IMSA
315-331-8205(F)
www.IMSAsafety.org

International Telecommunications Union (ITU)
Place des Nations
 Geneva 20, Geneva, CH-1211
Switzerland
+41 22 730 51 11
+41 22 733 7256(F)
www.itu.int

International VoIP Council
202-326-1743
www.voipcouncil.org

International Wireless Telecommunications Association (IWTA)
1150 18th St. NW, Suite 250
Washington, DC, 20036
202-331-7773
202-331-9062(F)
www.iwta.org

Internet Mail Consortium
127 Segre Place
Santa Cruz, CA, 95060
831-426-9827
831-426-7301(F)
www.imc.org

InterNIC
P.O. Box 1656
Herndon, VA, 22070
www.internic.net

Mid-America Cable Telecommunications Association (Mid-America)
P.O. Box 3306
Lawrence, KS, 66046
785-841-9241
785-841-4975(F)
www.midamericacable.com

Minnesota Telephone Association
30 East 7th Street
St. Paul, MN, 55101
651-291-7311
651-291-2795(F)
www.mnta.org

National Association of Broadcasters (NAB)
1717 N Street, NW
Washington, DC, 20036-2891
202-429-5300
202-429-4199(F)
www.nab.org

National Association of Radio and Telecommunications Engineers (NARTE)
P.O. Box 678
Medway, MA, 02053
508-533-8333
508-533-3815(F)
www.narte.org

National Association of Regulatory Utility Commissioners (NARUC)
1101 Vermont Avenue NW, Suite 200
Washington, DC, 20005
202-898-2200
202-898-2213(F)
www.naruc.org

National Association of State Telecommunications Directors (NASTD)
PO Box 11910 or 2760 Research Park Dr.
Lexington, KY, 40578-1910
859-244-8186
859-244-8001(F)
www.nastd.org

National Cable TV Association (NCTA)
1724 Massachusetts Ave., N.W.
Washington, DC, 20036
202-775-3669
202-775-3692(F)
www.ncta.com

National Emergency Number Association (NTCA)
422 Beecher Rd.
Columbus, OH, 43230
800-332-3911 or (614) 741-2080
(614) 933-0911(F)
www.nena9-1-1.org

National Exchange Carrier Association (NECA)
80 South Jefferson Rd.
Whippany, NJ, 07981-8597
973-884-8000 or 800-228-8597
973-884-8469(F)
www.neca.org

National Fire Protection Association (NFPA)
1 Batterymarch Park
Quincey, MA, 02169
617-770-3000
617-770-0700(F)
www.nfpa.org

National Institute of Standards and Technology (NIST)
100 Bureau DR
Gaithersburg, MD, 20899
301-975-2000
www.nist.gov

National Technical Information Service (NTIS)
U.S. Department of Commerce
Springfield, VA, 22161
703-605-6000
703-321-8547(F)
www.ntis.gov

National Telecommunications and Information Administration (NTIA)
U.S. Department of Commerce 1401 Constitution Ave. NW
Washington, DC, 20230
202-482-7002
www.ntia.doc.gov

National Telephone Co-op Association (NTCA)
4121 Wilson Blvd., Tenth Floor
Arlington, VA, 22203
703-351-2000
703-351-2001(F)
www.ntca.org

Network Professional Association (NPA)
195 South C St.
Tustin, CA, 92780
714-573-4780
714-669-9341(F)
www.npanet.org

Office of the Federal Register (OFR)
National Archives & Record Administration,
700 Pennsyvainia Ave. NW
Washington, DC, 20408
866-325-7208
202-523-6866(F)
www.nara.gov/fedreg

OFTEL
50 Ludgate Hill
London, EC4M 7JJ
England (UK)
44-020-7634-8700
44-020-7634-8845(F)
www.oftel.gov.uk

Optical Storage Technology Association (OSTA)
19925 Stevens Blvd.
Cupertino, CA, 95014
408-253-3695
408-253-9938(F)
www.osta.org

Organization for Promotion and Advancement of Small Telecom Companies (OPASTCO)
21 Dupont Circle, NW, Suite 700
Washington, DC, 20036
202-659-5990
202-659-4619(F)
www.opastco.org

Associations

Pacific Telecommunications Council (PTC)
2454 S. Beretania ST., Suite 302
Honolulu, HI, 96826
808-941-3789
808-944-4874(F)
www.ptc.org

PCI Industrial Computer Manufacturers Group (PICMG)
401 Edgewater Place, Suite 600
Wakefield, MA, 01880
781-246-9318
781-224-1239(F)
www.picmg.org

Personal Communications Industy Association (PCIA)
500 Montgomery St., Suite 700
Alexandria, VA, 22314-1561
703-739-0300
703-836-1608(F)
www.pcia.com

Portable Computer and Communications Association (PCCA)
P.O. Box 2460
Boulder Creek, CA, 95006
541-490-5140
419-831-4799(F)
www.pcca.org

Rural Cellular Association (RCA)
701 Brazos, Suite 320
Austin, TX, 78701
800-722-1872
512-472-1071(F)
www.rca-usa.org

Satellite Broadcasting & Communications Association (SBCA)
225 Reinekers Lane, Suite 600
Alexandria, VA, 22314
703-549-6990
703-549-7640(F)
www.sbca.com

Satellite Industry Association (SIA)
225 Reinekers Lane, Suite 600
Alexandria, VA, 22314
703-549-8697
703-549-9188(F)
www.sia.org

Small Business in Telecommunications (SBT)
1331 H St., NW Suite 500
Washington, DC, 20005
202-347-4511
202-347-8607(F)
www.sbthome.org

Society of Cable Telecommunications Engineers Inc. (SCTE)
140 Phillips Road
Exton, PA, 19341-1318
(800) 542-5040 or (610) 363-6888
610-363-5898(F)
www.scte.org

Society of Motion Pictures & Television Engineers (SMPTE)
595 W. Hartsdale Avenue
White Plains, NY, 10607-1824
914-761-1100
914-761-3115(F)
www.smpte.org

Telecommunications Industry Association (TIA)
2500 Wilson Blvd, Suite 300
Arlington, VA, 22201
703-907-7700
703-907-7727(F)
www.tiaonline.org

The Association for Telecommunications Professionals in Higher Education (ACUTA)
152 W. Zandale Drive, Suite 200
Lexington, KY, 40503
606-278-3338
606-278-3268(F)
www.acuta.org

The Computing Technology Industry Association (CompTIA)
1815 S. Myers Rd.
Oakbrook Terrace, IL, 60181
630-268-1818
630-268-1834(F)
www.comptia.org

The Consumer Electronics Association (CEA)
2500 Wilson Blvd.
Arlington, VA, 22201
703-907-7600
703-907-7675(F)
www.ce.org

The Electronic Frontier Foundation (EFF)
454 Sohotwell St.
San Francisco, CA, 94110-4832
415.436.9333
415.436.9993(F)
www.eff.org

The Open Group
44 Montgomery St. Ste. 960
San Francisco, CA, 94104
415-374-8280
415-374-8293(F)
www.opengroup.org

United States Telecom Association (USTA)
1401 H St. NW, Suite 600
Washington, DC, 20005-2164
202-326-7300
202-326-7333(F)
www.usta.org

United States Telecommunications Training Institute (USTTI)
1150 Connecticut Avenue, NW Suite 702
Washington, DC, 20036
USA
202.785.7373
202.785.1930(F)
www.ustti.org

United Telecom Counsel (UTC)
1901 Pennsylvania Avenue, NW 5th Floor
Washington, DC, 20006
202-872-0030
202-872-1331(F)
www.utc.org

Universal Wireless Communications Consortium (UWCC)
8302 159th Pl. NE
Redmond, WA, 98052
425-580-5031
www.uwcc.org

Wall Street Telecommunications Association (WSTA)
241 Maple Ave.
Red Bank, NJ, 07701
732-530-8808
731-530-0020(F)
www.wsta.org

Wireless Communications Association International (WCA)
1140 Connecticut Ave, NW Suite 810
Washington, DC, 20036
202-452-7823
202-452-0041(F)
www.wcai.com

Wireless Dealers Association (WDA)
9746 Tappenbeck Dr.
Houston, TX, 77055
800 624-6918 or 713 467-0077
800-820-2284(F)
www.wirelessdealers.com

Wireless Industry Association (WIA)
9746 Tappenbeck Drive
Houston, TX, 77055
800-624-6918 or 713-467-0077
800-820-2284(F)
www.wirelessindustry.com

Wireless LAN Alliance (WLANA)
P.O. Box 9097
San Jose, CA, 95157
650-352-4709
650-649-2305(F)
www.wlana.com

International Engineering Consortium (IEC)
549 West Randolph Street
Suite 600
Chicago, IL 60661-2208 USA
1-312-559-4100
1-312-559-4111(F)
www.iec.org

Associations

European Association for Standardizing Information and Communication Systems (ECMA)
114 Rue du Rhône
CH-1204 Geneva, Switzerland
http://www.ecma.ch/

European Telecommunications Standards Institute (ETSI)

650, route des Lucioles
06921 Sophia-Antipolis Cedex
FRANCE
33 (0)4 92 94 42 00
33 (0)4 93 65 47 16(F)

MIT Internet & Telecoms Convergence Consortium (MIT-ITC)
E40-234, 1 Amherst St.
Cambridge, MA 02139 USA
617-253-4138
617-253-7326(F)
http://itel.mit.edu/

Institute of Electrical and Electronics Engineers (IEEE)
1828 L Street, N.W., Suite 1202
Washington, D.C. 20036-5104
1 202-785-0017
1 202-785-0835(F)
www.ieee.org

Multiservice Switching Forum (MSF)
39355 California Street #307
Fremont, CA 94538
510-608-5922
510-608-5917(F)
http://www.msforum.org

International Multimedia Teleconferencing Consortium (IMTC)
Bishop Ranch 2
2694 Bishop Drive, Suite 275
San Ramon, CA 94583
1 925-275-6600
1 925-275-6691(F)
http://www.imtc.org/

Telecommunications Industry Association (TIA)
2500 Wilson Blvd., Suite 300
Arlington, VA 22201   USA
703-907-7700
703-907-7727
703-907-7776(F)
www.tiaonline.org

International Telecommunications Union (ITU)
ITU - Place des Nations
CH-1211 Geneva 20
Switzerland
4122 730 5115
4122 730 5595(F)
http://www.itu.int

TeleManagement Forum (TMF)
1201 Mt. Kemble Ave.
Morristown, NJ 07960-6628
1 973-425-1900
1 973-425-1515(F)
www.tmforum.org

The Internet Engineering Task Force
www.ietf.org

United States Internet Service Provider Association (USIPSA)
1330 Connecticut Avenue, NW
Washington, DC 20036
202-862-3816
202-261-0604(F)
http://www.cix.org

Association of Internet Professionals
4790 Irvine Boulevard, Suite 105-283
Irvine, CA 92620
866-AIP-9700
1-501-423-2248
http://www.association.org

International Internet Marketing Association
PO Box 4018
Vancouver Main
349 West Georgia Street
Vancouver, BC
V6B 3Z4
www.iimaonline.org

WA Internet Association
250 St Georges Terrace
Perth WA 6000
www.ix.waia.asn.au

Internet Service Providers' Consortium (ISP/C) OR Forum
1301 Shiloh Road, Suite 720
PO Box 1086
Kennesaw, GA 30144-8086
866-533-6990
678-819-1028(F)
www.ispc.org or http://www.ispf.com/

Internet SOCiety (ISOC)
1775 Wiehle Ave., Suite 102
Reston, VA 20190
703 326 9880
703 326 9881(F)
http://www.isoc.org/isoc/

Alliance for Global Internet Services (AGIS)
725 East 175 North
Lindon, UT 84042
Tel: 801-796-9311
http://agis.org

US Internet Industry Association (USIIA)
815 Connecticut Avenue, NW
Suite 620,
Washington, DC 20006
or
5810 Kingstowne Center Drive
Suite 120, PMB 212
Alexandria, VA 22315-5711
703-924-0006
703-924-4203(F)
http://www.usiia.org

Frame Relay Forum
39355 California St., Suite 307,
Fremont, CA 94538
510.608.5920
510.608.5917(F)
www.frforum.com/

CableLabs (Cable Television Laboratories, Inc.)
400 Centennial Parkway
Louisville, CO 80027-1266
303-661-9100
303-661-9199(F)
www.cablelabs.com

DSL Forum
39355 California Street, Suite 307
Fremont, CA 94538
510-608-5905
510-608-5917(F)
www.dslforum.org/

United States Telephone Associations (USTA)
1401 H Street, N.W., Suite 600
Washington, DC 20005-2164
202-326-7300
202-326-7333(F)
www.usta.org

ATM Forum
Presidio of San Francisco
P.O. Box 29920 (mail)
572B Ruger Street (surface)
San Francisco, CA 94129-0920
415-561-6275
415-561-6120(F)
www.atmforum.com

10 Gigabit Ethernet Alliance
1300 Bristol Street North, Suite 160
Newport Beach, CA 92660
949-250-7155
949-250-7159(F)

Association Management Solutions (AMS)
39355 California Street, Suite 307
Fremont, CA 94538
510-608-5900
510-608-5917(F)
www.amsl.com

ACCU
1330 Trinity Dr.
Menlo Park, CA 94025
650-233-9082
www.accu.org

Printed in the United States
74455LV00004B/457-460

9 781932 813340